Important Topics in
Organic
Chemistry

for

Class XII students
and all those preparing for competitive examinations like IIT, JEE, NEET, etc.

with

- Over 200 Selected Questions
- 1138 Multiple Choice Questions
- 712 Reasoning Type Questions
- Detailed Answers and Explanations

Important Topics in
Organic Chemistry

for

Class XII students
and all those preparing for competitive examinations like IIT, JEE, NEET, etc.

with

- **Over 200 Selected Questions**
- **1138 Multiple Choice Questions**
- **712 Reasoning Type Questions**
- **Detailed Answers and Explanations**

KC Goyal PhD

Former Reader
Department of Chemistry
Sri Guru Tegh Bahadur Khalsa College
University of Delhi
India

CBS

CBS Publishers & Distributors Pvt Ltd

New Delhi • Bengaluru • Chennai • Kochi • Kolkata • Mumbai
Hyderabad • Jharkhand • Nagpur • Patna • Pune • Uttarakhand

Important Topics in
Organic
Chemistry

ISBN: 978-93-86827-63-0

Copyright © Author and Publisher

First Edition: 2018

Published by Satish Kumar Jain and produced by Varun Jain for

CBS Publishers & Distributors Pvt Ltd

4819/XI Prahlad Street, 24 Ansari Road, Daryaganj, New Delhi 110 002, India.
Ph: 23289259, 23266861, 23266867 Website: www.cbspd.com
Fax: 011-23243014   e-mail: delhi@cbspd.com; cbspubs@airtelmail.in.
Corporate Office: 204 FIE, Industrial Area, Patparganj, Delhi 110 092

Ph: 4934 4934   Fax: 4934 4935   e-mail: publishing@cbspd.com; publicity@cbspd.com

Branches

- **Bengaluru:** Seema House 2975, 17th Cross, K.R. Road,
  Banasankari 2nd Stage, Bengaluru 560 070, Karnataka
  Ph: +91-80-26771678/79   Fax: +91-80-26771680   e-mail: bangalore@cbspd.com
- **Chennai:** 7, Subbaraya Street, Shenoy Nagar, Chennai 600 030, Tamil Nadu
  Ph: +91-44-26680620, 26681266   Fax: +91-44-42032115   e-mail: chennai@cbspd.com
- **Kochi:** Ashana House, No. 39/1904, AM Thomas Road, Valanjambalam,
  Ernakulam 682 016, Kochi, Kerala
  Ph: +91-484-4059061-65   Fax: +91-484-4059065   e-mail: kochi@cbspd.com
- **Kolkata:** 6/B, Ground Floor, Rameswar Shaw Road, Kolkata-700 014, West Bengal
  Ph: +91-33-22891126, 22891127, 22891128   e-mail: kolkata@cbspd.com
- **Mumbai:** 83-C, Dr E Moses Road, Worli, Mumbai-400018, Maharashtra
  Ph: +91-22-24902340/41   Fax: +91-22-24902342   e-mail: mumbai@cbspd.com

Representatives

- **Hyderabad** 0-9885175004   • **Jharkhand** 0-9811541605   • **Nagpur** 0-9021734563
- **Patna** 0-9334159340   • **Pune** 0-9623451994   • **Uttarakhand** 0-9716462459

Printed at Mudrak, Patparganj, Delhi, India

Preface

In the absence of a good and examination-oriented book of organic chemistry in a single unit, a majority of undergraduate students have to follow a number of books which is evident from my personal experience. Generally, students consider organic part of chemistry as a difficult subject and an endless material to memories. There are books available on the chemistry of compounds and on reactions in a greater detail but hardly any book is designed for competitive examinations. To overcome this difficulty, I have compiled this book, written in a systematic manner, to cater to the needs of students aspiring to appear in various competitive examinations like IIT, JEE, NEET, etc.

For better understanding, the book has been divided into four sections: Section I deals with various topics like definition, general organic chemistry, oxidising and reducing agents, tests and group tests, reagents, exceptions, single step reactions, polycyclic alkanes, dipole moments, aromaticity, organic intermediates, electron displacement effects, isomerism, hybridisation etc. Section II is divided into two parts: Part I contains about 100 questions pertaining to the text and Part II comprises five question papers and assorted questions along with answers and detailed explanations. Section III comprises 10 chapters on hydrocarbons, alkyl halides and aryl halides, alcohols and phenols, aldehydes and ketones, aliphatic and aromatic carboxylic acids, amines, biomolecules, chemistry in action, important points to remember, and IUPAC names. Section IV encompasses the chemistry of conversion and interconversion, with all possible routes and road maps for converting a compound to any other class of compound. Each part of the text is supplemented with MCQs, RTQs along with answers and explanations, and emphasis has been given on the reactions, particularly on the reactivity of alkenes and other related compounds.

I appreciate the support of my family members, my wife Mrs Vijaya Devi, daughter Dr Samta and son Prof Sunil Kumar, whose active support has made it possible to bring the book in the present form. I feel extremely grateful to Dr PP Rawat, retired Librarian, Lucknow, without whose help and efforts this book would not have seen the light of the day. I am equally indebted to the establishment of CBS Publishers & Distributors, New Delhi, particularly to Sh Satish Kumar Jain (Mataji), CMD; Sh Varun Jain, Director; Mr YN Arjuna, Senior Vice President Publishing, Editorial and Publicity; and Ms Sanjubala Tripathy, for making all possible efforts in the publication of this book.

KC Goyal

Contents

Section III

Section IV

Section I

Topics

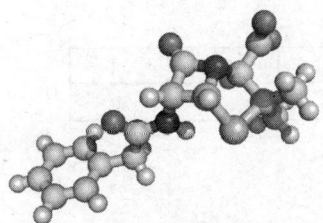

Definition of Organic Chemistry

INTRODUCTION

Organic chemistry may be defined as the chemistry of hydrocarbons (the compound of carbon and hydrogen) and their derivatives.

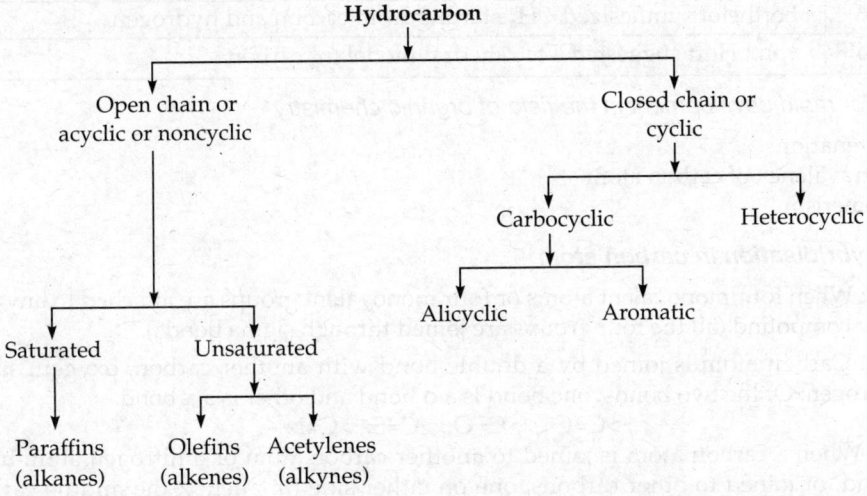

Alkanes are the saturated hydrocarbons in which the carbon is in tetrahedral form. The tetrahedral carbon is joined to four hydrogen atoms or four monovalent groups through bonds. The molecular composition of alkanes is C_nH_{2n+2}. Alkanes undergo free radical reactions only. When one or more hydrogen atoms resubstituted by monovalent groups or by a monovalent atom the alkanes are converted into various derivatives.

Alkenes or olefins have a molecular composition C_nH_{2n}. All alkenes contain only one pair of carbon joined through a double bond. The double bond consists of one σ bond and one π bond.

Acetylenes or alkynes have molecular composition C_nH_{2n-2}. All alkynes contain only one triple bond in between one pair of carbon atoms. The triple bond consists of one σ bond and two π bonds.

Electronegativity

F	O	N	Cl	Br	I	S	C	H
4	3.5	3	3	2.8	2.6	2.6	2.5	1

Classification of carbon

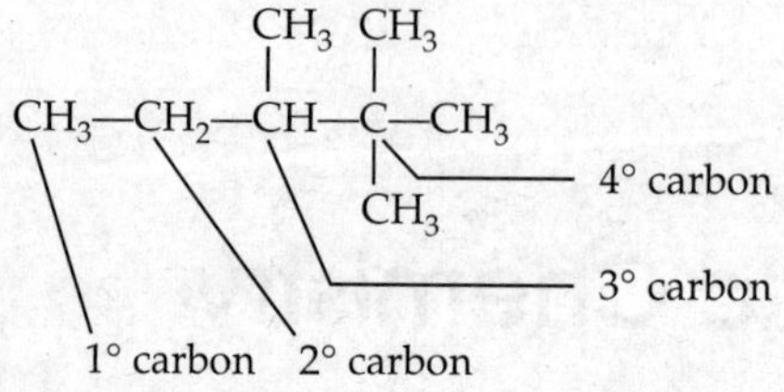

Some important dates in the development of organic chemistry

1828	Wholer prepared the first ever organic compound urea (NH_2CONH_2) from inorganic compounds by heating KCNO and NH_4Cl
1845	Kolbe synthesized CH_3COOH, starting from carbon and hydrogen, the elements present in the compound
1847	Simpson used chloroform $CHCl_3$ as the first anaesthetic in surgery
1856	Berthelot synthesized CH_4 starting from carbon and hydrogen
1874	van't Hoff suggested a tetrahedral model for carbon

Reasons for rapid development in the field of organic chemistry

  i. Catenation
 ii. Tetravalency of carbon atom
iii. Isomerism

State of hybridisation in carbon atom

  i. sp^3: When four monovalent atoms or four monovalent groups are attached to any carbon in any compound (all the four groups are joined through sigma bonds).
 ii. sp^2: Carbon atom is joined by a double bond with another carbon, oxygen, sulphur or nitrogen. Of the two bonds, one bond is a σ bond and other is a π bond.
>C=C<; >C=O; >C=S; >C=N—
iii. sp: When a carbon atom is joined to another carbon atom or a nitrogen atom by a triple bond, or joined to other carbons, one on either side (in allenes, the middle carbon is sp hybridised).
—C≡C—; —C≡N; >C=C=C<

Rules for IUPAC Nomenclature

  i. **Longest chain rule:** A longest chain (containing the functional group, as well as the C to C multiple bond) is selected.
 ii. **Lowest number rule:** After selecting the longest chain, the numbering is done in such a manner that it gives the lowest possible number to the substituents/side chain/locant (location of the substituent).
iii. **Lowest term in the set rule:** In case there are more than one substituent present in the molecule, then their positions are noted and sets are made of all the numbering. Then

these sets are compared 'term by term'. The preferred numbering is the one which have lowest term in the set, at first point of difference.

IUPAC Nomenclature

Preferential order of various functional groups in a polyfunctional compound
—SO_3H; COOH; $(CO)_2O$ anhydrides ; COOR esters; —COCl; —$CONH_2$
—CN; —CHO; >C=O; —OH; —O— ethers; >C=C<; —C≡C—

1. The isomers of butane

C_4H_{10} $CH_3CH_2CH_2CH_3$

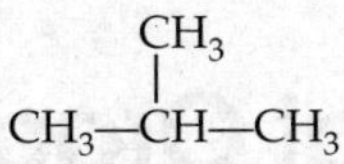

$$CH_3—\overset{\overset{\displaystyle CH_3}{|}}{CH}—CH_3$$

2. Isomers of pentane

C_5H_{12} $CH_3CH_2CH_2CH_2CH_3$

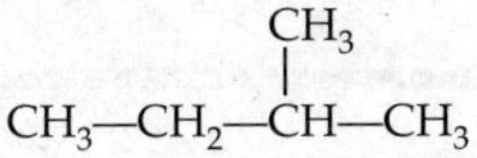

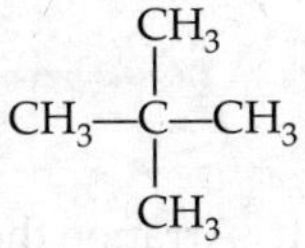

$$CH_3—CH_2—\overset{\overset{\displaystyle CH_3}{|}}{CH}—CH_3$$

$$CH_3—\overset{\overset{\displaystyle CH_3}{|}}{\underset{\underset{\displaystyle CH_3}{|}}{C}}—CH_3$$

3. Isomers of hexane

C_6H_{14} $CH_3H_2CH_2CH_2CH_2CH_3$

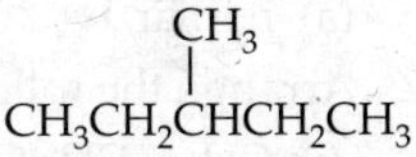

$$CH_3CH_2\overset{\overset{\displaystyle CH_3}{|}}{CH}CH_2CH_3$$
3 methyl pentane

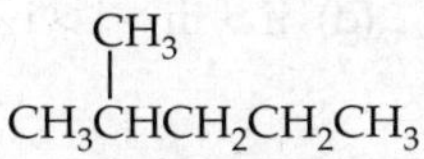

$$CH_3\overset{\overset{\displaystyle CH_3}{|}}{CH}CH_2CH_2CH_3$$
2-methyl pentane

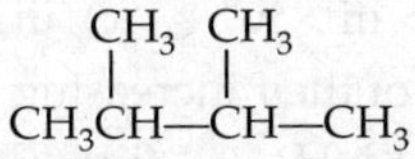

$$CH_3\overset{\overset{\displaystyle CH_3}{|}}{CH}—\overset{\overset{\displaystyle CH_3}{|}}{CH}—CH_3$$
2,3-dimethyl butane

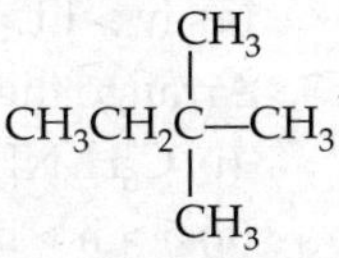

$$CH_3CH_2\overset{\overset{\displaystyle CH_3}{|}}{\underset{\underset{\displaystyle CH_3}{|}}{C}}—CH_3$$
2,2-dimethyl butane

4. Alkenes
    Isomers of pentene (C_5H_{10})

    $CH_3CH=CH—CH_2CH_3$
    butene-2

    $CH_3CH_2CH_2CH=CH_2$
    butene-1

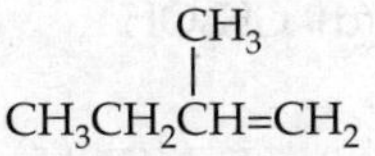

    $$CH_3CH_2\overset{\overset{\displaystyle CH_3}{|}}{C}H=CH_2$$
    2-methyl-1-butene

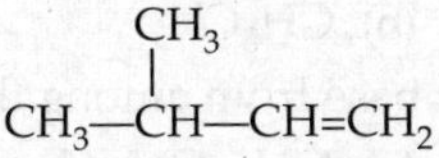

    $$CH_3—\overset{\overset{\displaystyle CH_3}{|}}{CH}—CH=CH_2$$
    3-methyl-1-butene

5. Alkynes
    Isomers of pentyne (C_5H_8)

    $CH≡C—CH_2CH_2CH_3$
    1-pentyne

    $CH_3—C≡C—CH_2CH_3$
    2-pentyne

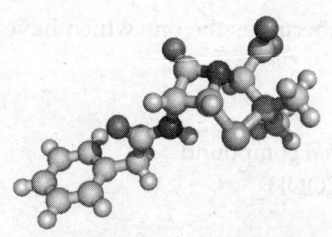

General Organic Chemistry

1. Arrange the following species in terms of their decreasing order (DO) of basicity.
   (i) OH^- (ii) $R-O$ (iii) NH_2^- (iv) Cl^-
   (a) i, ii, iii, iv   (b) iv, i, ii, iii   (c) iv, ii, i, iii   (d) iii, ii, i, iv

2. Arrange the following in terms of their increasing stability.
   (i) $CH_3-CH=CH_2$ (ii) $CH_3CH_2CH=CH$
   (iii) $(CH_3)_2CHCH=CH_2$ (iv) $(CH_3)_3CHC=CH_2$
   (a) iv > iii > ii > i   (b) i > ii > iii > iv   (c) iii > iv > ii > i   (d) ii > iii > i > iv

3. Arrange the following in order of their increasing **pKa** value.
   (i) $C_6H_5NH_2$ (ii) $(C_6H_5)_2NH$ (iii) $(C_6H_5)_3N$
   (a) i > ii > iii   (b) iii > i > ii   (c) i > iii > ii   (d) ii > iii > i

4. Arrange the following compounds in order of their increasing **pKa** value.
   (i) $C_6H_5CH_2COOH$ (ii) $CH_2=CHCH_2-COOH$
   (iii) CH_3-CH_2-COOH
   (a) i > ii > iii   (b) iii > ii > i   (c) ii > iii > i   (d) iii > i > ii

5. Which of the following can act both as an electrophile and a nucleophile?
   (a) CH_3NH_2 (b) CH_3Cl (c) CH_3CN (d) CH_3OH

6. Identify the strongest base from among the following compounds.
   (a) $CH_3-\overset{\mid}{N}-CH_3$ (b) $NH_2-\overset{\mid}{C}=NH$ (c) $C_6H_5-\overset{\mid}{N}-C_6H_5$ (d) CH_3NHCH_3
   $\qquad\; CH_3 \qquad\qquad\qquad NH_2 \qquad\qquad\qquad C_6H_5$

7. Arrange the following according to their decreasing order of nucleophilicity.

   (i) CH_3COO (ii) CH_3O (iii) CN (iv) $-\overset{\overset{O}{\|}}{\underset{\underset{O}{\|}}{S}O-O}$

   (a) i > ii > iii > iv   (b) iv > iii > ii > i   (c) ii > iii > i > iv   (d) iii > ii > i > iv

8. Which of the following compound will exhibit *d*-orbital resonance?

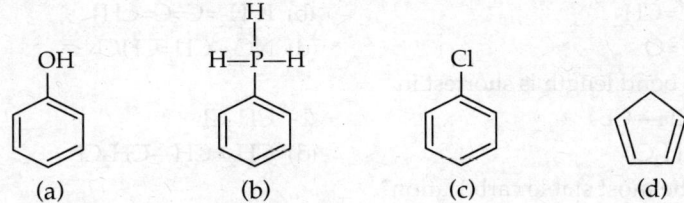

(a)　　　　　(b)　　　　　(c)　　　　　(d)

9. Which of the compounds will not show resonance?

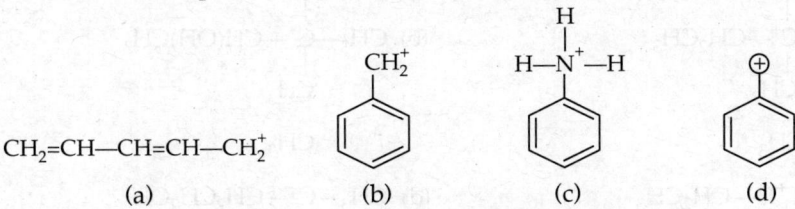

$CH_2=CH—CH=CH—CH_2^+$

(a)　　　　　(b)　　　　　(c)　　　　　(d)

10. Which of the compounds has the longest double bond?
    (a) $CH_2=CH—CH_3$　　　　　(b) $CH_2=C=—CH_2$
    (c) $(CH_3)_3—C—CH=CH_2$　　(d) $(CH_3)_2—C=CH_2$

11. The DO of the double bonds in the following.

(i)　　　　　(ii)　　　　　(iii)　　　　　(iv)

(a) i > ii > iii > iv　　(b) iv > ii > i > iii　　(c) iv > i > ii > iii　　(d) iii > ii > i > iv

12. Which one of the following compounds will operate all effects: inductive effect, mesomeric effect and hyperconjugation?

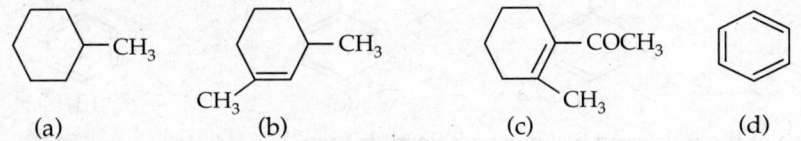

(a)　　　　　(b)　　　　　(c)　　　　　(d)

13. Which compound has the longest C—O bond?

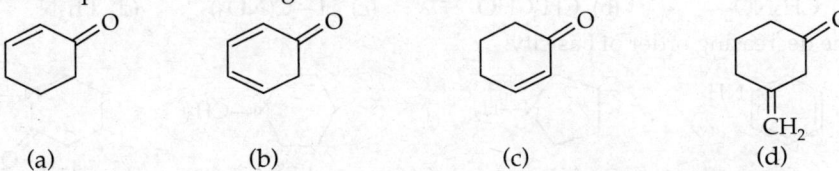

(a)　　　　　(b)　　　　　(c)　　　　　(d)

14. Arrange C—N bond length in the following compounds in decreasing order.

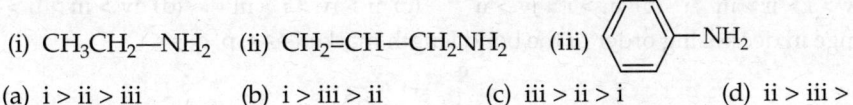

(i) $CH_3CH_2—NH_2$　(ii) $CH_2=CH—CH_2NH_2$　(iii) ⬡—NH_2

(a) i > ii > iii　　(b) i > iii > ii　　(c) iii > ii > i　　(d) ii > iii > i

15. Which one of the following is the most stabilised carbocation?

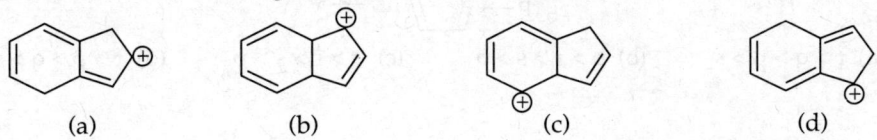

(a)　　　　　(b)　　　　　(c)　　　　　(d)

16. Which of the following is not planar?
    (a) $CH_2=C=CH_2$
    (b) $ICH_2=C=C=CH_2$
    (c) $CH_2=C=O$
    (d) $NC—CH=CHCN$

17. The C—Cl bond length is shortest in
    (a) $CH_2=CH—Cl$
    (b) CH_3Cl
    (c) $C_6H_5CH_2Cl$
    (d) $CH_2=CH—CH_2Cl$

18. Which is the most stable carbocation?

    (a) $CH_3—\overset{\overset{CH_3}{|}}{\underset{\underset{CH_3}{|}}{C^+}}—CH_2CH_3$

    (b) $CH_3—\overset{\overset{CH_3}{|}}{\underset{\underset{CH_3}{|}}{C^+}}—CH(OH)CH_3$

    (c) $HO—\overset{\overset{CH_3}{|}}{C^+}H—CH_2CH_3$

    (d) $CH_3—\overset{\overset{CH_3}{|}}{C^+}—CH_2CH_2OH$

19. Arrange the following carbanion in decreasing order of the stability.
    (i) $CH_2=\overset{\ominus}{CH}$
    (ii) $Ph—\overset{\ominus}{CH_2}$

    (iii) $CH_2=CH—\overset{\ominus}{CH_2}$
    (iv)

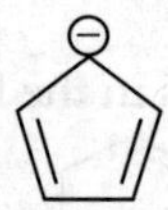

    (a) i > ii > iii > iv
    (b) iv > iii > ii > i
    (c) iii > iv > ii > i
    (d) iv > ii > i > iii

20. Which of the following is more stabilised by hyperconjugation?

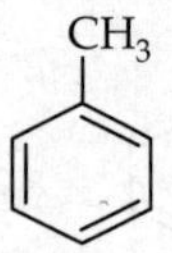

    (a) 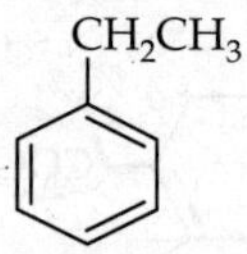   (b) 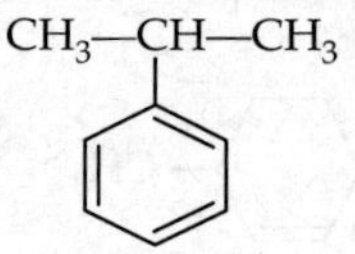   (c) 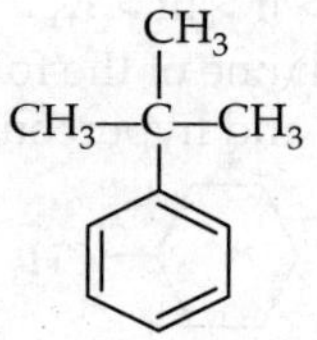   (d)

21. Which of the following has most acidic hydrogen?
    (a) CH_3NO_2
    (b) CH_3CHO
    (c) $H—C(NO_2)_3$
    (d) Ph_3N

22. The decreasing order of basicity:

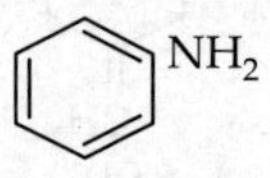

    (i) 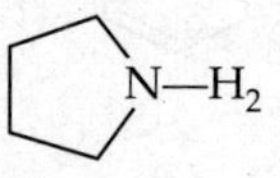   (ii) 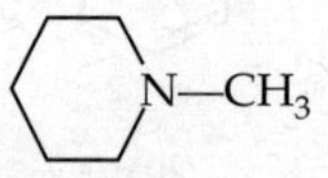   (iii) 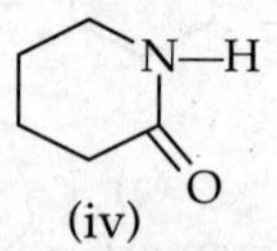   (iv)

    (a) iv > i > ii > iii
    (b) iii > i > iv > ii
    (c) ii > iv > i > iii
    (d) iv > iii > ii > i

23. Arrange in decreasing order of the bond length marked as s, p, q, s:

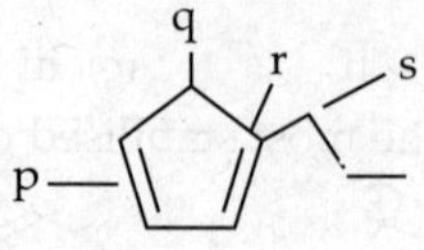

    (a) r > q > p > s
    (b) p > r > s > q
    (c) q > r > s > p
    (d) r > p > q > s

24. Arrange the following in increasing order of heat of combustion.

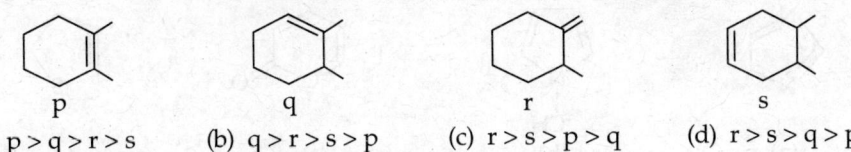

p   q   r   s

(a) p > q > r > s   (b) q > r > s > p   (c) r > s > p > q   (d) r > s > q > p

25. Which of the following does not have same C to C bond length?

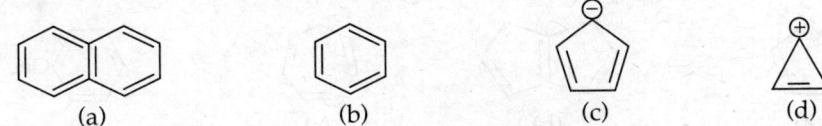

(a)   (b)   (c)   (d)

26. Which of the following compounds will show tautomerism?

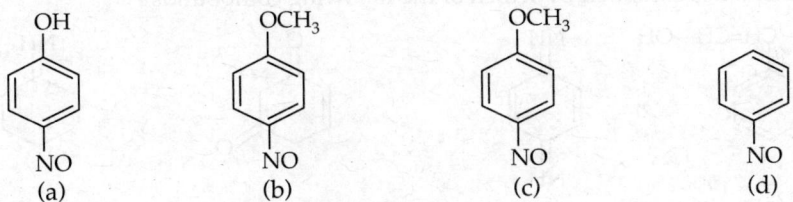

(a)   (b)   (c)   (d)

27. Arrange the following in DO of heat of hydrogenation?

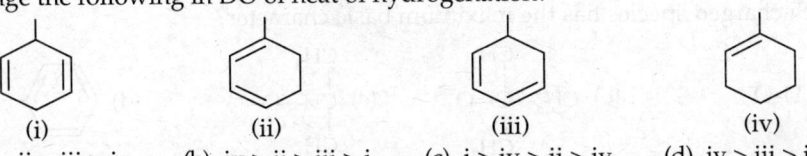

(i)   (ii)   (iii)   (iv)

(a) i > ii > iii > iv   (b) iv > ii > iii > i   (c) i > iv > ii > iv   (d) iv > iii > ii > i

28. Find out the correct order of energy required for heterolytic fission of the C—Cl bond.

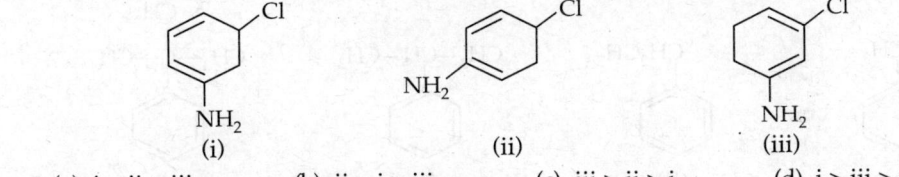

(i)   (ii)   (iii)

(a) i > ii > iii   (b) ii > i > iii   (c) iii > ii > i   (d) i > iii > ii

29. The correct order of heat of hydrogenation:

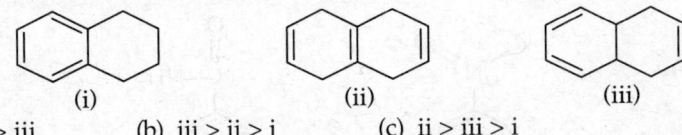

(i)   (ii)   (iii)

(a) i > ii > iii   (b) iii > ii > i   (c) ii > iii > i

30. Which of the following exhibit/exhibits dipole moment?

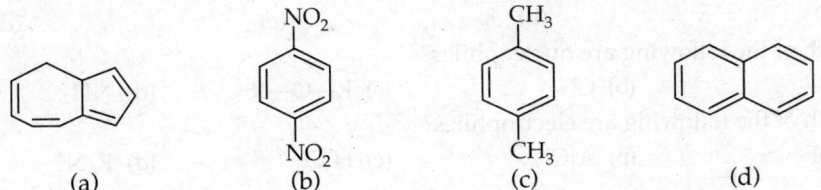

(a)   (b)   (c)   (d)

31. Which is an antiaromatic compound?

 (a) (b) (c) (d)

32. Which of the following compounds will not react with Na metal?

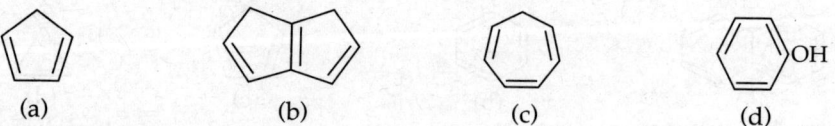

 (a) (b) (c) (d)

33. Tautomerism is exhibited by which of the following compounds?

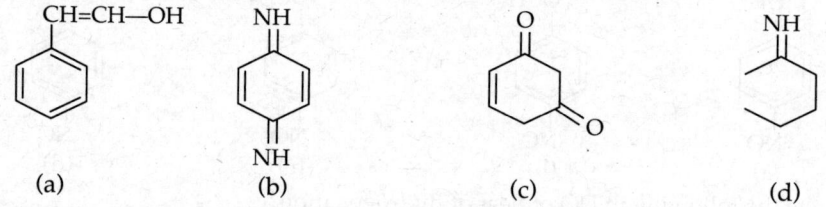

 (a) (b) (c) (d)

34. Which charged species has the maximum basic character?

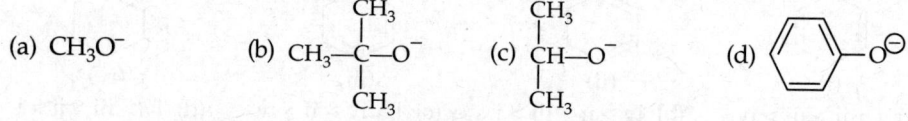

(a) CH_3O^- (b) (c) (d)

35. Which of the following has the maximum electron density?

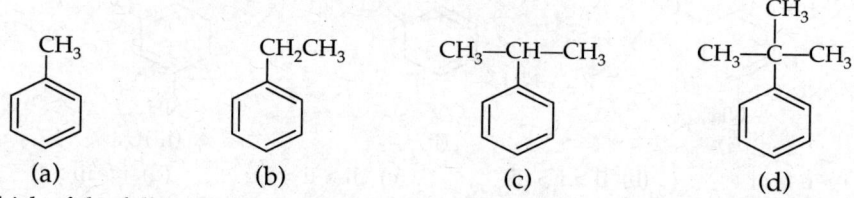

 (a) (b) (c) (d)

36. Which of the following has maximum electron density at o- and p-positions?

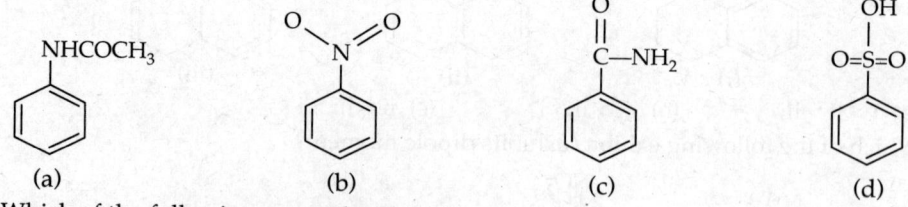

 (a) (b) (c) (d)

37. Which of the following are nucleophiles?
 (a) BF_3 (b) CN^- (c) R—O—R (d) NH_3

38. Which of the following are electrophiles?
 (a) BF_3 (b) $AlCl_3$ (c) H_3O (d) R_4N^+

39. Which is the most acidic hydrogen in the given compound?

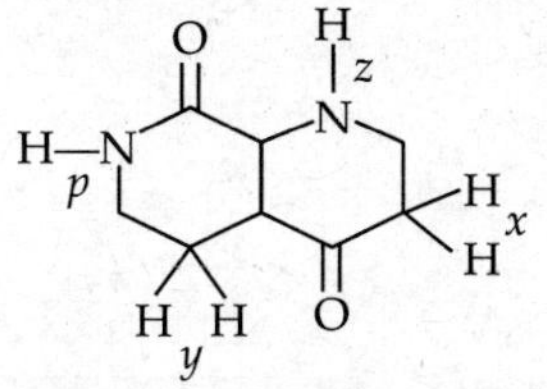

    (a) p (b) x (c) y (d) z

40. $PhCH_2NH_2 + CHCl_3/KOH \longrightarrow X \xrightarrow{H_2O/NaOH} Y$; Y is

    (a) PhCN   (b) $PhN\equiv C$ (c) $PhCH_2NH_2$ (d) $PhCH_2OH$

ANSWERS

1. (b)	2. (b)	3. (b)	4. (a)	5. (c)	6. (b)	7. (c)	8. (a)
9. (c)	10. (d)	11. (b)	12. (c)	13. (c)	14. (a)	15. (a)	16. (a)
17. (a)	18. (c)	19. (c)	20. (a)	21. (c)	22. (a)	23. (a)	24. (c)
25. (a)	26. (a)	27. (a)	28. (b)	29. (c)	30. (a), (b), (d)		31. (b)
32. (c)	33. (b), (c)	34. (b), (c), (d)		35. (a), (d)	36. (a)	37. (b), (c), (d)	
38. (a), (b)	39. (a)	40. (c)					

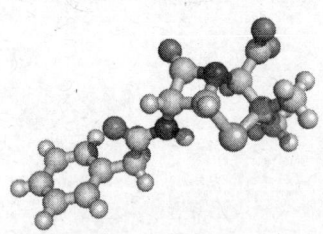

Oxidising and Reducing Agents

Oxidising Agents

1. Acidified $KMnO_4$ oxidises
    i. A primary alcoholic group (RCH_2OH) to an acid
    $$RCH_2OH \xrightarrow{\quad O \quad} RCHO \xrightarrow{\quad O \quad} RCOOH$$
    ii. A secondary alcoholic group (R_2CHOH) to a ketone
    $$R_2CHOH \longrightarrow R_2CO$$
    Ketone is the final product, in unsaturated compounds, the double and triple bonds are attacked.
    iii. Alkenes are oxidised to give a mixture of acids.
    $$RCH=CH_2 \longrightarrow RCOOH + HCOOH$$
    iv. Alkynes are oxidised to form a mixture of acids.
    $$CH \equiv CH \longrightarrow HCOOH + HCOOH$$
2. Dil. alk. $KMnO_4$ or Baeyer's reagent
    i. Alkenes are oxidised to glycols and hydroxylation takes place.

    $$\text{\Large$\underset{}{>}$C=C$\underset{}{<}$} + O + H_2O \longrightarrow \underset{\underset{cis\text{-glycol}}{OH \ OH}}{>C-C<}$$

    Other hydroxylating agents are:
    a. OsO_4(osmium tetroxide)/$NaHSO_3$: It hydroxylates alkenes to forms a *cis*-glycol
    b. Peroxy formic acid or any other peroxy acid: It hydroxylates alkenes to give a *trans*-glycol
    ii. Alkynes are oxidised to form an acid
    $$CH \equiv CH \longrightarrow \underset{\underset{\text{oxalic acid}}{COOH}}{\overset{COOH}{|}}$$

3. Strong alk. $KMnO_4$
   i. It attacks the double and triple bonds
      Alkenes are oxidised to form acid
      $$RCH=CH_2 \longrightarrow RCOOH + CO_2$$
      Alkynes are oxidised and a mixture is obtained
      $$RC\equiv CR' \longrightarrow RCOOH + R'COOH$$
   ii. Side chain oxidation: Any carbon containing group, joined to the benzene ring (through this carbon) is termed a side chain. On oxidation with Baeyer's reagent the whole of the side chain, irrespective of the length is oxidised to a —COOH, directly joined to the ring.

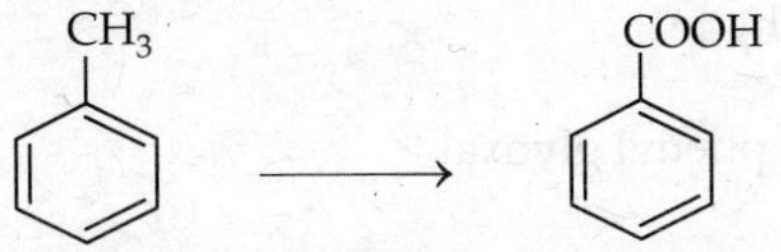

   toluene   benzoic acid

4. Acidified $K_2Cr_2O_7$ or H_2CrO_4 or $CrO_3 + H_2SO_4$
   i. A primary alcoholic group
      $$RCH_2OH \longrightarrow RCHO \longrightarrow RCOOH$$
      Carboxylic acid is the final product.
   ii. A secondary alcoholic group
      $$R_2CHOH \xrightarrow{\text{O}} R_2CO$$
      Ketone is formed, it is not oxidised further.
   iii. Unsaturated compounds
      Alkenes are oxidised breaking the double bond or triple bond to form a mixture of acids
      a. $CH_2=CH_2 \longrightarrow HCOOH + HCOOH \longrightarrow 2CO_2 + 2H_2O$
      b. $RCH=CH_2 \longrightarrow RCOOH + CO_2 + H_2O$
      c. $CH\equiv CH \longrightarrow HCOOH + HCOOH$

5. Periodic acid, HIO_4 oxidation: The glycolic compounds are decomposed and the two carbons holding the hydroxyl groups are separated and oxidised.
   i. $\underset{\overset{|}{HO}\ \ \overset{|}{OH}}{RCHCHR'} \longrightarrow RCHO + R'CHO$

   ii. $\underset{\overset{|}{OH}}{RCHCOR'} \longrightarrow RCOOH + R'CHO$

   iii. $RCOCOR' \longrightarrow RCOOH + R'COOH$

   Other oxidising agents, for vicinal and glycolic oxidation are sodium periodate $NaIO_4$, and lead tetra-acetate, $Pb(CH_3COO)_4$.

6. Chromyl chloride (CrO_2Cl_2) (used in ethereal solution)

   i.   toluene   benzyl alcohol   benzaldehyde

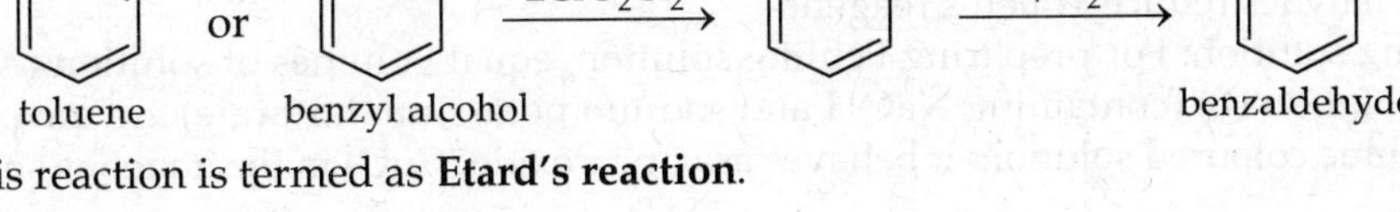

   This reaction is termed as **Etard's reaction**.

    ii. On treating toluene with chromium oxide and acetic anhydride, toluene is converted to form benzaldehyde.

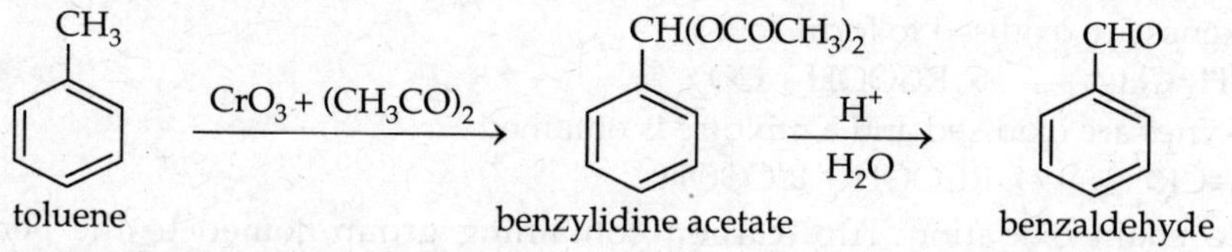

       toluene   benzylidine acetate   benzaldehyde

7. Selenium dioxide (SeO_2)

   i. Acetaldehyde is oxidised to form glyoxal

$$CH_3CHO \longrightarrow OHC—CHO$$

   ii. Acetone is oxidised to form pyruvic aldehyde

$$CH_3COCH_3 \longrightarrow CH_3COCHO$$

   iii. Phenyl methyl ketone is oxidised to give phenyl glyoxal

$$C_6H_5COCH_3 \longrightarrow C_6H_5COCHO$$

   iv. Acetylene is oxidised to form glyoxal

$$CH{\equiv}CH \longrightarrow OHC—CHO$$

   v. 1,2-diphenyl ethane is dehydrogenated to 1,2-diphenyl ethylene

$$C_6H_5CH_2CH_2C_6H_5 \longrightarrow C_6H_5CH{=}CHC_6H_5$$

8. PCC or pyridinium chlorochromate, $C_5H_5NHCrO_3Cl$

   It oxidises a 1° alcohol to an aldehyde; no further oxidation

$$RCH_2OH \longrightarrow RCHO$$

9. Sarret–Collin's reagent (CrO_3 + pyridine in CH_2Cl_2)

$$RCH_2OH \longrightarrow RCHO \text{ (no further oxidation)}$$

10. Ozonolysis: Ozone reacts with the compounds containing a double bond or a triple bond and forms ozonides.

   These ozonides on further hydrolysis are decomposed to give oxidised products.

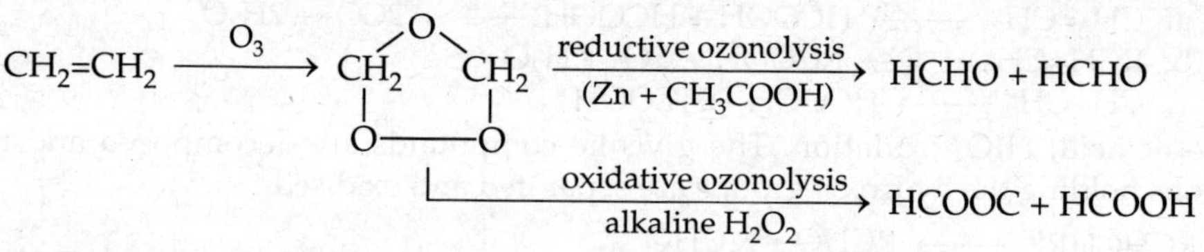

   Alkynes forms an ozonide, which on hydrolysis forms diketones

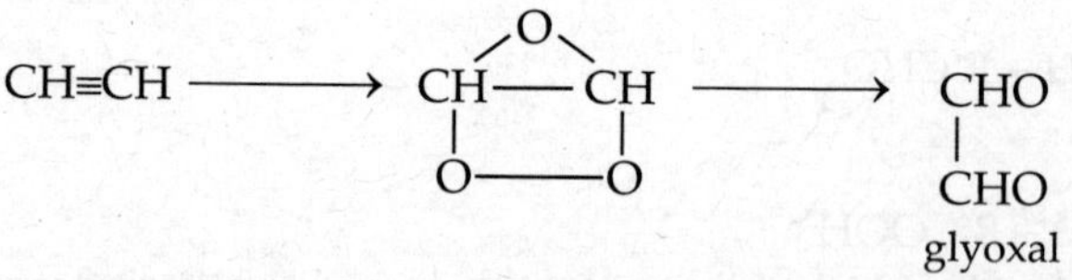

                                                glyoxal

11. Tollen's reagent: An aqueous solution of $AgNO_3$ is treated with excess of ammonia to get a clear, colourless solution. It behaves as Ag_2O in aqueous solution.

$$RCHO \text{ or } ArCHO + Ag_2O \longrightarrow RCOOH \text{ or } ArCOOH + 2Ag$$

                                                                       grey
                                                                     powder

   All aldehydes reduce Tollen's reagent.

12. Fehling solution: For preparing Fehling solution, equal volumes of solution (A) ($CuSO_4$), and solution (B) (containing NaOH and sodium potassium tartarate) are mixed to give a deep blue coloured solution. It behaves as cupric oxide (CuO) in the aqueous solution.

$RCHO + 2CuO \longrightarrow RCOOH + Cu_2O$

A red coloured powder of Cu_2O is formed

Aromatic nuclear substituted aldehydes do not reduce Fehling solution.

13. Sodium hypohalite, NaOI (I_2 + NaOH): It oxidises the compounds containing a terminal-$COCH_3$ or a $CH(OH)CH_3$ groups to form chloroform (CHI_3).

$$RCHCH_3 + NaOI \longrightarrow RCOOH + CHI_3$$
$$\overset{|}{OH}$$

$RCOCH_3 + NaOI \longrightarrow RCOOH + CHI_3$

The formation of CHI_3 is termed haloform reaction.

14. Waker's process or reaction: It involves oxidation of an alkene to an aldehyde.

$CH_2=CH_2 + PdCl_2 + H_2O \longrightarrow CH_3CHO + Pd + 2HCl$

15. Oxo process: In this process an alkene is oxidised to a higher aldehyde in presence of the catalyst, cobalt hydro tetra carbonyl at the high temperature and pressure.

$$CH_3CH=CH_2 \xrightarrow[\text{high temp. and pressure}]{CoH(CO)_4} CH_3CH_2CHO + (CH_3)_2CHCHO$$
$$\qquad\qquad\qquad\qquad\qquad\qquad\text{propionaldehyde} \quad \text{isopropionaldehyde}$$

16. Baeyer's villiger reaction: In presence of peroxy acid, ketone is oxidised to an ester.

$$RCOR' \xrightarrow{\text{peroxy acid}} RCOOR'$$

It involves the oxidation of a ketone, both aromatic as well as alicyclic ketones to esters (lactones), by means of peroxy benzoic acid, peroxy acetic acid, or trifluoro peroxyacetic acid.

17. Oppenauer oxidation: An oxidation process by which a secondary alcohol is oxidised to a ketone, by treating the secondary alcohol with the catalyst, alminum tertiary butoxide in presence of acetone.

$$\underset{\text{oxidised}}{R_2CHOH} + \underset{\text{reduced}}{CH_3COCH_3} \longrightarrow R_2CO + CH_3\overset{|}{C}HCH_3$$
$$\qquad\qquad\qquad\qquad\qquad\qquad\qquad\qquad\qquad OH$$

The method is useful for oxidising an unsaturated secondary alcohol to a ketone, as it does not affect the double bond.

18. Fenton's reagent: It is an aqueous solution of $FeSO_4$ or Fe $(OCOCH_3)_2$ containing H_2O_2. It acts as an oxidising agent.

$$Fe_2{}^+ + HO-OH \longrightarrow Fe_3{}^+ + OH + OH^-$$
$$\qquad\qquad\qquad\qquad\qquad\text{Hydroxyl}$$
$$\qquad\qquad\qquad\qquad\qquad\text{radical}$$

Hydroxyl radical oxidises some organic compounds:

i. Oxidation of aromatic ring: Benzene is hydroxylated to form phenol.

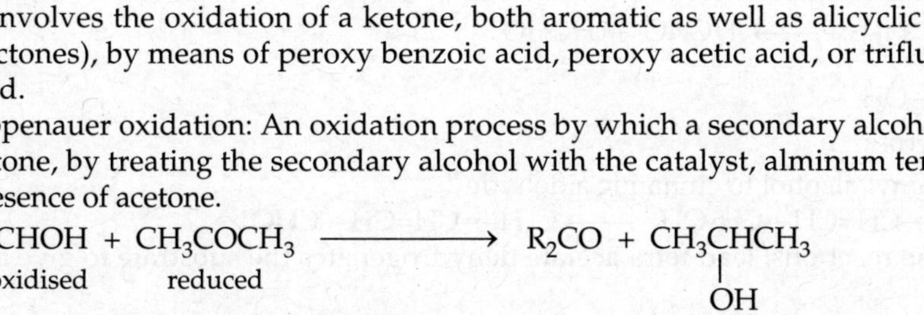

Yield of phenol is poor.

ii. Oxidation of tartaric acid: Fenton's reagent oxidises tartaric acid to dihydroxy fumaric acid.

$$
\begin{array}{ccc}
\text{CHOH—COOH} & & \text{HOOC—C—OH} \\
| & \longrightarrow & \| \\
\text{CHOH—COOH} & & \text{OH—C—COOH}
\end{array}
$$

   tartaric acid   dihydroxy fumaric acid

iii. Oxidation of glycol: Fenton's reagent oxidises glycol to glycolic aldehyde.

$$
\begin{array}{ccc}
\text{CH}_2\text{—OH} & & \text{CH}_2\text{—OH} \\
| & \longrightarrow & | \\
\text{CH}_2\text{—OH} & & \text{CHO}
\end{array}
$$

   glycol   glycolic aldehyde

19. Lead tetra acetate, $Pb(OCOCH_3)_4$: It oxidises the following compounds, due to its ability to dehydrogenate the molecule.

   i. 1° alcohol is oxidised to aldehyde.

   $$RCH_2OH + Pb(OCOCH_3)_4 \longrightarrow RCHO + Pb(OCOCH_3)_2 + 2CH_3COOH$$

   ii. 2° alcohol is oxidised to ketone.

   $$R_2CHOH \longrightarrow R_2CO$$

   iii. Glycol is oxidised to form formaldehyde

   $$
   \begin{array}{c}
   \text{CH}_2\text{—OH} \\
   | \\
   \text{CH}_2\text{—OH}
   \end{array}
   \longrightarrow \text{HCHO} + \text{HCHO}
   $$

   glycol

   iv. Cinnamyl alcohol to cinnamic aldehyde

   $$C_6H_5\text{—CH=CH—CH}_2\text{OH} \longrightarrow C_6H_5\text{—CH=CH—CHO}$$

   In all these reactions, lead tetra acetate dehydrogenates the substrate to give an oxidised product.

Oxidation of Compounds

1. Oxidation of primary alcohol, RCH_2OH

   i. $RCH_2OH \xrightarrow{\text{Acidified KMnO}_4 \text{ or K}_2\text{Cr}_2\text{O}_7} RCHO \longrightarrow RCOOH$

      a. Sarret–Collins reagent
      b. or PCC (CrO_3 + Pyridine + HCl)
      c. or Jone's reagent (CrO_3 + aqueous acetone)

   ii. $RCH_2OH \longrightarrow RCHO$

   iii. $RCH_2OH \xrightarrow{\text{Cu at 300°C}} RCHO + H_2$

2. A secondary alcoholic group, $[R_2CH(OH)]$

   i. $R_2CH(OH) \xrightarrow{\text{Acidified K}_2\text{Cr}_2\text{O}_7} RCOR$

   ii. $R_2CH(OH) \xrightarrow{\text{Jone's reagent (CrO}_3 \text{ + Aq. acetone)}} RCOR$

   iii. $R_2CH(OH) \xrightarrow{\text{Cu at 300°C}} RCOR + H_2$

3. A tertiary group $[R_3\text{—C(OH)}]$

   i. $R_3C(OH) \xrightarrow{\text{Conc. HNO}_3 \text{ or Cu at 300°C}} RCOR$

4. Oxidation of glycols: $RCH(OH)$—$(HO)HCR$

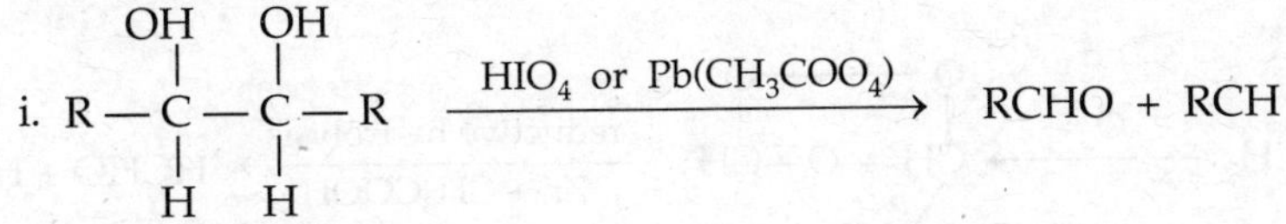

   i. $R—\overset{\displaystyle OH}{\underset{\displaystyle H}{C}}—\overset{\displaystyle OH}{\underset{\displaystyle H}{C}}—R \xrightarrow{\text{HIO}_4 \text{ or Pb(CH}_3\text{COO}_4)} RCHO + RCH$

   ii. $R—\overset{\displaystyle HO}{\underset{\displaystyle H}{C}}—\overset{\displaystyle OH}{\underset{\displaystyle H}{C}}—\overset{\displaystyle OH}{\underset{\displaystyle H}{C}}—R \xrightarrow{\text{HIO}_4 \text{ or Pb(CH}_3\text{COO}_4)} RCHO + HCOOH—RCHO$

   iii. $R—CO—CO—R \xrightarrow{\text{HIO}_4 \text{ or Pb(CH}_3\text{COO}_4)} RCOOH + RCOOH$

     (HIO_4 does not react or oxidise a double bond)

3. Aldehydes and ketones

   i. $RCHO \xrightarrow{\text{Tollen's reagent (Ag}_2\text{O)}} RCOOH + 2Ag$

   ii. $RCHO \xrightarrow{\text{Fehling solution (CuO)}} RCOOH + Cu_2O$

   iii. $CH_3—CHO \xrightarrow{\text{SeO}_2 \text{ (selenium dioxide)}} CHO—CHO$

   iv. $RCOCH_3 \xrightarrow{\text{NaOI (I}_2 + \text{NaOH)}} CHI_3 + RCOOH$

   v. $R—CHOH—CH_3 \xrightarrow{\text{Haloform reaction}} CHI_3 + RCOOH$

4. Aromatic compounds containing a side chain are oxidised to benzoic acid.

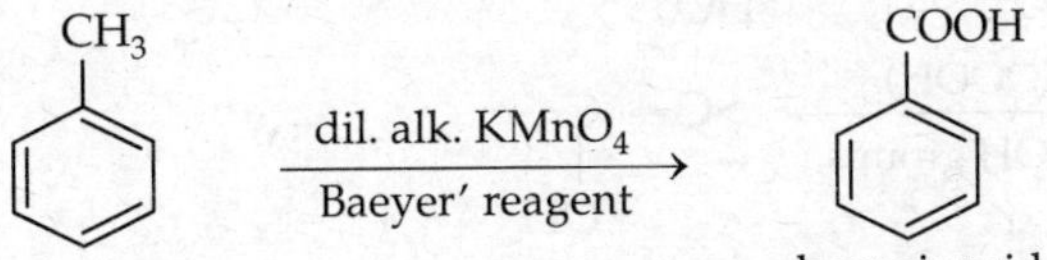

   i. Toluene is oxidised to benzaldehyde

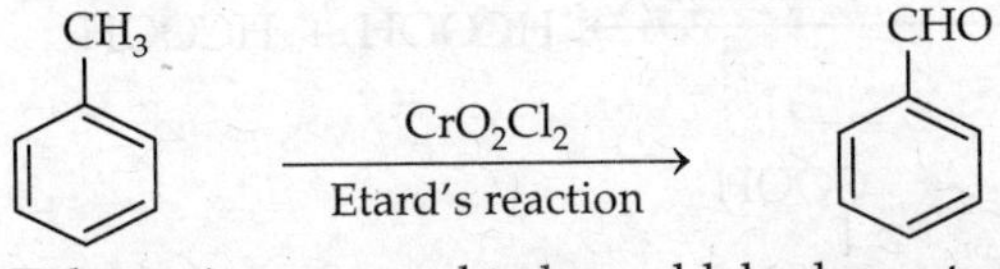

   ii. Toluene is converted to benzaldehyde, on treatment with CrO_3 and acetic anhydride

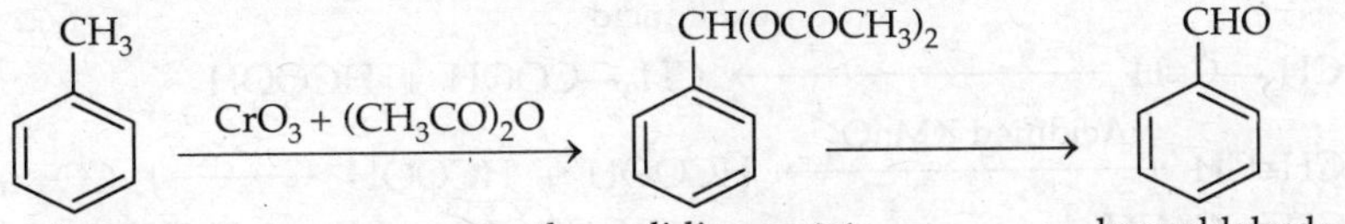

   iii. Oxidation with $V_2O_5 + O_2$

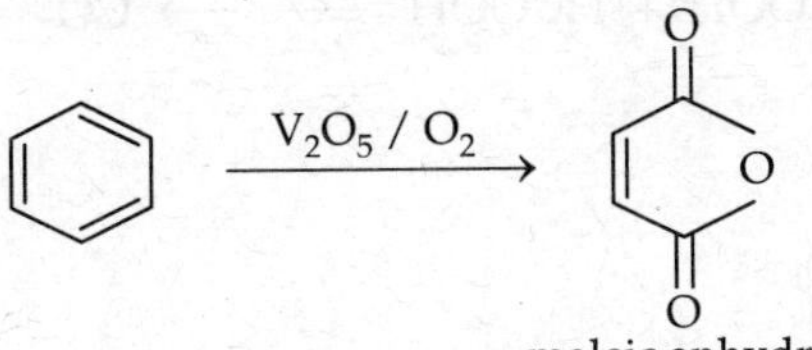

5. Alkenes and alkynes
   i. Ozonolysis

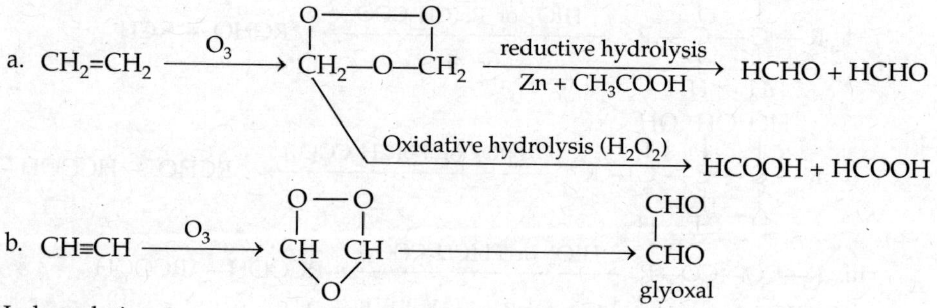

a. $CH_2=CH_2$ $\xrightarrow{O_3}$ $\underset{\underset{\displaystyle O}{|}}{CH_2}\text{—}O\text{—}\underset{\underset{\displaystyle O}{|}}{CH_2}$ $\xrightarrow[Zn + CH_3COOH]{\text{reductive hydrolysis}}$ HCHO + HCHO

$\xrightarrow{\text{Oxidative hydrolysis } (H_2O_2)}$ HCOOH + HCOOH

b. $CH{\equiv}CH$ $\xrightarrow{O_3}$ CH CH $\longrightarrow$ $\underset{\displaystyle CHO}{\overset{\displaystyle CHO}{|}}$
glyoxal

   ii. Hydroxylation
   Alkenes:

a. $CH_2=CH_2$ $\xrightarrow[\text{Baeyer's reagent}]{\text{dil. alk. } KMnO_4}$ $\underset{\displaystyle \text{cis-glycol}}{\overset{\overset{\displaystyle OH \quad OH}{|\quad\;\;|}}{CH_2\text{—}CH_2}}$

b. $CH_2=CH_2$ $\xrightarrow{\text{(i) } OsO_4 \text{ (ii) aq. } NaHSO_3}$ $\underset{\displaystyle \text{cis-glycol}}{\overset{\overset{\displaystyle OH \quad OH}{|\quad\;\;|}}{CH_2\text{—}CH_2}}$

c. $>C=C<$ $\xrightarrow[\text{trans-addition of –OH groups}]{\text{peroxy acid } (RCOOOH)}$ $\underset{\displaystyle \underset{\displaystyle \text{trans-glycol}}{OH}}{\overset{\displaystyle HO}{\underset{\displaystyle |}{\overset{\displaystyle |}{>C\text{—}C<}}}}$

d. $CH_2=CH_2$ $\xrightarrow{\text{hot } KMnO_4 \text{ or acidified } K_2Cr_2O_7}$ HCOOH + HCOOH
   Alkynes:

$CH{\equiv}CH$ $\xrightarrow{\text{dil. alk. } KMnO_4}$ $\underset{\displaystyle \text{oxalic acid}}{\overset{\overset{\displaystyle COOH}{|}}{COOH}}$

$CH_3\text{—}C{\equiv}H$ $\longrightarrow$ $CH_3\text{—}COOH$ + HCOOH

$CH{\equiv}CH$ $\xrightarrow{\text{Acidified } KMnO_4}$ HCOOH + HCOOH $\longrightarrow$ CO_2 + H_2O

$CH_3\text{—}C{\equiv}CH$ $\longrightarrow$ CH_3COOH + CO_2 + H_2O

$R\text{—}CH{\equiv}CH_2$ $\longrightarrow$ RCOOH + HCOOH

$CH{\equiv}CH$ $\longrightarrow$ HCOOH + HCOOH $\longrightarrow$ CO_2 + H_2O

2. Dil. alkaline $KMnO_4$ or Baeyer's reagent

i. $>C=C<$ $\longrightarrow$ $\underset{\displaystyle \text{cis-glycol}}{\overset{\overset{\displaystyle OH \quad OH}{|\quad\;\;|}}{>C\text{—}C<}}$

Reducing Agents

1. Lithium aluminium hydride (LiAlH$_4$)

Group			Reduced to
aldehyde	—CHO	$\longrightarrow$ —CH$_2$OH	primary alcohol
ketone	$>$CO	$\longrightarrow$ $>$CHOH	secondary alcohol
acid	—COOH	$\longrightarrow$ —CH$_2$OH	primary alcohol
ester	—COOR	$\longrightarrow$ —CH$_2$OH + ROH	mixture of primary alcohol and alcohol
acid chloride	—COCl	$\longrightarrow$ —CH$_2$OH	primary alcohol
amide	—CONH$_2$	$\longrightarrow$ —CH$_2$NH$_2$	primary amine
cyanide	—CN	$\longrightarrow$ —CH$_2$NH$_2$	primary amine
acid anhydride	(RCO)$_2$O	$\longrightarrow$ —CH$_2$OH	primary alcohol
aldoxime/ketoxime	$>$C=NOH	$\longrightarrow$ —CH$_2$NH$_2$	primary amine
nitro alkane	RNO$_2$	$\longrightarrow$ RNH$_2$	primary amine
nitro benzene	ArNO$_2$	$\longrightarrow$ ArNH=NHA	azo benzen
primary/secondary halide	RX	$\longrightarrow$ RH	corresponding alkane
tertiary halide	R$_3$CX	$\longrightarrow$ —C=C—	corresponding alkene
Azoxy benzene	Ar—N=N—Ar	$\longrightarrow$ RH	
Only 1° and 2°alkyl halides		$\longrightarrow$ RCH$_2$X/R$_2$CHX	alkanes
3° Alkyl halides	R$_3$C—X	$\longrightarrow$ R$_2$—C=CH$_2$	alkenes

LiAlH$_4$ does not reduce either a double bond or a triple bond.

Reduction by LiAlH$_4$ proceeds via reduction by hydride ion

LiAlH$_4$ $\longrightarrow$ Li$^+$ H$^-$ + AlH$_3$

Li$^+$ H$^-$ $\longrightarrow$ Li$^+$ + H^{-1}

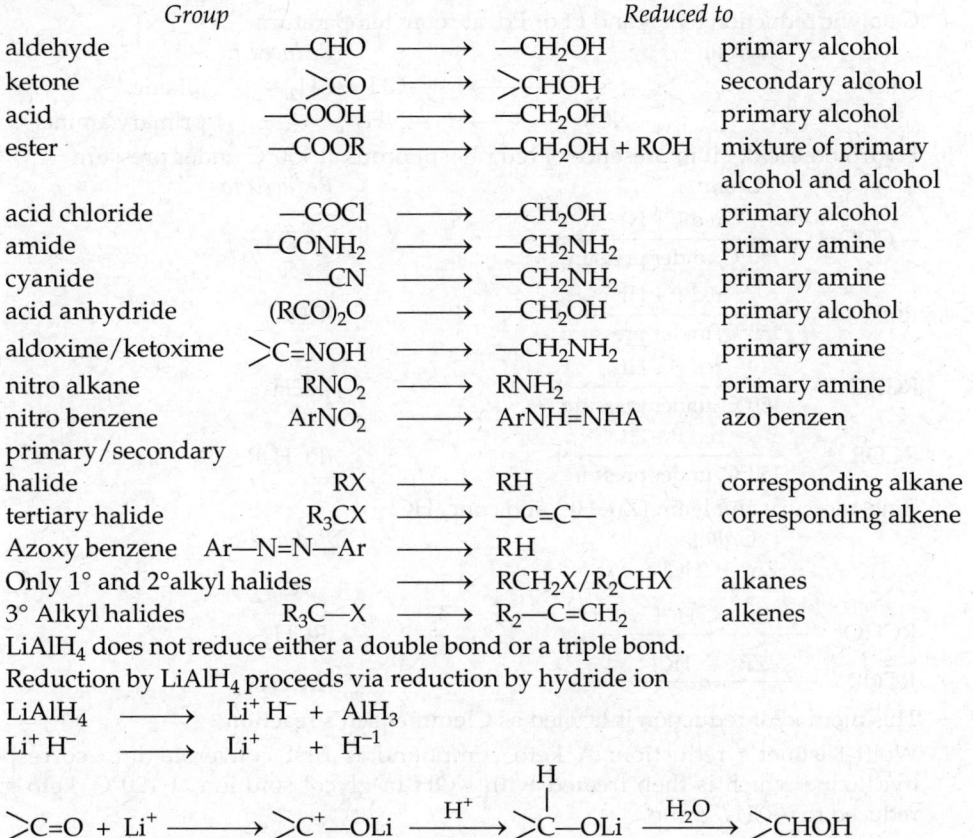

$$>C=O + Li^+ \longrightarrow >C^+\!\!-OLi \xrightarrow{H^+} >\!\!\overset{\overset{\displaystyle H}{|}}{C}\!\!-OLi \xrightarrow{H_2O} >CHOH$$

Reduction of either a double bond or a triple bond proceeds *via* electrophilic addition.

2. Sodium borohydride (NaBH$_4$)

Its reducing properties are like those of LiAlH$_4$, except it does not reduce nitro alkanes, RNO$_2$; nitro arenes, ArNO$_2$; carboxylic acids, RCOOH and esters RCOOR

Group			Reduced to
primary/secondary halide	RX	$\longrightarrow$ RH	corresponding alkane
aldehyde	—CHO	$\longrightarrow$ —CH$_2$OH	primary alcohol
ketone	$>$CO	$\longrightarrow$ $>$CHOH	secondary alcohol
acid chloride	—COCl	$\longrightarrow$ —CH$_2$OH	primary alcohol

3. Catalytic reduction by H$_2$ and Ni at 300°C

It is carried reducing the compound by hydrogen in the presence of Ni at 300 °C

Group			Reduced to
alkene	$>$C=C$<$	$\longrightarrow$ —CH$_2$—CH$_2$—	alkane
alkyne	—C≡C—	$\longrightarrow$ —CH$_2$—CH$_2$—	alkane
aldehyde	—CHO	$\longrightarrow$ —CH$_2$OH	primary alcohol

ketone	$>$CO $\longrightarrow$	$>$CHOH	secondary alcohol
acid	—COOH $\longrightarrow$	—CH$_2$OH	primary alcohol

4. Catalytic reduction by H$_2$ and Pt or Pd, at room temperature

	Group		*Reduced to*
alkene	$>$C=C$<$ $\longrightarrow$	—CH$_2$—CH$_2$—	alkane
nitro	—NO$_2$ $\longrightarrow$	—NH$_2$	primary amine

5. Hydroiodicacid, HI in presence of red phosphorous at 150°C under pressure

	Group		*Reduced to*
—COOH	$\xrightarrow[\text{150°C, under pressure}]{\text{red P + HI}}$		—CH$_3$
ROH	$\xrightarrow[\text{150°C, under pressure}]{\text{red P + HI}}$		RH
RCHO	$\xrightarrow[\text{150°C, under pressure}]{\text{red P + HI}}$		RCH$_3$
RCOR	$\xrightarrow[\text{150°C, under pressure}]{\text{red P + HI}}$		RCH$_2$R

6. Zinc–mercury amalgam (Zn–Hg) and conc. HCl

	Group	*Reduced to*
$>$CO	$\xrightarrow{\text{Zn + HCl}}$	$>$CH$_2$
RCHO	$\xrightarrow{\text{Zn + HCl}}$	RCH$_2$
RCOR	$\xrightarrow{\text{Zn + HCl}}$	RCH$_2$R

This method of reduction is termed as **Clemmensen's reaction**

7. Wolff-Kishner's reduction: A keto compound is first converted into corresponding hydrazine which is then treated with KOH in glycol solution at 190°C, keto group is reduced to —CH$_2$ group.

$>$C=O + NH$_2$NH$_2$ $\longrightarrow$ $>$C=N—NH$_2$ $\xrightarrow[\text{190°C}]{\text{KOH + glycol}}$ $>$CH$_2$

8. Sodium metal and ethyl alcohol or metal hydride
C$_2$H$_5$OH + Na $\longrightarrow$ C$_2$H$_5$ONa + H

	Group		*Reduced to*
isocyanide	—NC + 4H	$\longrightarrow$	—NHCH$_3$
cyanie	—CN + 4N	$\longrightarrow$	—CH$_2$—NH$_2$
amide	—CONH$_2$ + 4H	$\longrightarrow$	—CH$_2$—NH$_2$

The reduction of cyanide, isocyanide and amide by sodium metal and ethyl alcohol is known as **Mendius reaction**.

b. Reduction of 2-butyne by sodium and alcohol, leads to the formation of *trans*-butene.

CH$_3$C≡CCH$_3$ $\xrightarrow{\text{C}_2\text{H}_5\text{OH + Na}}$

$\begin{array}{c} \text{H}_3\text{C} \qquad \text{H} \\ \diagdown \;\;\;\; \diagup \\ \text{C=C} \\ \diagup \;\;\;\; \diagdown \\ \text{H} \qquad \text{CH}_3 \end{array}$

but-2-yne

trans-but-2-ene

Reduction of 2-butyne to *cis*- or *trans*-butene is termed **stereospecific reaction**.

c. Reduction of an aldehyde, a ketone or an ester, by sodium and alcohol is termed **Bouveault–Blanc reduction or reaction**.

	Group		*Reduced to*	
aldehyde	—CHO	$\longrightarrow$	—CH$_2$OH	(primary alcohol)
ketone	>CO	$\longrightarrow$	>CHOH	(secondary alcohol)
ester	—COOR	$\longrightarrow$	—CH$_2$OH + ROH	(mixture of primary alcohol)

9. Stannous chloride and conc. HCl

$$SnCl_2 + 2HCl \longrightarrow SnCl_4 + 4H$$

$$NO_2 \xrightarrow{SnCl_2 + HCl} —NH_2$$

Reduction of cyanides by stannous chloride and conc. HCl followed by hydrolysis leads to the formation of an aldehyde

$$\underset{\text{cyanide}}{—CN} \xrightarrow[H_2O]{SnCl_2 + HCl} \underset{\text{aldehyde}}{—CHO}$$

This reaction is termed **Stephen's reaction.**

10. Iron and HCl

$$Fe + 2HCl \longrightarrow FeCl_2 + 2H$$

$$\text{nitro} —NO_2 + 6H \longrightarrow —NH_2 \ (1° \text{ amine})$$

11. Zn dust + H$_2$O and NH$_4$Cl

$$Zn + 2H_2O \xrightarrow{NH_4Cl} Zn(OH)_2 + 2H$$

$$\underset{\text{nitrobenzene}}{C_6H_5NO_2} \xrightarrow[NH_4Cl]{Zn + H_2O} \underset{\text{N-phenyl hydroxyl amine}}{C_6H_5NHOH}$$

12. Lindler's catalyst: Hydrogen + Pd + BaSO$_4$, poisoned with sulphur or quinoline

    a. Reduction of 2-butyne gives a *cis*-2-butene

$$\underset{\text{but-2-yne}}{CH_3C≡CCH_3} \xrightarrow[Pd^- \ BaSO_4 + S]{\text{Lindlar's catalyst}} \underset{\textit{cis-but-2-ene}}{\overset{\displaystyle H \diagdown \quad \diagup H}{\underset{\displaystyle H_3C \diagup \quad \diagdown CH_3}{C=C}}}$$

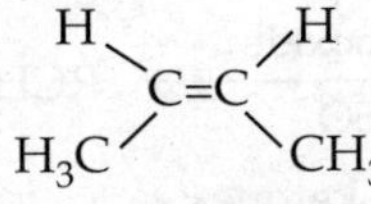

    b. Reduction of an acid chloride, RCOCl to an aldehyde by Lindler's catalyst is termed **Rosenmund's reaction.**

$$\underset{\text{acid chloride}}{RCOCl + H_2} \xrightarrow[Pd^- \ BaSO_4 + S]{\text{Lindlar's catalyst}} \underset{\text{aldehyde}}{RCHO}$$

13. Li or Na in presence of liquid ammonia

    Terminal alkenes are reduced by sodium and liquid NH$_3$, to corresponding alkanes. This reaction is termed **Birch reaction.**

$$C_2H_5OH + Na \longrightarrow C_2H_5ONa + H$$

$$RCH=CH_2 \xrightarrow{C_2H_5OH + Na} RCH_2CH_3$$

Reduction of Groups

1. An olefinic bond >C=C<

       Group   *Reduced to*

    a. $>C=C< \xrightarrow{H_2/Ni \text{ at } 300 \ °C} —CH_2—CH_2 \text{ (alkane)}$

    b. $>C=C< \xrightarrow{H_2/Pd \text{ or } Pt \text{ or Raney Ni}} —CH_2—CH_2 \text{ (alkane)}$

At room temperature

c. $>C=C< \xrightarrow[\text{(2) } CH_3COOH]{\text{(1) } BH_3} -CH_2-CH_2$ (alkane)

d. $R-CH=CH_2 \xrightarrow{Na + C_2H_5OH} R-CH_2-CH_3$ (alkane)

Terminal alkenes are reduced by sodium metal and alcohol to alkenes. This reaction is termed Birch reaction.

2. An acetylinic bond $-C≡C-$

   Group   *Reduced to*

   a. $-C≡C- \xrightarrow[\text{H}_2 + \text{Pd or Pt at room temp.}]{\text{H}_2 + \text{Ni at 300°C, or}} -CH_2-CH_2$ (alkane)

   b. $CH_3-C≡C-CH_3 \xrightarrow[\text{Pd + BaSO}_4 \text{ + sulphur}]{\text{Lindler's catalyst CH}_3\text{CH}_3}$ $\begin{array}{c} CH_3 \\ \diagdown \\ H \diagup \end{array} C=C \begin{array}{c} CH_3 \\ \diagup \\ \diagdown H \end{array}$

   2-butyne   *cis*-2-butene

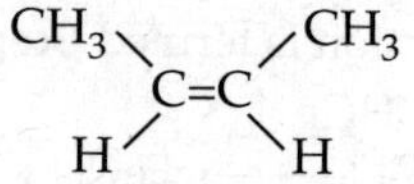

   c. $CH_3-C≡C-CH_3 \xrightarrow[\text{Li or Na + liquid NH}_3]{\text{Na + C}_2\text{H}_5\text{OH or}}$ $\begin{array}{c} CH_3 \\ \diagdown \\ H \diagup \end{array} C=C \begin{array}{c} H \\ \diagup \\ \diagdown CH_3 \end{array}$

   2-butyne   *trans*-2-butene

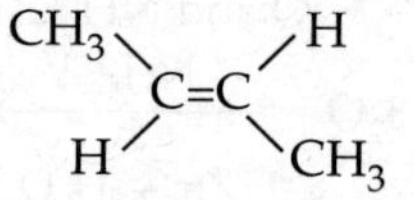

3. Alkyl halides and aryl halides

   $RX \xrightarrow[\text{LiAlH}_4]{\text{Zn couple + alcohol or}} RH$

   $RCH_2l \xrightarrow[\text{(CH}_3)_2\text{SO}]{\text{dimethyl sulphoxide}} RCHO$

4. Keto group in aldehydes ketones:

   a. $>C=O \xrightarrow{\text{Zn + Hg conc. HCl}} >CH_2$

   This reduction is known as **Clemmensen's reaction.**

   b. $>C=O \xrightarrow{\text{(1) NH}_2-\text{NH}_2/\text{(2) KOH + glycol at 150°C}} >CH_2$

   This reduction is known as **Wolff–Kishner reduction**

   c. $RCHO \xrightarrow[\text{LiAlH}_4 \text{ or NaBH}_4]{\text{H}_2/\text{Ni at 300°C}} RCH_2OH$

   aldehyde   1° alcohol

   d. $RCOR \xrightarrow{\hspace{2cm}} RCH_2R$

   ketone   alkane

5. A carboxylic—(COOH) group

   a. $RCOOH \xrightarrow[\text{Under pressure}]{\text{HI + red P/150°C}} RCH_3$

   alkane

   c. $RCOOH \xrightarrow{\text{LiAlH}_4 \text{ or NaBH}_4} RCH_2OH$

   1° alcohol

6. Nitro group—(NO_2) group

   a. $-NO_2 \xrightarrow{\text{Fe + HCl or SnCl}_2\text{ + HCl}} -NH_2$

   amino group

   b. $-NO_2 \xrightarrow{\text{Zn dust + H}_2\text{O/NH}_4\text{Cl}} -NHOH$

   hydroxyl amino group

   c. $C_6N_5NO_2 \longrightarrow C_6H_5-N=N-C_6H_5$

   azobenzene

7. A cyanide (—C≡N), isocyanide (—NC) and an amino (—$CONH_2$) group
   All these groups are reduced by Na + C_2H_5OH or $LiAlH_4$

   a. $-C\equiv N \longrightarrow -CH_2NH_2$

   1° amine

   b. $-NC \longrightarrow -NHCH_3$

   2° amine

   c. $-CONH_2 \longrightarrow -CH_2NH_2$

   1° amine

Reduction by sodium and alcohol is termed **Mendius reaction.**

MULTIPLE CHOICE QUESTIONS

1. $LiAlH_4$ is used as a/an
   - (a) oxidising agent
   - (b) reducing agent
   - (c) mordant
   - (d) water softener
2. The reduction of an alkyne to an alkene using $LiAlH_4$ results into
   - (a) *cis*-addition of hydrogen
   - (b) *trans*-addition of hydrogen
   - (c) a mixture obtained by addition of *cis*- and *trans*-addition, which are in equilibrium
   - (d) only a mixture of *cis* and *trans*
3. Cyclopentene on oxidation with alk. $KMnO_4$ gives
   - (a) cyclopentanol
   - (b) *trans*-1,2-cyclopentane diol
   - (c) *cis*-cyclopentane diol
   - (d) a mixture of *cis*- and *trans*- cyclopenane diol
4. Ethylene on treatment with 1% cold dil. $KMnO_4$ gives
   - (a) oxalic acid
   - (b) acetic acid
   - (c) glycerol
   - (d) ethylene glycol
5. The hydrocarbon that decolourises Baeyer's reagent, does not give any precipitate with amm. $AgNO_3$
   - (a) alkene
   - (b) alkane
   - (c) alkyne
   - (d) benzene
6. The most oxidising hydrocarbon is
   - (a) CO_2
   - (b) RCHO
   - (c) RCOOH
   - (d) RCOOOH

7. Reduction of 2-butyne with sodium in liquid ammonia gives predominantly
    (a) butane
    (b) *cis*-2-butene
    (c) *trans*-2-butene
    (d) none

8. Which will oxidise CH_3—CH=CH—CHO to CH_3—CH=CH—COOH
    (a) alk. $KMnO_4$
    (b) amm. $AgNO_3$
    (c) SeO_2
    (d) OsO_4

9. Acetylene and ethylene on reaction with dil. $KMnO_4$ will give
    (a) oxalic acid and formic acid
    (b) acetic acid and ethylene glycol
    (c) ethyl alcohol and ethylene glycol
    (d) oxalic acid and ethylene glycol

10. The reaction CH_3—C≡C—CH_3 + 2H gives a *trans*-2-butene, the reducing agent is
    (a) Linder's regent
    (b) ROH + Na
    (c) NH + Na
    (d) none

11. Alk. $KMnO_4$, is reacted with R—CH=CH—R, the compound formed is
    (a) RCHO
    (b) RCOOH
    (c) RCH(OH)—CH(OH)R
    (d) $CO_2 + H_2O$

12. Hydrogenation of the following compound is a result of

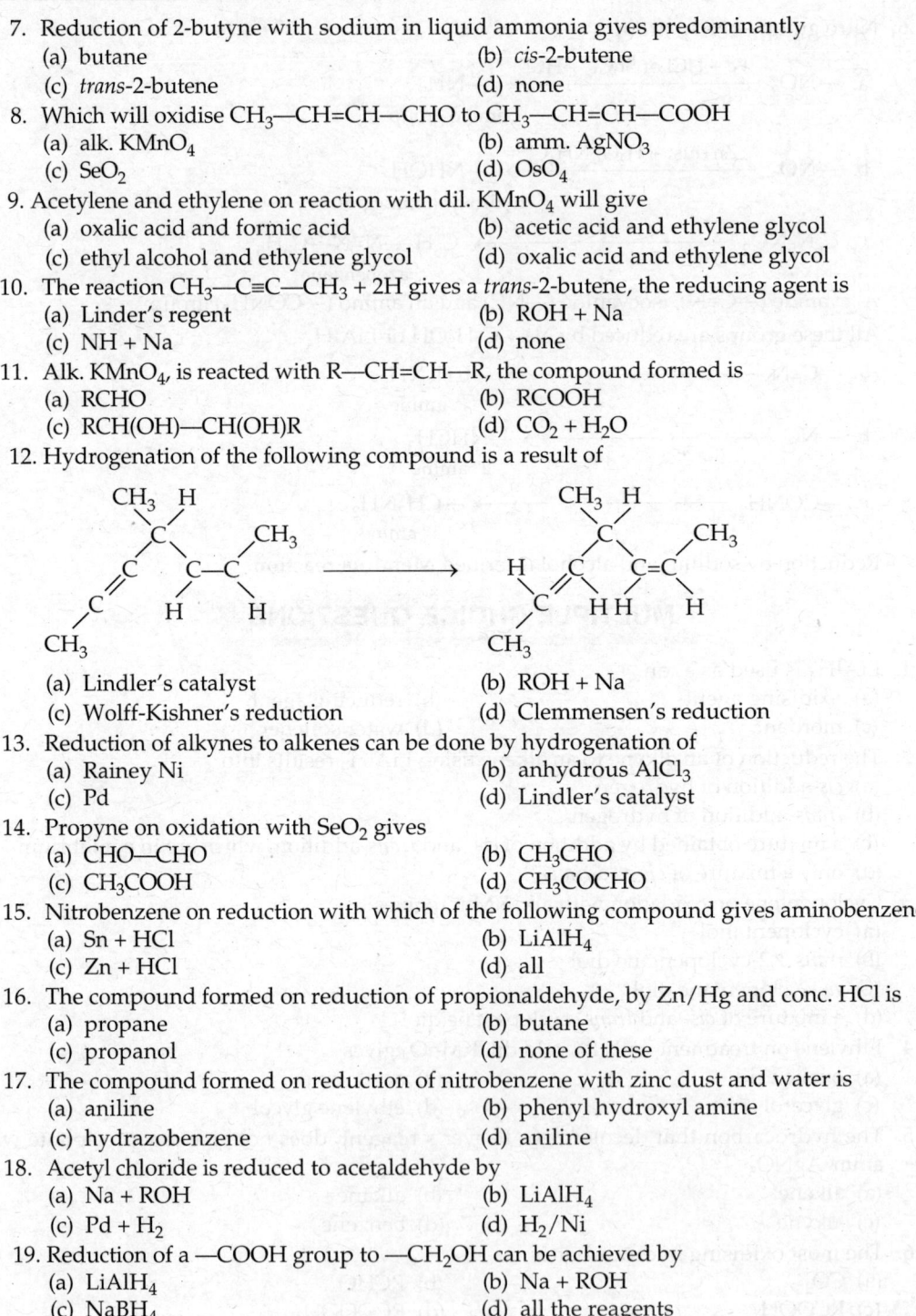

    (a) Lindler's catalyst
    (b) ROH + Na
    (c) Wolff-Kishner's reduction
    (d) Clemmensen's reduction

13. Reduction of alkynes to alkenes can be done by hydrogenation of
    (a) Rainey Ni
    (b) anhydrous $AlCl_3$
    (c) Pd
    (d) Lindler's catalyst

14. Propyne on oxidation with SeO_2 gives
    (a) CHO—CHO
    (b) CH_3CHO
    (c) CH_3COOH
    (d) CH_3COCHO

15. Nitrobenzene on reduction with which of the following compound gives aminobenzene
    (a) Sn + HCl
    (b) $LiAlH_4$
    (c) Zn + HCl
    (d) all

16. The compound formed on reduction of propionaldehyde, by Zn/Hg and conc. HCl is
    (a) propane
    (b) butane
    (c) propanol
    (d) none of these

17. The compound formed on reduction of nitrobenzene with zinc dust and water is
    (a) aniline
    (b) phenyl hydroxyl amine
    (c) hydrazobenzene
    (d) aniline

18. Acetyl chloride is reduced to acetaldehyde by
    (a) Na + ROH
    (b) $LiAlH_4$
    (c) Pd + H_2
    (d) H_2/Ni

19. Reduction of a —COOH group to —CH_2OH can be achieved by
    (a) $LiAlH_4$
    (b) Na + ROH
    (c) $NaBH_4$
    (d) all the reagents

20. Which reducing agent will convert the following cycloketone to cycloalkane
    (a) $P + I_2$
    (b) $NH_2—NH_2 + KOH$
    (c) $Na + ROH$
    (d) none

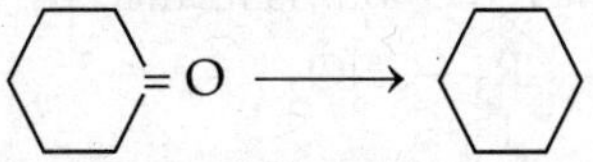

21. Fehling solution will be reduced by
    (a) sodium formate
    (b) soldium acetate
    (c) NaCl
    (d) $NaNO_3$

22. Which will oxidise butyne-2 to di-acetyl?
    (a) HNO_3
    (b) O_2
    (c) O_3
    (d) $KMnO_4$

23. 2-Methyl propene is isomeric with butene-1 by which of the following aspects they differ.
    (a) Baeyer's reagent
    (b) Tollen's reagent
    (c) Br_2 solution
    (d) $O_3/Zn + H_2O$

24. The order of relative strength of water, ammonia and ethyne
    (a) $H_2O > C_2H_4 > NH_3$
    (b) $C_2H_4 > H_2O > NH_3$
    (c) $C_2H_4 > NH_3 > H_2O$
    (d) $NH_3 > H_2O > C_2H_4$

25. The reducing agent used for reducing acetone to propane is
    (a) H_2SO_4
    (b) HCl
    (c) $P + I_2$
    (d) MnO_2

26. The reduction of phenol gives
    (a) cyclohexanol
    (b) benzene
    (c) benzyl alcohol
    (d) none

27. The following reaction can be brought about by

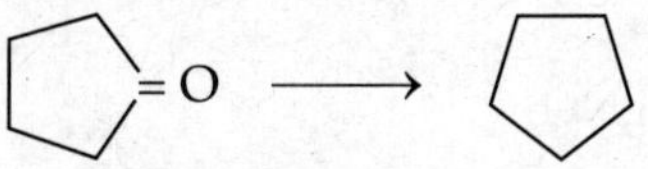

    (a) reduced $P + I_2$
    (b) Wolff-Kishner's reduction
    (c) Clemmensen's reduction
    (d) all

28. What will be the reduction products of $CH_3COOC_2H_5$, when reduced by $LiAlH_4$?
    (a) 5 molecules of ethanol
    (b) 3 molecules of ethanol
    (c) 2 molecules of alcohol
    (d) 4 molecules of alcohol

29. The reduction of carbonyl group to a $—CH_2$ group by $NH_2—NH_2$ and KOH is called
    (a) Clemmensen's reduction
    (b) Wolff–Kishner's reduction
    (c) Pondrof–Verley reduction
    (d) Wurtz reaction

30. Ethanol and isopropanol can be distinguished by
    (a) Lucas reagent
    (b) iodoform test
    (c) CAN test
    (d) none of these

31. A neutral compound gives red colour with ceric ammonium nitrate (CAN), the compound could be
    (a) an alcohol
    (b) an aldehyde
    (c) an ether
    (d) a ketone

32. Luca's reagent consists of a mixture of
    (a) conc. HCl + anhydrous $ZnCl_2$
    (b) conc. HCl + hydrated $ZnCl_2$
    (c) conc. $HNO_3 + ZnCl_2$
    (d) HCl + hydrated $ZnCl_2$

33. Luca's reagent is used to distinguish between
    (a) alcohols
    (b) aldehydes
    (c) amines
    (d) alkyl halides

34. In Victor Mayer's method of distinguishing alcohols, the red colour is formed by

$$RCH_2-CH-NO_2-HNa- \qquad R_2C-NO_2 \qquad R_3-CNO_2$$

$$| \qquad\qquad\qquad\qquad | \qquad\qquad\qquad |$$

NOH   NO   No reaction takes place

   (a)   (b)   (c)

ANSWERS

1. (b)
2. (a) *cis*-alkene

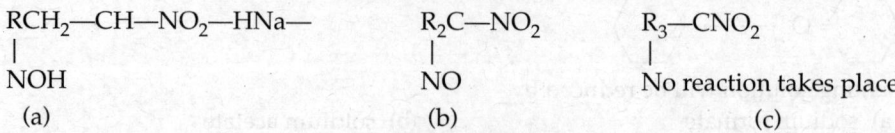

$$RC \equiv CR + H_2 \longrightarrow \begin{matrix} R \\ \diagdown \\ C=C \\ \diagup \quad \diagdown \\ H \qquad H \end{matrix}$$

3. (b) Cyclopentene forms cyclopentane 1,2-diol

4. (d) Ethylene is converted into ethylene glycol

$$CH_2=CH_2 + O + H_2O \longrightarrow \begin{matrix} CH_2-CH_2 \\ | \qquad | \\ HO \quad HO \end{matrix}$$

5. (a) Alkenes do not give any precipitate with amm. $AgNO_3$
6. (d) RCOOOH is the most oxidised product
7. (c) A *trans*-butene is obtained

$$CH_3C \equiv C-CH_3 + H_2 \longrightarrow \begin{matrix} CH_3 \qquad H \\ \diagdown \quad\quad \diagup \\ C=C \\ \diagup \qquad \diagdown \\ H \qquad CH_3 \end{matrix}$$

8. (b) Amm. $AgNO_3$ (Baeyer's reagent) will oxidise it to the corresponding acid

$$CH_3CH=CHCHO \longrightarrow CH_3CH=CH-COOH$$

9. (d) Oxalic acid and ethylene glycol are formed.

$$HC \equiv CH \longrightarrow \begin{matrix} COOH \\ | \\ COOH \end{matrix} \longrightarrow CH_2=CH_2 \longrightarrow \begin{matrix} CH_2-CH_2 \\ | \qquad | \\ HO \quad HO \end{matrix}$$

10. (a) *cis*-alkene is formed

$$CH_3C \equiv CCH_3 + 2H \longrightarrow \begin{matrix} CH_3 \qquad CH_3 \\ \diagdown \quad\quad \diagup \\ C=C \\ \diagup \qquad \diagdown \\ H \qquad H \end{matrix}$$

11. (c) Addition of hydroxyl groups takes place at the triple bond following *cis*-addition, giving rise to the formation of a product that contains a plane of symmetry, therefore the product is optically inactive.

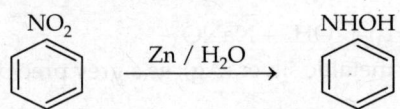

12. (a) Hydrogenation of the compound gives an optically inactive product, using Na/NH_3 as a reducing agent. The triple bond is reduced by the addition of hydrogen in *trans*-manner.

$$C\equiv C \xrightarrow{H_2} \underset{H}{\overset{H}{C=C}}$$

13. (d) A *cis*-compound is formed. No further reduction takes place, because Lindler's catalyst is poisoned.
14. (d) CH_3COCHO
15. (d) All
16. (a) propane
17. (b) Phenyl hydroxyl amine

$$\underset{}{\overset{NO_2}{\bigcirc}} \xrightarrow{Zn\,/\,H_2O} \underset{phenyl\ hydroxyl\ amine}{\overset{NHOH}{\bigcirc}}$$

18. (c) $RCOCl + H_2/Pd \longrightarrow RCHO$, Rosenmund reaction
19. (d) all these reagents
20. (b) $NH_2-NH_2 + KOH$
21. (a) sodium formate, HCOONa
22. (c) O_3; $CH_3-\underset{CH_3}{\overset{CH_3}{C}}=CH_2 + O_3 \longrightarrow CH_3COCH_3$

a nonreducing compound; $CH_3CH_2CH=CH_2 \longrightarrow CH_3CH_2CHO$ it is a reducing compound
23. (c) O_3
24. (a) The relative strength is of the order is $H_2O > C_2H_4 > NH_3$
25. (c) $P + I_2$
26. (a) cyclohexanol
27. (d) all
28. (b) 2 molecules of alcohol
29. (b) Wolff–Kishner's reduction
30. (a)
31. (a)
32. (a)
33. (a)
34. (a)

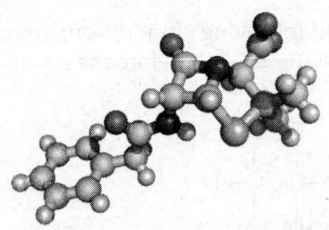

Tests and Group Tests, Rules and Reagents

Reagents and Solutions

i. **Tollen's reagent:** An ammoniacal solution of silver nitrate is termed Tollen's reagent. It is prepared by adding sodium hydroxide drop by drop to $AgNO_3$ solution followed by adding excess of ammonia, as a result a clear solution is obtained. It behaves as Ag_2O in the solution.

$$AgNO_3 + NaOH + 2NH_3 \longrightarrow [Ag(NH_3)_2]^+ OH^- + NaNO_3$$

Aldehydes reduce Tollen's reagent to form metallic silver (Ag), as a grey precipitate.

$$RCHO + Ag_2O \longrightarrow RCOOH + 2Ag$$

ii. **Fehling solution:** It is an alkaline solution containing a complex of copper tartarate, which is reduced by aldehydes to Cu_2O, a brick red coloured powder. The solution is prepared at the time of use only by mixing equal volumes of Fehling solution A (an aqueous solution of $CuSO_4$) and B (an aqueous solution of NaOH and sodium potassium tartarate or Rochelle salt) that result in the formation of a deep blue coloured clear solution.

Fehling solution (A)   Fehling solution (B)

Aqueous $CuSO_4$ solution $2NaOH + CH(OH)COONa$
                                            |
                                    $CH(OH)COOK$ (Rochelle salt)

On mixing equal volumes of solution of A + B

a. $CuSO_4 + 2NaOH \longrightarrow Cu(OH)_2 + Na_2SO_4$

$$Cu{\nearrow^{OH}_{\searrow OH}} \quad + \quad \begin{matrix} HOCH-COONa \\ | \\ HOCH-COOK \end{matrix} \quad \longrightarrow \quad Cu{\nearrow^{O-CHCOONa}_{\searrow O-CH-COOK}}$$

A deep blue coloured copper

Copper chelate compound behaves as CuO in aqueous solution.

b. $RCHO + 2CuO \longrightarrow RCOOH + Cu_2O$

On warming Fehling solution, with an aldehyde, a brick red powder of Cu_2O, or a thin film of metallic Cu is formed. Nuclear substituted aromatic aldehydes do not reduce Fehling solution conclusively.

iii. **Benedict's solution:** It is an alkaline solution of cupric ions, complexed with citrate ions.

It is prepared by mixing equal volumes of Benedict's solution $[Cu(OCOCH_3)_2]$ and NaOH + sodium citrate solution when a deep blue coloured solution is formed, it behaves as CuO.

On warming aldehydes or glucose with Benedict's solution, a brick red coloured powder of Cu_2O is deposited

$$RCHO + 2CuO \longrightarrow RCOOH + Cu_2O$$

This test is given by aliphatic aldehydes only.

iv. **Schiff's reagent:** It is an aqueous solution of magneta or fuschine or rosaniline hydrochloride (red coloured solution) decolourised by passing SO_2 gas through it. This colourless solution is known as Schiff's reagent. When schiff's reagent is treated with aldehydes, it becomes red again or its colour is restored again. Aromatic aldehydes restore the pink colour of Schiff's reagent. *Ketones* do not react with Schiff's reagent. However, acetone restores the pink colour of Schiff's reagent slowly.

v. **Luca's reagent:** It is a solution of $ZnCl_2$ dissolved in conc. HCl. It is used to distinguish between 1°, 2° and 3° alcohols.

a. 3° alcohol or tertiary alcohols produce a cloudiness within 1 minute, on mixing with Luca' reagent (due to the formation of alkyl halide).

$$ROH + HCl \longrightarrow RCl + H_2O$$

b. 2° alcohols or secondary alcohols produce cloudiness in 10–15 minutes on mixing with Luca's reagent

c. 1° alcohols or primary alcohols do not form cloudiness on mixing with Luca's reagent, even after 10–15 minutes.

vi. **Barfoed' s reagent:** It is a solution of copper acetate in acetic acid. It gives a red precipitate with monosaccharides. It behaves as CuO in the solution.

$$2CuO + Glucose \longrightarrow Cu_2O + gluconic\ acid$$

vii. **Millon's reagent:** It is a solution of mercuric nitrate $Hg(NO_3)_2$ and mercurous nitrate $Hg_2(NO_3)_2$ in conc. HNO_3. It is used for detecting proteins. On treating proteins with Millon's reagent, a white precipitate, turning red on heating is formed.

viii. **Fenton' reagent:** It is a solution of $FeSO_4$, containing a little of H_2O_2 used as an oxidising agent.

a. When glycerol is treated with Fenton's reagent, glycerol is oxidised to glyceric aldehyde and dihydroxy acetone.

$$
\begin{array}{lll}
CH_2OH & CHO & CH_2OH \\
| & | & | \\
CHOH \longrightarrow & CHOH & + & CO \\
| & | & | \\
CH_2OH & CH_2OH & CH_2OH \\
\text{glycerol} & \text{glyceric aldehyde} & \text{dihydroxy acetone}
\end{array}
$$

- Glyceraldehyde or glyceric aldehyde is used as a reference compound, for assigning absolute configuration to monosaccharides, such as D and L configurations.
- A mixture of glyceric aldehyde and dihydroxy acetone on warming in presence of weak alkali, condensed to form inactive glucose.

b. Lactic acid is oxidised to pyruvic acid.

$$
\begin{array}{ll}
CH_3CH(OH)COOH \longrightarrow & CH_3CO.COOH \\
\text{lactic acid} & \text{pyruvic acid}
\end{array}
$$

iv. **Sarret–Collin's reagent:** This reagent consists of chromium oxide CrO_3 and pyridine (C_5H_5N) dissolved in dichloromethylene CH_2Cl_2 as a solvent. This reagent oxidises a primary alcohol (aliphatic as well aromatic) to aldehyde, which do not oxidise further.

$$RCH_2OH \longrightarrow RCHO$$

x. **PCC, or pyridinium chloro-chromate**—$(C_5H_5N)Cr(OH)O_2Cl$:

This reagent also consists of CrO_3 + pyridine and HCl. It also oxidises a primary alcohol (aliphatic as well as aromatic) to an aldehyde, which are not oxidised further.

$$RCH_2OH \longrightarrow RCHO$$

xi. **Jone's Reagent:** It consists of a solution of chromic acid in an aqueous acetone. As the reagent is sufficiently mild, it oxidizes a primary alcohol to an aldehyde and a secondary alcohol to a ketone without rearranging or without oxidising the double bond, if present in the compound, but not further.

$$RCH_2OH \longrightarrow RCHO$$

$$R_2CHOH \longrightarrow RCOR$$

$$CH_3CH-CH=CH-CH-(OH) CH_3 \longrightarrow CH_3-CH-CH=CH-CO-CH_3$$

xii. **Lemieux reagent:** It is an aqueous solution of sodium periodate $NaIO_4$ containing a little of $KMnO_4$ (or osmium tetraoxide, OsO_4). The reagent is used for determining the position of a double bond and for preparing a keto compound. Its reaction involves the following steps:

a. First an alkene is oxidised to a *cis*-glycol
b. The *cis*-glycol is then cleaved by peroxide to either an aldehyde or a ketone
c. The aldehyde formed is oxidised to an acid by $KMnO_4$
d. In the case of a terminal alkene, it oxidised to HCHO
e. The reaction proceeds at room temperature

$$RCH=CH_2 \xrightarrow{KMnO_4} R-\overset{\displaystyle OH}{\underset{}{C}H}-\overset{\displaystyle OH}{\underset{}{C}H_2} \xrightarrow{NaIO_4} RCHO + HCHO$$

Instead of $KMnO_4$, osmium trraoxide OsO_4 has also been used and an aldehyde is formed, which is not oxidised further.

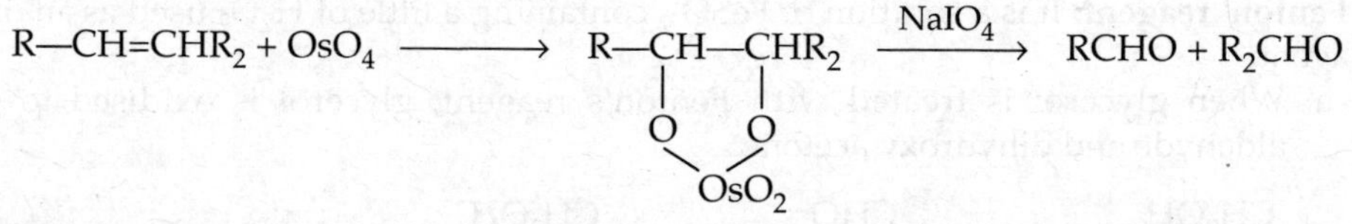

$$R-CH=CHR_2 + OsO_4 \longrightarrow R-\underset{O}{\overset{}{C}H}-\underset{O}{\overset{}{C}HR_2} \xrightarrow{NaIO_4} RCHO + R_2CHO$$
$$OsO_2$$

xiii. **Brady's reagent:** It is an alcoholic solution of 2,4-dinitrophenyl hydrazine. The carbonyl compounds form a yellow or an orange precipitate with Brady's reagent, which is insoluble in HCl.

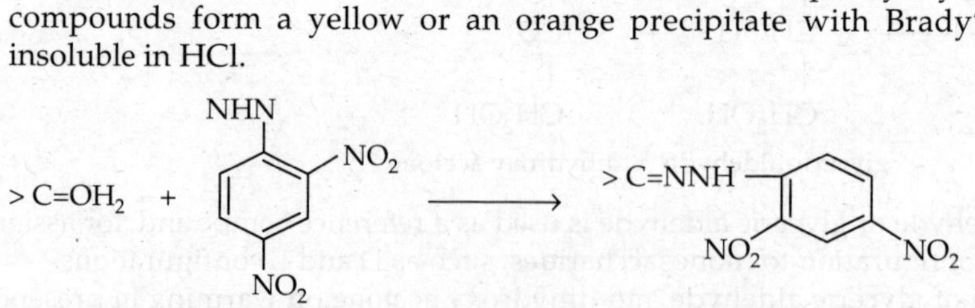

xiv. **Sanger's reagent:** 2,4-dinitro fluoro benzene is known as Sanger's reagent. This reagent used to detect amino acids, with which it reacts to produce yellow coloured dinitrophenyl amino acids.

2,4-dinitro fluoro benzene

xv. **Xanthoproteic test:** On warming a protein with conc. HNO_3 a yellow colour is produced. This test is given by all proteins, as the proteins contain an aromatic ring, which undergoes nitration, producing a yellow product.

Tests and Group Tests

i. **Beilstein's test:** The test is used to find out the presence of halogen in the organic compound. It is a qualitative test and does not indicate which member of the halogen family is present in the organic compound.

When the organic compound is heated, placed on the tip of a copper wire in the flame, appearance of a blue or green flame indicates the presence of a halogen in the organic compound.

Exception: Urea, thiourea (which do not contain halogen) also give this test.

ii. **Lassaigne's test:** This test is used for the detection of the extra elements present in the compound.

Procedure: Organic compound is fused with molten sodium metal and the fused mass is extracted with water. This extract is termed Lassaigne filtrate or LF, it used for detection of elements.

a. **Test for nitrogen:** 1 cc of LF is made alkaline with a few drops of NaOH solution. To this is added 2–3 ml of freshly prepared aqueous solution of $FeSO_4$ and the contents are boiled. After cooling, the contents are acidified with conc. HCl, followed by adding a few drops of a $FeCl_3$ solution. Appearance of a blue or a green colour, confirms the presence of nitrogen in the compound.

$$Na + C + N \longrightarrow NaCN$$
$$2NaCN + FeSO_4 \longrightarrow Fe(CN)_2 + Na_2SO_4$$
$$Fe(CN)_2 + 4NaCN \longrightarrow Na_4Fe(CN)_6$$
$$\underset{\text{sodium ferro cyanide}}{Na_4\,Fe(CN)_6} + 4FeCl_3 \longrightarrow \underset{\text{ferri ferro cyanide}}{Fe_4\,\{Fe(CN)_6\}_3} + 12NaCl$$
$$Na_4\,Fe(CN)_6 + FeCl_3 \longrightarrow \underset{\text{sodium ferri ferro cyanide}}{NaFe\{Fe(CN)_6\}_3} + 3NaCl$$

If sulphur is also present along with nitrogen in the organic compound, on adding $FeCl_3$ a blood red colour is formed.

$$Na + C + N + S \longrightarrow NaCNS$$
$$3NaCNS + FeCl_3 \longrightarrow \underset{\text{ferric thiocyanate (blood red colour)}}{Fe(CNS)_3} + 3NaCl$$

Formation of blood red colour indicates the presence of both nitrogen as well as sulphur.

b. **Test for sulphur:**
• 1 cc of LF + a few drops of sodium nitro prusside solution, formation of a pink colour confirms the presence of sulphur.

$$2Na + S \longrightarrow Na_2S$$
$$Na_2S + Na_2Fe(NO)(CN)_5 \longrightarrow \underset{\text{a purple coloured complex}}{Na_4Fe(NOS)(CN)_5}$$

- 1 cc of LF + CH_3COOH + $Pb(OCOCH_3)_2$

$$Na_2S + Pb(OCOCH_3)_2 \longrightarrow PbS + CH_3COONa$$

Formation of a black precipitate due to the formation of lead sulphide PbS, confirms the presence of sulphur in the compound.

c. **Test for halogens:**

1 cc LF + a few drops of conc. HNO_3 are boiled. After cooling on adding $AgNO_3$ solution, formation of a white or a whitish precipitate of silver halides indicates the presence of halogen.

$$Na + X \longrightarrow NaX$$
$$NaX + AgNO_3 \longrightarrow AgX + NaNO_3$$

- **Test for chlorine:** If in the above reaction, a white precipitate due to the formation of AgCl indicates the presence of chlorine, the white precipitate is soluble in ammonia.

$$NaCl + AgNO_3 \longrightarrow AgCl + NaNO_3$$
white precipitate
(soluble in ammonia)

$$AgCl + 2NH_3 \longrightarrow Ag(NH_3)_2\,Cl$$

- **Test for bromine:**

$$NaBr + AgNO_3 \longrightarrow AgBr + NaNO_3$$
straw coloured precipitate

This is partially soluble in ammonia.

- **Test for iodine:**

$$NaI + AgNO_3 \longrightarrow AgI + NaNO_3$$
yellow precipitate
(insoluble in ammonia)

iii. **Victor Meyer's test:** This test is used to distinguish between 1°, 2° and 3° alcohols. The procedure includes following steps

a. **First step:** Alcohols are first converted into alkyl iodides.

$$3ROH + PI_3 \longrightarrow RI + H_3PO_3$$

b. **Second step:** Alkyl iodides are converted into nitro alkanes by reacting with $AgNO_2$.

$$RI + AgNO_2 \longrightarrow R—NO_2 + AgI$$

c. **Third steps:** Nitro alkanes are then treated with HNO_2, when the nitro alkanes obtained from different alcohols react differently forming different derivatives.

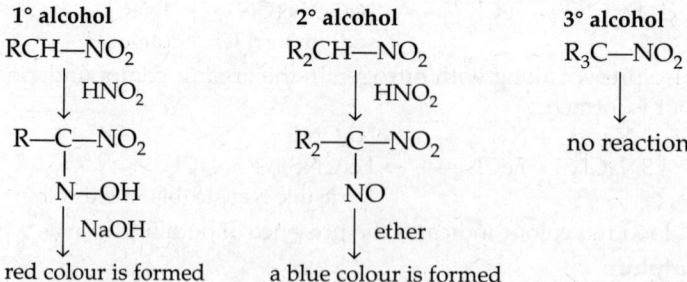

iv. **Liebermann's nitroso test:** Phenol on warming with conc. H_2SO_4 and $NaNO_2$ forms a brown colour, which soon changes to blue or a green colour. On diluting with water, the blue or green colour changes to brown, which changes to deep blue again on adding NaOH solution.

v. **Acrolein test:** Glycerol or glycerine, oil and fats, on warming the potassium bisulphate, $KHSO_4$ or conc. H_2SO_4 are dehydrated to form acrolein $CH_2=CH—CHO$, a sharp smelling liquid, which can easily be detected by its smell.

$$
\begin{array}{c}
CH_2OH \\
| \\
CHOH \\
| \\
CH_2OH
\end{array}
\xrightarrow[-2H_2O]{KHSO_4/H_2SO_4}
CH_2=CH—CHO
\uparrow KHSO_4/H_2SO_4
$$

$$
\begin{array}{c}
CH_2OCOR \\
| \\
CHOCOR \\
| \\
CH_2OCOR
\end{array}
\xrightarrow{KHSO_4/H_2SO_4}
\begin{array}{c}
CH_2OH \\
| \\
CHOH \\
| \\
CH_2OH
\end{array}
+ \quad 3RCOOH
$$

oils and fats   glycerine   fatty acids

v. **Dunston's test:** It is a characteristic test for glycerol. When phenolphthalein is added to cold dil. solution of borax, a pink colour is produced. On adding glycerol to it, the pink colour disappears, but on heating, the pink colour reappears. Oils and fats do not give this test between glycerol and oil and fats. Thus this test can be used to distinguish between oils or fats and glycerol.

vii. **Molische's test:** This test is used for detecting carbohydrates. To an aqueous solution of carbohydrate, are added a few drops of naphthalol, followed by adding conc. H_2SO_4 by the side of the test tube, a violet ring is produced at the junction of the two.

viii. **Carbylamine reaction:** When a primary amine (aliphatic or an aromatic) is warmed with chloroform and alcoholic KOH, it forms isocyanide or carbylamine, having a very offensive odour. It is used to detect a primary amino group in an organic compound.

$$RNH_2 + CHCl_3 + 3KOH \longrightarrow RNC + 3KCl + 3H_2O$$

Rules

i. **Huckel's rule:** If a compound is monocyclic and monoplaner and contains $(4n + 2)$ π-electrons, than it will exhibit aromatic character, i.e. will have unusual stability. Thus to be an aromatic compound, a molecule must have $2(n = 0)$, $6(n = 1)$, $10(n = 2)$ electrons. For benzene, $n = 1$, the molecule is a cloud of six π-electrons.

ii. **Markovnikov's rule:** It is an empirical rule. It relates to the addition of various molecules across the double bond or a triple bond in an unsaturated compound. This rule states that the negative part of the addendum (the molecule to be added), adds on to the carbon atom, that contains lesser number of hydrogen atoms. The addition occurs by a polar mechanism. In case of halogen acids, halogen atom is the negative part and hydrogen is positive part.

$$CH_3—CH=CH_2 + H–X \longrightarrow CH_3—CX—CH_3.$$
isopropyl halide

$$CH_3—C\equiv CH + H–X \longrightarrow CH_3—CX=CH_2 \longrightarrow CH_3—CX_2—CH_3$$

Mechanism

i. Electrophilic addition of H^+ to alkene

$$RCH=CH_2 + H^+ \longrightarrow R—CH^+—CH_3 + R—CH_2—CH_2^+$$
2° carbocation   1° carbocation

A 2°carbocation is more stable than a 1° carbocation, the final product is formed from 2° carbocation, 1° carbocation reverts back to the original compound.

iii. **Anti-Markovnikov's rule** or Kharasch reaction or peroxide effect:

The addition of HBr to an unsymmetrical alkene is abnormally effect in the presence of a peroxide, such as benzoyl peroxide. The addition takes place in the direction, opposite to that produced by Markovnikov's rule. The mechanism of the addition is a free radical reaction.

Mechanism

a. $(C_6H_5COO)_2 \longrightarrow 2C_6H_5\bullet + 2CO_2$
   benzoyl peroxide   phenyl free radical

b. $C_6H_5\bullet + H\overset{\frown}{\text{---}}Br \longrightarrow C_6H_6 + Br\bullet$
   bromine (free radical)

c. $R\text{---}CH=CH_2\text{---}$
   | Br• $RCHBrCH_2\bullet$ (less stable)
   | 1° alkyl free radical
   | RCH• $\text{---}CH_2Br$
   | Br•   2° alkyl free radical (more stable)

As a 2° free radical is more stable than a 1° free radical, the final product is formed from the 2° free radical.

So the formation of the final product is based on the formation of a more stable intermediate free radical.

$R\text{---}CH\bullet\text{---}CH_2Br + H\overset{\frown}{\text{---}}Br \longrightarrow RCH_2CH_2Br + Br\bullet$

iv. **Saytzeff's rule:** It is an empirical rule, which states that

a. In dehydration of a haloalkane the predominant product is the most substituted alkene, i.e. the one carrying largest number of alkyl groups, on the doubly bonded carbon atoms or

b. hydrogen is eliminated preferentially from the carbon atom, joined to the least number of hydrogen atoms.

$CH_3\text{---}CH=CH\text{---}CH_3 \rightleftharpoons CH_3\text{---}CHBr\text{---}CH_2\text{---}CH_3 \rightleftharpoons CH_3\text{---}CH_2\text{---}CH=CH_2$
Saytzeff's product   Hofmann's product
(major product)   (minor product)

v. **Hofmann's rule:** On dehydrohalogenation of alkyl halide, the least substituted product is the alkene containing smallest number of alkyl groups on doubly bonded carbon atoms is termed a Hofmann's product.

vi. **Popoff's rule:** According to this rule, when an unsymmetrical ketone is oxidised by 50% HNO_3, the carbon-carbon bond breaks in such a way that the smaller alkyl group preferentially goes with the carbonyl group, resulting in the formation of a mixture of carboxylic acids. The ―C―C―bond undergoes cleavage on either side of the carbonyl group, leading to the formation of two major and two minor products. Oxidation of 2-pentanone with 50% HNO_3 leads to the formation of four carboxylic acids.

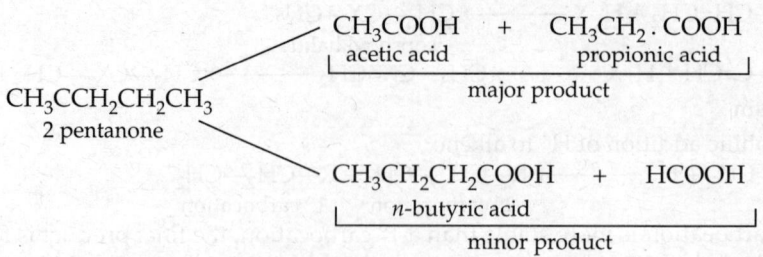

CH_3COOH + $CH_3CH_2 . COOH$
acetic acid   propionic acid
major product

$CH_3CCH_2CH_2CH_3$
2 pentanone

$CH_3CH_2CH_2COOH$ + $HCOOH$
n-butyric acid
minor product

MULTIPLE CHOICE QUESTIONS

1. Baeyer's reagent is
   (a) alk. $KMnO_4$ solution
   (b) acidified $KMnO_4$
   (c) neutral $KMnO_4$
   (d) aqueous bromine solution

2. The presence of unsaturation (olefinic or acetylinic) can be tested with
   (a) Schiff's reagent
   (b) Tollen's reagent
   (c) Fehling's solution
   (d) Baeyer's reagent

3. A mixture of ethane, ethylene and acetylene are treated with amm. $AgNO_3$, the gases which remain unreacted are
   (a) ethane and ethylene
   (b) ethane and ethyne
   (c) Ethene and ethyne
   (d) ethane only

4. Which one of the following can distinguish propyne from propene?
   (a) Br water
   (b) ammoniacal $AgNO_3$
   (c) aqueous $KMnO_4$
   (d) dil. H_2SO_4

5. The distinguishing test for triple bond containing acidic hydrogen is
   (a) Br_2 in CCl_4
   (b) alkaline $KMnO_4$
   (c) $Ag(NH_3)_2^+$
   (d) Luca's reagent

6. The compound most likely to decolourise a solution of $KMnO_4$ is
   (a) $CH_3—CH_3$
   (b) naphthalene
   (c) $CH_3CH=CHCH_2CH_3$
   (d) $(CH_3)_4C$

7. Which of the following compound does not decolourise Br_2 dissolved in CCl_4?
   (a) C_2H_2
   (b) C_3H_6
   (c) C_6H_6
   (d) C_2H_4

8. Alcoholic KOH is used for
   (a) dehydration
   (b) dehydrogenation
   (c) dehydrohalogenation
   (d) dehalogenation

9. Which reagent will quantitatively distinguish between butyne-2 and butyne-1?
   (a) Br_2
   (b) $Ag(NH_3)_2^+$
   (c) Cu_2^+
   (d) $KMnO_4$

10. Which of the following cannot be used to locate the position of the triple bond in the compound, $CH_3—C≡C—CH_3$?
    (a) Br_2
    (b) O_3
    (c) Cu_2^+
    (d) $KMnO_4$

11. Which reagent can be used to distinguish between propanol-1 and propanol-2?
    (a) Alkaline $KMnO_4$
    (b) acidified $K_2Cr_2O_7$
    (c) hot Cu
    (d) conc. H_2SO_4

12. The amide group is tested by heating the amide compound with
    (a) NaOH
    (b) H_2SO_4
    (c) treating it with Tollen's reagent
    (d) reaction with $C_6H_5SO_2Cl$

13. Which reagent can be used to distinguish between ethylene and acetylene?
    (a) aqueous alkaline $KMnO_4$
    (b) Cl_2 dissolved in CCl_4
    (c) ammoniacal Cu_2Cl_2
    (d) conc. H_2SO_4

14. Propene and propyne can be distinguished by
    (a) conc. H_2SO_4
    (b) Br_2 dissolve in CCl_4
    (c) dilute $KMnO_4$
    (d) ammoniacal $AgNO_3$

15. Baeyer's reagent is used for the detection of
    (a) a double bond
    (b) glucose
    (c) reduction
    (d) oxidation

16. Which cannot be used to locate the position of the triple bond in CH_3—$C \equiv C$—CH_3
    (a) Br_2
    (b) O_3
    (c) Cu^+
    (d) $KMnO_4$

17. 1–Butyne on reaction with hot al. $KMnO_4$ gives
    (a) $CH_3CH_2CH_2COOH$
    (b) $CH_3CH_2OH + CO_2$
    (c) CH_3CH_2COOH
    (d) $CH_3CH_2COOH + HCOOH$

18. CH_3—$C \equiv C$—CH_3 $\xrightarrow{X_2/H_2O + Zn}$ $CH_3-\overset{\overset{O}{||}}{C}-\overset{\overset{O}{||}}{C}-CH_3$

    In the above reaction X is
    (a) HNO_3
    (b) O_2
    (c) O_3
    (d) $KMnO_4$

19. A sample of $CHCl_3$ is tested before being used as an anesthetic. With _____
    (a) $AgNO_3$ solution
    (b) $AgNO_3$ solution after boiling with alcoholic KOH
    (c) Fehling solution
    (d) ammoniacal Cu_2Cl_2

20. Carbylamine test on the compound formed is
    (a) $C_6H_5C \equiv N$
    (b) $C_6H_5N \equiv C$
    (c) CH_3—O—$C \equiv N$
    (d) CH_3—$N = C = O$

21. Iodoform test is not given by
    (a) CH_3COCl
    (b) $CH_3COCH_2COC_2H_5$
    (c) CH_3CONH_2
    (d) all

22. Ethyl bromide and isopropyl chloride can be distinguished by
    (a) alcoholic $AgNO_3$
    (b) comparing their colours
    (c) burning compounds on a spatula
    (d) aqueous KOH

23. CCl_4 and $CHCl_3$ can be distinguished by
    (a) action of alcoholic KOH + RNH_2
    (b) RCN + KOH alcoholic
    (c) hydrolysis
    (d) burning in air

24. The alkyl halide which does not give a white precipitate with alcoholic KOH is
    (a) ethyl chloride
    (b) allyl chloride
    (c) isopropyl chloride
    (d) vinyl halide

25. Propanol-1 and propanol-2 can be best distinguished by
    (a) oxidation with alkaline $KMnO_4$ followed by reaction with Fehling's solution
    (b) oxidation with dichromate followed by reaction with Fehling's solution
    (c) oxidation by passing over hot Cu followed by reaction with Fehling's solution
    (d) oxidation with conc. H_2SO_4 followed by reaction with Fehling's solution

26. Which reagent cannot be used to distinguish between phenol and ethanol?
    (a) neutral $FeCl_3$
    (b) Na metal
    (c) an oxidizing agent
    (d) I_2 in presence of a base

27. Fenton's reagent is
    (a) $H_2O + FeSO_4$
    (b) $H_2O_2 + FeSO_4$
    (c) $H_2O_2 + ZnSO_4$
    (d) $NaOH + FeSO_4$

28. The reagent with which both aldehyde and ketone react easily is
    (a) Fehling solution
    (b) Grignard's reagent
    c) Schiff's reagent
    (d) Tollen's reagent

29. Fehling solution is used for the detection of
    (a) Ketonic group
    (b) alcoholic group
    (c) aldehydic group
    (d) carboxylic group

30. Formaldehyde can be distinguished from acetaldehyde by the use of
    (a) Schiff's reagent
    (b) Tollen's reagent
    (c) Fehling solution
    (d) NaOH + iodine

31. Methyl ketones are characterized by
    (a) Tollen's reagent
    (b) Iodoform test
    (c) Schiff's test
    (d) Benedict's reagent

32. When acetaldehyde is heated with Fehling solution a red precipitate is formed which consists of
    (a) Cu
    (b) $Cu + Cu_2O$
    (c) Cu_2O
    (d) Cu_2O

33. Fehling solution consists of
    (a) Acidified $CuSO_4$ solution
    (b) ammoniacal Cu_2Cl_2 solution
    (c) $CuSO_4$ + Rochelle's salt + NaOH
    (d) none of these

34. Schiff's reagent (magenta solution) is decolourised by
    (a) sulfurous acid
    (b) magenta solution decolourised by Cl_2
    (c) amm. $COCl_2$ solution
    (d) amm. $MnSO_4$ solution

35. How to differentiate between formic and acetic acids? Pick out the reaction.
    (a) acids liberate CO_2 when treated with $NaHCO_3$
    (b) form esters with alcohols
    (c) formic acid reduces ammoniacal $AgNO_3$ or Fehling solution or decolouise $KMnO_4$.

36. Oxalic acid may be distinguished from tartaric acid by
    (a) $NaHCO_3$
    (b) ammoniacal $AgNO_3$
    (c) litmus paper
    (d) phenolphthalein

37. Which test is used to differentiate 1°, 2° and 3° amines.
    (a) carbylamine test
    (b) iodoform test
    (c) Schiff's test
    (d) not defined

38. 1° and 2° amines are distinguished by
    (a) Br_2 + KOH
    (b) HCIO
    (c) HNO_2
    (d) NH_3

39. How primary, sec. and tertiary amines distinguished?
    (a) Hinsberg's reagent
    (b) Grignard's reagent
    (c) Fehling's solution
    (d) Tollen's reagent

40. A positive carbylamine test is not given by
    (a) *N,N*-dimethyl aniline
    (b) 2,4-dimethyl aniline
    (c) *N*-methyl-o-methyl aniline
    (d) *p*-methyl benzyl aniline

41. Which substance will evolve ammonia, when heated with NaOH?
    (a) ethyl amine
    (b) aniline
    (c) acetamide
    (d) acetoxime

42. Which of the following will give a dye test?
    (a) aniline
    (b) methyl amine
    (c) diphenyl amine
    (d) ethyl amine

43. Phenol and ethanol can be distinguished by
    (a) NaOH
    (b) PCl_5
    c) $FeCl_3$
    (d) benzoic acid

44. The reagent which will react with acetaldehyde and acetone?
    (a) Fehling solution
    (b) Grignard's reagent
    (c) Schiff's reagent
    (d) Tollen's reagent

45. Which does not give red precipitate with Fehling solution?
    (a) HCHO
    (b) CH_3CHO
    (c) Propionaldehyde
    (d) acetone

46. C_2H_5CHO and acetone can be distinguished by testing with
    (a) phenyl hydrazine
    (b) hydroxyl amine
    (c) Fehling solution
    (d) sodium bisulphite

47. The compound which will react with Luca's reagent, the fastest?
    (a) butane-1-ol
    (b) butane -2-ol
    (c) 2-methylpropan-1-ol
    (d) 2-methyl-2-propanol

48. Which of the following will not decolourise Br_2 water?
    (a) $CH_2=CH_2$
    (b) $(CH_3)_2C=C(CH_3)_2$
    (c) $CH\equiv CH$
    (d) none

49. Chlorobenzene and benzyl chloride can be distinguished by
    (a) treatment with KOH and then by $AgNO_3$ solution
    (b) Luca's reagent
    (c) decolourisation of Br_2 in CCl_4
    (d) orange colour with $CHCl_3/AlCl_3$

50. RCHO $\longrightarrow$ RCH_2OH, for this reaction the catalyst used is
    (a) Ni only
    (b) Pd
    (c) Pt only
    (d) any

51. Which one of the following compounds on reduction with $LiAl_4$, will give a product that will give a positive haloform test?
    (a) CH_3CH_2CHO
    (b) CH_3CH_2COOH
    (c) $C_2H_5{-}O{-}C_2H_5$
    (d) CH_3COCH_3

52. Ethane can be obtained from ethanol in one step only,
    (a) Na + Hg + water
    (b) Zn + Hg amalgam + conc. HCl
    (c) aluminum isopropoxide + isopropyl alcohol
    (d) $LiAl_4$ + ether

53. Collin's reagent converts
    (a) $>C=O \longrightarrow >CHOH$
    (b) $CHO \longrightarrow COOH$
    (c) $>CHOH \longrightarrow C=O$
    (d) $>CHOH \longrightarrow >C=O$

54. Formation of an alcohol from an aldehyde, is termed
    (a) reduction
    (b) oxidation
    (c) addition
    (d) substitution

55. Aliphatic aldehydes reduces Fehling's solution but benzaldehyde does not because
    (a) of a bulky ring and —CHO is the hinderer
    (b) of resonance, the oxidation of —CHO is difficult
    (c) —CHO group is in a cyclic structure
    (d) all of the above reasons

56. Fehling solution is used in the detection of
    (a) a ketonic group   (b) an alcoholic group
    (c) an aldehydic group   (d) an acid group

57. HCHO and CH_3CHO can be distinguished by the use of
    (a) Schiff's reagent   (b) Tollen's reagent
    (c) Fehling solution   (d) a haloform reaction

58. The reagent of choice for the selective reduction of a ketone in presence of an ester
    (a) $LiAl_4$ (b) $NaBH_4$
    (c) H_2 and Pd   (d) sodium in ethanol

59. The red precipitate obtained on heating Fehling solution with acetaldehyde consists of
    (a) Cu   (b) CuO
    (c) $Cu + CuO + Cu_2O$ (d) Cu_2O

60. Schiff's reagent is
    (a) magenta solution decolourised by H_2SO_3
    (b) magenta decolourise by SO_2 gas
    (c) ammonical $COCl_2$ solution
    (d) ammonical $MnSO_4$ solution

61. Acetaldehyde and acetone can be distinguished by treating
    (a) $NaHSO_3$ (b) NaCN
    (c) $NaI + I_2$ (d) $Ag(NH_3)_2^+$

62. The reaction of Tollen's reagent with acetaldehyde gives
    (a) CH_3OH (b) CH_3COOAg
    (c) silver mirror   (d) HCHO

63. Magenta is
    (a) alkaline phenolphthalein
    (b) red litmus
    (c) p-rosaniline hydrochloride or fuchsine
    (d) methyl red

64. C_2H_5CHO and CH_3COCH_3 can be distinguished by testing with
    (a) phenyl hydrazine   (b) Hydroxyl amine
    (c) Fehling solution   (d) sodium bisulphide

65. C_6H_5CHO and CH_3CHO can be distinguished by
    (a) Iodoform test   (b) 2,4-DNP test
    (c) NH_3 test   (d) Wolff-Kishner's reduction

66. HCHO and HCOOH can be distinguished by treating with
    (a) Tollen's reagent   (b) $NaHCO_3$
    (c) Fehling solution   (d) Benedict's solution

67. Benedict's solution provides
    a) Ag_+ (b) Cu_2^+ (c) Ba_2^+ (d) Li^+

68. Jone's reagent is
    (a) acid $KMnO_4$ (b) $K_2Cr_2O_7 + H_2SO_4$ or $CrO_3 + H_2SO_4$
    (c) alk. $KMnO_4$ (d) none

69. Schiff's reagent and Schiff's base are
    (a) same compounds
    (b) different compounds
    (c) physically same but chemically different

70. The reagent which can distinguish between acetophenone and benzophenone
    (a) 2,4-dinitro phenyl hydrazine
    (b) aqueous $NaHSO_3$
    (c) Benedict's solution
    (d) $I_2 + Na_2CO_3$

71. Which of the following will not give a red precipitate with Fehling solution
    (a) acetaldehyde
    (b) formalin
    (c) D-glucose
    (d) acetone

72. Sodium potassium tartrate is present in which one of the following?
    (a) Tollen's regent
    (b) Fehling solution
    (c) Benedicts solution
    (d) None

73. *p*-Chloro aniline and anilinium hydrochloride can be distinguished by
    (a) Sandmeyer's reaction
    (b) $NaHSO_3$
    (c) $AgNO_3$
    (d) carbylamine reaction

74. Which of the reducing agent will reduce?
    (a) Wurtz reaction
    (b) Wolff–Kishner's reaction
    (c) Na+ ethanol
    (d) $LiAlH_4$

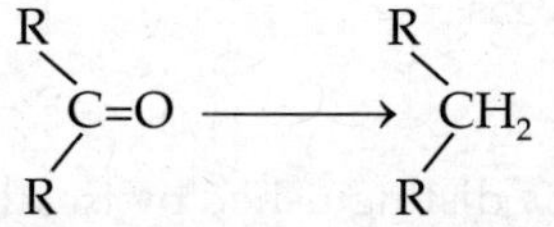

$$\begin{matrix} R \\ \diagdown \\ C=O \\ \diagup \\ R \end{matrix} \longrightarrow \begin{matrix} R \\ \diagdown \\ CH_2 \\ \diagup \\ R \end{matrix}$$

75. Which of the following reagent used to distinguish propyne-1 and propene?
    (a) bromine water
    (b) ammoniacal $AgNO_3$
    (c) aqueous $KMnO_4$
    (d) dil. H_2SO_4

76. Baeyer's reagent is used in the laboratory for
    (a) detection of a double bond
    (b) detection of glucose
    (c) reduction
    (d) oxidation

77. The compound which is most likely to decolourise $KMnO_4$ solution is
    (a) ethane
    (b) naphthalene
    (c) $CH_3—CH=CH—CH_2CH_3$
    (d) neopentane

78. Which of the following distinguishes butyne-1 and butyne-2?
    (a) Br_2
    (b) $Ag(NH_3)_2^+$
    (c) Cu_{2+}
    (d) $KMnO_4$

79. Lactic acid on oxidation with Fenton's reagent gives the major product
    (a) acetic acid
    (b) oxalic acid
    (c) pyruvic aid
    (d) none

80. On warming HCOOH with ammoniacal $AgNO_3$, the product formed is
    (a) Ag_2O
    (b) metallic silver
    (c) silver formate
    (d) HCHO

81. CH_3COOH and C_6H_5COOH can be distingushed by
    (a) flame test
    (b) solubility test
    (c) physical test

82. Cacodyl test is used for identification of
    (a) HCOOH   (b) CH_3COOH
    (c) oxalic acid   (d) alpha amino acid

83. p-amines and secondary amines are distinguished by
    (a) Br_2/KOH (b) HClO
    (c) HNO_2 (d) NH_3

84. 1°, 2° and 3° amines can be distinguished by
    (a) Hinsberg's reagent   (b) Grignard's reagent
    (c) Fehling solution   (d) Tollen's reagent

85. A 2°amine forms a yellow oil on treatment with HNO_2. On heating, they are converted back to form the secondary amine. The reaction is known as
    (a) Perkin's reaction   (b) Fries reaction
    (c) Liebermann's nitroso test   (d) Etard's reaction

86. Acetamide is treated separately with which of the following reagent to give methylamine?
    (a) PCl_5 (b) sodalime
    (c) $NaOH^+ Br_2$ (d) conc. H_2SO_4

87. The compound that will react fastest with Luca's regent at room temperature is
    (a) Butane-1-ol   (b) butane-2-ol
    (c) 2-methyl-propan-1-ol   (d) 2-methylpropan-2-ol

88. The reagent which will react with both acetaldehyde and acetone easily is
    (a) Fehling's reagent   (b) Grignard's reagent
    (c) Schiff's reagent   (d) Tollen's reagent

89. Baeyer's reagent is
    (a) alk. $KMnO_4$ (b) acidic $KMnO_4$
    (c) neutral $KMnO_4$ (d) aqueous Br_2 solution

90. Which will react faster with H_2 under catalytic conditions?

$$\underset{\text{(a)}}{\overset{R}{\underset{H}{C}}\!\!=\!\!\overset{R}{\underset{H}{C}}} \qquad \underset{\text{(b)}}{\overset{R}{\underset{R}{C}}\!\!=\!\!\overset{H}{\underset{H}{C}}} \qquad \underset{\text{(c)}}{\overset{R}{\underset{R}{C}}\!\!=\!\!\overset{R}{\underset{Ac}{C}}} \qquad \underset{\text{(d)}}{\overset{R}{\underset{R}{C}}\!\!=\!\!\overset{R}{\underset{R}{C}}}$$

91. A compound C_3H_6O gives no precipitate with 2,4-DNP, nor it reacts with Na metal is
    (a) CH_3CH_2CHO (b) CH_3COCH_3
    (c) $CH_2=CH—CH_2—OH$ (d) $CH_2=CH—O—CH_3$

92. In which case the reagent selected is not the correct one
    (a) Grignard's reagents elected for the presence of active hydrogen
    (b) Tollen's reagent for detecting the presence of unsaturation
    (c) Fehling solution for detecting glucose
    (d) $FeCl_3$ for detecting phenols

ANSWERS

1. (a)	2. (d)	3. (a)	4. Br	5. (c)	6. (c)	7. (c)	8. (c)
9. (b)	10. (a), (c), (d)		11. (c) hot Cu		12. (a)	13. (c) Cu_2Cl_2	
14. (d)	15. (a)	16. (a), (c), (d)		17. (d)	18. (c) O_3	19. (b)	20. (b)

21. (d) all   22. (a)   23. (a)   24. (d) resonance   25. (c)
26. (b) Na cannot be used as both react with Na metal
27. (b)   28. (b)   29. (c)   30. (d)   31. (b)   32. (d)   33. (c)   34. (a)
35. (c)   36. (b)   37. $1°$
38. (c) Primary amines form alcohol, sec amine gives nitroso amines
39. (a)
40. (d), (b) does not give this test because of steric hindrance
41. (c)   42. (a)   43. (c) $FeCl_3$
44. (b) Grignard's reagent
45. (d) acetone
46. (c) Fehling solution
47. (d) Tertiary alcohols react fastest with Luca's reagent
48. (b)
49. C—O bond breaks faster
49. (a) Benzyl chloride will give a white precipitate
50. any
51. (d)
52. (b) Clemmensen's reaction
53. (c) Collin's reagent (CrO_3 + pyridine) oxidises $2°$ alcohol to a ketone, and $1°$ alcohol to an aldehyde
54. (a) Reduction
55. (b) Resonance gives stability, it is difficult to oxidise a —CHO group when attached to a benzene ring
56. (c) a CHO group
57. (d) HCHO does not give a haloform test
58. (b) Reduction with $NaBH_4$ will convert it to form >CHOH, it will not affect the ester group
59. (d) Cu_2O
60. (b) magenta solution decolourised by SO_2 gas
61. (c) a haloform test, HCHO does not give a haloform test.
62. (c) a silver mirror
63. (c)   64. (c)   65. (a)
66. (b) HCOOH will evolve CO_2 when treated with $NaHCO_3$
67. (b)
68. (b) CrO_3 + H_2SO_4 it does not attack >C=C<
69. (b) Schiff's bases are the reaction products of –aromatic aldehyde + a primary amine $\longrightarrow$ aldemine
    —CHO$^+$—NH$_2$— $\longrightarrow$ —CH=N— + H$_2$O
    Schiff's reagent is an alcoholic solution of magenta, decolourised by SO_2
70. (d) $C_6H_5COCH_3$ will give haloform reaction
71. (d) acetone
72. (b) Fehling solution contains sod. potassium tartrate or Rochelle salt
73. (c) anilinium hydrochloride will form a white precipitate with $AgNO_3$
74. (b) reaction with NH_2 followed by treatment with KOH
75. (b)
76. (a) used for detection of unsaturation
77. (c)   78. (b)
79. (c) pyruvic acid, $CH_3CO.COOH$
80. (a) silver formate is formed
81. (a) Flame test-benzoic acid burns with a sooty flame

82. (b) it is a test given by acetic acid
83. (c) HNO_2 (1° amines)
84. (a) by Hinsberg's reagent
85. (c) a 2° amine forms a nitrosamine, (a yellow coloured oily compound)
86. (c)
87. (d) a tertiary alcohol
88. (b)   89. (a)   90. (d)   91. (d)   92. (b)

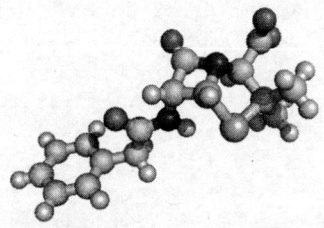

Reagents

REASONING TYPE QUESTIONS (RTQs)

1. Aldehydes reduce Tollen's reagent and Fehling solution but ketones do not. Explain.
2. Formaldehyde is a more powerful reducing compound as compared to the other aldehydes. Explain.
3. Give a suitable explanation, why the pink colour of decolorized Schiff's reagent is restored on treatment with an aldehyde?
4. An aldehyde and a ketone, both contain a keto group but can be distinguished by a chemical reaction. Explain
5. Both glycerol and oil and fats give acrolein test. Suggest a line by which the two can be distinguished.
6. Molische's test is used for testing a carbohydrate. How a monosaccharide can be identified. Name the reagent.
7. There are various methods for spot testing the alcohol. Discuss the one which is used by our police.
8. What reagents are used to convert a primary alcohol into an aldehyde? Give the names and the composition of the reagent.
9. Fehling solution is prepared by mixing solutions A and B, 'A' contain aqueous $CuSO_4$, 'B' contains, NaOH and sodium potassium tartrate. Explain the role of Rochelle salt in the preparation of Fehling solution.

ANSWERS

1. Aldehydes reduce Tollen's reagent because aldehydes contain a C—H bond which is easily oxidised. On the other hand, both the bonds of the >C=O group in ketones are connected to carbons, and under these conditions, the C to C σ-bond cannot be oxidised. This is why ketones do not reduce Tollen reagent.
2. In formaldehyde, both the bonds of the carbonyl group are joined to hydrogen atoms, which are oxidised easily, therefore HCHO is the most powerful reducing aldehyde.

$$\underset{H}{\overset{H}{>}}C=O + O \longrightarrow \underset{HO}{\overset{HO}{>}}C=O$$

3. Magenta or fuschine or rosaniline hydrochloride form a pink coloured solution, on passing sulphur dioxide gas SO_2 the pink colour is discharged. SO_2 being a reducing agent reduces to form a colourless compound. Hence the solution becomes colourless. When this colourless solution, termed Schiff's reagent, is treated with an aldehyde it is reduced and the reduced magenta is oxidised to the original compound, hence the Schiff's reagent restores its pink colour.

4. Both aldehyde and ketone form a yellow precipitate, with 2, 4-dinitro-phenyl hydrazine (or Brady's reagent) which is insoluble in HCl. Whereas aldehydes reduce Tollen's reagent and Fehling solution ketones do not.

5. All oils and fats as well as glycerol form acrolein, on heating with $KHSO_4$ or with H_2SO_4, yet glycerol gives Dunston's test, but oils and fats do not.

6. Molische's test is a confirmed test for all the carbohydrates. Only monosaccharides give Barfoed's reagent test.

7. Spot test for the detection of alcohol is one in the breath. The accused is asked to exhale the breath in a solution containing $K_2Cr_2O_7$ and H_2SO_4, if the solution turns green it means that the exhaled breath contains alcohol. It turns green because of the formation of Cr^{3+} ions.

8. $1°$ or a primary alcohol is oxidised to an aldehyde by a number of reagents.
    i. Sarret–Colliins reagent
       $$RCH_2OH \longrightarrow RCHO$$
    ii. PCC
    iii. Jone's reagent

9. CuO is an insoluble compound in water. Whenever it is formed it settles down as an insoluble compound formed. Rochelle salt reacts with Cu^{2+} ions to form a copper chelate compound, and therefore checks the CuO from pipetting out.

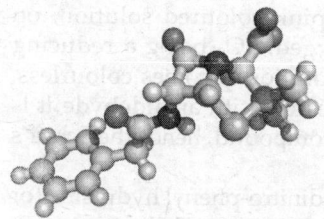

Exceptions

REASONING TYPE QUESTIONS (RTQs)

1. Nitrobenzene, phenyl cyanide, benzene, sulphonic acid, benzaldehyde, benzophenone, benzoic acid C_6H_5COOH, methyl benzoate, pyridine and aniline do not undergo Friedel–Craft reaction.
2. $(CH_3)_2CHCHO$, an aldehyde containing α-H undergoes Cannizzaro's reaction.
3. A 3° alkyl halide is converted into an alkene on reduction with $LiAlH_4$.
4. A compound of the following composition $(CH_3)C—O—CH_3$ is converted to methanol and $(CH_3)_3Cl$, on treatment with HI.
5. Benzaldehyde does not reduce Fehling solution.
6. Formic acid shows reducing properties like formaldehyde.
7. Methanol does not form dimethyl ether on heating with conc. H_2SO_4.
8. Wurtz reaction cannot be applied for the preparation of branch chain alkanes.
9. Aniline in conc. HCl gives a *m*-derivative.
10. Sodium benzoate forms *o*- and *p*-derivatives.
11. Tertiary butyl benzene does not form benzoic acid on oxidation with Baeyer's reagent and does not undergo a side chain oxidation with dilute alkaline $KMnO_4$.
12. Dehydrohalogenation of 2-bromobutane by a tertiary alcoholic solution of KOH gives 1-butene, a major product.
13. Terminal alkenes are reduced by sodium and alcohol (Birch reaction).
14. On heating, tetra methyl ammonium hydroxide, gives trimethyl amine and methanol, but alkene is not formed.
15. Aromatic ketones do not form sodium bisulphite addition compound.
16. Silver salt of a fatty acid does not undergo Borodine Hunsdiecker reaction, when treated with iodine, instead an ester is formed.
17. $LiAlH_4$ and $NaBH_4$ do not reduce double and triple bonds.
18. HI is a reducing compound but HBr and HCl are not.
19. Write a note on anti-Markovnikov's rule. HBr in presence of a peroxide adds to a double bond through anti-Markovnikov's rule but HCl and HI do not.
20. Alcohols are neutral to litmus paper but phenol is acidic.

ANSWERS

1. Some derivatives of benzene do not undergo Friedel–Craft reaction. This reaction is alkylation or acylation of benzene. Benzene or any aromatic containing benzene ring reacts with alkyl halide RX or with acyl chloride RCOCl, in presence of anhydrous $AlCl_3$ to form alkyl benzene or an acyl benzene. The role of catalyst, anhydrous $AlCl_3$ is to liberate an electrophile, which combines with benzene ring to form a π-complex, changing to a σ-complex and finally to form an alkyl benzene or acyl benzene. The various derivatives of benzene containing groups like $-NO_2$, $-CN$, $-SO_3H$, $-CHO$, $-COOH$, $-COCH_3$ or $-COOCH_3$, do not undergo Friedel–Craft reaction, because these groups deactivate the benzene ring to such an extent that electrophile is not able to form a π-complex.

$$R—Cl + AlCl_3 \longrightarrow R^+ + AlCl^-$$

$$(X = —NO_2, —CN, —SO_3H, —CHO, —COOH, —COCH_3 —COOCH_3)$$

Benzene derivative

Aniline and pyridine combine with electron deficient aluminium chloride $AlCl_3$ to form complexes, through pair of electrons, thus aniline and pyridine also do not undergo Friedel–Craft reaction.

$$C_6H_5NH_2 + AlCl_3 \longrightarrow C_6H_5NH_2^+ – AlCl_3^-$$
aniline–aluminum chloride complex

$$C_5H_5N + AlCl_3 \longrightarrow C_5H_5N^+ – AlCl_3^-$$
pyridine–aluminum chloride complex

So nitrobenzene, $C_6H_5NO_2$ is used as a solvent in Friedel–Craft reaction, since nitro benzene is steam volatile, it can be removed by steam distillation to recover the product.

2. Those aldehydes, which do not contain α-hydrogen atom, undergo Cannizzaro's reaction. In the presence of 50% aqueous solution of caustic soda, NaOH or potassium hydroxide KOH at room temperature two molecules of aldehyde, which do not contain α-hydrogen atom undergo self oxidation–reduction process to form a primary alcohol and an acid. Cannizzaro's reaction is shown by the aliphatic aldehydes, like formaldehyde HCHO, 2-methyl propionaldehyde $(CH_3)_3—C—CHO$ and all aromatic aldehydes Ar–CHO. α-hydrogen is the hydrogen, which is joined to the α-carbon atom.

$$2HCHO + NaOH \longrightarrow CH_3OH + HCOONa$$
methanol   sodium formate

$$(CH_3)_3—C—CHO + NaOH \longrightarrow (CH_3)_3—C—CH_2OH + (CH_3)_3—C—COONa$$
2-methyl propionaldehyde   2-methyl-1-propanol   2-methyl sodium propionate

In the same way benzaldehyde undergoes Cannizzaro's reaction to form benzyl alcohol, $C_6H_5CH_2OH$, and sodium benzoate, C_6H_5COONa.

$$2C_6H_5CHO + NaOH \longrightarrow C_6H_5CH_2OH + C_6H_5COONa$$

Mechanism

$$H-\overset{\overset{\displaystyle O}{\|}}{C}-H + OH^- \longrightarrow H-\overset{\overset{\displaystyle H}{|}}{\underset{\underset{\displaystyle OH}{|}}{C}}-O^- \longrightarrow H-C=O + H-\overset{\overset{\displaystyle H}{|}}{\underset{\underset{\displaystyle H}{|}}{C}}-O^\ominus \longrightarrow CH_3OH + HCOOH$$

When Cannizzaro's reaction is carried out with two different aldehydes which undergo Cannizzaro's reaction, individually, if one of them happens to be formaldehyde, it is the formaldehyde which is reduced to form methanol and the other aldehyde is oxidised to form an acid. This reaction is termed *crossed Cannizzaro's reaction*.

$$H-CHO + C_6H_5CHO + NaOH \longrightarrow \underset{reduced}{CH_3OH} + \underset{oxidised}{C_6H_5COONa}$$

In Cannizzaro's reaction, it is the two molecules which take part in the reaction, some compounds which contain two aldehyde group or a compound containing one aldehyde and one keto group also undergo reaction, known as the *internal Cannizzaro's reaction*.

$$\underset{\text{pyruvic aldehyde}}{CH_3COCHO} + NaOH \longrightarrow \underset{\text{lactic acid}}{CH_3CH(OH)\,COOH}$$

$$\underset{\underset{\displaystyle CHO}{|}}{CHO} + NaOH \longrightarrow \underset{\underset{\displaystyle COOH}{|}}{CH_2OH}$$

glyoxal   glycolic acid

All other aldehydes containing α-hydrogen atoms in the molecule, undergo 'aldol condensation'.

Acetaldehyde, CH_3—CHO, reacts with another molecule of acetaldehyde in presence of dil. alkali to form aldol. On heating aldol loses a molecule of water to form crotonaldehyde.

$$CH_3CHO + CH_3-CHO \xrightarrow{\text{dil. alkali/OH}^-} \underset{\text{aldol}}{CH_3-\overset{\overset{\displaystyle OH}{|}}{CH}-CH_2CHO} \longrightarrow \underset{\text{crotonaldehyde}}{CH_3-CH=CH-CHO}$$

In the case of taking two different aldehydes containing α hydrogen, two compounds are formed.

$$\underset{\text{acetaldehyde propionaldehyde}}{CH_3-CHO + CH_3-CH_2-CHO} \longrightarrow \underset{\substack{\text{2-methyl-3-hydroxy} \\ \text{butyraldehyde}}}{CH_3-\overset{\overset{\displaystyle OH}{|}}{CH}-\overset{\overset{\displaystyle CH_3}{|}}{CH}-CHO} + \underset{\text{3-hydroxy valeraldehyde}}{CH_3CH_2\overset{\overset{\displaystyle OH}{|}}{CH}-CH_2CHO}$$

2-methyl propionaldehyde or isobutyraldehyde $(CH_3)_2CH$—CHO contains an α-hydrogen in the molecule, it is expected to undergo aldol condensation to produce 2,2,4,-trimethyl-3-hydroxyl valeraldehyde.

$$(CH_3)_2-CH-CHO + (CH_3)_2-CH-CHO \longrightarrow CH_3-\overset{\overset{\displaystyle CH_3}{|}}{CH}-\overset{\overset{\displaystyle OH}{|}}{CH}-\overset{\overset{\displaystyle CH_3}{|}}{\underset{\underset{\displaystyle CH_3}{|}}{C}}-CHO$$

Isobutyraldehyde besides giving aldol condensation also undergoes Cannizzaro's reaction to form 2-methyl propyl alcohol and 2-methyl propionic acid

$$\underset{\text{isobutyraldehyde}}{(CH_3)_2-CH-CHO} + NaOH \longrightarrow \underset{\text{2-methyl propyl alcohol}}{(CH_3)_2CHCH_2OH} + \underset{\text{2-methyl propionic acid}}{(CH_3)_2CHCOOH}$$

So isobutyraldehyde is the only aldehyde which shows both the reactions, being an α-hydrogen containing aldehyde, undergoes aldol condensation and Cannizzaro's reaction.

3. The 3° alkyl halides CR_3—X, generally are dehydrohalogenated to form alkene, whenever they are treated with a strong nucleophilic reagent. On treating the 3° alkyl halides, CR_3—X, with alc. alk. KCN, etc. the tertiary alkyl halides lose a molecule of HX and are converted to form alkene lithium aluminium hydride $LiAlH_4$ dehydrohalogenates the tertiary alkyl halide and converts it into an alkene.

R_3C—X + $LiAlH_4$ ⟶ alkene

$LiAlH_4$ is a very powerful reducing agent. It reduces almost all the reducible compounds, but it does not reduce olefinic double bond and an acetylinic triple bond.

$LiAlH_4$ reduces various groups by providing hydride ion.

$LiAlH_4$ ⟶ AlH_3 + LiH

$$>C=O + Li^+ H^- \longrightarrow >\overset{\overset{\displaystyle H}{|}}{C}=O\ Li \xrightarrow{H_2O} >\overset{\overset{\displaystyle H}{|}}{C}-OH + LiOH$$

Alkenes undergo electrophilic addition only, therefore, $LiAlH_4$ does not reduce the alkene, formed from the tertiary alkyl halide.

4. Ethers are decomposed by hydrogen halides

R—O—R + HX ⟶ ROH + RX

The decreasing order of reactivity the halogen acids, on the ethers is in the order

HI > HBr > HCl

Because of low reactivity at the bond between oxygen and aromatic ring, an alkyl phenyl ether undergoes cleavage of the alkyl C=O bond yielding a phenol and alkyl halide.

C_6H_5—O—CH_3 + HI ⟶ C_6H_5OH + CH_3I
   anisole   phenol   methyl iodide

Cleavage involves nucleophilic attack by halide ion on the protonated ether, with displacement of weakly basic alcohol molecule.

HI ⟶ H^+ + I^-

$$(CH_3)_3C-O-CH_3 + H^+ \longrightarrow (CH_3)_3-\overset{\overset{\displaystyle H}{|}}{C}-O^+-CH_3 \xrightarrow{I^-} (CH_3)_3C-I + CH_3OH$$

[As tertiary alcohols are less acidic, or more basic as compared to methanol, which being the strongest acid, (a better leaving group) is formed]

5. Fehling solution is prepared by mixing an aqueous solution of copper sulphate as $CuSO_4$, sodium hydroxide NaOH, and Rochelle salt (sodium potassium tartrate). The Fehling solution behaves as cupric oxide, CuO in the aqueous solution (having a blue colour). The role of Rochelle salt is to stabilise it by forming a copper complex, also known as copper chelate complex so as to check the decomposition of copper salt to precipitate as CuO.

$CuSO_4$ + 2NaOH ⟶ $Cu(OH)_2$ + Na_2SO_4

$$Cu\begin{smallmatrix} \diagup OH \\ \diagdown OH \end{smallmatrix} + \begin{smallmatrix} HO-CHCOOK \\ HO-CHCOONa \end{smallmatrix} \longrightarrow Cu\begin{smallmatrix} \diagup O-CHCOOk \\ | \\ \diagdown O-CHCOONa \end{smallmatrix}$$

A copper chelate complex

In the absence of Rochelle salt, $Cu(OH)_2$ will decompose to form:

$Cu(OH)_2$ ⟶ CuO + H_2O

All aliphatic aldehydes reduce Fehling solution on heating. The blue colour disappears and there is formed a brick red coloured powder, due to the formation of cuprous oxide Cu_2O.

$$RCHO + 2CuO \longrightarrow RCOOH + Cu_2O$$

Benzaldehyde and other nuclear substituted aromatic aldehydes do not reduce Fehling solution.

6. Formic acid, shows reducing nature because it possesses an aldehydic group in the molecule.

```
      H                    OH
      |                    |
   H—C=O               H—C=O
   formaldehyde        formic acid
```

Therefore formic acid, shows all the reducing properties as shown by formaldehye.

7. Methanol, CH_3OH forms dimethyl sulphate on heating with conc. H_2SO_4 unlike ethanol which forms three compounds. On heating ethanol with conc. H_2SO_4 at 100°C, it forms ethyl hydrogen sulphate, $C_2H_5HSO_4$.

$$C_2H_5OH + H_2SO_4 \longrightarrow C_2H_5HSO_4 + H_2O$$

On heating ethyl hydrogen sulphate with excess of ethyl alcohol to a temperature 145°C, diethyl ether, $C_2H_5—O—C_2H_5$ is formed

$$C_2H_5HSO_4 + HOC_2H_5 \longrightarrow C_2H_5—O—C_2H_5 + H_2SO_4$$

On heating, ethyl hydrogen alone with excess of conc. H_2SO_4 ethylene is formed.

$$C_2H_5HSO_4 \longrightarrow CH_2=CH_2 + H_2SO_4$$

On heating, methanol with conc. H_2SO_4, to a temperature 170°C, dimethyl sulphate, $(CH_3)_2SO_4$ is the only compound formed.

$$2CH_3OH + H_2SO_4 \longrightarrow (CH_3)_2SO_4 + H_2O$$

8. Alkyl halides react with one another, in presence of sodium metal and ether, to form higher alkanes, containing double the number of carbon atoms present in the alkyl halide. This reaction is termed as Wurtz reaction.

$$R—X + 2Na + X—R \longrightarrow R—R + 2NaX$$

When two different alkyl halides are used, a mixture of three alkanes is formed.

i. $R—X + 2Na + R'—X \longrightarrow R—R' + 2\,NaX$
ii. $R—X + 2Na + RX \longrightarrow R—R + 2NaX$
iii. $R'—X + 2Na + R'X \longrightarrow R'R' + 2NaX$

However, Wurtz reaction has some limitations. These limitations are as follows:

i. The method is applicable only for 1° or primary alkyl halide, RCH_2X.
ii. The method can only be used for the preparative straight chain alkanes.

For the preparation of branched chain alkanes Corey–House synthesis is used.

Corey–House synthesis:

i. Alkyl halide, may be 1°, 2°, 3°. RX is treated with lithium metal, alkyl lithium RLi is formed,

$$R—X + 2\,Li \longrightarrow R\,Li + Li\,X$$

ii. Alkyl lithium is then treated with cuprous halide, CuX, when lithium dialkyl cuperate, R_2CuLi, is formed.

$$2RLi + CuX \longrightarrow R_2CuLi + LiX$$

iii. Finally, R_2CuLi, is treated with R'X, to prepare the alkane.

$$R_2CuLi + R'X \longrightarrow R—R' + RCu + LiX$$

9. Aniline, forms *o*- and *p*-derivatives. The amino group —NH$_2$ group is an activating group, it activates the *o*- and *p*-positions, and therefore forms *o*- and *p*-derivatives on converting it into a disubstituted compound, the anilinium ion.

anilinium ion

Amino group in presence of an acid acts as an *m*-directing group.

Aniline, when treated with hydrochloric acid, HCl, the amino group is converted to an anilinium salt. As a result, the —NH$_2$ group develops a +ve charge on the amino group it becomes NH$_3{}^+$; it now becomes electron deficient, and therefore it begins to acts as a deactivating group. The *o*- and *p*-positions are deactivated, leaving only *m*-position comparatively untouched, (*m*-position now are more electronegative), allowing the entering electrophile to join at *m*-positions.

10. The carboxylic group is a deactivating group. All *m*-directing groups deactivate the benzene ring, by withdrawing the electron density from *o*- and *p*-positions of the ring. This leaves the *m*-position little more electronegative. As the benzene ring undergoes electrophilic substitution, the entering group (the electrophile) goes to join at *m*-position (it may be noted that the densities at *m*-position do not increase, but as the densities decrease at the neighbouring carbon atoms, it becomes more negative).

Therefore, benzene containing *m*-directing groups undergo substitution slowly, as compared to *o*- and *p*-substitution, which proceeds very fast.

benzoic acid   sodium benzoate

On the other hand, in sodium benzoate, the —COO$^-$Na$^+$ group, behaves as an activating group, and acts as *o*- and *p*-directing group.

Being a salt, the sodium salt leaves the −ve charge on the COO$^-$ group, which becomes electron rich and increases the electron density on the *o*- and *p*-positions and therefore starts behaving as an activating group.

11. Side chain, is a group containing carbon through which it is joined to the benzene ring. On oxidation with dil alkaline KMnO$_4$ also known as Baeyer's reagent, the side chain is oxidised to a carboxylic group, directly joined to the benzene ring, irrespective of the length of the carbon chain.

For a side chain oxidation the carbon to which side is attached, must contain at least one hydrogen atom. This hydrogen is termed a benzylic hydrogen atom and the carbon is termed benzylic carbon. Tertiary butyl benzene is not oxidised to a —COOH group, though tertiary butyl group attached to benzene is a side chain.

benzylic carbon atom

$$CH_3 \overset{\overset{\displaystyle NH_2}{|}}{\underset{\underset{\displaystyle \bigcirc}{|}}{C}} CH_3$$

$\longrightarrow$ Tertiary butyl benzene, is not oxidised by dil. alk. $KMnO_4$

12. 2-Bromo butane, on dehydrohalogenation with alcoholic solution of tertiary butoxide forms 1-butene, instead of expected 2-butene. The reaction takes place in two stages:

   i. CH_3—$CHBr$—CH_2—CH_3 $\longrightarrow CH_3$—CH^+—CH_2—CH_3 + Br

   ii. Abstraction of the hydrogen is to be done by tertiary butoxide anion $(CH_3)_3C$—O^-, as it is a big ion, it is not able to approach the middle carbon, as per requirement by Saytzeff's rule, therefore, it abstracts one proton from the terminal carbon and forms a terminal alkene

   $(CH_3)_3$—C—O^- + H—CH_2—CH^+—CH_2—CH_3 $\longrightarrow CH_2{=}CH$—$CH_2$—$CH_3$

   1-butene

   + $(CH_3)_3$—C—OH

13. Alkenes are not reduced by sodium and alcohol, in general. However, terminal alkenes have been found to be reduced by sodium and alcohol. This reaction is termed Birch reaction.

   C_2H_5—OH + Na $\longrightarrow C_2H_5$—O^-Na^+ + H^+

   sodium ethoxide

   $RCH{=}CH_2$ + H_2 $\longrightarrow R\,CH_2$—CH_3

14. Tetra alkyl ammonium halides $R_4N^+X^-$ on treating with $AgOH$, form tetra alkyl ammonium hydroxide, $R_4N^+OH^-$

   $R_4N^+X^-$ + $AgOH$ $\longrightarrow R_4N^+OH^-$ + AgX

   On heating $R_4N^+OH^-$, it decomposes to form trialkyl amine R_3N, an alkene and a molecule of water.

   R_4N—OH $\longrightarrow R_3N$ + an alkene +H_2O

   These compounds are formed only when alkyl radical happens to be ethyl or a higher alkyl radical. On heating tetra methyl ammonium hydroxide, $(CH_3)_3NOH$, the compounds formed are trimethyl amine and water.

   $(CH_3)_3NOH$ $\longrightarrow (CH_3)_3N$ + CH_3OH

   Formation of methanol, can be explained as it requires at least two carbon atoms for the formation of an alkene. Since methyl contains only one carbon, and alkene formation requires at least 2 carbon atoms therefore methanol is formed.

15. Aromatic ketones, Ar—CO— R and Ar—CO—Ar, do not form sodium bisulphite addition compounds when treated with an aqueous solution of $NaHSO_3$ because of the steric hindrance offered by the aromatic ketones.

   Aliphatic aldehydes and ketones, form addition compound, on treatment with aqueous solution of $NaHSO_3$.

   $${>}C{=}O + NaHSO_3 \longrightarrow \overset{\overset{\displaystyle OH}{|}}{>}C{-}SO_3Na$$

   sodium bisulphate addition compound

These sodium bisulphite addition derivatives, are crystalline compounds, have sharp melting point and are used for their identification. These addition compounds, are hydrolysed easily, so this method can be used for their purification.

16. **Hunsdieker reaction:** On heating silver salt of a fatty acid, RCOOAg, with bromine or chlorine, the silver salts decompose to form alkyl bromide or an alkyl chloride.

$$RCOOAg + Br_2 \longrightarrow RBr + CO_2 + AgBr$$

However, on heating a silver salt of a fatty acid with iodine, an ester, RCOOR is formed.

$$2\ RCOOAg + I_2 \longrightarrow RCOOR + 2AgI + CO_2$$

This reaction is known as *Simonini Birnbaum reaction.*

17. Lithium aluminum hydride $LiAlH_4$ reduces all reducible groups, excepting an olefinic double and an acetylinic triple bond.

The reduction by $LiAlH_4$, involves reduction by a hydride ion (H^-).

$$Li\ Al\ H_4 \longrightarrow Al\ H_3 + Li\ H^-$$

$$\overset{+}{>}C=O^- + H^- \xrightarrow{Li^+} >C \overset{OLi^+}{\underset{H}{<}} \xrightarrow{H_2O} >CHOH$$

18. HI act as a reducing agent, but neither HBr nor HCl act as a reducing agent. The amount of energy required for decomposing one molecule into H and I atoms is ~70 kcal/mole. As this much energy is available in the system, HI breaks to form H and I atoms. Where as I combines with another iodine atom to form iodine molecule, the H reduce the compound. On the other hand, HCl requires a much higher amount of energy, which is not available in the system therefore HCl does not act as a reducing compound.

19. The anti-Markovnikov's rule is followed by HBr in presence of a peroxide, while adding to an unsymmetrical alkene, but HBr and HCl do not. Addition of HBr, across double bond in presence of a peroxide, is a free radical reaction. The amount of energy required for a molecule of HBr, to break into H + Br is around 83 kcal/mole, is available in the system. Taking the example of propylene, free radical Br• attacks the double bond, to produce two free radicals of different stabilities.

$$CH_3—CH=CH_2 \xrightarrow{Br^\bullet} CH_3CH^\bullet CH_2Br + CH_3CHBr—CH_2{}^\bullet$$

    propene   2° free radical   1° free radical

As a 2° free radical is more stable than a 1° free radical, the final product is formed from a 2° free radical (anti-Markovnikov's rule)

- The amount of energy required for HCl is around 103 kcal/mole, since this much energy is not available in the system therefore addition of HCl in the presence or absence, gives the same production, i.e. addition of HCl follows Markovnikov's rule.

- HI requires only 70 kcal/mole, breaks up to form a H and I, but iodine atom, so formed combine immediately with another iodine atom to form iodine molecule. For a free radical reaction, iodine atom is needed. This why, HI does not add across the double bond by anti-Markovnikov's rule.

20. Alcohols are neutral to litmus paper, but phenol is acidic in nature: Phenols are acidic because phenol form stable phenoxide ion, in aqueous solution. Phenol dissociate it self to release a proton and a phenoxide ion, which is stabilised by resonance.

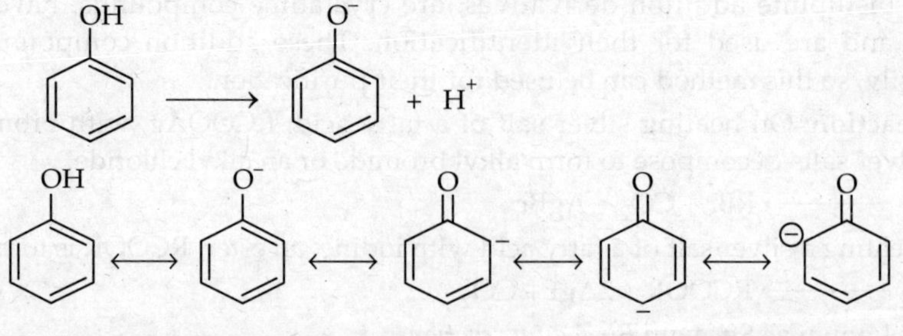

The −ve charge on the oxygen atom is delocalized throughout the benzene ring, there by effectively dispersed. This charge delocalisation is a stabilizing factor in phenoxide ion. On the other hand, the −ve charge is localised and concentrated on the oxygen atom of RO⁻. As a result alcohols are much weaker acids than phenols.

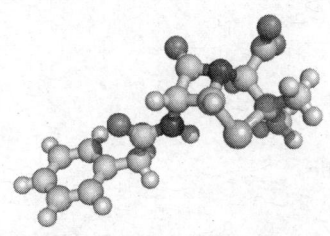

Single Step Reactions

REASONING TYPE QUESTIONS (RTQs)

1. An aldehyde to an ester

    RCHO $\longrightarrow$ RCOOR

2. A ketone to an ester

    RCOR $\longrightarrow$ RCOOR

3. A 1° alky halide to an aldehyde

    $RCH_2Cl \longrightarrow RCHO$

4. A monocarboxylic acid to a 1° alcohol

    $RCOOH \longrightarrow RCH_2OH$

5. A 1° amine to an alkyl isocyanide

    $RNH_2 \longrightarrow RNC$

6. A 1° alcohol to an aldehyde

    $RCH_2OH \longrightarrow RCHO$

7. Oxalic acid to formic acid

    $COOH_2 \longrightarrow HCOOH$

8. Glycerol to allyl alcohol

    $\underset{\overset{|}{OH}}{CH_2}-\underset{\overset{|}{OH}}{CH}.\underset{\overset{|}{OH}}{CH_2} \longrightarrow CH_2=CH-CH_2OH$

9. Glycerol to acrolein

    $\underset{\overset{|}{OH}}{CH_2}-\underset{\overset{|}{OH}}{CH}-\underset{\overset{|}{OH}}{CH_2} \longrightarrow CH_2=CH-CHO$

10. Glycol to formaldehyde

    $CH_2OH.CH_2OH \longrightarrow HCHO$

11. Glycol to acetaldehyde

    $CH_2OH.CH_2OH \longrightarrow CH_3CHO$

12. An alkene to an aldehyde

$$CH_2=CH_2 \longrightarrow CH_3CHO$$

13. An aldehyde to a methyl ketone

$$RCHO \longrightarrow RCOCH_3$$

14. A terminal alkyne to an alkyne with a triple bond in the middle

$$RCH_2C\equiv CH \longrightarrow RC\equiv C-CH_3$$

15. A carboxylic acid to an alkyl bromide

$$RCOOH \longrightarrow RBr$$

16. An alcohol to methyl ether

$$ROH \longrightarrow ROCH_3$$

17. A 1° amine to a 2° amine

$$RNH_2 \longrightarrow R_2NH$$

18. An alkyne to an alkene

$$RC\equiv CH \longrightarrow RCH=CH_2$$

19. A carboxylic acid to next higher homologue

$$RCOOH \longrightarrow RCH_2COOH$$

20. A carboxylic acid to 1° amine

$$RCOOH \longrightarrow RNH_2$$

21. An ether to a an ester

22. An ester to carbonyl compound

$$-RCOOR \longrightarrow RCOR$$

23. $n\ CH_3CH_2CHCHCH_3$ to $CH_3-CH-CH_3$

24. Glucose to gluconic acid

25. An amid to an amine

26. An alkene to a carboxylic acid.

ANSWERS

1. An aldehyde ($RCHO$) can be converted to an ester $RCOOR$, by heating aldehyde with aluminium ethoxide $Al(OC_2H_5)_3$.

$$2R-CHO \xrightarrow{Al\ (OC_2H_5)_3} R-CO-R$$

This reaction is known as *Tischenko's reaction*.

2. A ketone can be converted to form an ester by heating it with peroxy acid $RCOOOH$

$$RCOR + O \longrightarrow RCOOR$$

3. A 1° alkyl halide RCH_2Cl may be converted to form an aldehyde, by oxidizing with pyridinium chloro chromate, PCC, $(C_5H_5NH^+CrO_3Cl)$, in methylene dichloride.

$$R-CH_2Cl \xrightarrow{PCC/CH_2Cl_2} R-CHO + Cr_3^+$$

4. A monocarboxylic acid, $RCOOH$, can be reduced by lithium aluminium hydride, $LiAlH_4$ to a primary alcohol, RCH_2OH.

$$RCOOH \xrightarrow{LiAlH_4} RCH_2OH$$

Mechanism

$$LiAlH_4 \longrightarrow AlH_3 + Li^+H^-$$

$$>C=O + Li^+H^- \xrightarrow{\hspace{1cm}} >\overset{\overset{\displaystyle H}{|}}{C}-O^- Li^+ \xrightarrow{H^+ (H_2O)} >\overset{\overset{\displaystyle H}{|}}{C}-OH + LiOH$$

5. A 1° or a primary alcohol RCH_2—OH is converted to form an aldehyde, RCHO on oxidation by pyridinium chloro chromate or PCC, $(C_5H_5N^+HCr_3Cl)$. In this case the oxidation stops at the aldehydic stage and the aldehyde formed is not oxidised further.

$$RCH_2—OH \xrightarrow{\hspace{2cm}} RCHO$$

6. Oxalic acid $(COOH)_2$ is converted to formic acid (HCOOH) by heating with glycerol to a temperature of 110°C.

$$
\begin{array}{l}
CH_2OH \\
| \\
CH_2OH \\
| \\
CH_2OH \\
\text{glycerol}
\end{array}
+
\begin{array}{l}
COOH \\
| \\
COOH \\
\text{oxalic} \\
\text{acid}
\end{array}
\xrightarrow{110°C}
\begin{array}{l}
CH_2OCO—COOH \\
| \\
CHOH \\
| \\
CH_2OH \\
\text{glyceryl mono-} \\
\text{formate}
\end{array}
\xrightarrow{110°C}
HCOOH +
\begin{array}{l}
CH_2OH \\
\text{formic} \\
\text{acid}
\end{array}
\begin{array}{l}
CH_2OH \\
| \\
CH_2OH \\
| \\
CH_2OH
\end{array}
$$

8. Allyl alcohol $CH_2=CH—CH_2OH$ is obtained by heating oxalic acid with glycerol to a temperature 260°C.

$$
\begin{array}{l}
CH_2OH \\
| \\
CH_2OH \\
| \\
CH_2OH \\
\text{glycerol}
\end{array}
+
\begin{array}{l}
COOH \\
| \\
COOH \\
\text{oxalic} \\
\text{acid}
\end{array}
\xrightarrow{260°C}
\begin{array}{l}
CH_2OCO \\
| \\
CHOCO \\
| \\
CH_2OH \\
\text{glyceryl dioxalin}
\end{array}
\xrightarrow{H_2O}
\begin{array}{l}
CH_2 \\
\| \\
CH \\
| \\
CH_2OH \\
\text{allyl alcohol}
\end{array}
$$

9. Acrolein, $CH_2=CH—CHO$ is prepared from glycerol, by heating it with either $KHSO_4$ or conc. H_2SO_4.

$$
\begin{array}{l}
CH_2OH \\
| \\
CHOH \\
| \\
CH_2OH
\end{array}
+ KHSO_4 \xrightarrow{\hspace{2cm}}
\begin{array}{l}
CH_2 \\
\| \\
CH \\
| \\
CHO
\end{array}
$$

10. Glycol is subjected to periodic oxidation with HIO_4, when the two —OH groups, attached to two adjacent carbons undergo oxidation with cleavage of carbon–carbon bonds.

$$
\begin{array}{l}
CH_2OH \\
| \\
CH_2OH
\end{array}
\xrightarrow{HIO_4} HCHO + HCHO
$$

11. On heating glycol with conc. H_2SO_4 it is dehydrated and rearranges to form acetaldehyde

$$
\begin{array}{l}
CH_2OH \\
| \\
CH_2OH
\end{array}
\xrightarrow{\hspace{2cm}} CH_3CHO
$$

12. An alkene may be converted into an aldehyde, by the following two processes.

    i. Oxo process: An alkene is heated with $CO + H_2$ in the presence of a catalyst, when it is converted to form aldehydes.

    $$CH_3{-}CH{=}CH_2 + CO + H_2 \xrightarrow{Co(CO)_4} CH_3CH_2CH_2\,CHO + (CH_3)_2\,CH\,CHO$$

        propene   butyraldehyde
            iso butyraldehyde

    ii. Waker's reaction

    $$CH_2{=}CH_2 + PdCl_2 + H_2O \xrightarrow{\hspace{3cm}} CH_3CHO + Pd + 2HCl$$

13. An aldehyde may be converted to form a methyl alkyl ketone, by treating with diazomethane CH_2N_2.

    $$RCHO + CH_2N_2 \xrightarrow{\hspace{2cm}} RCOCH_3 + N_2$$

14. A terminal alkyne may be converted into an alkyne, having the triple bond in the middle by heating it with $NaNH_2$ to a temperature 170 °C, when the terminal triple bond moves to the middle in the molecule.

    $$R{-}CH_2{-}Ca{\equiv}CH \xrightarrow{\hspace{2cm}} R{-}Ca{\equiv}C\,CH_3$$

15. A carboxylic acid may be converted into an alkyl bromide by treating the silver salt of the acid with bromine.

    $$RCOOH \xrightarrow{\hspace{1cm}} RCOOAg + Br_2 \xrightarrow{\hspace{1cm}} RBr + AgBr + CO_2$$

    This reaction is termed *Borodine–Hunsdieker reaction*.

16. An alcohol may be converted to form methyl ether by treating it with diazomethane, CH_2N_2.

    $$R{-}OH + CH_2N_2 \xrightarrow{\hspace{2cm}} R{-}O{-}CH_3$$

17. A primary amine may be converted into a secondary amine, by treating it with diazomethane, CH_2N_2.

    $$RNH_2 + CH_2N_2 \xrightarrow{\hspace{2cm}} RNHCH_3$$

18. An alkyne into an alkene, reduction with two different reducing agents.

    i. By sodium and alcohol, the hydrogen adds in the alkyne in the *trans*-manner.

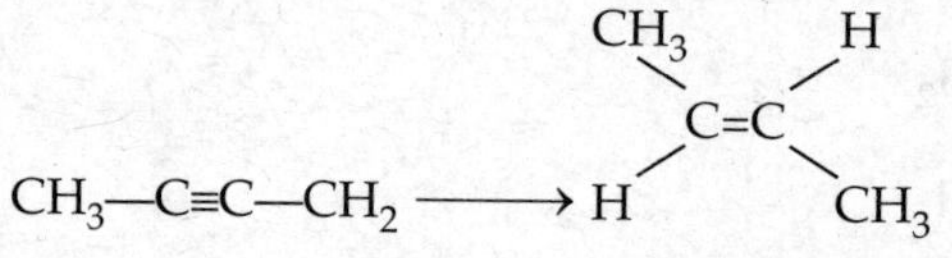

    $$CH_3{-}C{\equiv}C{-}CH_2 \longrightarrow \underset{\text{\textit{trans}-alkene}}{\begin{array}{c} CH_3 \qquad H \\ \diagdown \; / \\ C{=}C \\ / \qquad \diagdown \\ H \qquad CH_3 \end{array}}$$

    ii. By Lindlar's catalyst and hydrogen, when a *cis*-addition takes place:

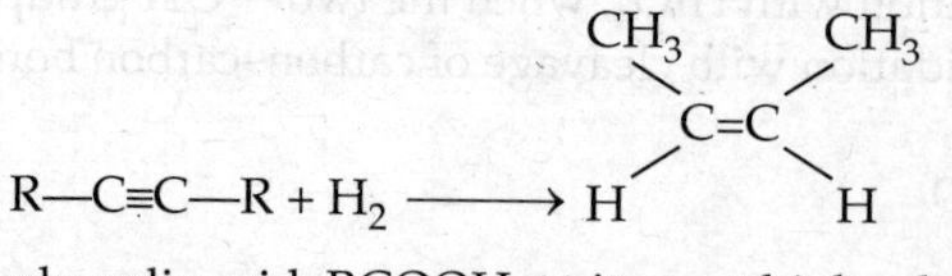

    $$R{-}C{\equiv}C{-}R + H_2 \longrightarrow \begin{array}{c} CH_3 \qquad CH_3 \\ \diagdown \; / \\ C{=}C \\ / \qquad \diagdown \\ H \qquad H \end{array}$$

19. A carboxylic acid, RCOOH, to its next higher homologue (in two steps):

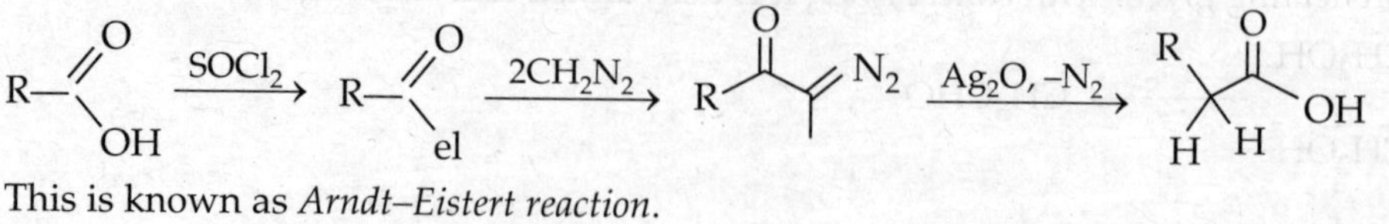

    This is known as *Arndt–Eistert reaction*.

20. A RCOOH is treated with hydrazoic acid N_3H to form a primary amine.

$$RCOOH + N_3H \longrightarrow RNH_2 + CO_2 + N_2$$

21. An ether R—O—R is converted to an ester RCOOR by heating it to a temperature of 150 °C/500 atms. in presence of BF_3 as a catalyst.

$$R—O—R \xrightarrow{\text{BF}_3} RCOOR$$

22. An ester RCOOR to a ketone RCOR by heating it with $(CH_3)_2SO$.

$$RCOOR + (CH_3)_2SO \longrightarrow RCOR + (CH_3)_2SO_2$$

23. An n-butane $CH_3—CH_2—CH_2—CH_3$ to iso butane, $CH_3—CH—(CH)_3—CH_3$ by heating it with anhydrous $AlCl_3$.

$$CH_3—CH_2—CH_2—CH_3 \longrightarrow CH_3—CH(H_3)—CH_3$$

24. Glucose to gluconic acid by oxidizing it with Tollen's reagent.

$$
\begin{array}{ccc}
\text{CHO} & & \text{COOH} \\
| & \xrightarrow{\text{Ag}_2\text{O}} & | \\
(\text{CHOH})_4 & & (\text{CHOH})_4 \\
| & & | \\
\text{CH}_2\text{OH} & & \text{CH}_2\text{OH} \\
\text{glucose} & & \text{gluconic acid}
\end{array}
$$

25. There are two methods to convert an amide to an amine
    i. by Hofmann's bromamide reaction (amine obtained contains one carbon less)

    $$RCONH_2 + Br_2 + 4\,KOH \longrightarrow RNH_2 + K_2CO_3 + 2KBr + H_2O$$

    ii. Reduction of amide by sodium and alcohol or by $LiAlH_4$ (the amine obtained contains same number of carbon atoms).

    $$RCONH_2 + 2H_2 \longrightarrow RCH_2NH_2$$

26. An alkene can be converted to a carboxylic acid by heating the alkene with CO and H_2O, in presence of H_3PO_4.

$$RC=CH_2 + CO + H_2O \xrightarrow{\text{H}_3\text{PO}_4} R—CH_2—COOH$$

This reaction is termed as *Koch synthesis*.

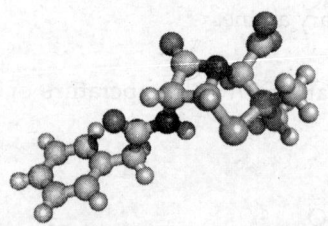

Polycyclic Alkanes

Salient Features

These are the cyclic hydrocarbons and their derivatives, which have rings with two or more carbon atoms in common. Compounds of this type are named systematically by attaching the prefix "bicyclo" to the name of open chain hydrocarbon having the same total number of carbon atoms as in the ring. The size of the two rings is specified by the number of carbon atoms in each of the three linkages which connect the two atoms which are the ring junctions.

For example: Decalin has total 10 carbon atoms in the rings, as in the case of decane, the two rings are joined at carbon number 5 and carbon number 6; each ring contains 4 carbon atoms and there is no other carbon in between carbon number 5 and 6 (C-5 and C-6 are termed as ring junctions). Therefore, the molecule has only two bridges.

$$\begin{array}{ccc}
& {}^1CH_2 & & {}^{1'}CH_2 & \\
CH_2{}^2 & & CH & & {}^{2'}CH_2 \\
\mid_3 & & \mid & & \mid_{3'} \\
CH_2 & & CH & & CH_2 \\
& {}_4CH_2 & & {}_{4'}CH_2 &
\end{array}$$

In naming the cycloalkane, the numbers are enclosed in the square brackets, after the prefix "bicyclo" and before the name of the hydrocarbon. The numbers are listed in order of decreasing magnitude and properly separated by periods, not commas as decalin is named as bicyclo[4.4.0] decane.

Example 1

- The total number of carbon in the compound = 4 (as in the case of butane)
- Two rings are joined at carbon 1 and 3, which may be termed ring junctions. Therefore, the compound has two bridges.

$$\begin{array}{ccc}
& {}^1CH_2 & \\
{}^4CH_2 & {}_3\mid & {}^2CH_2 \\
& CH_2 &
\end{array}$$

- There is no other carbon in between carbon number 1 and 3. Therefore the compound may be named as bicyclo[1.1.0] butane.

Example 2
- The compound has total number of carbon equal to 6 (as in the case of hexane).
- The two rings are joined at carbon number 1 and 4 (carbon number 1 and 4 are ring junctions). Therefore, there are three bridges.
- The larger ring contains two carbons in the ring, the smaller contains only one carbon in the ring. The third ring also contains one carbon. Therefore, compound may be termed as bicyclo[2.1.1] hexane.

$$
\begin{array}{c}
\diagup {}^{1}CH_2 \diagdown \\
CH_2 \quad | \\
|^{6} \quad {}^{2} \\
|_{5} \quad CH_2 \quad {}_{3}CH_2 \\
CH_2 \quad | \\
\diagdown {}_{4}CH_2 \diagup
\end{array}
$$

Example 3
- The total number of carbon atoms are 8 (as in the case of octane).
- The 3 rings are formed involving carbon number 1 and 5.

$$
\begin{array}{c}
\diagup {}^{1}CH_2 \diagdown \\
CH_2 \quad | \\
|^{8} \quad {}^{2} \\
|_{7} \quad CH_2 \\
CH_2 \quad | \quad {}^{3}CH_2 \\
|_{6} \quad {}_{4}CH_2 \diagup \\
CH_2 \quad | \\
\diagdown {}_{5}CH_2
\end{array}
$$

- The larger ring contains 3 carbon atoms, and the other two, rings contain 2 and 1 carbon each, respectively.
- The compound may be named as bicyclo [3.2.1] octane.

For cycloalkanes, *containing nitrogen* in the ring, a prefix "aza" is used, to indicate the presence of nitrogen. For the purpose of determining the size of the ring, nitrogen is taken as the carbon, and if possible, is given lower number.

Example 4
- The number of carbon atoms, including the nitrogen, is equal to 8 (as is present in octane). nitrogen 1 and carbon 6 are ring junctions and three rings are formed.
- Three rings or three bridges are formed.
- The larger ring contain 3 carbon, other small ring contains 2 carbon atoms, the third ring contains only one atom.

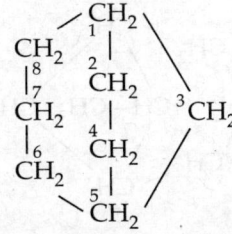

- The compound may be named as 1-aza-bicyclo [3:2:1] octane.

Similarly, for the compounds *containing oxygen,* being a part of the ring, a prefix 'oxa' is used, to indicate the position of oxygen in it and is given lower possible number.

Example 5

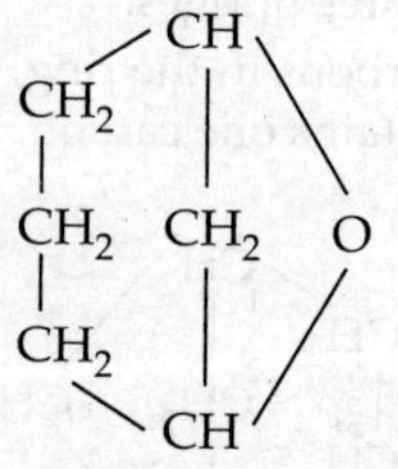

$$
\begin{array}{c}
CH \\
CH_2 \quad | \\
| \quad | \\
CH_2 \quad CH_2 \quad O \\
| \quad | \\
CH_2 \quad | \\
CH
\end{array}
$$

- The compound may be named as 1-oxa- bicyclo [3.1.1] heptane.

For compound *containing substituted groups,* like chlorine or a methyl group, the compound is named by applying the usual IUPAC rules.

- The compound may be named as 1-chloro-7-methyl-bicyclo [3.1.1]heptane.

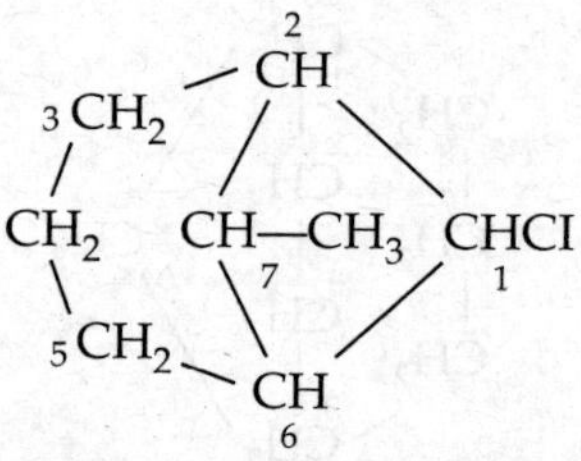

$$
\begin{array}{c}
\overset{2}{CH} \\
{}_3CH_2 \\
{}_4CH_2 \quad CH-CH_3 \quad CHCl \\
\qquad \quad 7 \qquad \quad 1 \\
{}_5CH_2 \\
\overset{CH}{_6}
\end{array}
$$

Spiro compounds: These compounds are **bicyclic, but have only one carbon atom common to both the rings.**

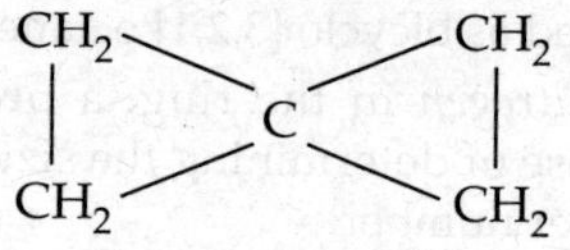

$$
\begin{array}{c}
CH_2 \qquad\qquad CH_2 \\
| \qquad C \qquad | \\
CH_2 \qquad\qquad CH_2
\end{array}
$$

- The compound is named as spiro[2.2]pentane

Bridged Cyclokanes

Name the following Structures

1. 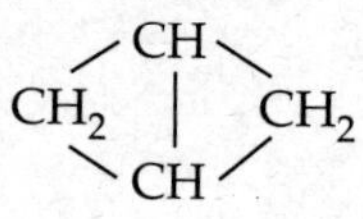
$$
\begin{array}{c}
CH \\
CH_2 \quad | \quad CH_2 \\
CH
\end{array}
$$

2.
$$
\begin{array}{c}
CH \\
CH_2 \quad | \quad CH_2 \\
| \quad CH_2 \quad | \\
CH_2 \qquad CH_2 \\
CH
\end{array}
$$

3.
$$
\begin{array}{c}
CH \\
CH_2 \quad | \quad CH_2 \\
| \quad CH_2 \quad | \\
CH_2 \qquad CH \\
CH_2
\end{array}
$$

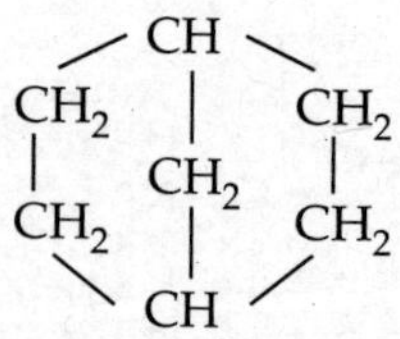

4. 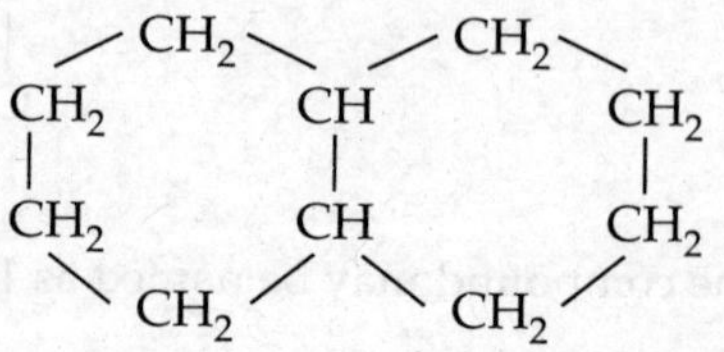
$$
\begin{array}{c}
CH_2 \qquad CH_2 \\
CH_2 \quad CH \quad CH_2 \\
| \qquad\qquad | \\
CH_2 \quad CH \quad CH_2 \\
CH_2 \qquad CH_2
\end{array}
$$

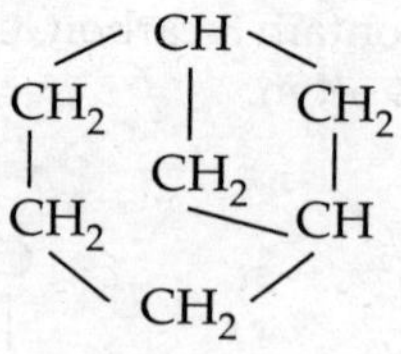

5.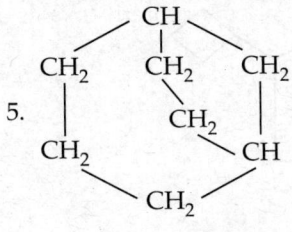

6.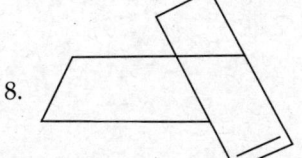

7.

8.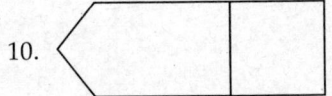

9.

10.

11.

12.

13.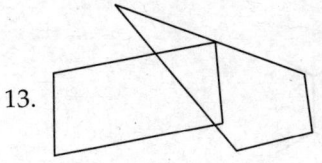

14.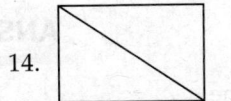

15.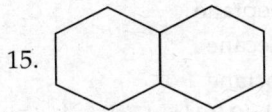

16.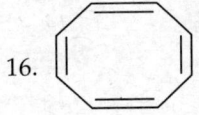

17.

18.

19.

20.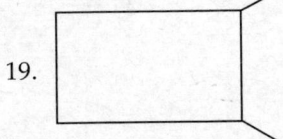

21.

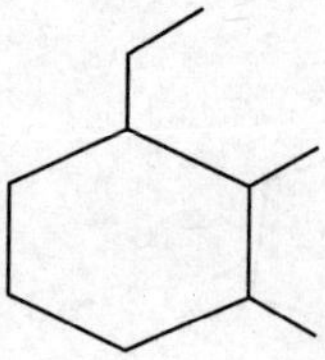

22.

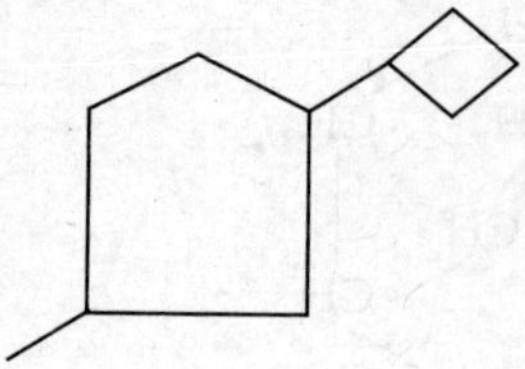

ANSWERS

1. Bicyclo[1.1.0] butane
2. Bicyclo[2.2.1] heptane
3. Bicyclo[3.3.1] heptane
4. Bicyclo[4.4.2] decane
5. Bicyclo[3.2.1.] octane
6. 1-Chloro bi cyclo[3.2.1] octane
7. Bicyclo[2.2.1] heptane
8. Bicyclo[2.2.2] 1-octene
9. Bicyclo[2.2.1] heptane
10. Bicyclo[2.2.1] heptane
11. Bicyclo[4.2.0] octane
12. Bicyclo[3.2.2] Nonane
13. Bicyclo[3.2.1] octane
14. Bicyclo[1.1.0] butane
15. Bicyclo[4.4.0] decane
16. Cyclo-1,3,5,7-octatetraene
17. Cyclobutene
18. 1-ethyl cyclopropane
19. 1,2-Dimethyl cyclo butane
20. 1-isopropyl-3,3-dimethyl cyclo pentane
21. 1-Iodo-2-methyl-3-ethyl cyclohexane
22. 1-Cyclobutyl-3-mehyl cyclopentane

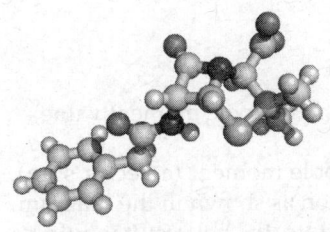

Dipole Moment

Polarity of Bonds

Organic compounds are the compounds of covalency. Two atoms joined by a covalent bond, share electrons; their nuclei are held together by the same electron cloud. But in most cases, the two nuclei do not share the electron clouds equally; the electron cloud is denser about one atom than the other. One end of the bond is denser as compared to the other end. One end is thus slightly negative and the other end is relatively positive: that is, there is a negative pole and a positive pole. Such a bond is said to be a polar bond or that the bond possess polarity.

Polarity is indicated by using the symbols $\delta+$ and $\delta-$ which indicates partial +ve and partial −ve charge respectively. A bond is expected to be a polar bond, if it is joined by the atoms of different electronegativities. Greater is the difference in the electronegativities, the more polar the bond will be.

Example:

i. $H^+ \!-\! F$

ii. 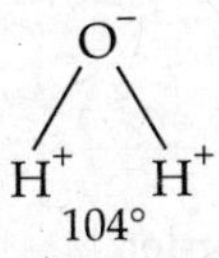
$$
\begin{array}{c}
O^- \\
/ \ \backslash \\
H^+ \quad H^+ \\
104°
\end{array}
$$

iii. 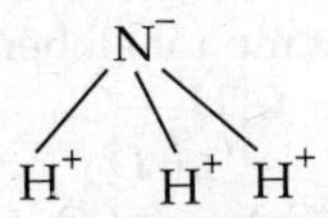
$$
\begin{array}{c}
N^- \\
/ \ | \ \backslash \\
H^+ \quad H^+ \ H^+
\end{array}
$$

Electronegativity: $F > O > Cl, N > Br > C > H$
                        $4 \quad 3.5 \quad 3 \quad 3 \quad 2.8 \quad 2.5 \quad 1$

Polarity of Molecules

A molecule is polar, if the center of negative charge does not coincide with the center of positive charge. Such a molecule will have a dipole; two equal and opposite charges separated in space. A dipole is indicated by the arrow points from positive to negative. The molecule possesses a dipole moment μ which is equal to the magnitude of the charge, e multiplied by distance d between the center of charges.

$$
\begin{array}{ccccc}
\mu & = & e & \times & d \\
\text{dipole moment} & & \text{charge} & & \text{distance} \\
\text{(debye units, D)} & & \text{(esu)} & & \text{(cm)}
\end{array}
$$

Example

Hydrogen fluoride (H–F)

$\mu = 1.82$ D. It is a small molecule, but d is small e is large

(F has very high negativity; pulls the electrons strongly, hence a high dipole moment value).

Some dipole moment values:

As nitrogen in ammonia molecule is sp^3 hybridized, the net dipole moment (a vector sum), resulting from the three individual moments would be in the direction as shown in the diagram. In the case of water molecule, since the oxygen is in sp^3 hybridized state, the net result would be from two individual moments in the direction as shown in the figure.

The dipole moment, therefore, can be described as the vector sum of all individual moments of all the bonds present in the molecule, including any lone pair present in the molecule.

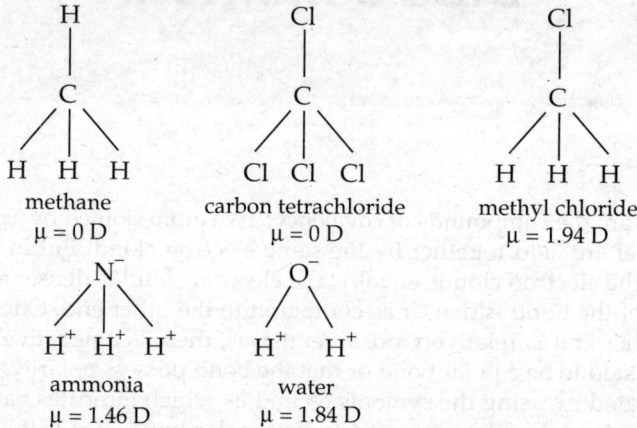

methane
$\mu = 0$ D

carbon tetrachloride
$\mu = 0$ D

methyl chloride
$\mu = 1.94$ D

ammonia
$\mu = 1.46$ D

water
$\mu = 1.84$ D

The dipole moments of methyl halides are in the decreasing order:

$$CH_3—Cl > CH_3—F > CH_3—Br > CH_3—I$$
$$\mu = 1.94\ D \quad 1.82\ D \quad\quad 1.79\ D \quad\quad 1.64\ D$$

The order of the dipole moment of the polychloro derivatives of methane:

$$CH_3Cl > CH_2Cl_2 > CHCl_3 > CCl_4$$

The decreasing order of the dipole moments of monohalobenzenes

$$C_6H_5—I > C_6H_5—Br > C_6H_5—Cl > C_6H_5—F$$
$$\mu = 1.4\ D \quad\quad 1.5\ D \quad\quad\quad 1.6\ D \quad\quad\quad 1.5\ D$$

The decreasing order of the dipole moment of CO_2 and SO_2 in the following order:

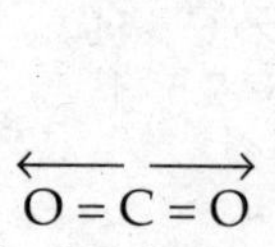

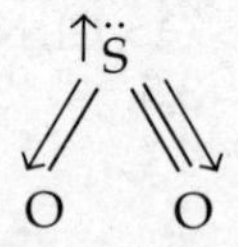

$$\overset{\longleftarrow}{O} = C = \overset{\longrightarrow}{O} \qquad\qquad \underset{O\quad\quad O}{\overset{\uparrow \ddot{S}}{\diagup\!\diagup \ \ \diagdown\!\diagdown}}$$

Dipole moments of the *trans*-derivatives of ethenes will be zero; *cis*-compounds will show some moment:

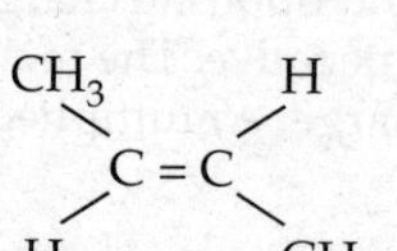

$\mu = $ zero

trans-2-butene

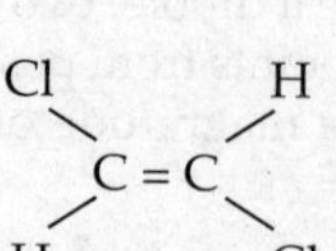

$\mu = $ zero

trans-1,2-dichloro ethane

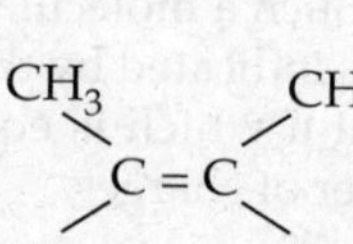

$\mu = $ somewhere

cis-compounds

Similarly 1,4-derivatives of benzene, having same group will show zero dipole moment; while 1, 2 and 1, 3 derivatives will show some dipole moment.

1, 3-disubstituted $\mu = 0$ 1, 2-disubstituted
                                                          μ = somewhere more than
                                                          m-disubstituted derivative

This property of dipole moment has been used to determine the relative positions of the groups in the disubstituted benzenes.

For example: The dipole value for 1,4-dichlorobenzene is zero; the value for 1,3-dichloro-benzene (m-dichlorobenzene) is lower (having angle of 120°) than 1,2-dichlorobenzene (o-dichlorobenzene) which has a higher value (having an angle of 60°).

o-dichlorobenzene   m-dichlorobenzene   p-dichlorobenzene
1,2-dichlorobenzene   1,3-dichlorobenzene   1,4-dichlorobenzene
(angle = 60°)   (angle = 120°)   (angle = 180°)
$\mu = 2.3$ D $\mu = 1.48$ D μ = zero D

Inertness of chlorine in vinyl chloride and chlorobenzene: The dipole moment of vinyl chloride and chlorobenzene has been found to be unusually very small. Organic halogen compounds are polar compounds; the displacement of electrons towards the more electronegative atom should make the halogen atom more electronegative and a carbon relatively more electropositive.

The dipole moment values for alkyl halides is in the range of 2.02 D to 2.15 D. The mobile electrons of the benzene ring and the C to C double bond should be particularly easy to displace towards the halogen atom, with the result, the dipole moment value of vinyl chloride and chlorobenzene is expected to be higher as compared to alkyl halides.

In the cases of the vinyl chloride, the dipole moment value has been found to be only 1.4 D and in case of chlorobenzene, the dipole moment is 1.7 D. On account of resonance, the hybrid structures formed in case of vinyl chloride and chlorobenzene, oppose the usual displacement of the electrons toward the halogen atom. Although, there is still a net displacement toward the halogen atom in vinyl chloride and chloro benzene, it is less than in the case of alkyl halides.

The resonating structure of chlorobenzene:

I   II   III   IV   V

As there is a partial double bond in between C—Cl bond (III, IV and V), there is a slight positive charge on chlorine and a negative charge on the carbon atom.

The resonating structure of vinyl chloride:

$CH_2=CH-Cl \longleftrightarrow CH_2{}^--CH=Cl^+$

The carbon to which the chlorine is joined is sp^2 hybridized. It is more electronegative as compared to sp^3 hybridized carbon and it will not release electrons to chlorine. Thus –Cl is firmly attached and will require a large amount of energy to react with any compound. Therefore, chlorobenzene (or bromobenzene) or vinyl chloride do not undergo usual displacement reactions.

MULTIPLE CHOICE QUESTIONS

1. Which molecule will have some dipole moment value?

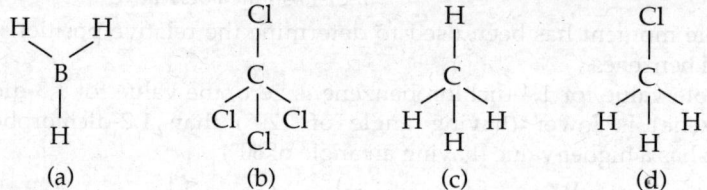

    (a)   (b)   (c)   (d)

2. Dipole moment is zero for:
    (a) ClF   (b) PCl_3 (c) SiF_4 (d) $CFCl_3$

3. If a molecule MX_3 has a zero dipole moment, the σ-orbitals used by M (atomic number 21) are:
    (a) A pure p (b) sp hybridised
    (c) sp^2 hybridised   (d) sp^3 hybridised

4. The dipole moment (in Debye units) value of m-dibromobenzene is 1.54 D, what will be the value of o-dibromobenzene?
    (a) zero   (b) 2.67
    (c) 0.62   (d) more than 1.54 D

5. The dipole moment value of chlorobenzene is the same as that of:
    (a) o–dichlorobenzene   (b) m-dichlorobenzene
    (c) p-dichlorobenzene   (d) 1, 3, 5-trichlorobenzene

6. Which of the following has a zero dipole moment value?
    (a) CO_2 (b) SO_2
    (c) p-dibromobenzene   (d) chlorobenzene

7. The dipole moment of NF_3 is much less than that of NH_3, because:
    (a) the two molecules do not have the same geometry
    (b) fluorine is more electronegative as compared to nitrogen, whereas hydrogen is less electronegative as compared to nitrogen
    (c) unshared pair of electron is not present in NF_3
    (d) no suitable answer

8. Which of the following will have a zero dipole moment?
    (a) *cis*-2-butene   (b) *trans*-2-butene
    (c) 1-butene   (d) *trans*-2-pentene

9. Which of the following will have a zero dipole moment?
    (a) 1, 1-dichloro ethylene   (b) *cis*-1, 2-dichloro ethylene
    (c) *trans*-1, 2-dichloroethylene   (d) per bromobutane

10. The compound having no dipole moment is:
    (a) CH_3Cl (b) CCl_4 (c) CH_2Cl_2 (d) $CHCl_3$

11. Which of the following has a zero dipole moment?

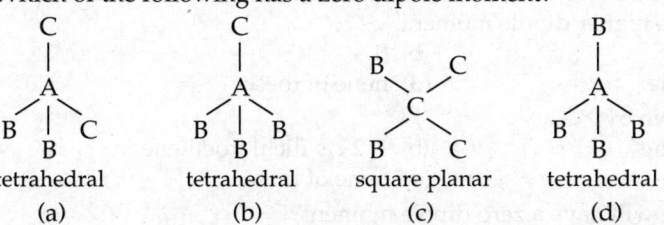

    tetrahedral   tetrahedral   square planar   tetrahedral

      (a)   (b)   (c)   (d)

12. Which one of the following has dipole moment?
    (a) CCl_4
    (b) *cis*-2-butene
    (c) *trans*-2-butene
    (d) 2-methylpropane

13. Which of the following compound will have highest dipole moment?
    (a) NH_3
    (b) PH_3
    (c) AsH_3
    (d) SbH_3

14. Which of these compounds have a net dipole moment?

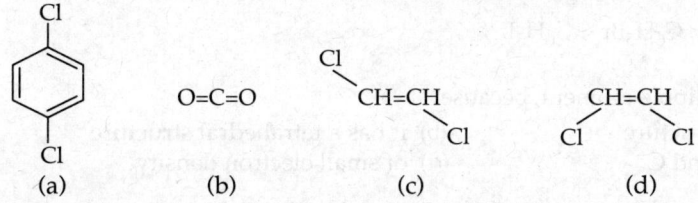

    (a)   (b)   (c)   (d)

15. Dipole moment of p-nitro aniline, as compared to that of aniline (X) and nitrobenzene (Y) will be
    (a) greater than (X) and (Y)
    (b) smaller than (X) and (Y)
    (c) Greater than (X), but smaller than (Y)
    (d) equal to zero

16. In defining the dipole moment, mark the correct statement.
    (a) It is the product of q and d, where q is the magnitude of particle charge and d is the distance between of the opposite charges.
    (b) the unit for dipole moment is debye
    (c) all are correct

17. Mark the correct order of the alkyl halides
    (a) $CH_3F > CH_3Cl > CH_3Br > CH_3I$
    (b) $CH_3Cl > CH_3F > CH_3Br > CH_3I$
    (c) $CH_3I > CH_3Br > CH_3Cl > CH_3F$
    (d) none of these

18. The dipole moments of CO_2 and SO_2 are in the order of:
    (a) $CO_2 > SO_2$
    (b) $SO_2 > CO_2$
    (c) $CO_2 = SO_2$
    (d) none of these

19. The correct order of the dipole moment of CH_3Cl, CH_2Cl_2, $CHCl_3$ and CCl_4.
    (a) $CH_3Cl > CHCl_3 > CH_2Cl_2 > CCl_4$
    (b) $CH_3Cl > CH_2Cl_2 > CHCl_3 > CCl_4$
    (c) $CH_2Cl > CHCl_3 > CH_3Cl > CCl_4$
    (d) $CCl_4 > CHCl_3 > CH_2Cl_2 > CH_3Cl$

20. Out of the two compounds (A) 1, 2-dibromobenzene and (B) 1, 4-di bromobenzene, the compound which has a higher dipole moment.
    (a) A
    (b) B
    (c) have the same value
    (d) none of these

21. Dipole moment is shown by:
    (a) 1, 4-dichlorobenzene
    (b) 1, 2-*cis*-dichloroethene
    (c) 1, 2 dichloroethene
    (d) none of these

22. Which of the following will have a zero dipole moment?
    (a) 1, 1-dichloroethylene
    (b) *cis*-1, 2 dichloroethylene
    (c) *trans* 1, 2-dichloroethylene
    (d) none of these

23. Which of the following will have zero dipole moment?
    (a) *cis*-2-butene
    (b) *trans*-2-butene
    (c) 1-butene
    (d) 2-mehyl-1-propene

24. Arrange the following in the decreasing order of dipole moment.
    (a) $C_6H_5Cl > C_6H_5I > C_6H_5Br > C_6H_5F$
    (b) the reverse of (a)
    (c) $C_6H_5F < C_6H_5Cl < C_6H_5Br < C_6H_5I$
    (d) none of these

25. CCl_4 does have any dipole moment, because
    (a) it has a planar structure
    (b) it has a tetrahedral structure
    (c) small size of Cl and C
    (d) of small electron density

ANSWERS

1. (d)
2. (c) SiF_4 is sp^3 hybridised
3. (c)
4. (d) it has more than 1.54 D, because in o-compound the angle is more acute
5. (c) the value of C—Cl, $\mu = 1.2$ D
6. (c) $\mu = \mu_1 + \mu_2\ \mu_1\mu_2 \cos = \alpha, \mu_1 = \mu_2, \alpha$ incomplete
   The p-dibromobenzene will have $\mu = 0$, because the angle $\mu = 180°$
7. (c) In F—N—F, the lone pair on nitrogen contributes opposite dipole moment to N—F̈
   whereas in NH_3, the lone pair contributes for additive dipole moment
8. (b) 9. (b) 10. (d) 11. (d) 12. (c) 13. (a)
14. (d) the value of C—Cl M = 1.2 D

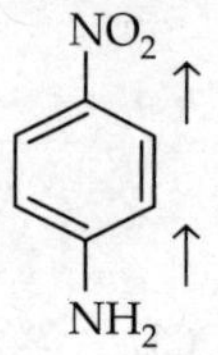

15. (a) 16. (d) 17. (b) 18. (b) 19. (a) 20. (a) 21. (b) 22. (c)
23. (b) 24. (b) 25. (b)

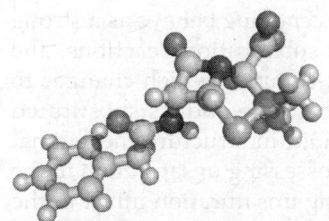

Aromaticity and Nonbenzenoid Aromatic Rings

Salient Features

Benzene and its derivatives are highly unsaturated compounds. These contain three alternate double bonds in the molecule. These are expected to form addition compounds, like olefins and should respond to positive colour test. In spite of being highly unsaturated, benzene and its derivatives undergo substitution reaction, rather than forming addition compounds, keeping the unsaturation intact in the molecule. This characteristic property of forming substituted compounds been characterised as aromatic character or the compound possessing aromaticity. The aromatic character or aromaticity observed in benzene and its derivatives has been attributed to the following factors.

  i. Presence of a cyclic ring.
 ii. Monoplanar structure of the ring, i.e. all atoms in ring being on the same plane.
iii. The number of π-electrons in the ring follow Huckel's rule, i.e. $(4n + 2)$ π-electrons, where the value of $n = 0, 1, 2, 3$, etc. and the number of electrons in the ring should be 2, when the value of $n = 0$; it should be 6, when the value of $n = 1$; it should be 10, when the value of $n = 3$ and so on.

Some aromatic rings:

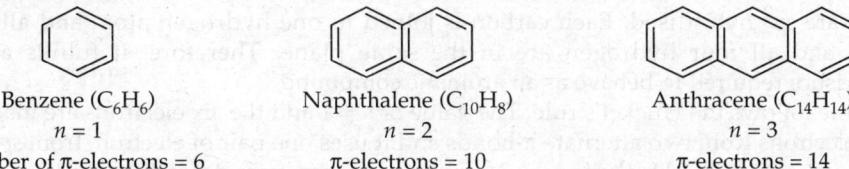

Benzene (C$_6$H$_6$)	Naphthalene (C$_{10}$H$_8$)	Anthracene (C$_{14}$H$_{14}$)
$n = 1$	$n = 2$	$n = 3$
number of π-electrons = 6	π-electrons = 10	π-electrons = 14

In benzene, the 6π-electrons or in naphthalene 10π electrons or in anthracene, 14π-electrons are delocalised and are present in the form of clouds spread over and below the ring, sandwiching the flat ring from top and below, in between which the benzene ring lies.

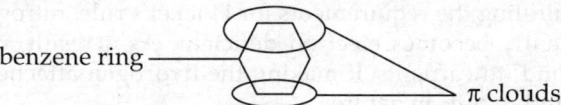

benzene ring —— π clouds

The delocalization of electrons, makes the benzene ring more stable and make all C to C bond lengths equal, i.e. 1.39 Å (instead of 1.54 Å for a single bond and 1.34 Å for a C to C double bond).

An uniform spread of π-clouds, apparently results in making the benzene ring behave as a strong negative center. Since all reactions in benzene are electrophilic substitution reactions, the electrophile gets embedded through π-clouds to form initially a π-complex which changes to σ-complex and finally forming a substituted derivative. Besides benzene and its substituted derivatives there are several other compounds, though not benzenoid in structure (hexagonal carbocyclic), yet they behave as aromatic compounds, i.e. inspite of possessing unsaturation in the cyclic ring, these compounds undergo substitution reactions, keeping unsaturation intact in the molecule. Some of these examples are: nonbenzenoid carbocyclic aromatic hydrocarbons–Azulene, $C_{10}H_8$, a bicyclic hydrocarbon, consists of two cyclic rings, fused together. One ring is a seven membered ring, fused with a five membered carbocyclic ring in 1,2-positions. Seven membered ring contains three double bond and five membered ring contains two double bonds. One double bond is common in the both the rings. All the carbons are sp^2 hybridised, so the molecule is planar and all the atoms of the ring are in the same plane.

Seven membered ring may be regarded as having a seven electron system, while the five membered ring may be regarded having five electrons. If one electron is transferred from seven membered ring to the five membered ring, both will contain, six electrons each, fulfilling Huckel's rule. In doing so, the seven membered ring become electron deficient acquiring a positive charge and the five membered ring becomes electron rich and therefore it becomes electronegatively charged.

It undergoes electrophilic substitution generally in five membered ring.

iv. Five membered heterocyclic aromatic rings:

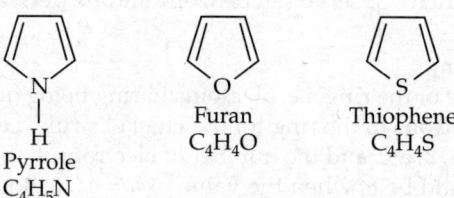

<table>
<tr><td></td><td></td><td></td></tr>
</table>

N | H Pyrrole C_4H_5N	O Furan C_4H_4O	S Thiophene C_4H_4S

Pyrrole: A five membered heterocyclic ring containing four carbons and one nitrogen in the ring. The molecule has two alternate double bonds (or two π-bonds). All four carbons and nitrogen are sp^2 hybridised. Each carbon is joined to one hydrogen atom and all the carbon, nitrogen and all four hydrogen are in the same plane. Therefore, it fulfills all the three characteristics required to behave as an aromatic compound.

Pyrrole follows the Huckel's rule. The value of $n = 1$ and the six electrons are made available, by four electrons from two alternate π-bonds and it uses one pair of electron, from sp^2 hybridised nitrogen, as required by Huckel's rule. These six π-electrons are delocalised and are spread in the form of π-clouds over and below the flat pyrrole ring. Pyrrole therefore behaves as an aromatic compound. It undergoes substitution, preferably in position 2 and 5. In case these positions are not available for substitution, then substitution takes place in positions 3 and 4.

In this process of fulfilling the requirements for Huckel's rule, nitrogen loses its lone pair of electrons and consequently becomes electron deficient. As a result, nitrogen pulls bonding electrons of N ← :H bond, towards itself making the hydrogen attached to nitrogen acidic or making N—H bond slightly acidic in nature.

Pyrrole is not only nonbasic in nature, although it contains = NH group, it is slightly acidic in nature. It reacts with alkali and sodium metal to form sodium salt.

This can be explained in terms of electron densities. Each carbon contributes one electron each, whereas nitrogen contributes a pair of electrons. This group of six electrons is delocalised, over the entire pyrrole ring, consisting of five atoms only, four carbon and one nitrogen. So the electron density on each atom of the ring, comes to $6/5 = 1.2$. Nitrogen contributes two electrons but gets share of 1.2 electron density, this makes nitrogen electron deficient.

$$H_2O + \underset{\underset{Na^+}{N}}{\boxed{}} \xleftarrow{\quad NaOH \quad} \underset{\underset{H}{N}}{\boxed{}} \xrightarrow{\quad Na \quad} \underset{\underset{Na^+}{N}}{\boxed{}} + \frac{1}{2}H\uparrow$$

pyrrole

Furan: It is a five membered ring, consisting of four carbon and one oxygen atom. All atoms are in sp^2 hybridised state. The lone pair of oxygen is in unhybridised *p*-orbital, perpendicular to the plane in which the three hybridised orbitals lie. The furan molecule contains two alternate double bonds. All atoms of furan lie in the same plane. The value of $n = 1$. In order to fulfill the requirements as per Huckel's rule, the four electrons of two π-bonds, along with two electrons of the lone pair of oxygen, furan is able to raise six π-electrons, which in the form of π-clouds are spread over the entire furan ring to behave as an aromatic compound. The substitution takes place at position 2 or 2 and 5, if not available then in positions 3 and 4.

$$\underset{\underset{1}{O}}{\overset{4\overbrace{}3}{\boxed{}}}\,{}^2_{5}$$

furan

Thiophene: Like furan, it is a five membered ring consisting of four carbons and one sulphur atom. All four carbons are sp^2 hybridised, including sulphur. The lone pair of sulphur is in unhybridised p-orbital, being perpendicular to the three hybridised orbitals. As a result, the molecule is flat, with all the atoms lying in the same plane. The value of $n = 1$.

It follows the Huckel's rule. It requires six π-electrons, four π-electrons are contributed by two π-bonds, and two electrons are contributed by sulphur from its lone pair, thus making six electrons in all.

It fulfills all the three requirements, required for a monocyclic flat ring to be aromatic therefore thiophene behaves as an aromatic compound. It undergoes substitution, keeping the unsaturation intact, in the molecule.

$$\underset{\underset{1}{\ddot{S}}}{\overset{4\overbrace{}3}{\boxed{}}}\,{}^2_{5}$$

thiophene

Nonbenzenoid heterocyclic aromatic compound:

Pyridine (C_5H_5N): It is heterocyclic, and is a strongly basic compound. The ring consists of five carbons and one nitrogen. All carbon and the nitrogen are in sp^2 hybridised state and lie in the same plane. The molecule contains three alternate double bonds in the ring. The lone pair of electrons is retained by nitrogen, on account of which pyridine is basic in nature.

As pyridine behaves as a strongly basic compound, it undergoes substitution in the ring, both by electrophilic as well as nucleophilic substitution process. Whereas, the electrophilic substitution takes place in position 3 or 3 and 5 positions, the nucleophilic substitution takes place in 2 and 6 positions both.

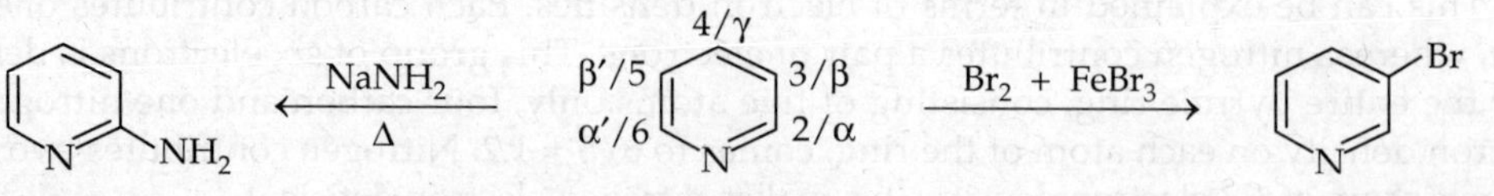

α-amino derivative   pyridine   β-bromo derivative

Nonbenzenoid nonaromatic rings: Some carbocyclic hydrocarbons, containing three carbon membered ring, like cyclopropene C_3H_4 or having a five membered ring like cyclopentadiene and cycloheptatriene C_7H_8 having a seven membered ring, do not fulfill some conditions, as required to behave as aromatic compounds. Therefore, these compounds do not behave as aromatic compounds.

Cyclopropene, (C_3H_4): It is a carbocyclic hydrocarbon, having a three membered ring, containing a double bond in the molecule, can be converted into a nonbenzenoid ring, fulfilling all the conditions, as required for the compounds to behave as aromatic compounds.

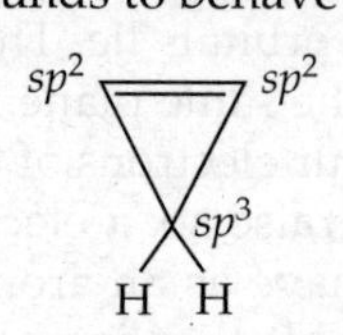

Cyclopropene is a carbocyclic compound, containing a double bond, or two electrons, in the ring, as required by the Huckel's rule, but the ring is not planar, i.e. all the atoms are not in the same plane, as one of the carbon is sp^3 hybridised.

However, cyclopropene can be converted into an aromatic ring in case one hydrogen from sp^3 hybridised carbon, is removed as a hydride ion, (H^-), converting cyclopropene into cyclopropenyl cation.

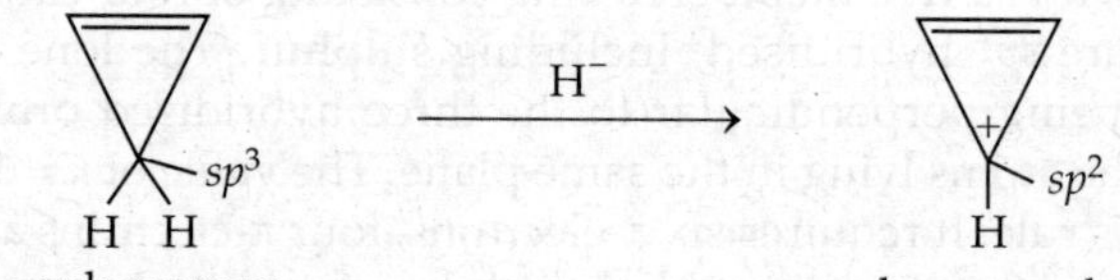

cyclopropene   cyclopropenyl cation
(not a planar ring)   (a monoplanar ring)

As cyclopropenyl cation fulfills all the three conditions, required to behave as an aromatic ring, it behaves as an aromatic compound and undergoes substitution keeping the unsaturation, intact in the cation.

Cyclopropenyl cation has been prepared from 3-chlorocyclopropene, by treating it with antimony pentachloride, $SbCl_5$.

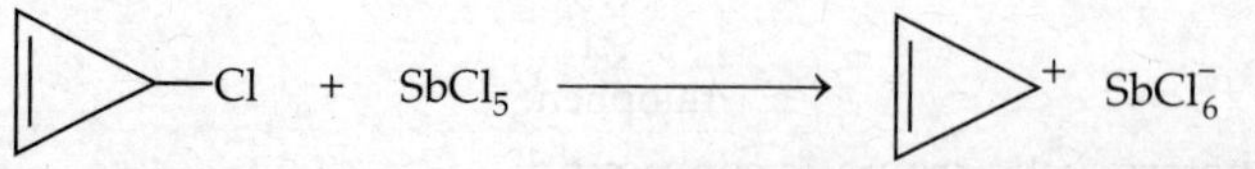

3-chlorocyclopropene   cyclopropenyl hexachloroantimonate

Cyclopentadiene (C_5H_6): It is a five membered carbocyclic compound having a five membered ring, with two alternate double bonds. It has four carbons in the sp^2 hybridised state and the fifth carbon in sp^3 hybrdised form.

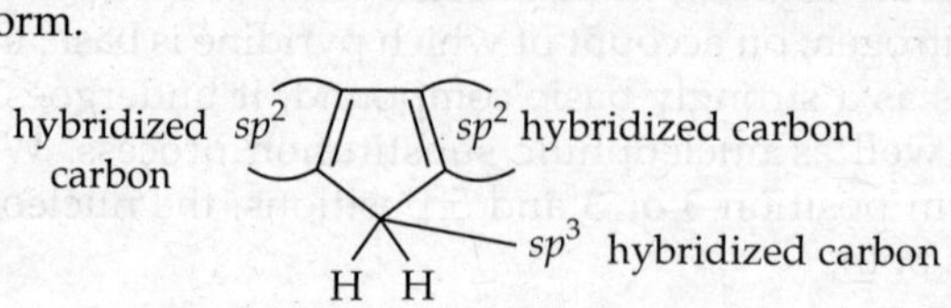

hybridized sp^2 carbon sp^2 hybridized carbon

sp^3 hybridized carbon

Cyclopentadiene has two double bonds, or four electrons in the ring. The ring is not mono-planar as one of the carbon is in sp^3 hybridised form. Cyclopentadiene can be converted into an aromatic compound, if one of the hydrogen is removed from sp^3 hybridised carbon atom as a proton (H^+), leaving its bonded pair of electrons on the carbon. This carbon, after losing a proton, is converted into a sp^2 hybridised form, in which the pair of electrons is in unhybridised p-orbital form, perpendicular to the plane in which the hybridised orbitals lie, converting cylopentadiene into cyclopentadienyl anion.

Cyclopentadienyl anion (C_5H_5) fulfills all the three conditions, required for the anion to behave as aromatic ring. It is cyclic, monoplanar and contains six electrons, as required by Huckel's rule.

Cyclopentadienyl anion has been prepared by treating cyclopentadiene with potassium metal. It reacts with metal halide to form metal cyclopentadienyls. The reaction is carried out by treating metal halide with sodium salt of cyclopentadiene in the tetrahydrofuran, as a solvent. One of the most discussed compounds, namely ferrocene or dicyclopentadienyl iron, (C_5H_5)$_2$ Fe, has been prepared by treating sodium salt of cyclopentadiene with ferrous chloride in tetra-hydrofuran as solvent. It reacts with metal halide to form metal cyclopentadienyls. The reaction is carried out by treating metal halide with sodium salt of cyclopentadiene in the tetrahydrofuran as a solvent. One of the most discussed compound, namely ferrocene or dicyclopentadienyl iron (C_5H_5)$_2$ Fe, has been prepared by treating sodium salt of cyclopentadiene with ferrous chloride in tetrahydrofuran as solvent.

cyclopentadienyl anion

sodium cyclopentadiene

In ferrocene, the two cyclopentadienyl rings lie in parallel planes with the iron atom, placed symmetrically between two rings. In ferrocene, entire ring is bonded uniformly with the ferrous atom. Bonding occurs by the overlap of the sextet of π-electrons of the ring with d–orbitals of the metal atom. These compounds are usually termed π-complex molecule.

Cycloheptatriene or tropylidene (C_7H_8): It is a seven membered cyclohydrocarbon, containing three alternate double bonds in the ring. Out of seven carbons, six carbons are sp^2 hybridised and the seventh carbon is sp^3 hybridised. Therefore, the ring is not planar, i.e. all the atoms of the ring do not lie in the same plane. Since cycloheptatriene does not fulfill all the three criteria, required for the compound to behave as an aromatic compound, cycloheptatriene, behaves as typical olefinic compound. It has ($4n + 3$) π-electron system in the ring.

Cycloheptatriene can be converted into an aromatic ring, in case a single π-electron is lost from the antibonding MO. By the loss of a single π-electron from the antibonding molecular orbital, it is converted into 'tropylium cation' which has all the atoms of the ring, lying in the same plane. 'Tropylium cation' behaves as an aromatic ring.

Tropylium cation $C_7H_7^+$ contains $(4n + 2)$ π-electrons in the system. It has been prepared from tropylium bromide by heating.

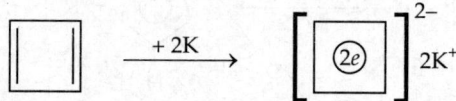

Br

$$\xrightarrow[-HBr]{\Delta}$$

tropylium cation

Antiaromatic nonbenzenoid ring: Conjugated monocyclic polyenes, containing $(4n)$ π-electrons system, where $n = 1, 2, 3$, etc. are generally called annulenes. These behave as olefinic compounds. For example:

1. Cyclobutadiene-1,3
2. Cyclooctatetraene

1. *Cyclobutadiene 1,3*: The molecule has two alternate double bonds in the ring. All the carbons are in the state of sp^2 hybridised form, and the ring is monoplaner. However, the molecule contains only two double bonds or the number of π-electrons is only four. The compound does not behave as an aromatic compound.

   However, cyclobutadiene-1,3, can be converted into an aromatic ring by treating it with two equivalent of potassium metal, which supplies two electrons raising the number of electrons to be six in the ring. By the gain of two electrons, cyclobutadiene molecule acquires six electrons and fulfills Huckel's rule. Therefore, dipotassium salt of cyclo-butadiene $(C_4H_4)_2K^+$, behaves as an aromatic compound.

$$\square \xrightarrow{+\ 2K} \left[\ \fbox{2e}\ \right]^{2-} 2K^+$$

2. *Cyclo-octatetraene* (C_8H_8): It is a cyclic hydrocarbons containing eight carbons in the ring and it is highly unsaturated. It contains four alternate double bonds in the molecule or four π-bonds or 8-electrons. All the eight carbons are in the state of sp^2 hybridised form. It does not possess a closed shell planar configuration in the ground state, being a $(4n)$ π-electron molecule. Therefore, it is highly unstable as well as highly reactive. It exhibits all the reactions of a typical unsaturated compound.

   Cyclo-octatetraene molecule can be converted into an aromatic ring. On treating it with two equivalent of potassium metal, the ring gains two electrons, raising the number of π-electrons in the ring to be $8 + 2 = 10$. Therefore, dipotassium salt of cyclo-octatetraene, $(C_8H_8)_2\ 2K^+$, behaves as an aromatic compound. This gain of two electrons (one to each nonbonding MO), converts the cyclo-octatetraene molecule into cyclo-octatetraenyl dianion having a closed shell of $(4n + 2)$ π-electrons in the ring.

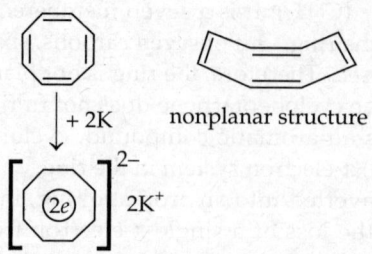

nonplanar structure

+ 2K

$$\left[\ \fbox{2e}\ \right]^{2-} 2K^+$$

MULTIPLE CHOICE QUESTIONS

1. The hybrid orbitals used by benzene belong to
   a) sp^3 (b) sp^2
   (c) sp (d) $sp^2 d^3$
2. According to Huckel's rule, monocyclic compounds will show aromaticity, when
   (a) it has 4π electrons   (b) it has no electrons
   (c) it has $4 + 2\pi$ electrons   (d) it has $(4n + 2)\pi$ electrons
3. Carbon in benzene is sp^2 hybridised and the bond angle is 120°, the shape of the molecule is
   (a) Linear   (b) planar
   (c) pyramidal   (d) planar hexagonal
4. According to Huckel, the number of electrons in anthracene are
   (a) 12   (b) 14
   (c) 10   (d) 20
5. The overlapping in benzene in carbon to carbon orbitals of type
   (a) p-p (b) sp-sp
   (c) sp^2-sp^2 (d) sp^3-sp^3
6. The correct structure of benzene was proposed by
   (a) Faraday   (b) Huckel
   (c) Wohler   (d) none of these
7. Which of the monocyclic species is nonaromatic in character?

       (a) 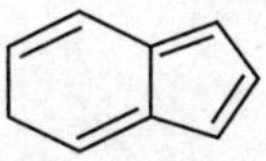   (b)    (c)    (d)

8. Which of the following would be aromatic?

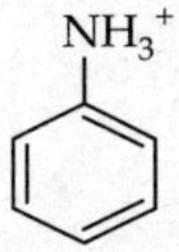

                      NH_3^+ NH_3^+

                      (a) 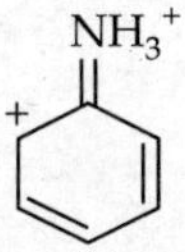   (b)

9. The given cyclic compound belongs to which class of compounds.

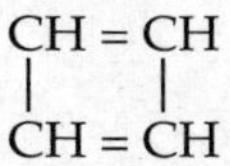

$$CH = CH$$
$$|\qquad\quad|$$
$$CH = CH$$

   (a) polyene   (b) annulene
   (c) aromatic compound   (d) heterocyclic compound
10. The given compound is not aromatic, but it can be converted into an aromatic compound.

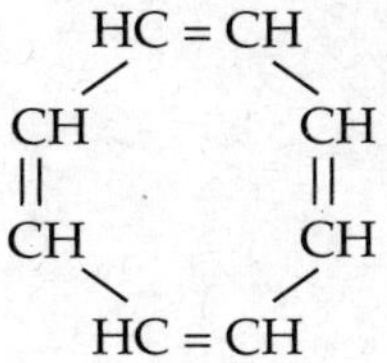

$$HC = CH$$
$$CH \qquad\qquad CH$$
$$||\qquad\qquad\quad ||$$
$$CH \qquad\qquad CH$$
$$HC = CH$$

(a) by reduction
(b) by treating it with two equivalent of potassium metal
(c) by treating it with two equivalent of Cu metal
(d) none

ANSWERS

1. (b)   2. (d)   3. (b)   4. (b)   5. (c)   6. (b)   7. (d)   8. (a)
9. (b)   10. (b)

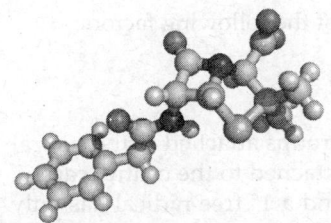

Organic Intermediates

Reaction Intermediates or Organic Radicals

Organic compounds are the compounds of covalency. These compounds do not ionise. However, these compounds may undergo a homolytic or a heterolytic fission, leading to the formation of reaction intermediates. The reaction intermediates are the species, formed during the various reactions, which the organic compounds undergo: homolytic or heterolytic fission. These are immediately consumed and therefore nonisolable. These species are charged and have incomplete octet.

- **Homolytic fission:** When a carbon to carbon (-C-C-) σ-bond undergoes a homolytic fission, it leads to the formation of a highly reactive species, known as free radicals.

$$-C : C- \xrightarrow{\text{homolytic fission}} -C \bullet + -C \bullet$$
$$\text{free radicals}$$

- **Heterolytic fission:** When a C to C (-C : C-) bond undergoes a heterolytic fission, it gives two charged ions, a positively charged ion known as carbonium ion or a carbocation and a negatively charged ion, known as carbanion.

$$-C : C- \xrightarrow{\text{heterolytic fission}} -C^+ + -C:^-$$
$$\text{carbocation} \quad \text{carbanion}$$

- **Free radicals**: These are formed during a homolytic fission of a σ-bond. These are not isolable and are consumed during the reaction as soon as they are formed. These are highly reactive.

A free radical has the following characteristics:

  i. The number of the electrons, around the central atom are only seven or an odd number of electrons.

  ii. The species is neutral, as there is no charge on it.

  iii. The species is sp^2 hybridised, i.e. it is triangular in shape. The odd electron is in unhybridised p-orbital, perpendicular to the plane in which the three hybridised orbitals lie.

  iv. It is paramagnetic in nature.

Stabilities of the free radicals: A free radical is stable on account of the following factors:

i. + I effect
ii. Hyperconjugation
iii. Resonance

i. **+I Effect:** An alkyl free radical may have one or more alkyl groups attached to the central carbon atom. A 3° alkyl free radical has three alkyl groups attached to the central carbon atom, a 2°alkyl free radical has two alkyl groups attached and a 1° free radical has only one alkyl group or no group is attached to the central carbon atom. A 3°alkyl radical is more stable, than a 2° free radical which in turn is more stable than a 1° free radical, on account of +I effect of alkyl groups (three alkyl groups in the case of a 3° free radical, two alkyl groups in the case of a 2° free radical and only one alkyl groups in the case of 1° free radical attached to the central carbon atom. More the +I effect of alkyl groups, more the free radical becomes stable + I effect of alkyl groups increases the electron density around the electron deficient carbon atom, is stabilised further. Hence, the stabilities of the free radical is in the decreasing order as follows.

$$
\begin{array}{cccc}
R_2 & R_2 & H & \\
| & | & | & \\
R_1-C & > \quad R_1-C & > \quad R_1-C & > \quad CH_3 \\
| & | & | & \\
R_3 & H & H & \\
3° & 2° & 1° & 1° \text{ methyl}
\end{array}
$$

By +I effect of alkyl groups, the electron density around the central carbon increases, though it does not become equivalent to the eight electrons or equivalent to an octet. Therefore, the free radicals are stabilised by +I effect of the alkyl free radicals. More is the number of alkyl groups more stable is the species. Hence the order

$3° > 2° > 1° >$ methyl

ii. **Hyperconjugation:** Another factor aiding to the stability of the free radicals is the hyper-conjugation. Greater is the hyperconjugation or canonical forms of a free radical, greater is the stability of the free radicals. Since hyperconjugation is not possible in the case of a methyl free radical, hence its least stable.

Canonical forms in the case of an ethyl free radical

$$
\begin{array}{cccc}
H \ H & H \ H & \bullet H \ H & H \ H \\
| \ | & | \ | & \ \ | \ | & | \ | \\
H \ \ C{=}C & \longleftrightarrow \quad H{-}C{=}C & \longleftrightarrow \quad H{-}C{=}C & \longleftrightarrow \quad H{-}C{-}C\bullet \\
| \ | & | \ | & | \ | & | \ | \\
H \ H & \bullet H \ H & H \ H & H \ H \\
\ \ \text{I} & \text{II} & \text{III} & \text{IV}
\end{array}
$$

As four canonical forms are possible in the case of ethyl free radical, it is more stable than a methyl free radicals. One methyl group forms three canonical forms therefore more the number of methyl groups attached to electron deficient carbon, the more stable it is.

Thus, isopropyl free radical (a 2° free radical), can form seven canonical forms. In the same way, a tertiary butyl free radical can form ten canonical forms. Hence, the decreasing order of stabilities of alkyl free radicals, on the basis of hyperconjugation, further supports the order:

$3° > 2° > 1° > CH_3\bullet$

iii. **Resonance:** The third factor involved in the stabilities of free radicals is the resonance, wherever it is possible. In the case of allyl free radical, the allylic species is stabilised by resonance.

$$CH_2=CH—CH_2\bullet \longleftrightarrow \bullet CH_2—CH=CH_2 \longleftrightarrow \overset{\underset{\displaystyle \bullet}{\rule{2cm}{0.4pt}}}{CH_2—CH—CH_2}$$

allyl free radical   resonance hybrid

The alkyl free radicals do not undergo resonance, but allylic free radicals do undergo resonance stabilisation. Hence, allyl free radical is more stable than a 3° free radical. Hence the order:

$$CH_2=CH—CH_2\bullet > R_3C\bullet > R_2CH\bullet > RCH_2\bullet > CH_3\bullet$$

allyl free radical   3°   2°   1°   methyl

Aryl free radical contain one or more phenyl groups attached to benzylic carbon holding odd electron on it, or phenyl groups attached to electron deficient carbon atom. A benzyl free radical $C_6H_5CH_2 \bullet$ is a 1° free radical. It is even more stable than a 3° alkyl free radical since it is stabilised by resonance and it is capable of forming three resonating structures.

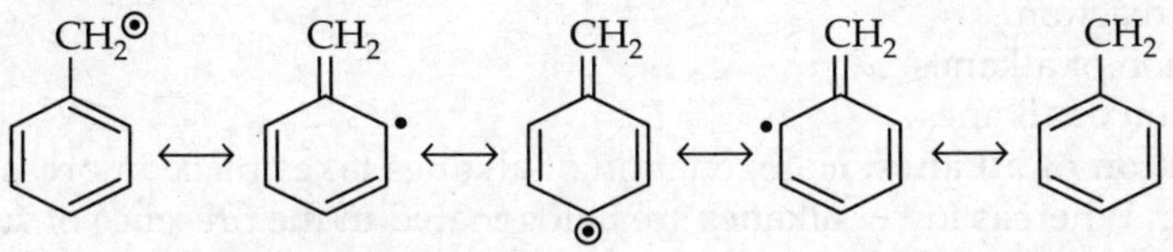

A resonance stabilised structure is more stable as compared to the induced structure. Any charge on benzylic carbon, whether an odd electron or a +ve charge or a –ve charge, it is stabilised by forming a large number of resonating structures. Resonating structure signifies the stability of a species.

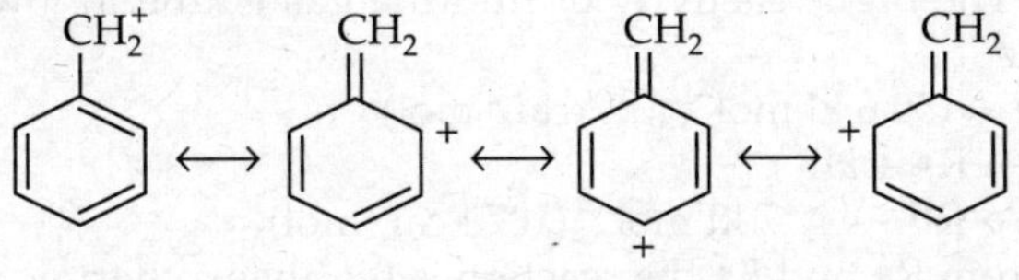

Benzyl carbocation

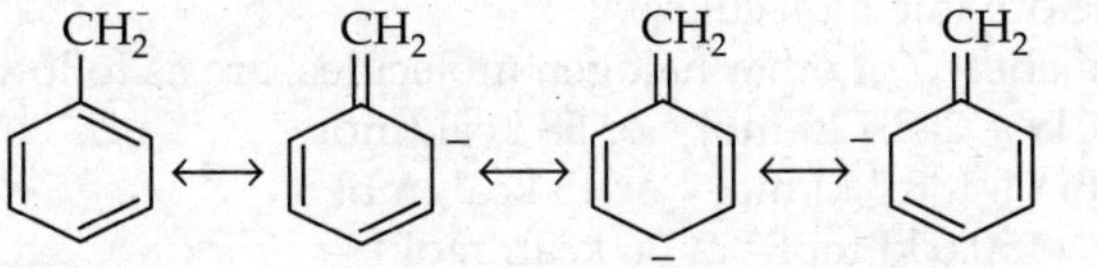

Benzyl carbanion

A phenyl ring forms three resonating structures, therefore benzyl free radical, $C_6H_5CH_2$— is capable of forming 3 + 1 = 4 resonating structures; diphenyl methyl free radical, $(C_6H_5)_2CH$— is capable of forming 6 + 1 = 7 resonating structures; similarly triphenyl methyl free radical $(C_6H_5)_3C$ is capable of forming 9 + 1 = 10 resonating structures.

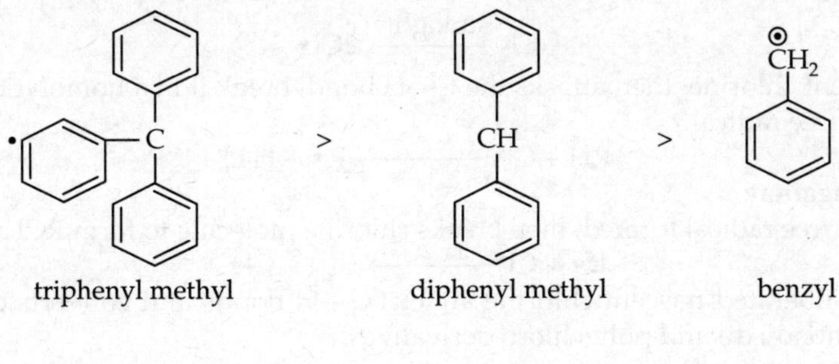

     triphenyl methyl   diphenyl methyl   benzyl

Since, resonance stabilised aryl free radical structures than inductive induced structures, the overall stabilities of organic free radicals may be formulated as follows:

$(C_6H_5)_3$—$C\bullet$ > $(C_6H_5)_2CH\bullet$ > $C_6H_5CH_2\bullet$ > CH_2=CH—$CH_2\bullet$ > $R_3C\bullet$ > $R_2CH\bullet$ > $RCH_2\bullet$ > $CH_3\bullet$

resonance stabilised free radicals   +I effect stabilised structures

The following reactions involve the formation of free radicals as the reaction intermediate, and follow the free radical mechanism.

   i. Halogenation of alkanes
   ii. Kolbe's electrolytic synthesis
  iii. Wurtz reaction
  iv. Borodine–Hunsdiecker reaction
   v. Sandmeyer's reaction
  vi. Kharasch reaction
 vii. Chlorination of alkanes
viii. Bromination of alkanes

   i. **Halogenation of alkanes:** Halogenation of alkanes takes place in presence of sunlight or on heating. Whereas lower alkanes are halogenated in the presence of sunlight, the higher ones are halogenated on heating. The decreasing order of the reactivity of halogens is in the order of:

$$Cl_2 > Br_2 > I_2$$

Alkanes cannot be fluorinated with fluorine, as many alkanes either burn or explode when reacted with it. The high reactivity of fluorine gas is due to low dissociation energy of fluorine molecule.

$F_2 \longrightarrow 2F\bullet + 150.6$ kJ mol^{-1} (36 kcal/ mol)
$R H + F\bullet \longrightarrow R\bullet + HF$
$R\bullet + F\bullet \longrightarrow RF + 447.7$ kJ mol^{-1} (107 kcal/mol)

In forming HF from $R\bullet$ and $F\bullet$ the reaction is highly exothermic and the energy evolved, (83 kcal/mol) is large enough to break a C to C single bond, a σ-bond, thereby resulting in the fission of the organic molecules.

The dissociation energy of other halogen molecules, are as follows:

$Cl_2 \longrightarrow 2Cl\bullet + 242.7$ kJ mol^{-1} or 58 kcal/mol
$Br_2 \longrightarrow 2Br\bullet + 188.3$ kJ mol^{-1} or 45 kcal/mol
$I_2 \longrightarrow 2I\bullet + 150.6$ kJ mol^{-1} or 36 kcal/mol

A free radical reaction involves three steps:

   i. Initiation
   ii. Propagation
  iii. Termination

   a. **Initiation:**
Cl_2 (or halogen), breaks in the homolytic fission to form nascent chlorine.

$$Cl_2 \xrightarrow{\text{sunlight}} 2Cl\bullet$$

Nascent chlorine, then attacks the C—H bond, break it in a homolytic fission, forming alkyl free radical.

$$RH + Cl\bullet \longrightarrow R\bullet + HCl$$

   b. **Propagation:**
Alkyl free radical formed, then breaks chlorine molecule to form RCl and liberates $Cl\bullet$.

$$R\bullet + Cl_2 \longrightarrow RCl + Cl\bullet$$

Thus liberated nascent chlorine attacks C—H bonds and converts organic molecule into mono-, di- and poly-chloro derivatives.

c. **Termination:**
   When the conditions are no longer favourable for the reaction to go on, it comes to an end, by joining free radicals available in the reaction mixture.

$$Cl\bullet + Cl\bullet \longrightarrow Cl_2$$
$$R\bullet + R\bullet \longrightarrow R—R$$

One of the property of free radical is to undergo dispropotionation, i.e. to undergo hydrogenation, one of the molecule acquiring hydrogen, at the expense of the other molecule.

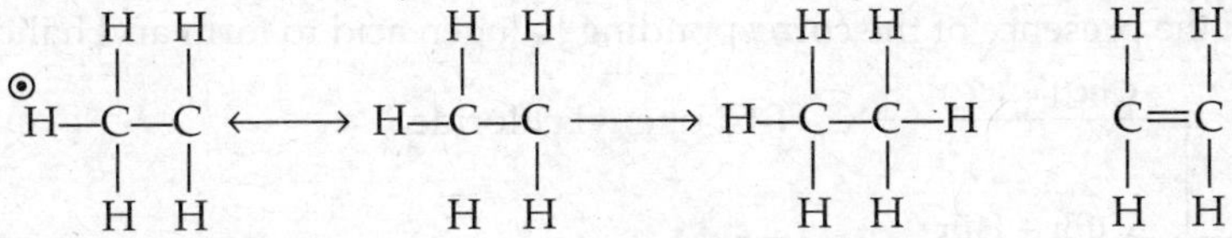

| ethyl free radicals | | ethane | ethylene |

ii. **Kolbe's electrolytic synthesis:** When an aqueous solution of sodium salt of a fatty acid is electrolyzed, a higher alkane, containing even number of carbon atoms is obtained.
   Sodium salt of a fatty acid, being an electrolyte, dissociates on its own.

$$CH_3COONa \rightleftharpoons CH_3COO^- + Na^+$$

On switching the current, the ions move towards oppositely charged electrodes, and are discharged, either by losing or gaining the electrons.

At anode

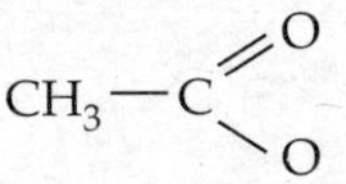

$$CH_3—C{\overset{\displaystyle O}{\underset{\displaystyle O}{\big<}}}$$

$CH_3OO—CNa$

$+e$
CH_3COO

$CH_3\bullet + \downarrow CO_2$

$CH_3\bullet + CH_3^-\bullet \longrightarrow C_2H_6$

At cathode

Na^+

$\downarrow +e$

$$Na + H_2O \longrightarrow NaOH + H$$

Alkanes prepared by this method, contains even number of carbon atoms. CH_4 cannot be prepared by this method.

iii. **Wurtz Reaction:** In this method, alkyl halide is reacted with sodium metal, in ether to give higher alkanes. Formation of an alkyl free radical is as an intermediate free radical.

$$R : X + Na \longrightarrow R\bullet + NaX$$
$$R\bullet — + R\bullet \longrightarrow R—R$$
$$\text{A higher alkane}$$

iv. **Borodine–Hunsdiecker Reaction:** Silver salt of a fatty acid is decomposed with Br_2, in carbon tetrachloride, to prepare alkyl bromide.

$$RCOONa + Br_2 \longrightarrow RBr + CO_2 + NaBr$$

The reaction with Cl_2 is identical, however, the yield of alkyl chloride is poor, on account of formation of various side products.
The halaids formed are of the following decreased order:
1° Alkyl halide > 2° alkyl halide > 3° alkyl halide
Mechanism

$$RCOOAg + Br_2 \longrightarrow RCOOBr + AgBr + CO_2$$
$$RCOO : Br \longrightarrow RCOO\bullet + Br$$

$$RCOO\bullet \longrightarrow R\bullet + CO_2$$
$$R\bullet + Br\bullet \longrightarrow RBr$$
$$R\bullet + Br_2 \longrightarrow RBr + Br\bullet$$
$$R\bullet + RCOO:Br \longrightarrow RBr + CO_2$$

Unlike bromine, iodine reacts with RCOOAg to form an ester.

$$2RCOOAg + I_2 \longrightarrow RCOOR + 2AgI + CO_2$$

This reaction is known as *Simonini–Birnbaum* reaction.

v. **Sandmeyer's Reaction:** The reaction involves an aryl diazonium halide with cuprous halide in the presence of the corresponding halogen acid to form aryl halide.

$$C_6H_5N_2Cl$$

- $\xrightarrow{CuCl + HCl}$ C_6H_5Cl phenyl chloride
- $\xrightarrow{CuBr + HBr}$ C_6H_5Br phenyl bromide
- $\xrightarrow{CuI_2 + KI}$ C_6H_5I phenyl iodide
- $\xrightarrow{Cu_2 + (CN)_2}$ C_6H_5CN phenyl cyanide

Mechanism

$$Cu_2Cl_2 \longrightarrow 2Cu^+ + 2Cl^-$$
$$Cu^+ \longrightarrow Cu^{2+} + e$$
$$C_6H_5N_2^+ + e \longrightarrow C_6H_5{}^\bullet + N_2$$
$$C_6H_5{}^\bullet + Cl^- + Cu^{2+} \longrightarrow C_6H_5Cl + Cu^+$$

or

$$CuCl + Cl^- \longrightarrow CuCl_2^-$$
$$C_6H_5N_2^+ + e + CuCl_2^- \longrightarrow C_6H_5{}^\bullet + N_2 + CuCl_2$$
$$C_6H_5{}^\bullet + CuCl_2 \longrightarrow C_6H_5Cl + CuCl$$

vi. **Kharasch reaction or peroxide effect or anti–Markovnikov's rule:** HBr adds on to an unsymmetrical alkene, in the presence of a peroxide or a peroxy acid (but not H_2O_2) to give an addition product, opposed to what is expected by Markovnikov's rule.

$$(C_6H_5COO)_2 \longrightarrow 2C_6H_5{}^\bullet + 2CO_2$$
$$HBr + C_6H_5 \longrightarrow C_6H_6 + Br^\bullet + H^\bullet$$

$$CH_2{=}CH{-}CH_3 + Br^\bullet \longrightarrow CH_2Br{-}{}^\bullet CH{-}CH_3 \xrightarrow{HBr} CH_2Br{-}CH_2{-}CH_3$$

propene   *n*-propyl bromide

vii. **Chlorination of alkanes** (Allylic chlorination): Propene, on chlorination with chlorine at high temperature (300 °C) forms allyl chloride.

$$Cl_2 \xrightarrow{high\ temp.} 2Cl^\bullet$$
$$CH_2{=}CH{-}CH_3 + Cl^\bullet \longrightarrow CH_2{=}CH{-}CH_2{}^\bullet + HCl$$
$$CH_2{=}CH{-}CH_2{}^\bullet + Cl^\bullet \longrightarrow CH_2{=}CH{-}CH_2Cl$$

propene   allyl chloride

At high temperature chlorine does not add to double bonds.

viii. **Bromination of alkanes** (Allylic bromination): Unsymmetrical alkenes are brominated by *n*-bromo succinimide.

$$\begin{array}{c} CH_2{-}CO \\ | \qquad\qquad\ \diagdown \\ \qquad\qquad\qquad N{-}Br \\ | \qquad\qquad\ \diagup \\ CH_2{-}CO \end{array}$$

$$CH_2=CH-CH_3 \xrightarrow{n\text{-}BS} CH_2=CH-CH_2\,Br + (CH_2-CO)_2NH$$

allyl bromide

Carbonium ion or carbocation

These are positively charged, electron deficient species, which have the following characteristic:

i. These species are electron deficient species, have only six electron (in the valence shell) around the central carbon atom.

ii. These are sp^2 hybridised ions and their shape is triangular planar.

iii. Being positively charged, these species act as electrophile.

These carbocation are formed during the following set of reactions:

a. By a heterolytic fission or an unusual cleavage of a covalent bond as in the case of dehydrogenation of alkyl halides by alkali.

$$-C : Cl \longrightarrow -C^+ + Cl^-$$

b. By dehydrogenation of alcohols.

$$-C : OH \longrightarrow -C^+ \text{ ion of halogen acids} + H_2O$$

c. In Friedel–Craft reaction

$$R-X + AlCl_3 \longrightarrow R^+ + AlCl_4^-$$

d. By protonation of alkenes, i.e. addition of halogen acids to alkenes.

$$CH_3CH=CH_3 + H^+ \longrightarrow CH_3CH^+-CH_3 \xrightarrow{X^+} CH_3CHX-CH_3$$

e. By hydration of alkenes, [BH_3, (borohydration)]

By reaction of alkenes with ($Hg(OOCCH_3)_2$), i.e. oxymercuration demercuration

$$R-CH=CH_2 + BH_3 \longrightarrow R-CH^+-CH_2(BH_2) \longrightarrow RCH_2-CH_2OH$$

(BH_3 joins at the double bond as $BH_2^+ + H^-$)

Alkyl carbocations may contain three alkyl groups, attached to the central carbon atom, R_3C, these are termed as 3°carbocations. Alkyl carbocation may contain 2 alky groups attached to the central carbon atom, R_2CH, which are termed 2°carbocation. Likewise, carbocation may contain one alkyl group joined to the central carbon atom, RCH_2, this is termed 1°carbocation.

As alkyl groups are electron repelling or electron donating, or have a +I effect, a 3°carbocation will be more stable than a 2°carbocation, which in turn may be more stable than a 1°carbocation.

$$R_3C^+ > R_2CH^+ > RCH_2^+ > CH_3^+$$
3°   2°   1°

· As aryl carbocations may contain 3 phenyl groups attached to the central carbon atom, i.e. triphenylmethyl, $(C_6H_5)_3C$ is termed a 3°carbocation. Those which contain two phenyl groups attached to the central atom, diphenylmethyl $(C_6H_5)_2CH$—may be termed a 2° carbocation. In the same way benzyl carbocation, $C_6H_5CH_2$ may be termed 1° aryl carbocation.

triphenylmethyl   diphenylmethyl   benzyl carbocation

The stability of carbocations may be explained on the basis of the following phenomenon:
- Inductive effect
- Hyperconjugation
- Resonance
- **Inductive effect:** Alkyl radicals are electron repelling or electron donating groups, greater is the dispersal and consequently more stable. Therefore, a 3° carbocation is more stable than a 2° or 1° carbocation. So a methyl group will be least stable. Hence, the stability of the carbonium ions is based on + I effect of the alkyl groups, attached to the electron deficient carbon. More are the alkyl group, more is +I effect.
- **Hyperconjugation:** It leads to the formation of more conjugative structures, greater is the number of conjugative structures or canonical forms, greater will be the hyperconjugative energy and greater will be the the the stability of the carbonium ion.

$$H\overset{H}{\underset{H}{\overset{|}{\underset{|}{C}}}}-\overset{H}{\underset{H}{\overset{|}{\underset{|}{C^+}}}} \longleftrightarrow H\overset{H^+}{\underset{H}{\overset{|}{\underset{|}{C}}}}=\overset{H}{\underset{H}{\overset{|}{\underset{|}{C}}}} \longleftrightarrow H\overset{+}{\underset{H}{C}}\overset{H}{\underset{H}{=C}} \longleftrightarrow H\overset{H}{\underset{H^+}{\overset{|}{\underset{|}{C}}}}=\overset{H}{\underset{H}{\overset{|}{\underset{|}{C}}}}$$

Three canonical structures are possible for each methyl CH_3– group, therefore with the addition of each —CH_3 group, three canonical structures are added. In case of isopropyl carbonium ion (a 2° carbonium ion) with two more methyl groups on the central carbon atom, it is able to form 7 canonical structures. Therefore, it is more stable than ethyl carbonium ion (a 1° carbocation). Likewise tertiary butyl carbonium ion, with 10 canonical forms is more stable than isopropyl carbocation.

$$\begin{matrix} CH_3 \\ \\ CH_3 \end{matrix}\!\!\!\!\!\!\!\!\searrow_{\nearrow}\!\! CH^+ \text{ isopropyl carbocation}$$

- **Resonance:** The carbonium ions are more stabilised by resonance. These stabilised resonating structures are even more stable than alkyl carbocations, which are stabilised by +I effect of alkyl group (do not undergo resonance stabilisation).

$$CH_2{=}CH{-}CH_2{}^+ \longleftrightarrow CH_2{}^+{-}CH{=}CH_2 \longleftrightarrow \overset{-+-}{CH_2{-}CH{-}CH_2}$$

aryl carbocation-resonating structures

Aryl carbocation, contain one or more phenyl groups attached to the central carbon atom, having a +ve charge on the benzylic carbon atom. A benzylic carbocation is a 1° carbocation and it is stabilised by resonance, and it is more stable than a 3° alkyl carbocation. Any charge on the benzylic carbon is accommodated, by resonance, by forming a large number of resonating structures. In the case of benzylic carbon atom, containing one phenyl group, three resonating structures are possible adding 4 reasonating structures in all.

In diphenyl carbocation, $(C_6H_5)_2CH^+$, being a 2° carbocation, six resonating structure are formed, making seven structures in all. It is even more stable than benzyl carbocation.

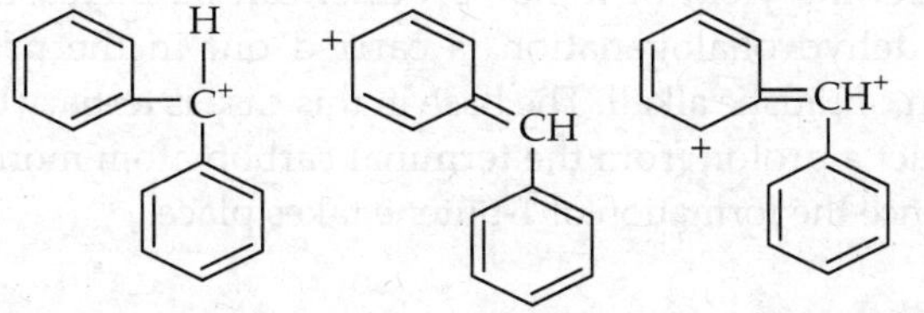

In triphenylmethyl carbocation, $(C_6H_5)_3C^+$ a 3° carbocation, containing three phenyl ring, nine structures are formed, making in all 10 resonating structures.

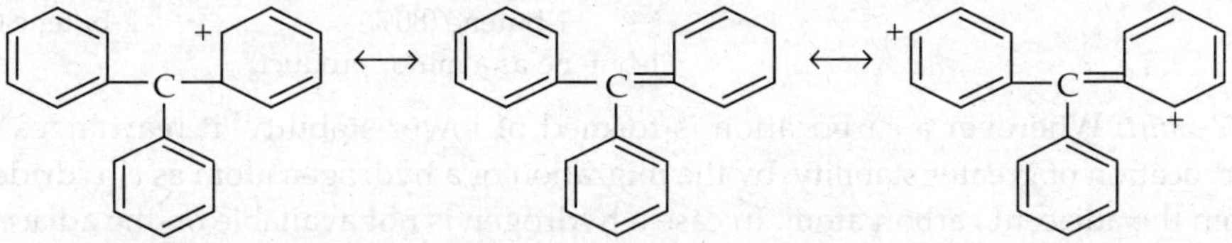

It is more stable than diphenylmethylc carbocation a 2° carbocation. Hence, the decreasing order of the stabilities of aryl carbocations are:

$$(C_6H_5)_3C^+ \quad > \quad (C_6H_5)_2CH^+ \quad > \quad C_6H_5CH_2^+$$

triphenyl methyl   diphenyl methyl   benzyl

As these aryl carbocations are more stable than alkyl carbocations, the over all stability of aryl and alkyl carbocations, in the decreasing order is as follows:

$$(C_6H_5)_3C^+ > (C_6H_5)_2CH^+ > \text{tropylium ion} > C_6H_5CH_2^+ > R_3C > R_2CH^+ > RCH_2^+ > CH_3^+$$

Tropylium ion is capable of forming seven resonating structures, like diphenyl methyl carbocation, but it is less stable than diphenyl methyl carbocation.

   i. Dehydrohalogenation of alkyl halides
  ii. Dehydration of alcohols
 iii. Protonation of alkenes
 iv. Alkylation of benzene

i. **Dehydrohalogenation of alkyl halides by alcoholic potash:** Alkyl halides are dehydrohalogenated by alcoholic caustic alkali or sodamide, $NaNH_2$ to form alkenes and carbocations are formed as intermediates.

$$CH_3-CH_2\overset{\overset{\displaystyle Cl}{|}}{C}H-CH_3 \longrightarrow CH_3-CH_2-\overset{+}{C}H-CH_3 \longrightarrow$$

$$CH_3-CH_2-CH=CH_2 \;+\; CH_3-CH=CH-CH_3$$

1-butene 20%   2-butene 80%

(Hofmann's product)   (Saytzeff's product)

In forming 2-butene, a proton is lost from the adjacent carbon atom, containing lesser number of hydrogen atoms, to give a major product, as per *Saytzeff's rule*. This is the major product. Alkene, which is formed as a major product, contains more alkyl groups, attached to the doubly bonded carbon atoms. 1-butene is formed as major product, when a proton is lost from form adjacent carbon atom, containing larger number of hydrogen atoms. It is known as Hofmann's rule.

In the above case, the yield of minor product can be increased, to become the major product, if the dehydrohalogenation is carried out in the presence of tertiary butyl alcoholic solution of caustic alkali. The base in this case is tertiary butoxide ion, $(CH_3)_3CO-$, which can abstract a proton from the terminal carbon atom more easily, than the middle carbon atom. Hence the formation of 1-butene takes place.

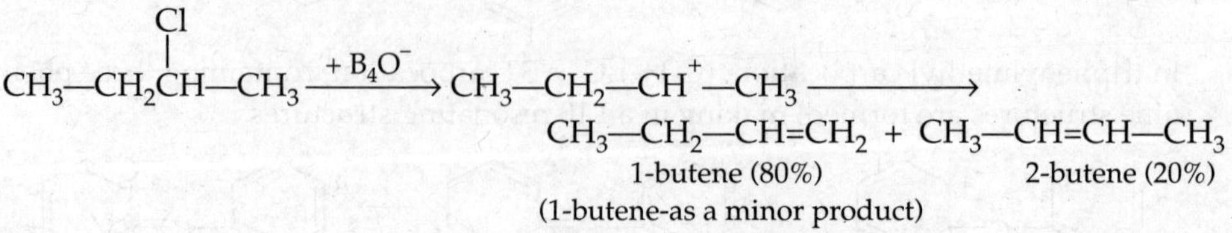

$$CH_3-CH_2\overset{\overset{\displaystyle Cl}{|}}{CH}-CH_3 \xrightarrow{+\,B_4O^-} CH_3-CH_2-\overset{+}{CH}-CH_3 \longrightarrow$$

$$CH_3-CH_2-CH=CH_2 \;+\; CH_3-CH=CH-CH_3$$
$$\text{1-butene (80\%)} \qquad\qquad \text{2-butene (20\%)}$$
$$\text{(1-butene-as a minor product)}$$

1, 2-Shift: Wherever a carbocation is formed of lower stability, it rearranges to form a carbocation of greater stability, by the migration of a hydrogen atom as a hydride ion, (H^-) from the adjacent carbon atom. In case a hydrogen is not available on the adjacent carbon atom, then a methyl group migrates as a methyl ion, (CH_3^-) from the adjacent carbon atom, to convert the carbocation into a more stable carbocation. In case neither of the two are available, then no migration takes place. This is known as a **1, 2-shift**.

Examples:

a.
$$CH_3-\overset{\overset{\displaystyle CH_3}{|}}{CH}-CH_2CH_2Cl \xrightarrow{\text{aq. alkali}} CH_3-\overset{\overset{\displaystyle CH_3}{|}}{CH}-CH_2\overset{+}{CH_2} \xrightarrow[\text{shift}]{\text{1,2 hydride}}$$
1-chloro-3-methyl butane

$$CH_3-\overset{\overset{\displaystyle CH_3}{|}}{CH}-\overset{+}{CH}CH_3 \longrightarrow CH_3-\overset{\overset{\displaystyle CH_3}{|}}{C}=CHCH_3 \;+\; CH_3-\overset{\overset{\displaystyle CH_3}{|}}{CH}-CH=CH_2$$
$$\text{2-methyl-2-butene} \qquad \text{3-methyl-1-butene}$$

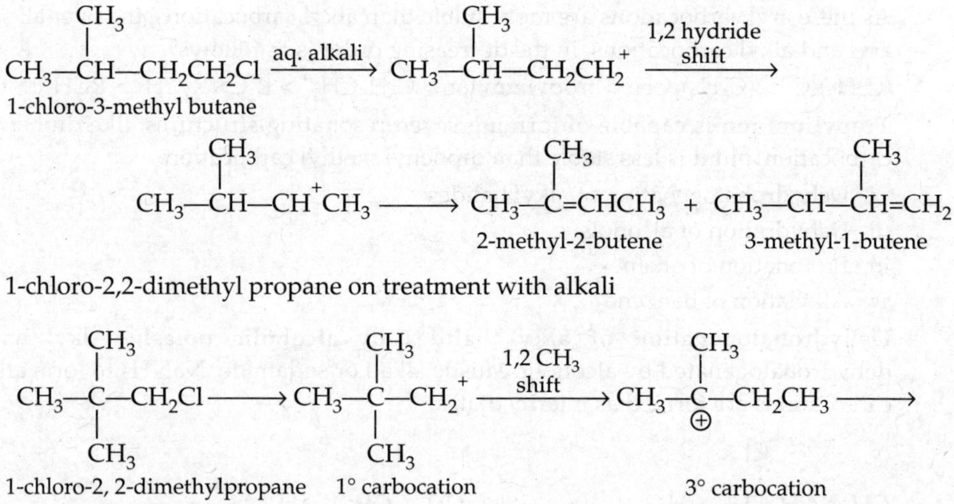

b. 1-chloro-2,2-dimethyl propane on treatment with alkali

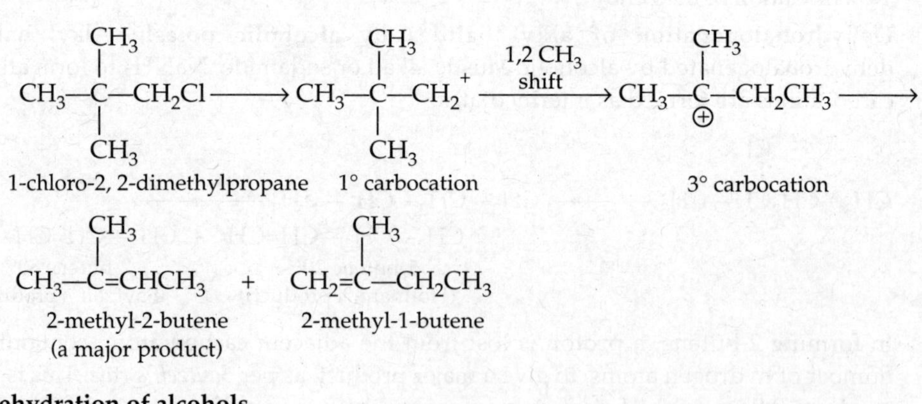

$$CH_3-\overset{\overset{\displaystyle CH_3}{|}}{\underset{\underset{\displaystyle CH_3}{|}}{C}}-CH_2Cl \longrightarrow CH_3-\overset{\overset{\displaystyle CH_3}{|}}{\underset{\underset{\displaystyle CH_3}{|}}{C}}-\overset{+}{CH_2} \xrightarrow[\text{shift}]{\text{1,2 }CH_3} CH_3-\overset{\overset{\displaystyle CH_3}{|}}{\underset{\underset{\displaystyle \oplus}{|}}{C}}-CH_2CH_3 \longrightarrow$$
$$\text{1-chloro-2, 2-dimethylpropane} \quad \text{1° carbocation} \qquad\qquad \text{3° carbocation}$$

$$CH_3-\overset{\overset{\displaystyle CH_3}{|}}{C}=CHCH_3 \;+\; CH_2=\overset{\overset{\displaystyle CH_3}{|}}{C}-CH_2CH_3$$
$$\text{2-methyl-2-butene} \qquad \text{2-methyl-1-butene}$$
$$\text{(a major product)}$$

ii. **Dehydration of alcohols**

Alcohols are dehydrated by conc. H_2SO_4 or H_3PO_4 to form alkenes. Carbocations are formed as intermediate ions, which rearrange, if possible to form carbocation of greater stability, before forming alkene/alkenes as final product, following Saytzeff's rule .

a.

$$CH_3-\underset{\underset{\textstyle CH_3}{|}}{CH}-CH_2-CH_2OH \longrightarrow CH_3-\underset{\underset{\textstyle CH_3}{|}}{CH}-CH_2-\overset{+}{CH_2} \xrightarrow[\text{shift}]{\text{1,2 hydride}} CH_3-\underset{\underset{\textstyle CH_3}{|}}{CH}-\overset{\oplus}{CH}-CH_3$$

3-methyl butanol-1   1° carbocation   3° carbocation

$$CH_3-\underset{\underset{\textstyle CH_3}{|}}{C}-CHCH_3 \qquad + \qquad CH_3-\underset{\underset{\textstyle CH_3}{|}}{CH}-CH=CH_2 \qquad \longleftarrow$$

2-methyl-2-butene   1-methyl-2-butene
a major product   a minor product

b.

$$CH_3-\underset{\underset{\textstyle CH_3}{|}}{\overset{\overset{\textstyle CH_3}{|}}{C}}-CH_2-OH \longrightarrow CH_3-\underset{\underset{\textstyle CH_3}{|}}{\overset{\overset{\textstyle CH_3}{|}}{C}}-\overset{+}{CH_2} \xrightarrow[\text{shift}]{\text{1,2 }CH_3} CH_3-\underset{\underset{\textstyle CH_3}{|}}{\overset{\overset{\textstyle \oplus}{|}}{C}}-CH_2-CH_3$$

3-methyl butanol-1   1° carbocation   3° carbocation

$$CH_3-\underset{\underset{\textstyle CH_3}{|}}{C}=CHCH_3 \qquad + \qquad CH_2=\underset{\underset{\textstyle CH_3}{|}}{C}-CH_2CH_3 \qquad \longleftarrow$$

2-methyl-2-butene   2-methy butene-1
a major product   a minor product

iii. Protonation of alkenes

Alkenes react with halogen acids, to form alkyl halides RX or on reaction with conc. H_2SO_4 to form alkyl hydrogen sulphate. The addition follows Markovnikov's rule. A proton from acid adds to the doubly bonded carbon, as per Markovnikov's rule, to give an intermediate carbocation. This carbocation ion may rearrange to form a more stable carbocation of higher-stability, if possible, before converting to the final product.

Examples:

a.

$$CH_3-\underset{\underset{\textstyle CH_3}{|}}{CH}-CH=CH_2 \xrightarrow{H^+} CH_3-\underset{\underset{\textstyle CH_3}{|}}{CH}-\overset{+}{CH}-CH_3 \xrightarrow[\text{shift}]{\text{1,2 hydride}} CH_3-\underset{\underset{\textstyle CH_3}{|}}{\overset{+}{C}}-CH_2CH_3$$

3-methy-1-butene   2° carbocation   3° carbocation

$$\downarrow CH_3 + Cl^- \qquad\qquad\qquad \downarrow CH_3 + Cl^-$$

$$CH_3-\underset{\underset{\textstyle Cl}{|}}{CH}-CH-CH_2 \qquad\qquad CH_3-\underset{\underset{\textstyle Cl}{|}}{C}-CH_2CH_3$$

2-chloro-3-methyl butane   2-chloro-2-methyl butane

b.

$$CH_3-\underset{\underset{\textstyle CH_3}{|}}{\overset{\overset{\textstyle CH_3}{|}}{C}}-CH=CH_2 \xrightarrow{H^+} CH_3-\underset{\underset{\textstyle CH_3}{|}}{\overset{\overset{\textstyle CH_3}{|}}{C}}-\overset{+}{CH}-CH_3 \xrightarrow[\text{shift}]{\text{1,2 }CH_3} CH_3-\underset{\underset{\textstyle CH_3}{|}}{\overset{+}{C}}-CHCH_3$$

3,3-dimethyl butene-1   2° carbocation   3° carbocation

$$\downarrow CH_3 + Cl^- \qquad\qquad\qquad \downarrow CH_3 + Cl^-$$

$$CH_3-\underset{\underset{\textstyle CH_3}{|}}{C}-\underset{\underset{\textstyle Cl}{|}}{CH}-CH_2 \qquad\qquad CH_3-\underset{\underset{\textstyle Cl}{|}}{C}-\underset{\underset{\textstyle CH_3}{|}}{CH}-CH_2$$

2-chloro-3,3-dimethyl butane   2-chloro-2,3-dimethyl butane

iv. **Alkylation of benzene (Friedel–Craft) Reaction's:** Aromatic compounds are alkylated in the ring by alkyl halides, in the presence of anhydrous $AlCl_3$. Anhydrous $AlCl_3$, helps to release alkyl carbocation, which alkylates benzene ring.

$$RX + AlCl_3 \longrightarrow R^+ + AlCl_4^-$$

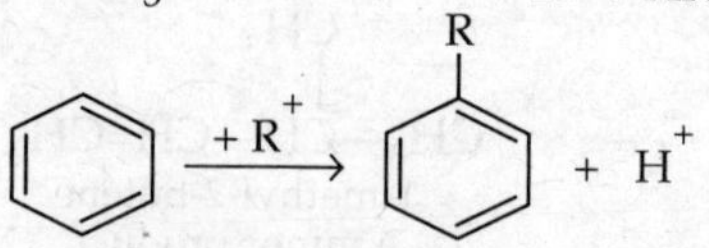

Carbocation formed during alkylating may rearrange to form carbocation of greater stability before alkylating the benzene ring.

Examples:

a. $CH_3—CH_2—CH_2—Cl + AlCl_3 \longrightarrow CH_3—CH_2—CH_2^+ + AlCl_4^-$

b. $CH_3—CH_2—CH_2^+ \longrightarrow CH_3—CH^+—CH_3$

1° carbocation 2° carbocation

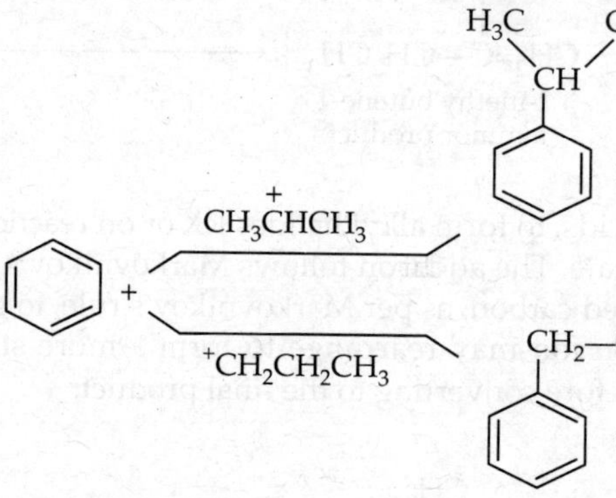

isopropyl benzene

n-propyl benzene

c. $CH_3CH(CH_3)CH_2Cl \longrightarrow CH_3CH(CH_3)CH_2^+ \longrightarrow CH_3C^+(CH_3)CH_3 \longrightarrow$

isobutyl chloride 1° carbocation 3° carbcation

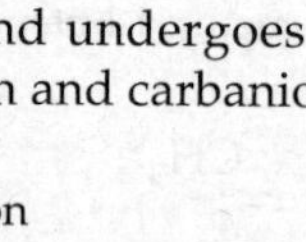

tertiarybutyl benzene

Carbocation or carbonium ion and carbanion: When a C to C s-bond undergoes a heterolytic fission it leads to the formation of charged species, namely carbocation and carbanion.

$$—C : C— \longrightarrow —C^+ \quad C^-$$

σ-bond carbocation carbanion

Carbanions

Carbanion are negatively charged species, having a complete octet around central carbon atom. Carbanion have the following characteristics.

i. These are negatively charged species, having a complete octet around central carbon atom.
ii. These species are sp^2 hybridised, and are tetrahedral in nature. One corner of the tetrahedron, is occupied by a pair of electrons, giving it a negative charge.
iii. These act as a nucleophile.

As the octet around central carbon is complete, the presence of any alkyl group on the central carbon, intensifies the negative charge on the central negatively charged carbon atom, will destabilize the it. hence the most stable carbanion will be methyl group and the decreasing order of the various carbanion will be:

$$CH_3:^- > CH_3CH_2:^- > (CH_3)_2CH^- > (CH_3)_3C:^- \text{ or}$$

$$CH_3:^- > RCH_2:^- > R_2CH:^- > R_3C:^-$$

methyl   1°   2°   3° carbanion

Aryl carbanion may contain one or more aryl groups, attached to the central carbon atom holding a negative charge, with the complete octet. Since, the negative charge on the benzylic carbon is accommodated in the benzene ring through resonance, making it more stable, than even a methyl carbanion. Benzyl carbanion can form three resonating structures, making it 3 + 1 = 4 in all.

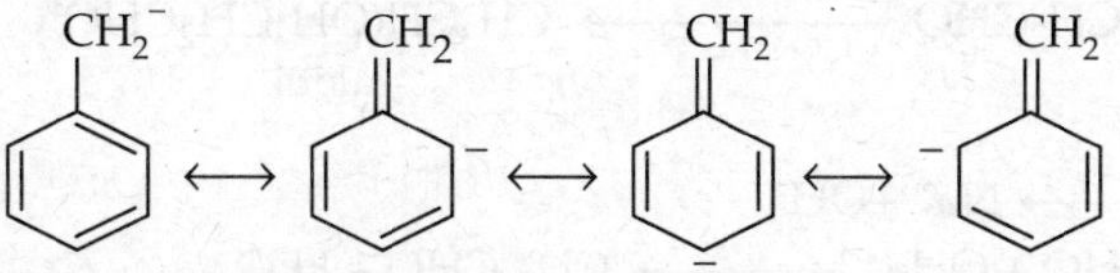

In the case of diphenyl methyl carbanion, $(C_6H_5)_2CH:^-$ with two phenyl rings, the number of resonating structures possible, comes to 6 + 1 = 7, and in the case of triphenyl methyl carbanion, $(C_6H_5)_3C:^-$ the number comes to 9 + 1 = 10.

$$(C_6H_5)_3C:^- > (C_6H_5)_2CH:^- > C_6H_5CH_2:^-$$

3°   2°   1°

More is the number of resonating structures, more it becomes stable. Hence the decreasing order of stabilities of the carbanion is:

$$(C_6H5)_3C:^- > (C_6H_5)_2CH:^- > C_6H_5CH_2:^- > CH_3:^- > RCH_2:^- > R_2CH:^- > R_3C:^-$$

Stability of the carbanion is influenced by the following two factors.

- **Inductive effect:** The groups with +I effect will always destabilize the carbanion and the groups with –I effect will always stabilise the carbanion.

G $\longrightarrow$ –C:^- + I effect   G $\longleftarrow$ –C:^- –I effect

Electron releasing group, intensify the negative charge on the carbon having –ve charge on the carbon and destabilize it. On the other hand, the electron releasing groups disperse –ve charge and therefore stabilise it,

$$CH_3:^- > RCH_2:^- > R_2CH:^- > R_3C:^-$$

methyl   1°   2°   3°

- **Resonance:** Carbanions may be stabilised by resonance and these carbanions are more stable than the ones, stabilised by +I effect. Greater is the number of resonating structures more is the stability.

Examples:

Allyl carbanion: its resonating structures are

$$CH_2=CH—CH_2:^- \longleftrightarrow {}^-CH_2—CH=CH_2 \longleftrightarrow CH_2—CH—CH_2$$

Benzyl carbanion

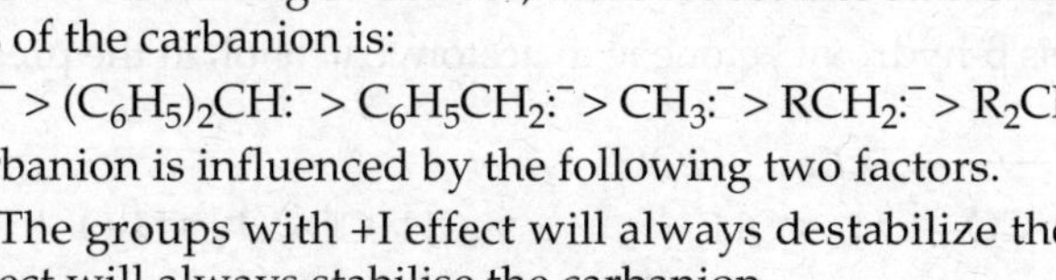

Cyclopentadienyl carbanion

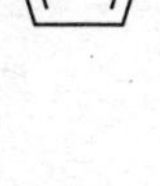

In benzyl carbanion, the 6 electrons are delocalised over the six carbon atoms, while in cyclopentadienyl carbanion, 6 electrons are delocalised over five carbon atoms, so the dispersal is more in benzylic carbanion than in cyclopentadienyl carbanion. The greater is the dispersal, more is the stability. Hence benzylic carbanion is more stable than cyclopentadienyl carbanion.

Carbanions are formed as reaction intermediates, in the following reactions.

Aldol condensation: In aldol condensation, one molecule of aldehyde (or ketone) condenses with another molecule of aldehyde (or ketone), containing α hydrogen, in the presence of a dil. base (Na_2CO_3, $Ba(OH)_2$ or dil. NaOH), to form, β-hydroxy ketone.

Examples:

i. Acetaldehyde forms hydroxy butyraldehyde or aldol

$$CH_3{-}CHO + HCH_2CHO \xrightarrow{NaOH} \underset{\text{aldol}}{CH_3CH(OH)CH_2CHO}$$

Mechanism

a. $NaOH \longrightarrow Na^+ + OH^-$

b. $H : CH_2{-}CHO + OH{-} \longrightarrow :CH_2^-CHO + H_2O$

c. $CH_3{-}\overset{O}{\overset{\|}{C}}{-}H + :CH_2CHO \longrightarrow CH_3{-}\underset{H}{\overset{OH}{C}}{-}CH_2{-}CHO$

B-hydroxy butyralehyde

ii. Acetone forms β-hydroxy ketone ie di acetonyl alcohol, in the presence of $Ba(OH)_2$.

Mechanism

a. $Ba(OH)_2 \longrightarrow Ba_2^+ + 2OH$

b. $H : CH_2{-}CO{-}CH_3 + OH^- \longrightarrow :CH_2\,COCH_3 + H_2O$

$$(CH_3)_2C{=}O + H{-}CH_2COCH_3 \longrightarrow (CH_3)_2C(OH){-}CH_2COCH_3 \longrightarrow (CH_3)_2C{=}CHCOCH_3$$

diacetonyl alcohol mesityl oxide

iii. **Perkin's Reaction:** Benzaldehyde condenses with acetic anhydride, in presence of fused sodium acetate to form cinnamic acid.

Mechanism

a. $CH_3COONa \rightleftharpoons CH_3COO^- + Na^+$

b. $H :CH_2CO.O.CO.CH_3 + CH_3COO^- \longrightarrow {}^-CH_2{-}CO.O.COCH_3 + CH_3COOH$

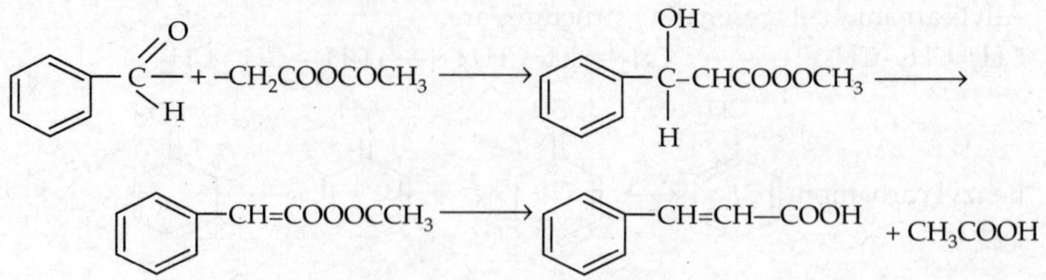

cinnamic acid

MULTIPLE CHOICE QUESTIONS

1. The most stable carbocation is:

   (a) $C_6H_5-\overset{+}{\underset{\underset{C_6H_5}{|}}{C}}-C_6H_5$

   (b) $C_6H_5-\overset{+}{\underset{\underset{CH_3}{|}}{C}}-CH_3$

   (c) $CH_3-\overset{+}{\underset{\underset{CH_3}{|}}{C}}H$

   (d) $CH_3-\overset{+}{\underset{\underset{CH_3}{|}}{C}}-CH_3$

2. Removal of a hydride ion from methane molecule will give a
   - (a) methyl radical
   - (b) carbocation
   - (c) carbanion
   - (d) methyl group

3. Heterolysis of a carbon–chlorine bond produces
   - (a) two free radicals
   - (b) two carbonium ions
   - (c) two carbanion
   - (d) one cation and one anion

4. Which of the following contains only three pairs of electrons?
   - (a) carbanion
   - (b) carbocation
   - (c) free radical
   - (d) none

5. The compound in which C* uses its sp^3 hybridised orbitals for bond formation is
   - (a) HCOOH
   - (b) $H_2N-CO-NH_2$
   - (c) $(CH_3)_3C-OH$
   - (d) CH_3CHO

6. Which is the strongest nucleophile?
   - (a) Br^-
   - (b) OH^-
   - (c) CN^-
   - (d) CH_3-O^-

7. Which in neither is a nucleophile nor an electrophile?
   - (a) X^-
   - (b) N^+O_2
   - (c) NH_3
   - (d) NH_4^+

8. Which of the following is an electrophilic reagent?
   - (a) RO–
   - (b) BF_3
   - (c) NH_3
   - (d) ROH

9. A solution of (+)2-chloro-2-phenyl ethane in toluene slowly racemises in presence of a small amount of $SbCl_5$ due to the formation of
   - (a) carbanion
   - (b) carbene
   - (c) free radical
   - (d) a carbocation

10. The structure that has a +ve charge on oxygen atom is

    $H-\overset{+}{\underset{\underset{H}{|}}{O}}-H$ CH_3-O-CH_3 $CH_3-O:$ $H-O-O-H$

      (a)   (b)   (c)   (d)

11. Which is the active species used for nitration?
    - (a) NO_2
    - (b) NO^+
    - (c) NO_2^+
    - (d) NO_3^+

12. The chief reaction product of a reaction in between *n*-butane and Br_2 at 130 °C is
    - (a) $CH_3CH_2CH_2CH_2Br$
    - (b) $CH_3CH_2CHBrCH_3$

    (c) $CH_3-\overset{\overset{\displaystyle Br}{|}}{\underset{\underset{CH_3}{|}}{C}}-CH_3$

    (d) $CH_3-\underset{\underset{CH_3}{|}}{CH}-CH_2Br$

13. The correct decreasing order of basic nature (Lewis bases):
    (a) $CH_3CH_2^- > CH_2=CH- > CH\equiv C- > OH^-$
    (b) $CH_3CH_2^- > CH\equiv C^- > CH_2=CH^- > OH^-$
    (c) $CH_3CH_2^- > OH^- > CH\equiv C^- > CH_2=CH^-$
    (d) $OH^- > CH\equiv C^- > CH_2=CH^- > CH_3-CH_2^-$

14. Which of the following is a nucleophile?
    (a) BF_3 (b) $FeCl_3$
    (c) $ZnCl_2$ (d) C_2H_5MgBr

15. The decreasing order of stability of the free radicals.
    (i) $C\bullet H_3$ (a) (ii) $C_2\bullet H_5$ (b) (iii) $(CH_3)_2C\bullet H$ (c) (iv) $(CH_3)_3C\bullet$ (d) is
    (a) (i) > (ii) > (iii) > (iv)   (b) (iv) > (iii) > (ii) > (i)
    (c) (ii) > (i) > (iii) > (iv)   (d) (iii) > (iv) > (i) > (ii)

16. In which of the following species, the central atom is negatively charged?
    (a) carbonium ion   (b) carbanion
    (c) carbocation   (d) free radicals

17. Which of the following contains only three pair of electrons?
    (a) carbanion   (b) carbonium ion
    (c) free radical   (d) none

18. Which of the following species is paramagnetic?
    (a) carbonium ion   (b) carbanion
    (c) a free radical   (d) all

19. Carbanion initiates:
    (a) addition reactions   (b) substitution reactions
    (c) both a and b   (d) none

20. Which of the following is a free radical?
    (a) Cl^+ (b) Cl^-
    (c) $Cl\bullet$ (d) NO_2

21. Electrophiles are
    (a) electrons loving   (b) electrons hating
    (c) nucleus loving   (d) nucleus hating

22. Nucleophiles are
    (a) nucleus loving   (b) nucleus hating
    (c) electron loving   (d) electron hating

23. Which of the following is an electrophilic reagent?
    (a) RO^- (b) BF_3
    (c) NH_3 (d) ROH free radical

24. A heterolytic cleavage gives:
    (a) cationic   (b) anionic
    (c) both a and b   (d) free radical

25. Which of the following does not acts as a nucleophile?
    (a) CH_3NH_2 (b) RO^-
    (c) $AlCl_3$ (d) CH_3MgBr

26. Which of the following have identical bond order?
    (a) CN^- (b) O_2
    (c) NO_2 (d) CN^+

27. Pick out the nonisoelectronic structure from the following:
    (a) CH_3^+ (b) H_3O^+
    (c) NH_3 (d) CH_3^-

(i) (a) and (b)   (ii) (a) and (c)   (iii) (c) and (d)   (iv) (b), (c) and (d)

28. An example of electrophilic addition:
    (a) $CH_2=CH_2 + Br_2$— ⟶ Br—CH_2—CH_2Br
    (b) $CH_3HO + CH_3MgBr$
    (c) $RCl + OH$ ⟶ $ROH + Cl^-$
    (d) none

29. Which is the decreasing order of the following bases:
    $OH^-, NH_2^-, HC≡CH^-, CH_3CH_2^-$
    (a) $C_2H_5^- > NH_2^- > CH≡C^- > OH^-$
    (b) $HC≡C^- > C_2H_5^- > NH_2^- OH^- ≡$
    (c) $CH_3CH_2^- > NH_2^- > HC^- ≡C^- > OH^- ≡$
    (d) $NH_2^- > CH≡C^- > OH^- > C_2H_5^-$

30. The compound which is not a Lewis acid is
    (a) BF_3 (b) $AlCl_3$ (c) $BeCl_2$ (d) $SnCl_2$

31. The most stable carbonium ion:
    (a) p—NO_2—$C_6H_4CH_2^+$ (b) $C_6H_5CH_2^+$
    (c) p—Cl—$C_6H_4CH_2^+$ (d) pCH_3O—$C_6H_4CH_2^+$

32. Decreasing order of stability of the following compounds:
    (i) CH_3–CH^+CH_3 (ii) $CH_3CH^+OCH_3$(2)
    (c) CH_3–CH^+COCH_3 (3)
    (a) i > ii > iii   (b) iii > ii > i
    (c) ii > iii > i   (d) iii > i > ii

33. One of the following set, behaves as an electrophile as well as a nucleophile is
    (a) Cl^- and H_2O (b) HF and HBr
    (c) HCHO and CH_3CN (d) H_2 and CH_4

34. Formation of ether takes place through the formation of
    CH_3CH_2—$OH + H +$ ⟶ C_2H_5—O—C_2H_5
    $\qquad\qquad\qquad\qquad\qquad\qquad\qquad |$
    $\qquad\qquad\qquad\qquad\qquad\qquad\qquad H$
    (a) $CH_3CH_2CH_2^+$ (b) $CH_3CH_2^+$
    (c) CH_3CH_2—^+O—CH_2CH_3 (d) none

35. The halogen atom can easily be replaced by nucleophile is
    (a) OR—   (b) SH—   (c) CN—   (d) all

36. +M effect will be due to
    (a) —OCH_3 (b) —OH   (c) —NH_2 (d) —Cl

37. –M effect will be due to
    (a) —NO_2 (b) CHO   (c) —COOH   (d) all

38. In which of the following, 1, 2-shift or a hydride shift will not take place?

    (a) CH_3—$\overset{+}{C}$—CH_2OH
    $\qquad\qquad |$
    $\qquad\quad CH_3$

    (b) CH_3—$\overset{CH_3}{\underset{CH_3}{C}}$—$\overset{+}{C}H$—$CH_3$

    (c) CH_3—$\overset{}{\underset{CH_3}{C}}$—$CH_2^+$

    (d) CH_3—$\overset{C_6H_5}{\underset{CH_3}{C}}$—$C_+H$—$CH_3$

39. The compound which gives the most stable carbonium ion on dehydration is

    (a) CH_3—$CH(CH_3)$—CH_2OH

    (b) CH_3—$\underset{\underset{CH_3}{|}}{\overset{\overset{CH_3}{|}}{C}}$—$OH$

    (c) $CH_3CH_2CH_2CH_2OH$

    (d) $CH_3\overset{\overset{CH_3}{|}}{C}HCH_2CH_3$

40. The compound which will readily react with NaOH to form methanol?
    (a) $(CH_3)_4N^+\,OH^-$
    (b) CH_3—O—CH_3
    (c) $(CH_3)_4S^+\,I^-$
    (d) $(CH_3)_3C$—Cl

41. Dehydration of alcohols involves
    (a) carbonium ion
    (b) carbanion
    (c) a free radical
    (d) carbene

42. A salt solution is treated with $CHCl_3$ and shaken with Cl_2 water. The salt solution becomes violet, salt solution contains
    (a) NO_2^-
    (b) NO_3^-
    (c) I^-
    (d) Br^-

43. In S_N1 reactions, the first step is the formation of
    (a) free radical
    (b) carbanion
    (c) carboniumion
    (d) final product

44. Addition of HI on a double bond of propene yields isopropyl iodide, a major product, the addition proceeds through
    (a) a more stable carbonium ion
    (b) more stable carbanion
    (c) a stable free radical
    (d) homolysis

45. Electrophiles are characteristic
    (a) Lewis bases
    (b) Lewis acids
    (c) amphoteric
    (d) none

46. Which of the following is an electrophile?
    (a) CH_3OH
    (b) NH_3
    (c) BCl_3
    (d) $AlCl_4^-$

47. Which of the following behaves as an electrophile?
    (a) RNH_2
    (b) $:CH_2$
    (c) H_2O
    (d) $:CN$

48. A nucleophile must necessarily have
    (a) one unpaired electron
    (b) two lone pairs of electrons
    (c) an overall +ve charge
    (d) tendency to donate electron pair

ANSWERS

1. (a) Due to resonance a 3° carbocation is more stable
2. (b) $CH_4 \longrightarrow CH_3^+ + H^-$ (methyl carbocation)
3. (d) $C : Cl \longrightarrow C^+ + Cl^-$
4. (b) Carbocation
5. (c) C in $(CH_3)_3\,C$—OH

6. (b) The weaker the acid, stronger is the nucleophile

$HI > HBr > HCl > HCN > H_2O > C_2H_5OH$

$HCl > RCOOH > ROH > HCN > H_2O > NH_3 > C_2H_2 > C_2H_4 > CH_3-CH_3$

7. (d) NH_4^+ is neither an electrophile nor a nucleophile

(in the same way $Me_3 N^+$ is also neither an electrophile nor a nucleophile

As these species do not have an electron pair for donation nor these cannot accommodate a pair of electron, as all shells of N are fully filled.

8. (b) BF_3

9. (d) $SbCl_5$ will abstract a Cl to form $SbCl_6^-$ and will convert 2-chloro-2-phenl ethane into a carbocation

$$C_6H_5-\underset{\underset{Cl}{|}}{CH}-CH_3 \xrightarrow{SbCl_5} C_6H_5CH^+ CH_3 + SbCl_6^-$$

10. (a) H_3O^+

11. (c) NO_2^+ produced in the reaction-

$2H_2SO_4 \longrightarrow 2H^+ + 2HSO_4^-$

$2H^+ HNO_3^+ \longrightarrow H_3N^+O_3$

$H_3N^+O_3 \longrightarrow NO_2^+ + H_3^+O$

12. (b) $CH_3CH_2CHBrCH_3$; 2° H is more reactive 1° H

13. (a) $CH_3CH_2^- > CH_2=CH^- > CH\equiv C-> OH-$

$H_2O > C_2H_2 > C_2H_4 > CH_3CH_3$

14. (d) C_2H_5MgBr act as a nucleophile and an electrophile

15. (b) $(CH_3)_3C\bullet > (CH_3)_2C\bullet H > C\bullet_2H_5 > C\bullet H_3$

16. (b) A carbanion is negatively charged

17. (b) A carbonium ion contains only 3 pairs of electrons

18. (c) A free radical contains odd number of electron C, Cl•

Only a free radical has an unpaired electron

19. (c) Electrophiles are electron deficient specie, they can share an electron pair, with a carbanion and a called Lewis acids; carbanion initiate both addition and substitution process

20. (c) Cl•   21. (a)

22. (a) nucleophiles are electron rich species, they would like to join an electron deficient specie or carbonium ion

23. (b)   24. both (a) and (b)

25. (d)   26. (a), (c)   27. (iv)   28. (a)   29. (a)   30. (c) $BeCl_2$

31. (b) due to resonance

32. (a) —OCH_3 and $COCH_3$ are electron withdrawing, –; CH_3 is electron donating

33. (b), (c)   34. (b), (c)   35. (d) all   36. (d) all   37. (d)   38. (a)   39. (b)

40. (a) $(CH_3)_4N-OH \longrightarrow CH_3OH + (CH_3)_3N$

41. (a)   42. (c)   43. (b)   44. (c) R—OH $\longrightarrow$ R^+ + OH—, carbonium ion

45. (b)   46. (c)   47. (b)   48. (d)

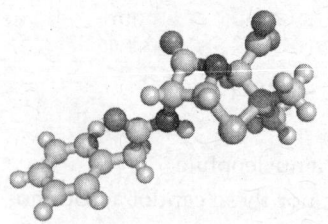

Electron Displacement Effect

There are four types of electronic displacement mechanisms observed in organic molecules. These are i. Inductive effect, ii. Electromeric effect, iii. Mesomeric effect, iv. Hyper-conjugative effect.

i. Inductive effect: When two atoms like carbon and hydrogen are joined by a covalent bond, it is understood that the two atoms share the pair of electron equally in between them. When a carbon atom forms a covalent bond with another atom of high electronegativity, the electron pair is shifted towards the more electronegative atom, making the bond polar in nature. As a result of this minor displacement of the bonding electron pair, there develops a small fraction of charges on both the bonded atoms (indicated by = δ^+ and δ^-).

$C - H$ bond   a standard bond, nonpolar in nature
Electronegativity   —   2.5   2.1
$C^{\delta+} - Cl^{\delta-}$ a polar bond
Electronegativity   —   2.5   3

This polarisation thus induced in the molecule is of permanent in nature. This permanent effect caused by a small displacement of the bonding pair of electrons, from a less electronegative atom towards the more electronegative atom, caused by difference in their electronegativities, introduces polarity in the molecule and is called as negative inductive effect or I-effect. It depends upon the 'intrinsic' tendency of a substituent group to release or withdraw electrons based on its electronegativity, either through the molecular chain or through the space. The effects weaken steadily with increasing distance from the substituents.

In an organic compound, when a highly electron withdrawing group 'X' is attached at the terminal carbon atom of a chain, it withdraws electrons from the α-carbon atom, this produces a small +ve charge on the α-carbon atom. This in turn withdraws electrons from the β-carbon atom. This causes the β-carbon to acquire a little +ve charge which would be smaller than that of the charge on the α-carbon atom. This charge is relayed throughout the chain till it reaches the last carbon atom of the chain, with a much decreased magnitude.

$$C^{\delta\delta\delta+} - C^{\delta\delta+} - C^{\delta\delta+} - C^{\delta+} - X^{\delta-}$$
$$\quad\delta \qquad\quad \gamma \qquad\quad \beta \qquad\quad \alpha$$

Actually it dies down completely, after the third or the fourth carbon atom of the chain.

As the various elements are likely to substitute for hydrogen in an organic compound, all are more electronegative as compared to hydrogen, so that almost all the groups exert electron withdrawing inductive effect.

Since the inductive effect is shown by the molecules containing single bond, the effect is permanent in nature. Generally this effect is shown by the compounds which have electron withdrawing or an electron donating groups linked to the terminal carbon atom.

The electron withdrawing groups are represented by the symbol = '–I'

The electron donating groups are represented by the symbol = '+I'

The following **groups**, when present in a chain exert an **electron withdrawing effect in the decreasing order.**

$- 'I' = -NO_2 > -CN > -F > -COOH > -Cl > -Br > -I > -OH > -OCH_3 > -C_6H_5$

The **groups, with +I effect are given in decreasing order**

$+I \text{ effect} = (CH_3)_3-C > (CH_3)_2-C > CH_3CH_2- > CH_3-$

In general, the inductive effect of the alkyl group is in the order of $3° > 2° > 1°$.

ii. Electromeric effect: The compounds containing, a double or triple bond in the molecule show this effect. When an electrophile attacks either a double or a triple bond, a polarisation takes place in the molecule and a pair of electron is completely transferred from one atom of the double or the triple bond to the other atom. An olefinic double bond consists of one α-bond and one π-bond. It is the electrons of the π-bond, being mobile, that shifts to either side depending upon form which side the electrophile (E) approaches

$$\overset{\overset{E^+}{|}}{>C} = C< \longrightarrow \overset{\overset{E}{|}}{>C} - C^+ \longrightarrow >C^+ - \overset{\overset{E}{|}}{C<} \longrightarrow \overset{\overset{E^+}{|}}{>C} = C<$$

The atom, thus acquires a negative charge to which the electron pair has been transferred, and the other becomes positively charged.

Electrophilic reagent:

$$A = B \longrightarrow A^+ - B^-$$

It is purely a temporarily attack and remains in the play only in the presence of the electrophilic reagent. It disappears once the attacking electrophile is removed and the polarised molecule reverts to its original electronic state.

In the presence of an electrophile

$$A = B \rightleftharpoons A^+ - B^-$$

On removing the electrophile

Alkenes or the compounds containing an olefinic double bond, undergo electrophilic addition reaction only. The effect that produces a temporary polarisation in the molecule, at the seat of a multiple bond, by the shift of a pair of electrons, from one atom to the other atom, under the influence of an electrophilic reagent, is called as electromeric effect. In a symmetrical molecule, the electron shift may take place on either side, depends upon from which side the electrophile approaches and attacks the double bond.

$$\overset{\overset{E^+}{|}}{>C} = C< \longrightarrow \overset{\overset{E}{|}}{>C} - C^+ \qquad >C^+ - \overset{\overset{E}{|}}{C<} \qquad \overset{\overset{E^+}{|}}{>C} = C<$$

In the case of acetylinic compounds a triple bond consists of one α-bond and two π-bonds.

$$-C \equiv C- \qquad -C-C-$$

Acetylene undergoes both types of addition reactions, electrophilic as well as nucleophilic addition process.

Electrophilic addition reactions: The following compounds add across the triple bond by an electrophilic addition process, i.e. the electrophile adds first, followed by the addition of the nucleophile.

- $H_2SO_4 \longrightarrow H^+ + HSO_4^-$
- $BH_3 \longrightarrow BH_2^+ + H^-$
- $Br_2 \longrightarrow Br^+ + Br^-$ $—C \equiv C \longrightarrow —C = C^+ \longrightarrow —C = C$
- $HX \longrightarrow H^+ + X^-$ ($E^+ = H^+, BH_2^+, Br^+, H^+$)

The following compounds add across the triple bond by the nucleophilic addition process, i.e. the nucleophile adds first, followed by the addition of the electrophile.

- $CH_3COOH \longrightarrow CH_3COO^- + H^+$ (in presence of Hg_2^+ ions)
- $CH_3OK \longrightarrow CH_3O^- + K^+$ (in the presence of potassium methoxide)
- $H_2O \longrightarrow H^+ + OH^-$ ($H_2SO_4 + Hg_2^+$ ions)$—C \equiv C— \longrightarrow —C = C^- \longrightarrow —C = C—$
- $HCN \longrightarrow H^+ + CN^-$

  In presence of $Ba(CN)_2$ ($Nu– = CH_3COO^-, CH_3O^-, OH^-, CN^-$)

Difference between electromeric and inductive effects

1. The effect is shown by molecules containing a—C to C—single bond.	1. Shown by the substances containing a double bond, (—C=C—) or a triple bond.
2. Takes place under the influence of a substituent group (electron withdrawing or an electron releasing) attached to the terminal carbon atom.	2. Takes place only when the molecule is exposed to attack by an electrophile.
3. Polarity is caused by the shift of electron pair from one atom to the other atom.	3. A complete transfer of the electron pair from one atom to other, joined by a multiple bond.
4. The charge developed is very small and is shown by δ +ve and δ –ve, depending upon +I or –I effects of the substituent group.	4. The charge acquired by the atom, gaining a pair of electron is +1, and the other atom gets –1.
5. Permanent effect depending upon the structure of the substrate group.	5. A temporary effect, till the attacking electrophile is around; effect disappears with the removal of the electrophile.

iii. Mesomeric effect or conjugative effect: It may be defined as a case of electron displacement, in the molecule, which have an extended chain with conjugate π-bonds, (having an alternate σ-bond and π-bonds). The electron displacement is relayed through π-electrons of multiple bond, in the carbon chain of the molecule. This causes a permanent polarisation in the molecule. The π-electrons get delocalized as a consequence of this mesomeric effect, resulting in the formation of a number of resonating structures of the molecule. This leads to a greater magnitude of mesomeric effect, than the corresponding inductive effect, operates in the molecule, based on the electronegativities of the bonded atoms. This may be also be defined as the polarity produced in the molecule, as a result of an interaction between two π-bonds or in between a π-bond and a lone pair of electron, when these happen to be in a conjugated system. This effect is transmitted alongwith the carbon chain of the molecule, as I the inductive effect.

If a carbonyl group is in conjugation with a carbon to carbon double bond (—C=C—), the polarisation is transmitted through π-electrons.

$$CH_3—CH=CH—\underset{\underset{H}{|}}{C}=O \longrightarrow CH_3—\overset{+}{CH}—CH=C—O^-$$

The π-electrons are delocalized as a consequence of mesomeric effect, forming a number of resonating structures of the molecule. This leads to a greater magnitude of the mesomeric effect than the corresponding inductive effect, for a given difference of electronegativities of the bonded atoms. Mesomeric effect may be positive or negative. Those atoms or groups, which lose electrons towards the carbon atom are said to have a positive M effect (+M). Those atoms or groups which withdraw electrons from carbon atom, are said to be have a negative M effect (–M) .

The atoms or groups, having a **+ M effect** are:

—Cl, —Br, —I, —NH_2, —NR_2, —OH, —OCH_3

All these atoms have a pair of electrons, in conjugation with an attached unsaturated system which is used for enhancing the degree of delocalization, bringing a greater degree of stability to the molecule. The +M effect of the halogen (—Cl or —Br), in vinyl halide or in aryl halide, explains the low reactivity effect.

$$CH_2{=}CH{-}Br \longrightarrow \overset{-}{C}H_2{-}CH = Br^+$$

vinyl bromide

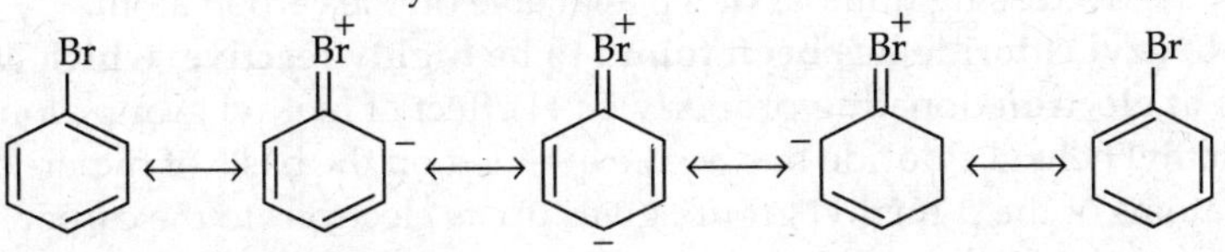

bromobenzene

The groups that have a –M effect are:

—NO_2, —C ≡ N, —C=O, —SO_3H

All these groups, have –M effect and are highly electronegative, like oxygen or nitrogen, functioning as a sink for electrons.

iv. Hyperconjugation: It is also referred as 'no bond resonance', it is used to explain the unusual high reactivity of some aromatic compounds, as well as stabilities of some carbocation and organic free radicals. When a carbon atom containing at least one hydrogen atom, (C—H bond) is attached to an unsaturated carbon atom, the electrons of C—H bond, become less localised, by entering into partial conjugation, with the attached unsaturated system, i.e. σ-, π-conjugation.

$$H{-}\underset{H}{\overset{H}{C}}{-}CH{=}CH_2 < \longrightarrow H{-}\underset{H}{\overset{H}{C}}{-}CH = CH_2 < \longrightarrow {-}H{-}\underset{H}{\overset{H}{C}}{-}C = H{-}CH_2 <$$

This type of conjugation between a single and a multiple bond, is termed as hyperconjugation, and this effect is permanent. The delocalization, is through overlapping between a p-orbital and a σ-orbital. Because of this overlapping, each pair of electrons, is not limited to binding together just two atoms, i.e. the doubly bonded carbon and hydrogen but to a certain extent helps bind together all four atoms. Delocalisation of this kind, involving σ-bond orbitals with π-bond orbitals is called 'hyperconjugation'.

The three hydrogen atoms in the propene molecule contribute to this effect.

$$H{-}\underset{H}{\overset{H}{C}}{-}CH{=}CH_2 < \longrightarrow H{-}\underset{H}{\overset{H}{C}}{-}CH = CH_2 < \longrightarrow {-}H{-}\underset{H}{\overset{H}{C}}{-}C = H{-}CH_2 <$$

The hydrogen atoms are not free; the effect is to increase the ionic character of the C—H bond, the electrons of the C—H bond, become partially delocalized, through conjugation. This is further proved on the basis of the bond length.

A single —C to C— bond in the case of an alkane, has a bond length = 1.54 Å

The bond length as measured in the case of a conjugated system = 1.46 Å

$$CH_3—CH_3CH_3—CH = CH_2$$
$$\text{1.54 Å} \qquad\qquad \text{1.46 Å}$$

The inductive effect of the alkyl groups (+R) is as follows:

$(CH_3)_3—C—$ > $(CH_3)_2—CH—$ > $CH_3—CH_2—$ > $CH_3—$

The hyperconjugative effect has been found to be in the reverse order

$CH_3—$ > $CH_3—CH_2—$ > $(CH_3)_2CH—$ > $(CH_3)_3C—$
3-H   no-H

It is based on the decreasing number of 'H' available on the carbon atom.

a. p-Methyl benzyl chloride has been found to be highly reactive, which cannot be explai-ned on the basis of electron donating property or +I effect of methyl group alone. The abnormal reactivity of p-methyl benzyl chloride has been explained on the basis of hyper- conjugation. One of the hydrogen atoms of the p-methyl group, gives up its electrons to the carbon atom to which it was attached, but remains in the vicinity of the carbon atom of the methyl group.

This release of hydrogen atom, as a proton, makes the carbon, having an unshared pair of electrons, which is used to convert the benzenoid ring to a quinonoid ring, with the valences of each atom in the quinonoid ring is fulfilled, and setting chlorine free as an anion. This explains the abnormal high reactivity of chlorine.

b. The **formation of 1,2-dibromopropane from allyl bromide, when it reacts with HBr may be explained as follows:**

$$CH_2{=}CH—CH_2Br \xrightarrow{H^+} CH_2{=}CH—\overset{-}{C}H—Br< \xrightarrow{H^+}$$

$$\overset{-}{C}H_2{-}\overset{+}{C}H{=}CH—Br \xrightarrow{+\,HBr} CH_3—CHBr—CH_2Br$$

The inductive effect of bromine would tend to give 1,3-diromo propane, however the hyper-conjugative bond is stronger, hence the formation of 1,2 dibromo propane may be explained.

Hyperconjugative structure of a species defines its stability. For example, three structures may be written for ethyl carbocation:

The stability of carbonium ions decreases as one moves down from tertiary butylcarbocation to methyl carbocation.

$$(CH_3)_3—C^+ > (CH_3)_2—CH^+ > CH_3—CH_2{}^+ > CH_3{}^+$$

The stability is based on the formation of various resonating structures.

c. The relative stability of the various classes of carbonium ions may be explained on the basis of no bond resonance structures that can be written for them. Such structures are arrived, by shifting the bonding electrons from an adjacent C—H bond to the electron deficient carbon atom. In this way, the +ve charge originally on the electron deficient carbon atom is dispersed to the hydrogen atom. This manner of electron release by assuming no bond resonance character in the adjacent C—H bond is called hyperconjugation or no-bond resonance.

The resonating structures for tertiary butyl carbocation are nine, six for isopropyl carbocation and only three for ethyl carbocation.

d. Stabilities of free radicals: The greater stability of the tertiary butyl free radical is based on theformation of nine hyperconjugative structures. The decreasing order of stabilities of free radicals is the same as in the case of carbonium ions:

$(CH_3)_3$—C— $\longrightarrow$ $(CH_3)_2$CH— $\longrightarrow$ CH_3—CH_2— $\longrightarrow$ CH_3—

tertiary butyl   isopropyl   ethyl   methyl free radicals

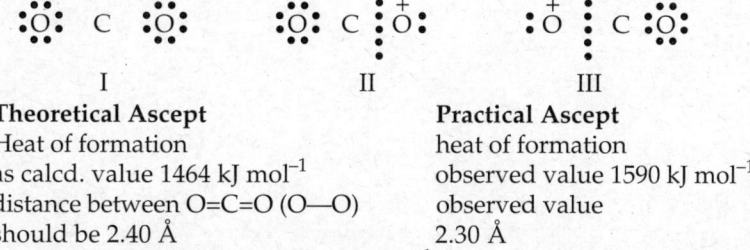

Nine such hyperconjugated structures are formed in the case of tertiary butyl free radical; in the case of isopropyl free radical six, in the case of ethyl three conjugative structure are formed, explains the decreasing order of stabilities of the free radicals.

e. Resonance: There are many compounds, which cannot be represented by a single electronic structure, which may explain, satisfactorily, all the properties shown by the compound.

For example, the electronic structure of carbon dioxide may be represented by at least three possible structures, as shown

$$\ddot{O}: \quad C \quad :\ddot{O} \qquad\qquad :\ddot{\overset{\_}{O}}: \quad C \quad :\overset{+}{O}: \qquad\qquad :\overset{+}{O} \quad C \quad :\overset{\_}{\ddot{O}}:$$

I   II   III

Theoretical Ascept	**Practical Ascept**
Heat of formation	heat of formation
as calcd. value 1464 kJ mol^{-1}	observed value 1590 kJ mol^{-1}
distance between O=C=O (O—O)	observed value
should be 2.40 Å	2.30 Å

In other words, carbon dioxide requires 125 kJ mol^{-1} more energy, than the expected energy to break it into its elements, i.e. carbon dioxide is more stable than the anticipated structure, O=C=O. The difference between the enthalpy of formation of the actual compound (the observed value), and that of the resonating structure of the lowest internal energy (obtained by calculation) is called resonance energy. As none of these structures account for its observed values, therefore it led to the idea that the true structure of CO_2 is the one that none of these single structure represent. The actual structure lies in between these and each one of these structure contribute towards the actual structure that cannot be represented by any single structure. When several structures contribute, can be assumed to contribute towards the actual or a true structure of a molecule, but none of these can represent it completely, the molecule is referred to as having a resonance hybrid structure, and the phenomenon is known as *resonance.*

Mechanism

a. the nuclei in each structure are same or nearly same; b. the number of unpaired electrons is the same in each structure, c. and all the contributing hybrid structure are of the same or of nearly same stability. The concept of hybridisation of two or more conventional valence bond structures, to yield a new stable structure, with more stability or having less energy, is termed as *resonance*.

Criteria of resonance:

• Heat of formation of a resonance hybrid is exceptionally high. Heat of formation of carbon dioxide, observed is higher than a calculated value by 126 kJ mol^{-1}(1590 kJ mol^{-1} –1464 kJ mol^{-1}). This amount of energy, i.e. the resonance energy is lost, during the formation of carbon dioxide.

• Shortening of bond length: In the resonating structure, the bond lengths are generally shorter, than in any of the canonical structures.

• Stabilisation: Larger is the value of resonance energy; the smaller is the internal energy of the resonance hybrid, greater is the stability of the resonating structure.

Benzene is a resonance hybrid of many contributing structures. It may be represented by the following structures.

The resonance energy for benzene is = 144.4 kJ mol^{-1}or 36 kcal/mol. All the six C to C— bond lengths are equal= 1.39 Å, instead of having an alternate double bond (bond length = 1.34 Å) and a single bond, (bond length = 154 Å).

It is clear, that no single classical structure, (with localised bonds), completely describes the properties of the molecule fully, and to describe the molecule fully, several structures must be written. None of these structures is a unique true description of the molecule, but each structure contributes towards the behavior of the molecule. By using the concept of resonance, the actual molecule can be expressed as a combination or a hybrid of the contributing structures. These contributing structures are called as canonical or resonating structures. In fact, none of these canonical forms have any real or separate existence.

It may be therefore, be said that only the hybrid structure is the actual structure and all other structures are fictitious or imaginary.

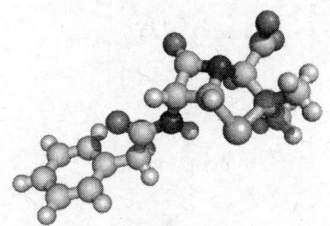

Isomerism

In organic chemistry, there are many compounds that have same molecular formula, but differ in their physical and chemical properties; these compounds are known as *isomers* and the phenomenon as *isomerism*. Isomers contain the same number of same kind of atoms, but the atoms are attached to one another in different ways. Isomers are different compounds, because they have different molecular structures. The differences in molecular structures, gives rise to a difference in properties.

Broadly, isomerism may be divided in the following two broad classes on the basis of their structures.

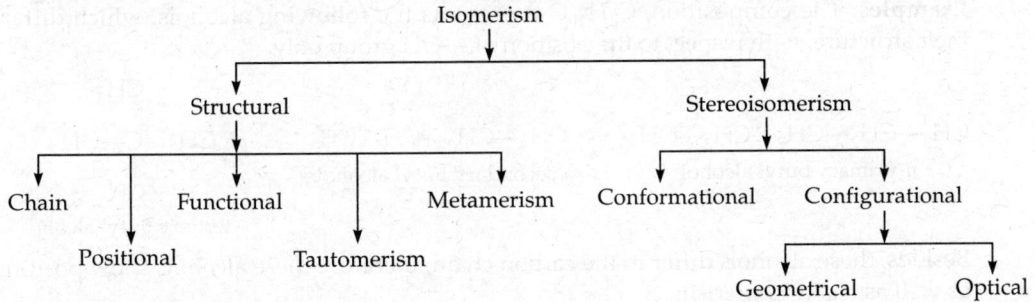

The two broad classes of isomerism are:

i. **Structural isomerism:** When the isomerism is simply due to difference in the arrangement of atoms within the molecule without any reference to space. It may be explained on the basis of tetravalency of carbon taken in the same plane. It is further in divided into 5 classes.

   a. **Chain or nuclear isomerism:** These isomers differ only in the structure of carbon chain. In alkanes, butane onwards, every alkane exists in a number of isomers.

Examples:

- Butane exists in two isomeric forms, namely *n*-butane and isobutane.

    $CH_3CH_2—CH_2—CH_3$ and $CH_3—CH—(CH_3)—CH_3$
        n-butane   isobutane

- Pentane exists in three isomeric forms *n*-pentane, isopentane, and neopentane

$$CH_3—CH_2—CH_2—CH_2—CH_2;$$
n-pentane

$$CH_3—CH_2—\overset{\overset{\displaystyle CH_3}{|}}{CH}—CH_3;$$
isopentane

$$CH_3—\overset{\overset{\displaystyle CH_3}{|}}{\underset{\underset{\displaystyle CH_3}{|}}{C}}—CH_3$$
neopentane

All these compounds differ only in the carbon chains.

ii. **Functional isomerism:** These isomers have same molecular composition, but differ in the functional group, present in the isomers. There are several groups of compounds which have same molecular compositions, but have different functional groups.

Examples: Alcohols and ethers have same molecular composition, i.e. $C_nH_{2n}O$, but differ in the functional groups.

$C_4H_{10}O$: both ethers and alcohols may be represented by:

$$CH_3CH_2CH_2CH_2—OH;$$
n-butyl alcohol

$$CH_3—CH_2—O—CH_2—CH_3$$
diethyl ether

The composition, $C_4H_{10}O$ refers to various ethers, which may be formulated out of this formula:

$$CH_3—CH_2—O—CH_2—CH_3;$$
diethyl ether

$$CH_3—CH_2—CH_2—O—CH_3;$$
methyl *n*-propyl ether

$$CH_3—\overset{\overset{\displaystyle CH_3}{|}}{CH}\overset{\displaystyle O}{\diagdown}CH_3$$
methyl isopropyl ether

This property is known as '*metamerism*'.

iii. **Position isomerism:** In these isomers, the functional group is the same, but they differ in their positions.

Examples: The composition, $C_4H_{10}O$ represents the following alcohols, which differ in their structure, with respect to the position of —OH group only.

$$CH_3—CH_2—CH_2—CH_2—OH;$$
n-primary butyl alcohol

$$CH_3—CH_2—\overset{\overset{\displaystyle OH}{|}}{CH}—CH_3;$$
secondary butyl alcohol

$$CH_3—\overset{\overset{\displaystyle CH_3}{|}}{\underset{\underset{\displaystyle CH_3}{|}}{C}}—OH$$
tertiary butyl alcohol

Besides, these alcohols differ in the carbon chain, therefore these alcohols show positional as well as chain isomerism.

iv. **Metamerism:** It is the isomerism shown by the members of the same homologous series, due to the difference in the nature of alkyl groups attached, on either side of a divalent functional group.

Examples: The formula, $C_4H_{10}O$ represents alcohols and ethers both, (showing both position and functional isomerism).

The formula, $C_4H_{10}O$, may also represent by three isomeric ethers.

$$CH_3—CH_2—O—CH_2—CH_3; \quad CH_3—O—CH_2—CH_2—CH_3; \quad CH_3—O—CH(CH_3)_2$$
diethyl ether   methyl *n*-propyl ether   methyl isopropyl ether

These ethers differ in the nature of the alkyl chain attached on either side of the functional group, i.e. oxygen.

The other example are ketones and secondary amines.

$C_5H_{10}O-;$ $CH_3—CH_2—CO—CH_2—CH_3;$ $CH_3CO—CH_2—CH_2—CH_3;$
ketones   diethyl ketone   methyl *n*-propyl ketone

$CH_3—CO—CH—(CH_3)_2$
methyl isopropyl ketone

These ketones differ in the alkyl chains attached on either side of the divalent keto group as functional group.

Secondary amines: $C_4H_{10}N$

$CH_3—CH_2—NH—CH_2—CH_3;$ $CH_3—NH—CH_2—CH_2—CH_3;$ $CH_3—NH—CH(CH_3)_2$
diethyl amine   methyl *n*-propyl amine   methyl isopropyl amine

Similarly, these secondary amines differ in the alkyl chains attached on either side of the divalent secondary amino group.

v. **Tautomerism:** When two structural isomers are usually interconvertible and exist in dynamic equilibrium. They are called as *tautomers* and the phenomenon is known as tautomerism. It is a special type of functional isomerism in which the isomers are in equilibrium with each other.

Example: Ethyl acetoacetate, $CH_3COCH_2COOC_2H_5$ is an equilibrium mixture of a keto form (93%), and an enol form (7%) (enol form means that there is a double bond or an 'ene' and a hydroxyl group).

$$\underset{\text{keto form}}{CH_3—\overset{\overset{O}{\|}}{C}—CH_2—COOC_2H_5} \rightleftharpoons \underset{\text{enol form}}{CH_3—\overset{\overset{OH}{|}}{C}=CH—COOC_2H_5}$$

As a matter of fact, the same compound shows the chemical and physical properties of both the groups, namely a keto group and an enol group.

Properties of keto and an enol form:

$$CH_3—\overset{\overset{O}{\|}}{C}—CH_2—COOC_2H_5 \rightleftharpoons CH_3—\overset{\overset{OH}{|}}{C}=CH—COOC_2H_5$$

• Long colourless needles, mp –39°C
• Does not give any colour with $FeCl_3$
• Does not decolourise Br_2 water

• Colourless oil, bp –78°C
• Gives violet colour with $FeCl_3$
• Decolourise yellow colour of bromine water

Dynamic equilibrium means that when a reagent reacts with keto form, the enol form changes to the keto form till all the ester has reacted as a keto compound.

In the same way, when a reagent reacts with the enol form, the keto form changes to the enol form, till all the ester has reacted as an enol compound.

vi. **Stereoisomerism:** It may be defined as the phenomenon, possessed by some compounds, which can be explained only on the basis of tetrahedral nature of carbon, or by taking the tetravalency of the carbon, which are directed towards the four corners of the regular tetrahedron, inclined at an angle of 109°28.

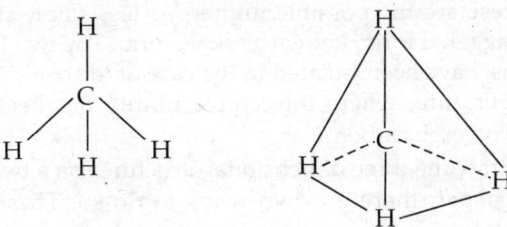

The isomers, which have same structure, but differ in spatial arrangement of atoms or groups or both, and have different configuration (arrangement of several atoms in space), are said to be *stereoisomers*.

Stereoisomerism is further divide into following two groups.

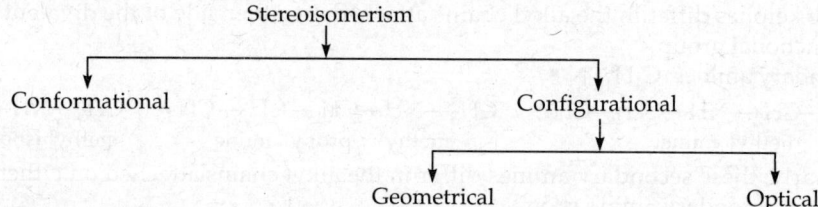

a. **Conformational isomerism:** It is also termed rotational isomerism. A carbon to carbon single bond is free to rotate about its axis, it can give rise to the formation of various structures, known as conformers, by rotation about a σ-bond with spatial arrangements of the atoms and groups. Since the energy involved in the conversion of one form to the other form is very small, these forms convert into each other readily and ultimately into the most stable form, the staggered form.

A molecule of ethane, containing two carbon atoms attached by a single bond, may have the following forms possible; however it finally converts into the staggered form, being most stable.

| eclipsed | skew form | staggered form |

Eclipsed conformational form: The three hydrogen atoms attached to the front carbon, are exactly in front of the three hydrogen attached to the back carbon.

Staggered conformational form: In the staggered form all the six hydrogen atoms are clearly seen and are symmetrically situated at the maximum possible distance (when seen from front the side of the ball and the stick model).

In the staggered form, the hydrogen atoms are as far apart as possible, minimizing the repulsion between hydrogen atom, called as nonbonded interaction (inter nuclear distance = 3.1 Å), and in the eclipsed form, the nonbonded interaction is large (inter nuclear distance = 2.29 Å). This is the reason for the staggered conformation to be more stable as compared with the eclipse form.

Skew forms: These are the possible numerous less stable structures, in between the eclipsed and staggered form, known as skew form. The two forms namely eclipsed and staggered forms, have been isolated in the case of the compound, $CHBr_2—CHBr_2$, at a very low temperature, where interconvertibility has been checked by the size of bromine atoms on each carbon.

In order to draw the three dimensional structures on a two dimensional surface, i.e. on a plane of paper, there are two ways to do so. These methods are known as *projection formulae*.

- *Sawhorse projection formula*: Looking at the molecule through the nearest carbon, two extreme structures are possible, for ethane molecule. One structure, as shown earlier in which three hydrogen atoms attached, on the nearest carbon overlap the three hydrogen atoms; on the other carbon. Such a structure is termed as *eclipsed* form. This eclipsed form is unstable, because the two hydrogen atoms are close enough for a mutual repulsion (internuclear distance between the two (H—H) = 2.29 Å.

  The other structure is that in which the three hydrogen atoms on one carbon are at the maximum distance from those hydrogen on the other carbon atom. Since the repulsion is minimum, this will be more stable. This form is termed as *staggered form*, (internuclear distance between the two (H—H) = 3.1 Å.

  In between these forms there are a large number of possible structures, termed as *skew forms*.

  H—H = 2.29 Å   H—H distance = 3.1 Å

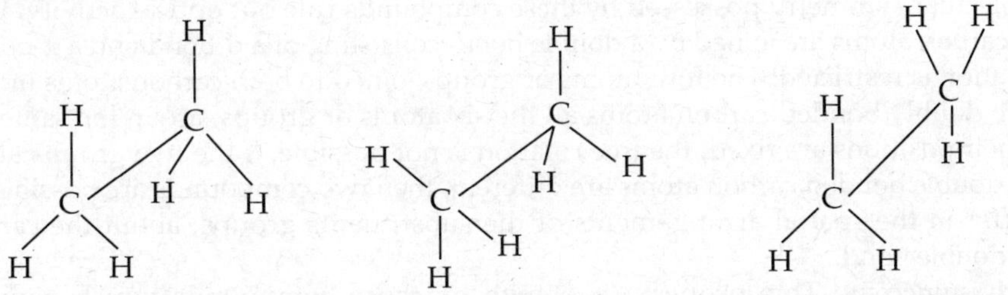

  eclipsed   skew   staggered form

- *Newman projection formula*: In Newman's projection formula, the C—C bond is seen end on. The nearest carbon is represented by a dot, while the three bonds linking the three hydrogen atoms attached to this carbon, are represented by three lines, radiating from this point. The carbon situated farther away, is represented by a circle, and three bonds are represented by three lines radiating from the circle. The eclipsed and staggered forms for ethane molecule, are given below.

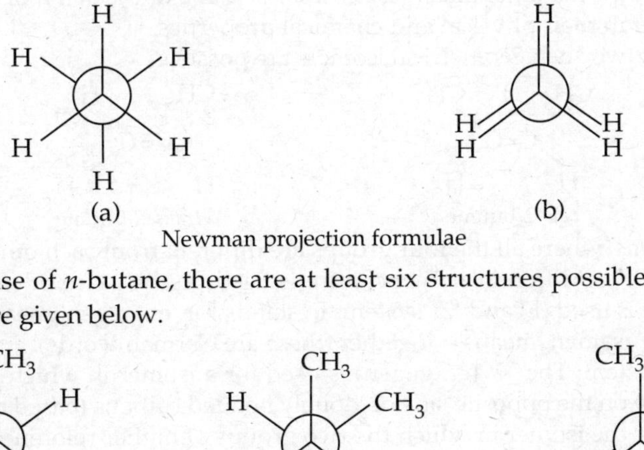

  (a)   (b)

Newman projection formulae

Taking the case of *n*-butane, there are at least six structures possible. Out of which only three forms are given below.

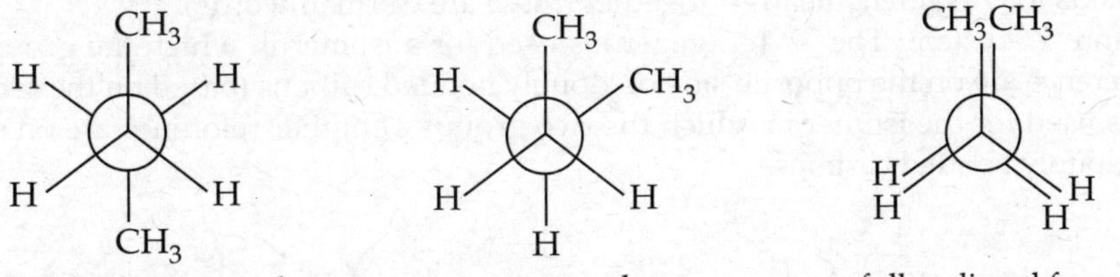

completely staggered   staggered   fully eclipsed form

By rotating at 60°, C—C bond eclipsed forms are possible.

b. **Configurational isomerism:** These are identical looking structures of the same compound, which are nonsuperimposable on each other. These may be termed *stereoisomers*, being mirror images of each other. They have the same structural formula, but differ in the arrangements of the various atoms and groups in space.

Diastereoisomerism

i. *Geometrical isomerism or a cis- and trans-isomerism:* This is characterized by the compounds, which have same structure but have different configuration are also classified as diastereoisomers, because these are not mirror image of each other.

The molecular symmetry possessed by these compounds rule out optical activity. When a pair of carbon atoms are joined by a double bond, consisting of a σ-bond and a π-bond, the free rotation is restricted. The four atoms or groups joined to both carbon atoms including both the doubly bonded carbon atoms all the six atoms or groups, are in the same plane. Since their positions are fixed, the free rotation is not possible. If the two groups attached to each double bonded carbon atoms are different, then two compounds are possible. They will differ in the spatial arrangements of the substituents groups, about the carbon to carbon double bond.

ii. *Optical isomerism:* This property is shown by those compounds which have same structure but differ in configurations and because of their molecular 'asymmetry' these compounds are optically active.

The compound, in which the two identical groups are close to each other or are on the same side of the carbon to carbon double bond, is termed *cis*-compound. The other isomer in which the two identical groups are on the opposite side of the doubly bonded carbon atoms is termed a *trans*-isomer.

These two compounds, namely *cis*- and *trans*-compounds are known as geometrical isomers and the phenomenon as *geometrical isomerism* or *cis- and trans-isomerim*. These two isomers differ in their physical and chemical properties.

For 2-butene, two such geometrical isomers are possible.

For the systems, where all the four groups are different from each other, like AB-C = C-DC or AC-C=C-BD, a different nomenclature is used. In the place of *cis*- and *trans*- nomenclature, there is used 'E' and 'Z' system 'E' stands for 'entgegen' means = opposite and 'Z' = stands for zusamen, means = together (these are German words).

'E' and 'Z' system: The 'E' (entgegen) is used for a isomer in which the groups of higher preference are on the opposite side of doubly bonded carbons (based on the sequence rule). 'Z' is used for the isomer in which the two groups of higher reference are on same side of the doubly bonded carbons.

Examples:

- For the compound having the composition -I-Cl – Br-F- preference based on the atomic number as per the sequence rule is

I > Br > Cl > F
53   35   17   9

I   F
$\diagdown$ $\diagup$
C=C
$\diagup$ $\diagdown$
Cl   Br

E isomer

I   Br
$\diagdown$ $\diagup$
C=C
$\diagup$ $\diagdown$
Cl   F

Z isomer

- Preference as per the sequence rule

—COOH > —C≡N > CH_2—CH_3 > —CH_3

COOH   CN
$\diagdown$ $\diagup$
C = C
$\diagup$ $\diagdown$
CH_3CH_2 CH_3

Z isomer

CH_3CH_2 CN
$\diagdown$ $\diagup$
C = C
$\diagup$ $\diagdown$
COOH CH_3

E isomer

Auwer's Skita rule: For predicting *cis-* and *trans*-isomers, there is Auwer's Skita rule. As per this rule, these can be distinguished from each other by comparing the following properties:

mp –	*cis*-isomer has	lower mp	than	*trans*-isomer
stability –	*cis*-isomer is	less stable	than	*trans*-isomer
density –	*cis*-isomer has	greater	than	*trans*-isomer
refractive index –	*cis*-isomer has	more	than	*trans*-isomer
solubility in water –	*cis*-isomer has	greater	than	*trans*-isomer
dipole moment –	*cis*-isomer has	more	than	*trans*-isomer
heat of combustion –	*cis*-isomer has	greater	than	*trans*-isomer
dissociation constant –	*cis*-isomer has	greater	than	*trans*-isomer
bp –	*cis*-isomer has	higher bps	than	*trans*-isomer

cis-isomer is usually is more labile.

Optical isomerism possessed by certain compounds, on passing a monochromatic polarized light through their aqueous solution, they rotate the plane of the polarized light either to the right or to the left. This phenomenon is termed optical isomerism and the compound is said to be optically active.

Monochromatic light: According to the wave theory, an ordinary ray of light, consists of waves vibrating in all plane, perpendicular to the direction of propagation. The light from sodium lamp consists of rays vibrating in a single plane is termed a monochromatic light.

Plane Polarised light (PPL): On passing the light from sodium lamp, (or a monochromatic light) through a crystal of iceland spar (or calcite, or crystalline $CaCO_3$), the light splits up in two rays, each ray vibrating in one plane only. Each of the two ray formed has different refractive indices. By taking a combination of iceland spar crystals, suitably cut for this purpose (or a Nicol prism) on passing through a monochromatic light, one of the two rays suffers total internal reflection, while the other ray passes through, is termed a polarised light, vibrating in one plane only. This ray is termed plane polarised light. Suppose the vibrations of the polarised light are taking place, in the horizontal plane, then the vertical plane is termed the plane of the polarized light. When another Nicol prism is placed in the path of the plane polarised light, the light passes with full intensity, provided the axis of the second prism is parallel to the axis of the first prism. If the axis of the second prism is rotated through an angle of 90° or at the right angle to that of first, the light is totally cut off due to total internal reflection. However, in the various intermediate positions of the second 'Nicol' prism, the light passes partially. Whereas, the first prism is called a *polariser*, the second prism is termed as *analyser*. On placing an aqueous solution of sugar in between the two, namely polariser and analyser, the analyser is to be rotated, either to the right or to the left, to see the light with full intensity.

+O+	|\ |	|	|	|\ |
Source of light	polariser	sample	analyser	eye

The optically active compounds exist in three isomeric forms, with one asymmetric carbon atom.

1. The isomers which rotate the plane of the polarised light to the right, are known as dextro-rotatory compound (Latin, dexter = right) or direction (+) –form.
2. Those which rotate the plane of the polarised light to the left are known as laevo-rotatory compounds (Latin laevo = left) or direction (–) –form.
3. An inactive form, which does not rotate the plane of the polarised light at all, is termed as a *racemic mixture*, (Latin, racemic = mixture of equal amounts). This mixture consists of an equal amount of a dextro and laevo forms of an optically active compound.

(Note: Whenever an organic compound is prepared in the laboratory, there obtained a racemic mixture. On the other hand, if an optical active compound is found in nature, it is either a dextro or a laevo rotatory compound. Both the forms of an optical active organic compound are not found in nature.)

Optically active substances: Besides quartz (a crystalline form of SiO_2), which has been found to exist in two forms, other optically active substances like benzyl, $NaClO_4$, $ZnSO_4$, N_2H_4 and barium formate $(HCOO)_2Ba$, have been found to be optical active in crystalline form. As soon as the crystals are fused, vapourised or dissolved in a solvent, the optical activity is lost.

A compound that have a carbon atom whose all the four valences are joined by four different atoms or groups or both, is termed an *asymmetric carbon atom*. An asymmetric carbon does not have a plane of symmetry or a center of symmetry; hence the molecule is termed as a '*chiral*' molecule. Such a compound, exists in two forms, which are not superimposable on each other. The two forms are the mirror image of each other (superimposable means that the four groups joined at the carbon atom, cannot cover the four groups on its mirror image, only three overlap but not all the four groups). These two forms are termed as 'enantiomers' (enantio = opposite; meros = part). Enantiomers resemble each other in their physical and chemical properties, except that they rotate the plane polarised in opposite direction.

Lactic acid, $(CH_3–CH(OH)—COOH)$ exists in two isomeric optically active forms. These two forms are mirror image of each other and are not superimposable on each other. They are termed as enantiomers.

$$\begin{array}{ccc} \text{COOH} & & \text{COOH} \\ | & | & | \\ \text{H—O—C—H} & & \text{H—C—O—H} \\ | & | & | \\ \text{CH}_3 & & \text{CH}_3 \end{array}$$

Plane of mirror

Mirror image of Lactic acid

Tartaric acid: HOOC CH(OH) CH(OH)COOH

The acid is a dibasic acid and there are two similar asymmetric carbon atoms in the molecule. Tartaric acid exists in four isomeric forms. These are:

1. d-Tartaric acid 2. l-Tartaric acid 3. a racemic tartaric acid 4. meso-tartaric acid

```
        COOH                    COOH                    COOH
    H ──┼── OH              HO ──┼── H              HO ──┼── H
   HO ──┼── H               H ──┼── OH             HO ──┼── H
        COOH                    COOH                    COOH
```

d-tartaric acid	l-tartaric acid	meso-tartaric acid
mp. 170 °C	170 °C	143 °C

1. Dextro-tartaric acid rotates the plane of the polarised light to the right, and it can be assumed that the upper half rotates the plane of polarised light to the right, and the lower half strengthen the rotation to the right. The d-tartaric acid does not have a plane of symmetry. The molecule cannot be divided into two equal halves, by a line.

2. Laevo-tartaric acid is a mirror image of d-tartaric acid, and rotates the plane of the polarised light to the left. The rotation due to upper half and the lower half is super-imposed. The laevo-isomer does not have a plane of symmetry, like that of the d-tartaric form.

3. Racemic tartaric acid or 'dl' form: It is a mixture of d- and l-tartaric acids, and contains equal amounts of both d-and l- forms. The racemic form is optically inactive. However, the racemic form can be separated into the d- and l- forms. This method is known as resolution. Racemic variety is also termed 'externally compensated molecule', because of the presence of the other isomer of opposite rotation, racemic variety is optically inactive.

4. Meso tartaric acid: It is also optically inactive, as the rotation of upper half is opposite to that of the lower half. Since the molecule can be divided into two halves, the molecule therefore possesses 'a plane of symmetry'. A meso variety is also termed 'internally compensated molecule'. This cannot be resolved.

Besides, there is another element, which also make the compound optically inactive if they possess 'a center of symmetry'. Where as d- or l- tartaric acids are mirror image of each other, and are termed enantiomers, the meso tartaric acid and d- or l- form are termed 'diastereoisomers'.

A center of symmetry is an imaginary point in the molecule such that, if an imaginary line is drawn from any group in the molecule and then it is extended to an equal distance beyond this point, it meets the mirror image of the original group. For example, the substituted, diketo-piperazines, contain two asymmetric carbon atoms, but these compounds are still optically inactive, because they contain a center of symmetry in the molecule, as in *trans*-1,4-dimethyl-diketo piperazine.

```
  CH₃   CO ──── NH   CH₃            CH₃   CO ──── NH    H
    \  /            \  /              \  /            \  /
     C               C                 C               C
    /  \            /  \              /  \            /  \
   H   NH ──── CO   H.             H   NH ──── CO   CH₃
          cis-                              trans-
```

In the case of the *trans*- derivatives, if a line drawn from methyl group on carbon number 1, to the center of symmetry and extended beyond this point, by an equal distance, it meets carbon number 4, (mirror image of methyl group), and therefore the *trans*-isomer is optically inactive. On the other hand, the *cis*-isomer is optically active, because it does not contain center of symmetry in the molecule. An organic compound that contains two dissimilar asymmetric carbon atoms in the molecule can give four possible stereoisomers.

Thus, 2-bromo-3-chloro butane, CH_3—CHBr—CHCl—CH_3 may be written as:

```
      CH3              CH3              CH3              CH3
       |                |                |                |
  H—C—Br           Br—C—H           Br—C—H           H—C—Br
       |                |                |                |
  H—C—Cl           Cl—C—H            H—C—H           Cl—C—H
       |                |                |                |
      CH3              CH3              CH3              CH3
       I                II               III              IV
```

The forms I and II are optical enantiomers, (object and its mirror image), in the same way forms III and IV are also, optical enantiomers. These two pairs of enantiomers will also form two racemic forms or 'dl' forms. The forms I and III are not the mirror images of each other, similarly II and IV are also not enantiomers, although these are isomers of the same compound. Therefore I and III, (or II and IV) forms are termed diastereoisomers which are not mirror image of each other, but are optical isomers of the same compound.

Stereo Chemical Notations: The various optical active compounds, are classified as belonging to either 'D' or 'L' series, a convenient way to designate the configuration of a molecule. Capital 'D' and capital 'L' are neither substitute for small 'd' or small 'l' nor they indicate the optical rotation of the organic compound.

For this purpose, glyceric aldehyde has been regarded as the simplest sugar. Glyceric aldehyde can be written in the following two ways:

```
        CHO                    CHO
         |                      |
    H—C—OH                 OH—C—H
         |                      |
      CH2OH                  CH2OH
       I   'D'                II   'L'
```

Structure 1, has the —OH group on the right side, this has been assigned a capital 'D' and the other structure has been assigned as capital 'L'.

In writing any structure, if second last —OH group is on the right hand side, it assigned as 'D' and in case if the second last —OH, on the left hand side, it given as 'L' compound, irrespective of the fact whether it dextro-rotatory or a laevo-rotatory. The 'D' and 'L' refers to the absolute configuration, (a three dimensional structure) and sign (–) and (+) refer to the rotation of the plane polarised light which is a physical property.

Example:

```
        CHO                    COOH
         |                      |
    H—C—OH                 H—C—OH
         |                      |
      CH2OH                  CH2OH
  D-(+) glyceric aldehyde    D-(–) lactic acid
```

R and S notation of optical isomers: 'Cahn–Ingold–Prelog' proposed a scheme of representation of optical isomer, on plane of paper, by using prefix 'R', (R = Rectus means right handed) and 'S', (S = sinister means left handed), for optically active compounds.

The assignment is based on the nature of the groups, attached to the asymmetric carbon atom. The nature of the groups is determined by knowing the priority of the groups, based on the sequence rule.

Rules:
- Looking at the tetrahedral arrangement of the groups about the carbon from the side remote from the substituents of lowest priority, if the sequence, A → B → C, are traced clockwise, is said to have the 'R' configuration
- In case A, B, C are traced in anti-clockwise direction, then the isomer is said to have a 'S' configuration or the molecule is visualized in such a way that the group of the lowest priority is directly away from us; the remaining three groups then, looking like the three spokes of a wheel, are then traced; in case, they are traced in a clockwise fashion, the isomer is said to have the 'R' configuration. In case if traced in the anticlockwise fashion, the isomer is aid to have 'S' configuration. The racemic modification, or a 'dl' mixture is termed as 'R S' mixture. For a given compound, the specific rotation is not known, it is therefore to be determined experimentally, while writing the configuration of an optically active isomer, if any of the two groups are interchanged, it is converted into its enantiomer if interchange is done, then the original isomer is obtained.

To determine and designate, the configuration of a chiral carbon atom, the double interchange is done to place the group of the lowest priority, either to the bottom or to the top of the perspective or the Fischer projection formula. For R-glyceraldehyde, suppose the 'H' is on the horizontal line, and it has to be brought to the bottom, as per priority rule.

First 'H' is exchanged with 'CH$_2$—OH' group, R–glyceraldehyde is converted into its enantiomer. By doing another exchange of the —OH group with CHO group, the original configuration or R-glyceraldehyde is obtained.

$$\underset{\text{CH}_2\text{OH}}{\overset{\text{CHO}}{H-C-OH}} \qquad \underset{H}{\overset{\text{CHO}}{CH_2OH-C-OH}} \qquad \underset{H}{\overset{\text{CHO}}{CH_2OH-C-OH}} \qquad \underset{H}{\overset{\text{OH}}{CH_2OH-C-CHO}}$$

first exchange between H and CH$_2$OH   second exchange between OH and CHO

The sequence of the group has, 'R' configuration: OH > CHO > CH$_2$OH > H

Mutarotation: Sugars are optically active compounds. They are dextro-rotatory compounds. An aqueous solution of glucose, shows an initial rotation value = 113°, final value = 52.5°.

This phenomenon of initial rotation changing to a different value is termed as *mutarotation*.

Predicting the number of optical isomers:
a. When optically active compound has no symmetry:
   Number of optically active isomers $a = 2^n$ [n = number of chiral carbon atoms]
   Number of racemic forms $r = a/2$
   Number of meso forms $m = $ zero
b. When molecule has a symmetry:
   Number of optically active isomers $a = 2^{n-1}$
   Number of racemic forms $r = a/2$
   Number of meso forms $m = n^{(n/2-1)}$

Sequence rule:
Priority of the groups arranged in the decreasing order:
—I > —Br > —Cl > —SH > —F > —OCOR > —OR > —OH > —NO$_2$ > —NHCOR > —NR$_2$ > —NHR > —NH$_2$ > —CCl$_3$ > —COCl > —COOR > —COOH > —CONH$_2$ > —COR > —CHO > —CH$_2$OH > —CN > —C$_6$H$_5$ > —CR$_3$ > —CHR$_2$ > —CH$_2$R > —CH$_3$ > —D > —H

MULTIPLE CHOICE QUESTIONS

1. The pair of structures, given below represent

$$
\underset{\overset{|}{CH_3}}{H-\underset{\overset{|}{CH_3}}{C}-Cl} \qquad \underset{\overset{|}{CH_3}}{H-\underset{\overset{|}{CH_2Cl}}{C}-H}
$$

   (a) position isomers
   (b) functional isomers
   (c) metamers
   (d) tautomers

2. Stereoisomerism can be exhibited by compounds possessing
   (a) one chiral center
   (b) two chiral centers
   (c) unsymmetrical molecule
   (d) all of these

3. Optically active compound among the following is
   (a) 2-ethylbutanol-1
   (b) n-butanol
   (c) 2,2-dimethyl butanol
   (d) 2-methyl butanol-1

4. Which one of the following pair represents stereoisomerism?
   (a) geometrical isomerism and conformational isomerism
   (b) geometrical isomerism and optical isomerism
   (c) optical isomerism and geometrical isomerism
   (d) optical isomerism and metamerism

5. Which of the following does not show resonance?
   a) CO_2
   (b) benzene
   (c) nitro methane
   (d) propane

6. The structure 2-methylpent-3-enoic acid shows

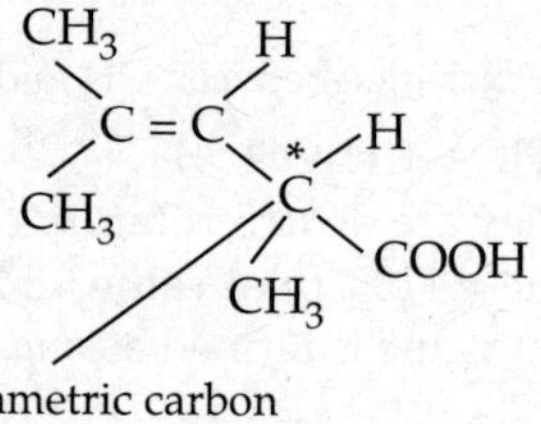

$$
\underset{CH_3}{\overset{CH_3}{\diagdown}}C=C\underset{\overset{\ast}{C}\diagup\underset{CH_3}{\diagup}\diagdown COOH}{\diagup}\overset{H}{\diagup}\diagdown H
$$

   asymmetric carbon
   (a) optical isomerism
   (b) geometrical isomerism
   (c) both (a) and (b)
   (d) none

7. The isomers which can be interconverted into through rotation around a single bond, are
   (a) conformers
   (b) diastereo isomers
   (c) enantiomers
   (d) positional isomers

8. Which one of the following compounds exhibits geometrical isomerism?
   (a) C_2H_5Br
   (b) $(CH)_2(COOH)_2$
   (c) CH_3CHO
   (d) $CH_2(COOH)_2$

9. Position isomerism is shown by
   (a) o-, m and p-nitro phenol
   (b) dimethyl ether and ethanol
   (c) pentane-2 one and pentane-3-one
   (d) CH_3CHO and CH_3COCH_3

10. Which is incorrect about enantiomorphism?
    (a) they rotate the plane of polarised light in different direction
    (b) they have mostly identical physical properties

(c) they have same configuration

(d) they have different physical properties

11. Glucose and fructose are
 (a) chain isomers
 (b) position isomers
 (c) functional isomers
 (d) optical isomers

12. The isomerism exhibited by acetic acid and methyl formate is
 (a) functional isomerism
 (b) chain
 (c) geometrical
 (d) optical

13. How can an E isomer be changed to a Z isomer or vice versa?
 (a) Switching groups on both the doubly bonded carbon atoms
 (b) Switching groups on the same carbon
 (c) Substituting groups of high priority by higher derivative
 (d) none

14. The tautomer that is less stable is
 (a) anion form
 (b) cation form
 (c) labile form
 (d) all

15. Ethyl acetoacetate ($CH_3COCH_2COOC_2H_5$), exhibits
 (a) optical isomerism
 (b) geometrical isomerism
 (c) tautomerism
 (d) enantio isomerism

16. Which one is a chiral molecule?
 (a) CH_3Cl
 (b) CH_2Cl_2
 (c) $CHBr_3$
 (d) CH—Cl Br I

17. Chiral molecules are those, which are
 (a) non-superimposable on its mirror image
 (b) are superimposable on its mirror image
 (c) show geometrical isomerism
 (d) are unstable molecule

18. An optical active compound is
 (a) 1-bromo butane
 (b) 2-bromo butane
 (c) 1-bromo-2-methyl propane
 (d) 2-bromo-2-methyl propane

19. A compound contains 2 dissimilar asymmetric carbons atoms. The number of optical isomers are
 (a) 2
 (b) 3
 (c) 4
 (d) 5

20. Which of the following is optically active compound?
 (a) *n*-propanol
 (b) 2-chlorobutane
 (c) *n*-butane
 (d) 3-hydroxypentane

21. Which of the following may exist as enantiomorphs?
 (a) CH_3—CH—COOH
         |
        CH_3
 (b) $CH_3CH=CH—CH_2CH_3$
 (c) CH_3—CH—CH_3
         |
        NH_2
 (d) CH_3CH_2—CH—CH_3
               |
             Br

22. Dichloro ethylene shows (CHCl=CClH)
 (a) geometrical isomerism
 (b) position isomerism
 (c) both a and b
 (d) none

23. Which of the following structure are/is superimposable?

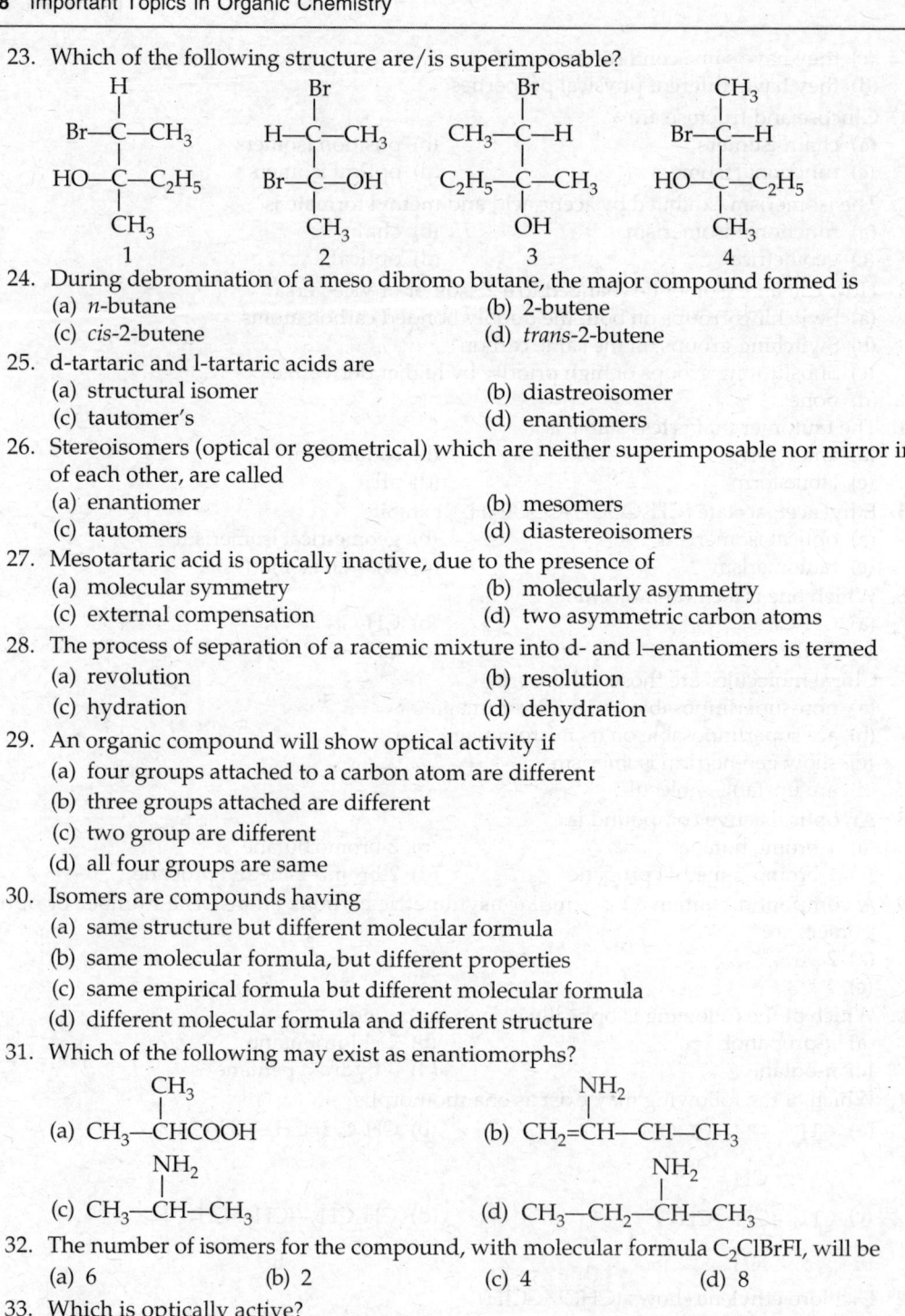

       1   2   3   4

24. During debromination of a meso dibromo butane, the major compound formed is
    (a) n-butane
    (b) 2-butene
    (c) *cis*-2-butene
    (d) *trans*-2-butene

25. d-tartaric and l-tartaric acids are
    (a) structural isomer
    (b) diastreoisomer
    (c) tautomer's
    (d) enantiomers

26. Stereoisomers (optical or geometrical) which are neither superimposable nor mirror image of each other, are called
    (a) enantiomer
    (b) mesomers
    (c) tautomers
    (d) diastereoisomers

27. Mesotartaric acid is optically inactive, due to the presence of
    (a) molecular symmetry
    (b) molecularly asymmetry
    (c) external compensation
    (d) two asymmetric carbon atoms

28. The process of separation of a racemic mixture into d- and l–enantiomers is termed
    (a) revolution
    (b) resolution
    (c) hydration
    (d) dehydration

29. An organic compound will show optical activity if
    (a) four groups attached to a carbon atom are different
    (b) three groups attached are different
    (c) two group are different
    (d) all four groups are same

30. Isomers are compounds having
    (a) same structure but different molecular formula
    (b) same molecular formula, but different properties
    (c) same empirical formula but different molecular formula
    (d) different molecular formula and different structure

31. Which of the following may exist as enantiomorphs?

    (a) CH_3—$\overset{\overset{\displaystyle CH_3}{|}}{CH}COOH$
    (b) CH_2=CH—$\overset{\overset{\displaystyle NH_2}{|}}{CH}$—$CH_3$

    (c) CH_3—$\overset{\overset{\displaystyle NH_2}{|}}{CH}$—$CH_3$
    (d) CH_3—CH_2—$\overset{\overset{\displaystyle NH_2}{|}}{CH}$—$CH_3$

32. The number of isomers for the compound, with molecular formula $C_2ClBrFI$, will be
    (a) 6
    (b) 2
    (c) 4
    (d) 8

33. Which is optically active?
    (a) α-chloro propionic acid
    (b) β-chloro propionic acid
    (c) isobutyric acid
    (d) propionic acid

34. *n*-propyl alcohol and isopropyl alcohol are
    (a) position isomer   (b) chain isomers
    (c) tautomers   (d) geometrical isomers
35. The maximum number of stereoisomers possible for 3-hydroxy-2-methyl butanic acid
    (a) 1   (b) 2
    (c) 3   (d) 4
36. Resonance arises from a
    (a) migration of atoms   (b) migration of protons
    (c) delocalisation of sigma electron   (d) delocalisation of π-electrons
37. Geometrical isomerism is caused by
    (a) restricted rotation around C=C bond
    (b) by the presence of one asymmetric carbon
    (c) due to different groups attached to the same functional group
38. Which of the following is optically active?
    (a) *n*-propanol   (b) 2-chloro butane
    (c) *n*-butanol   (d) 3-hydroxy pentane
39. If the bond order (number of bonds) increases
    (a) bond length increases
    (b) bond energy decreases
    (c) bond length decreases and bond energy increases
    (d) bond length increases and bond energy decreases
40. Sometimes the behavior of a compound is explained by assuming that it exists in two or more possible structures. This phenomenon is termed
    (a) isomerism   (b) resonance
    (c) muta rotation   (d) allotropism
41. Geometrical isomerism can exist only when the molecule
    (a) has a plane of symmetry
    (b) has a center of symmetry
    (c) has two different groups attached to the doubly bonded carbon atoms
    (d) rotates the plane of polarised light to a particular direction
42. Chiral molecules are those; which
    (a) are not superimposable on their mirror image
    (b) are superimposable on their mirror images
    (c) show geometrical isomerism
    (d) are unstable
43. If the bond has a zero% ionic character then the bond is
    (a) partially ionic   (b) partially covalent
    (c) purely covalent   (d) purely ionic
44. In the following structures, which two forms are staggered form of ethane?

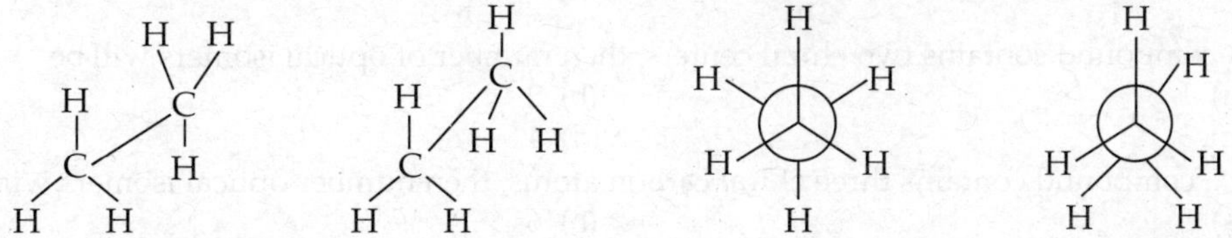

    (a)   (b)   (c)   (d)

45. Which of the following compounds are optically active?
    (a) Quartz   (b) benzil
    (c) ZnSO$_4$ (d) All of these
46. An alkane, which is optically active will contain number of carbon atoms
    (a) 5   (b) 6
    (c) 7   (d) 8
47. The two isomers given below are

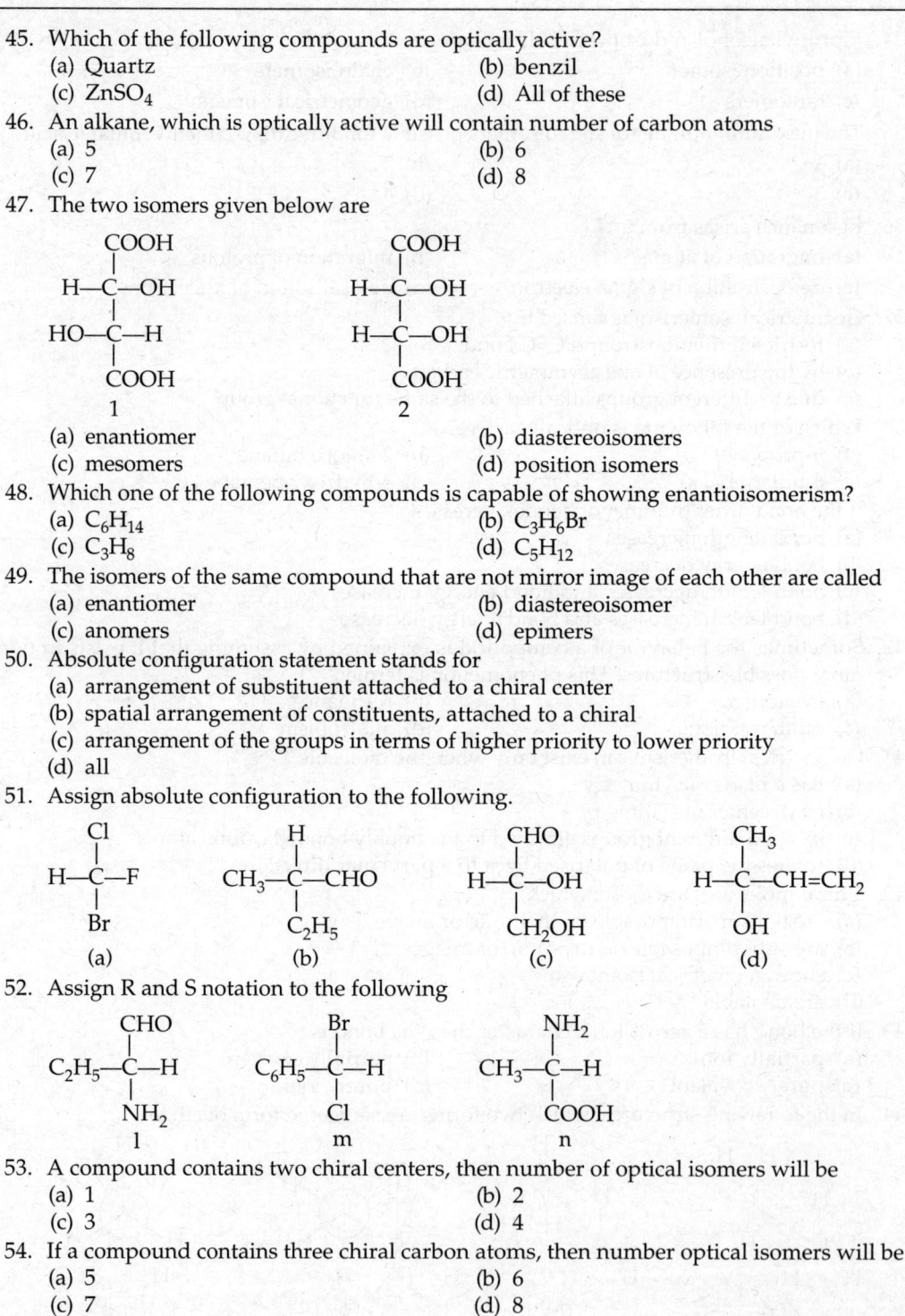

    (a) enantiomer   (b) diastereoisomers
    (c) mesomers   (d) position isomers
48. Which one of the following compounds is capable of showing enantioisomerism?
    (a) C$_6$H$_{14}$ (b) C$_3$H$_6$Br
    (c) C$_3$H$_8$ (d) C$_5$H$_{12}$
49. The isomers of the same compound that are not mirror image of each other are called
    (a) enantiomer   (b) diastereoisomer
    (c) anomers   (d) epimers
50. Absolute configuration statement stands for
    (a) arrangement of substituent attached to a chiral center
    (b) spatial arrangement of constituents, attached to a chiral
    (c) arrangement of the groups in terms of higher priority to lower priority
    (d) all
51. Assign absolute configuration to the following.

52. Assign R and S notation to the following

53. A compound contains two chiral centers, then number of optical isomers will be
    (a) 1   (b) 2
    (c) 3   (d) 4
54. If a compound contains three chiral carbon atoms, then number optical isomers will be
    (a) 5   (b) 6
    (c) 7   (d) 8

Conformational Isomerism

55. The arrangement of atoms that characterises a particular isomer is termed
    (a) configuration
    (b) conformation
    (c) geometrical
    (d) none

56. The dihedral angle between the hydrogen atoms of the methyl groups is
    (a) 120°
    (b) 180°
    (c) 90°
    (d) 60°

57. The least energetic conformational isomer of cyclohexane is
    (a) Boat form
    (b) chair form
    (c) half chair form
    (d) twisted form

58. Correct order of the stabilities of different configurational isomers of butane is
    (a) Staggered > gauche > partially eclipsed > fully eclipsed
    (b) Staggered > fully eclipsed > partially eclipsed > gauche
    (c) Gauche > staggered > partially eclipsed > fully eclipsed
    (d) none

59. Isomers that can be interconverted into each other, by rotation about a single bond are
    (a) conformers
    (b) diastereoisomers
    (c) enantiomers
    (d) positon isomers

60. Which is optically active?
    (a) α-chloro propionic acid
    (b) β-chloro propionic acid
    (c) iso butyric acid
    (d) propionic acid

61. Why the rotation about a single bond is not completely free
    (a) There exist a small repulsion between the electrons of the adjacent carbons
    (b) hydrogen is bulky, therefore it does not allow free rotation
    (c) C—H bond generates torsional strain
    (d) none

62. Conformation in molecules is due to
    (a) rotation about a single bond
    (b) change in the direction of light
    (c) structural changes
    (d) restricted rotation about the double bond

63. The number of conformational isomers of ethane will have
    (a) low energy
    (b) high energy
    (c) same energy
    (d) each have different energy

64. Conformers of ethane:
    (a) low energy
    (b) high energy
    (c) same energy
    (d) each have different energy

65. Which structural features are necessary for a compound to exhibit conformational isomerism?
    (a) Four atoms or groups to be joined to a carbon
    (b) Four atoms or the groups sequentially joined through bond only
    (c) The molecule should possess a tetra hedral structure
    (d) all

Geometrical Isomerism (Questions)

66. Stereoisomerism can be exhibited by compounds possessing
    (a) one chiral center
    (b) two chiral centers

(c) unsymmetrically substituted doubly bonded carbon atoms

(d) all of these

67. Geometrical isomerism can exist only when the molecule

(a) has a center of symmetry

(b) has a plane of symmetry

(c) has two different groups attached to both the carbon atoms attached by a double bond

(d) rotate the plane of polarised light in a particular direction.

68. Which of the following compounds will show *cis*-, *trans*-isomerism.

(a) $X_2C = CY_2$ (b) $XYC = CZ_2$

(c) $X_2C = CX_2$ (d) $YXC = CXY$

69. The number of geometrical isomers in the case of the compound will be
$CH_3—CH=CH—CH=CH—C_2H_5$

(a) 4 (b) 3

(c) 2 (d) 5

70. Maleic acid and fumaric acids are pair of

(a) tautomers (b) geometrical isomers

(c) chain isomers (d) functional isomers

71. Geometrical isomerism is exhibited by the compounds having which of the following group?

(a) $>C=C<$ (b) $>C=N—$

(c) $—N=N—$ (d) all of these

72. Geometrical isomerism is caused by

(a) different groups attached to the same functional group

(b) Swing of a hydrogen atom in between two polyvalent atoms

(c) Presence of an asymmetric carbon atom

(d) restricted rotation about a carbon to carbon double bond $>C=C$.

73. Which among the following is likely to show geometrical isomerism?

(a) $CH_3—CH=CH_2$ (b) $CH_3—CH=NOH$

(c) $(CH_3)_2—C=C—Cl(CH_3)$ (d) $CH_2=CH—CH=CH_2$

74. Mark correct statement about fumaric acid $—HOOC—HC=CH—COOH$

(a) has a zero dipole moment

(b) exhibit geometrical isomerism

(c) has a tendency to undergo dehydrogenation

(d) does not respond to Bayer's test

75. Which of the following will exhibit geometrical isomerism?

(a) 2-butene (b) 2-butyne

(c) 2-butanol (d) butanal

76. Which alkenes do not show geometrical isomerism or where geometrical isomerism disappears?

(a) When an alkene is converted into an alkyne

(b) As soon as the two groups attached to the doubly bonded carbon atom become identical

(c) When an alkene is hydrogenated to form alkane

(d) All the above

77. Assign E/Z configuration each of the following

(a) 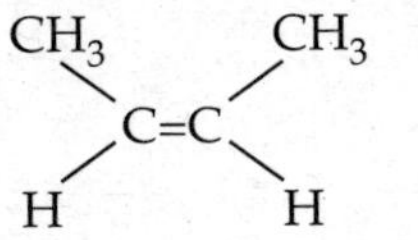
$$CH_3 \quad CH_3$$
$$\backslash C=C \diagup$$
$$H \qquad H$$

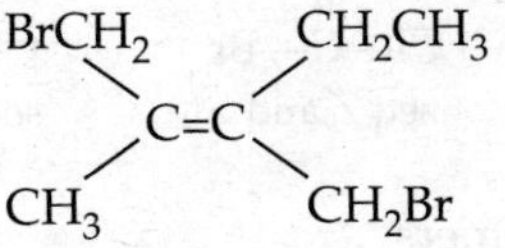

(b)
$$BrCH_2 \quad CH_2CH_3$$
$$\backslash C=C \diagup$$
$$CH_3 \qquad CH_2Br$$

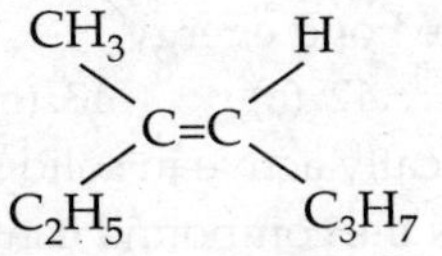

(c)
$$CH_3 \quad Br$$
$$\backslash C=C \diagup$$
$$CH_2OH \qquad C_3H_7$$

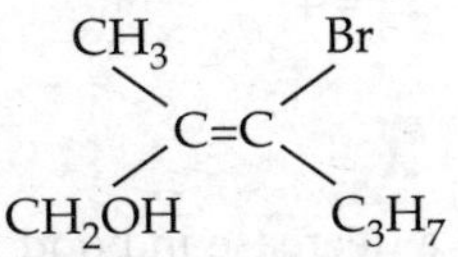

(d) 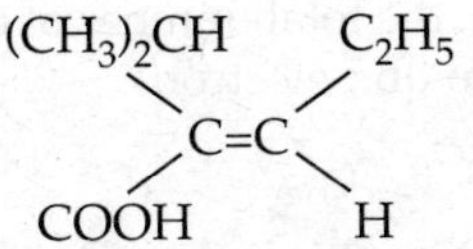
$$(CH_3)_2CH \quad C_2H_5$$
$$\backslash C=C \diagup$$
$$COOH \qquad H$$

78. (a) 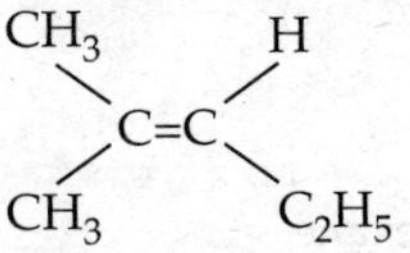
$$CH_3 \quad H$$
$$\backslash C=C \diagup$$
$$CH_3 \qquad C_2H_5$$

(b)
$$CH_3 \quad H$$
$$\backslash C=C \diagup$$
$$C_2H_5 \qquad C_3H_7$$

(c)
$$Cl \quad Br$$
$$\backslash C=C \diagup$$
$$H \qquad F$$

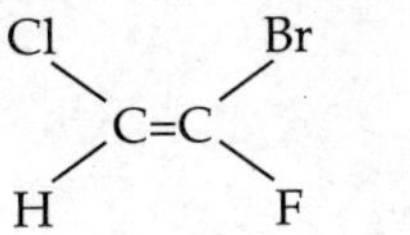

(d) 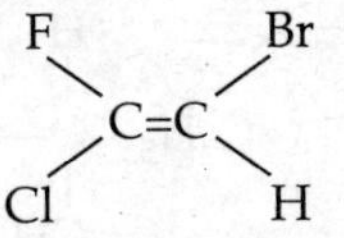
$$F \quad Br$$
$$\backslash C=C \diagup$$
$$Cl \qquad H$$

ANSWERS

1. (a) These are position isomers; CH_3—CH—Cl—CH_3 and CH_2Cl—CH_2—CH_3
2. (d) all   3. (d)   4. (c)   5. (d) Propane
6. (a) one asymmetric carbon is present
7. (d) Conformers
8. (b) $(CH)_2(COOH)_2$ maleic acid and fumaric acid
9. (a) o, m and p are position isomers
10. (c) the configuration of enantiomers are different, e.g.

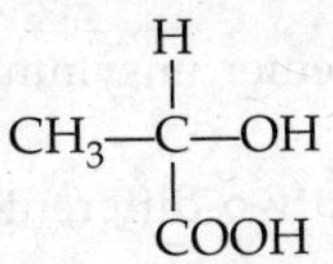

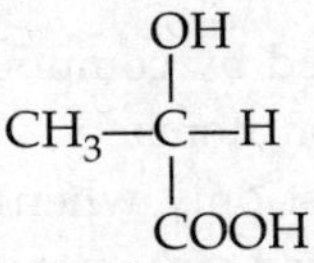

$$CH_3—\overset{\overset{\displaystyle H}{|}}{\underset{\underset{\displaystyle COOH}{|}}{C}}—OH \qquad\qquad CH_3—\overset{\overset{\displaystyle OH}{|}}{\underset{\underset{\displaystyle COOH}{|}}{C}}—H$$

11. (c) They are functional isomers
12. (a)   13. (b)   14. (c)   15. (c)   16. (d)
17. (a) a chiral molecule should not contain any kind of symmetry
18. (b) 2-bromobutane, CH_3—CH_2—CHBr CH_3
19. (c) $a = 2^n$, where n is the number of dissimilar carbon atoms and 'a' is the number of optically active carbon atoms
20. (b) 2-chlorobutane   21. (d)   22. (a)   23. (d) (1 and 3 are identical)
24. *trans*-2-butene   25. (d) the pair is enantiomers
26. (d) it is the definition of diastereoisomers
27. (a) molecular symmetry; internally compensated molecule
28. (b)   29. (a)   30. (b)   31. (d)
32. 6 isomers

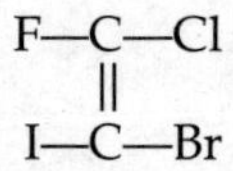

$$
\begin{array}{ccc}
\text{F—C—Cl} & \text{F—C—I} & \text{F—C—Br} \\
\| & \| & \| \\
\text{I—C—Br} & \text{Cl—C—Br} & \text{Cl—C—I} \\
\text{seq Z and E} & \text{seq Z and E} & \text{seq Z and E}
\end{array}
$$

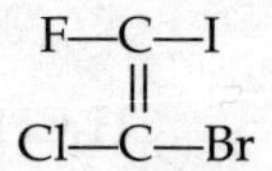

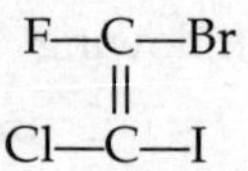

33. (b)
34. (a) position isomers
35. (d) 4 isomers, i.e. the total number of optical isomers $= 2^n$ or $= 2^2 = 4$
36. (d) delocalisation of π-electrons
37. (a)
38. (b)
39. (c) an increase in multiplicity of bond (e.g. N≡N) brings in a decrease in bond length and increases in the bond energy
40. (b)   41. (c)   42. (a)   43. (c)   44. (a) and (c)
45. (d) These are optically active in solid state only
46. (c) 7 carbon atoms the compound can be named as 3-methyl hexane

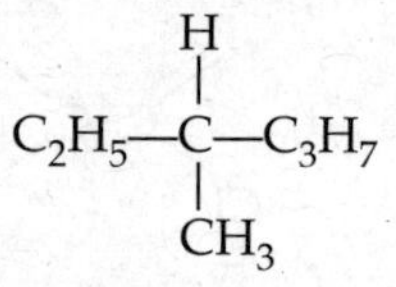

$$
\begin{array}{c}
\text{H} \\
| \\
\text{C}_2\text{H}_5\text{—C—C}_3\text{H}_7 \\
| \\
\text{CH}_3
\end{array}
$$

47. (b)   48. (b)   49. (b)   50. (d)
51. a = S; b = R; c = R; d = R
52. l = R; m = R; n = S
53. (d) 4   54. (d) 8

Answers to Questions on Conformational

55. (b)   56. (b) 180°   57. (b)   58. (a)   59. (a)   60. (b)   61. (c)
62. (a)   63. (c)   64. (d)   65. all

Answers to Questions on Geometrical Isomerism

66. (d) Stereoisomerism can be exhibited by compounds possessing center unsymmetrically substituted doubly bonded carbon atoms.
67. (c) Geometrical isomerism can exist only when the molecule has two different groups attached to both the doubly bonded carbon atoms.
68. (d) YX—C=C—XY will permit *cis- trans-* isomerism.

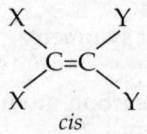
cis

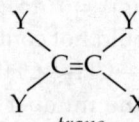
trans

69. (a) Four isomers are possible.
70. (b) Maleic acid and fumaric acids are geometrical isomers.

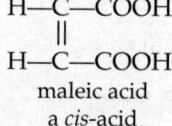

$$
\begin{array}{c}
\text{H—C—COOH} \\
\| \\
\text{H—C—COOH} \\
\text{maleic acid} \\
\text{a } cis\text{-acid}
\end{array}
\qquad
\begin{array}{c}
\text{HOOC—C—H} \\
\| \\
\text{H—C—COOH} \\
\text{fumaric acid} \\
\text{a } trans\text{-acid}
\end{array}
$$

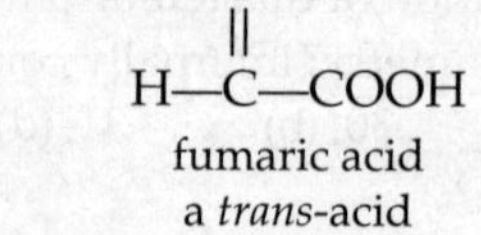

71. (d) All compounds possessing >C=C<, >C=N, —N=N—groups in the molecule
72. (d) Restricted rotation about a >C=C<.
73. (b) CH_3—CH=NOH will show geometrical isomerism

$$\underset{anti}{\overset{\displaystyle CH_3 \qquad O-H}{\underset{H}{\diagdown}C=N\diagup}}\qquad\qquad\underset{syn}{\overset{\displaystyle CH_3}{\underset{H}{\diagdown}C=N\diagdown}_{O-H}}$$

74. (a) Fumaric acid has a zero dipole moment, it is a *trans* acid

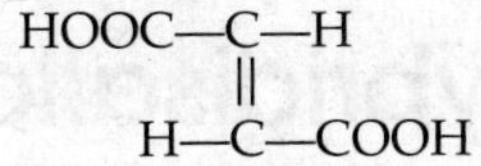

$$\begin{array}{c} HOOC-C-H \\ \| \\ H-C-COOH \end{array}$$

fumaric acid
a *trans*-acid

75. (b) 2-Butene will show geometrical isomerism

$$\underset{cis\text{-}2\text{-butene}}{\overset{\displaystyle CH_3 \qquad CH_3}{\underset{H \qquad\ \ H}{\diagdown}C=C\diagup}}\qquad\qquad\underset{trans\text{-}2\text{-butene}}{\overset{\displaystyle CH_3 \qquad H}{\underset{H \qquad\ \ CH_3}{\diagdown}C=C\diagup}}$$

76. (d)
77. a = Z; b = E; c = E; d = E
78. (b)

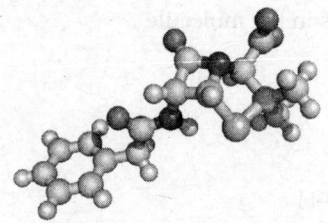

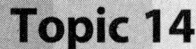

Hybridisation

It may be defined as the blending of orbitals of different energies and creating an equal number of orbitals, having almost an equal amount of energy; the process is termed *'process of hybridisation.'*

Salient Features

- The number of bonds (both σ- and π- bonds) remains same in a compound and its tautomer
- When a sp^3–sp^3 bond changes to a sp^2–sp^2 bond, the total number of bonds decreases by one only.
- Similarly when a sp^2–sp^2 bond changes to a sp–sp bond, the total number of bonds decreases by one only
- In other words, bond sp^3–sp^3 change to a sp–sp bond the total number bonds decreases by two.

Hybridization in nitrogen

The atomic number of nitrogen is seven. It has five electrons in the valence shell (outermost shell). It also undergoes sp^3, sp^2 and sp hybridisation like carbon.

$2p^3 \uparrow \uparrow \uparrow$ $2sp^3 \uparrow \downarrow \uparrow \uparrow \uparrow$
$2s^2 \uparrow \downarrow$
$1s^2 \uparrow \downarrow$ $1s^2 \uparrow \downarrow$

ground state   excited state (with same energy taken)

The $2sp^3$ orbitals of nitrogen are directed towards the four corners of a tetrahedron. Since nitrogen has one electron more than that of carbon, as a result, one of the orbitals is completely filled, and cannot take part in bond formation, and nitrogen shows a valency of three, as in ammonia, NH_3.

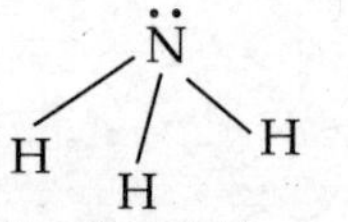

$\ddot{N}$
H   H
H
Angle = 107°
bonding in ammonia

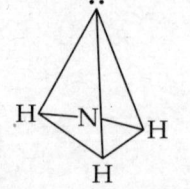

tetrahedral model of ammonia

All the three, nitrogen–hydrogen bonds formed are σ- bonds. These bonds are directed at an angle of 107°. In 'sp^3' hybridised form, the nitrogen is tetrahedral in shape; its three corners are occupied by three hydrogen atoms and the fourth corner is occupied by lone pair. Nitrogen also undergoes 'sp^2' hybridisation. Whenever nitrogen is bonded to two other atoms or groups, it uses its two 'sp^2' orbitals, plus the unhybridised orbital to form bonds. For example, in imine (and also in azo-compounds), nitrogen as well as carbon are in sp^2 hybridised state or the nitrogen is joined by a double bond to the other atom.

Orbital picture of RCH=NH (aldemine)

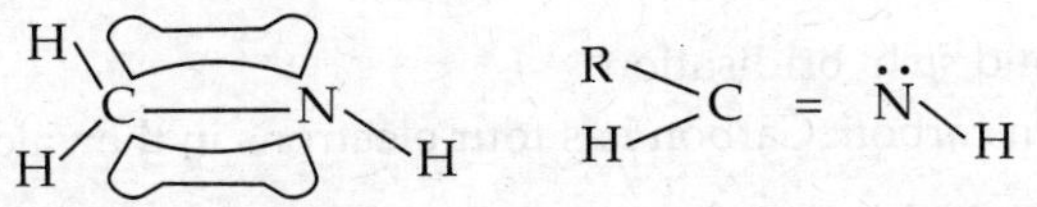

pi bonds

Bonding in aldemine; both carbon and nitrogen are sp^2 hybridised

Nitrogen also undergoes 'sp' hybridisation. Whenever nitrogen is bonded to one other atom or a group, it uses its 'sp' orbital plus its two unhybridised orbital $2p$ orbital, ($2p_x$ and $2p_y$ orbital) to form bonds or nitrogen is joined by a triple bond with another atom. For example, in H—C≡N, nitrogen is in 'sp' hybridised state.

$$H—C≡N$$

bonding in hydrogen cyanide

Hybridisation in oxygen

The atomic number of oxygen is eight. It has six electrons in the valence shell (outermost shell).

$2p^3$ ↑↓ ↑ ↑ $2sp^3$ ↑↓ ↑↓ ↑↑

$2s^2$ ↑ ↓

$1s^2$ ↑ ↓ $1s^2$ ↑ ↓

ground state   excited state (with same energy taken)

Oxygen is tetrahedral in shape. Its four sp^3 orbitals are directed towards the four corners of a tetrahedron. As two of its orbital are completely filled, it can accommodate only two electrons. Therefore, oxygen shows a valency of only two, (the filled orbitals cannot take part in the bond formation), in the compounds when oxygen is attached to two other atoms or groups.

The case of water molecule, the two hydrogen atoms are joined by two sp^3 hybridised orbitals to oxygen through σ-bonds and are inclined at an angle of 104°.

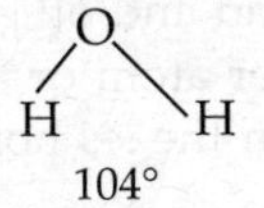

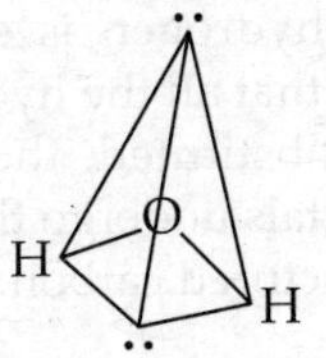

bonding in water

Oxygen undergoes sp^2 hybridisation, in the compounds when oxygen atom is attached to one another atom or a group (as in the case of aldehydes or ketone), it uses its sp^2 orbitals plus the unhybridised 'p_z' orbital to form the compound, or in the compounds when oxygen is joined to the other atom by a double bond (as in aldehydes or ketone).

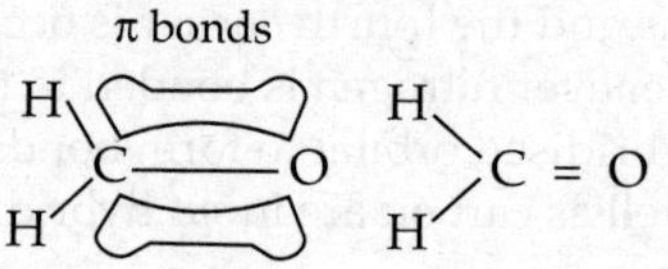

π bonds

bonding in formaldehyde, [HCHO or $H_2C = O$]

Hybridisation in carbon

It undergoes sp^3, sp^2 and sp hybridisation.

a. sp^3 hybridisation in carbon: Carbon has four electrons in the valence shell.

$2p^2$ ↑ ↑
$2s^2$ ↑ ↓
$1s^2$ ↑ ↓

ground state

As the $2s^2$ orbital is completely filled, carbon, therefore can accommodate only two electrons, in the half filled valence shell, therefore in such a situation, carbon can only accommodate two electrons, showing a valency of two. However, carbon shows a valency of four, in all the organic compounds, which are quite stable. It is possible only when $2s^2$ orbitals take part in the process of raising the valency to four. By taking some energy from the surroundings, one of the $2s^2$ electron jumps to the vacant 'p' orbital, making the half-filled orbitals to four.

$2p^3$ ↑ ↑ ↑
$2s^1$ ↑
$1s^2$ ↑ ↓

excited state

In this manner, carbon is able to create four half-filled orbitals, which can accommodate four electrons, showing a valency of four. However, this model of carbon contains two types of orbitals, one $2s^1$, spherical shaped orbital and three $2p^2$, perpendicular orbitals, inclined with each other, at an angle of 90°.

Such a model of carbon, in the excited state will have four half-filled orbitals or four unpaired electrons, showing a valency of four. The four atoms or groups, joined to carbon atom, will have two types of σ- bonds, one will be *s-s* bond and other three will be *s-p* bonds. When any of the three hydrogen is substituted by another atom or a group, the derivative will be different as compared to the derivative formed, when the *s-s* bonded hydrogen is substituted by the same group.

It has been found, that whichever hydrogen is substituted by a particular group, the compound obtained is the same. It means that all the hydrogen are joined with the carbon by the same manner, whichever hydrogen is substituted, the derivative obtained is the same. It is possible only when all four half- filled orbitals undergo the process of hybridisation, involving $2s^1$ and $2p^3$ orbitals, forming a tetrahedral structured carbon.

$2p^3$ ↑ ↑ ↑
$2s^1$
$1s^2$ ↑ ↓

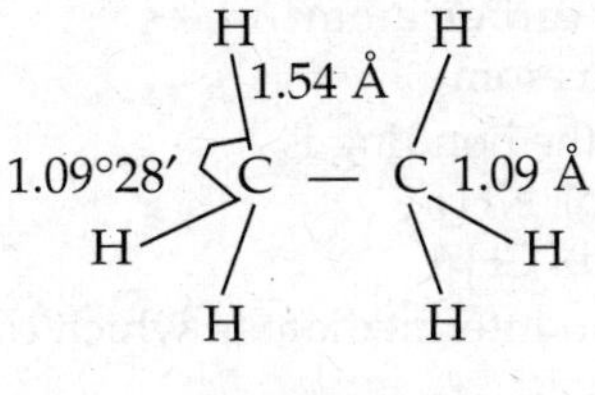

bonding in ethane 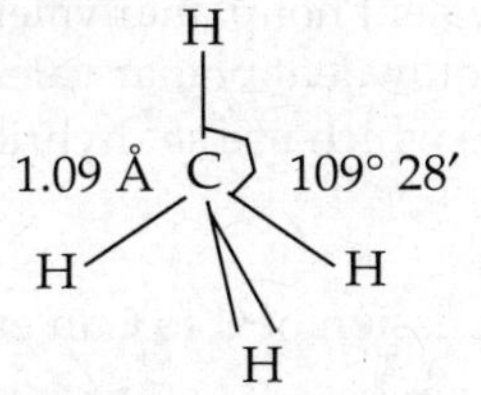   bonding in methane

In this process of hybridisation, it gives out more energy than it has taken, to promote one $2s$ electron to the vacant $2p$ orbital. Thus, carbon undergoes $2sp^3$ hybridisation; all the four half-filled orbitals, are inclined with each other at an angle of 109° 28', forming a regular tetrahedron.

  b. sp^2 Hybridisation

    Carbon also undergoes sp^2 hybridisation, when out of four sp^3 orbitals, only three orbitals, $2s^1$ and $2p^2$ half-filled orbitals, pool up their energies and create three orbitals, inclined with each other at an angle of 120° leaving one of the $2p$ orbital in the unhybridised state, which is at an angle of 90° to the three hybridised orbitals.

Such a carbon atom, when forms a σ-bond with another carbon atom, through a linear overlapping of the hybridised $2sp^2$, one from each sp^2 hybridised orbitals and a π-bond, through sideways overlapping of $2p$ unhybridised orbitals, one from each sp^2 hybridised carbon atom.

$2p^1 \uparrow$

$2sp^2 \uparrow \uparrow \uparrow$

$1s^2 \uparrow \downarrow$

bonding in ethylene   π-bond formation in ethylene

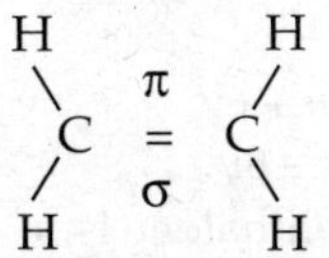

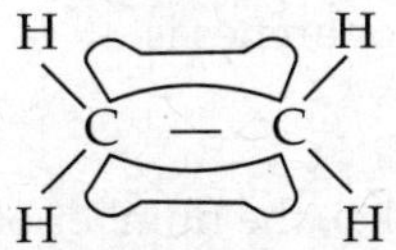

  c. sp hybridisation

    Carbon also undergoes sp hybridisation, when out of four $2sp^3$ hybridised orbitals, only two orbitals, $2s^1$ and $2p^1$, are half-filled orbitals pool up their energies and create two orbitals, inclined at an angle of 180°, leaving two $2p$ orbitals unhybridised, which are at an angle of 90° to the hybridised orbitals.

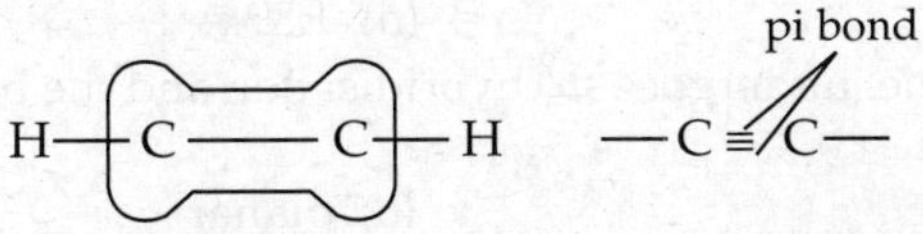

acetylene

MULTIPLE CHOICE QUESTIONS

1. Concept of hybridisation is essential to explain
   (a) the tetravalency in carbon compound
   (b) the 4π-electrons of carbon atom

      (c) 4 equivalent nonplanar valency bonds of carbon atom

      (d) 4 nonequivalent planar valency of carbon atom

2. The species which use sp^2 hybrid orbitals in the bonding is

    (a) PH_3 (b) NH_3

    (c) CH_3^+ (d) CH_4

3. Homolytic fission of C to C in ethane gives an intermediate in which carbon is hybridised to

    (a) sp^3 (b) sp^2

    (c) sp (d) $sp^2 d^2$

4. The hybridisation of carbon in diamond, graphite and acetylene is in the order of

    (a) sp^3, sp^2, sp (b) sp^2, sp^3, sp

    (c) sp, sp^2, sp^3 (d) sp^2, sp, sp^3

5. The bond energy for catenation next to carbon is

    (a) N   (b) S

    (c) Si   (d) P

6. In benzene there are

    (a) 2π-electrons   (b) 6π-electrons

    (c) 8π-electrons   (d) 3π-electrons

7. The benzene molecules contains

    (a) 6 sp^2 hybridised carbons   (b) 3 sp^2 hybridised carbons

    (c) 6 sp^3 hybridised carbons   (d) 3 sp^3 hybridised carbons

8. According to Huckel, the monocyclic rings will show aromaticity, when

    (a) it has 4 π electrons   (b) it has no electrons

    (c) it has 4 + 2 π electrons   (d) it has $(4n + 2)$ π electrons

9. General formula for arenes is

    (a) $C_nH_{2n + 6}$ (b) $Cn\ H_2n + 6y$

    (c) C_nH_2n (d) $C_nH_2n - 6y$

10. According to Huckel, the number of π electrons in naphthalene is

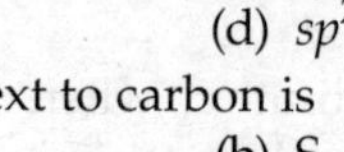

Naphthalene

    (a) 8   (b) 10

    (c) 4   (d) 12 π electrons

11. The C to C bond in alkane is

    (a) 1.54 Å   (b) 1.20 Å

    (c) 1.48 Å   (d) 1.39 Å

12. The carbon in benzene, undergoes sp^2 hybridisation and the bond angle is 120°, the shape of the molecule is

    (a) linear   (b) planar

    (c) pyramidal   (d) planar hexagonal

13. Cyclopentadienyl anion is aromatic due to the presence of

    (a) 6 π-electrons   (b) 10 π-electrons

    (c) 4 π-electrons   (d) 12 π-electrons

14. Allyl isocyanide contains——and——bonds

    (a) 9 σ and 3 π bonds   (b) 9 σ and 9 π bonds

    (c) 3 σ and 4 π bonds   (d) 5 σ and 7 π bonds

15. The bond between carbon number 1 and carbon number 2, in the compound

$$N\equiv C—CH=CH_2$$

    involves hybridisation as
    (a) sp^2-sp^2
    (b) sp^3-sp
    (c) sp-sp^2
    (d) sp-sp

16. The central oxygen atom in ether is
    (a) sp
    (b) sp^2
    (c) sp^3
    (d) sp^3d^2

17. Hybridisation in carbon in carbonyl group is
    (a) sp
    (b) sp^2
    (c) sp^3
    (d) none

18. The enolic form of acetone of contains $CH_2=\overset{\overset{\displaystyle O—H}{|}}{C}—CH_3$
    (a) 9σ, 1π, 2 lone pairs
    (b) 8σ, 2π and 2 lone pairs
    (c) 10σ, 1π and one lone pair
    (d) 9σ, 2π bonds and 1 lone pair

19. The carbon in carboxylic acid is
    (a) sp
    (b) sp^2
    (c) sp^3
    (d) none

20. The nitrogen in amine is
    (a) sp^3
    (b) sp^2
    (c) sp
    (d) none (amines are tetrahedral)

21. The overlapping in benzene involves
    (a) sp-p
    (b) sp-sp
    (c) sp^2-sp^2
    (d) sp^3-sp^3

22. Which hybrid orbitals in carbon form the compound: $CH_3—CH=CH—CH_2CH_3$?
    (a) sp^2-sp^3
    (b) only sp^3
    (c) sp-sp^3
    (d) sp-sp^2

23. Hypercojugation is
    (a) σ-π conjugation
    (b) noticed due to delocalisation of σ and π electrons
    (c) no bond resonance
    (d) all

24. Overlapping of which of the following orbitals would be maximum to form the strongest covalent bond?
    (a) $1s$-$2s$ σ
    (b) $1s$-$2p$ σ
    (c) $2p$-$2p$ π
    (d) $2p$-$2p$ σ

25. Among the following orbital bonds, the angle is minimum in between
    (a) sp^3-sp^3 bonds
    (b) p_x and p_y orbitals
    (c) H—O—H in water
    (d) sp-sp bonds

26. The s-character of sp, sp^2 and sp^3 hybrid orbitals follow the order
    (a) $sp > sp^2 > sp^3$
    (b) $sp^3 < sp^2 < sp$
    (c) $sp^3 > sp > sp^2$
    (d) sp has 50% sp^2 has 33% and sp^3 has 25% s character.

27. Lateral overlapping of orbitals forms
    (a) σ bond
    (b) π bond
    (c) both σ and π bond character
    (d) none
28. The number of π-bonds present in propyne
    (a) 2
    (b) 3
    (c) 1
    (d) 4
29. Homolytic fission of C—C bonds in ethane gives an intermediate in which carbon is
    (a) sp^3
    (b) sp^2
    (c) sp
    (d) sp^2d^2
30. How many types of σ- and π- bonds are present in tetracyanoethane?

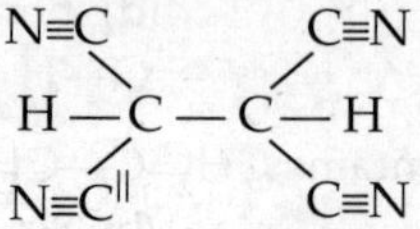

$$N\equiv C \qquad\qquad C\equiv N$$
$$H-C-C-H$$
$$N\equiv C^{\parallel} \qquad\qquad C\equiv N$$

    (a) 9 each
    (b) 8-σ and 10-π
    (c) 11 σ and 8-π
    (d) none
31. Among the following, which one has more than one kind of the hybridisation?
    (a) CH_3—CH_2—CH_2—CH_3
    (b) CH_3—CH=CH—CH_3
    (c) CH_2=CH—$C\equiv CH$
    (d) $CH\equiv CH$
32. In piperidine, the nitrogen atom involves hybridisation is

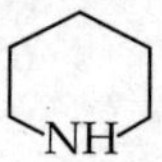

    (a) sp
    (b) sp^2
    (c) sp^3
    (d) sp^2d^2
33. Select the compound which has only one π bond
    (a) $CH\equiv CH$
    (b) CH_2=$CHCHO$
    (c) CH_3—CH=CH_2
    (d) CH_3CH=$CHCOOH$
34. The cylindrical shape of alkyne is due to
    (a) three π-bonds
    (b) three σ-bonds
    (c) two σ and 1 π bond
    (d) 1 σ and two π bonds
35. The hybridisation of boron in diborane is
    (a) sp
    (b) sp^2
    (c) sp^3
    (d) sp^3d
36. The structure and hybridisation of Si $(CH_3)_4$ is
    (a) sp^3; tetrahedral
    (b) sp^2; pyramidal
    (c) sp; hexagonal
    (d) none
37. The total number of σ bonds in CH_2=CH—CH =CH—$C\equiv CH$
    (a) 4 π and 11 σ
    (b) 4 π and 10 σ
    (c) 3 π and 10 σ
    (d) 5 π and 11 σ
38. The hybridisation in I_3^- is
    (a) sp^3
    (b) sp^3d
    (c) sp^3d^2
    (d) dsp^2
39. The geometry associated with hybridisation sp^3d is
    (a) pentagonal pyramidal
    (b) pentagonal planar
    (c) tetragonal pyramidal
    (d) trigonal pyramidal geometry

40. The type of hybridisation for the inter-bond angle 178° is
    (a) sp^3 (b) sp^2
    (c) sp (d) sp^2d
41. In one of the following systems, the central atom uses sp^2 hybrid orbitals
    (a) NO_2^+ (b) NO_2^-
    (c) C_2H_2 (d) CH_4
42. Number of σ and π bonds in benzaldehyde
    (a) 4-π and 13-σ bonds   (b) 4-π and 8σ bonds
    (c) 4-π and 14-σ bonds   (d) 8-π and 14-σ bonds
43. In which of the following, the central atom does not utilise sp^2 hybrid orbitals in its bonding?
    (a) BeF_3^- (b) OH^+_3
    (c) NH_2^- (d) NF_3
44. CO_2 has——hybridisation
    (a) sp (b) sp^2
    (c) sp^3 (d) none
45. The number of σ and π bonds in 1-butene-3-yne CH_2=CH—C≡CH are
    (a) 10 σ-Bonds 3π-bonds   (b) 9-σ-bonds 3π-bond
    (c) 13-σ-bonds 3-π-bond   (d) 7 σ-bonds 3π-bond
46. π-molecular orbital can result from
    (a) p-p orbital   (b) sp^2 orbitals
    (c) s orbitals   (d) sp-orbitals
47. Which of the following will have least hindered rotation about the C—C bond?
    (a) ethane   (b) ethylene
    (c) acetylene   (d) hexa chloro ethane
48. Which of the following statement is not correct?
    (a) > C=C< is made up of 5σ and 1π bonds
    (b) a σ bond is stronger than a π bond
    (c) a σ bond can exist independently without a π bond
    (d) a double bond is stronger than a single bond
49. The enolic form of acetone CH_2=C—CH_3 contains
    $$\underset{\text{H—O}}{\big|}$$
    (a) 9σ and 1π bond   (b) 8σ and 2π bond
    (c) 7σ + 3π bond   (d) none
50. The amide group contains
    (a) only σ bonds   (b) contains both σ and π bonds
    (c) dative Bonds   (d) 2σ bonds and π bonds
51. The hybridisation in SO_2 is
    (a) sp^3 (b) sp^2
    (c) sp (d) sp^2d
52. The bond angle in H—C—H in ethylene is
    (a) 120°   (b) 180°
    (c) 90°   (d) 109°
53. In CH_3CH_2OH, the bond that undergoes heterolytic fission most readily is
    (a) C—C   (b) C—O
    (c) C—H   (d) O—H

54. The bonds present in between two carbon atoms in ethylene are
    (a) 3σ and 1π bond
    (b) 1σ and 1π bond
    (c) 1π and 3σ bond
    (d) 1π and 5σ bonds

55. A methyl free radical involves
    (a) sp
    (b) sp^2
    (c) sp^3
    (d) sp^3d

56. The structural formula of a compound is CH_3—$CH=C=CH_2$; the type of hybridisation of the four carbon atoms is
    (a) $sp^3 sp^2 sp\ sp^2$
    (b) $sp^3 sp^2 sp^2 sp^2$
    (c) $sp^3 sp^2 sp^2 sp^3$
    (d) $sp^3 sp\ sp^2$

57. The orbitals used by carbon in CH_2Cl_2 is
    (a) $4sp^2$
    (b) $4sp^3$
    (c) $4sp$
    (d) one s and 3p-orbitals

58. Which statement is not true?
    (a) acetylene has a linear structure
    (b) alkynes undergo electrophilic addition not nucleophilic addition
    (c) alkenes show geometrical isomerism
    (d) there is sp^3 hybridisation in propane

59. The ratio of σ to π-bond in benzene is
    (a) $2 : 4$
    (b) $4 : 1$
    (c) $6 : 9$
    (d) $8 : 2$

60. The species which use sp^2 hybrid orbitals in its bonding is:
    (a) PH_3
    (b) NH_3
    (c) CH_3^+
    (d) CH_4

61. Homolytic fission of a C—C bond gives an intermediate, in which carbon is
    (a) sp^3
    (b) sp^2
    (c) sp
    (d) sp^2d

62. Hybridisation involves
    (a) addition of electron pair
    (b) mixing of electron pair and redistribution
    (c) removal of electron
    (d) separation of orbitals

63. A triple bond in ethyne is made up of
    (a) 3σ bonds
    (b) 1σ and 2-π
    (c) 3-π
    (d) 2σ and 1-π bonds

64. The bonds which are present in propyne are
    (a) 3σ and 3π
    (b) 1σ and 1π
    (c) 1σ and 3π
    (d) 6σ and 2π bonds

65. The bond angle in sp^3, sp^2, sp hybridised orbital are respectively
    (a) $109° 28'$; $120°$ and $180°$
    (b) $120°$, $109°.28'$, $180°$
    (c) $180°$, $109°.28'$, $120°$
    (d) $120°$, $180°$, $109°.28'$

66. The energy in π-bond in kcal is about
    (a) 36
    (b) 50
    (c) 74
    (d) 140

67. A compound has a structural formula CH_3—$CH=C=CH_2$. The type of hybridisation at the four-carbon atoms is
    (a) sp^3, sp^2, sp and sp^2
    (b) sp^3, sp^2, sp^2, sp^2
    (c) all sp^2
    (d) sp^3, sp, sp, sp^2

68. Propadiene, C_3H_4 molecule contains
    (a) $2sp^2$, $1sp$
    (b) $2sp^2$ and $2sp$
    (c) $1sp^2$ and $3sp$
    (d) none

69. The number of σ and π bonds present in CH_3—CH=CH_2 are
    (a) 9σ and 2π   (b) 8σ and 1π
    (c) 6σ and 4π   (d) none
70. Which electronic configuration would exhibit the lowest ionisation energy?
    (a) $1s^2$ (b) $1s^2, 2s^2, 2p^6, 3s^1$
    (c) $1s^2, 2s^2, 2p^2$ (d) $1s^2, 2s^2, 2p^3$
71. The molecule in which the distance between the two adjacent carbon atoms is largest?
    (a) ethane   (b) ethene
    (c) ethyne   (d) benzene
72. The compound, 1,2-butadiene (CH_2=C=$CHCH_3$) has
    (a) only sp hybridised carbon atoms (b) has only sp^2
    (c) has sp and sp^2 (d) has sp, sp^2 and sp^3 carbon atoms
73. The double bond in ethylene is formed by the overlapping of which of the following orbitals of each carbon.
    (a) two sp^1 hybridised orbitals   (b) 2 sp^2 orbitals
    (c) two p-orbitals   (d) one sp^2 and one p orbitals
74. Among the following compounds one has the central atom with sp^2 hybridisation is
    (a) H_2CO_3 (b) SiF_5
    (c) BF_3 (d) $HClO$
75. The central atom having sp^3 hybridisation is
    (a) PCl_3 (b) SO_3
    (c) NO_3^- (d) BF_3^-
76. Which one of the following has sp^2 hybridised structure?
    (a) CO_2 (b) SO_2
    (c) N_2O (d) CO
77. A sp^3 hybrid contains
    (a) ¼ s character   (b) ½ s
    (c) 2/3 s (d) 3/4 s character
78. How many σ and π bonds are present in: CH_2=CH—CH=CH—CH_3
    (a) 10, 4   (b) 11, 3
    (c) 12, 2   (d) 13, 1
79. Which of the following have ionic, covalent and a coordinate bond?
    (a) C_2H_6 (b) NH_4Cl
    (c) HCl (d) $AlCl_3$
80. Unique hydrogen bridging is present in
    (a) diborane B_2H_6 (b) C_2H_6
    (c) Si_2H_6 (d) Ge_2H_6
81. A triple bond is made up of
    (a) 3 σ   (b) 1 σ and 2-π
    (c) 3 π   (d) 2σ and 1π bonds
82. The number of sp^3 hybridised carbons in 2-butyne are
    (a) 2   (b) 3
    (c) 4   (d) 1
83. The hybrid orbitals at C-2 and C-3 from left to right in the compound, CH_3—CH=CH—Cl are
    (a) sp^3-sp (b) sp^2-sp
    (c) sp^2-sp^2 (d) sp-sp

84. Only two monochloro derivatives are possible for
    (a) n-pentane
    (b) 2,4-dimethyl pentane
    (c) benzene
    (d) 2-methyl propane
85. The monosodium salt of acetylene, on treatment with dry CO_2 forms
    (a) $CH \equiv C{-}COOH$
    (b) $CH \equiv C{-}COONa$
    (c) $CH \equiv C\ COOH$
    (d) none of these
86. HCHO and acetylene react in presence of copper acetylide as a catalyst to form
    (a) 2-butyne-1,4-diol
    (b) 1-butne-1,4-diol
    (c) 2-butyne-1,2-diol
    (d) none
87. The greatest strain involved in cycloalkane when the bond angle is
    (a) 60°
    (b) 90°
    (c) 120°
    (d) 108°
88. When the hybridisation state of a carbon changes from sp^3 to sp^2 and finally changes to sp, the angle between the hybridised orbital
    (a) is not effected
    (b) increases progressively
    (c) decreases considerably
    (d) decreases slowly

ANSWERS

1. (c)
2. (c) CH_3^+ has a planar structure
3. (b) CH_3^+
4. (a) $sp^3\ sp^2\ sp$
5. (c) The bond energy of catenation is in the order of
    $$C > Si > S > P$$
6. (b)　　　7. (a)　　　8. (d)　　　9. (d)
10. (b) 10 π electron　　11. (a)　　12. (d)
13. (a) 6 π-electrons　　14. (a)　　15. (c) $sp\text{-}sp^2$
16. (c) sp^3　　　17. sp^2
18. (a)
19. (b) sp^2
20. (a) sp^3 hybridised
21. (c) $sp^2\text{-}sp^2$
22. (a) Carbon number 1, 3 and 5 are sp^3 and 2 and 3 are sp^2
23. (d) all
24. (b) The more closely related orbitals show more overlapping; the more closer is the nucleus, the more is the overlapping
25. (b) $(p_x \text{-} p_y)$ 90° angle
26. (a) and (b)
27. (b) lateral overlapping produces a π bond
28. (a)
29. (b) it produces free radical sp^2 hybridised
30. (c)　　　31. (b) and (c)　　　32. (c)　　　33. (c)　　　34. (d)
35. (c) B_2H_6 has boron in sp^3 hybridised state
36. (a) silicon is sp^3 hybridised; the structure is tetrahedral
37. (a)
38. (b) I_3^- has sp^3d hybridisation, with three lone pairs in central I　$I \bullet I \bullet I$
39. (d) sp^2d hybridisation leads to trigonal pyramidal geometry

40. (c) the *sp-sp* hybrid orbitals are at 180° to each other
41. (b) NO_2^- has the geometry and possess sp^2 hybridisation :O⁻N=O
42. (c) benzaldehyde has 14 σ and 4 π bonds
43. (a) BeF_3^-
44. (a) it is linear molecule
45. (d)
46. (a) π-orbitals are formed by the overlap of *p-p* orbitals
47. (a) least hindered rotation means, free rotation, along sigma bond is free.
48. (a) >C=C< is composed of 5σ and 1π-bond
49. (a) 9-σ and one -π bond
50. (b)   51. (b)   52. (a)   53. (b)   54. (b)   55. (b) sp^2
56. (a) sp^3, sp^2, sp, sp^2
57. (b)
58. (b) alkyne undergoes nucleophilic addition
59. (b) Benzene contains 12σ- and 3-π-bonds; the ratio is 4 : 1
60. (c) CH_3^+
61. (b) sp^2
62. (b) mixing of electrons and redistribution
63. (b)   64. (d)   65. (a)   66. (b)   67. (a)   68. (a)   69. (b)   70. (b)
71. (a)   72. (d)   73. (d)   74. (a)   75. (a)   76. (c)   77. (a)   78. (c)
79. (b)   80. (a)   81. (b)   82. (a)   83. (c)   84. (d)   85. (b)   86. (c)
87. (a)   88. (b)

Section II

Question Papers

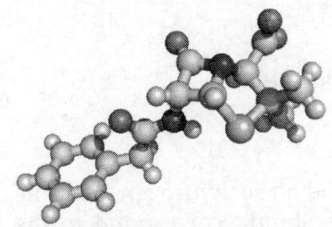

Part I

Paper 1

REASONING TYPE QUESTIONS (RTQs)

1. Benzene molecule contains π cloud in the molecule, besides the fact that in benzene molecule, all the six C to C bond are equal, i.e. 1.39 Å (a value in between that of a single bond 1.54 Å and a double bond 1.34 Å). What is the proof?

2. Explain the difference between conformation and configuration.

3. Name the compound that reacts almost with all the organic compounds.

4. Name the reducing agent that reduces all the functional groups except the olefinic double and acetylinic triple bonds.

5. Explain the difference between polarity and polarisability.

6. An aqueous solution of glucose shows the phenomenon of mutarotation. Explain.

7. Glucose shows some properties of aldehydes, it reduces Fehling solution and Tollen's reagent, but it does not restore the pink colour of Schiff's reagent, nor it forms an addition compound with $NaHSO_3$.

8. A *cis*-alkene may be converted into a *trans*-alkene and vice-versa. How?

9. Explain the nature of the intermediates formed in S_N1 and S_N2 mechanism in case of nucleophilic, i.e. substitution in alkyl halides.

10. Cyclopentanone may be converted into cyclohexanone, when it is treated with a reactant, name the reactant.

11. Picric acid is a stronger acid, though it does not contain a carboxylic group. Justify

12. Which group is easily reduced in vinyl acetylene, $CH_2=CH—C\equiv CH$, a double bond or a triple, when reduced by a metal and hydrogen.

13. Explain, why C_6H_6 molecule undergoes only electrophilic substitution reactions and not nucleophilic substitution reactions.

14. What is the difference between 'inter' and 'intra' molecular rearrangements? Explain with suitable examples.

15. Glucose and fructose react with phenyl hydrazine ($C_6H_5NHNH_2$) to form the same derivative namely osazone. Explain with diagram.

16. Name the compounds those do not undergo Friedel–Craft reaction. Explain why?

17. Anthranilic acid (o-amino benzoic acid) does not exist as zwitterion or a dipolar ion, but sulphanilic acid (p-amino benzene sulphonic acid) exists, explain.

18. Does benzyne exist free in nature?
19. Explain the term 'aromaticity'. What are its requirements?
20. Explain the difference between 'anomeric' and 'epimeric' carbon atoms.

ANSWERS

1. **Benzene** is a flat molecule. All atoms, namely all six carbon and all six hydrogen atoms lie in the same plane. The thickness of the benzene molecule should correspond to the diameter of the carbon atom. However, the **thickness of the benzene molecule** has been found to be **more than the diameter of the carbon atom**. All this proves that the benzene ring is engulfed from the above and from below in by the π-clouds, which make the benzene molecule thicker.

2. When models of various organic compounds, with arrangements of atoms or groups, are considered in space, the concept of conformation and configuration, surface for discussion.

    i. **Conformation:** The different arrangements of atoms in space, that results in from the rotation of atoms or groups, about the C-C single bond axis, are called conformation or conformational isomers. Since, the energy involved in the rotation of one form to another form is very small, these forms convert into each other and ultimately to the most stable form, i.e. the staggered form.

    A molecule of ethane may have the following forms possible; however it finally converts into a staggered form, being the most stable.

229 ppm   310 ppm

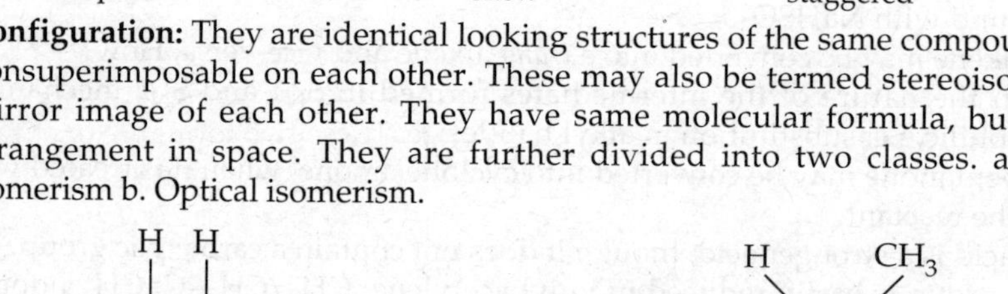

eclipsed   skew   staggered

    ii. **Configuration:** They are identical looking structures of the same compound, which are nonsuperimposable on each other. These may also be termed stereoisomers, and are mirror image of each other. They have same molecular formula, but differ in the arrangement in space. They are further divided into two classes. a. Geometrical isomerism b. Optical isomerism.

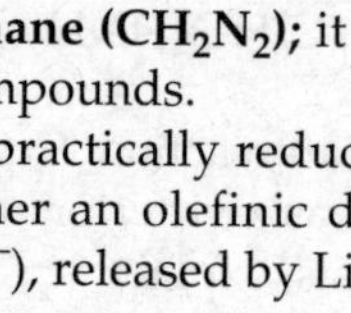

cis-2-butene   *trans*-2-butene
Geometrical isomerism   Optical isomerism

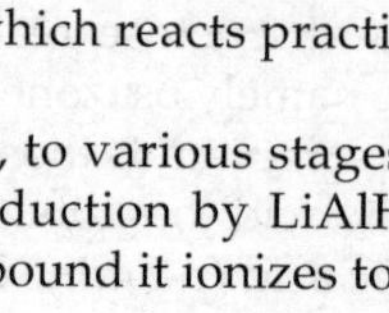

3. **Diazomethane (CH_2N_2);** it is an organic compound, which reacts practically with all the organic compounds.

4. **$LiAlH_4$;** it practically reduces all the reducible groups, to various stages, but it does not reduce either an olefinic double or a triple bond. Reduction by $LiAlH_4$ is brought by hydride (H^-), released by Li^+H^-, Li^+H^- is an ionic compound it ionizes to form Li^+ and H^- ion.

$$\text{LiH} \longrightarrow \text{Li}^+ + \text{H}^-$$

$$>C = O + \text{Li} \longrightarrow C^+\text{OLi} \xrightarrow{\text{H}^-} \underset{\text{H}}{\overset{\text{OLi}}{C}} \xrightarrow{\text{H}_2\text{O}} \underset{\text{H}}{\overset{\text{OLi}}{C}}$$

On the other hand, reduction of a double or triple bond takes place by an electrophilic attack by a proton, on one of the more electronegative carbon.

5. **Polarity is based on electronegativity** of the atoms joined by a single bond. The bonding electrons are partially shifted towards more electronegative atom. This makes the bond polar in nature.

For example: C—Cl bond is polar, with chlorine being more electronegative (2.53)

$$-C - Cl \longrightarrow C^{\delta+} - Ci^{\delta-}$$

Polarisability is a phenomenon, associated with the atoms, having a high atomic number, like iodine (atomic number = 53), which has electronegativity almost equal to that of carbon atom, yet the bond between carbon and iodine becomes more polar. This is because the electrons from the outermost orbit of iodine atom, being away from the nucleus are able to tilt towards the approaching electrophile,

$$\underset{2.5 \qquad 2.6}{C-I \longrightarrow C^{\delta+} \quad I^{\delta+}}$$

The decreasing order of reactivity of halides:

C—I > C—Br > C—Cl

6. **Mutarotation:** The change in initial specific rotation value of an aqueous solution of a reducible sugar (i.e. an aqueous solution of glucose) to a constant value is termed mutarotation. Glucose is a cyclic compound, having an oxide ring in the molecule. The oxide ring consists of five carbon and one oxygen atom, known as pyranose ring. On account of the ring, glucose exists as a mixture of two cyclic forms namely α-glucose and β-glucose.

   Ordinary glucose is a mixture of isomeric forms of glucose, namely α-glucose, mp = 140 °C, having a specific rotation value = (+)112° and β-glucose, mp 148–50 °C, having a specific rotation value = (+)19°.

On dissolving glucose in water, the two forms of glucose, start converting into one another, till a new equilibrium is established, where the amount of the two forms are different from that of the glucose in the solid state. These two forms are in equilibrium with each other through an open aldehydic form of glucose, where α-form is ~64% and the β-form is ~30% and the open form is 5%.

α-Glucose ⇌ (open aldehydic form) ⇌ β-Glucose

The phenomenon of mutarotation has been regarded as a proof that the sugar has a ring structure. The ring in the case of glucose consists of five carbons and one oxygen, known as a pyranose ring and in the case of fructose the ring consists of four carbon and one oxygen, known as furanose ring.

Fructose also shows mutarotation, but its two forms have not been isolated.

7. An aqueous solution of glucose, after attaining the final value of optical rotation, i.e. +52.5, contains two cyclic forms of glucose (glucose = ~65% and glucose = ~30%), in an equilibrium with an open chain aldehydic form of glucose (~5%). Those reagents which would react, with even such a low quantity of open chain aldehydic form of glucose, reacts with this form and show reducing properties of glucose.

It is this aldehydic form of **glucose which shows some of the reducing properties** like reduction of Fehling solution and Tollen's reagent. Those reagents, for which a higher concentration of aldehydic form is required, do not react with an aqueous solution of glucose. Thus, glucose solution fails to react with $NaHSO_3$ solution nor it restores the pink colour of Schiff's reagent.

8. **A *cis*-alkene** may be converted into a ***trans*-alkene or vice-versa**, by heating the alkene with iodine. Iodine adds in a *trans*-manner and the conversion may be brought by heating the alkene with iodine, either to a high temperature or through exposer of light.

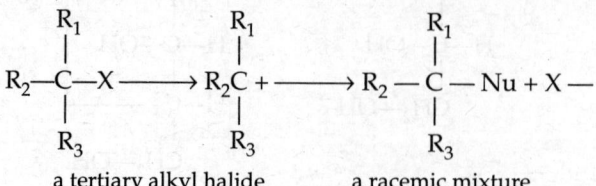

9. The **intermediate in S_N1 mechanism:**
The shape of the intermediate is sp^2 hybridised form, it is triangular in shape, having all the three groups in the same plane with the +ve charge on the central carbon atom.

The molecule may attack the intermediate either from the top or from the bottom side, to form the derivative. Since an attack is possible from either side, it results in the formation of a racemic mixture (a mixture containing equal quantity of both d- and l- forms). A racemic mixture is optically inactive.

The intermediate in the S_N2 mechanism:

i. The transition state in S_N2 mechanism has five groups attached to the central carbon atom.

ii. The attacking nucleophile "Nu", attacks the alkyl halide from the side opposite to X, forming a half bond in the transition state. At the same time, the bond of the leaving group, i.e. the "X" is half broken. As the attack takes place from the back side, the molecule undergoes inversion, and the derivative has the optical rotation, opposite to that of the alkyl halide. Whereas the tertiary alkyl halides, R_3—C—, undergo nucleophilic attack by S_N1 mechanism, a primary alkyl halide, RCH_2X, undergoes nucleophilic substitution by S_N2 mechanism. A secondary alkyl halide undergoes nucleophilic substitution partly by S_N1 and S_N2 mechanism.

$$R_2 \overset{\overset{\displaystyle R_1}{|}}{\underset{\underset{\displaystyle R_3}{|}}{C}} {-}X \longrightarrow R_2 \overset{\overset{\displaystyle R_1}{|}}{\underset{\underset{\displaystyle R_3}{|}}{C}} + \longrightarrow R_2 {-} \overset{\overset{\displaystyle R_1}{|}}{\underset{\underset{\displaystyle R_3}{|}}{C}} {-} Nu + X{-}$$

a tertiary alkyl halide   a racemic mixture

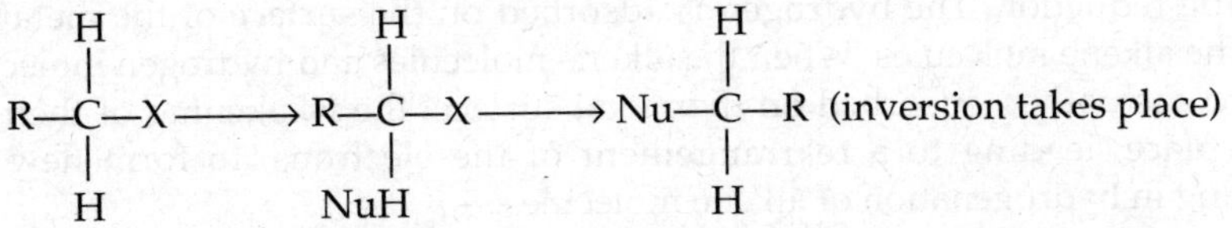

$$\underset{\substack{| \\ H \\ \text{a primary alkyl halide}}}{\overset{\substack{H \\ |}}{R-C-X}} \longrightarrow \underset{\substack{| \\ NuH}}{\overset{\substack{H \\ |}}{R-C-X}} \longrightarrow \underset{\substack{| \\ H}}{\overset{\substack{H \\ |}}{Nu-C-R}} \text{ (inversion takes place)}$$

a derivative

10. Cyclopentanone undergoes a ring expansion, to form cyclohexanone and cycloheptanone, when it reacts with diazo methane, CH_2N_2.

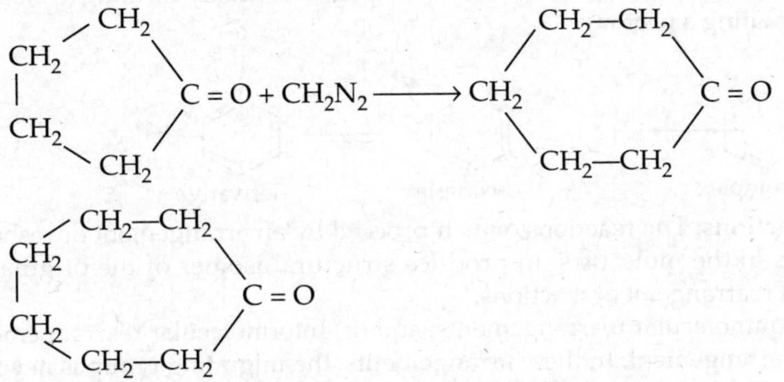

$$\begin{array}{c} CH_2 \\ | \\ CH_2 \\ | \\ CH_2 \end{array}\!\!\!\!\!\!\!\! \begin{array}{c} CH_2 \\ \diagdown \\ \diagup \\ CH_2 \end{array}\!\!\!\! C=O + CH_2N_2 \longrightarrow \begin{array}{c} CH_2 \\ \diagup \\ CH_2 \\ \diagdown \\ CH_2-CH_2 \end{array}\!\!\!\! \begin{array}{c} CH_2-CH_2 \\ \diagdown \\ \diagup \\ CH_2-CH_2 \end{array}\!\!\!\! C=O$$

$$\begin{array}{c} CH_2 \\ | \\ CH_2 \end{array}\!\!\!\! \begin{array}{c} CH_2-CH_2 \\ \diagup \\ \diagdown \\ CH_2-CH_2 \end{array}\!\!\!\! C=O$$

11. **Picric acid** is a 2,4,6-trinitro phenol. Though it does not contain a carboxylic acid group, yet it behaves as a very strong acid, it liberates CO_2 from $NaHCO_3$ while other phenols do not liberate CO_2 from $NaHCO_3$. The three NO_2 groups in o- and p- positions exert a very strong deactivating effect on —OH group which release a proton from —OH group and stays ionized.

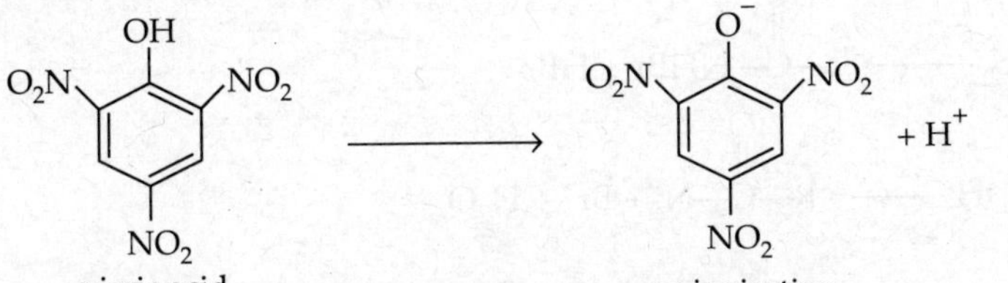

picric acid   on ionisation

a. The reduction of a triple bond in preference to a double bond, takes place, if it is in conjugation with a double bond.

$$CH_2=CH-C\equiv CH \longrightarrow \underset{\text{1,3-butadiene}}{CH_2=CH-CH=CH_2}$$

b. When the two are not in conjugation with each other, it is the double bond, that is more reactive than the triple bond in an electrophilic addition.

$$CH_2=CH-CH_2C\equiv CH+HBr \longrightarrow CH_3-CHBr-CH_2-C\equiv CH$$

The addition of hydrogen, across the double bond, takes place in the *cis*- manner, i.e. the addition take place from the same side.

$$CH_3-CH=CH_2+H_2 \longrightarrow \underset{\substack{| \ | \\ H \ H}}{CH_3-CH=CH_2} \longrightarrow CH_3CH_2CH_3$$

The reduction of a double bond is carried out with H_2, in presence of platinum, at room temperature or in presence of a heated nickel or palladium known as catalytic reduction. The hydrogen is adsorbed on the surface of the metal, along with the alkene molecules. When the alkene molecules and hydrogen molecules lie next to each other, adsorbed on the metal surface, the weakening of their bonds takes place, leading to a rearrangement of the electrons, to form new bonds resulting in hydrogenation of alkene molecules.

13. A benzene molecule is a flat molecule, engulfed from top to bottom, by π-clouds which imparts a strong negative charge on the molecule. Any attack by a nucleophile on the benzene ring will be repulsed by it. On the other hand, an attack by an electrophile will be facilitated, to form a π-complex changing to a σ-complex and finally forming a substituted derivative, by eliminating a proton.

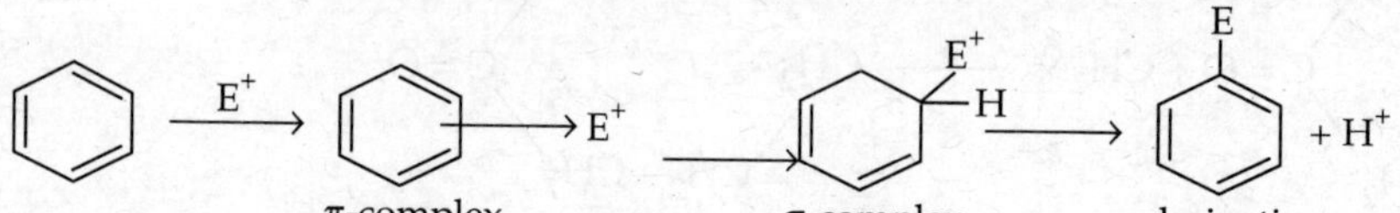

π-complex σ-complex   derivative

14. **Rearrangement reactions:** The reactions, which proceed by an arrangement or reshuffling of atoms or groups in the molecules, to produce structural isomer of the original compounds, are termed rearrangement reactions.

These may be (i) Intramolecular rearrangements, and (ii) Intermolecular rearrangements

i. Intramolecular arrangement: In these arrangements, the migrating group is never fully detach form the system, during the process of migration.

Example: In Hofmann's bromamide reaction (or degradation), in which, an amide is converted into a corresponding amine, by treating with Br_2 and KOH, an intermediate formed, namely, an unstable acyl nitrene, rearranges intra molecularly to give iso-cyanate.

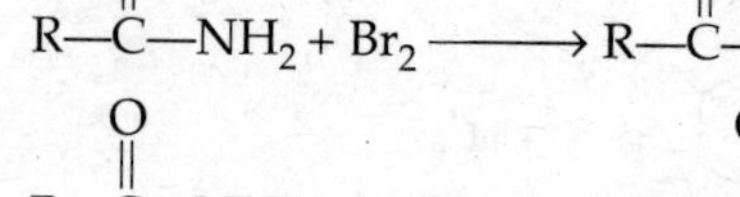

ii. Intermolecular rearrangement: The rearrangements in which the migrating group, completely detaches from the molecule and later joins or gets attached at some site of the molecule, producing, thereby a structural isomer of the original substance.

Example: Diazo amino benzene, on warming with dil. HCl, rearranges to form *p*-amino azo benzene.

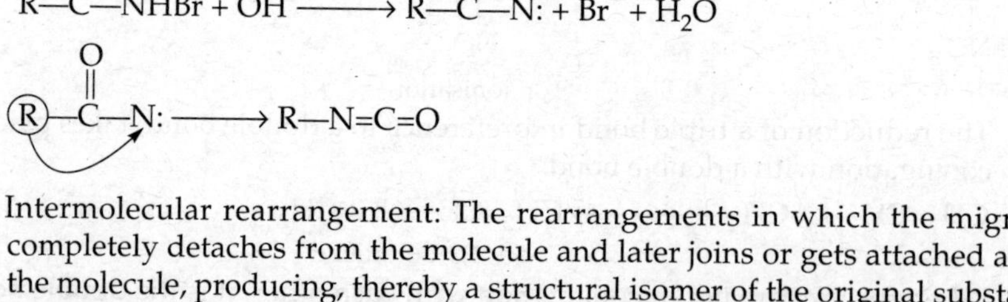

15. **Glucose and Fructose**, both react with an excess of phenyl hydrazine to form the same compound namely **osazone**. As the reaction with phenyl hydrazine is carried out in aqueous solution, it is the free aldehydic form of glucose that reacts with $C_6H_5NHNH_2$ to form osazone. **Glucose** reacts with one molecule of $C_6H_5NHNH_2$, to form glucose phenyl hydrazone, which further reacts with more $C_6H_5NHNH_2$, to yield glucosazone, through a number of intermediate equilibrium stages. The second—CHOH is converted to a >C=O group, though $C_6H_5NHNH_2$ is not an oxidising reagent. This >C=O group then further reacts with more $C_6H_5NHNH_2$, to form osazone finally.

Fructose also reacts with $C_6H_5NHNH_2$ to form fructose phenyl hydrazone. This reacts with more of phenyl hydrazine to convert the —CH_2OH into a —CHO group, which reacts with more phenyl hydrazine to form finally osazone. The reaction stops at this stage because of formation of a cyclic compound, involving a hydrogen bonding.

The configuration of glucose and fructose is the same from C-3 to C-6 Therefore, both glucose and fructose form same osazone.

```
1        CHO + H₂NNHC₃H₅    CH = NNHC₆H₅        CH = NHNC₆H₅
         |                  |                   |
2      H—C—OH               HCOH                C = NHNC₆H₅
         |                  |          H₂NHC₆H₅ |
3-5   (H—C OH)₃  ────────→ (HC—OH)₃  ────────→ (HCOH)₃
         |                  |          excess   |
6        CH₃OH              CH₂OH               CH₂OH
       glucose        glucose phenyl hydrozone  osazone
```

```
CH₂OH                 CH₂OH              CH = N—NHC₆H₅
|                     |                  |
C=O + H₂NNHC₆H₅       C=N—NHC₆H₅         C = N—NHC₆H₅
|                     |                  |
(CHOH)₃  ────────→   (CHOH)₃  ────────→ (CHOH)₃
|                     |      excess      |
CH₂OH                 CH₂OH   phenyl     CH₂OH
                              hydrazine
fructose        fructose phenyl hydrozone  osazone
```

16. The following aromatic compounds do not undergo Friedel–Craft reaction.

    where X = NO_2, COOH, CHO, $COCH_3$, COOR, SO_3H, CN

    These groups deactivate the benzene ring to such an extent that attacking electrophile does not form a complex with the benzene ring. Aniline does not undergo Friedel–Craft reaction, as it forms π-complex with it. Pyridine does undergo Friedel–Craft reaction as the lone pair on the π nitrogen, co-ordinate with the $AlCl_3$ (a Lewis acid).

17. **Anthranilic acid or *o*-amino benzoic acid,** does not exist as a zwitter ion, because the —NH_2 group in the *o*-position, is deactivated to such an extent, that it is incapable of forming a salt with proton, released by the —COOH group.

    On the other hand *p*-amino benzene sulphonic acid, or sulphanilic acid, exists as a zwitterion, being a derivative of mineral acid, H_2SO_4 ionises fully.

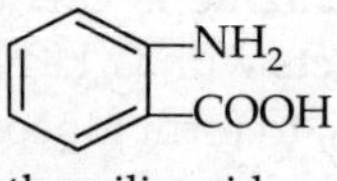

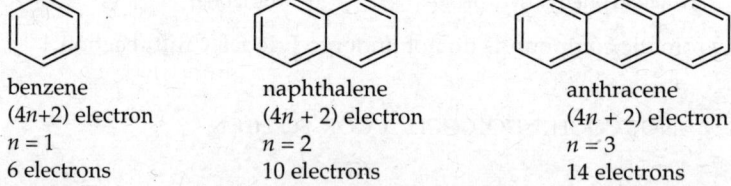

anthranilic acid

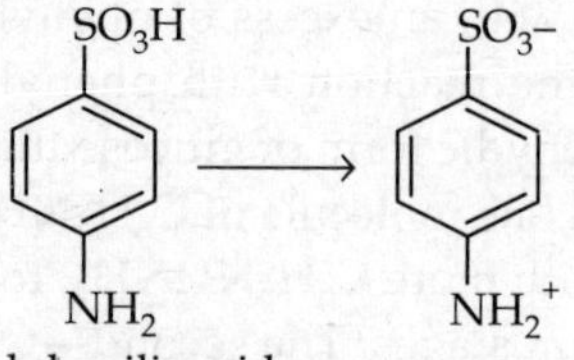

sulphanilic acid

18. Benzyne does not exist free in nature, because it is unstable in free state. It is formed, when chlorobenzene reacts with sodamide, $NaNH_2$.

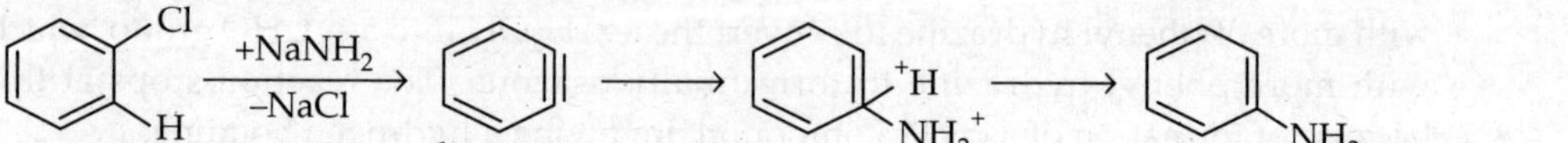

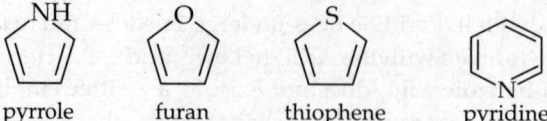

benzyne

Sodamide $NaNH_2$, abstracts the chloride ion from chlorobenzene, converting it into benzyne. An amino group, then attacks and joins the ring, next to the position vacated by the chloride ion, and converts it into aniline.

19. **Aromaticity:** It may be defined as a property possessed by cyclic, monoplanar unsaturated rings to undergo substitution reaction, rather than addition keeping the unsaturation intact in the molecule. For a cyclic compound to behave as an aromatic compound, it must fulfill the following conditions:

1. The compound must be cyclic in nature, and be monoplanar, i.e. all the atoms of the ring should be in the same plane.

2. Each atom of the ring must be in the sp^2 hybridised state, i.e. all the atoms of the ring must have a p-orbital. These orbitals must be parallel to each to each other, so that a continuous overlap by 'p' orbitals or π electron cloud is possible around the ring.

3. The cyclic molecular orbital (electron cloud), formed by overlap of p-orbitals, must contain $(4n + 2)$ π-electrons, where $n = 0, 1, 2, 3$ etc, known as Huckel's rule. The compound must follow Huckel's rule.

Some of the benzenoid aromatic rings:

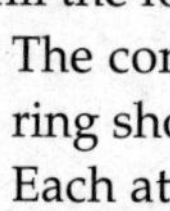

benzene
$(4n+2)$ electron
$n = 1$
6 electrons

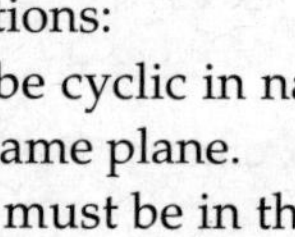

naphthalene
$(4n + 2)$ electron
$n = 2$
10 electrons

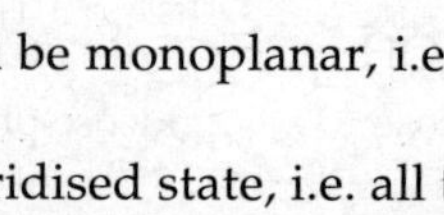

anthracene
$(4n + 2)$ electron
$n = 3$
14 electrons

Some of the nonbenzenoid aromatic rings, each ring contains 6π-electrons

pyrrole   furan   thiophene   pyridine

The following rings are not aromatic in character, as these do not fulfill all the conditions, required for the compound to behave as aromatic compound. These rings are termed nonaromatic rings.

Nonaromatic rings:

i. Cyclopropene

- It is cyclic in nature.

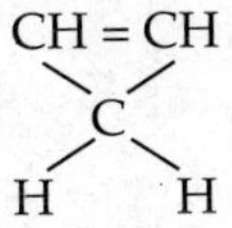

$$CH = CH$$
$$\diagdown \diagup$$
$$C$$
$$\diagup \diagdown$$
$$H \qquad H$$

- It has 2 electrons, as required by Huckel's rule.
- But the third condition is not fulfilled, as the ring is nonplanar. One of the carbon is in sp^3 hybridised form. However, cyclo propene may be converted into an aromatic compound, if one hydrogen from sp^3 hybridised carbon be removed as a hydride ion. It can be done by treating cyclopropene with tri phenyl methyl chloride $((C_6H_5)_3$—C—Cl) in liquid SO_2. The tri phenyl methyl carbocation formed in the reaction between $(C_6H_5)_3$C—Cl and SO_2, abstracts a hydride from the cyclopropene and converts it into cyclopropenyl carbocation, which will behave as an aromatic compound.

$$(H_6H_5)_3C\text{—}Cl + SO_2 \longrightarrow (C_6H_5)_3C^+ + SO_2Cl$$

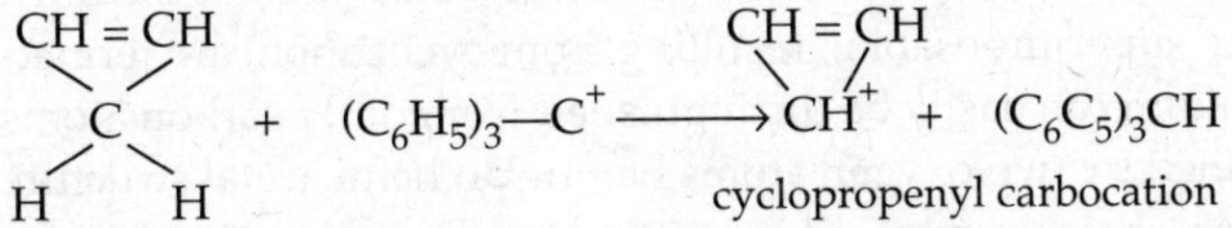

$$CH = CH \qquad\qquad\qquad CH = CH$$
$$\diagdown \diagup \qquad\qquad\qquad \diagdown \diagup$$
$$C \quad + \quad (C_6H_5)_3\text{—}C^+ \longrightarrow CH^+ \quad + \quad (C_6C_5)_3CH$$
$$\diagup \diagdown$$
$$H \qquad H \qquad\qquad\qquad\qquad \text{cyclopropenyl carbocation}$$

ii. Cycloheptatriene, C_7H_8

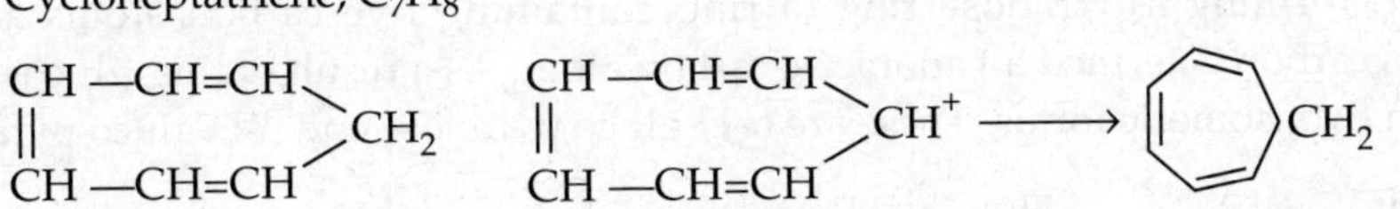

$$CH\text{—}CH=CH \diagdown \qquad CH\text{—}CH=CH \diagdown \qquad\qquad \diagup \diagdown$$
$$\| \qquad\qquad\quad CH_2 \quad \| \qquad\qquad\quad CH^+ \longrightarrow \qquad CH_2$$
$$CH\text{—}CH=CH \diagup \qquad CH\text{—}CH=CH \diagup$$

- It is cyclic in nature.
- The number of electrons in the ring are = 6, it fulfills Huckel's rule.
- However, the ring is not planar, as one of the carbon is in sp^3 hybridised form. Cycloheptatriene ring may be converted into an aromatic ring, in case one hydrogen is removed from sp^3 hybridised carbon, i.e. from, —CH_2, by treating with triphenyl methyl carbonium ion. The cycloheptatrienyl carbocation will behave as an aromatic ring.

Antiaromatic rings: These rings have the $(4n + 2)$ number of π-electrons, as required by Huckel's rule.

i. Cyclobutadiene (C_4H_4)

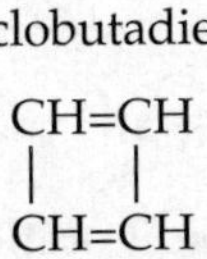

$$CH=CH$$
$$| \qquad |$$
$$CH=CH$$

- It is cyclic in nature
- It is monoplanar
- However, the number of π-electrons is only 4, so it does not fulfill the Huckel's rule. But it can be converted into an aromatic ring, by treating it with two equivalent of sodium metal thereby converting into a disodium salt of cyclobutadiene. Two sodium

atoms donate one electron each, raising the number of π-electrons to six, thus making the ring aromatic in nature.

ii. Cyclo-octatetraene (C_8H_8)

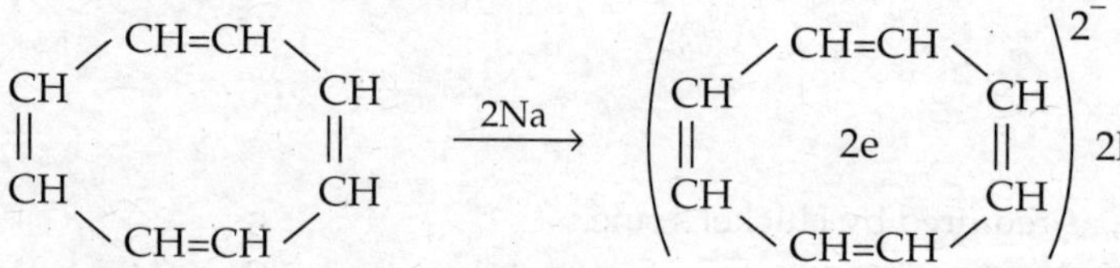

$$
\begin{array}{c}
\text{CH} \overset{\displaystyle\diagup\, \text{CH=CH} \,\diagdown}{\underset{\displaystyle\diagdown\, \text{CH=CH} \,\diagup}{\underset{\displaystyle\|}{\underset{\displaystyle\text{CH}}{}}}} \text{CH}
\end{array}
\xrightarrow{\;2\text{Na}\;}
\left(
\begin{array}{c}
\text{CH} \overset{\displaystyle\diagup\, \text{CH=CH} \,\diagdown}{\underset{\displaystyle\diagdown\, \text{CH=CH} \,\diagup}{}} \text{CH} \quad 2e
\end{array}
\right)^{2-}
2\text{Na}^{+}
$$

- The ring is cyclic in nature
- The ring is monoplanar
- The number of π-electrons is only 8. Therefore, it is not an aromatic ring. However, by treating it with two equivalent of sodium metal, it is converted into a disodium salt of cyclo-octatetraene. Two sodium atoms donate two electrons making the number of π-electrons to be 10, thus fulfilling the Huckel's rule.

Therefore, the disodium salt of cyclo-octatetraene will behave as an aromatic.

20. In carbohydrate chemistry, the diastereo isomers (those which are not the mirror image of each other nor they are superimposable), resulting from cyclization, are termed "anomer". The anomeric carbon atom can easily be distinguished from other carbon atoms in the ring by the fact that it is joined by two oxygen atoms having an hemi acetal structure.

Glucose is not an open chain compound, but it is a cyclic compound, involving carbon numbered 1, forming a pyranose ring (a ring containing five carbon atoms and one oxygen. This carbon is termed an anomeric carbon atom. As a result of this, glucose molecule exists in two isomeric forms. These are α-D-gluco pyranose and β-D-gluco pyranose

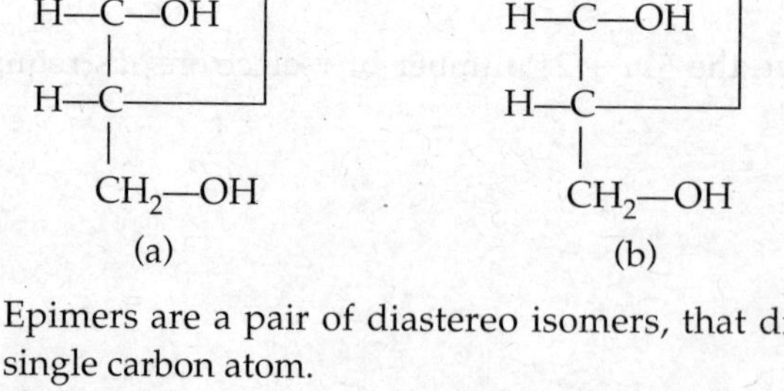

(a)   (b)

Epimers are a pair of diastereo isomers, that differ in only in the configuration about a single carbon atom.

Example: D-glucose and D-mannose differ only in configuration about one asymmetric carbon atom, i.e. carbon number 2.

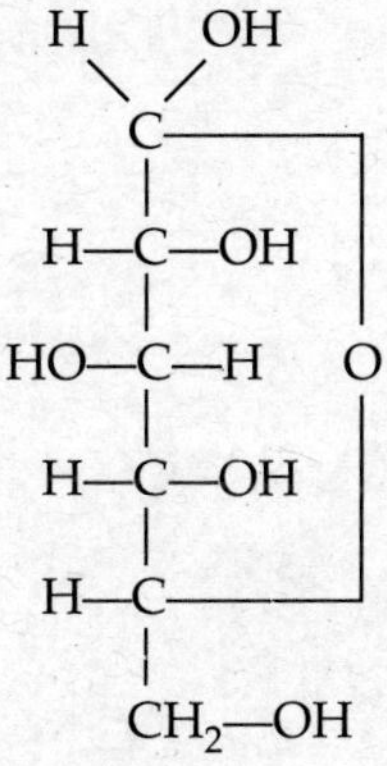

```
        H   OH                      H   OH
         \ /                         \ /
          C                           C
          |                           |
       H—C—OH                    HO—C—H
          |                           |
      HO—C—H    O                HO—C—H    O
          |                           |
       H—C—OH                     H—C—OH
          |                           |
       H—C                         H—C
          |                           |
       CH₂—OH                      CH₂—OH
      α-D-glucose                  α-D-mannose
```

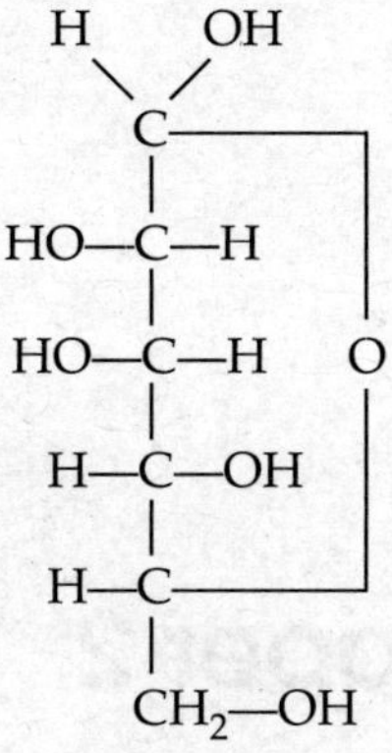

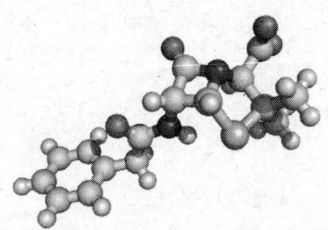

Paper 2

REASONING TYPE QUESTIONS (RTQs)

1. What do you understand by the term "sequence rule"? Explain with example.
2. What do you understand by term "asymmetric synthesis"? Explain with example.
3. Explain the term "stereospecific reduction" of alkynes with detailed explanation.
4. What do you understand by the term 'plane of symmetry'? Explain with suitable example, the terms used (i) enantiomers and (ii) diastereoisomers.
5. What are the requirements for a geometrical isomerism or a *cis-* and *trans-*isomerism to take place? Discuss the requirements in details.
6. Discuss 'absolute configuration' and 'specification of the configuration' by using the "R" and "S" system, which is used for asymmetric molecules.
7. Explain the term 'chirality'.
8. What do you understand by the term 'tautomerism'? Discuss.
9. Give suitable examples of aromatic nucleophilic substitution reactions.
10. In which type of isomerism, the projection formulae are used? Explain.
11. Explain the terms 'racemisation' and Walden inversion.
12. What do you understand by the term 'transition state'? Explain with suitable examples.
13. How will you define the terms 'electrophile' and 'nucleophile'?
14. Explain the terms 'octane number' and 'cetane number'?
15. What do you understand by the terms "electromeric" and "mesomeric" effect? Explain.
16. Discuss what do you understand by the term "orientation" with respect to substitutions in mono substituted benzene.
17. What are the requirements for optical isomerism to take place? Discuss.
18. What do you understand by the term *trans-*esterification? Explain.
19. Explain the term "resolution".
20. What do you understand by "free radical reactions"? Discuss various reactions.

ANSWERS

1. **Sequence rule:** The atoms or groups joined to the asymmetric carbon atom are arranged in the order of **decreasing priority**, based on their atomic number.

Rules:

i. Of the various atoms, attached to the asymmetric carbon atom, the one with the highest atomic number will have the highest priority.

Element	I > Br > Cl > S > F > O > N > C > H
At. number	53 35 17 16 9 8 7 6 1

A ready reference for the most common groups and atoms is given below, from highest priority to the lowest.

—I, —Br , —Cl, —SH, —F, —OCOR, —OR, —OH, —NO$_2$, —NHCOR, —NR$_2$, —NH$_2$, —CCl$_3$, —COCl, —COOR, —CONH$_2$, —COR, —CHO, —CN, —CH$_2$OH, —C$_6$H$_5$, —CR$_3$, —CHR$_2$, —CH$_2$R, —CH$_3$, —H,

ii. In case, the atoms attached to the **central carbon** atom are the same, then the priority is determined by going to the next atom away from the carbon atom. If it does not help in determining the priority then next atom attached is considered the carbon atom is to be considered for determining the priority, and so on, till an atom with different atomic number determines the priority.

iii. When an atom is attached directly to the carbon atom, having a triple bond, it is considered to be attached by three like atoms of the same atomic number. Similarly, an atom, attached by a double bond, is regarded as being attached by two like atoms of the same atomic number.

2. **Asymmetric synthesis:** When a compound, containing an asymmetric carbon atom, is synthesized by ordinary methods, the product is a racemic mixture, which is optically inactive containing an equal amounts of both the optical isomers, namely d- and l-isomers. However, if the synthesis is carried out in the presence of (or under the influence of) some optically active compound it results in the formation of either one of the isomer or a mixture containing one of the optically active isomer in dominating amount. The process by which an optically active compound (or an asymmetric compound) is synthesized, from a symmetric compound, to yield either a dextro or laevo isomer, is termed **asymmetric synthesis**.

Example:

(i) By reduction: By usual reducing agents, pyruvic acid gives a racemic mixture. On the other hand, when pyruvic acid is reduced by yeast, only laevo lactic acid is obtained. In the same way, pyruvic acid $CH_3CO.COOH$, is first converted into an ester, (–) menthyl pyruvate, with (–) menthol alcohol, $C_{10}H_{19}$—OH. This on reduction gives (–) menthyl-l-lactate, $CH_3CH(OH)COOC_{10}H_{19}$ in excess and (–) menthyl-d-lactate. This upon hydrolysis gives (–) lactic acid in excess and small amount of (+) lactic acid.

$$CH_3CO.COOH + C_{10}H_{19}OH \longrightarrow CH_3CO.COOC_{10}H_{19} \longrightarrow CH_3CH(OH)COOC_{10}H_{19}$$

Pyruvic acid   (–)menthol   (–)methyl pyruvate   (–)menthyl lactate
                                                        (+)menthyl lactate

$$\longrightarrow CH_3CH(OH)COOH + C_{10}H_{19}—OH$$

Excess of (–) lactic acid + menthol
+small amount of (+) lactic acid

3. **Stereospecific or stereoselective reduction:** Alkynes on reduction in the presence of finely divided platinum or Raney Nickel, add up two molecule of H$_2$, first forming alkene and finally alkane. The catalytic reduction of disubstituted alkynes with **Lindlar's catalyst, (palladium mounted on BaSO$_4$, poisoned with quinoline) forms a *cis*-alkene, exclusively.**

$$R—C \equiv C—R + H_2 \xrightarrow{Pd + BaSO_4} \begin{array}{c} R \\ \\ H \end{array} C=C \begin{array}{c} R \\ \\ H \end{array} \quad \textit{cis}\text{-compound}$$

On the other hand, when reduced in presence of **sodium or lithium metal and liquid ammonia**, predominantly a *trans*-alkene is formed.

$$R—C \equiv C—R + H_2 \xrightarrow{Na + liq. NH_3} \begin{array}{c} R \\ \\ H \end{array} C=C \begin{array}{c} H \\ \\ R \end{array} \quad \textit{trans}\text{-compound}$$

Such reactions, which produce only one of the isomer exclusively are termed **stereo-specific or stereoselective reduction reactions**.

4. **Plane of symmetry:** A plane, which divides a molecule into two equal halves is known as plane of symmetry. An object, lacking a plane of symmetry is called **dissymmetric or chiral**. A symmetric object is referred to as **achiral**.
   Tartaric acid: The compound has two asymmetric carbon atoms, in the molecule, and therefore exists in **four isomeric** forms. These are:
   I. d-tartaric acid II. l-tartaric acid III. racemic tartaric acid IV. meso tartaric acid.

COOH	COOH	COOH
H—C—OH	HO—C—H	HO—C—H
HO—C—H	H—C—H	HO—C—H
COOH	COOH	COOH
(+) acid	(−) acid	meso
m.p. = 170 °C	m.p. = 170 °C	m.p. = 143 °C
d-tartaric acid	l-tartaric acid	meso tartaric acid
I	II	III

a racemic mixture
(+−) acid
m.p. 206 °C
optically inactive

IV

Optically active, possess plane of symmetry, mirror image of each other termed enantiomers

Optically inactive, can be resolved, are not superimposable on each other termed disymmetric or chiral

Whenever an optically active compound is prepared in the lab, a **racemic mixture is always obtained** (an optically inactive mixture containing an equal amount of both d- and l-isomers).

- **Enantiomers:** The optically active compounds, existing in two optically active forms, and are mirror image of each other. These are not superimposable on each other.
- **Diastereoisomers:** d-tartaric acid (or l-tartaric acid) and meso tartaric acid are the isomers of the same compound, but they are not mirror image of each other.

5. **For geometrical isomerism to occur, following are the requirements:**
   A pair of carbon atoms be joined by a double bond
   The groups attached on either carbon atoms, should be different from each other.
   Example: abC = Cab
   The restricted rotation about the C to C double bond, is responsible for this phenomenon. The two compounds are not superimposable on each other.
   The isomer, in which the **identical or similar groups are close to each other**, with respect to the double bond, is termed and *cis*-**isomer** or compound the isomer, in which the **identical or similar groups are on the opposite side of a double bond or are farther apart**, is termed a *trans*-**isomer**.

$$\underset{\substack{\diagup \\ B}}{\overset{\substack{A \\ \diagdown}}{}} C=C \underset{\substack{\diagdown \\ B}}{\overset{\substack{A \\ \diagup}}{}}$$

cis-isomer

$$\underset{\substack{\diagdown \\ B}}{\overset{\substack{A \\ \diagup}}{}} C=C \underset{\substack{\diagup \\ A}}{\overset{\substack{B \\ \diagdown}}{}}$$

trans-isomer

For the systems, where all the four groups attached are different from each other like ABCDC$_2$ or A, B-C = C-C, D, a different nomenclature is used. The symbols "E" and "Z" are used in place of *cis*- and *trans*-nomenclature. **"E" stands for entgegen (a German word) = opposite and "Z" stands for zusamen (a German word) meaning together.**

An isomer, in which the atoms or groups of higher preference, based on the **sequence rule,** are close to each other (or are on the same side), is termed a "Z" isomer; and the other one that contains the two groups of higher preference on the opposite side or are farther apart, is termed as an "E" isomer.

Examples:

1. Preference is based on   I > Br > Cl > F
      atomic number.   53   35   17   9

                               F > Br
                               I > Cl

$$\underset{\substack{\diagup \\ Cl}}{\overset{\substack{I \\ \diagdown}}{}} C=C \underset{\substack{\diagdown \\ Br}}{\overset{\substack{F \\ \diagup}}{}}$$

"E" isomer

$$\underset{\substack{\diagup \\ Cl}}{\overset{\substack{I \\ \diagdown}}{}} C=C \underset{\substack{\diagdown \\ F}}{\overset{\substack{Br \\ \diagup}}{}}$$

"Z" isomer

2. Preference is based on sequence rule

     —COOH > —CN > CH$_3$CH$_2$– > CH$_3$–
     —COOH > CH$_3$CH$_2$–
     —CN   > CH$_3$–

$$\underset{\substack{\diagup \\ CH_3-CH_2}}{\overset{\substack{HOOC \\ \diagdown}}{}} C=C \underset{\substack{\diagdown \\ CH_3}}{\overset{\substack{C=N \\ |}}{}}$$

"Z" isomer

$$\underset{\substack{\diagup \\ HOOC}}{\overset{\substack{CH_3-CH_2 \\ \diagdown}}{}} C=C \underset{\substack{\diagdown \\ CH_3}}{\overset{\substack{C-N \\ \diagup}}{}}$$

"E" isomer

6. **R and S system for asymmetric molecule:** The actual three dimensional arrangement of groups in an asymmetric molecule is called **absolute configuration.** This type of configuration, can be specified by using the "R" and "S" system.

   The **"R" and "S"** can also be called as **Cahn–Ingold–Prelog system**, after the names of the inventor. The four groups attached to the asymmetric carbon atoms are arranged in a decreasing order of priority, (1.2.3.4) by applying the 'sequence rule'. The asymmetric compound is then viewed by holding in such a way, that the group with lowest priority, is held away from the observer or lies on the opposite side. The three groups (1 > 2 > 3), which are close to the observer, are then traced from highest priority to the lowest; if the tracing is **a clockwise direction, the compound is assigned a "R" configuration (R = rectus or right).**

   Similarly, if the tracing from a group of highest priority to a lowest priority follows in an **anticlockwise manner, the compound is assigned a "S" configuration (S = sinister or left)**

   Example:

   Priority A > B > C > D

(4) D — C $\overset{\displaystyle A\ (1)}{\underset{\displaystyle \underset{(3)}{C}}{\overset{\displaystyle |}{\diagup}}}$ B (2)

$\overset{\displaystyle A\ (1)}{\underset{\displaystyle \underset{(3)}{C}\ \ \ \ B\ (2)}{C}}$

A > B > C > D   Clockwise = R

7. **Chiral or an asymmetric carbon atom:** A carbon atom, which is joined by **four different atoms or groups, or by both atoms and groups,** is called an asymmetric carbon atom and the compound as an "asymmetric molecule". **Such a molecule lacks a plane of symmetry and the molecule is without any symmetry. A dis-symmetric or a chiral molecule is often used for such asymmetric molecules.**

$$CH_3 - \overset{\displaystyle C-Br}{\underset{\displaystyle COOH}{\overset{\displaystyle |}{\underset{\displaystyle |}{C}}}} - CH_2CH_3 \qquad\qquad CH_3 - \overset{\displaystyle C_2H_2}{\underset{\displaystyle COOH}{\overset{\displaystyle |}{\underset{\displaystyle |}{C}}}} - CH_2CH_3$$

asymmetric carbon

8. **When two structural isomers are mutually interconvertible and exist in dynamic equilibrium are termed tautomers and the phenomenon as tautomerism.** It is a special type of isomerism, in which the isomers exist in equilibrium with each other.

For Example: Ethyl acetoacetate, $CH_3COCH_2COOC_2H_5$, exists in an equilibrium mixture of two forms, a keto form and an enol form at room temperature.

$$CH_3COCH_2COOC_2H_5 \rightleftharpoons CH_3\overset{\displaystyle HO}{\overset{\displaystyle |}{C}}=CHCOOC_2H_5$$

a keto form (93%)   an enol form (7%)

As a matter of fact, the same compound shows the chemical and physical properties of both the groups, namely of the keto group as well as of the enol group (enol group means an ene + ol; a double bond + OH group).

Properties of keto group	Properties of enol group
$CH_3COCH_2COOC_2H_5 \rightleftharpoons CH_2\overset{\displaystyle HO}{\overset{\displaystyle \mid}{C}}=CHCOOC_2H_5$	
1. Long colourless needles, mp = –39 °C	1. Colourless liquid bp = – 78 °C
2. No colour with $FeCl_3$	2. Gives a red colour with $FeCl_3$
3. Does not decolourise Br_2 water	3. Decolourises yellow colour of Br_2 water.

Dynamic equilibrium means that when a reagent reacts with the keto form, Enol form changes to keto form, till all ester has reacted as keto form. In the same way, if the reagent reacts with the enol form, the keto form change to the enol form till all the keto has reacted in enol form.

9. **Aromatic nucleophilic substitution:** Benzene as such does not undergo nucleophilic substitution reactions. However, substituted derivatives containing group like —NO_2 does undergo nucleophilic substitution reactions.

Examples:

   i. When nitrobenzene, $C_6H_5NO_2$ is fused with solid KOH, it is converted into o-nitro phenol. A hydroxyl group, —OH gets substituted in the benzene nucleus, in the o-position to the —NO_2 group. The hydroxyl, —OH group is a nucleophilic group.

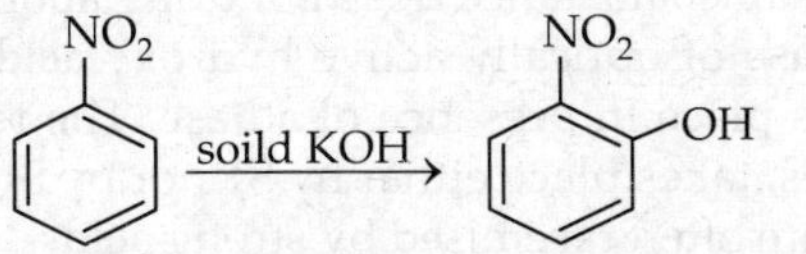

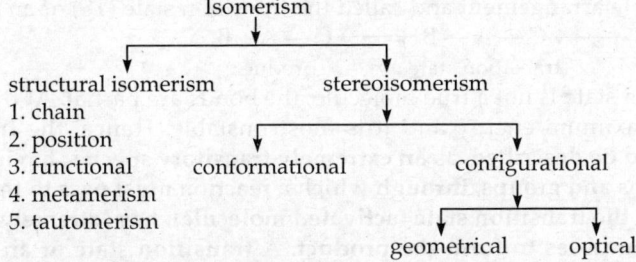

ii. When dinitrobenzene is treated with potassium. Ferro cyanide, $K_4Fe(CN)_6$ a cyano-group, —CN, enters in between the two—NO_2 groups to form 1,3-dinitro-2-cyano benzene.

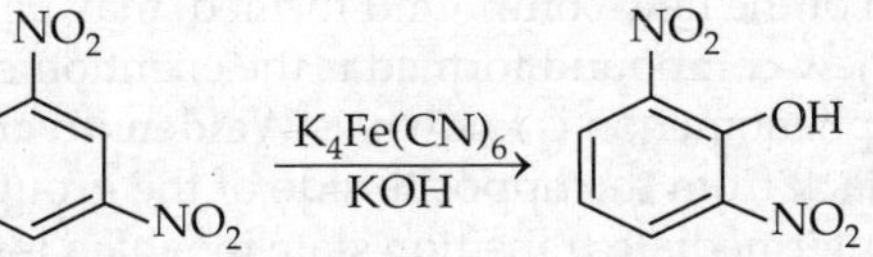

iii. On heating pyridine with solid KOH, at a temperature ~320°C, pyridine forms 2-pyridinol or 2-hydroxy pyridine. In the same way, on heating pyridine with sodamide $NaNH_2$, pyridine is coverted into 2-amnio pyridine.

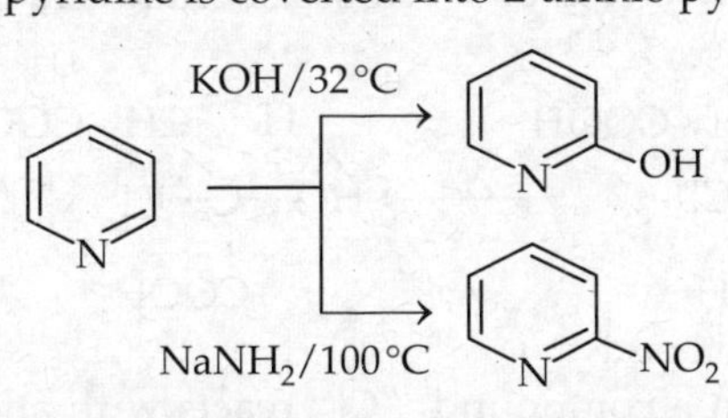

10. **Isomerism:** Isomers are the compounds, with **same molecular formula**, but **differ in their physical** and **chemical properties**, and the phenomenon is termed isomerism.
    Isomerism is divided in two main classes:

Isomerism
- structural isomerism
  1. chain
  2. position
  3. functional
  4. metamerism
  5. tautomerism
- stereoisomerism
  - conformational
  - configurational
    - geometrical
    - optical

Projection formulae: It is a methodology to represent a three dimensional structure on a two dimensional surface. It is used in *conformational isomersim.*
There are three ways to draw structures:
1. **Sawhorse projection formulae**
2. **Fischer projection**
3. **Newman projection formulae**

11. **Racemisation and Walden inversion**
    i. **Racemisation:** The process of converting an optically compound, (+) or (–) into a racemic mixture, is known as racemisation. A (+) or (–) form of almost all optically active, are racemised, under the influence of **heat, light or chemical reactions.** In this process, one half of the active compound changes to the isomer of the opposite rotation, resulting in the formation of a racemic mixture. It has been observed that in the process of racemisation of different optically active compound, undergoing the process of

racemisation, the same substance does not bring about change in all the compounds. For example, in case of optically active hydroxy acids, like lactic and tartaric acids, racemisation takes place in presence of a base. The racemisation in case of optically active alkyl halides, takes place either by S_N1 or by S_N2 mechanism. Similarly, asymmetrical alcohols are often racemised by strong acids.

Therefore, it may be suggested that an intermediate species, having sp^2 hybridised structure, i.e. a triangular plane species with all the groups attached to it.

ii. **Walden inversion:** When a group, attached to an asymmetric carbon atom, is replaced, the configuration of the new compound formed, may be opposite to that of the original compound. The new compound formed is the enantiomer (mirror image) of the original compound. This phenomenon is known as **Walden inversion.** The attacking group may approach and attack from the opposite side of the group to be replaced, leading to the formation of an intermediate transition state in which three groups and the asymmetric carbon may lie in the same plane followed by separation of the group to be replaced. **The change is like an umbrella, being blown inside out.**

For example, the conversion of (+) chloro succinic acid to (−) malic acid, is an example of Walden inversion.

$$
\begin{array}{ccc}
HO^- \longrightarrow \underset{\substack{| \\ COOH}}{\overset{\substack{H \quad CH_2-COOH \\ \diagdown \diagup}}{C}}-Cl & \longrightarrow HO\text{-----}\underset{\substack{| \\ COOH}}{\overset{\substack{H \quad CH_2-COOH \\ \delta^- \diagdown \diagup \delta^-}}{C}}\text{-----}Cl & \longrightarrow \underset{\substack{| \\ COOH}}{\overset{\substack{CH_2-COOH \\ |}}{HO-C-H}}
\end{array}
$$

(+) chloro-succinic acid   (−) malic acid

12. **Transition state:** When a compound, "C", reacts with another compound "A–B", it is expected to approach the compound from direction remote from "B" (proper alignment). While "C" draws closer to "A–B", and starts pulling away "A", until a stage is reached, when "C" and "B" are losely attached to "A", approximately equidistance from it. This is the **least stable arrangement** and called the transition state (TS) or an activated complex.

$$
\underset{\text{reactant}}{C - + A - B} \longrightarrow \underset{\text{transition state}}{C - A - B} \longrightarrow \underset{\text{product}}{C - A + B}
$$

The transition state is not a true molecule, the bonds are partial. At this stage, the system possesses maximum energy and it is most unstable. **Hence, the transition state of a system, could be described, as an extremely transitory species, having specific arrangement of atoms and groups, through which a reaction must pass to form the products. In other words, the transition state (activated molecule), has extremely short life time and at once decomposes to form the product. A transition state or an activated complex refers to an imaginary molecule that has not been isolated.**

A reaction, that proceeds through the formation of an intermediate (being most unstable) must pass through two energy barriers, one for the conversion of the reactant to the intermediate and the other to the conversion products.

13. i. **Electrophile: A reagent, that can accept a pair of electrons in a reaction, is called an electrophile.** The meaning of an electrophile, is that it is electron-loving. It attacks the region of high electron density (or a negative center), in the substrate molecule. These are the species which carry either a positive charge or are electron deficient, and are capable of accepting a pair of electrons. For example:

$H,^+ Br,^+ Cl,^+ I,^+ NO,^+ NO_2,^+ R_3C,^+ RCO,^+ SO_3H,^+ CH_3^+$

Some of electron deficient compounds, having an incomplete octet, also act as an electrophile, for example:

$BF_3,\ AlCl_3,\ SO_2,\ FeCl_3,\ SnCl_4,\ :CH_2$

ii. **Nucleophile:** A reagent that can donate a pair of electron, in a reaction, either to an electrophile or to an electron deficient compound. These species either have a negative charge or are electron rich, which can act as donors of electron pair. Nucleophiles, are nucleus loving. These attack the region of low electron density (or a positive center), in the substrate molecule. They have a negative charge on them (including carbanions) or are neutral molecules.

Charged species = Cl^-, Br^-, I^-, CN^-, OR^-, $RCOO^-$, CH_3^-, SH^-, $H-C \equiv C^-$

Some compound, having lone pairs of electrons but are neutral, also act a nucleophiles.

Neutral species:

NH_3, RNH_2, R_2NH, R_3N, $H_2O:$, ROH, ROR

14. **Octane number:** Octane number of a particular gasoline is the percentage of iso-octane in a mixture of iso-octane and heptane, which gives equivalent knock performance as that of gasoline.

   • **Gasoline is a complex mixture containing C4 to C10 of hydrocarbons.** The branched chain hydrocarbons have been found to give a smooth performance in a combustion engine.

   • **2,2,4-trimethyl pentane or iso-octane** has no tendency to knock and it has been assigned the octane number = 100, on the other hand, heptane, has a great tendency to knock, and it has en assigned the octane number = zero

$$CH_3-CH_2-CH_2-CH_2-CH_2-CH_2-CH_3$$

n-heptane

2,2,4-trimethylpentane or iso-octane

Cetane number: It is a number given to diesel fuel sample, which is a measure of its performance in a **compression engine. It is the volume of percentage hexadecane (cetane), in a blend with 1-methyl naphthalene, that gives the same performance in a standard compression engine as a fuel sample under the same operating conditions.**

Hexadecane has been given an arbitrary number 100 and 1-methyl naphthalene zero

1-methyl napthalene (cetane value = zero)

$$CH_3-(CH_2)_{14}-CH_3$$

hexadecane (cetane value = 100)

The diesel engine charges air into a chamber and compresses it during intake. Liquid fuel is then injected into the compressed chamber, where combustion occurs generating energy, required for the locomotive.

Fuel, with cetane number greater than 45 are required for good performance in a compression engine.

15. **Electromeric and mesomeric effect**

   i. **Electromeric effect:** The temporary shift of π-electrons to one of the atoms, joined by a multiple bond, at the time of attack by a reagent, is known as electromeric effect.

      • Owing to the shift of electrons, from one atom to the other atom in a multiple bonded compound, a polarity is produced in the compound, when it is attacked by a reagent.

When a double bond or a triple bond is exposed to an attack by an electrophile (E) the two π-electrons, which form the π–bond are completely transferred to the other atom. The electromeric effect may be represented as follows:

$$A = B^- \xrightarrow{E} A^- \longrightarrow B^+$$

In case of C=C double bond, in ethylene molecule, **the double bond is made up of one σ-bond and one π-bond.** Since the π-electrons are held loosely and are quite exposed under the influence of positively charged electrophile, the symmetry of the molecular orbital are completely disturbed. As the two π-electrons migrate to one of the carbon atom, making it negatively charged while the other carbon acquires a +ve charge.

- **The electromeric effect is a temporary effect as it takes place only at the time of attack by an electrophile (or in presence of a reagent).**

ii. **Mesomeric effect:** It may be defined as **a case of electron displacement in the molecule causing a permanent polarisation.** This electron displacement is relayed **through π-electrons of the multiple bond in the carbon chain of the molecule.** Unlike the inductive effect, which operates in the molecule having σ-bonds, the mesomeric effect operates in those system, **which have an extended chain with conjugate π-bonds (alternate π- and σ- bonds).** This is referred to as **conjugate effect.**

- This may also be defined as the polarity produced in the molecule, as result of an **interaction between two π-bonds or a π-bond and a lone pair of electrons,** when this happens, it is regarded as to be **a conjugate system. The effect is transmitted along the carbon chain of the molecule, in a similar way as is transmitted in inductive effect.**

- If the carbonyl group is conjugated with a carbon to carbon double bond (C=C), the polarisation will be transmitted further via π-electrons.

$$CH_3-CH=CH-C{\overset{O}{\underset{H}{\lessgtr}}} \longleftrightarrow CH_3-CH^+-CH=C{\overset{O^-}{\underset{H}{\lessgtr}}}$$

The π-electrons are delocalised as a consequence of mesomeric effect, giving a number of resonating structures of the molecule. This leads to a greater magnitude of the mesomeric effect, than the corresponding inductive effect, for a given difference of electronegativities of the bonded atoms.

The mesomeric effect may be +ve or a –ve. Those atoms or groups which lose their electrons towards carbon atoms, are said to have **+M effect,** such groups are **Cl, Br, I, NH$_2$, OH, OCH$_3$** and those atoms or groups which draw electrons from a carbon are said to have **–M effect,** e.g.

$$NO_2, CN, >C=O$$

16. **Orientation:** To determine the relative position of the groups, attached in the benzene ring, is termed as orientation. Thus the process of finding the relative positions of the various groups, attached in the benzene ring of an unknown derivative is termed **orientation.**

There are three methods to find out the relative positions in the benzene ring.

- **Korner's absolute method:** The **disubstituted derivatives** are converted into **trisubstituted derivatives** and the number of trisubstituted derivatives obtained, in each case determine the relative positions of the two groups in the disubstituted derivative.

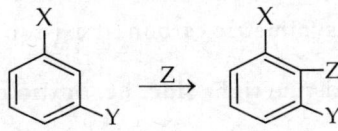

o-disubstituted derivatives

two differernt

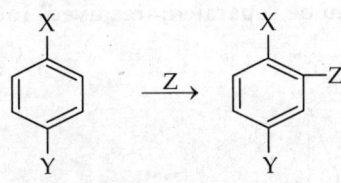

m-disubstituted derivatives

three differernt

p-disubstituted derivatives

only one trisubstituted derivative

a. **ortho disubstituted** derivative gives **two trisubstitute derivative**.
b. A **m-disubstituted** derivative forms **three trisubstituted derivatives**.
c. A **p-disubstituted** derivative forms **only one trisubstitute derivative**.

- **Relative method:** In this method, the compound with relatively unknown positions is converted into another compound with known positions (provided the groups substituted occupy the same position or exchange position with the incoming group).
- **Dipole moment measurement method:** In certain cases by determining the dipole moment of the compounds (with atoms or simple groups), may help in determining the relative positions.
  When two groups are same in the compound the dipoles is maximum, in the case of an o-disubstituted compound. In the case of a p-disubstituted derivative it will be minimum or it could be zero. In the case of a m-di substituted derivative it would be found to be in between the two.

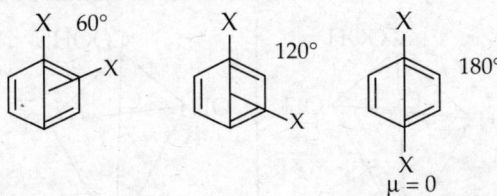

17. **Optical isomerism:** It is a type of stereoisomerism. The outstanding features of optical isomerism is that they could rotate the plane-polarised light **(PPL)**. This property is referred to as optical isomerism or optical activity.

A requirement of optical activity to occur, is that the compound must contain four different atoms or groups or both attached to the central carbon, or it should contain an asymmetric carbon atom. Such a molecule lacks a plane of symmetry, and such a molecule is termed a dissymmetric or a chiral molecule.

- The necessary condition for a molecule to exhibit optical isomerism, is that it should be dissymmetric, i.e. the molecule should not be superimposable on its mirror image. The non-superimposable mirror image of a chiral molecule, are called 'enantiomers'. They represent, two optical isomers; the one that rotate the PPL to the right is termed as 'dextro' or (+) isomer and the other isomer that rotate the PPL to the left is termed as 'laevo' or (−) isomer.

When an optical isomer **contains only one asymmetric carbon**, it exist in **three isomeric forms**.

i. Optical isomer that rotates the PPL towards the **right side, i.e. dextro or d-or (+) isomer.**

ii. Optical isomer that rotates the PPL to **the left side, i.e. laevo or l-or (−) isomer.**

iii. A **racemic form** that is optically inactive, being a mixture of an equal amount of d- and l- forms; it can be resolved, or it can be separated, (resolved) into the two forms.

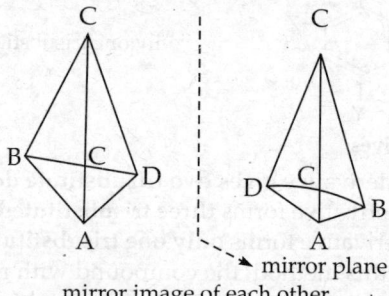

mirror image of each other

On the other hand when an optical isomer contains two asymmetric carbon atoms, it exists in four isomeric forms.

i. A **dextro rotatory** or (+) form

ii. A **laevo rotatory** or (−) form

iii. A **racemic mixture** containing an equal amount **of d- and l-forms; it is optically inactive is termed externally compensated molecule, however it can be resolved.**

iv. A **Meso form,** which is optically inactive, due to the presence of two asymmetric carbon atoms in the molecule having the opposite rotation power to each other. This is termed as an internally compensated molecule. **It cannot be resolved.**

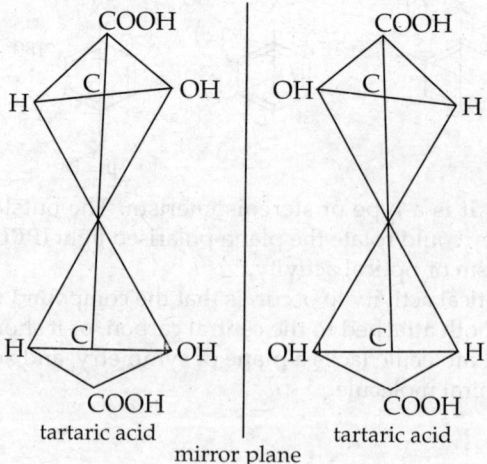

tartaric acid   tartaric acid

mirror plane

18. **Transesterificaion:** The formation of a new ester, by mutual exchange or transfer of the alcohol residue is termed as *trans*-esterification. The reaction is catalyzed by acids like H_2SO_4 or HCl or bases. The most effective catalyst is a trace of sodium salt of the reacting alcohol. A base catalyzed *trans*-esterification is preferred, to an acid catalyzed reaction, since in the former case, the reaction proceeds to completion.
An ester, RCOOR′ can react with an alcohol R″ OH, to give a new ester RCOOR″.
$$RCOOR' + R''OH \longrightarrow RCOOR'' + R'OH$$
a new ester

In this case, the alcohol residue (R′O) has been replaced by the alcohol residane, (R″O) of the reacting alcohol.
Trans-esterification is a reversible process and it is **pushed to completion by using a large excess of the reacting alcohol.** It is generally used for the preparation of **an ester of a higher alcohol** from that of a lower alcohol, since the lower alcohol produced in the reaction, can be removed by distillation.

19. **Resolution:** When an optically active compound is synthesized, it **produces an optically inactive mixture, consisting of both (+) and (–) isomers, in equal amount.** Such mixtures are called racemic mixture or a racemate.
The process of separation of racemic mixture into optically active components (+) and (–) is known as resolution.
There are various methods of separation:
   - **Chemical method:** As optical isomer of the same compound resemble with each other, therefore this method becomes possible. In this method, the racemic mixture is made to combine with another, optically active compound ("S") and the derivatives formed from both the isomers differ in their solubilities. A fractional crystallization helps to separate the two derivatives, one with (+) and the other with (–).
   Reaction mixture + optically active compound ("S"), two derivatives are formed
   (+) isomer "S" and
   (–) isomer "S"
   Upon hydrolysis, the derivatives decompose to generate the isomers
   (+) isomer "S" $\longrightarrow$ (+) isomer + "S"
   (–) isomer "S" $\longrightarrow$ (–) isomer + "S"
   By removing the substance "S", which has been used for the purpose, the two optically isomers, i.e. the two (+) and (–) can be isolated.
   - **Biological separation:** In this method, the racemic mixture is treated with certain moulds which decomposes one of the two isomers leaving behind the other isomer, which can easily be isolated.
   - **Physical separation:** It is applicable only in those cases where the crystals are sufficiently large enough to pick them and separate.

20. **Free radical reactions: A reaction in which the attacking reagent is a free radical termed free radical** reactions. Free radicals may be defined as a species with one or more unpaired electrons. These are formed when a covalent bond undergoes a homolytic fission.
   The various free radical reactions are given below:
   i. **Halogenation of alkanes:** Alkanes react with chlorine in the presence of sunlight or a diffused light, or at a temperature 300–400°C, to form chloro substituted products.

In the process of chlorination of alkanes, the following steps are involved.

a. **Chain initiation steps:** Chlorine undergoes a homolytic fission to form free chlorine.

$$C : Cl \xrightarrow{\text{UV light}/300-400\,°C} Cl\bullet + Cl\bullet$$

b. **Chain propagation step:** Chlorine free radical, then attack the alkane molecules to form chloro derivatives.

$$H{-}CH_3 + Cl^\bullet \longrightarrow CH_3^\bullet + HCl$$

CH_3 free radical then attacks another molecule of chlorine to form mono chloro-methane and liberates a Cl free radical.

$$CH_3^\bullet + Cl : Cl \longrightarrow CH_3Cl + Cl^\bullet$$

The steps (a) and (b) are repeated again and again, till no more energy is available for the reaction to proceed further.

c. **Chain termination step:** The chain reaction comes to a halt, and the free radicals combine with another free radical to form molecules.

$$Cl^\bullet + Cl^\bullet \longrightarrow Cl_2$$
$$CH_3^\bullet + CH_3^\bullet \longrightarrow C_2H_6$$

ii. **Wurtz reaction or Wurtz synthesis:** When an alkyl halide is treated with sodium metal in presence of diethyl ether as a solvent two molecules react with sodium metal to form a high alkane. The reaction proceeds via formation of an alkyl free radical.

a. Alkyl halide reacts with sodium metal to form an alkyl free radical

$$RX + Na \longrightarrow R^\bullet + NaX$$

b. The alkyl free radicals combine to form a higher alkane

$$R^\bullet + R^\bullet \longrightarrow R : R$$

iii. **Kolbe's electrolytic synthesis:** On electrolysis, a concentrated aqueous solution of a fatty acid of a higher alkane is obtained.

$$RCOONa \longrightarrow RCOO^- + Na^+$$

On switching the current these ions formed, discharge at their respective electrodes.

at anode	at cathode
$CH_3COONa \longrightarrow$	$CH_3COO^- + Na^+$
CH_3COO^-	
	$Na^+ + e^-$
$CH_3{-}COO + e$	
	$NaOH + H^+$
$CH_3 + CH_3^-$	

iv. **Addition of HBr to an unsymmetrical alkene, in the presence of a peroxide:** The reaction is also known as **Kharasch reaction, or a peroxide effect or anti-Markovnikov's reaction.**

The reaction involves an attack by a bromine free radical, the peroxide decomposes to form a free radical which reacts with HBr to form a $Br^\bullet$ free radical.

A peroxide decomposes to give an alkoxy free radical.

$$RO^\bullet + HBr \longrightarrow ROH + Br^\bullet$$
$$CH_3{-}CH{=}CH_2 + Br^\bullet \longrightarrow CH_3{-}CH^+ CH_2Br + CH_3{-}CHBr{-}CH_2^+$$

2° carbocation   1° carbocation

As a 2°carbocation is more stable as compared to a 1°carbocation, the final product is formed from the 2°carbocation.

$$CH_3CH^+CH_2Br + H + Br \longrightarrow CH_3CH_2CH_2Br + Br^\bullet$$

HCl and HI do not form anti-Markovnikov's product, when they are treated with an unsymmetrical alkene in the presence of a peroxide.

v. **Sandmeyer's reaction:** Aryl diazonium chloride react with HX, in the presence of cuprous salt of the same acid, the aryl diazonium salt is converted to form corresponding aryl halide.

$$C_6H_5N_2Cl \xrightarrow{\quad CuX + HX \quad} C_6H_5X + N_2$$

A free radical mechanism is involved in Sandmeyer's reaction.

$C_6H_5N_2Cl \longrightarrow C_6H_5N_2^+ + Cl^-$

$C_6H_5N_2^+ + Cu^+ \longrightarrow C_6H_5^\bullet + N_2 + Cu_2^+$

$C_6H_5^\bullet + Cl^- \longrightarrow C_6H_5Cl + e$

$Cu_2^+ + e \longrightarrow Cu^+$

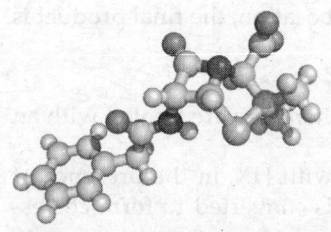

Paper 3

REASONING TYPE QUESTIONS (RTQs)

1. i. Why do HCl and HI do not form anti-Markovnikov's product, when treated with an unsymmetrical alkene? Explain.
   ii. Discuss anti-Markovnikov's rule, as it is observed in the case of unsymmetrical alkenes.
2. Are their compounds, which do not contain an asymmetrical carbon in the molecule, yet show optical activity? Discuss.
3. What are the species formed when a C to C single bond undergoes a homolytic fission? Discuss their properties.
4. When a C to C single bond undergoes a heterolytic fission, what are the species formed? Discuss their properties.
5. Explain the term 'inductive effect'.
6. Explain the terms 'bond' energy and 'bond dissociation energy'.
7. Explain the term 'isomerisation' with suitable examples.
8. Explain the term 'hybridisation' as it happens in the case of carbon, nitrogen and oxygen.
9. What are the types of bonds involved in organic compound? Discuss.
10. Arrange in decreasing order, the bond lengths observed in the case of an alkane, an olefinic bond and an acetylinic bond. Also give reasons.
11. Arrange an alkane, alkene and alkyne in decreasing order of their acidic character of the hydrogen atoms.
12. Explain the reasons behind this order, as observed in Q No 11.
13. Arrange the various hydrogenated compounds in the decreasing order of their acidic character, where the hydrogen is attached to oxygen, nitrogen and carbon atoms.
14. In various reactions to which the organic compounds are subjected to it is observed that a σ-bond is not broken. Explain.
15. A suggestion made by vant Hoff and Le Bel in the year 1884, led to an important development in the organic chemistry. What was this suggestion and how it changed some basic concepts in organic chemistry.
16. What was the first organic compound prepared in the laboratory, and who prepared it? Name the chemist and also the starting material used?
17. What are the shapes of the following compounds? Arrange them in the decreasing order of their bond angles. The compounds are methane, ammonia, hydrogen sulphide and water.

18. When methane is treated with chlorine and bromine, forms chloro or bromo substituted derivatives of methane, but in the case of methane with iodine, some oxidising agent is used. Explain why?
19. On alkylation of benzene with n-propyl chloride, by Friedel–Craft reaction, the compound obtained is n-isopropyl benzene. Suggest how n-propyl benzene may be prepared, starting from benzene.
20. Fluorine is more reactive than other halogens, yet the fluoro-compounds of alkanes, alkenes and alkynes are not reactive. Explain.

ANSWERS

1. **Peroxide effect or Kharasch reaction or anti-Markovnikov's rule:** When halogen acids react with an unsymmetrical alkene in presence of a peroxide, the addition of HCl and HI is not affected and the product is the one as is formed in the absence of a peroxide. On the other hand, the addition of HBr to an unsymmetrical atom is affected and the product formed is a compound opposite to that is obtained in the absence of a peroxide.

$$CH_3—CH=CH_2 + HBr \xrightarrow{\text{peroxide}} CH_3—CH_2—CH_2Br$$
$$\text{propene} \qquad\qquad\qquad\qquad\qquad \text{n-propyl bromide}$$

The energy required for the homolytic fission is made available by the decomposition of the peroxide.

$$R—O:O—R \longrightarrow R—O\cdot + R—O\cdot \text{ energy}$$
$$\text{organic peroxide} \qquad \text{alkoxy free radicals}$$

i. The amount of energy required for a homolytic fission of HCl = 10^3 kcal/mol.

$$H:Cl \xrightarrow{10^3 \text{ kcal/mol}} H\cdot + Cl\cdot$$

As this much energy is not available in the system, therefore, HCl does not undergo a homolytic fission. As a result, HCl does not form an anti-Markovnikov's product, when it reacts with an unsymmetrical alkene. The reaction proceeds by an addition of a H^+ ion.

HCl ionizes to form a H^+ and a Cl^- ion.

The compound formed is isopropyl chloride, a Markovnikov's product.

$$CH_3—CH=CH_2 \xrightarrow{H^+} CH_3—\overset{+}{C}H—CH_3 \xrightarrow{Cl^-} CH_3CH—CH_3$$
$$\text{propene} \qquad\qquad \text{a 2° carbocation} \qquad\qquad \text{isopropyl chloride}$$

ii. The energy required for a homolytic fission of HBr = 87 kcal/mol.
Since this much of energy is available in the system, HBr undergoes homolytic fission, and produces a H atom and a Br atom.

$$H:Br \xrightarrow{87 \text{ kcal/mol}} H\cdot + Br\cdot$$

The Br· atom then attacks either of the doubly bonded carbon atoms either to give at 2° free radicals (a) or 1° free radical (b).

a. $CH_3—CH=CH_2 + Br \longrightarrow CH_3—\overset{\bullet}{C}H—CH_2Br$ **(2° degree free radical)**

b. $CH_3—CH=CH_2 + Br \longrightarrow CH_3—CHBr—\overset{\bullet}{C}H_2$ **(1° degree free radical)**

As a 2° free radical is more stable as compared to a 1° free radical, both may be formed but the reaction proceeds further with the more stable 2° free radical.

$$CH_3—\overset{\bullet}{C}H—CH_2Br + HBr \longrightarrow CH_3—CH_2—CH_2Br + Br\cdot$$
$$\text{n-bromopropane}$$

c. The energy required for the homolytic fission of HI is 71 kcal/mol.

$$H \overset{\bullet}{\underset{\bullet}{\,}} I \xrightarrow{\text{71 kcal/mol}} H\bullet + I\bullet$$

Since this much energy is available in the system, HI beaks up to form a hydrogen and an iodine atom. As soon as iodine atom is liberated, it combines immediately with another atom of iodine, to form molecular iodine.

$$I\bullet + I\bullet \xrightarrow{\hspace{3cm}} I_2$$

iodide atoms   molecular iodine

As iodine atom is not available, for attacking the unsymmetrical alkene, therefore HI also does not form an anti-Markovnikov's product.

Like HCl, HI reacts with an unsymmetrical alkene and give a Markovnikov's product.

$$CH_3\!-\!CH=CH_2 \xrightarrow{\text{HI}} CH_3\!-\!CHI\!-\!CH_3$$

propene   isopropyliodide

2. **Optical activity without asymmetric carbon is exhibited by:**
   i. Allene derivatives or propadiene derivatives
   ii. Biphenyl derivatives
   iii. Optically active cycloalkene
   i. **Allene derivatives:** Some allene derivatives, for example, 1, 3-diphenyl propadiene, exhibits optical isomerism. Allene is a highly unsaturated compound, which contains the central carbon atoms as 'sp^2' hybridised. The central carbon atom forms two sp-sp^2 σ-bonds.

$\longmapsto$ Central carbon atom sp hybridised

$CH_2=C=CH_2$

$\downarrow$ $\downarrow$ allene contains two π-bonds, and
two σ-bonds

sp^2 hybridised carbon atoms

The central carbon atom also has two 'p' orbitals, which are mutually perpendicular. These form σ- bonds with 'p' orbitals, on the other terminal carbon atom. As a result, the substituents at the end of the molecule are in a plane, which is **perpendicular to that of the substituents at the other end of the carbon atom**, so that compounds exist in two isomeric forms, which are non-superimposable mirror images and are optically active.

mirror plane

C_6H_5 H   H C_6H_5

C=C=C   C=C=C

H C_6H_5 C_6H_5 H

1, 3-diphenylpropadiene

   ii. **Biphenyl derivatives:** In biphenyl, the two phenyl rings are joined to each other by a single bond at positions 1 and 1' positions. The two phenyl rings are in the same plane or the two rings are coplanar. When positions 2 and 2' (as well as 6 and 6') are occupied by big groups like —SO_3H or COOH, the free rotation about the single is restricted. As a result, the two rings are forced to incline with each other at right angle.

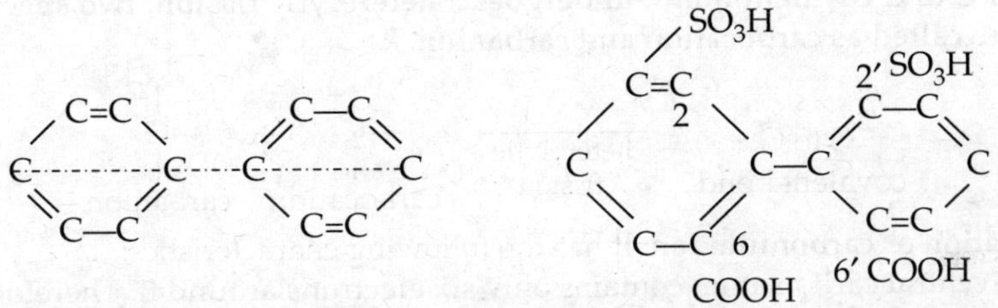

The two rings are not in the same plane. A mirror image of such a molecule will not be superimposable on each other. Therefore, the molecule becomes "chiral".

iii. **Optically active cycloalkene:** A *trans*-cyclo octene has been found to exist in two optically active forms, (though the *cis*-form does not show optical activity). On account of the olefinic bond in the *trans*-cyclo octane, it assumes a nonplanar configuration and as a result it exists in two optically active forms.

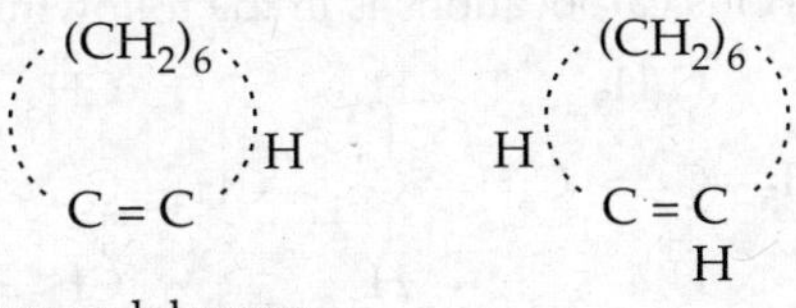

Trans-cyclohexene

3. When a C to C single bond undergoes a homolytic fission, two species are formed, both contain an even number of electrons, seven electrons each.

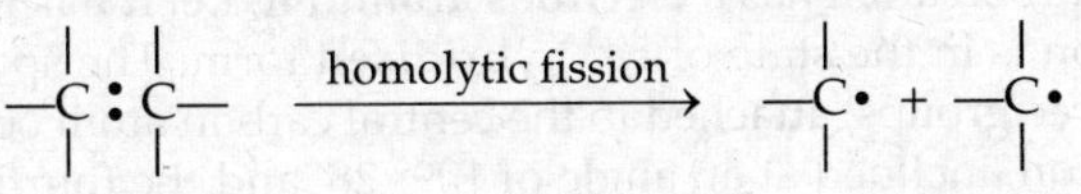

$$ -\overset{|}{\underset{|}{C}} : \overset{|}{\underset{|}{C}}- \xrightarrow{\text{homolytic fission}} -\overset{|}{\underset{|}{C}}\bullet \ + \ -\overset{|}{\underset{|}{C}}\bullet $$

free radicals

The following are the characteristics of the free radicals:

i. **Each radical are in the state of sp^2 hybridised form, i.e. there are 7 electrons around** each species and the three groups attached to the central carbon atom are in the same planes, inclined at an angle of 120°, while the odd electron lies perpendicular to the plane in which the three groups lie.

ii. Each radical is associated with a large amount of energy, therefore they are **highly reactive and are highly unstable**.

iii. When these free radicals come in contact with any substance, these free radicals abstract an atom from these substances, convert itself into a compound and set the reaction in motion.

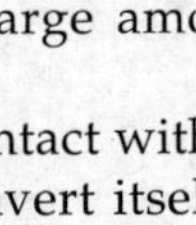

$$ R'-\overset{R''}{\underset{R'''}{C}}\bullet $$

The stabilities of free radicals is in the following order:

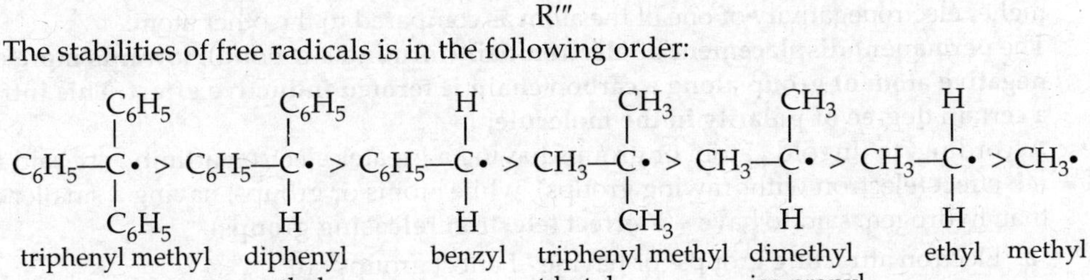

$$ C_6H_5-\overset{C_6H_5}{\underset{C_6H_5}{C}}\bullet > C_6H_5-\overset{C_6H_5}{\underset{H}{C}}\bullet > C_6H_5-\overset{H}{\underset{H}{C}}\bullet > CH_3-\overset{CH_3}{\underset{CH_3}{C}}\bullet > CH_3-\overset{CH_3}{\underset{H}{C}}\bullet > CH_3-\overset{H}{\underset{H}{C}}\bullet > CH_3\bullet $$

triphenyl methyl   diphenyl methyl   benzyl   triphenyl methyl t-butyl   dimethyl iso propyl   ethyl methyl

4. **When a C to C covalent bond, undergoes a heterolytic fission, two species are formed; these are called as carbocation and carbanion.**

$$\text{C to C} \atop \text{covalent bond} \quad \xrightarrow[\text{fission}]{\text{heterolytic}} \quad \overset{|}{\underset{|}{-C}}^{+} \quad + \quad \overset{|}{\underset{|}{-C}}^{-}$$

$$\text{carbocation} \quad \text{carbanion}$$

i. **Carbocation or carbonium ion:** It has the following characteristics:

  a. The central carbon atom contains only **six electrons around** it. Therefore, it is **electron deficient**.

  b. **The carbocation is in the state of sp^2 hybridised form.** The three groups attached to the central carbon atom are inclined, to each other at an angle of $120°$.

  c. Being electron deficient, it readily combines with any substance that can provide it a pair of electrons.

  d. Therefore it acts as an **electrophile.**

  e. The stability of various carbocations is in the following order.

$$C_6H_5-\overset{C_6H_5}{\underset{C_6H_5}{\overset{|}{\underset{|}{C}}}}^{+} > C_6H_5-\overset{C_6H_5}{\underset{H}{\overset{|}{\underset{|}{C}}}}^{+} > C_6H_5-\overset{H}{\underset{H}{\overset{|}{\underset{|}{C}}}}^{+} > CH_3-\overset{CH_3}{\underset{CH_3}{\overset{|}{\underset{|}{C}}}}^{+} > CH_3-\overset{CH_3}{\underset{H}{\overset{|}{\underset{|}{C}}}}^{+} > CH_3-\overset{H}{\underset{H}{\overset{|}{\underset{|}{C}}}}^{+} > CH_3^{+}$$

triphenyl methyl diphenyl methyl benzyl ter. butyl isopropyl ethyl methyl

ii. **Carbanion:** It has the following characteristics:

  a. The central carbon atom has **8 electrons around it, i.e. it has its octet complete.**

  b. The carbanion is in the state of sp^3 **hybridised** form. The species is in the tetrahedral form. The three groups, attached to the central carbon atom occupy the three corners of the tetrahedron, inclined at an angle of $109°.28'$ and the fourth bond is being occupied the by the lone pair of electron.

  c. Therefore, it acts as a **nucleophile.** It readily combines with any species, which is electron deficient.

  d. The stabilities of the carbanion are as follows.

$$C_6H_5-\overset{C_6H_5}{\underset{C_6H_5}{\overset{|}{\underset{|}{C}}}}{\colon} > C_6H_5-\overset{C_6H_5}{\underset{H}{\overset{|}{\underset{|}{C}}}}{\colon} > C_6H_5-\overset{H}{\underset{H}{\overset{|}{\underset{|}{C}}}}{\colon} > CH_3-\overset{H}{\underset{H}{\overset{|}{\underset{|}{C}}}}{\colon} > CH_3-\overset{H}{\underset{H}{\overset{|}{\underset{|}{C}}}}{\colon} > CH_3-\overset{CH_3}{\underset{H}{\overset{|}{\underset{|}{C}}}}{\colon} > CH_3-\overset{CH_3}{\underset{CH_3}{\overset{|}{\underset{|}{C}}}}{\colon}$$

triphenyl methyl diphenyl methyl benzyl methyl ethyl isopropyl t-butyl

5. **Inductive effect: It involves σ-electrons.** The electrons, which forms a covalent bond, are seldom shared equally in between two atoms, because of the **difference in their electron negativity.** The inductive effect refers to the polarity produced in a molecule as a result of higher electronegativity of one of the atom as compared to the other atom.

The **permanent displacement of electron pair (forming a σ-bond), towards more electro-negative atom or group along a carbon chain is termed inductive effect. This introduces a certain degree of polarity in the molecule.**

According to 'Ingold', atom or groups having a greater electron affinity, are said to have $-'I'$ effect (electron withdrawing groups), while atoms or groups, having a smaller affinity than hydrogen, said to have $+'I'$ **effect (electron releasing groups).**

  a. Electron attracting groups- or having –I effect groups are:

   $-NO_2$, $-F$, $-Cl$, $-Br$, $-I$, $-OH$,$- C_6H_5$

b. Electron releasing groups or groups having a +I effect are:

$(CH_3)_3$—C—, $(CH_3)_2$CH—, CH_3CH_2— CH_3—

Consider the carbon–chlorine bond. The chlorine atom is more electronegative with respect to the carbon atom.

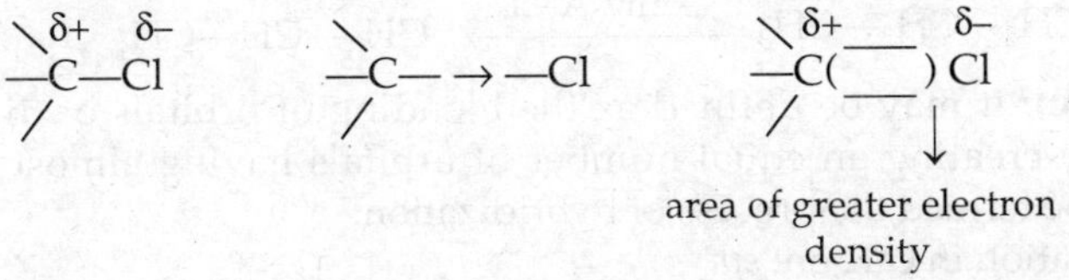

$$\overset{\delta+}{\underset{/}{\overset{\backslash}{-}}}\overset{\delta-}{C-Cl} \qquad \overset{\backslash}{\underset{/}{-}}C- \rightarrow -Cl \qquad \overset{\delta+}{\underset{/}{\overset{\backslash}{-}}}\overset{\delta-}{C(\underline{\quad})\,Cl}$$

area of greater electron
density

An inductive effect is not confined to the polarisation of only one bond, it is transmitted along the chain of carbon atoms although it tends to be insignificant, beyond the second carbon atom of the chain.

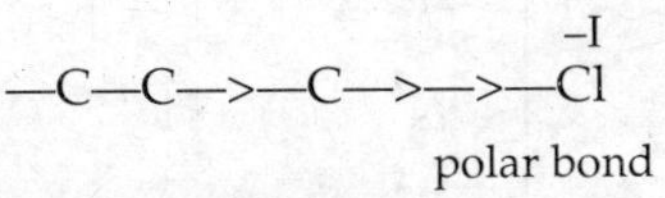

$$\overset{\qquad\qquad\qquad -I}{—C—C—>—C—>—>—Cl}$$

polar bond

6. **Bond dissociation energy and bond energy:**

   i. **Bond energy:** It is an average value of the dissociation energies of a given bond in a series of different dissociating species.

   For example:

   In the case of CH_4, the first dissociation energy in a C—H dissociation is 102 kcal, whereas the second dissociation value is different, i.e. only 87 kcal. In the same way, it is different in the case of third and fourth dissociation the values are given below.

   CH_3—H $\longrightarrow$ $CH_3H \bullet + H \bullet$ $\Delta H = 102$ kcal/mole

   CH_2—H $\longrightarrow$ $CH_2 \bullet$ $+ H \bullet$ $\Delta H = 87$ kcal/mole

   CH—H $\longrightarrow$ CH$\bullet$ $+ H \bullet$ $\Delta H = 123$ kcal/mole

   C—H $\longrightarrow$ C(g)$\bullet$ $+ H \bullet$ $\Delta H = 81$ kcal/mole

   ii. **Bond dissociation energy:** It is defined as the amount of energy required to break a bond in a molecule.

   - The value depends upon the type of bond as well as the structural environment in which the bond is situated in the compound.
   - It relates to the endothermic homolysis of a covalent bond represented by ΔH. It is a positive quantity.
   - The reverse of this reaction is termed bond formation energy, represented as ΔH.

   Some values are given below.

   $\overset{|\quad|}{\underset{|\quad|}{—C—C—}}$ single bond $\longrightarrow$ 83 kcal/mole

   $>C=C<$ double bond $\longrightarrow$ 146 kcal/mole

   $C\equiv C$ triple bond $\longrightarrow$ 200 kcal/mole

   The bond formation is accompanied by a decrease in the energy of the system. It means that this much energy is lost into the separation of the bonded molecule into atoms. It means that this much energy must be supplied to separate the bonded atoms in a (6.625×10^{23}) molecule.

7. **Isomerisation:** The conversion of an organic compound into its isomer is termed **isomerisation.**

On treating a normal alkane with anhydrous $AlCl_3$ in presence of HCl at a temperature of 25°C, it is converted into its isomer. For example. On heating n-butane with anhydrous $AlCl_3$ at a temperature of 25°C, it is converted into isobutane.

$$CH_3—CH_2—CH_2—CH_2—CH_3 \xrightarrow{\text{Anhy. } AlCl_3} CH_3—\overset{\overset{\textstyle CH_3}{|}}{CH}—CH_3$$

8. **Hybridisation:** It may be defined as the blending of orbitals of different energies and subsequently creating an equal number of orbitals having almost an equal amount of energy may be termed the process of hybridization.

   i. Hybridisation in **carbon:** sp^3

   Carbon has 4 electrons in the valence shell.

These are:

$$
\begin{array}{lll}
\uparrow\,| & 2p^2 & \uparrow \quad \uparrow \\
E\,| & 2s^2 & \uparrow\downarrow \\
| & 1s^2 & \uparrow\downarrow
\end{array}
$$

Energy diagram of carbon

The $2s^2$ orbital is completely filled therefore can accommodate only two electrons into the half filled valency shell, therefore in such a situation, carbon will show a valency of only two. However, the carbon compounds show a valency of 4 in all organic compounds and all the compounds are quite stable. It is possible only when the $2s^2$ orbitals take part in the process of raising the valency to 4. By taking some energy from the surroundings, one of the $2s^2$ orbitals jumps to the vacant p-orbitals, making it as in the exited state:

$$
\begin{array}{lll}
\uparrow\,| & 2p^3 & \uparrow \quad \uparrow \quad \uparrow \\
E\,| & 2s^1 & \uparrow \qquad \text{in excited state} \\
| & 1s^2 & \uparrow\downarrow
\end{array}
$$

In this manner, carbon is able to create 4 half filled orbitals, which can show a valency of 4. However, this contains two types of orbitals, one $2s^1$ or a spherical shaped orbital and other $2p^3$ or the three perpendicular orbitals, inclined with each other at an angle of 90°. Such a model of carbon in the exited state will have 4 unpaired electrons, showing a valency of 4. The four atoms or groups joined to the four valencies of carbon will have two types of σ-bonds, one will be a s–s bond, and the other will be 3 sp bonds. When any of three hydrogen of the sp bonds are substituted by any atom or a group, the derivatives will be different as compared to the derivative formed when the s–s bond is substituted by the same group. It has been found that whichever H is substituted by a particular group, the compound is the same. It means that all hydrogens are joined with the carbon in the same manner, whichever hydrogen is substituted, the compound obtained is the same. It is possible when all the four half filled orbital undergo the process of hybridization, creating four identical orbitals. It can be concluded, therefore, that all 4 orbitals undergo the process of hybridisation, involving $2s^1$ and $2p^3$ orbitals and forming a tetrahedral structure.

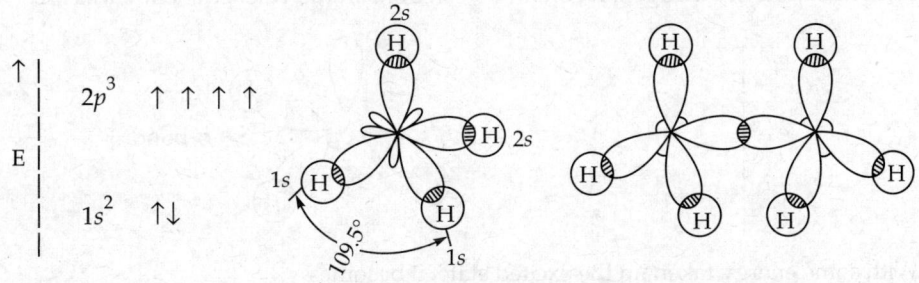

↑ | 2p^3 ↑ ↑ ↑ ↑

E |

  | 1s^2 ↑↓

Methane   Ethane

In the process of hybridisation, gives out more energy than it had taken to promote one $2s^1$ orbital to the vacant $2p^3$ orbital. Thus, carbon undergoes sp^3 hybridisation, all the 4 half filed orbitals, inclined with each other at an angle of 109°.28′, forming a regular tetrahedral structure.

- sp^2 **hybridisation**

  Carbon also undergoes sp^2 hybridisation, when out of the four sp^3 orbitals, only three orbitals, ($2s^1$ and $2p^2$ half filled) pool up their energies and creat three orbitals, inclined to each other with angle of 120°, leaving one of the '$2p$' orbitals at an angle of 90° to the three sp^2 hybridised orbitals.

  Such a carbon, when forms a bond with another carbon atom, besides forming a σ-bond, the two unhybridised 'p' orbitals, one from each sp^2 hybridised carbons, overlap each other in the lateral fashion and forms a π-bond.

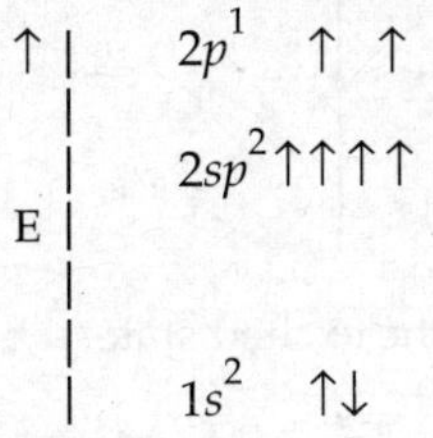

  ↑ | $2p^1$ ↑ ↑

    | $2sp^2$ ↑↑↑↑

  E |

    | 1s^2 ↑↓

- *sp* **hybridisation**

  Carbon undergoes *sp* hybridisation also when out of four sp^3 orbitals, only two orbitals, $2s^1$ and $2s^1$ half filled orbitals, pool up their energies and creat two orbitals, inclined at an angle of 180°, leaving two of $2p$ orbitals at an angle of 90° to the two hybridisation orbit.

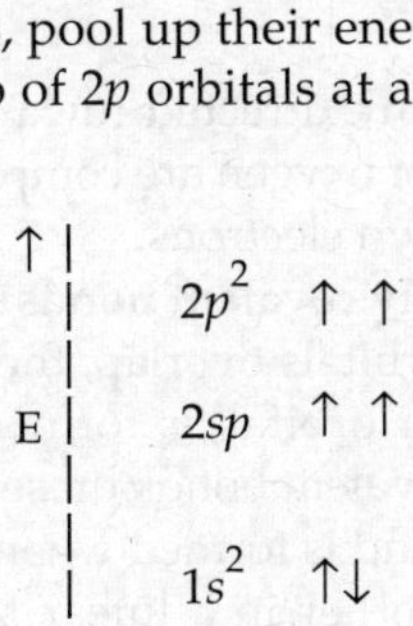

  ↑ | $2p^2$ ↑ ↑

    | $2sp$ ↑ ↑

  E |

    | 1s^2 ↑↓

Such a carbon, when forms a bond with another *sp* hybridised carbon, besides forming a bonding, the other two unhybridised orbitals overlap each other in a lateral fashion, forming two π-bonds.

ii. **Hybridisation in Nitrogen:** It contains 5 electrons in the valence shell. These are:

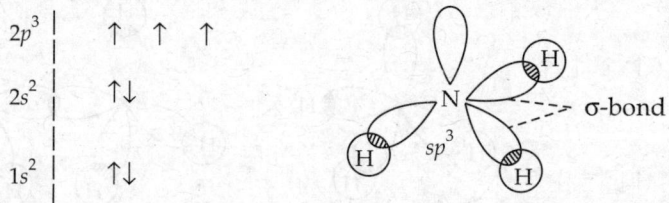

$2p^3$ ↑ ↑ ↑

$2s^2$ ↑↓

$1s^2$ ↑↓

σ-bond

With some energy taken, in the excited state, it becomes

$2sp^3$ ↑↓ ↑ ↑ ↑

$1s^2$ ↓↑

The sp^3 orbitals of nitrogen are directed towards the four corners of a regular tetrahedron. Since nitrogen has one more electron than carbon, as a result one of the orbital of nitrogen is compltely filled and cannot take part in bond formation, this is why nitrogen shows a valency of three as in the molecule of NH_3. All the three bonds formed are σ-bonds. The bonds in the molecule are directed at an angle of 107°.

iii. **Hybridisation in Oxygen:** It has 8 electrons in the valence shell.

a. These are:

$2p^4$ ↓↑ ↑ ↑

$2s^2$ ↑↓

$1s^2$ ↓↑

b. With some energy taken in the excited state, it becomes:

$2sp^3$ ↓↑ ↑↓ ↑ ↑

$1s^2$ ↑↓

The sp^3 orbitals of oxygen are directed towards the four corners of a regular tetrahedron. As the two orbitals of oxygen are completely filled, it is the other two half filled orbitals that accommodate two electrons.

9. **Organic compounds contain only covalent bonds.** These are formed by the overlapping of atomic orbitals. When atomic orbitals overlap, they share the same region in space and a new orbital, called a 'molecular orbital' is formed. The molecular orbital is a 'covalent bond'. There are two types of covalent bonds present in organic compounds.

i. **Sigma (σ) bond:** A sigma bond is formed when two atomic orbitals of a sp^3 hybridized carbon atom overlap each other in a linear fashion or an end to end overlapping, giving out maximum energy resulting in the formation of a strong single bond. A sp^3 hybridized carbon forms a σ-bond, when it combines with another sp^3 hybridized carbon atom.

When organic compounds are subjected to various reactions, generally a σ-bond, (in carbon-to carbon) is not broken, as the amount of energy required for breaking the single σ bond is not available in the system. On the other hand, a σ-bond formed in between an sp^3 orbital of carbon and a s-orbital of oxygen, nitrogen or halogen is broken, when organic compounds are subjected to various reactions.

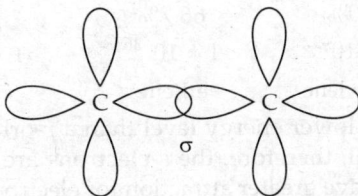

ii. **pi (π) Bond:** A sp^2 hybridized carbon atom contains three orbitals inclined to each other at an angle of 120° and are in the same plane. The remaining 4th orbital remains undisturbed, and it remains in the perpendicular position to the three hybridized orbitals. Such a sp^2 hybridized carbon atom, forms two bonds when it combines with other sp^2 hybridized carbon atom. One σ-bond is formed, by the overlap of one sp^2 hybridized orbital of a carbon atom with the sp^2 hybridized orbital of the other carbon atom, through a linear overlapping (resulting in the formation of a σ-bond). The unhybridized 4th orbital of each sp^2 hybridized carbon atom, overlap each other in a lateral fashion, or in a side by side fashion, resulting in the formation of a weaker bond, known as π-bond. Since the two unhybridized orbitals, come close enough, to overlap each other, the bond distance between the two carbons nuclei, joined by a π-bond (a C to C double bond) is much smaller than the one joined by a σ-bond alone (a C to C single bond).

The bond distance for a C to C single bond (σ-bond) = 1.54 Å units
The bond distance for a C to C double bond (π-bond) = 1.34 Å units

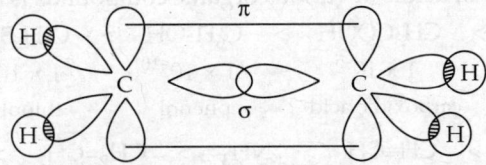

In the same manner a sp hybridized carbon atom, that contains two unhybridized orbitals forms two π-bonds when it combines with another sp hybridized carbon atom, besides forming a σ-bond, resulting in the formation of a (C to C) triple bond. The bond distance for a C to C triple bond is much smaller than that of either a C to C single bond and that of a C to C double bond. A C to C triple bond, consists of one σ bond and two π bonds.

The bond distance for a C to C triple bond = 1.2 Å units.

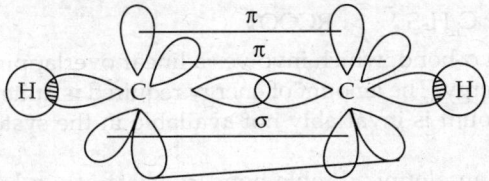

10. The decreasing order of the acidic character of the 'hydrogen attached to a sp^3, sp^2 and sp hybridized carbon are in the following order.

$$CH \equiv CH \quad > \quad CH_2{=}CH_2 \quad > \quad CH_3{-}CH_3$$

	sp	sp^2	sp^3
s-character	50%	33.3%	25%
p-character	50%	66.7%	75%
K_a values	1×10^{-22}	1×10^{-36}	1×10^{-42}
	acetylene	ethene	ethane

Reason: As s-orbitals lie at a lower energy level than a p-orbital, the s-electrons are close to nucleus, as compared to p-orbital, therefore, the s-electrons are held more tightly. As a result, sp-hybridized carbon of alkyne, have greater attraction for electrons so the electron pair forming a bond between carbon and hydrogen is displaced towards carbon, making it more acidic in nature as compared to alkenes and alkanes.

The corresponding Lewis bases have the following decreasing order of basic character:

$$CH_3{-}CH_2\overset{..}{}{}^- \; > \; CH_2{=}CH\overset{..}{}{}^- \; > \; CH{\equiv}CH\overset{..}{}{}^-$$

11. The decreasing order of the bond length in the case of a σ-bond, an ethylinic bond and an acetylinic bond is as follows:

$$CH_3{-}CH_3 \quad > \quad CH_2{=}CH_2 \quad > \quad CH{\equiv}CH$$

bond distance	1.54 Å	1.34 Å	1.2 Å

12. In the formation of a π-bond, the p-electrons have to come closer, so that a side by side overlapping may take place. This brings the nuclei of the two participating carbons close enough, hence the bond length of an alkene (a C to C double bond) is smaller than that of an alkane, (a C to C single bond).

    Similarly, in the formation of a triple bond in between two carbon atoms, that consists of one σ-bond and two π bonds, the two carbon atoms move still closer as a result the bond length in case of a triple bond is smallest in the series of alkane, alkene and alkyne.

13. The decreasing order of acidic of various organic compounds is as follows character:

$$HCl \; > \; CH_3COOH \; > \; C_6H_5OH \; > \; C_6H_5SH \; > \; H_2O$$

K_a values	1×10^8	1×10^{-5}	1×10^{-10}	1×10^{-11}	1×10^{-14}
		carboxylic acid	phenol	thiophenol	water

$$ROH \; > \; CH{\equiv}CH \; > \; NH_3 \; > \; CH_2{=}CH_2 \; > \; CH_3{-}CH_3$$

K_a values	1×10^{-18}	1×10^{-22}	1×10^{-35}	1×10^{-36}	1×10^{-40}
	alcohol	alkyne	ammonia	alkene	alkane

   • The decreasing order of basic character of their corresponding Lewis bases is as follows.

$$CH_3{-}CH_2\overset{..}{}{}^- > CH_2{=}CH\overset{..}{}{}^- > NH_2\overset{..}{}{}^- > CH{\equiv}C\overset{..}{}{}^- > R{-}O\overset{..}{}{}^- >$$

$$C_6H_5S\overset{..}{}{}^- > C_6H_5S\overset{..}{}{}^- > RCOO\overset{..}{}{}^-$$

14. In the formation of a σ-bond, which involves a linear overlapping, resulting in giving out a large amount of energy. The amount of energy required is quite large to break the σ-bond. Since this much amount is invariably not available in the system, therefore a σ-bond is generally not broken.

    On the other hand, an olefinic double bond consists of a σ-bond and a π bond. In the formation of a π-bond, on account of sideways overlapping, the amount of energy given out is not that large as in the case of a σ-bond. Therefore, a π-bond requires much less energy to break it. This is why olefinic and acetylinic bonds, are easily broken whenever these compounds are subjected to various reactions. Generally the olefinic and acetylinic compounds, undergo addition reactions, and are easily converted into derivatives of alkanes.

$$H_2C \overset{\pi}{\underset{\sigma}{=}} CH_2 \qquad\qquad H-C \overset{\pi}{\underset{\sigma}{\equiv}} C-H$$

Bond angle between C—H bond = 120°;   Bond angle, between C—H bond = 180°;
bond distance —C=C –1.34 Å   bond distance C≡C— 1.20 Å

15. V' Hoff and Le Bel suggested in the year 1884, that the four valencies of carbon may not be in the same plane, but these may be directed in the space. This gave the concept of a tetrahedral nature of the carbon atom. They made this suggestion in order to explain the phenomenon of optical activity, observed in the organic compounds.

16. Friedrich Wohler, a German chemist prepared an organic compound, (1828) in the laboratory, namely, urea, (NH_2—CO—NH_2) for the first time by heating a mixture of lead cyanate ($Pb (CNO)_2$) with ammonium hydroxide.

$$Pb (CNO)_2 + 2 NH_4OH \longrightarrow Pb (OH)_2 + 2 NH_4CNO$$

$$\underset{\text{ammonium cyanate}}{NH_4CNO} \overset{\Delta}{\longrightarrow} \underset{\text{urea}}{NH_2CONH_2}$$

17. The decreasing order with respect to bond angles in the compounds namely methane, ammonia, water and sulphurated hydrogen is as follows:
Oxygen in H_2O; carbon in methane; nitrogen in ammonia and sulphur in hydrogen sulphide, are in the tetrahedral form. In the molecule of water, oxygen atom has two lone pairs of electrons; in the molecule of ammonia, nitrogen has one lone pair of electrons and in the H_2S, the sulphur atom has two lone pairs of electrons, all being in the tetrahedral shape.

Bond angles	110°	109.28°	107°	104°	92.5°
		methane	ammonia		hydrogen sulphide

18. A reaction of methane, either chlorine or bromine, leads to the formation of chloromethanes or bromomethanes, besides the formation of HCl or HBr.

$$CH_4 + X_2 \longrightarrow CH_3X + HX \ (X = Cl \text{ or } Br)$$
$$+ CH_2X_2 + CHX_3 + CX_4 + HX$$

As HCl or HBr are not reducing agents, the reaction is irreversible.
On the other hand, a reaction of methane with iodine gives iodomethanes and hydrogen iodide.

$$CH_4 + I_2 \rightleftharpoons CH_3I + HI$$

As HI is a powerful reducing agent, the products formed react back to form the starting reactants again. In other words, the reaction, being reversible does not lead to the formation of iodomethanes. In order to lead the reactions to completion, an oxidizing reagent is used, to oxidize HI formed, to iodine again. The oxidizing agents used may be HIO_3, HNO_3 etc.

$$2\,HI + O \longrightarrow H_2O + I_2$$

19. In **Friedel–Craft reaction** (or F–C reaction) alkylation of benzene with *n*-propyl chloride, $CH_3CH_2CH_2Cl$, leads to formation of isopropyl benzene.

Mechanism

$$\underset{\text{\textit{n}-propyl carbocation}}{CH_3{-}CH_2{-}CH_2Cl} \longrightarrow \underset{1° \text{ carbocation}}{CH_3{-}CH_2{-}CH_2{}^+} + AlCl_4$$

n-propyl carbocation, being a 1° carbocation rearranges to a 2° carbocation, being more stable than a 1° carbocation, before it reacts with benzene.

$$\underset{\substack{\text{\textit{n}-propyl-}\\1° \text{ carbocation}}}{CH_3{-}CH_2{-}CH_2{}^+} \longrightarrow \underset{\substack{\text{\textit{n}-propyl carbocation}\\2° \text{ carbocation}}}{CH_3{-}CH^+{-}CH_3}$$

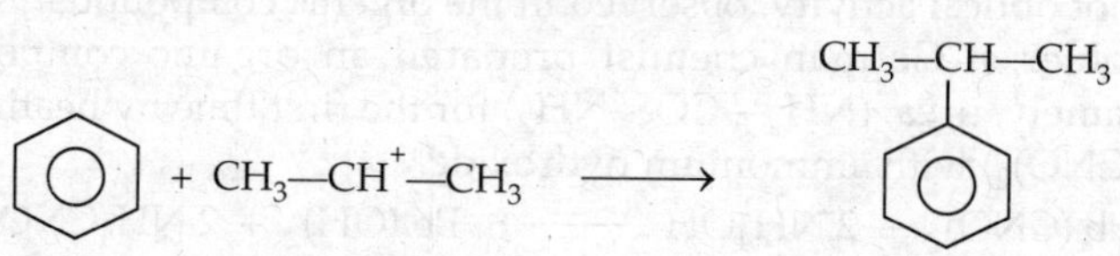

benzene + CH$_3$—CH$^+$—CH$_3$ (isopropyl-) $\longrightarrow$ isopropyl benzene

In order to prepare *n*-propyl benzene, a F–C reaction of benzene is carried with *n*-propionyl chloride ($CH_3{-}CH_2{-}CO{-}Cl$) in the presence of anhydrous $AlCl_3$ to obtain propionyl benzene. The –CO group is then reduced either by Zn/Hg amalgam (Clemmensen's reduction) or with hydrazine in the presence of NaOH (Wolf—Kishner's reduction) to convert into a –CH$_2$ group.

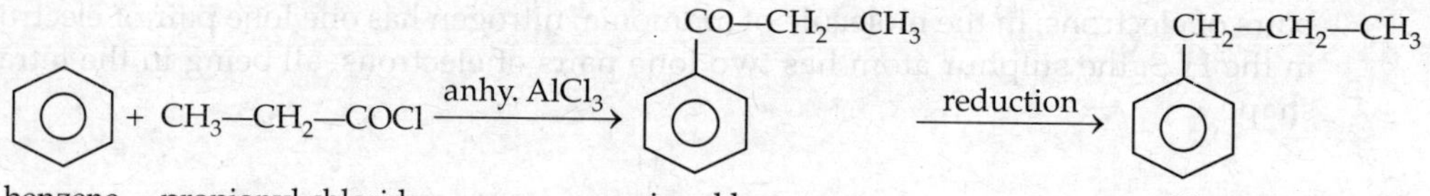

benzene + CH$_3$—CH$_2$—COCl $\xrightarrow{\text{anhy. AlCl}_3}$ propionyl benzene (CO—CH$_2$—CH$_3$) $\xrightarrow{\text{reduction}}$ *n*-propyl benzene (CH$_2$—CH$_2$—CH$_3$)

20. The electronegativities of various halogens are as follows.

Fluorine	>	chlorine	>	bromine	>	iodine
4		3		2.8		2.6

On account of highest electronegativity of fluorine, it is expected for the alkyl fluoride to be much more reactive than other halides. However, it is observed that fluorides are totally inert and nonreactive. The inertness in alkyl fluorides, has been explained on the basis of back bonding. Owing to **back bonding, fluorine is firmly bound to the carbon** and therefore, it is **totally inert and nonreactive**.

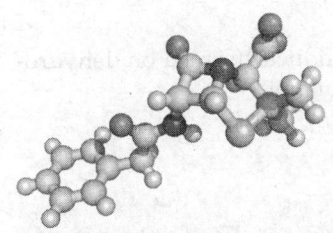

Paper 4

REASONING TYPE QUESTIONS (RTQs)

1. What do you understand by the term 'hyperconjugation'? Explain with suitable example.
2. Explain the effect of 'hydrogen bonding' on the properties of the compounds, which show hydrogen bonding.
3. The concept of delocalization is used in the case on benzene molecule. Explain.
4. Ozonolysis is used to determine the structure and the position of double bond in the unsaturated compounds. Explain with suitable examples.
5. Discuss the various methods of hydration of alkenes and alkynes. What are the reagents that can be used for hydration of alkenes and alkynes?
6. How does 'hydration' differs from the term 'hydrolysis', as in both the cases a water molecule is involved.
7. What do you understand by the term 'hydroxylation'? What are the various reagents that can be used for hydroxylation of alkenes?
8. Discuss various reagents that are use for dehydration of alcohols.
9. Write a note on hydroboration–oxidation reaction. Name the products which are obtained on the hydrolysis of the products.
10. It is possible to subject alkynes to partial hydrogenation. What reagents are there, that can be used for partial hydrogenation of alkynes.
11. Explain, why alkenes are more reactive than alkanes. Discuss the types of reactions, which alkanes undergo. How these reactions differ from the reactions, which alkenes undergo?
12. What are the types of dienes? Discuss, the types, with suitable examples.
13. Name the products that are obtained when 1,3-butadiene reacts with (i) HCl, (ii) Bromine and (iii) H_2O. Discuss the mechanism as well.
14. Write a note on Diel–Alder reaction.
15. Name the product that is obtained on reactions of dihalides with (i) aqueous KOH (ii) alcoholic KOH. Name reagents that can be used for dehalogenation of vic-dihalides.
16. Explain why acetylene and 1-alkynes are acidic in nature?
17. Write a note on the conformations of cyclohexane. Discuss the stabilities of various forms of cyclohexanes.

18. Discuss 1,2-shift methide shift or hydride shift.
19. Discuss Saytzeff's rule. Discuss the formation of different products, formed on dehydro-halogenation of 2-chlorobutane as well as dehydration of 2-butanol.
20. Discuss Markovnikov's rule.

ANSWERS

1. **Hyperconjugation is also referred to as 'no bond resonance':** It is used to explain the high reactivity of some aromatic halogen compounds and the stabilities of carbocations. When a carbon atom, containing at least to one hydrogen atom, is attached to an unsaturated carbon atom or at least one unshared orbital a delocalization of electrons takes place. The delocalization is through overlapping between a p-orbital and a σ-bond. Because of overlapping, each pair of electrons is not limited to binding together with just two atoms—the doubly bonded carbon or carbon and hydrogen, but to a certain extent, helps bind together all four atoms. Delocalization of this kind, involving σ-bond orbitals is called 'hyperconjugation'.

   i. p-methyl benzyl chloride has been found to be much more reactive than it can be explained on the basis of +I effect of p-methyl group. This has been explained on the basis of no bond resonance structures that can be written for p-methyl benzyl chloride. One hydrogen atom of p-methyl group, gives up its electrons to the carbon atom, but remains within the sphere (vicinity) of the methyl group. This release of hydrogen, as a proton, makes the carbon having an unshared pair of electrons, which is used to convert the benzenoid ring to a quinonoid ring, setting chlorine free as an anion. As a –vely charged Cl atom is much more reactive than a covalently bonded chlorine, which explains the hyperactivity of p-methyl benzyl chloride.

   ii. The relative stability of various classes of carbocations, may be explained by the number of no-bond resonance structures, that can be written for them. Such structures are obtained by shifting the binding electrons from an adjacent —C—H bond to the electron deficient carbon. In this way, the +ve charge, originally on carbon atom, is dispersed to the hydrogen atom and electrons release by assuming no-bond character in the adjacent —C—H bond is called hyperconjugation or a no-bond resonance. Hyperconjugation defines the stability of a species. For example, for an ethyl carbo-cation, the three hyperconjugation structures may be written as below.

ethyl carbocation

$$\begin{array}{ccc}
\overset{\displaystyle H+}{\underset{\displaystyle H}{\overset{|}{\underset{|}{H-C}}} \overset{\displaystyle H}{\underset{\displaystyle H}{\overset{|}{\underset{|}{= C}}}} & \longleftrightarrow & \overset{\displaystyle H}{\underset{\displaystyle H}{\overset{|}{\underset{|}{H+ C}}}} \overset{\displaystyle H}{\underset{\displaystyle H}{\overset{|}{\underset{|}{= C}}}} & \longleftrightarrow & \overset{\displaystyle H}{\underset{\displaystyle H+}{\overset{|}{\underset{|}{H-C}}}} \overset{\displaystyle H}{\underset{\displaystyle H}{\overset{|}{\underset{|}{= C}}}}
\end{array}$$

2. Hydrogen bonding is an attractive force, which occurs in those compounds whose molecules contain either –OH bond or –NH bond (as it occurs in water, alcohols, acids, amines and amides). The —O—H bond is highly polar. Oxygen is more electronegative (electronegativity = 3.5), as compared to hydrogen (electronegativity 1), and therefore, oxygen pulls the bonding electrons towards itself as a result, oxygen aquires a little negative charge (δ^-) and hydrogen acquires a small positive charge (δ^+). Adjacent molecules of the compound, containing –OH groups, will be attracted by virtue of the opposite charges to each other. This force of attraction is due to hydrogen bond. Usually a hydrogen bond is represented by a dotted line.

$$\cdots\cdots O \cdots\cdots H \cdots\cdots O \cdots\cdots H \cdots\cdots O \cdots\cdots H \cdots\cdots$$

3. The concept of delocalization has been used to explain the unusual properties as exhibited by benzene molecule. Being a highly unsaturated compound, it contains three alternate double bonds in the molecule and it is expected to undergo only addition reactions. However, it undergoes not only substitution reactions, under normal conditions, also keeps the double bond intact in the molecule.

   This unusual behavior, as exhibited by benzene, has been explained on the basis of delocalization of π-electrons in the benzene ring. That all the six π-electrons, leave their respective carbon atoms, make a pool of six π-electrons, which is in the form of π-clouds, engulfs the molecule from the top and bottom.

   One of the proofs for delocalization, is provided by the bond length of C to C in the benzene ring. The bond length for all carbon to carbon is the same and it is equal to 1.39 Å. This value is more than the bond length as observed in the case of double bond (C to C double bond length = 1.34 Å) and it is less than the bond length for a C to C single bond (C to C single bond length = 1.54 Å).

   The second proof is provided by the thickness of the benzene ring. As the molecule is flat, all the six carbon atoms and all the six hydrogen atoms, lie in the same plane, the thickness of the benzene ring should be equal to the diameter of the carbon atom. However, the thickness has been found to be more than the diameter of the carbon atom, this proves that the benzene ring engulfed the π-clouds from top and bottom.

4. **Ozonolysis:** Alkenes and alkynes react with ozone (O_3) to form ozonides. One molecule of O_3 reacts with one double bond and the ozonide formed, contains one oxygen atom, inserted in between the two carbon atoms, originally joined by the double bond.

$$\underset{\text{alkene}}{\overset{\displaystyle H \quad R}{R-\overset{|}{C}=\overset{|}{C}-R}} + \underset{\text{ozone}}{O_3} \longrightarrow \underset{\text{ozonide}}{\overset{\displaystyle O \text{-----} O}{R-\overset{|}{C}H-O-\overset{|}{C}-R}} \xrightarrow{\text{hydrolysis}} \underset{\text{aldehyde}}{R-CH=O} + \underset{\text{ketone}}{\overset{\displaystyle R}{R-\overset{|}{C}=O}}$$

keto compounds

On hydrolysis, the ozonide molecule decomposes to form the keto compounds, with each carbon, originally joined by a double bond, now separates as oxidized carbon. From the analysis of the keto compounds, formed upon the hydrolysis of the ozonide, not only the position of the double bond, but structure of alkene can also be established by joining the oxidized carbon atoms.

Alkynes also react with ozone to form ozonides. One molecule of O_3 reacts with one triple bond and the ozonide formed has both carbon atoms joined by the oxygen to form keto structure, the third oxygen forms an oxide ring, (an epoxide or forms an oxirane ring), with the two carbons originally joined by a triple bond.

On hydrolysis, ozonide is converted into a diketo compound, which contains both the carbons, originally joined by a triple bond, now joined by a single bond.

$$
\underset{\text{alkyne}}{\overset{\displaystyle \text{H}\quad\text{R}}{\underset{|\qquad|}{\text{R—C≡C—R}}}} + \underset{\text{ozone}}{O_3} \longrightarrow \underset{\substack{\text{ozonide}\\\text{epoxide or an oxirane}}}{\overset{\displaystyle O\text{--}O}{\underset{O}{\text{R—C—C—R}}}} \xrightarrow[\text{H}_2\text{O}]{\text{hydrolysis}} \underset{\text{a diketo compound}}{\overset{\displaystyle O\quad O}{\underset{O}{\text{R—C—C—R}}}} + H_2O_2
$$

(oxiranes or epoxides are the compounds which contains an oxide ring with two carbon and one oxygen).

5. **Hydration:** It may also be termed a process of addition of a molecule water. Alkenes react may be hydrated to form alcohols by any of the following methods:

a. Alkenes with a molecule of water, to form alcohol, generally in the presence of a strong acid (usually H_2SO_4).

$$
{>}C{=}C{<} + H_2O \xrightarrow{\text{H}_2\text{SO}_4} \underset{\text{H}\quad\text{OH}}{\text{—C—C—}}
$$

Mechanism

i. Sulphuric acid, adds to the alkene molecule, following Markovnikov's rule to from a hydrogen sulphate derivative.

$$
\underset{\text{R—C=C—R}}{\overset{\displaystyle\text{R}\quad\text{H}}{}} \xrightarrow{\text{H}^+,\ \text{HSO}_4} \underset{\text{HSO}_4\quad\text{H}}{\overset{\displaystyle\text{R}\qquad\text{H}}{\text{R—C——C—R}}}
$$

ii. On hydrolyzing, the hydrogen sulphate derivatives (by boiling with excess water) converted into an alcohol.

$$
\underset{\text{HSO}_4\quad\text{H}}{\overset{\displaystyle\text{R}\qquad\text{H}}{\text{R—C——C—R}}} + \xrightarrow{\text{H}_2\text{O}} \underset{\underset{\text{alcohol}}{\text{OH}\quad\text{H}}}{\overset{\displaystyle\text{R}\qquad\text{H}}{\text{R—C——C—R}}} + H_2SO_4
$$

b. *Hydroboration: Oxidation of alkenes.* This is a two step reaction:

i. Alkenes are treated with diborane (B_2H_6) to form a trialkyl boranes. In this reaction, diborane, reacts as BH_3, adds to the double bond, to form a trialkyl borane.

$$
\underset{\text{alkene}}{3RCH{=}CH_2} + \underset{\text{borane}}{BH_3} \longrightarrow \underset{\text{trialkyl borane}}{(RCH_2{—}CH_2)_3B}
$$

A molecule of BH_3 adds as BH_2^+ and H^-, although, BH_3 itself is an electron deficient compound, yet the addition takes place as BH_2^+ and H^-.

BH_2^+ adds to the carbon having less number of hydrogen and hydrogen as a negative part (H^-), adds at the carbon having larger number of hydrogen atoms. Thus, BH_3 adds as per Markovnikov's rule, when one looks on the charges these ions have. On the other hand, this addition may be termed anti-Markovnikov's addition, if the is simply taken to be the addition of hydrogen and BH_2 without considering the charges on the ions.

On hydrolyzing the trialkylborane, with aqueous alkaline solution of H_2O_2, alcohol is obtained. In the case of terminal alkenes, their trialkylborane derivatives on hydrolysis, give primary alcohols.

$$(R—CH_2—CH_2)_3\,B\ +\ H_2O_2 \xrightarrow[\text{OH}^-]{\text{H}_2\text{O}} \underset{\text{primary alcohol}}{3R—CH_2—CH_2—OH\ +\ H_3BO_3}$$

In a nutshell, the addition may simply as addition of a molecule of water across the double bond, following anti-Markovnikov's rule.

$$\underset{\substack{|\ \ \ \ |\\ H\ \ OH}}{R—CH=CH_2}\ +\ (H_2O) \longrightarrow RCH_2—CH_2—OH$$

Hydration of alkynes

Alkynes may be hydrated by H_2SO_4 in the presence $HgSO_4$ as a catalyst. The products are either aldehydes or ketones.

$$R—C\equiv CH \xrightarrow[\text{H}_2\text{SO}_2]{\text{H}_2\text{O/Hg}_2^+} \underset{\substack{|\\ \text{O—H}}}{R—CH=CH_2} \longrightarrow \underset{\text{ketone}}{R—\overset{\overset{\text{O}}{\|}}{C}—CH_3}$$

$$\underset{\text{O—H}^-\ \text{H}^+}{}$$

$$\text{unstable, rearranges itself}$$

$$CH\equiv CH \xrightarrow{\text{H}_2\text{O/Hg}_2^+} \underset{\substack{\text{vinyl alcohol}\\ \text{unstable, rearranges to form}\\ \text{either an aldehyde or a ketone}}}{CH_2=CH—OH} \longrightarrow \underset{\text{acetaldehyde}}{CH_3—CHO}$$

6. **Hydration** may be defined as a process of addition of a molecule of water to the reacting compound. The reaction may be carried out in the presence of some catalyst or a reagent, which may help the reacting compound to form an intermediate compound, which then reacts with the water molecule. Alkenes are hydrated in the presence of H_2SO_4 to form alcohol. In the same manner, alkenes are also hydrated to form alcohols, by hydroboration–oxidation process.

$$R—CH=CH_2 \xrightarrow{\text{H}_2\text{SO}_4/\text{H}_2\text{O}} \underset{\substack{\text{OH}\\|}}{R—CH—CH_3}$$

$$R—CH=CH_2 \xrightarrow[\text{OH}^-]{\text{B}_2\text{H}_6/\text{H}_2\text{O}} R—CH_2—CH_2OH$$

Alkynes react with H_2SO_4 with $HgSO_4$ as a catalyst, to give either aldehydes or ketones.

$$R—C\equiv CH \xrightarrow[\text{H}_2\text{SO}_4]{\text{H}_2\text{O/HgSO}_4} R—CO—CH_3$$

Hydrolysis may be defined as a process, in which a reaction is carried out in the presence of water, taken as a solvent.

i. Esters are hydrolyzed to form alcohol and acid, when treated with H_2O in presence of a little catalyst.

$$R{-}COOR' + H_2O \xrightarrow[H_2SO_4]{H^+} RCOOH + R'OH$$

ester $\qquad\qquad\qquad\qquad$ a carboxylic acid $\quad$ alcohol

ii. Cyanides on treating with water in the presence of a suitable catalyst are hydrolyzed to give carboxylic acids.

$$R{-}C{\equiv}N + 2H_2O \xrightarrow{\text{acid catalyst}} R{-}COOH + NH_3$$

alkyl cyanide $\qquad\qquad\qquad\qquad$ carboxylic acid

iii. Isocyanides are hydrolyzed to form primary amines and formic acid.

$$R{-}NC + H_2O \xrightarrow{OH^-} RNH_2 + HCOOH$$

alkyl isocyanide $\qquad\qquad\qquad$ a primary amine $\quad$ formic acid

iv. Alkyl halides are hydrolyzed by alkaline aqueous solutions, to form corresponding alcohols.

$$R{-}X + H_2O \xrightarrow{OH^-} R{-}OH + X$$

alkyl halides

7. **Hydroxylation:** Alkenes are hydroxylated to form 1,2-dihydroxy compounds also known as glycols. There are three reagents, which are used for the hydroxylation of alkenes.

i. A dil. aqueous solution of $KMnO_4$: On treating an alkene with a dil. aqueous solution of $KMnO_4$, alkene adds two –OH groups across the double bonds on the adjacent carbon atoms in a *cis*-manner, to form a *cis*-glycol.

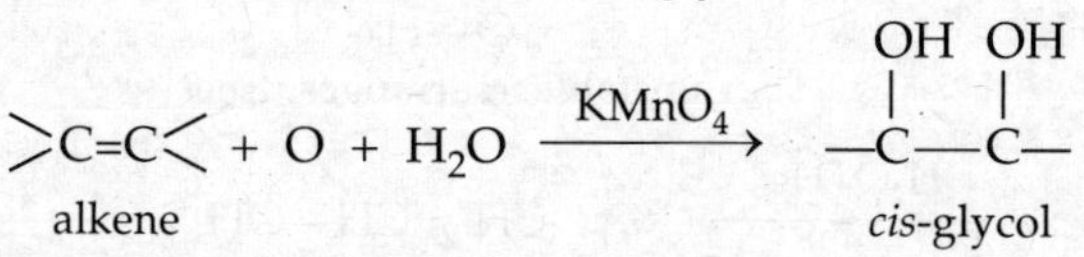

$$>C{=}C< + O + H_2O \xrightarrow{KMnO_4} \underset{\textit{cis}\text{-glycol}}{\overset{\text{OH OH}}{-C-C-}}$$

alkene

ii. Alkene are treated with osmium tetroxides (OsO_4) to give a complex, a osimate ester, which on further mild reduction with sodium sulphite (Na_2SO_3) is hydrolyzed to from a *cis*-glycol.

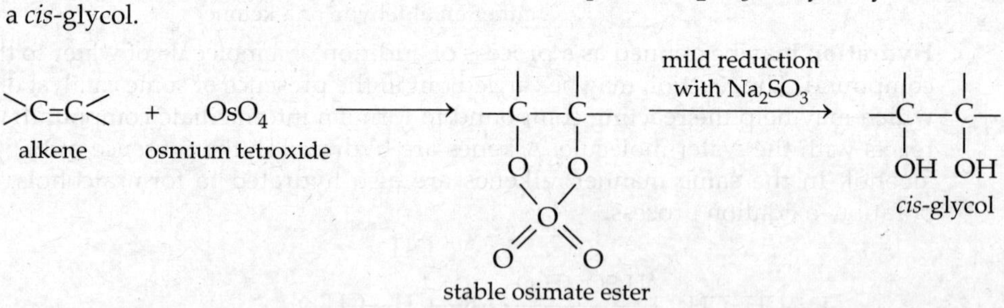

$$>C{=}C< + OsO_4 \longrightarrow \text{(stable osimate ester)} \xrightarrow[\text{with Na}_2\text{SO}_3]{\text{mild reduction}} \underset{\text{OH OH}}{-C-C-}$$

alkene $\quad$ osmium tetroxide $\qquad\qquad\qquad\qquad\qquad\qquad\qquad\qquad$ *cis*-glycol

stable osimate ester

iii. Alkenes also form *trans-glycols*, on treatment with hydrogen peroxide (H_2O_2) in formic acid (HCOOH).

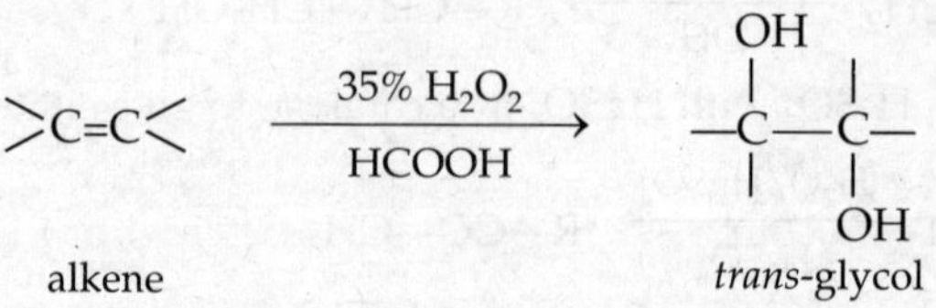

$$>C{=}C< \xrightarrow[\text{HCOOH}]{35\% \ H_2O_2} \underset{\text{OH}}{\overset{\text{OH}}{-C-C-}}$$

alkene $\qquad\qquad\qquad\qquad\qquad\qquad\qquad\qquad$ *trans*-glycol

8. **Dehydration of alcohols:** Alcohols may be dehydrated to form alkenes and ethers.

i. *Dehydration of alcohols to alkenes*: Alcohols, on heating with conc. H_2SO_4 at 170 °C, lose a molecule of water to give alkenes.

$$R-CH_2-CH_2-OH \xrightarrow[170\,°C]{conc.\ H_2SO_4} R-CH=CH_2$$

alcohol   alkene

Dehydration of alcohols to form alkenes is in the following order.

$$\begin{array}{ccccc} & R & & R & \\ & | & & | & \\ R-C-OH & > & R-C-OH & > & RCH_2-OH \\ & | & & | & \\ & R & & H & \\ 3° & > & 2° & > & 1° \end{array}$$

Note: Secondary and tertiary alcohols, containing four or more carbon atoms to give a mixture of two alkenes. For example:

$$\underset{\text{2-butanol}}{CH_3-CH_2CH-CH_3} \;\; (OH) \xrightarrow[-H_2O]{conc.\ H_2SO_4} \underset{\substack{\text{1-butene}\\\text{(minor)}}}{CH_3-CH_2-CH=CH_2} + \underset{\substack{\text{2-butene}\\\text{(major)}}}{CH_3-CH=CH-CH_3}$$

The formation of alkene in major quantity, is governed by the *Saytzeff's rule*. According to this rule, alkene produced in greater quantity is the one that contains the higher number of alkyl groups in the molecule.

ii. Alcohols may also be dehydrated, by passing the vapours of alcohol over heated alumina (Al_2O_3) kept at 350 °C. For example:

$$CH_3-CH_2-OH \xrightarrow[350\,°C]{Al_2O_3} CH_2=CH_2 + H_2O$$

ethene

iii. *Dehydration of alcohols to ethers*:

- On heating excess of ethyl alcohol, with conc. H_2SO_4, at 140 °C forms, two molecules of ethyl alcohol together, with elimination of one molecule of water, to form diethyl ether.

$$CH_3-CH_2-OH + HO-CH_2-CH_3 \xrightarrow[140\,°C]{conc.\ H_2SO_4} CH_3-CH_2-O-CH_2-CH_3 + H_2O$$

diethyl ether

- Alcohols can be dehydrated to form ethers, by passing the vapours of alcohol over heated alumina (Al_2O_3) kept at 200 °C. For example:

$$2C_2H_5-OH \xrightarrow[200\,°C]{Al_2O_3} C_2H_5-O-C_2H_5 + H_2O$$

diethyl ether

9. **Hydroboration–oxidation reaction:** Alkenes react with diborane (B_2H_6) to form an addition compound B_2H_6 which adds across the double bond as borane, BH_3.

B_2H_6 itself is an electron deficient compound, yet it adds on the double bond as BH_2^+ and H^- to form an addition compound. BH_2^+ adds to the carbon atom, having lesser number of hydrogen atoms and H^- adds to the carbon having larger number of hydrogen atoms.

On looking at the charges, the species undergoes addition following Markovnikov's rule. On the other hand, the addition may be considered as adding by anti-Markovnikov's, rule, if the addition is considered as the adding of 'H' and 'BH_2', without the charges on them.

$$3R\text{—}CH=CH_2 + BH_3 \longrightarrow (R\text{—}CH_2\text{—}CH_2)_3B$$
alkene   trialkyl borane

The addition compound formed with alkaline aqueous hydrogen peroxide, the trialkyl borane is hydrolyzed to give alcohol.

$$(R\text{—}CH_2\text{—}CH_2)_3B + H_2O \longrightarrow 3R\text{—}CH_2\text{—}CH_2OH + H_3BO_3$$

In brief, the addition may simply be taken as the addition of a molecule of water across the double bond.

$$R\text{—}CH=CH_2 + (H_2O) \longrightarrow R\text{—}CH_2\text{—}CH_2\text{—}OH$$
$\quad\quad\;\;|\quad\;\;|$
$\quad\quad\;\;H\quad OH$

The terminal alkenes form trialkyl boranes and primary alcohols on hydrolysis Therefore, this method is used to synthesize primary alcohols.

10. Alkynes, can be hydrogenated partially to form alkenes, when alkynes are reduced with Lindler's catalyst. The reduction stops at alkene stage.
Lindler's catalyst consists of Pd metal, mounted on $BaSO_4$ is poisoned by quinoline.

$$CH\equiv CH + H_2 \xrightarrow[\substack{quinoline \\ (Lindler's\ catalyst)}]{Pd + BaSO_4} CH_2=CH_2$$
alkyne   alkene

On catalytic hydrogenation, in presence of a catalyst (Ni) at a temperature 250–300 °C, Rainey nickel, or finely divided Pt or Pd, alkynes add up two molecules of H_2 to form alkanes. Alkenes are formed as intermediate compounds, however alkenes cannot be isolated.

$$CH\equiv CH + H_2 \xrightarrow[250\,°C\text{–}300\,°C]{Ni} CH_2=CH_2 \xrightarrow[250\,°C\text{–}300\,°C]{Ni/H_2} CH_3\text{—}CH_3$$
alkyne   alkene   alkane

11. Alkenes are more reactive than alkanes, because alkenes contain one double bond in the molecule.
Alkanes: *Alkanes undergo only substitution reactions.*
There are two types of atoms present in alkanes.

i. The C to C single bond, known as the σ-bond, formed by the overlapping of sp^3 hybridized orbitals of the two participating carbon atoms. The σ-bond (formed by the overlapping of sp^3-sp^3 orbitals) is not broken, under the conditions, alkanes are subjected to various reactions.

ii. The other atoms present in alkanes are hydrogens attached to the carbons by the σ-bonds (formed by the overlap of sp^3-s orbitals).

iii. The hydrogen atoms undergo substitution, through halogens by a free radical mechanism to form halogen substituted alkanes.

iv. Hydrogen in alkanes can also be substituted by sulphonic acid group (–SO_3H) when subjected to a prolonged reaction with hot, fuming sulphuric acid. In the alkanes, one hydrogen atom is substituted by a –SO_3H group. This process is termed 'sulphonation'.

$$R\text{—}H + H_2SO_4 \xrightarrow[fuming\ H_2SO_4]{\Delta} R\text{—}SO_3H + H_2O$$
alkane   alkyl sulphonic acid

Hydrogen in alkanes can also be substituted by a nitro group, by heating the vapours of alkanes mixed with vapours of conc. HNO_3 at 400–500 °C. One hydrogen in alkane is substituted by a nitro group (NO_2).

$$R-H + HNO_3 \xrightarrow{400-500\,°C} R-NO_2 + H_2O$$

alkane   nitroalkane

Since the reaction is carried out at a very high temperature, rupture of carbon to carbon bonds occurs during the process. Thus, on nitration ethane gives a mixture of nitro-methane and nitroethane.

$$CH_3-CH_3 + HNO_3 \xrightarrow{450°C} CH_3-CH_2-NO_2 + CH_3-NO_2$$

ethane   nitroethane   nitromethane

Alkenes: Alkenes contain one double bond in the molecule. The double bond consists of one σ-bond and one π-bond. In the addition properties shown by alkenes, it is the π-bond that breaks easily in the reaction conditions, which facilitates addition.

Majority of properties shown by alkenes are *addition properties*. The double bond breaks and the reacting compound adds to the doubly bonded carbon atoms.

i. The addition across the double bond, follows Markovnikov's rule.

ii. The addition across takes place by electrophilic mechanism.

$$R-CH=CH_2 + \overset{+}{H}-\overset{-}{B} \longrightarrow R-\overset{+}{C}H-CH_3 \longrightarrow R-CHB-CH_3$$

alkene   carbocation

12. **Dienes:** Alkenes, containing two double bonds in the molecule are called dienes or 'alkadienes'.

The alkadienes are of the three types:

i. The dienes, containing two double bonds in the molecule, are separated by more than one single bond, are termed nonconjugated dienes.

ii. The dienes, containing two double bonds in the molecule are separated by only one single bond are termed conjugated dienes.

iii. The dienes, containing two double bonds in the molecule are adjacent to each other are termed cumulated dienes.

$$>C=\overset{|}{C}-\overset{|}{C}-\overset{|}{C}=C< \qquad >C=\overset{|}{C}-\overset{|}{C}=C< \qquad >C=C=C<$$

nonconjugated   conjugated   cumulative compound

The most important class of dienes is conjugated dienes. The properties of the conjugated dienes are different from those of other dienes.

13. **Reactions of 1,3- butadiene, with (i) HCl (ii) Br$_2$ and (iii) water.**

i. Addition of HCl

1, 3-butadiene, $CH_2=CH-CH=CH_2$ is a conjugated diene. It reacts with one molecule of HCl, to form a mixture of 1, 2-addition and 1, 4-addition compounds. At low temperature 1, 2-addition is preferred, whereas at high temperature it prefers to form 1, 4-addition compounds.

Mechanism

a. 1, 2 butadiene adds on a proton to form allylic carbonium ion, which is stabilized by resonance.

$$CH_2=CH-CH_2 + \overset{+}{H} \longrightarrow CH_3-\overset{+}{C}H-CH=CH_2$$

1, 3-butadiene   allylic carbocation

$$CH_3=CH-CH=\overset{+}{C}H_2 \longleftrightarrow CH_3-\overset{+}{C}H=CH-CH_2 \longleftrightarrow \overset{\,\,\bullet\bullet\bullet\,\,\,\,\bullet\bullet\bullet}{CH_2-CH-\overset{H}{CH}-CH_2}$$

I   II   III

The energy of activation for carbonium ion (III) is higher for 1, 4-addition than for 1, 2-addition, therefore 1, 4-addition is favoured at high temperature and 1, 2-addition at low temperature.

b. The anion Cl^- adds to either carbon number 2 or number 4 to give 1, 2- or 1, 4-addition compounds.

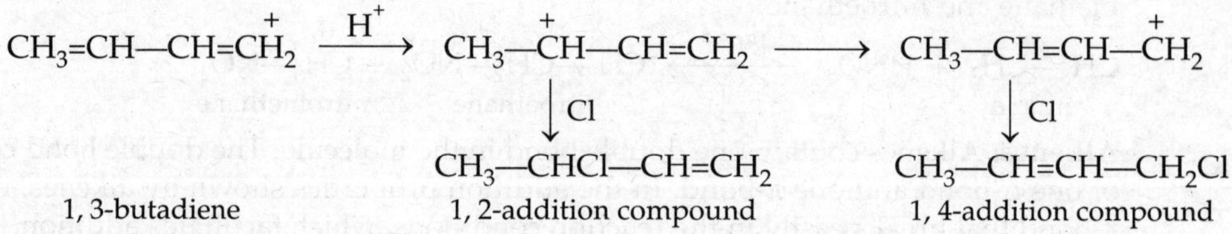

$$CH_2=CH-CH=CH_2 \xrightarrow{H^+} CH_3-\overset{+}{C}H-CH=CH_2 \longrightarrow CH_3-CH=CH-\overset{+}{C}H_2$$

$$\downarrow Cl \qquad\qquad\qquad \downarrow Cl$$

$$CH_3-CHCl-CH=CH_2 \qquad\qquad CH_3-CH=CH-CH_2Cl$$

1, 3-butadiene   1, 2-addition compound   1, 4-addition compound

ii. Addition of bromine

The addition of Br_2 to 1, 3-butadiene follows the same pattern as in the case of HCl. Mechanism is also the same.

In a reaction between Br_2 (CCl_4) and 1, 3-butadiene, both the addition compounds, namely 1, 2-addition product and 1, 4-addition compounds are formed. Whereas 1, 2-addition is favoured at low temperature and 1, 4-addition compound is formed at a high temperature.

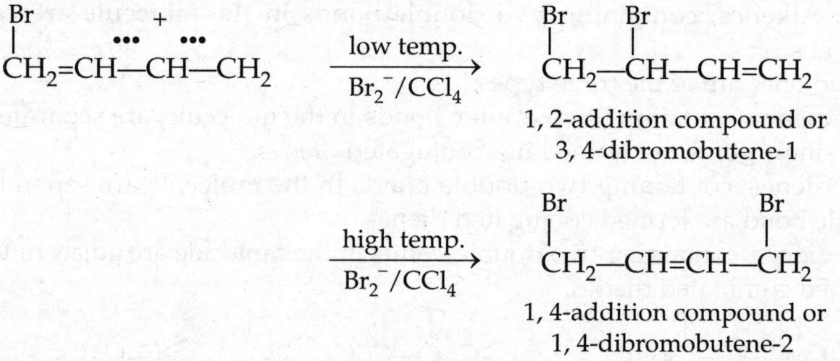

$$\overset{Br}{\underset{|}{C}}H_2=CH-CH-CH_2 \xrightarrow[Br_2^-/CCl_4]{\text{low temp.}} \overset{Br}{\underset{|}{C}}H_2-\overset{Br}{\underset{|}{C}}H-CH=CH_2$$

1, 2-addition compound or
3, 4-dibromobutene-1

$$\xrightarrow[Br_2^-/CCl_4]{\text{high temp.}} \overset{Br}{\underset{|}{C}}H_2-CH=CH-\overset{Br}{\underset{|}{C}}H_2$$

1, 4-addition compound or
1, 4-dibromobutene-2

iii. Addition of hydrogen

Hydrogen adds to 1,3-butadiene to form both the addition compounds namely 1, 2 and 1, 4-addition compounds, in the presence of a catalyst.

$$CH_2=CH-CH=CH_2 \xrightarrow{H_2/\text{catalyst}} CH_3-CH_2-CH=CH_2 + CH_3-CH=CH-CH_3$$

butene-1   butene-2

14. **Diel–Alder reaction:** The Diel–Alder reaction, involves in the addition of a suitable vinyl derivative (called dienophile) to the two ends of a conjugated diene system, in the absence of a catalyst, the product formed is a cyclic compound, invariably a six membered ring is formed and the product is termed an adduct.

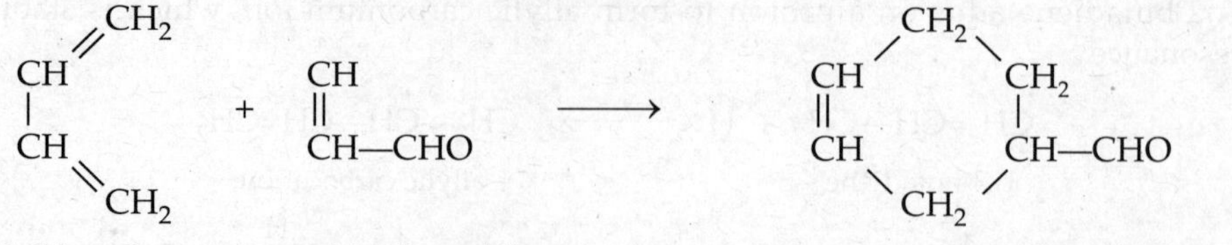

1, 3-butadiene   acrolein dienophile   1, 2, 3, 6-tetra hydro benzaldehyde

This reaction can be used to prepare derivatives of benzene and six membered rings.

15. **Reaction of dihalides:** The dihalogen derivative alkanes may be divided into three categories:

   i. *Geminal dihalides:* These dihalides, contain two halogen atoms attached to the same carbon atom.

   CH_3—CH_2—$CHCl_2$ (gem-dihalides)
   1, 1-dichloropropane

   ii. *Vicinal dihalides* or *vic-dihalides:* These dihalides contain two halogens attached at the adjacent carbon atoms.

   CH_3CHCl—CH_2Cl (vic-dihalide)
   1, 2-dichlorpropane

   iii. *Other dihalides,* which may contain two halogen atoms attached on the two different carbon atoms, which are neither geminal nor vicinal.

   CH_2Cl—CH_2—CH_2Cl
   1, 3-dichloropropane

a. Reaction with aqueous KOH

   i. *Gem-dihalides:* On treatment with aqueous KOH, aldehyde or ketone is obtained.

   CH_3—CH_2—$CHCl_2$ + 2KOH $\longrightarrow$ CH_3—CH_2—$CH(OH)_2$ $\longrightarrow$ CH_3—CH_2—CHO + H_2O
   1, 1-dichloropropane (aqueous)   unstable   propionaldehyde

   CH_3—CCl_2—CH_3 + KOH $\longrightarrow$ CH_3—$C(OH)_2CH_3$ $\longrightarrow$ CH_3—CO—CH_3 + H_2O
   2, 2-dichlopropane (aqueous)   unstable   acetone

   ii. *Vic-dihalides:* 1,2-Dihydroxy compounds or glycols are obtained.

   CH_3—CHCl—CH_2Cl + 2KOH $\longrightarrow$ CH_3—CH (OH)—$CH_2(OH)$ + 2KCl
   1, 2-dichloropropane (aqueous)   1, 2-dihydroxy propane (glycol)

   iii. *Other dihalides:* With aqueous KOH, both chlorines are substituted by –OH groups.

   CH_2Cl—CH_2—CH_2Cl + 2KOH $\longrightarrow$ CH_2 (OH)—CH_2—$CH_2(OH)$ + 2KCl
   1, 3-dichloropropane (aqueous)   1, 3-dihydroxy propane

b. Reaction with alcoholic KOH

   i. *Gem-dihalides:* Dehydrohalogenation takes place, leads to the formation of an alkyne.

$$
\begin{array}{c}
\quad\; H \;\; Cl \\
\quad\; | \;\;\; | \\
CH_3\text{—}C\text{—}CH \;\; + \;\; 2KOH \longrightarrow CH_3\text{—}C{\equiv}CH + 2KCl + 2H_2O \\
\quad\; | \;\;\; | \\
\quad\; H \;\; Cl
\end{array}
$$

   1, 1-dichloro propane (alco.)   propyne

   ii. *Vic-dihalides:* Dehydrohalogenation takes place and an alkyne is formed.

$$
\begin{array}{c}
\quad\; H \;\; Cl \\
\quad\; | \;\;\; | \\
CH_3\text{—}C\text{—}CH \;\; + \;\; 2KOH \longrightarrow CH_3\text{—}C{\equiv}CH + 2KCl + 2H_2O \\
\quad\; | \;\;\; | \\
\quad\; Cl \;\; H
\end{array}
$$

   1, 2-dichloropropane (alco.)   propyne

   iii. *Dehydrohalogenation*

   CH_2Cl—CH_2—CH_2—Cl + 2KOH $\longrightarrow$ $CH_2{=}C{=}CH_2$ + 2KCl
   1, 3-dichloro propane   (alco.)   propadiene

   CH_2Cl —CH_2—CH_2—CH_2Cl + KOH $\longrightarrow$ $CH_2{=}CH$—$CH{=}CH_2$ + 2KCl
   1, 4-dichlorobutane   (alco.)   1,3-butadiene

c. Dehalogenation

i. *Vic-dihalides and gem-dihalides*, on treatment with zinc dust in methanol are dehalogenated to give alkenes.

$$CH_3—CHBr—CH_2Br + Zn \xrightarrow[\Delta]{methanol} CH_3—CH=CH_2 + ZnBr_2$$

1, 2-dibromo propane   propane
vic-dihalide

$$CH_3—CH_2—CHBr_2 + Zn \xrightarrow[\Delta]{methanol} CH_3—CH=CH_2 + ZnBr_2$$

1, 1-dibromopropane   propene
gem-dihalide

ii. 1, 3- to 1, 6-dihalides, on treatment with Zn dust in methanol give cycloalkanes.

$$CH_2 \Big\langle {CH_2Br \atop CH_2Br} + Zn \xrightarrow{methanol} CH_2 \Big\langle {CH_2 \atop |} {CH_2} \quad ZnBr_2$$

1, 3-dibromopropane   cyclopropane

16. **Terminal alkynes are more acidic than either alkenes or alkanes:**
1-alkynes are acidic in nature, because of presence of a triple bond in the molecule. The two π-bonds, present in the molecule engulf the molecule in the form of π-clouds, forming a barrel type cover over the two carbons, originally joined by a triple bond.
In forming π-clouds, the electrons move away from the two carbon atoms making them more electron deficient, in addition to fact that on account of each carbon being in 'sp' hybridized state, the bonding electrons are more tightly held by each carbon atom, the bonding electrons of the C–H bond, are already displaced towards the carbon atom. As a result, it becomes easier to remove hydrogen as a proton.

17. **Cyclohexane:**
The cyclohexane ring consists of six methylene groups (–CH$_2$–) joined together to form a cyclic compound. The compound is as stable as any other alkane. Taking into account the valencies of the tetrahedral carbon, the compound should be highly unstable. As against the normal angle of –C to C– single bond of 109°.28', the angle comes out to be 120° in case the molecule of cyclohexane is considered to be a planer ring.

In the year 1918, 'Sachse and Mohr', suggested that such rings can become strain free, if all the ring carbons are not forced into one plane. In other words, the rings can be multiplaner in shape. Such rings not only will have a normal angle between –C to C– single bond, the ring will be free from any strain, and will be as stable as any other alkane. This was termed as 'Sachse and Mohr theory of strainless rings'. Therefore, a molecule of cyclohexane can exist in two non-planer puckered conformations, both of which are free from strain, as the angle between the carbon atoms remains normal, i.e. 109°.28'. They are called as 'chair' and 'boat' forms.

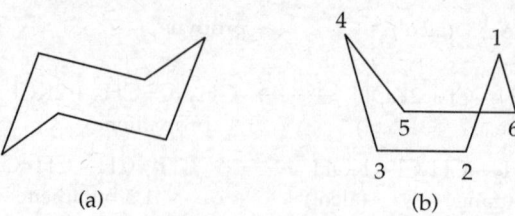

(a)   (b)

18. **1, 2-Shift or a methide shift and hydride shift:**

   i. *A Methide (CH_3^-) shift*

   A 3° carbocation is more stable than a 2° carbocation, which in turn, is more stable than a 1° carbocation.

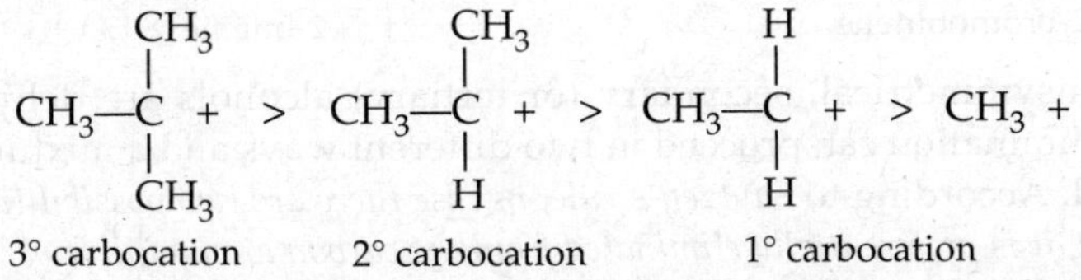

$$CH_3\text{--}\overset{\overset{\displaystyle CH_3}{|}}{\underset{\underset{\displaystyle CH_3}{|}}{C}}\overset{+}{{}} \;>\; CH_3\text{--}\overset{\overset{\displaystyle CH_3}{|}}{\underset{\underset{\displaystyle H}{|}}{C}}\overset{+}{{}} \;>\; CH_3\text{--}\overset{\overset{\displaystyle H}{|}}{\underset{\underset{\displaystyle H}{|}}{C}}\overset{+}{{}} \;>\; CH_3 +$$

           3° carbocation   2° carbocation   1° carbocation

   Wherever, a corbocation of a lower stability is formed in the reaction, as an intermediate species, it rearranges to the carbocation of a greater stability, before it react to form the derivative. The reactions that proceed through the formation of a carbocation are always *susceptible to undergo rearrangement.*
   For example:

   Dehydration of 2-hydroxy-3,3-dimethyl butane, $(CH_3)_3\text{--}\overset{\overset{\displaystyle OH}{|}}{C}\text{--}CH\text{--}CH_3$ by conc. H_2SO_4.

$$(CH_3)_2C\text{--}\underset{\underset{\displaystyle CH_3}{|}}{\overset{\overset{\displaystyle OH}{|}}{C}H}\text{--}CH_3 \xrightarrow[-H_2O]{H^+} (CH_3)_2C\text{--}\underset{\underset{\displaystyle CH_3}{|}}{\overset{\overset{\displaystyle OH}{|}}{C}H}\text{--}CH_3 \longrightarrow (CH_3)_2\overset{\overset{\displaystyle +}{}}{C}\text{--}\underset{\underset{\displaystyle CH_3}{|}}{C}H\text{--}CH_3 \xrightarrow{-H^+}$$

   2-hydroxy-3, 3-dimethyl-   carbon-2   carbon-3
          butane   2° carbocation $\longrightarrow$ 3° carbocation
                               rearranges

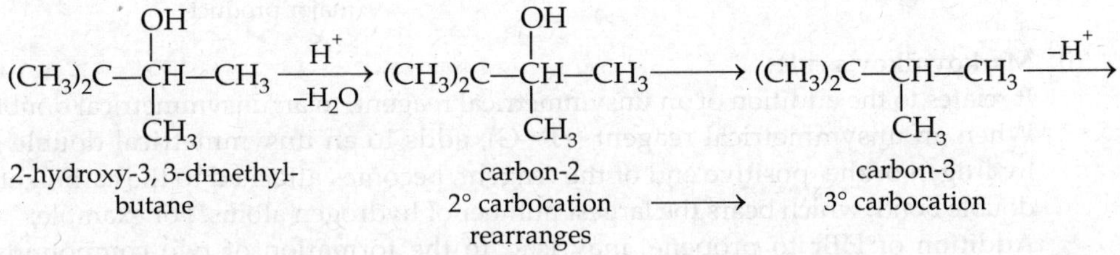

   The 2° carbocation is converted to a 3° carbocation, by the shift of a $-CH_3$ group, from carbon 2 to carbon 3, along with the pair of the bonding electrons, as a –vely charged species, termed as the methide group ($-CH_3$).

   ii. *A hydride shift*:
   Dehydration of 1-butanol

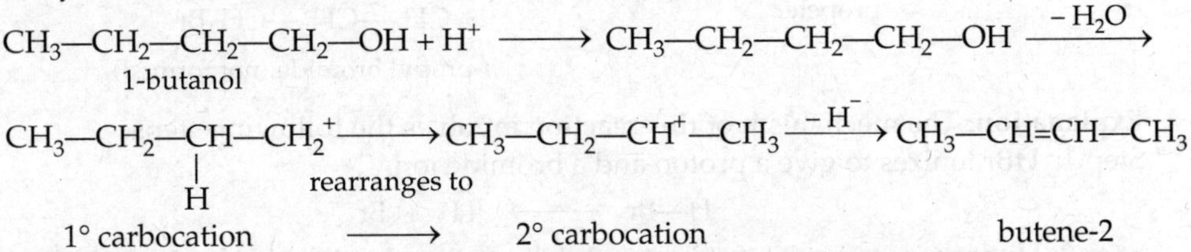

$$CH_3\text{--}CH_2\text{--}CH_2\text{--}CH_2\text{--}OH + H^+ \longrightarrow CH_3\text{--}CH_2\text{--}CH_2\text{--}CH_2\text{--}OH \xrightarrow{-H_2O}$$
           1-butanol

$$CH_3\text{--}CH_2\text{--}\underset{\underset{\displaystyle H}{|}}{C}H\text{--}CH_2^+ \longrightarrow CH_3\text{--}CH_2\text{--}\overset{+}{C}H\text{--}CH_3 \xrightarrow{-H^-} CH_3\text{--}CH{=}CH\text{--}CH_3$$
                   rearranges to
     1° carbocation $\longrightarrow$ 2° carbocation   butene-2

   1° carbocation is converted to a more stable 2° carbocation by the shift of a hydrogen atom along with the pair of bonding electrons as a hydride ion, (H^-) from carbon 1 to carbon 2.

19. **Saytzeff's rule:**

   a. When alkyl halides are dehydrohalogenated, alkenes are formed. If the dehydrohalogenation of an alkyl halide can yield more than one alkene, then according to *Saytzeff's rule, the main product is the most highly substituted alkene.* For example, two alkenes are possible, when 2-bromobutane is heated with alcoholic KOH.

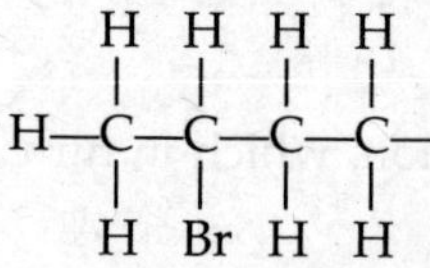

$$H-\underset{\underset{H}{|}}{\overset{\overset{H}{|}}{C}}-\underset{\underset{Br}{|}}{\overset{\overset{H}{|}}{C}}-\underset{\underset{H}{|}}{\overset{\overset{H}{|}}{C}}-\underset{\underset{H}{|}}{\overset{\overset{H}{|}}{C}}-H \longleftrightarrow$$

2-bromobutene

$CH_2=CH-CH_2CH_3$ 1-butene (20%)

$\xrightarrow{OH^-}$ $CH_3-CH=CH-CH_3$

2-butene (80%)

b. When unsymmetrical, secondary (or tertiary) alcohols are dehydrated with conc. H_2SO_4 elimination can proceed in two different ways and a mixture of two alkenes is obtained. According to *Saytzeff's rule, in case there are two possibilities of elimination of a hydrogen, it is preferentially eliminated from the carbon atom with fewer number of hydrogen atoms.* In other words, the major product will be the alkene, with the larger number of alkyl groups, attached to the doubly bonded carbon atoms.

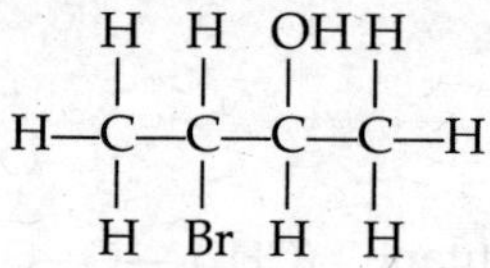

$$H-\underset{\underset{H}{|}}{\overset{\overset{H}{|}}{C}}-\underset{\underset{Br}{|}}{\overset{\overset{H}{|}}{C}}-\underset{\underset{H}{|}}{\overset{\overset{OH}{|}}{C}}-\underset{\underset{H}{|}}{\overset{\overset{H}{|}}{C}}-H \longrightarrow$$

2-butanol

$CH_2=CH-CH_2-CH_3$
1-butene (20%)
(minor product)

$CH_3-CH=CH-CH_3$
2-butene (80%)
(major product)

20. **Markovnikov's rule:**

It relates to the addition of an unsymmetrical reagent, to an unsymmetrical double bond. When an unsymmetrical reagent (H—G), adds to an unsymmetrical double bond, the hydrogen or the positive end of the reagent, becomes attached to the carbon atom of the double bond, which bears the largest number of hydrogen atoms. For example:

Addition of HBr to propene, may lead to the formation of two compounds, namely *n*-propyl bromide and isopropyl bromide, however *only isopropyl bromide is formed.*

$CH_3-CH=CH_2$ $\xrightarrow{HBr}$

propene

$CH_3-CHBr-CH_3$
isopropyl bromide (formed)

$CH_3-CH_2-CH_3Br$
n-propyl bromide (not formed)

Explanation: The mechanism of this reaction involves the following steps.

Step 1: HBr ionizes to give a proton and a bromide ion.

$$H-Br \longrightarrow H^+ + Br^-$$

Step 2: Hydrogen or a proton (electrophile), attacks the double bond, to give a more stable carbonium ion or a carbocation.

$$CH_3-CH=CH_2 + H^+ \longrightarrow$$

$CH_3-CH^+-CH_3$
a 2° carbocation

$CH_3-CH_2-CH_2^+$
a 1° carbocation

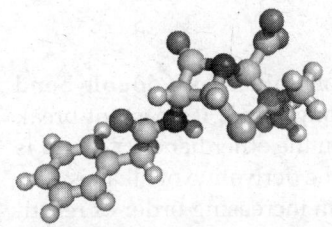

Paper 5

REASONING TYPE QUESTIONS (RTQs)

1. Discuss the stabilities of alkenes. Arrange them in the decreasing order of their stabilities.
2. How a cyclohexane ring can be opened to give the corresponding open chain alkane. Also check out the steps by which the *n*-hexane be converted into cyclohexane ring again.
3. Discuss Arndt–Eistert synthesis or reaction. What are its unique utility, which other synthetic methods cannot give?
4. What is Beilstein test? What it is used for? Discuss its chemistry.
5. What are the various components of pyroligneous acid? How these can be separated? Give chemical reactions in support to your suggestions.
6. Give the definition of fermentation. Discuss the various examples where the process of fermentation is used to manufacture the required products.
7. Glycol is poisonous, when consumed but glycerol is not poisonous, when consumed. Discuss the chemistry involved.
8. What are oxonium salts? Discuss their formation and stability.
9. Explain why alkanes are inert and relatively unreactive.
10. Explain the utility of the process of 'pyrolysis' used in chemistry. How catalytic cracking is used in alkane chemistry.
11. How does propene reacts with Cl_2 at 500°C? Name the product and suggest a mechanism for the reaction.
12. Write a short note on Baeyer's strain theory.
13. Explain, why the carbon–carbon bonds in cyclopropane are weaker than the carbon–carbon bonds in propane.
14. Write a short note on (i) tautomerism, (ii) resonance.
15. Write a short note on the stability of conjugated dienes.
16. Give the evidence that chlorination of methane, involves a free radical mechanism.
17. Explain, why formic acid behaves as a reducing compound, but other fatty acids do not?
18. Name the compounds, which neutralize dil. acidic or alkaline $KMnO_4$ solution. Explain on the basis of chemical reactions involved.
19. Methanol and ethanol are both monohydroxy alcohols, yet methanol is poisonous, when consumed, but ethanol is not. Explain.
20. Name the substances the would inhibit the chlorination of methane, proceeding by a free radical mechanism.

ANSWERS

1. **Alkenes are hydrocarbons, containing one pair of carbon atom joined by a double bond consisting of one σ-bond and one π-bond, σ-bond is a strong bond**, it does not break under the conditions to which alkenes may be subjected to. On the other hand **a π-bond is a weak bond**, that opens easily and adds the reagent to form the derivative of alkanes.

   The decreasing order of relative stabilities of the alkenes or an increasing order of reactivities of alkenes is as follows-

$$R_2C=CR_2 > R_2C=CHR > R_2C=CH_2 > RCH=CHR > RH=CH_2 > CH_2=CH_2$$

   A common test for unsaturation is the decolourisation of yellow colour of bromine water, which is shown by all alkenes except $R_2C=CR_2$, which does not decolourise bromine water.

2. **A cyclohexane ring may be opened and to be closed by the following steps.**

   opening of the ring

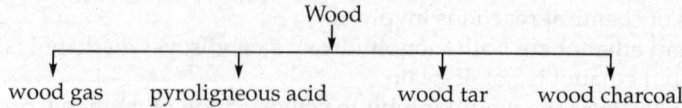

   cyclohexane   bromoderi   cyclohaxane   hexane-1,6-di al

   closure of ring

   n-haxene   benzene   cyclohaxene

   (n-haxene → Pt/Al$_2$O$_3$, 500 C, 15 atm → benzene → catalytic red H$_2$/Nl → cyclohaxene)

3. **Arndt–Eistert reaction:** It is a reaction that can give immediate higher homologue of any fatty acid or a mono carboxylic acid in a two step reaction procedure. An acid is treated with thionyl chloride, to give the corresponding acid chloride, which on reaction with diazomethane (CH_2N_2) gives a diazoketone, which on warming with water converted to a ketene, then add a molecule of water to form the next higher fatty acid.

$$RCOOH + SOCl_2 \longrightarrow RCOCl \longrightarrow RCOCHN_2$$

   acid chloride   diazoketone

$$\xrightarrow{CH_2N_2H_2O/Ag_2O} RCH=C=O \xrightarrow{H_2O} RCH_2COOH$$

   ketene

4. **Beilstein test:** The test is used to detect the presence of halogen in the organic compound. The compound is heated, placed on the tip of copper wire in the naked flame. Upon heating for a while, the halogen present in the organic compound is converted into volatile copper halide which imparts a blue or green colour to the flame. Though the test is sensitive, but it is not reliable. Substances like urea, which does not contain any halogen, also colours the flame green.

5. **Pyroligneous acid (pyro = heat; ligneous = of wood):** When wood is heated to ~400°C, in iron retorts in the absence of air (destructive distillation) the products obtained are:

   Wood

   wood gas   pyroligneous acid   wood tar   wood charcoal

Pyroligneous acid is an aqueous distillate, in addition to water it contains

$CH_3OH = 3\%$	$CH_3COCH_3 = 0.5\%$	$CH_3COOH = 10\%$
methanol	acetone	acetic acid
BP $= 65\,°C$	$56\,°C$	$118\,°C$

Separation of the constituent: The pyroligneous acid first treated with milk of lime, $Ca(OH)_2$, when CH_3OH and CH_3COCH_3 distil over and collected. They are separated by a fractional distillation, as their boiling points are very close to each other. Acetic acid remains in the retort, as a solid, calcium acetate $(Ca(OCOCH_3)_2)$ is distilled with conc. H_2SO_4, when CH_3COOH distils over at $118\,°C$.

$$CH_3COOH + Ca(OH)_2 \longrightarrow Ca(OCOCH_3)_2 + H_2O$$
$$Ca(OCOCH_3)_2 + H_2SO_4 \longrightarrow CH_3COOH + CaSO_4$$

6. **Fermentation:** It may be defined as the slow decomposition of big molecules of certain organic compounds, into simple ones under the influence of nonliving, complexes substances, known as ferments.

   i. **Alcoholic fermentation:** Conversion of sugar into ethanol, by **yeast**, is called as alcoholic fermentation.

   Method
   - Dil. solution of cane sugar is treated with yeast, an enzyme, **invertase** present in it, hydrolyses it to form a molecule of glucose and a molecule of fructose

$$C_{12}H_{22}O_{11} + H_2O \xrightarrow{\text{invertase}} C_6H_{12}O_6 + C_6H_{12}O_6$$

   canesugar   glucose   fructose

   - Glucose is then acted upon by another enzyme, '**zymase**', present in yeast, break it into ethanol and CO_2.

$$C_6H_{12}O_6 \longrightarrow C_2H_5OH + 2CO_2$$

   (fructose also ferments to form alcohol, but the process is quite slow)

   ii. **Acetone:** It can also be prepared by fermentation of starch and molasses.

   iii. **Glycerol or glycerine:** Glycerol can also be prepared by the fermentation of molasses and sugar, in the presence of Na_2SO_3 (sodium sulphite).

$$C_6H_{12}O_6 \xrightarrow{Na_2SO_3} CH_3CHO + CH_2OH\!-\!CHOH\!-\!CH_2OH$$

   Acetaldehyde glycerol upon hydrolysis with water, forms alkaline solution, which favours formation of glycerol. As high as ~25% glycerol has been obtained by this process.

   iv. **Quick vinegar process:** Acetic acid is manufactured by this process, from fermented liquor, containing ~12-15% of ethanol, by aerobic-oxidation, in presence of air, by mycoderma acetii.

7. **Glycol is poisonous**, when consumed orally, because, in the system, it is oxidised to oxalic acid, which is poisonous in nature.

$$\begin{array}{c} CH_2OH \\ | \\ CH_2OH \end{array} \xrightarrow{\text{in human body}} \begin{array}{c} COOH \\ | \\ COOH \end{array}$$

   glycol   oxalic acid

On the other hand, glycerol is present in the oils and fats, which are consumed in large quantities by humans. In the human body, oils and fats are hydrolysed to glycerol, and monocarboxylic acids.

$$
\begin{array}{l}
CH_2OCOR_1 \\
| \\
CHOCOR_2 \\
| \\
CH_2OCOR_3
\end{array}
\quad\longrightarrow\quad
\begin{array}{l}
CH_2OH \\
| \\
CHOH \\
| \\
CH_2OH
\end{array}
\quad + \quad 3R'COOH \quad (R' = R_1, R_2, R_3)
$$

oil and facts   glycerol

8. **Oxonium salts:** Ethers are weakly basic in nature, just as alcohols. These react with H_2SO_4 or HCl to form oxonium salts.

$$
C_2H_5{-}O{-}C_2H_5 + H^+ + HSO_4^- \longrightarrow (C_2H_5{-}\overset{\overset{\displaystyle H}{|}}{O}{}^+{-}C_2H_5)\,HSO_4^-
$$

oxonium salt of diethyl ether

Oxonium salts are stable and soluble in highly conc. acidic solution. On treatment with water, these salts are decomposed to regenerate ether, because water is a stronger base as compared to ether.

9. **Nonreactive nature of alkanes:** Alkanes are nonreactive as these compounds undergo only a few reactions, unlike other hydrocarbons like alkenes, alkyne and other aromatic compounds. Alkanes do not react with alkali, acids, oxidising agents at room temperature. **This is because the C—C and C—H bonds are nonpolar**

10. **Pyrolysis or Cracking:** The decomposition of a compound, under the influence of heat is termed pyrolysis (pyre = fire; lysis = loosening).
The process, when applied to alkane is known as cracking. When alkanes are heated to a temperature 500–800 °C) in the absence of air, a thermal decomposition occurs. The large alkane molecules break down in a mixture of smaller and lower weight alkanes, alkenes and hydrogen.

    The process of cracking can be carried at a lower temperature also in the presence of a finely divided catalyst consisting of 'silica-alumina' and the process is termed 'catalytic cracking' which involves a free radical mechanism.

11. A reaction **of propene with Cl_2** at elevated temperature of ~500 °C follows a *free radical mechanism* and the compound formed is 3-chloropropene or allyl chloride, $CH_2Cl{-}CH = CH_2$

$$
\underset{\text{propene}}{CH_3{-}CH = CH_2} + Cl_2 \xrightarrow{500\,°C} \underset{\text{allyl chloride}}{CH_2Cl{-}CH = CH_2}
$$

12. **Baeyer's strain theory:** *Stabilities of cycloalkanes*–cycloalkanes are as stable as open chain alkanes. Cyclohexane onwards, the cycloalkanes assume a **puckered structure,** which result in an angle of 109°.28′ in between C to C. The larger rings prepared by Ruziicka, containing as many as 32 carbon atoms, were found to be stable as any other alkanes.

13. **The cyclopropane ring is a monoplaner ring:** The three carbons occupy three corners of an equilateral triangle, and cyclopropane has a bond angle of 60°. In forming a mono-planer ring, each carbon to deviate from normal angle of 109°.28′, to form an angle of 60°, causing a strain in the molecule. This may be called as '**angle strain**'. If the strain in the ring be considered to be equal to the 'angle strain', then the angle strain may be calculated as $\frac{1}{2}$ (109°.28′ – 60°) = 24°.64′. It is the angle by which each of the bond involved have to be pulled to form an angle of 60°.

As cyclopropane has maximum strain in the molecule, therefore it is most unstable. The ring opens with slight provocation, releasing the strain in the molecule.

In the same way, angle strain may be calculated for other rings.

	bond angle	angle strain
Cyclobutane	$90°$	$9°44'$ $[(1/2)\ 109°.28' - 90° = 9°.44']$
Cyclopentane	$108°$	$0.44'$ $[(1/2)\ 109°.28' - 108° = 0.44']$
Cyclohexane	$120°$	$-5°16$ $[(1/2)\ 109°.28 - 120° = -5°.16']$

14. **a. Tautomerism** is used for those compounds which give the reactions of two different groups present, while the compound remains the same. Acetoacetic ester or ethyl acetoacetate ($CH_3COCH_2COOC_2H_5$) is one such example. On account of migration of a hydrogen from one polyvalent atom to another polyvalent atom in the same molecule, the compound behaves as a keto compound as well as an enol (ene + ol) compound.

$$CH_3COCH_2COOC_2H_5 \rightleftharpoons CH_3C(OH) = CHCOOC_2H_5$$

     a keto form   an enol form

b. Resonance is used for certain compounds which cannot be represented by a single structure. In other words, a single structure cannot explain the properties, shown by the compound. For example: benzene shows the properties of a saturated compounds as it undergoes substitution reactions, though the molecule is highly unsaturated, and contains three double bonds in the molecule. It has been assumed that benzene is a resonance hybrid of the following structures.

15. **Stabilities of dienes:** Dienes are highly reactive compounds, on account of the presence of two double bonds in the molecule.

The stabilities of the dienes are as follows.

$$CH_2 = CH - CH = CH_2 > CH_2 = CH - CH_2 - CH = CH_2 > CH_2 = C = CH_2$$

    1,3-butadiene   1,4 pentadiene   propadiene or allene

    Conjugated diene   isolated dienes   cumulative diene

The relative stability of allenes reflects an extra strain, as a result of one carbon forming two double bonds.

16. There is **no reaction between Cl_2 and CH_4 in the dark.** However, the two molecules react within the presence of either sunlight or on heating. Methane reacts with Cl_2 by the free radical mechanism and the proof is provided by treating the reacting mixture with a catalytic amount of dibenzoyl peroxide ($C_6H_5COO)_2$.

$$Cl : Cl \xrightarrow{\text{sunlight } \Delta} Cl\bullet + Cl\bullet$$

$$CH_4 + Cl \longrightarrow CH_3Cl + CH_2Cl_2 \text{ etc}$$

Dibenzoyl peroxide break up in absence of light to form a phenyl free radical, which reacts with chlorine molecule to form atomic chorine. Atomic chlorine so formed reacts with methane to form chloromethane.

$$(C_6H_5COO)_2 \xrightarrow{\text{dark}/25°C} 2C_6H_5COO\bullet \longrightarrow C_6H_5\bullet + CO_2$$

$$C_6H_5\bullet + Cl : Cl \longrightarrow C_6H_5Cl + Cl\bullet$$

17. **Formic acid, HCOOH:** Formic acid is a powerful reducing agent. It reduces mercuric chloride ($HgCl_2$) to mercurous chloride (Hg_2Cl_2).

$$2HgCl_2 + HCOOH \longrightarrow Hg_2Cl_2 + 2HCl + CO_2$$

In formic acid, the C—H bond is easily oxidised to C—OH group with this formic acid is converted into a dihydroxy compound in which two —OH groups are joined to the same carbon. As such compounds are unstable; lose a molecule of water and finally form H_2O and CO_2.

$$H-C{\overset{O}{\underset{OH}{}}} + O \longrightarrow HO-C{\overset{O}{\underset{OH}{}}} \longrightarrow CO_2 + H_2O$$

Other fatty acids do not contain a $-C=O$, group in the molecule, therefore all other fatty acid do not behave as reducing compounds.

18. **i. An aqueous solution of $KMnO_4$ is neutralized by (i) HCOOH (ii) Oxalic acid.**

   $2KMnO_4 + 3H_2SO_4 \longrightarrow K_2SO_4 + 2MnSO_4 + 3H_2O + 5O$

   a. $HCOOH + O \longrightarrow H_2O + CO_2$

   b. $\begin{matrix} COOH \\ | \\ COOH \end{matrix} + O \longrightarrow 2CO_2 + H_2O$

ii. An aqueous dilute alkaline solution of $KMnO_4$ is also neutralised by compounds containing olefinic double bonds and triple bonds.

   $2KMnO_4 \longrightarrow 2KOH + 2MnO_2 + 3O$

   a. Olefins are hydroxylated to form glycols.

   $$>C=C< + O + H_2O \longrightarrow >\overset{HO}{\underset{}{C}} - \overset{OH}{\underset{}{C}}<$$
   $$\text{glycol}$$

   b. Acetylene is oxidised oxalic acid; other alkynes are oxidised to form a mixture of fatty acids.

   $$HC{\equiv}CH + 4O \longrightarrow \begin{matrix} COOH \\ | \\ COOH \end{matrix} \quad \text{oxalic acid}$$

   $$R-C{\equiv}C-R' + 4O \longrightarrow RCOOH + R'COOH$$
   $$\text{mixture of fatty acids}$$

19. **Methanol is oxidised to formaldehyde (HCHO) in human body which is a powerful reducing agent, it reduces the blood, coagulates** and causes the arteries to block, eventually causes the death.

   $$CH_3OH + O \longrightarrow HCHO$$

   Ethyl alcohol, on the other hand is good when taken in small quantity. However, prolonged and excessive intake, may cause a permanent damage to liver.

20. The substances are termed **inhibitors** which would cause a reaction, proceeding by a free radical mechanism to bring termination to a reaction are termed **"radical traps"**. In chlorination of methane, **oxygen acts as an inhibitor**. Oxygen combines readily with methyl free radicals to form comparatively stable peroxymethyl, which terminates the chain reaction.

   $$CH_3{\bullet} + O_2 \longrightarrow CH_3-O-O{\bullet}$$
   methyl free radical $\qquad\qquad$ peroxy methyl free radical

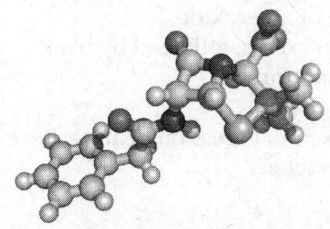

Part II

Assorted Questions

MULTIPLE CHOICE QUESTIONS

1. Name the products formed, when an aqueous solution of sodium acetate is electrolyzed.
   (a) alkane   (b) alkene   (c) alkyne   (d) all of these

2. Which drying agent will be used for drying ethyl alcohol?
   (a) conc. H_2SO_4 (b) P_2O_5 (c) $CaCl_2$ (d) CaO

3. Name the alkene that does not decolourise Br_2 water.

4. Anthranilic acid does not exist as a zwitterion.

5. Nitrobenzene does not undergo Friedel–Craft reaction. Which are the other compounds that also do not undergo F–C reaction.

6. Identify the positions which have maximum electron density in m-amino phenol.

7. Which reagent will be used to introduce a *n*-propyl group in the benzene ring.
   (a) *n*-propyl chloride   (b) isopropyl chloride
   (c) propionyl chloride

8. Name the alkene which will be formed in major quantity, when 2-bromobutane is treated with potassium tertiary butoxide.

9. Name the electrophile generated in nitration, chlorination and alkylation of a benzene ring.

10. Which of the following has higher heat of hydrogenation?
    (i) $R_2C = CR_2$ (ii) $CH_2 = CH_2$

11. The compound formed when β hydroxy propenoic acid is heated.

12. Reaction of iodine on *cis*-2-butene, followed by heating

13. Chlorination of propene at 500 °C.

14. Action of heat on $(CH_3)_4OH$. Name the product formed.

15. Action of conc. H_2SO_4 on methanol.

16. Identify the products, 'X', 'Y' and 'Z' in the following set of reactions.

$$R-C{\equiv}C\,H \xrightarrow{BH_3} X \xrightarrow{\text{alkaline hydrolysis}} Y \xrightarrow{CH_3COOH} Z$$

17. Name the products formed, by the reaction of $NaNH_2$ on 2-butyne at 170 °C.

18. Name the strongest Lewis base: NH_2^-, OH^-, $C_2H_5^-$, $C_2H_5O^-$
    Arrange these in the decreasing order of their basic character.

19. Name the product formed in a reaction between CH_3CHO and HCN, followed by hydrolysis. Discuss the stereochemistry of the product formed.

20. Reaction product from cinnamic acid and HBr in the presence of a peroxide.
21. Complete the following reaction-$CH_2 (NH_2) COOH + NaOH$ in excess, followed by heat.
22. Name the product formed on nitration of aniline by nitrating mixture.
23. What is the composition of constant boiling alcohol?
24. The product formed in a reaction between glycerol and HI, taken in excess on heating.
25. Arrange the following in the decreasing order of the basic character.

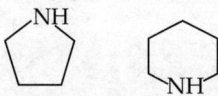

$R_2NH > RNH_2 > C_6H_5NH_2$
26. Product formed when acetamide is heated with NaOH solution.
27. What is the relation in between CH_3COCH_3 and $CH_3—C(OH)CH_3$?
28. Reaction of R_3CX with $LiAlH_4$ or with KCN.
29. Glyoxal is heated with NaOH solution.
30. Which chloro compound is more reactive?
    (a) $C_6H_5CH_2Cl$; C_6H_5Cl ; o-$C_6H_4 (CH_3) Cl$
    (b) $CH_2Cl—CH=CH_2$; $CH_2=CHCl$; $CH≡C—Cl$; CH_3Cl
31. Which has higher dipole moment value?
    $CH_3F, CH_3Cl, CH_3Br, CH_3I$
32. When benzene is methylated with CH_3Cl, in the presence of anhydrous $AlCl_3$, the product formed is (a) toluene (b) polymethyl benzene (c) dimethyl benzene
33. How chlorine in C_6H_5Cl can be made more reactive?
34. Why o-nitrophenol is steam volatile but p-nitrophenol is not?
35. Product formed when diethyl ether reacts with chlorine in the presence of sunlight.
36. Phenol does not liberate CO_2, when reacts with $NaHCO_3$ but acids do.
37. Sodium metal can be preserved in ether but not in ethanol.
38. What is the function of $ZnCl_2$ in Luca's reagent?
39. Both glycerol and fats or oils give acrolein test, so how to distinguish between them.
40. Bond angle in ether is
    (a) 110°   (b) 106°.5'   (c) 105°   (d) none
41. The products formed on heating anisole with HI.
42. How to distinguish between methanol and ethanol.
43. Product formed when acetamide reacts with P_2O_5.
44. Pure HCN does not react with CH_3CHO.
45. *cis*- and *trans*-isomerism is shown by
    (a) alkenes   (b) aldoximes
    (c) hydrazone derivatives   (d) all
46. α- and β-glucose are
    (a) epimers   (b) anomers   (c) tautomers   (d) all
47. Monochlorination of sodium benzoate leads to the formation of
    (a) an o-disubstituted derivative   (b) a p-disubstituted derivative
    (c) a m-disubstituted derivative   (d) both (i) and (ii).
48. Compound obtained on heating malonic acid is
    (a) an anhydride   (b) an acid   (c) a hydrocarbon   (d) none
49. What are the number of optical isomers in the case of tartaric acid?
    (a) 2   (b) 3   (c) 4   (d) more than 4
50. The bond length of sodium formate in between carbon and oxygen are
    (a) 1.20 Å and 1.40 Å   (b) 1.27 Å and 1.27Å
    (c) 1.40 Å and 1.20 Å   (d) none.

51. Which acid is stronger than the other?
    (a) $CH_2=CH-COOH$ (b) $CH\equiv CH-COOH$
    (c) none
52. Compound formed on distillation of a calcium salt of adipic acid.
    (a) cyclopentanone   (b) cyclohexanone
    (c) cyclohexane   (d) none
53. Formic acid is a resonance hybrid of how many structures?
    (a) 1   (b) 2   (c) 3   (d) more than 3.
54. Acidified $KMnO_4$ is decolourised by
    (a) formic acid   (b) oxalic acid
    (c) acetic acid   (d) both by (a) and (b).
55. RCN and RNC are more polar than RX, because of
    (a) van der Wall's forces   (b) dipole-dipole attraction
    (c) hydrogen bonding   (d) none
56. IUPAC name for the compound CH_2-CH_2 is
    $$\overset{NH}{\underset{CH_2-CH_2}{\diagup\diagdown}}$$
    (a) aziridine   (b) aza cyclopropane
    (c) both (a) and (b)   (d) none
57. The number of possible isomers formed in a reaction between 2-pentyne and HBr.
    (a) 2   (b) 3   (c) 1   (d) none.
58. Glycerol on oxidation with Fenton's reagent forms a compound that is used as a reference compound in "absolute configuration".
    (a) dihydroxy acetone   (b) glyceric aldehyde
    (c) glyceric acid   (d) none of these
59. Alkyl isocyanide on reduction forms a compound that possesses reducing properties. The compound is
    (a) alkyl amines   (b) methanoic acid
    (c) HCN   (d) none of these
60. On fusing nitrobenzene with solid KOH, the compounds formed are
    (a) m-nitrophenol   (b) o- and p- nitrophenols
    (c) phenol   (d) none
61. The calcium salt of α-dibasic acids on heating produces compounds containing
    (a) same number of carbon atoms   (b) one carbon less
    (c) two carbons less   (d) none of these
62. Name the alkane, with minimum molecular wt, showing optical isomerism.
    (a) 3-methyl hexane   (b) octane
    (c) isomer of heptane   (d) none
63. An alkane, with molecular wt 72 on monochlorination, gives only one single mono-chlorinated alkane, it is
    (a) pentane   (b) isomer of pentane
    (c) neopentane   (d) none of these
64. Which of the following contain more acidic hydrogen?
    (a) alkane   (b) alkene
    (c) alkyne   (d) benzene
65. Conjugation in dienes can be detected by heating it with
    (a) maleic anhydride   (b) acrolein
    (c) an enophile   (d) all of these

66. Which of these are/is optically active in the crystalline form?
    (a) $NaClO_4$ (b) benzyl   (c) $ZnSO_4$ (d) N_2H_4
    (e) $(HCOO)_2Ca$ (f) quartz   (g) all

67. Oxidation of neo butyl benzene by Bayer's reagent gives
    (a) benzoic acid   (b) benzophenone
    (c) benzene   (d) it is not oxidised

68. Acylation of benzene, by Friedel–Craft reaction, gives
    (a) a mono acyl derivative   (b) a diacyl derivative
    (c) a ploy acyl derivative   (d) none

69. What is the effect of adding oxygen, in a reaction undergoing a free radical reaction, like chlorination of methane, in the presence of sunlight.

70. The product formed is a/an

    $R-CH=N-OH + P_2O_5 \xrightarrow{\Delta} X:$

    (a) amide   (b) amine   (c) chloride   (d) none

71. What is the effect of adding diphenyl amine, $(C_6H_5)_2NH$ or catechol o-$C_6H_4(OH)_2$, in a reaction between an unsymmetrical alkene, reacting with HBr, in the presence of a peroxide.
    (a) it becomes faster   (b) it becomes slower
    (c) no effect   (d) addition follows Markovnikov's rule

72. Compare the K_a1 and K_a2 values in the case of maleic acid and fumaric acids.

73. An alkene, with no unsymmetrical carbon atom, but optically active.

74. The number of different bromo substituted products possible, when ethane is allowed to react with bromine in the presence of sunlight.

75. The number of structural isomers possible (open and cyclic) for an organic compound with mol. formula C_5H_{12}.

76. Chiral molecule are those, which are
    (a) nonsuperimposable on their mirror images
    (b) superimposable on their mirror images
    (c) show geometrical isomerism
    (d) unstable molecules

77. The most stable carbonium ion is
    (a) $CH_3-CH_2-CH^+-CH_3$ (b) $CH_3-CH^+-CH_3$
    (c) $CH_3-CH_2-CH_2-CH_2^+$ (d) CH_3^+

78. Among the following orbitals the angle is minimum in between
    (a) sp^3 bonds,   (b) p_x-p_y orbitals,
    (c) H–O–H,   (d) sp bonds.

79. Which of the following is an electrophile?
    (a) RO—   (b) BF_3 (c) NH_3 (d) R—OH

80. The compound with highest bp
    (a) n-hexane   (b) n-pentane
    (c) 2-methyl pentane   (d) 2,2-dimethylpropane

81. In which of the following, the distance between two adjacent carbon atoms is the largest?
    (a) C_2H_6 (b) C_2H_4 (c) C_6H_6 (d) C_2H_2

82. Which among the following would give acetone on ozonolysis?
    (a) 2-butene   (b) 1-butene
    (c) 2-methyl-1-propene   (d) propene

83. Toluene is subjected to chlorination in the presence of sunlight and heat followed by treatment with NaOH, what would be the most likely compound formed?

84. The correct order of the solubility of the alcohols in water would be
    (a) $3° > 2° > 1°$ (b) $1° > 2° > 3°$
    (c) $3° > 1° > 2°$ (d) none
85. Phenol is
    (a) a weaker base than ammonia   (b) stronger than carbonic acid
    (c) weaker than carbonic acid   (d) none
86. Between 1-nitrophenol and salicylaldehyde, the solubility in a base is
    (a) nil in both the cases   (b) higher in case of 1-nitrophenol
    (c) higher in case of salicylaldehyde   (d) equal for both
87. Which of the following has a lowest bp?
    (a) phenol   (b) o-nitrophenol
    (c) m-nitrophenol   (d) p-nitrophenol
88. Carboxylic reacts with diazomethane to form
    (a) an amine   (b) an ester
    (c) alcohol   (d) an amide
89. What is formed on heating glutaric acid?
90. Ethylene glycol on oxidation with periodic acid gives
    (a) oxalic acid   (b) glyoxal
    (c) glyollic acid   (d) formaldehyde
91. Aldol condensation between which of the following two compounds, followed by dehydration, gives methyl vinyl ketone.
    (a) $HCHO + CH_3COCH_3$ (b) $HCHO + CH_3CHO$
    (c) $CH_3CHO + CH_3CHO$ (d) $CH_3COCH_3 + CH_3COCH_3$
92. Which is considered to be more basic?
    (a) aniline   (b) methyl amine
    (c) ethyl amine   (d) hydroxyl amine
93. Strongest base in acidic medium is
    (a) chlorobenzene   (b) nitro benzene
    (c) aniline   (d) phenol
94. Which of the following forms the most stable carbonium ion, when attacked by an electrophile?
    (a) acetamide   (b) aniline
    (c) N,N-dimethylaniline,   (d) anisole
95. $p\text{-}NH_2\text{—}C_6H_4CHO$ does not show typical nucleophilic addition reaction, when treated with HCN.
96. Name the electrophile formed and involved in Reimer–Tiemann reaction.
97. Which of the following species is formed in Hofmann's bromamide reaction?
    (a) CH_3NCO (b) $CH_3CONHBr$
    (c) CH_3CONBr_2 (d) none of these.
98. How an amino group in aniline can be made a m-directing?
99. How a —COOH group in benzoic acid can be made an o- and p-directing group?
100. A group attached to the benzene ring, that acts as an electron withdrawing group from o- and p-positions but acts as an electron donating group from m-position.
101. Write the reaction in which a —NO_2 group, attached to the benzene ring acts as m-directing?
102. Optically active diphenyl derivatives do not contain a chiral carbon atom. Explain.
103. Give an example of a bicyclic nonbenzenoid aromatic ring.

104. i. 2-butyne is reduced by two different reducing reagents to form 2-butene, which are not identical. Name the reducing agents. Explain, why in one case the compound formed is different than in the other?

    ii. What is the product formed when 2-butyne is treated with oxygen in the presence of Cu_2^+ ions.

105. 2-butene is hydroxylated by two different hydroxylating reagents, give stereoisomeric butane-2,3-diol.

106. The product formed by the reaction between CH_3F and Na or Mg in presence of diethyl ether.

107. Ethane is chlorinated with excess of chlorine in the presence of sunlight.

    (a) How many chloro derivatives are possible?

    (b) How many pairs of isomers are formed?

108. Isopentane is monochlorinated.

    (a) How many isomers are possible?

    (b) On distillation how many fractions are obtained?

109. What happens when, an ozonide is treated with dimethyl sulphide?

110. What is the relation between an acid and its conjugate base?

111. Which of the two has a higher dipole moment: pentyne-1 and pentyne-2?

112. What is the number of isomers in case of C_2 Br, Cl, F, I?

113. Which is stronger bond, C—D or C—H?

114. What compound is formed when vinyl chloride, $CH_2=CH—Cl$, is treated with HI?

115. $C_6H_5CH_2—CH=CH_2 + HBr \longrightarrow$ complete the equation.

116. What is the relative heat of hydrogenation, in case of $CH\equiv CH$, C=O, $CH_2=CH_2$ and C_6H_6? Arrange them in the decreasing order.

117. Which will be most exothermic?

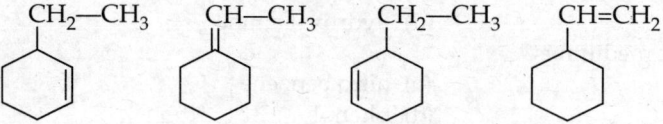

118. Write a short note on (i) borazine (ii) fullerene

ANSWERS

1. Kolbe's electrolytic reaction: a free radical reaction; ethane, an alkane is formed.
Sodium acetate (CH_3COONa) is an electrolyte, dissociates to form sodium ion and acetate ion. On switching the current, Na^+ ions move toward cathode, get an electron each and converted into sodium metal, attack water to form NaOH and hydrogen, which escapes as a gas. CH_3COO^- ions move to anode, lose an electron and converted into acetate free radicals. $CH_3COO\bullet$, free radicals, being unstable, decompose to form CO_2 and methyl free radicals ($CH_3\bullet$). Finally $CH_3\bullet$ free radicals combine to form ethane (C_2H_6).

$CH_3COONa \longrightarrow CH_3COO^- + Na^+$

At cathode

$Na^+ + e \longrightarrow Na;\ Na + H_2O \longrightarrow NaOH + H\uparrow$

At anode

$CH_3COO^- \longrightarrow CH_3COO\bullet + e\ ;\ CH_3 : COO\bullet \longrightarrow CH_3\bullet + CO_2$

$CH_3\bullet + CH_3\bullet \longrightarrow C_2H_6$

2. (d) CaO.

3. $(CH_3)_2C=C(CH_3)_2$

4. It is due to –I effect of the —NH_2 group.

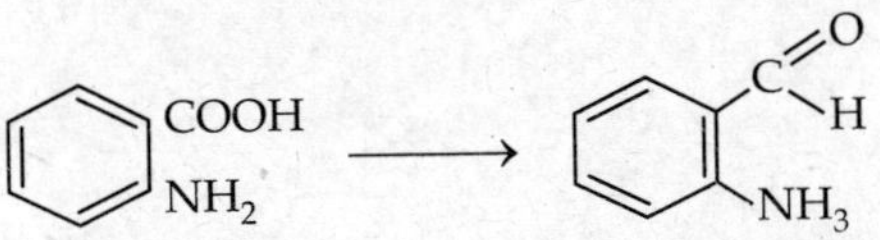

anthranilic acid

Availability of lone pair decreases on nitrogen.

5. C_6H_5X, X=CN, NO_2, SO_3H, CHO, COOH, $COCH_3$, $COOC_2H_5$ and —NH_2.
—NH_2 acts as a Lewis base and thus combines with $AlCl_3$. All other groups make the benzene ring electron deficient; hence a π-complex is not formed.

6. The positions that have a maximum electron density in m-aminophenol are ortho and meta positions.

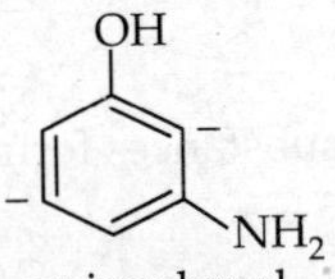

m–aminophenol

7. Propionyl chloride (CH_3—CH_2—COCl) will be used to introduce n-propyl group to the benzene ring in presence of anhydrous $AlCl_3$.

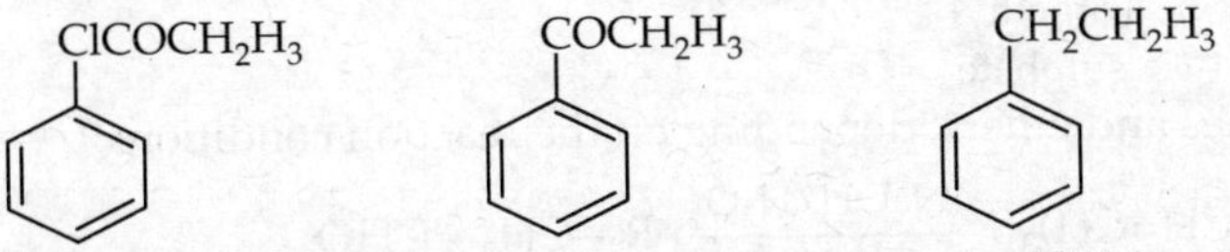

8. 1-butene will be formed as a major product. As $(CH_3)_3$—CO^-Na^+, being a big species, is not able to reach the middle carbon, hence the formation of 1-butene, as a major product.

9. Electrophile in nitration $\longrightarrow$ H—$\overset{\overset{\displaystyle H}{|}}{O^+}$—N=O $\longrightarrow$ NO_2^+

   nitronium ion

   Electrophile in chlorination $\longrightarrow$ Cl^+ H_2SO_4 $\longrightarrow$ H^+ HSO_4^-

   chloride ion

   Electrophile in sulphonation $\longrightarrow$ H—$\overset{\overset{\displaystyle O}{|}}{\underset{\underset{\displaystyle H}{|}}{O^+}}$—$\overset{\overset{\displaystyle O}{|}}{\underset{\underset{\displaystyle O}{|}}{S}}$=O $\longrightarrow$ $\overset{\overset{\displaystyle O}{||}}{\underset{\underset{\displaystyle O}{||}}{—S}}$=O

   sulphur trioxide

   Electrophile in alkylation $\longrightarrow$ R—X $\xrightarrow{AlCl_3}$ R^+ + $AlCl_4^-$

10. R_2—C=C—R_2 –28.6 kcal/mole (average value = 28.6 kcal/mol)
    CH_2=CH_2 32.6 kcal/mole

11. Action of heat on β-hydroxy propenoic acid

    $\underset{\underset{\displaystyle OH}{|}}{CH_2}$—$CH_2$—COOH $\xrightarrow{\Delta}$ CH_2=CH—COOH + H_2O

    β-hydroxy propionic acid   propenoic acid

12.

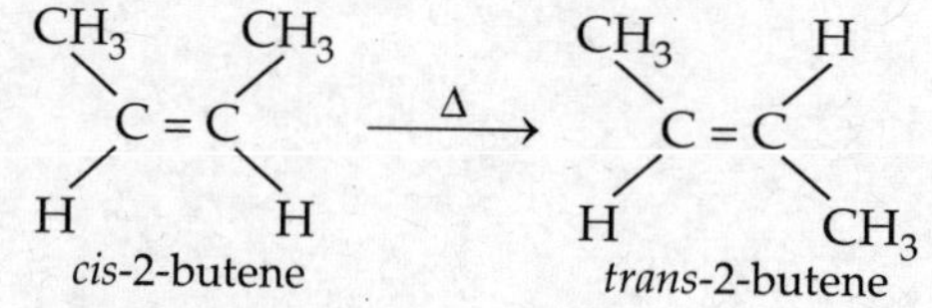

$$CH_3 \quad CH_3 \qquad\qquad CH_3 \quad H$$
$$\underset{H}{\overset{}{}}C=C\underset{H}{\overset{}{}} \xrightarrow{\Delta} \underset{H}{\overset{}{}}C=C\underset{CH_3}{\overset{}{}}$$

cis-2-butene   *trans*-2-butene

Iodine adds in *trans* manner, across the double bond on an alkene and on heating, iodine molecule is eliminated to give an isomer of alkene.

$$Cl_2 \xrightarrow{500\,°C} 2Cl\bullet$$

13. $CH_3-CH=CH_2 + Cl_2 \xrightarrow{500\,°C} ClCH_2-CH=CH_2 + HCl$

$Cl\bullet$ attacks the $-CH_3$ group, keeping the double bond intact, being a free radical reaction.

14. $(CH_3)_4\,NOH \longrightarrow CH_3OH + (CH_3)_3N$

tetraethylammonium   methanol   trimethyl amine
     hydroxide

In the case of higher alkyl groups, for example, $(C_2H_5)_4NOH$, the substances formed are an alkene, trialkylamine and water molecule.

$(C_2H_5)_4\,NOH \longrightarrow CH_2=CH_2 + (C_2H_5)_3N + H_2O$

     ethylene   triethylamine

15. $2CH_3OH + H_2SO_4 \xrightarrow{heat} (CH_3)_2SO_4 + 2H_2O$

     Dimethyl
     sulphate

Other alcohols, give alkenes and ethers, depending on the reaction conditions.

16. $3R-C\equiv CH \xrightarrow{BH_3} (R-CH=CH)_3B \xrightarrow{NaOH/H_2O} R-CH_2-CHO$

   alkyne   aldehyde

$\xrightarrow{CH_3COOH} R-CH=CH_2$

                    alkene

This reaction may be another method, to convert alkyne to alkene

17. $CH_3-C\equiv C-CH_3 \xrightarrow{NaNH_2/170\,°C} CH_3-CH_2-C\equiv CH$

  butyne-2   butyne-1

The double bond shifts from the middle position to a terminal position

18. $C_2H_5^- > NH_2^- > C_2H_5O^- > OH^-$

19. $CH_3CHO \xrightarrow{HCN} CH_3CH(OH)CN \xrightarrow{H_2O/H^+} CH_3CH(OH)COOH$

                                               lactic acid
                                         (optically active)

20. $C_6H_5CH=CH-COOH \xrightarrow{HBr} C_6H_5CH_2CHBr-COOH$

  cinnamic acid   β-phenyl-α bromo propionic acid

21. $CH_2(NH_2)COOH + NaOH \xrightarrow{\Delta} CH_2(OH)COONa \longrightarrow CH_3NH_2$

22. $-NH_2 + H^+ \longrightarrow NH_3^+$; it becomes m-directing as NH_3^+ becomes electron deficient

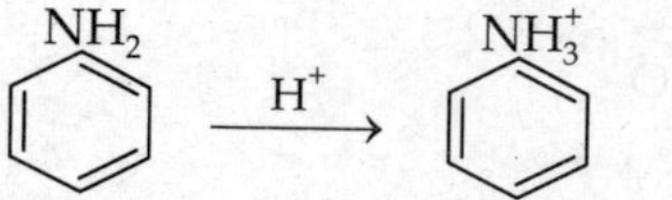

$$\underset{\text{aniline}}{\overset{NH_2}{\bigodot}} \xrightarrow{H^+} \underset{\text{anilinium ion}}{\overset{NH_3^+}{\bigodot}}$$

23. Azeotropic distillation

bp	alcohol	H_2O	C_6H_6
64.8°C	18.4%	7.4%	74.1%
68.2°C	32.4%	nil	67.8%

constant boiling alcohol

bp	78.15%	4.13%

24.

$$\begin{array}{ccccc}
CH_2OH + HI & & CH_2 & & CH_3 \\
| & & | & & | \\
CHOH + HI & \xrightarrow[\text{excess}]{HI} & CH & \xrightarrow{HI} & CHI \\
| & & | & & | \\
CH_2OH + HI & & CH_2I & & CH_3
\end{array}$$

25.

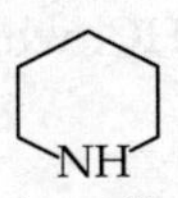

> 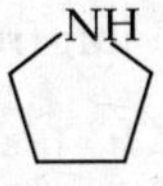 > R_2NH > RNH_2 > $C_6H_5NH_2$

   piperidine   pyrrolidine

K_b values

-2×10^{-3} 1×10^{-3} 6×10^{-4} 4.7×10^{-4} 4.2×10^{-10}

26. $CH_3CONH_2 + NaOH \xrightarrow{\Delta} NH_3 + CH_3COONa$

27. These are tautomers of acetone.

28. $R_3CX + LiAlH_4 \longrightarrow$ alkene + HX

29. $\begin{array}{l} CHO \\ | \\ CHO \end{array}$ + NaOH $\longrightarrow CH_2(OH)COOH$; an example of Cannizzaro's reaction

   glyoxal   glycolic acid

30. (a) $C_6H_5CH_2Cl$ > C_6H_5Cl > $o\text{-}C_6H_4(CH_3)Cl$

   (b) $\underset{sp^3}{CH_2Cl-CH=CH_2}$ > CH_3Cl > $\underset{sp^2}{CH_2=CHCl}$ > $\underset{sp}{CH\equiv C-Cl}$

31. CH_3Cl > CH_3F > CH_3Br > CH_3I

   $\mu = 1.91$ 1.82   1.79   1.65

32. toluene   m-xylene

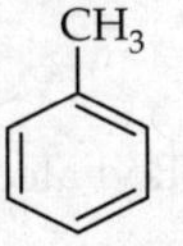

and

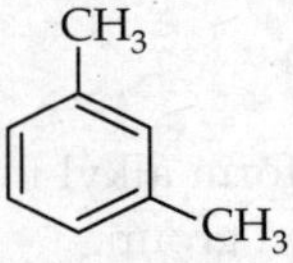

   thermodynamically more stable

33. By introducing an election withdrawing group, e.g. —NO_2 group, o- in p-and m–positions, —Cl becomes more reactive.

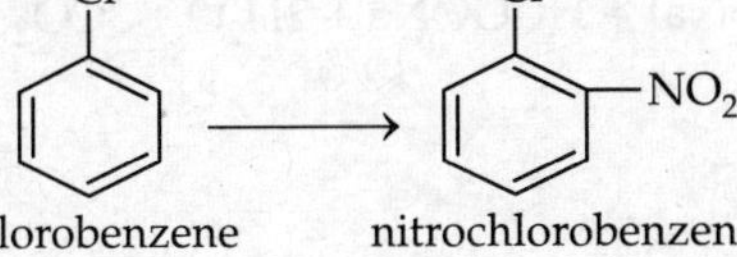

$\longrightarrow$

   chlorobenzene   nitrochlorobenzene

34. o-nitrophenol has an intramolecular hydrogen bonding.

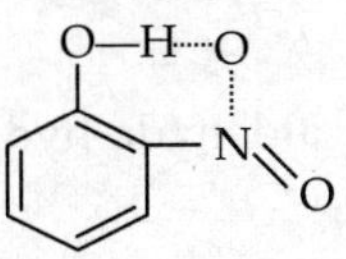

35. CCl_3—CCl_2—O—CCl_2—CCl_3
    perchloro diethyl ether

36. $C_6H_5OH + NaOH \longrightarrow C_6H_5ONa + H_2CO_3$
    $pK_a = 10$ stronger base $pK_a = 6$
    weak acid   strong acid

    Phenol does not liberate CO_2 from $NaHCO_3$, because in a reaction with alkali, it forms a stronger acid and a stronger base.

37. Ethanol reacts with sodium metal to form sodium ethoxide (C_2H_5ONa). Therefore, sodium metal cannot be preserved in ethyl alcohol.

    $C_2H_5OH + Na \longrightarrow C_2H_5O^-Na^+ + H$

38. The function of $ZnCl_2$ in the Luca's reagent is to liberate H^+ ion from HCl which will result in the formation of the R^+ radical.

    $ZnCl_2 + HCl \longrightarrow ZnCl_3^- + H^+$

$$ROH + H^+ \longrightarrow R\overset{\overset{\displaystyle H}{|}}{—}O^+—H \longrightarrow R^+ + H_2O$$
    alcohol   alkyl
                                    carbonium ion

39. Duncan's test for glycerol
    Phenolphthalein + borax $\longrightarrow$ a pink colour $\longrightarrow$ on adding glycerol, pink colour disappears, which reappears on heating again.

40. Bond angles

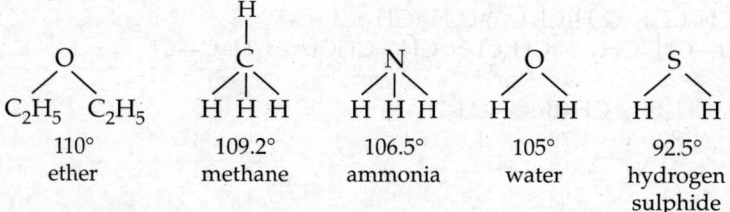

          110°   109.2°   106.5°   105°   92.5°
          ether   methane   ammonia   water   hydrogen
                                                    sulphide

41. C_2H_5—O—CH_3 + HI $\longrightarrow C_2H_5OH + CH_3I$
    Ethers are decomposed when heated with HI, to form alkyl iodide and alkyl alcohol. In unsymmetrical ethers, iodine goes with smaller alkyl group.

42. Ethyl alcohol gives a haloform test, but methanol does not give haloform test, and on warming ethyl alcohol with iodine and Na_2CO_3, a yellow precipitate of iodoform (CHI_3) separates.

    $CH_3CH_2OH + 4I_2 + 3Na_2CO_3 \longrightarrow CHI_3 + 5NaI + HCOONa + 2H_2O + 3CO_2$

43. $CH_3CONH_2 \xrightarrow[-H_2O]{P_2O_5} CH_3CN$
    methyl cyanide
    or acetonitrile

44. HCN is a weak acid, it does not ionise in the presence of water

    $HCN + H_2O \longrightarrow H_3O^+ + CN^-$

45. (a) Alkenes exist in *cis* and *trans* forms; aldehyde-oximes and aldehyde-hydrazone derivatives exist in syn- and anti-forms.

A–C=C with A, B substituents *cis-* and *trans-* **alkenes**; C=N oximes *syn* and *anti*; N=N hydazone derivatives *syn* and *anti*

$$\underset{B}{\overset{A}{\diagdown}}C=C\underset{B}{\overset{A}{\diagup}} \quad \underset{B}{\overset{A}{\diagdown}}C=C\underset{A}{\overset{B}{\diagup}} \quad \underset{R}{\overset{}{\diagdown}}C=N\diagdown_{OH} \quad \underset{R}{\overset{OH}{\diagup}}C=N \quad \underset{R}{\overset{}{\diagdown}}N=N\diagdown_{OH} \quad \underset{R}{\overset{OH}{\diagup}}N=N$$

$\qquad$ *cis-* $\qquad$ *trans-* $\qquad$ syn $\qquad$ anti $\qquad$ syn $\qquad$ anti

$\qquad$ alkenes $\qquad\qquad\qquad$ oximes $\qquad\qquad$ hydazone derivatives

46. (b) Glucose is a cyclic compound, and it exists in two isomeric forms. The two forms are α-glucose and β-glucose. These two isomers differ only in the configuration at carbon number 1 and are termed anomers.

$$\begin{array}{cc} \overset{H\quad OH}{\diagdown\diagup} & \overset{HO\quad H}{\diagdown\diagup} \\ C{-}\!\!\!\!-\!\!\!\!\underset{|1}{} & C{-}\!\!\!\!-\!\!\!\!\underset{|1}{} \\ (CHOH)_3\ O & (CHOH)_3\ O \\ |2 & |2 \\ CH{-}\!\!\!\! & CH{-}\!\!\!\! \\ |3 & |3 \\ \underset{4}{CH_2OH} & _4CH_2OH \\ \alpha\text{-glucose} & \beta\text{-glucose} \end{array}$$

47. (c) Benzoic acid (C_6H_5—COOH) forms m-derivatives, as —COOH group deactivates the benzene ring. However, sodium salt in —COONa group activates the benzene ring, and forms o- and p-derivatives, when reacts with Cl_2. Na^+

benzoic acid $\qquad$ sodium benzoate

48. (b) $CH_2(COOH)_2 \longrightarrow CH_3COOH + CO_2$; all gem-dibasic acids on heating, converted into monocarboxylic acids.

49. (c) Tartaric acid exists in four isomeric forms; these are dextro, laevo, racemic and meso tartaric acids.

$$\begin{array}{lll} (HO)-CH-COOH & HOOC-CH-(OH) & (OH)CH-COOH \quad \text{an equal} \\ \quad\quad\quad\ | & \quad\quad\quad\ | & \quad\quad\quad\ | \\ HOOC-CH-(OH) & (OH)-CH-COOH & (OH)CH-COOH \quad \text{mixture} \\ \text{dextro- and laevo- isomers} & \text{meso forms} & \text{d and l forms} \\ & & \quad\quad\quad \text{racemic} \end{array}$$

50. (b) Sodium formate shows the bond length of C=O and C—OH are identical is both are equal to 1.27Å and 1.27Å.

$$H-C\overset{\displaystyle O}{\underset{OH}{\diagup}} \qquad H-C\overset{\displaystyle O}{\underset{O^-Na^+}{\diagup}}$$

$\qquad$ Formic acid $\qquad$ Sodium formate

Bond length

C = O	1.21Å	C—O	1.27Å
C—O	1.43Å	C—O	1.27Å

51. (b) $CH{\equiv}C-COOH \ > \ CH_2{=}CH-COOH$

Prop-2-ynic acid is stronger than prop-2-enoic acid.

52. (a)

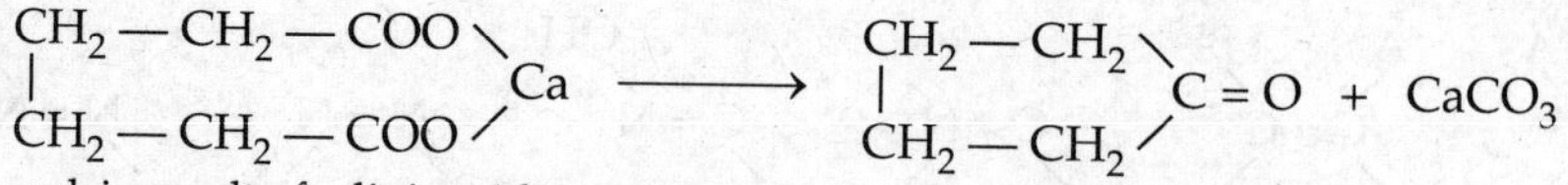

$$\begin{array}{c} CH_2 - CH_2 - COO \\ | \qquad\qquad\qquad\quad Ca \\ CH_2 - CH_2 - COO \end{array} \longrightarrow \begin{array}{c} CH_2 - CH_2 \\ | \qquad\qquad C = O \\ CH_2 - CH_2 \end{array} + CaCO_3$$

calcium salt of adipic acid   cyclopentanone

53. (c) Formic acid is a resonance hybrid of three structures.

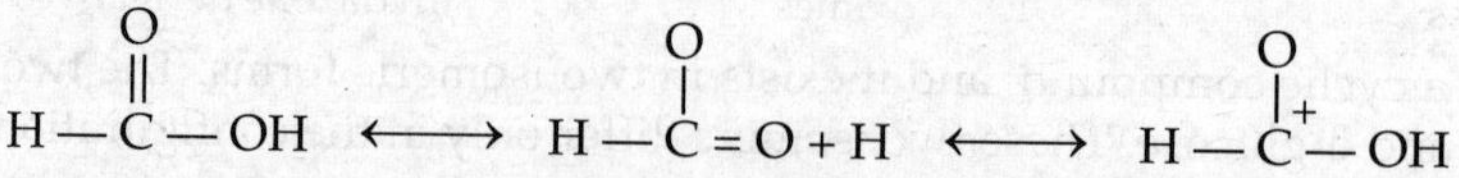

54. (b) Formic acid and oxalic acids decolourise acidified $KMnO_4$

$$2KMnO_4 + 3H_2SO_4 \longrightarrow K_2SO_4 + 2MnSO_4 + 3H_2O + 5O$$

$$HCOOH + O \longrightarrow H_2O + CO_2 \quad COOH + O \longrightarrow 2CO_2 + H_2O + COOH$$

Formic acid   Oxalic acid

55. (b) $R - C\equiv N$ and $R - N\equiv C$ are more polar, as compared to alkyl halides, because of dipole-dipole attraction.

56. (c) IUPAC name for the compound is aziridine and aza cyclopropane

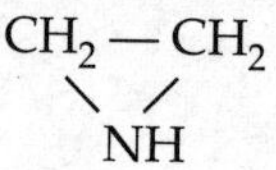

$$\begin{array}{c} CH_2 - CH_2 \\ \quad \backslash \quad / \\ \quad NH \end{array}$$

57. (b) $CH_3 - C\equiv C - CH_2 - CH_3 \xrightarrow{HBr}$ two forms

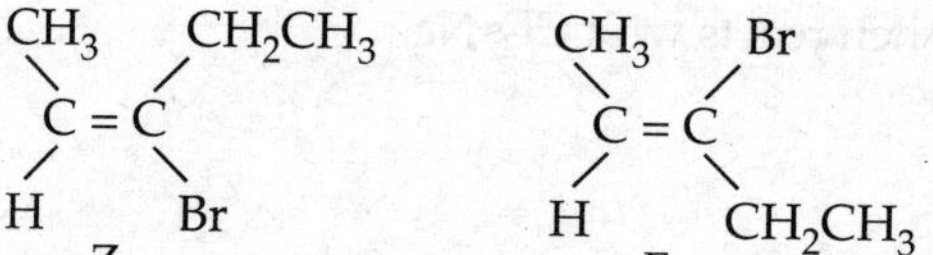

Z   E

58. (b) Fenton's reagent

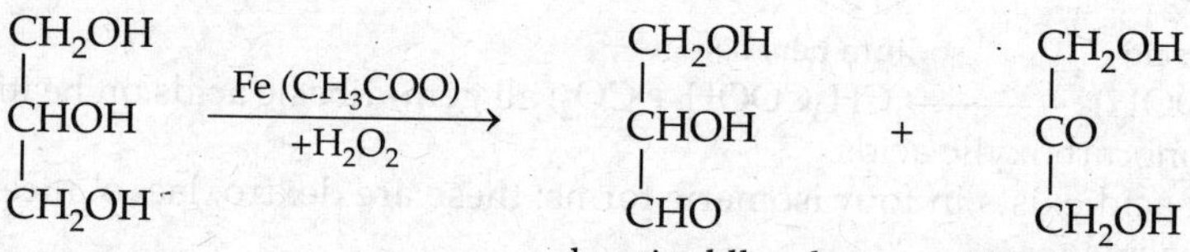

glyceric aldheyde   dihydroxy acetone

Fenton's reagent, ferric acetate containing a small amount of H_2O_2 oxidises glycerol to form glyceric aldehyde and dihydroxy acetone. Glyceric aldehyde is used as a reference compound in "absolute configuration".

59. (b) $RNC + 2H_2O \longrightarrow RNH_2 + HCOOH$

Formic acid possesses reducing properties.

60. (b) Nitrobenzene

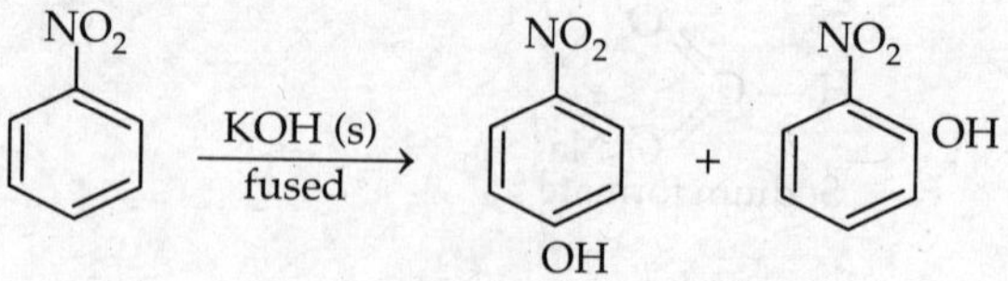

p-nitro + o-phenol

61. (b) On heating calcium salts of α-dibasic acids, cyclic ketones are obtained, containing one carbon less. For example, on heating calcium salt of succinic acid, cyclo-propanone is formed.

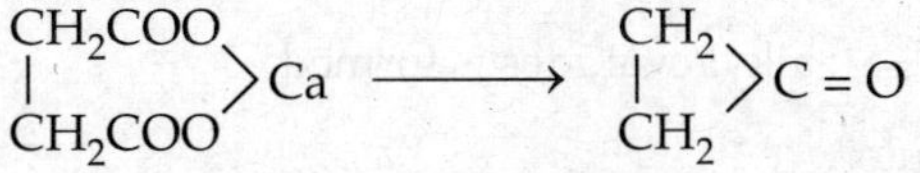

$$\begin{matrix} CH_2COO \\ | \\ CH_2COO \end{matrix} \rangle Ca \longrightarrow \begin{matrix} CH_2 \\ | \\ CH_2 \end{matrix} \rangle C = O$$

62. (a) $CH_3 — CH_2 — CH_2 — \underset{\underset{H}{|}}{\overset{\overset{CH_3}{|}}{C}} — CH_2 — CH_3$ 3-methyl hexane, will show optical activity.

63. (c) Neopentane; $CH_3 — \underset{\underset{CH_3}{|}}{C} — CH_3,$ on chlorination neopentane will give only one mono-

    chlorinated derivative.

64. (c) $CH{\equiv}CH \quad > \quad CH_2 = CH_2 \quad > \quad CH_3 — CH_3$
    K_a value
    $1 \times 10^{-22} \qquad 1 \times 10^{-36} \qquad 1 \times 10^{-42}$

    The acetylinic hydrogen will be more acidic in nature.

65. (c) Conjugated dienes form cyclic compounds or form adducts, when they are treated with an enophile. This reaction is termed as Diels–Alder reaction. All of these reagents, undergo Diels–Alder reaction, react with butadiene-1,3 to form adducts.

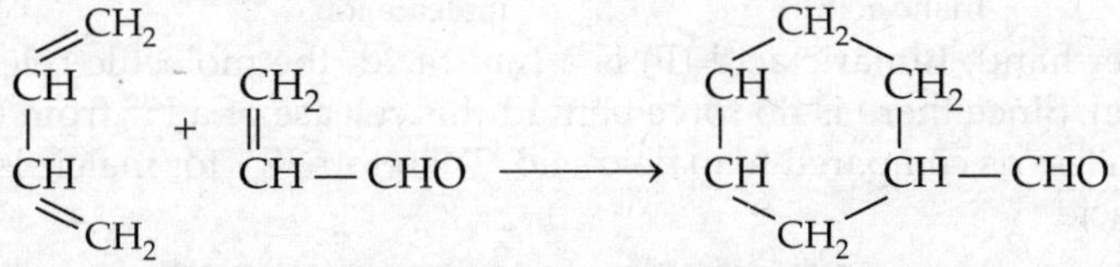

    $$\begin{matrix} & \overset{CH_2}{\|} \\ & CH \\ & | \\ & CH \\ & \underset{CH_2}{\|} \end{matrix} \quad + \quad \begin{matrix} CH_2 \\ \| \\ CH — CHO \end{matrix} \quad \longrightarrow \quad \begin{matrix} CH_2 \\ CH \quad CH_2 \\ | \qquad | \\ CH \quad CH — CHO \\ CH_2 \end{matrix}$$

    butadiene-1,3 acrolein 1, 2, 3, 6-tetra hydro benzaldehyde

66. (g) All of these substances show optical activity in the crystalline form.

67. (d) It is not oxidised. Though the neo butyl group is a side chain, yet it is not oxidised by alk. $KMnO_4$ to a COOH, because it does not have the α-hydrogen, i.e. no hydrogen attached to the central carbon of the neo butyl group.

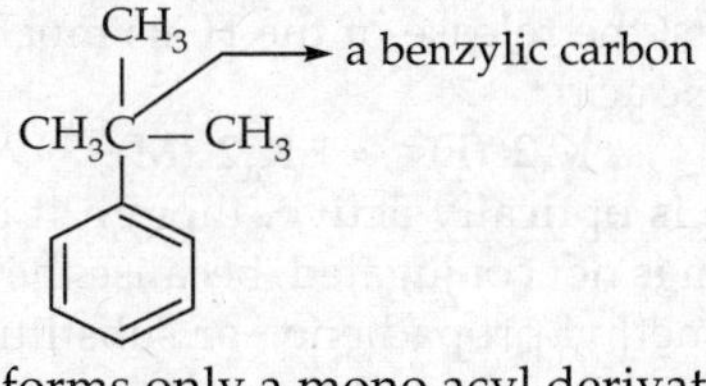

    $$CH_3\underset{}{\overset{\overset{CH_3}{|}}{C}} — CH_3 \longrightarrow \text{a benzylic carbon}$$

68. (a) Acylation of benzene forms only a mono acyl derivative, i.e. acetyl benzene or phenyl methyl ketone.

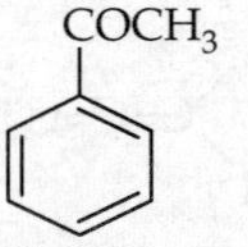

    $COCH_3$

69. On adding oxygen in the reaction undergoing a free radical reaction, the reaction becomes slower. An oxygen molecule, combines with the $CH_3\bullet$ free radical to for a methyl per oxide. This results in the decrease of number of methyl free radical, hence the reaction slowdown.

    $CH_3\bullet \qquad + \qquad O—O \longrightarrow CH_3—O—O\bullet$
    methyl free radical $\qquad\qquad$ oxygen $\qquad\qquad$ methyl peroxide

70. (d) R—CH=N—OH $\xrightarrow[\Delta]{P_2O_5}$ R—C≡N; alkyl cyanides is formed.

71. (d) Addition follows Markovnikov's rule.

$(C_6H_5)_2NH \longrightarrow$ capable of forming a free radical $(C_6H_5)_2 N\bullet$.

$2\,(C_6H_5)_2N\bullet \longrightarrow (C_6H_5)_2N—N\,(C_6H_5)_2$

It will combine with a $Br\bullet$ free radical, and will effect the course of the reaction. As a result, the reaction will proceed by a ionic mechanism, i.e. it will follow the Markovnikov's rule.

72. Maleic acid (M) is a *cis*-dibasic acid. A free rotation is restricted on account of the double in between the two carbon atoms. The molecule has two carboxylic groups close to each other. As there is mutual repulsion between the two COOH groups, the molecule releases one H^+, forms a monocarboxylate ion which is stabilised by the formation of a hydrogen bonding in between the —COOH and COO—.

$$
\begin{array}{l}
CH—COOH \\
\parallel \\
CH—COOH
\end{array}
\longrightarrow
\begin{array}{l}
CH—C{\overset{\displaystyle /\!/O}{}}{}—O \\
\parallel \\
CH—C—O^{-} \\
{\overset{}{\searrow}}O
\end{array}
\;\; H + H^{+}
$$

malic acid   maleate ion

On the other hand, fumaric acid (F) is a *trans* acid, the molecule releases H^+, to form a fumerate ion. Since there is no force behind this release of a H^+ from the fumaric acid, it ionise much less as compared to maleic acid. Therefore K_a1 for maleic is more than the K_a1 of fumaric acid

$$
\begin{array}{l}
CH—COOH \\
\parallel \\
HOOC—CH
\end{array}
\longrightarrow
\begin{array}{l}
CH—COO \\
\parallel \\
HOOC—CH
\end{array}
\longrightarrow
\begin{array}{l}
CH—COO— \\
\parallel \\
—OC—CH
\end{array}
$$

Fumric acid   Fumerate ion   Fumerate di-ion

$$K_a1\,(M) \;>\; K_a1\,(F) \quad (M = \text{maleic acid})$$

In second ionisation, the fumerate ion ionise much more, where as in maleate ion the H^+ is held by two oxygen atoms, the release of the H^+ is much less. Therefore K_a2 for fumaric acid is more for than maleic acid.

$$K_a2\,(F) \;>\; K_a2\,(M)$$

73. 1,3-dimethyl propadiene is optically active, though it does not contain an asymmetric carbon atom. The π-system is not conjugated, because the two α-bonds are at right angle to each other. Therefore, dimethyl propadiene, or substituted allenes, have no asymmetric carbon atom in the molecule, but have asymmetric molecule, i.e. they have no plane of symmetry.

$$
\begin{array}{c}
CH_3 H \\
\searrow \nearrow \\
C=C=C \\
\nearrow \searrow \\
H CH_3
\end{array}
$$

74. When ethane reacts with bromine in the presence of sunlight, nine isomers are formed

$\underset{1}{CH_3—CH_2Br};\ \underset{2}{CH_3CHBr_2};\ \underset{3}{CH_3—CBr_3};\ \underset{4}{CH_2Br—CH_2Br};\ \underset{5}{CHBr_2—CH_2Br},$

$\underset{6}{CH_2Br—CBr_3};\ \underset{7}{CBr_3—CH_2Br};\ \underset{8}{CBr_3—CHBr_2};\ \underset{9}{CBr_3—CBr_3}$

None of these bromo derivatives are optically active, therefore the number of fraction will be 9 only, i.e. equal to the number derivatives.

75. The number of possible structural isomers, with mol. formula C_5H_{12} = 3.

$$CH_3-CH_2-CH_2-CH_2-CH_3;$$

$$CH_3-CH_2-\overset{\overset{\displaystyle CH_3}{|}}{CH}-CH_3;$$

$$CH_3-\overset{\overset{\displaystyle CH_3}{|}}{\underset{\underset{\displaystyle CH_3}{|}}{C}}-CH_3$$

n-pentane	isopentane	neopentane

76. (a) Chiral molecules are those, which are non-superimposable on its mirror image. For example, d-lactic and l-lactic acids are not superimposable on each other; therefore these may be termed chiral molecules.

77. (b) A 2° carbocation is more stable than a 1° carbocation. The list contains two 2° carbocations out of which $CH_3-CH^+-CH_3$ is more stable than the other carbocation.

78. (b) There exists minimum bond angle between p_x and p_y orbital.
    i. In sp_3 bonds angle = 109°.28′
    ii. In p_x and p_y bonds angle = 90°
    iii. In H_2O, bond angle = 105°
    iv. In sp bonds angle = 180°.

79. (b) BF_3 is an electrophile, as number of electrons around boron is only six.

$$\text{F} \ : \ \text{B} \ : \ \text{F}$$
$$\overset{..}{\text{F}}$$

80. (a) *n*-hexane will have the highest bp on account of van der Waal's forces.

81. (a) Alkanes have the largest bond distance between the two adjacent carbon atoms. The bond distance between other molecules is as follows.
    In alkanes bond distance between the two carbons = 1.54 Å
    In alkenes bond distance between the two carbons = 1.34 Å
    In alkynes bond distance between the two carbons = 1.20 Å
    In benzene bond distance between the two carbons = 1.34 Å

82. (c) $CH_3-\overset{\overset{\displaystyle CH_3}{|}}{CH}=CH_2 + O_3 \longrightarrow CH_3-\overset{\overset{\displaystyle CH_3}{|}}{\underset{\underset{\displaystyle O}{|}}{CH}}-O-\overset{\underset{\underset{\displaystyle O}{|}}{}}{CH_2} \xrightarrow{HO} CH_3CO-CH_3 + HCHO + H_2O_2$

83. On chlorination, toluene will form $C_6H_5CCl_3$, which on treatment with NaOH forms C_6H_5COOH. On chlorination of toluene in the presence of sunlight and heat, the side chain reacts with chlorine to form benzotrichloride. The reaction proceeds through a free radical reaction.
    i. $C_6H_5CH_3 \ + \ 3Cl_2 \longrightarrow C_6H_5CCl_3 + 3HCl$
    ii. $C_6H_5CCl_3 \ + \ 3NaOH \longrightarrow C_6H_5COOH + 3NaCl + H_2O$

84. (a) 3° alcohol will be much more soluble in water as compared to the 2° alcohol. 1° Alcohol will be least soluble. The reason behind this is that a 3° alcohol, is almost a round molecule; a 2° alcohol will occupy a larger space and in the same way a 1° alcohol will occupy much more space, being a linear molecule.
    The solubility is based on the formation of hydrogen bonding.

85. (c) Phenol is a weaker acid than H_2CO_3.
    $C_6H_5OH = pK_a = 10$; $H_2CO_3 = pK_a = 6$
    (higher K_a value or a lower pK_a value means a stronger acid).

86. (b) p-nitrophenol is more soluble, because the molecule has a hydrogen bonding and it is a round molecule.

87. (b) o-nitrophenol has lower bp, because the molecule has intramolecular H-bonding.

88. (b) Diazomethane (CH_2N_2) reacts with acid to form an ester.
    $$RCOOH + CH_2N_2 \longrightarrow RCOOCH_3$$

89.

$$\underset{\text{glutaric acid}}{CH_2 \Big\langle \begin{array}{c} CH_2-COOH \\ CH_2-COOH \end{array}} \xrightarrow{\text{heat}} \underset{\text{glutaric anhydride}}{CH_2 \Big\langle \begin{array}{c} CH_2-CO \\ CH_2-CO \end{array} \Big\rangle O}$$

90. $HCHO;$ $\underset{CH_2-OH}{\overset{CH_2-OH}{|}} \xrightarrow{HIO_4} 2HCHO$

    Periodic acid on oxidation of a vic-glycol breaks the C to C bond and oxidise them.

91. (a) $HCHO + H_2CHCOCH_3 \longrightarrow CH_2=CHCOCH_3$; vinyl methyl ketone.

92. (c) K_a values: $CH_3NH_2 = 10.62$; $C_2H_5NH_2 = 10.63$.

93. (c) In acid medium aniline reacts with acid to form anilinium salt, other compounds do not react with acid.
    $$C_6H_5NH_2 + H^+ \longrightarrow C_6H_5NH_3 + Cl^-.$$

94. (b) Aniline and dimethyl aniline react with an electrophile to form a carbonium ion.

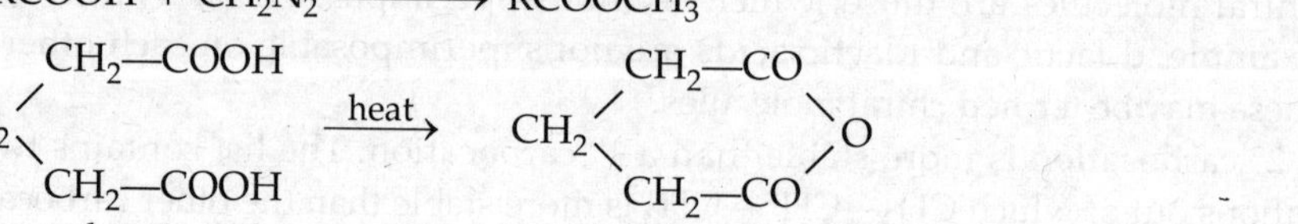

aniline   carbocation   dimethyl aniline   carbocation

On attack by an electrophile, the molecule is converted into a π- complex first, which then changes to form a σ-complex. The σ-complex has a +ve charge on the nitrogen atom. The carbocation formed from dimethyl aniline is more stable, on account of the two methyl groups attached to the nitrogen.

95. p-amino benzaldehyde exists as an ionic compound and CN^- does not adds to the C=O group, because nitrogen is not electron deficient.

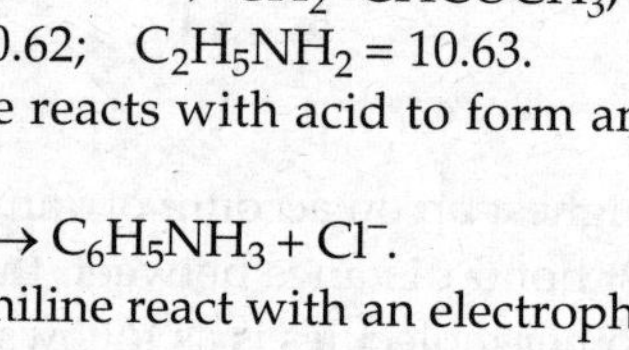

96. In Reimer–Tiemann reaction, phenol reacts with chloroform, in presence of caustic alkali to form salicylaldehyde,
    $$\underset{\text{chloroform}}{H-CCl_3} + KOH \longrightarrow \underset{\substack{\text{dichloro-} \\ \text{carbene}}}{:CCl_2} + KCl + H_2O$$

:CCl_2^- Dichlorocarbene is the attacking reagent or the electrophile on phenol molecule.

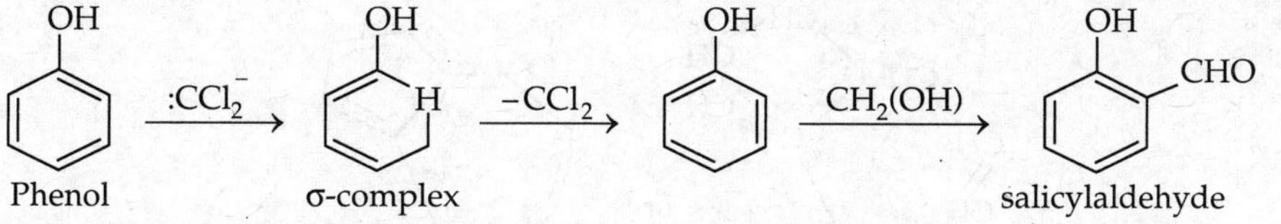

97. Hofmann's bromamide reaction is used to prepare an amine from the corresponding amide by treating the amide with Br_2 and a caustic alkali.

$$CH_3CONH_2 \xrightarrow{Br_2} CH_3CONHBr \xrightarrow{KOH} CH_3NCO \xrightarrow{2KOH} CH_3NH_2$$

   acetamide   *n*-bromamide   methyl isocyanate   methyl amine

   CH_3NCO and $CH_3CONHBr$ are the two intermediates formed in Hofmann's reaction.

98. In the acidic medium, the amino group is protonated to form an anilium ion, which acquires a +ve charge; it deactivates the benzene ring, and becomes m-directing group.

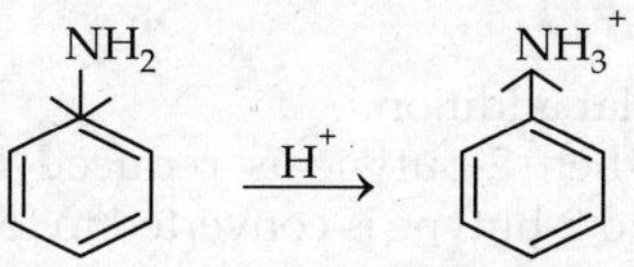

99. —COOH group acts as a deactivating group and therefore acts as an m-directing group. On converting —COOH group to sodium carboxylate (—COONa) group, the group activates and as a result it begins to behave as an o- and p-directing group.

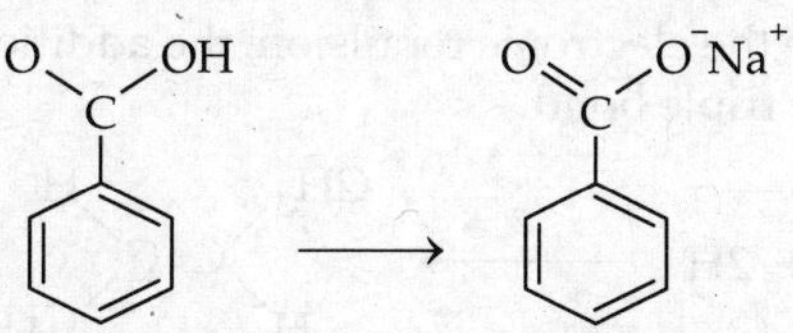

100. A methoxy group (—OCH_3) acts as an activating group or a o- and p-directing group, but from the m-position it the —OCH_3 group acts as deactivating group.

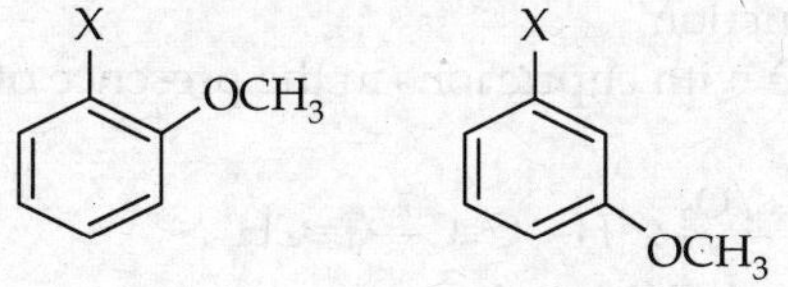

101. Nitrobenzene on fusion with solid KOH, gives a mixture of o- and p- nitrophenols.
    In this case, the attacking group is a nucleophile, i.e. —OH^- group, and it attacks the electron deficient positions, created by the deactivation of —NO_2 group.
    [See answer to Q. No. 60].

102. Diphenyl derivatives are optically active, because the mirror images of these diphenyl derivatives is not superimposable on each other. The two rings are not on the same plane, at right angle to each other. Hence these are non-imposable.

103. Azulene is a bicyclic compound, that behaves as a nonbenzenoid aromatic compound.

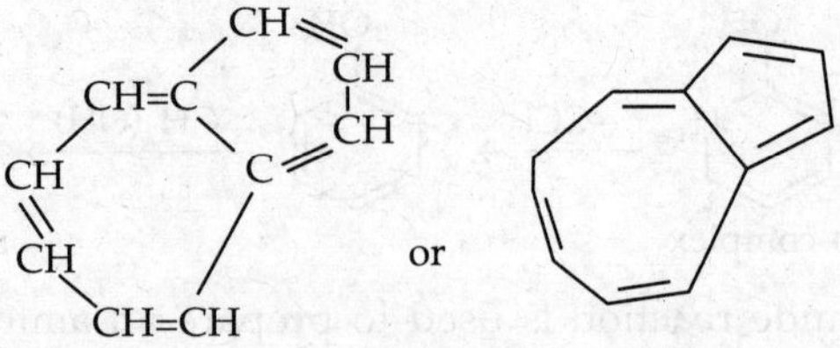

104. i. When 2-butyne is reduced by $Pd/BaSO_4$, it is reduced to form 2-*cis*-butene; a *trans*-addition takes place. The catalyst absorbs the alkyne molecule as well as hydrogen, from the same side, to form *cis*-2-butene, (the catalyst Pd-BaSO$_4$, does not absorb alkene).

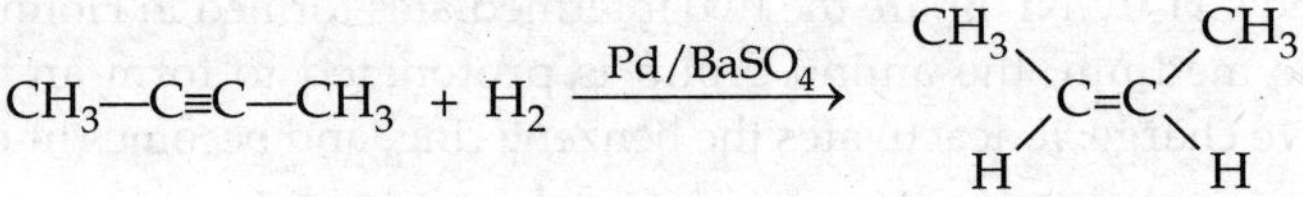

$$CH_3-C\equiv C-CH_3 + H_2 \xrightarrow{Pd/BaSO_4}$$

cis-2-butene

It can be termed molecular addition.
On the other hand, when 2-butyne is reduced by sodium metal and ammonia, reduction takes place and 2-butyne is converted to form a *trans*-2-butene.

$$2NH_3 + 2Na\bullet \longrightarrow 2Na\,NH_2 + 2e$$

$$CH_3-C\equiv C-CH_3 + 2e \longrightarrow CH_3-\overset{\bullet\,\bullet}{C}=\overset{\bullet\,\bullet}{C}-CH_3$$

In order to minimize the electronic repulsion, the addition of electrons takes place in the opposite faces of the triple bond.

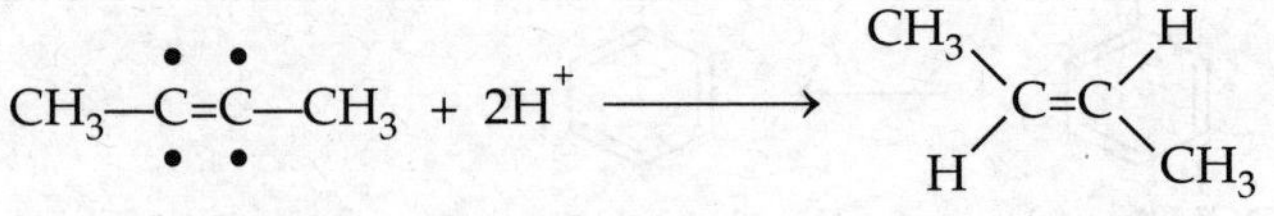

$$CH_3-\overset{\bullet\,\bullet}{C}=\overset{\bullet\,\bullet}{C}-CH_3 + 2H^+ \longrightarrow$$

trans-2-butene

Since the reduction produces only one of the two compounds, it is often termed "stereosphecific reduction"

ii. On heating acetylene with cupric ions in the presence of oxygen the compound formed is 1,3- butadine.

$$2H-C\equiv C-H \xrightarrow{Cu_2^+/O_2} H-C\equiv C-C\equiv CH$$

This reaction is termed *Erlington's coupling*.

$$2CH_3-C\equiv C-H \xrightarrow{Cu_2^+/O_2} CH_3-C\equiv C-C\equiv C-CH_3$$
propyne   1,4-hexa-diene

105. 2-butene is hydroxylated by (i) alkaline $KMnO_4$; (ii) $HCOOH + H_2O_2$

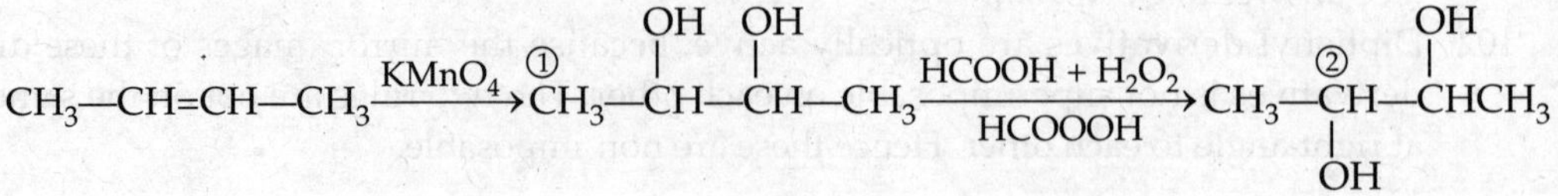

$$CH_3-CH=CH-CH_3 \xrightarrow[\text{①}]{KMnO_4} \underset{\text{cis-glycol}}{CH_3-\underset{OH}{\underset{|}{CH}}-\underset{OH}{\underset{|}{CH}}-CH_3} \xrightarrow[HCOOOH]{HCOOH + H_2O_2} \underset{\text{trans-glycol}}{CH_3-\underset{|}{\underset{OH}{CH}}-\underset{OH}{\underset{|}{CH}}CH_3}$$

2-butene   cis-glycol   peroxy formic   trans-glycol
                                            acid

106. Alkyl fluorides do not react with either sodium or magnesium metals.

107. In all 9 derivatives are formed (see answer to Q. No. 74).

108. a. On monochlorination of isopentane, the number of monochlorinated derivatives formed are 7.

$$CH_3-CH_2-\underset{\underset{CH_3}{|}}{CH}-CH_3 \xrightarrow{Cl_2} \underset{I}{CH_2Cl-CH_2-\underset{\underset{CH_3}{|}}{CH}-CH_3} + \underset{II}{CH_3-CHCl-\underset{\underset{CH_3}{|}}{CH}-CH_3}$$

$$\underset{III}{CH_3-CH_2-\underset{\underset{Cl}{|}}{\overset{\overset{CH_3}{|}}{C}}-CH_3} \qquad \underset{IV}{CH_3-CH_2-\underset{\underset{CH_3}{|}}{CH}-CH_2Cl}$$

The compounds I, II and IV are optically active and therefore, these monochloro isopentanes will exist in both d and l-forms. As compound number III is not optically active, the number of isomers should be 6 + 1 = 7.

b. On distillation, the number of fractions will be only 4, because the bp of the d-and-l fractions of I, II and IV will be the same.

109. Dimethyl sulphide, $(CH_3)_2S$, acts as a reducing agent, the ozonide is converted to form the corresponding keto compounds.

$$\underset{\underset{\text{ozonide}}{O\text{--------}O}}{>C-O-C<} + (CH_3)_2S \longrightarrow >C=O + >C=O + (CH_3)_2SO$$

keto compounds

110. Stronger is the acid, weaker is its conjugate base.

111. Pentyne-1 has a higher dipole moment, as compared to pentyne-2.
$$\underset{\text{pentyne-1}}{CH_3-CH_2-CH_2-C\equiv CH} > \underset{\text{pentyne-2}}{CH_3-CH_2-C\equiv C-CH_3}$$

112. 6 Isomers are possible.

$$\underset{I}{\overset{Br}{\underset{Cl}{>}}C=C\overset{F}{\underset{I}{<}}} ; \underset{II}{\overset{Br}{\underset{Cl}{>}}C=C\overset{I}{\underset{F}{<}}} ; \underset{III}{\overset{Br}{\underset{F}{>}}C=C\overset{Cl}{\underset{I}{<}}} ; \underset{IV}{\overset{Br}{\underset{F}{>}}C=C\overset{I}{\underset{Cl}{<}}} ; \underset{V}{\overset{Br}{\underset{Cl}{>}}C=C\overset{F}{\underset{I}{<}}} ; \underset{VI}{\overset{Br}{\underset{I}{>}}C=C\overset{Cl}{\underset{F}{<}}}$$

113. C—D bond is stronger than —C—H bond.

114. $CH_2 = CH-\ddot{\underset{\cdot\cdot}{Cl}} \colon \xrightarrow{HI} \overset{+}{CH_2}-\overset{-}{CH}-\ddot{\underset{\cdot\cdot}{Cl}} \colon \longrightarrow CH_2I-CH_2Cl$
2-iodo-1-chloro ethane

On account of the I effect of the —Cl group, the expected compound should be 2-iodo-1-chloro ethane. However, the compound formed is CH_3—$CH(Cl)I$ namely 1-chloro-1-iodo ethane. It is the resonance effect, which dominates, and vinyl chloride reacts as CH_2^{-}—$CH = Cl^{+}$.

$$CH_2=CH-\ddot{\underset{\cdot\cdot}{Cl}} \colon \longrightarrow \overset{-}{CH_2}-CH=\overset{+}{\underset{\cdot\cdot}{Cl}} \colon \xrightarrow{HI} CH_3-CH(Cl)\,I$$

115. The H^+ adds to the 3-phenyl propene-1, C_6H_5—CH_2—$CH=CH_2$, to give a carbocation C_6H_5—CH_2—CH^+—CH_3

$$HBr \longrightarrow H^+ + Br^-$$

$$C_6H_5CH_2-CH=CH_2 + H^+ \longrightarrow C_6H_5CH_2-CH^+-CH_3$$

The carbocation formed is now stabilised by the shift of a proton as a hydride ion, (H⁻), to the adjacent carbon atom, and the benzylic carbon atom, now acquires a +ve charge. Br⁻ joins at this carbocation, to form the bromo derivative, namely 1-bromo-1 phenyl propane.

CH₂—CH=CH₂   CH₂—CH⁺CH₃   CH⁺—CH₂CH₃   CHBrCH₂CH₃

⬡   ⟶   ⬡   ⟶   ⬡   ⟶   ⬡

1-bromo 1-1pheynyl propane

116. a. Heat of the reaction: It is defined as the amount of heat evolved or absorbed in the reaction concerned, at constant pressure, represented by the symbol $= \Delta H$.

In those reactions, where heat is evolved, the reaction is termed exothermic reaction, represented by the negative sign $= -\Delta H$.

In those reaction, where heat is absorbed, the reaction is termed as endothermic reaction, and it is represented by the positive sign, symbol $= \Delta H$.

b. Heat of combustion: The amount of evolved as 1 g. mol. of the compound is oxidised form the products to the highest oxidation state, at the constant temperature and pressure.

Heat of combustion of ethane, ethylene and acetylene are listed below.

C_2H_2 (g) + 5/2 O_2 (g) $\longrightarrow$ 2CO_2 (g) + H_2O (l)   H = –311 kcal

C_2H_4 (g) + 3O_2 (g) $\longrightarrow$ 2CO_2 (g) + 2H_2O (l)   H = –337 kcal

C_2H_6 (g) + 7/2 O_2 (g) $\longrightarrow$ 2CO_2 (g) + 3H_2O (l)   H = –373 kcal

The higher heat of acetylene flame is possible, in spite of the lower heat of combustion of acetylene as compared with those of ethylene and ethane, because the heat capacity of the products is small; less amount of water formed and less heat of reaction is used to bring the combustion products up to the flame temperature.

c. Heat of hydrogenation may be defined as the amount of heat evolved when a gram mol of the unsaturated compound is saturated by reacting with hydrogen.

(alkenes = the more highly substituted the double bond of alkenes amount of heat evolved is much less; the *trans* isomer evolves much less heat as compared to that of *cis* isomers).

Heat of hydrogenation of alkenes and alkynes, ketones and benzene are as follows.

CH_3—C ≡ C = 69.3 kcal

CH_2 = CH_2 = 30 kcal

>C = O   = 57 kcal

C_6H_6 = 49 kcal

Reduction of the compounds containing following groups:

—C ≡ C—> C = C > C = O > C_6H_6

i. All these compounds are reduced by metal/H_2.

ii. $LiAlH_4$ reduces only keto compound, but does not reduce a double and a tripe bond.

iii. Lindler's catalyst reduces benzene and conjugated dienes, but does not reduce either a double or a triple bond.

iv. An acetylinic compound is reduced more easily, as compared to the other compounds listed.

117. Benzene molecule has three alternate double bonds in the molecule.
The heat of combustion for the benzene molecule
$$C_6H_6(g) + 15/2\,O_2(g) \longrightarrow 6CO_2(g) + 3H_2O(g)$$
calculated = 827 kcal/mol.
experimental value = 789 kcal/mol.

As the experimental value is much less as compared to the calculated value, (827–789 kcal/mol) by 38 kcal/mol is termed stabilization energy or resonance energy, in other words, benzene contains much less energy as compared to the calculated value.

The heat of hydrogenation values, calculated for the molecule containing three alternate double bonds in the molecule.
As calculated = 85.8 kcal/mol.

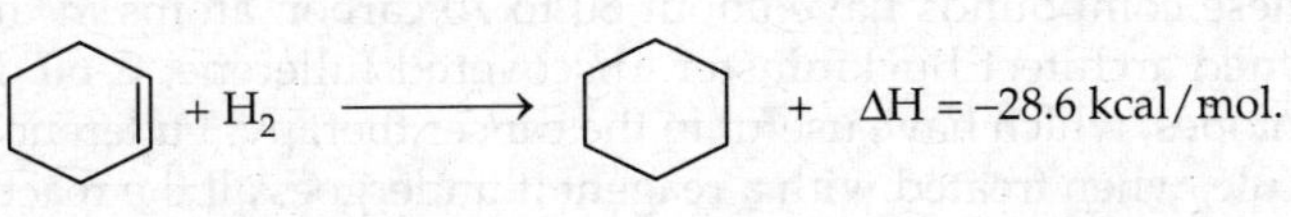

cyclohexene

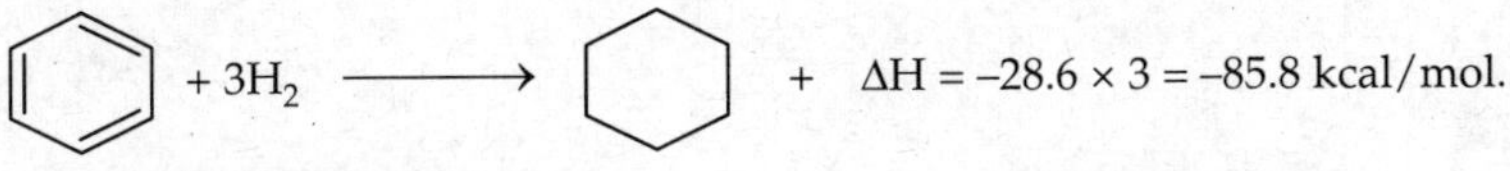

1,3,5-cyclohexatriene/or
benzene

However, the experimental value = 49.8 kcal/mol

In fact, a benzene molecule evolves only 49.8 kcal/mol or in other words, benzene molecule contains 85.8 – 49.8 kcal/mol = 36 kcal/mol less. This is termed resonance energy. Therefore, benzene molecule should not be represented, as 1,3,5-cyclohexa-triene, or it should be regarded as a deactivated molecule (the three double bonds are deactivated) by an amount = 36 kcal/mol.

In case an energy = 36 kcal/mol is provided, the deactivated double bonds will become active and the activated benzene molecule is then expected to behave as benzene 1,3,5-cycohexatriene. Benzene molecule, therefore is represented as a molecule, in which the electrons are delocalized, forming electron clouds over and below the benzene ring.

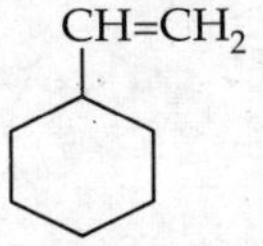

Most exothermic is the least substituted alkene.

2-cyclohexyl ethene (iv), is the least substituted cycloalkanes derivative and is exothermic in nature.

118. i. Borazine or borazole $B_3N_3H_6$
It is prepared by heating diborane (B_2H_6) with ammonia at 200°C. It is a liquid with bp = 55°C.
All nitrogen-boron bonds have same bond length. The boron–nitrogen the bond angle is 120°, as in the case of benzene. Benzene is often termed as inorganic benzene.

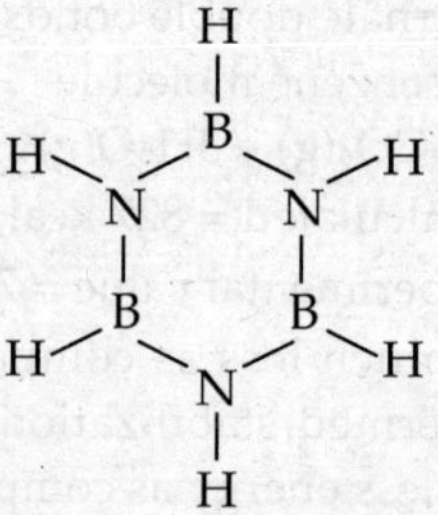

$$
\begin{array}{c}
\text{H} \\
| \\
\text{B} \\
\text{H}-\text{N} \quad \text{N}-\text{H} \\
| \quad\quad | \\
\text{H}-\text{B} \quad \text{B}-\text{H} \\
\text{N} \\
| \\
\text{H}
\end{array}
$$

Borazine has a flat cyclic structure, it decomposes slowly in presence of air water acids and on heating. Boron and nitrogen are in sp^2 hybridised state. Each nitrogen has two nonbonding electrons in the p-orbital, making it to a total 6 p-delocalised electrons.

ii. Fullerenes: These compounds have about 60 to 70 carbon atoms in the molecule. In 1985, a renowned architect Buckminster, discovered fullerene, C-60. It encapsulates radioactive isotopes, which have useful in the cancer therapy. Fullerene molecule adds just one molecule, when treated with a reagent it undergoes all the reactions associated with electron deficient alkenes.

Section III

Chemistry of Hydrocarbons and Their Derivatives

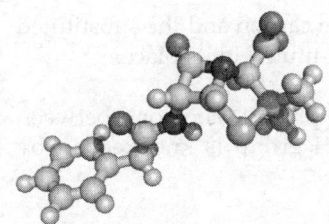

Chapter 1

Hydrocarbons

Salient Features

1. When the molecule is unsymmetrical (optically active), the number of meso compound formed is zero.
2. *Heterolysis* is favoured in acid or base catalyzed reactions.
3. *Homolysis* is favoured in gas phase reactions, occurring at a high temperature or in presence of sunlight.
4. Fractional distillation cannot be carried out for a mixture of substances, which form a constant boiling mixture.
5. Stronger the base, stronger will be the nucleophile but poor leaving group.
   H_2O is a weak base, a poor nucleophile, but stronger leaving group.
6. *Nucleophilicity* is the ability to react with an electrophile; basicity is the ability to remove a proton from an acid.
7. Weak base and medium to high polarity favours substitution; strong base and medium to low polarity favours elimination.

$$CH_3-\underset{\underset{CH_3}{|}}{\overset{\overset{CH_3}{|}}{C}}-Cl \quad \begin{array}{c} \xrightarrow{\text{aq. KOH}} \quad (CH_3)_3C\,OH \\[2mm] \xrightarrow{\text{alc. KOH}} \quad \underset{CH_3}{\overset{CH_3}{\diagdown}}CH=CH_2 \end{array}$$

8. The presence of groups like CN^-, $C_2H_5O^-$, NH_2^- favours elimination in S_N1 reaction, whereas S_N2 is exothermic in nature.

9. Benzylic hydrogen is a must for side chain oxidation. In other words, the carbon through which a side chain is attached to the ring must contain a hydrogen atom, e.g. tertiary butyl benzene is not oxidised by alk. $KMnO_4$ to form benzoic acid, because there is no hydrogen on benzylic carbon.

$$CH_3-\underset{\underset{\text{(benzylic carbon)}}{|}}{\overset{\overset{CH_3}{|}}{C}}-CH_3$$

benzylic carbon

10. **IPSO Reactions:** Are those reactions in which the bond between carbon and the substituted group is broken (or knocked off) by the reagent and it gets substituted in its place.
    Example:
    i. Salicylic acid, on reaction with Br_2 forms 2, 4, 6-tribromo phenol. The bond between carbon and —COOH group is broken and the —COOH group is knocked off by bromine.

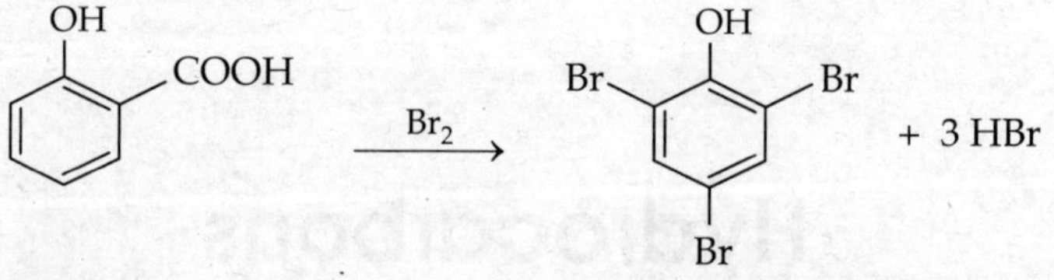

    salicylic acid   2, 4, 6-tribromophenol
                                      (—COOH group is knocked off)

    ii. On treating, sulphanilic acid (p-aminobenzene sulphonic acid) with bromine, 2, 4, 6-tribromoaniline. The bond between carbon and SO_3H group is broken by bromine.

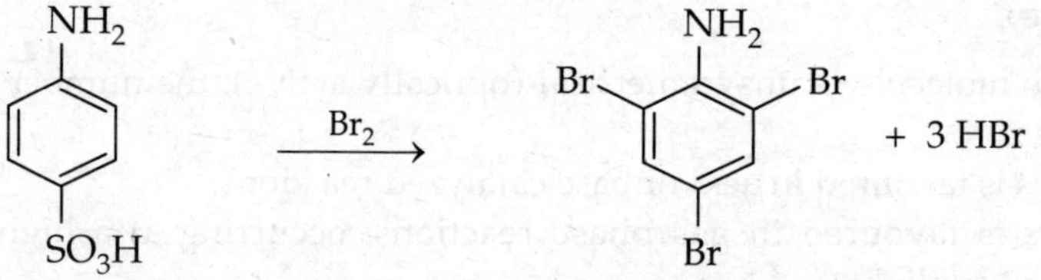

    sulphanilic acid   2, 4, 6-tribromophenol

11. Octane number of toluene = 104.
12. The stability of structural isomers, generally increases with increasing branching.
13. C_6H_5–X derivatives of benzene do not undergo Freidel–Craft reaction, where X = –NO_2, –CN, –NH_2, –SO_3H, –CHO, –COOH, –$COOCH_3$ and pyridine (C_5H_5N).
14. *cis-hydroxylation* of a double bond, can be carried out by treating the alkene with silver acetate, CH_3COOAg in presence of I_2 in aqueous acetic acid.

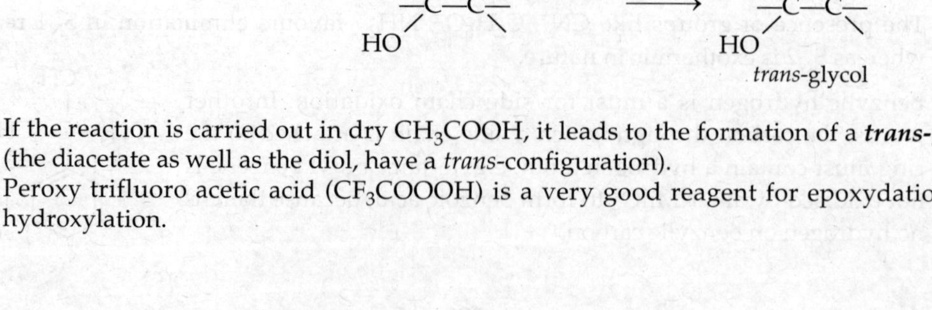

If the reaction is carried out in dry CH_3COOH, it leads to the formation of a ***trans*-glycol** (the diacetate as well as the diol, have a *trans*-configuration).
Peroxy trifluoro acetic acid (CF_3COOOH) is a very good reagent for epoxydation and hydroxylation.

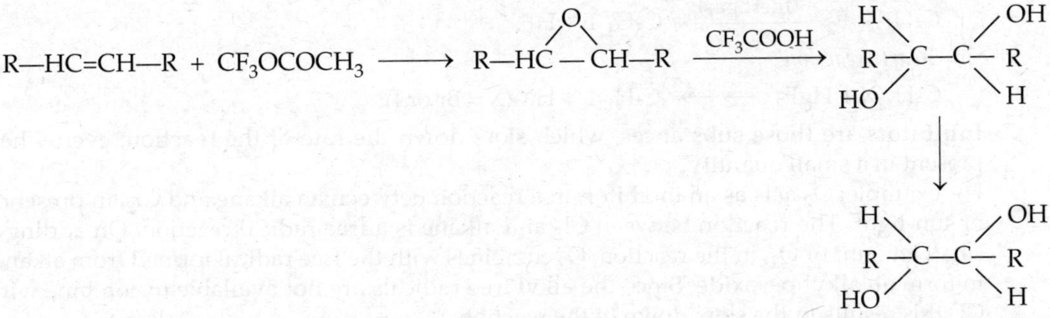

$$R-HC=CH-R + CF_3OCOCH_3 \longrightarrow R-HC\overset{O}{\underset{}{\diagdown}}CH-R \xrightarrow{CF_3COOH}$$

$$\begin{array}{c} H \diagdown \quad \diagup OH \\ R-C-C-R \\ HO \diagup \quad \diagdown H \end{array}$$

$$\downarrow$$

$$\begin{array}{c} H \diagdown \quad \diagup OH \\ R-C-C-R \\ HO \diagup \quad \diagdown H \end{array}$$

trans-glycol

14. RDX: Research developed explosive.
    The compound can be named as 1,3,5–trinitro-2,4,6 trihydro triazine.

ALKANES

Salient Features

1. A good quality gasoline, used in motor engines, have octane number 80 or more.
2. For aeroplanes, gasoline has 100 or a higher octane number.
3. *n*-heptane has an octane number zero and 2,2,3 trimethyl pentane (or triptane) has an octane number 124.

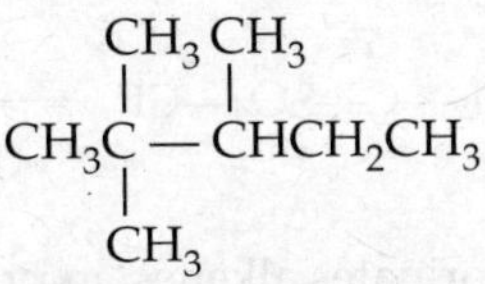

$$\begin{array}{c} \quad\quad CH_3\ CH_3 \\ \quad\quad |\quad\ | \\ CH_3C-CHCH_2CH_3 \\ \quad | \\ \quad CH_3 \end{array}$$

2,2,3-trimethyl pentane (or triptane)

Octane number of a gas is defined as the % of iso-octane present in a fuel under examination having same performance as an imaginary mixture of iso-octane and *n*-heptane. A gasoline with higher octane number is a better fuel.

It may be defined as the number, which gives the % by volume of iso-octane in the mixture of iso-octane and *n*-heptane, that has the same antiknock quality as the fuel under examination.

4. Best known antiknock compound is TEL, $(CH_3)_3$ Pb (trimethyl lead).
5. Cetane (decahexane, $C_{16}H_{34}$), has a rating of 100, α-methyl naphthalene ($C_{10}H_7-CH_3$) has a rating of zero.

$$CH_3(CH_2)_{14}-CH_3$$

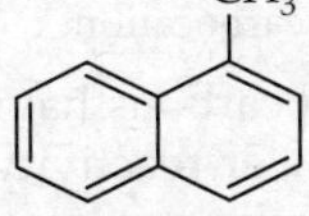

cetane   α-methyl naphthalene

Cetane number of the fuel is defined as the % of cetane (*n*-hexadecane) present in an imaginary mixture of *n*-hexadecane and α-methyl naphthalene.

6. Fluoroalkanes can be prepared by treating with fluorine (diluted with an inert gas) or by action of mercuric fluoride (HgF_2) on bromo (RBr) or iodo alkanes (RI).

     i. $C_5H_{12} + F_2 \xrightarrow{\text{(Inert gas)}} C_5H_{11}F + HF$

    ii. *Swartz reaction*

       $C_5H_{11}X + HgF_2 \longrightarrow C_5H_{11}F + HX \ (X = Br \text{ or } I)$

7. **Inhibitors** are those substances, which slow down the rate of the reaction, even when present in a small quantity.

   For example: O_2 acts as an inhibitor, in a reaction between an alkane and Cl_2, in presence of sun-light. The reaction between Cl_2 and alkane is a free radical reaction. On adding a small amount of O_2, in the reaction, O_2 combines with the free radical formed from alkane, to form an alkyl peroxide. Since the alkyl free radicals are not available to combine with $Cl^{•}$ this results in the slow down of the reaction.

     i. $Cl\!:\!Cl \xrightarrow{\text{sunlight}} 2Cl^{•}$

    ii. $R\!:\!H + Cl \longrightarrow R^{•} + HCl$

    iii. $R^{•} + O_2 \longrightarrow R{-}O{-}O^{•}$
                      alkyl peroxide

8. The dissociation energy of a $-\overset{|}{\underset{|}{C}}{-}H$ bond is 98.8 kcal/mol during pyrolysis.

9. $Cl_3{-}C{-}SO_2Cl$, trichloro methyl sulphonyl chloride, chlorinates only 2° and 3° hydrogen, (but not 1° hydrogen).

$$CH_3{-}\overset{\overset{\displaystyle CH_3}{|}}{CH}{-}CH_3 + Cl_3{-}C{-}SO_2{-}Cl \longrightarrow CH_3{-}\overset{\overset{\displaystyle CH_3}{|}}{\underset{\underset{\displaystyle Cl}{|}}{C}}{-}CH_3$$

10. Sulphuryl chloride, SO_2Cl_2 chlorinates alkenes, in presence of peroxide or sunlight.

$$CR_2{=}CH_2 + SO_2Cl_2 \xrightarrow{\text{peroxide/light}} CR_2{=}CH{-}Cl + SO_2 + H{-}Cl$$

11. The migration of anions, in 1, 2 shift is in the decreasing order: $C_6H_5{-} > CH_3 > H^-$. The groups migrate as phenyl anion ($C_6H_5^-$), methide, (CH_3^-) or a hydride ion (H^-) to convert a carbocation of a lower stability to a carbocation of higher stability, before the formation of the final product.

$$CH_3{-}\overset{\overset{\displaystyle CH_3}{|}}{\underset{\underset{\displaystyle Cl}{|}}{C}}{-}CH_2Cl \xrightarrow{\text{aq. KOH}} CH_3{-}\overset{\overset{\displaystyle CH_3}{|}}{\underset{\underset{\displaystyle Cl}{|}}{\overset{+}{C}}}{-}CH_2 \longrightarrow CH_3{-}\overset{\overset{\displaystyle CH_3}{|}}{\overset{+}{C}}{-}CH_2CH_3 \xrightarrow{OH^-} CH_3{-}\overset{\overset{\displaystyle CH_3}{|}}{\underset{\underset{\displaystyle OH}{|}}{C}}{-}CH_2CH_3$$

                 1° carbocation   3° carbocation   2-methyl-2-butanol

    Open chain saturated hydrocarbons having general formula C_nH_{2n+2} are called alkanes or paraffin (on account of their lower reactivity). In alkanes, the carbon is sp^3 hybridised and the shape of the molecule is *tetrahedral* in nature.

    The bond distance between the C to C is 1.54 Å. The amount of energy required to break C to C single bond (or a σ-bond) in a homolytic fission equal to ~85 kcal/mole.

$$>\!\overset{|}{C} : \overset{|}{C}\!< \longrightarrow \overset{|}{C}{•} + >\!\overset{|}{C}{•}$$

                 bond   Free radicals

Alkanes, when arrange in the order of their increasing molecular mass, constitute a group of closely related group of compounds is called a homologous series.

Alkanes	Name	Difference	Molecular composition
CH_4	Methane		
C_2H_6	Ethane	CH_2	
C_3H_8	Propane	CH_2	C_nH_{2n+2}
C_4H_{10}	Butane	CH_2	
C_5H_{12}	Pentane	CH_2	
C_6H_{14}	Hexane	CH_2	

1. A homologous series of alkanes (or any other closely related group of compounds), have the following characteristics. These can be governed by the general formula C_nH_{2n+2}
2. A succeeding member differs from the preceding member, by a CH_2 group or by a methylene group.
3. These can be prepared by the general methods of preparation.
4. Their physical properties and chemical behavior are similar.
5. These show a regular gradation in their physical and chemical behaviour.

Method of Preparation

1. *Decarboxylation of fatty acids*: By heating carboxylic acid with excess of sodalime or by heating a sodium salt of a fatty acid with sodalime (an equivalent mixture of NaOH and lime)

   For preparation of methane:
   $$CH_3COONa + NaOH \longrightarrow CH_4 + Na_2CO_3$$
   Sodium acetate

   For preparation of any alkane: All alkanes may be prepared by heating the sodium salt of the carboxylic acid with sodalime.
   $$RCOONa + NaOH \longrightarrow RH + Na_2CO_3$$

2. *By reduction of an alkyl halides*: RX Alkyl halides may be reduced to corresponding alkane, by reduction with various reducing agents as given below.

   a. Zn–Cu couple and ethyl alcohol:
   $$Zn + C_2H_5OH \longrightarrow Zn(OC_2H_5)_2 + 2H$$
   $$RX + 2H \longrightarrow RH + HX$$

   b. Zn +acetic acid or Zn + 2HCl
   $$Zn + 2CH_3COOH \longrightarrow Zn(OCOCH_3)_2 + 2H$$
   $$Zn + 2HCl \longrightarrow Zn(Cl)_2 + 2H$$
   $$RX + 2H \longrightarrow RH + HX$$

   c. Zn + 2NaOH
   $$Zn + 2NaOH \longrightarrow Na_2ZnO_2 + 2H\uparrow$$

   d. Sodium and ethanol
   $$C_2H_5OH + Na \longrightarrow C_2H_5ONa + H$$

   e. Lithium aluminum hydride
   $LiAlH_4$ or $NaBH_4$
   $$LiAlH_4 \longrightarrow LiH + AlH_3$$
   $$NaBH_4 \longrightarrow NaH + BH_3$$
   $$RX + LiH \longrightarrow RH + LiX$$

   For example: Ethane may be prepared from C_2H_5X by reducing with any of the reducing agents
   $$C_2H_5X + 2H \longrightarrow C_2H_6 + HX$$

Reduction with LiAlH₄; reduces only a primary RH_2X and a secondary alkyl halide R_2CHX to the corresponding alkane.

$$RCH_2X + LiH \longrightarrow RCH_3 + LiX$$

However, the tertiary alkyl halide R_3CX give mainly an alkene

$$R_3—CX + LiH \longrightarrow \text{an alkene} + HX$$

Reduction with NaBH₄; reduces only 2° and a 3° alkyl halides to corresponding alkane but does not reduce 1° alkyl halides.

3. *By reducing alkyl iodide with HI*: In presence of red phosphorous in a sealed tube at 150 °C

$$RI + HI \longrightarrow RH + I_2$$

4. *From Grignard's reagent (RMgX)*: On treating RMgX, with any compound containing atom, active hydrogen, like H_2O, alcohol, ammonia, primary and secondary amines, alkanes are obtained.

$$RMgX + H_2O \longrightarrow RH + MgX(OH)$$
$$RMgX + HOR \longrightarrow RH$$
$$RMgX + NH_3 \longrightarrow RH$$
$$RMgX + RNH_2 \longrightarrow RH$$
$$RMgX + R_2NH \longrightarrow RH$$

5. *From alkyl halides by Wurtz reaction*: On treating an ethereal solution of alkyl halide, preferably bromide, with sodium metal, linear higher alkanes, containing even number of carbon atoms are formed (other metals can also be used like Ag, Cu, etc. in place of sodium).

$$CH_3X + Na + CH_3X \longrightarrow CH_3—CH_3 + 2NaX$$
$$\text{or} \quad RX + 2Na + XR \longrightarrow —R—R + 2NaX$$

Only 1° alkyl halides would undergo Wurtz reaction, leading to the formation of linear alkanes containing even number of carbon atoms. By taking a mixture of two different alkyl halide, an alkane, containing odd number of carbon may be prepared.

Mechanism: Formation of higher alkane by Wurtz reaction involves, both a free radical as well as organometallic compound formation mechanism.

- **Free radical mechanism:** A free radical is formed, when alkyl halide reacts with sodium metal, this free radical then reacts with other same free radical to form molecule of alkane.

$$R : X + Na \longrightarrow R\,Na + R\bullet$$
$$R\bullet + R\bullet \longrightarrow R : R$$

- **Organometallic compound formation:** An organometallic compound is formed in a reaction between sodium metal and alkyl halide as an intermediate, which on further reaction with another molecule of alkyl halide, forms a higher molecule of alkane.

$$R—X + 2Na \longrightarrow R^-Na^+ + NaX; \quad R^-Na^+ + RX \longrightarrow R : R + NaX$$

Methane cannot be prepared by this method.

Tertiary halides undergo dehydrohalogenation leading to the formation of an alkene, on treatment with sodium metal.

$$R_3—CX + Na \longrightarrow \text{alkene} + NaX$$

The methods discussed so far are for the preparation of linear alkanes, for the preparation of branched chain alkanes, other methods are used.

Corey–House synthesis: The reagent used in this reaction is known as alkyl lithium cuperate, R_2CuLi.

$$R''X + 2Li \longrightarrow R''Li + LiX \quad (R'' \text{ a sec. or tertiary alkyl radical})$$
$$2R''Li + CuI \longrightarrow R''_2CuLi$$

On treating this reagent with R'X, the desired compound may be prepared.

$$R''_2CuLi + RX \longrightarrow R''—R + RCu + LiX$$

6. *By Kolbe's electrolytic synthesis*: In this process, an aqueous solution of a sodium salt of a fatty acid is electrolyzed, when alkane is formed. The sodium salt of a fatty acid is an electrolyte so it dissociates on its own in an aqueous solution to form sodium ion and carboxylate ion. On switching the current, Na^+ ions are liberated as sodium metal at cathode, by taking an electron from the cathode, attacks water to form NaOH and hydrogen. The carboxylic ions give up an electron, converts into a free radical, which decomposes to form a free radical and CO_2. Two such free radicals combine to form a molecule of higher alkane.

Methane cannot be prepared by this method. By taking a mixture of sodium salt of two different acids, an alkane containing odd number of carbon may also be prepared. The mechanism involved is a free radical mechanism.

Reaction at anode   *Reaction at cathode*

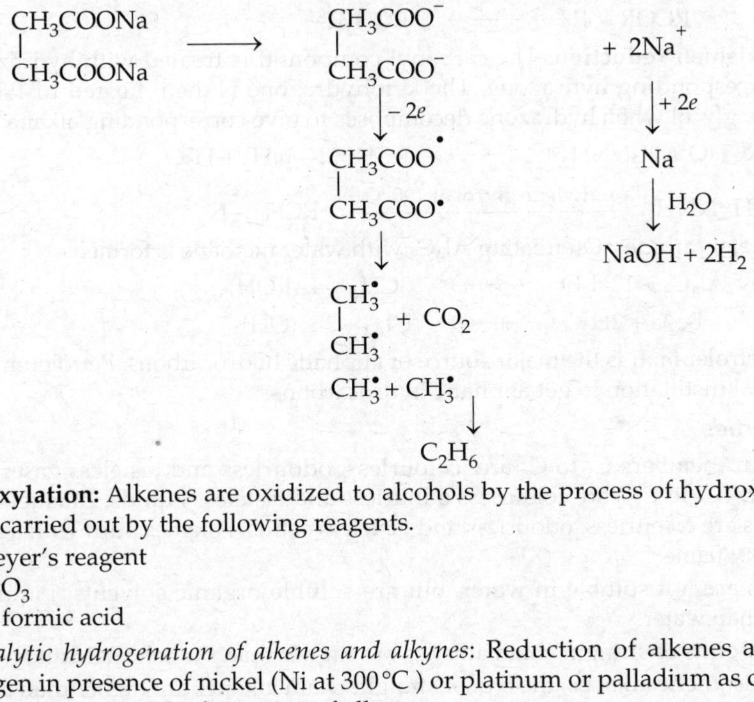

Hydroxylation: Alkenes are oxidized to alcohols by the process of hydroxylation, which can be carried out by the following reagents.

1. Baeyer's reagent
2. By O_3
3. By formic acid

7. *By catalytic hydrogenation of alkenes and alkynes*: Reduction of alkenes and alkynes by hydrogen in presence of nickel (Ni at 300 °C) or platinum or palladium as catalyst at room temperature leads to the formation of alkanes.

$$CH_2=CH_2 + H_2 \longrightarrow CH_3—CH_3$$
ethene (an alkene)   ethane

$$CH \equiv CH + 2H_2 \longrightarrow CH_3CH_3$$
ethyne (an alkyne)

This process of hydrogenation of alkene and alkynes is known as **Sabatier and Sanderens reduction**. The reduction of alkenes and alkyne is an exothermic reaction except methane. Terminal alkenes may be reduced by sodium and alcohols, and the reaction is known as Birch reaction.

$$R—CH=CH_2 + 2H \longrightarrow RCH_2CH_3$$

8. *By reduction of alcohols, aldehydes, ketones and carboxylic acids:* Alcohols, aldehydes, ketones and carboxylic acids may be reduced to corresponding alkanes, by heating in a sealed tube to 150°C with excess of HI and red phosphorous.

$$\text{Red P + HI /150°C /}\Delta$$

$$ROH + 2HI \longrightarrow RH + I_2 + H_2O$$
$$RCHO + 4HI \longrightarrow RCH_3 + 2H_2O$$
$$RCOR + 4HI \longrightarrow RCH_2R + 2H_2O$$
$$RCOOH + 4HI \longrightarrow RCH_3 + 2H_2O$$

9. By **reduction of aldehydes and ketones** by either by **Clemmensen's or by Wolf–Kishner reaction**

   Clemmensen's reaction: The carbonyl compound is reduced with Zn/Hg amalgam and conc. HCl.

$$RCHO + 4H \xrightarrow{\text{Zn/Hg + conc. HCl}} RCH_3 + H_2O$$
$$RCOR + 4H \longrightarrow RCH_2R$$

   Wolf–Kishner reduction: The carbonyl compound is treated with hydrazine to convert into corresponding hydrazone. The ketohydrazone is then heated to 190°C in boiling ethylene glycol when hydrazone decomposes to give corresponding alkane.

$$RCHO + H_2NNH_2 \longrightarrow RCH=N—NH_2 + H_2O$$
$$RCH=NNH_2 \xrightarrow{\text{ethylene glycol at 190°C}} RCH_3 + N_2$$

10. From metal carbides: On treating Al_4C_3 with water methane is formed.

$$Al_4C_3 + 12H_2O \longrightarrow 3CH_4 + 4Al(OH)_3$$
$$Be_2C + 2H_2O \longrightarrow CH_4 + 2Be(OH)_2$$

11. From petroleum: It is the major source of aliphatic hydrocarbons. Petroleum is subjected to fractional distillation to get aliphatic hydrocarbons.

Physical poperties

a. First four members C_1 to C_4 are colourless, odourless and tasteless gases, next thirteen members (C_5 to C_{17}) are colourless odourless and tasteless liquids. Alkanes containing C_{18} onwards are colourless, odourless and tasteless solids. Alkanes burn with a non-luminous, non-sooty flame.

b. Alkanes are not soluble in water, but are soluble organic solvents. Liquid alkanes are lighter than water.

c. Boiling points of straight chain alkanes are higher than branch chain alkanes.
   Among isomeric alkanes, the b.p. of straight chain alkanes are higher than branched chain alkanes. As branching increases, the b.p. decreases further. For example: pentane exists in three isomeric forms

$$CH_3CH_2CH_2CH_2CH_3 \qquad CH_3-\overset{\overset{\displaystyle CH_3}{|}}{C}HCH_2CH_3 \qquad CH_3-\overset{\overset{\displaystyle CH_3}{|}}{\underset{\underset{\displaystyle CH_3}{|}}{C}}HCH_2CH_3$$

n-pentane	isopentane	neopentane
bp = 36°C	= 28°C	= 9°C

It is on account of van der Waal's forces, which are directly proportional to the surface area (these attractive forces are easily overcome by thermal energy). As branching increases, the surface decreases, leading to decrease in van der Waal's forces. Intramolecular forces or van der Waal's forces in a crystal depends not only upon size but also upon how well they fit into crystal lattice.

The dipole moment of all **alkane,** whether straight chain or branched chain, the **dipole moment is zero.** As soon as one hydrogen is substituted by another atom or a group (other than hydrogen), the dipole moment increases immediately. Greater is the value of dipole moment, higher is the b.p..

Chemical properties

All the alkanes are inert, being saturated in nature. Alkanes do not react with alkali, bases or reducing agents. However, alkanes undergo some substitution reactions, with highly reactive substances like halogens or react under vigorous conditions.

1. *Halogenation*: Halogenation of alkanes, is a substitution reaction with one or more hydrogen atoms substituted by halogen atoms in presence of sunlight or UV or heat (a temperature 250–400 °C) or a catalyst.

   Chlorination Methane and other alkanes react with Cl_2 (or Br_2 or I_2), in presence of a diffused light or heat (250–400 °C) to form halogen substituted products. Methane does not react with Cl_2 in dark, but reacts in presence of diffused light to form chlorine substituted derivatives of methane.

$$CH_4 + Cl_2 \xrightarrow{\text{sunlight}} CH_3Cl + HCl$$
methyl
chloride

$$CH_3Cl + Cl_2 \longrightarrow CH_2Cl_2 + HCl$$
methylene
chloride

$$CH_2Cl_2 + Cl_2 \longrightarrow CHCl_3 + HCl$$
chloroform or
trichloro methane

$$CHCl_3 + Cl_2 \longrightarrow CCl_4 + HCl$$
carbon
tetrachloride

Higher alkanes react with Cl_2 at a higher temperature in a similar manner. Bromine and iodine react with alkanes in the same manner as chlorination, but the reaction is not so vigorous. The decreasing order of reactivity of halogens is in the following order.

$$Cl_2 > Br_2 > I_2$$

Iodination is a slow reaction and is reversible.

$$RH + I_2 \rightleftharpoons RI + HI$$

However, iodination may be carried out in the presence of any oxidising agent, like HNO_3, HIO_3 etc, which oxidize HI to iodine, sending the reaction towards the formation of product.

The ease with which various hydrogens are substituted are of the following order.
3° carbon > 2° carbon > 1° carbon.

Mechanism:

Halogenation of alkanes is a free radical reaction. A free radical reaction involves the three steps:

- Chain initiation step
- Chain propagating step
- Chain terminating step

i. **Chain initiating step:** The energy from sunlight or UV or heat, cleaves the chlorine molecule in a homolytic fission into nascent chlorine, which is associated with a large amount of energy.

$$Cl : Cl \longrightarrow 2Cl^{\bullet}$$

ii. **Chain propagating step:** Chlorine so liberated, attack the C—H bond to break, it undergo homolytic fission to form HCl and an alkyl free radical.

$$—C—H + \bullet Cl \longrightarrow —C^{\bullet} + HCl$$

Alkyl free radical then attacks another Cl_2 molecule to form another C—Cl bond, at the same time liberating another Cl free radical, which attacks another C—H bond to convert it into a Cl—bond. The reaction goes on till the energy is available in he system. In this manner the reaction goes on till the energy is available in the system.

$$—C^{\bullet} + Cl : Cl \longrightarrow —C^{\bullet}—C—Cl + Cl^{\bullet} + HCl$$

$$\underset{\underset{H}{|}}{—C}—H + Cl \longrightarrow \underset{\underset{H}{|}}{—C^{\bullet}} + HCl$$

$$\underset{\underset{H}{|}}{—C^{\bullet}—} + Cl : Cl \longrightarrow \underset{\underset{H}{|}}{—C}—Cl + Cl^{\bullet} \text{ and so on}$$

When no more energy is available or the reaction comes to an end.

iii. **Chain terminating step:** This reaction terminates when free radials combine with each other.

$$\underset{\underset{H}{|} \; \underset{H}{|}}{\overset{\overset{H}{|} \; \overset{H}{|}}{H—C—C—H}} + \underset{\underset{H}{|} \; \underset{H}{|}}{\overset{\overset{H}{|} \; \overset{H}{|}}{H—C—C—H}} \longrightarrow CH_2=CH_2 + CH_3—CH_3$$

Chain terminating step may also involve disproportion. In another words, the free radical can also undergo disproportion, one alkyl free radical can gain a hydrogen at the expense of other which loses a hydrogen. In alkanes the decreasing order of substitution of hydrogen, based on the formation of more stable intermediate free radical have the following decreasing order of stability.

$$\underset{\underset{R}{|}}{\overset{\overset{R}{|}}{R—C\bullet}} > \underset{\underset{H}{|}}{\overset{\overset{R}{|}}{R—C\bullet}} > \underset{\underset{H}{|}}{\overset{\overset{H}{|}}{R—C\bullet}} > \underset{\underset{H}{|}}{\overset{\overset{H}{|}}{H—C\bullet}}$$

$$3° \qquad\qquad 2° \qquad\qquad 1° \qquad\qquad \text{methyl}$$
$$\text{free radical}$$

Chlorination of 2° and 3° hydrogen may be carried by treating the alkane with trichloro methyl sulphonyl chloride (CCl_3SO_2Cl) as the chlorinating reagent. This reagent does not chlorinate 1° hydrogen. Fluorination, chlorination, and bromination are exothermic reactions, whereas iodination is an endothermic reaction.

Effect of inhibitors in a free reaction: Inhibitors are those substances which retard the rate of reaction, even when present in a small amount.

On adding a small amount of oxygen, in a reaction undergoing a free radical reaction, the process slows down. On adding a small amount of O_2 in the process of halogenation, the process of chlorination is slowed down, depending upon the amount of oxygen added.

$$\begin{array}{c} H \\ \diagdown \\ H - C \bullet \\ \diagup \\ H \end{array} + \; O\!:\!O \;\longrightarrow\; CH_3-O-O\bullet$$

As CH_3O-O is much more reactive than a CH_3 free radical, it prevents the formation of CH_3Cl.

a. **Fluorination of alkanes:** The direct fluorination of alkanes leads to the decomposition of alkanes, as the reaction is highly exothermic. However, when fluorine is diluted with nitrogen gas it fluorinates alkanes. Alkyl fluorides can also be prepared by treating alkyl halides with AgF, AsF_3, SbF_3 etc.

$$RX + AgF \;\longrightarrow\; RF + AgX$$

This reaction is known as *Swartz reaction*.

b. **Sulphonation:** Alkane react with hot conc. H_2SO_4 or oleum at $400\,^{\circ}C$ to form alkyl sulphonic acid.

$$RH + H_2SO_4 \xrightarrow{\;400\,^{\circ}C\;} RSO_3H + H_2O$$

The relative rate or a decreasing order of substitution of hydrogen by $-SO_3H$ group is as follows:

Decreasing order $3^{\circ} > 2^{\circ} > 1^{\circ}$

c. **Nitration:** Alkanes react with conc. or fuming HNO_3 to form nitro derivatives, preferably in the vapour phase.

$$RH + HNO_3 \xrightarrow{\;150\text{–}475\,^{\circ}C\;} RNO_2 + H_2O$$

Vapour phase nitration may also cleave a C—C bond and may result in the formation of a mixture of nitro alkanes.

For example: $CH_3CH_2CH_3 \xrightarrow{\;\text{vapour phase}\;} CH_3CH_2NO_2 + CH_3CH_2CH_2NO_2$
$$+ \; CH_3NO_2CHNO_2$$

d. **Isomerization:** The process of conversion of one isomer into the other isomer is known as isomerization.

$$CH_3CH_2CH_2CH_2CH_3 \xrightarrow{\;\text{anhyd. AlCl}_3\;} CH_3\overset{\displaystyle CH_3}{\overset{\displaystyle |}{C}}HCH_2CH_3$$

e. **Thermal decomposition or cracking or pyrolysis:** High boiling alkanes are converted into low boiling alkanes when dropped on hot surface, like heated catalyst (oxides of Mo, V, Cr, Pt or Al_2O_3) or on passing vapours of the high boiling alkanes through a hot metal tube ($500\text{–}700\,^{\circ}C$) where alkane molecules break up. This process enables conversion of high boiling fraction of petroleum into low boiling gasoline.

f. **Aromatisation or hydroforming or catalytic reforming or platforming:** *n*-alkanes containing 6 or more carbon atoms undergo dehydrogenation on passing its vapour through a catalyst followed by cyclisation and are converted into benzene and their homologues. The catalyst consists of either oxides of Mo, V, Cr or Pt or Al_2O_3, kept at a temperature $400\text{–}500\,^{\circ}C$. Alkanes first undergo dehydrogenation, followed by cyclisation and isomerization, if possible.

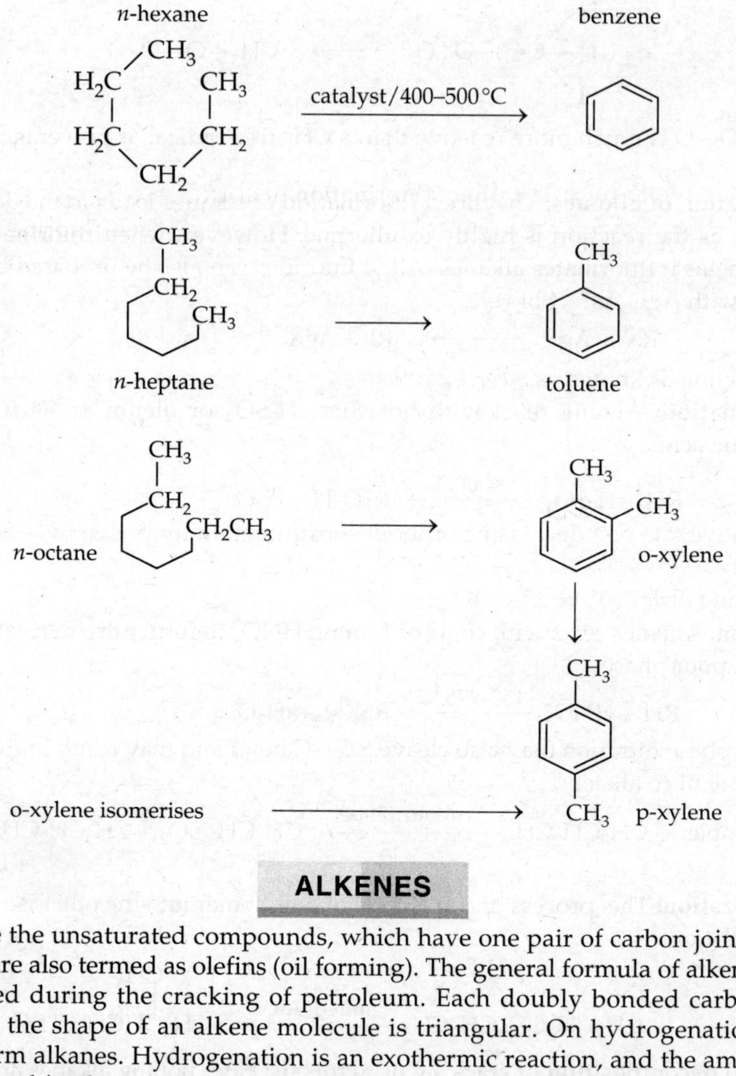

n-hexane → benzene (catalyst/400–500 °C)

n-heptane → toluene

n-octane → o-xylene

o-xylene isomerises → p-xylene

ALKENES

Alkenes are the unsaturated compounds, which have one pair of carbon joined by a double bond. Alkenes are also termed as olefins (oil forming). The general formula of alkene is C_nH_{2n} and these are formed during the cracking of petroleum. Each doubly bonded carbon atom is sp^2 hybridised and the shape of an alkene molecule is triangular. On hydrogenation, alkenes are converted to form alkanes. Hydrogenation is an exothermic reaction, and the amount of energy released = 28.6 kcal/mole.

Nomenclature: Alkenes are named after the name of corresponding name of alkane, by changing or substituting the suffix "ane" by "ylene" in the common nomenclature and by substituting the suffix "ane" by "ene" in the IUPAC nomenclature. The series is known as alkene series. The longest carbon chain, i.e. the molecule containing the double bond is chosen as the parent alkene and the position of the double bond side chain is indicated by the lowest number.

Olefin	Corresponding alkane	Common name	IUPAC name
$CH_2=CH_2$	$CH_3—CH_3$	ethylene	ethene
$CH_3—CH=CH_2$	$CH_3CH_2CH_3$	propylene	propene
$CH_3CH_2CH=CH_2$	$CH_3CH_2CH_2CH_3$	butylene-1	butene-1
$CH_3CH=CHCH_3$	$CH_3H_2CH_2CH_3$	butylene-2	butene-2

Methods of Preparation

1. *By dehydrohalogenation of alkyl halides:* When alkyl halide are heated with alc. KOH or caustic potash, they lose a molecule of hydrogen halide and are converted to form alkenes.

$$H—\overset{\overset{H}{|}}{C}—\overset{\overset{|}{H}}{C}—Br + \text{alcoholic KOH} \xrightarrow{\Delta} CH_2=CH_2 + KBr + H_2O$$

Ethyl bromide

In elimination of a molecule of HBr, the bromide ion is eliminated from the adjacent carbon and a proton from the other carbon atom. This type of elimination is termed *trans* or β-elimination.

Dehydrohalogenation of 2-bromobutane: By treaing 2-bromobutane with alc. KOH, a mixture of 2-butylene is obtained as a major product and other compound 1-butylene is obtained as a minor product.

$$CH_3—\overset{\overset{Br}{|}}{C}HCH_2—CH_3 \longrightarrow CH_3—\overset{+}{\underset{\overset{|}{H}}{C}}CH_2CH_3 \longrightarrow$$

$$CH_3—CH=CH—CH_3$$
80% Saytzeff's product
$$+$$
$$CH_3CH_2CH=CH_2$$
20% Hoffmann's product

The formation of two products can be explained on the basis of Saytzeff's rule.

Saytzeff's rule: According to this rule, the major products formed, on dehydrohalogenation of an alkyl halide by alco. KOH is the one which contains more alkyl groups, joined to the doubly bonded carbon atoms.

$$CH_3CH=CHCH_3 \qquad CH_3CH_2CH=CH_2$$
(contains two alkyl groups) (contains only one alkyl group)

In other words, the hydrogen is lost from the less substituted carbon atom in the formation of a double bond. The relative rate of formation stablities of the alkenes is in the following order:

Therefore,

$$R_2C=CR_2 > R_2C=CHR > R_2C=CH_2 > RCH=CHR > RCH=CH_2 > CH_2=CH_2$$

Therefore, the more stable is the alkene, more easily, it is formed. The dehydrohalogenation of an alkyl halide, by alc. caustic potash, and the decreasing order of reactivity of the alkyl halide is as below.

$$R—\overset{\overset{R}{|}}{\underset{\underset{R}{|}}{C}}X > R—\overset{\overset{R}{|}}{\underset{\underset{H}{|}}{C}}—X > R—\overset{\overset{H}{|}}{\underset{\underset{H}{|}}{C}}—X$$
$$3° \qquad\qquad 2° \qquad\qquad 1°$$

Alkenes on hydrogenation forms alkanes, which is an exothermic reaction. The more stable is the alkene, less is the heat of hydrogenation. For ethene the value = 32.8 kcal/mole. This value goes on decreasing as more and more of ethylene is alkylated. For $R_2C = CR_2$, the value = 28.6 kcal/mole.

2. *Dehydration of alcohols*: Alcohols are dehydrated by conc. H_2SO_4.

   a. Ethyl alcohol is heated with conc. H_2SO_4 at 150-170°C, loses a molecule of water to form alkene.

- $C_2H_5OH \xrightarrow{100°C} C_2H_5HSO_4 + H_2O$
  excess of
  ethyl alcohol

- $C_2H_5HSO_4 + HO—C_2H_5 \xrightarrow{H_2SO_4/145°C} C_2H_5—O—C_2H_5 + H_2O$

- $C_2H_5HSO_4 \xrightarrow{\text{conc. } H_2SO_4/170°C} CH_2 = CH_2$

Mechanism:

i.

$$\underset{\underset{H}{|}}{\overset{\overset{H}{|}}{H—C}}—\underset{\underset{SO_4H}{|}}{\overset{\overset{H}{|}}{C}}—H \longrightarrow CH_2=CH_2 + H_2SO_4$$

H_3PO_4 or P_2O_5 or Al_2O_3 (at 300°C), may also be used for dehydration of alcohols, in place of conc. sulphuric acid.

Dehydration of secondary and tertiary alcohols may be carried by using dilute H_2SO_4.

$$CH_3CH_2CH_2CH_2OH \xrightarrow{75\% \ 100°C} CH_3CH=CHCH_3 + CH_3CH_2CH=CH_2$$

$$\underset{\underset{CH_3}{|}}{\overset{\overset{OH}{|}}{CH_3CHCH_2CH_3}} \xrightarrow{50\%} CH_3CH=CHCH_3$$
2-butene

$$\underset{\underset{CH_3}{|}}{\overset{\overset{CH_3}{|}}{CH_3—C—OH}} \xrightarrow{20\%} (CH_3)_2C=CH_2$$

The ease of dehydrogenation of alcohols is in the decreasing order:
$$3° > 2° > 1°$$

ii. *Formation of butene-1 and-2, from n-butyl alcohols*: The formation may be explained on the basis of formation of an intermediate 1° carbocation, which rearrange to form a 2° carbocation, before forming butene.

$$H_2SO_4 \longrightarrow H^+ + HSO_4^-$$

$$\underset{n\text{-butyl alcohol}}{CH_3CH_2CH_2CH_2OH} + H^+ \longrightarrow \underset{\substack{\text{protonated} \\ \text{butyl alcohol}}}{CH_3CH_2CH_2CH_2\overset{+}{O}—H} \underset{-H_2O}{\xrightarrow{\hspace{1cm}}}$$

$$\underset{1° \text{ carbocation}}{CH_3CH_2CH_2CH_2^+} \xrightarrow{\text{rearrange}} \underset{2° \text{ carbocation}}{CH_3CH_2\overset{+}{C}HCH_3} \longrightarrow CH_3CH=CH—CH_3 + CH_3CH_2CH=CH_2$$

Methyl alcohol on heating with conc. H_2SO_4 forms dimethyl sulphate

$$CH_3OH + H_2SO_4 \longrightarrow \underset{\substack{\text{Methyl hydrogen} \\ \text{sulphate}}}{CH_3HSO_4} \longrightarrow \underset{\text{dimethyl sulphate}}{(CH_3)_2SO_4}$$

3. *By dehalogenation of vicinal or a geminal-di halides*: Alkenes may be prepared by heating dihalides with Zn dust and methanol. Dehalogenation generally occurs in a *trans*-manner. It is stereospecific.

$$CH_3CHBr-CH_2Br \xrightarrow{Zn + CH_3OH/\Delta} CH_3CH=CH_2 + ZnCl_2$$

1, 1-dibromopropane a gem-di bromide 1, 2 dibromo propane– a vic-di bromide.

$$CH_3CH_2CHBr_2 \xrightarrow[\Delta]{Zn/CH_3OH} CH_3CH=CH_2$$

If sodium/ether is used in place of Zn/methanol, an ethereal solution of propylidine bromide gives 3-hexene.

$$CH_3CH_2CHBr_2 + 4Na + Br_2HCCH_2CH_3 \longrightarrow CH_3-CH_2-CH=CH-CH_2-CH_3$$

4. *By heating a tetra alkyl quarternary ammonium halide, $R_4N^+X^-$*: The salt is first treated with AgOH to convert it into corresponding hydroxide, which on heating gives an alkene, besides the formation of a tertiary amine.

$$R_4N^+X^- + AgOH \longrightarrow R_4NOH \longrightarrow Alkene + R_3N + H_2O$$

$$\underset{Base}{(C_2H_5)_4NX} \longrightarrow (C_2H_5)_4NOH \longrightarrow CH_2=CH_2 + (C_2H_5)_3N + H_2O$$

$$(CH_3)_4NOH \longrightarrow CH_3OH + (CH_3)_3N$$

5. *By controlled and selective reduction of alkynes*: Alkynes can be reduced to corresponding alkene, using selective reducing agents.

   a. By using Lindlar's catalyst: Lindlar's catalyst, consists of a mixture of Pd mounted on $BaSO_4$, containing a small amount of sulphur or quinoline.

   $$CH_3-C\equiv C-CH_3 + H_2 \longrightarrow \underset{cis\text{-compound}}{\overset{\displaystyle CH_3 \qquad CH_3}{\underset{\displaystyle H \qquad\quad H}{C=C}}}$$

   Out of *cis*- and *trans*- isomers the *cis*-isomer is obtained in a large quantity.

   b. By using Na + ROH or Li or Na in presence of NH_3 as a reducing agent butyne-2 is reduced to form a *trans*-2-butene

   $$CH_3-C\equiv C-CH_3 \xrightarrow[Li/Na + NH_3]{Na + ROH} \underset{trans\text{-isomer}}{\overset{\displaystyle CH_3 \qquad H}{\underset{\displaystyle H \qquad\quad CH_3}{C=C}}}$$

   If in the process of reduction, where out of two possible stereoisomer, only one is obtained, then it is termed as *stereo specific reduction*.

6. *Kolbe's electrolytic synthesis*: Ethylene may also be prepared by electrolysing an aqueous solution of sodium succinate.

	at Anode	*at Cathode*
CH_2COONa	CH_2COO^-	
|	|	$+ 2Na^+$
CH_2COONa	CH_2COO^-	
	$\downarrow -2e$	$\downarrow +2e$
	$CH_2COO^{\bullet}$	
	|	$2\,Na$
	$CH_2COO^{\bullet}$	
	$\downarrow$	$\downarrow H_2O$
	CH_2	
	$\|\| \quad + CO_2$	$NaOH + H_2$
	CH_2	

Physical properties

Alkenes containing C_2 to C_4, are colourless, odourless and tasteless gases, C_5 to C_{15} are colourless, odourless liquids and still higher ones are waxy solids. Alkenes show geometrical isomerism when each of the doubly bonded carbon atom is attached to two different monovalent groups or atoms.

$$CH_3 \diagdown \atop H \diagup C=C \diagup CH_3 \atop \diagdown H \qquad CH_3 \diagdown \atop H \diagup C=C \diagup H \atop \diagdown CH_3$$

cis-   butene-2   *trans-*

In a pair of *cis-* and *trans-*isomers, the *cis-*isomer has a higher boiling point a higher density, higher refractive index, higher dipole moment value, higher dissociation constant (if acids), but lower melting point, if solids and have a lower stability than the *trans-*isomer.

Alkenes show a regular gradation in the physical and chemical properties

Combustion: Alkenes burn with a luminous and sooty flame. This can be explained on the basis of incomplete oxidation of carbon contents of alkenes (alkenes have a higher carbon contents). During combustion, some carbon particles are liberated in elemental form, which obstruct the flame, turns it luminous and finally deposit as soot.

Colour test:
a. Alkenes decolourise the orange colour of bromine water or (5% solution of bromine in CCl_4). The Br_2 molecule adds on the double bond to form addition compound.
    When no more bromine is left in the solution then the solution becomes colourless.

$$>C=C< + Br_2 \longrightarrow \underset{\underset{Br}{|}}{\overset{\overset{Br}{|}}{C-C}} < \quad \text{bromine adds on the double bond in a } \textit{trans}\text{-manner}$$

b. Alkenes decolourise pink colour of the acidified $KMnO_4$. $KMnO_4$ is an oxidising agent, liberates oxygen which with molecule of water hydroxylate alkene and converts it into diol.

$$>C=C< + O + H_2O \longrightarrow >\underset{\underset{HO}{|}}{C} - \underset{\underset{OH}{|}}{C} <$$

Chemical properties

Alkenes are the unsaturated hydrocarbons, containing one pair of carbon joined by a double bond consisting of one σ-bond and one π-bond and are more reactive than alkanes. Alkenes undergo addition reactions, however under special conditions alkenes may also undergo substitution reactions, as well.

1. The high reactivity of alkenes is due to the presence of a double bond, i.e. due to the presence of two electrons, forming a σ-bond, between a pair of carbon atoms. The bond readily undergoes electromeric change at the time of requirement of the attack by a nucleophile.

$$>C=C< \longrightarrow >C=C \xrightarrow{E^+} >C^+-C \xrightarrow{E^+} -\underset{\underset{}{|}}{\overset{\overset{Nu}{|}}{C}}-\underset{\underset{}{|}}{\overset{\overset{E}{|}}{C}}-$$

The electrophile join at the newly created negative centre, converting it into a charged species, i.e. a carbocation or a carbonium ion as an intermediate. More readily, the carbocation is formed, the faster is the reaction. Alkenes that give rise to the formation of a more stable carbocation, would undergo addition faster.

2. The *addition reaction*, in alkene, takes place across the double bond (more electronegative) facilitating the addition of electropositive part on it.

$$CH_3 \longrightarrow CH=CH_2 \xrightarrow{H^+} CH_3-CH^+-CH_3 \xrightarrow{X^-} CH_3-CHX-CH_3 + CH_3-CH_2-CH_2^+$$
$$\quad\quad\quad\quad\quad\quad\quad\quad\quad \text{2° carbocation} \quad\quad\quad\quad\quad\quad\quad\quad\quad\quad\quad\quad \text{1° carbocation}$$

As there is a possibility of formation of both 1° and 2° carbocation, the 2° carbocation being more stable than the 1° carbocation, therefore the final product is formed from the 2° carbocation (1° carbocation reverts back to the starting alkene. In alkenes, where the allylic group exerts more-I effect, CH_3, on the double bond, follows the anti-Markovnikov's rule.

$$HX \longrightarrow H^+ + X^-$$

- $CCl_3CH=CH_2 \longrightarrow CCl_3-CH=CH_2 \xrightarrow{H^+} CCl_3CH_2-CH_2^+ \xrightarrow{X^-} CCl_3CH_2CH_2X$

- $CH_2^+CH-COOH \longrightarrow CH_2=CH\,COOH \xrightarrow{H^+} {}^+CH_2CH_2COOH \xrightarrow{X^-}$
$$\quad\quad\quad\quad\quad\quad\quad\quad\quad\quad\quad\quad\quad\quad\quad\quad\quad\quad\quad\quad\quad\quad\quad\quad\quad\quad\quad XCH_2-CH_2-COOH$$

3. **Hydrogenation or addition of hydrogen:** All alkenes react with H_2 in presence of a catalyst heated at 300°C in presence of Ni or Pt or Pd at room temperature.
H_2/Ni at 300°C Pt or Pd at room temperature.

$$>C=C< \longrightarrow >CH-CH< \,|\, >C=C< + H_2 \longrightarrow >CH-CH<$$

Hydrogenation of alkane is an exothermic reaction. The average value of heat of hydrogenate = 28.6 kcal/mole. The process of hydrogenation of alkene is termed **Sabatiers and Senderens reaction**.

The relative rate of hydrogenation is as below.

$$CH_2=CH_2 > RCH=CH_2 > RHC=CHR > R_2C=CH_2 > R_2C=CH_2 > R_2C=CR_2$$

Terminal alkenes are reduce by Na + ROH. The reaction is known as *Birch reaction*.

$$Na + ROH \longrightarrow RO^-Na^+ + H$$
$$RCH=CH_2 + 2H \longrightarrow RCH_2CH_3$$

$LiAlH_4$ and $NaBH_4$ do not reduce either an olefinic double bond ($>C=C<$) or an acetylinic triple bond ($-C\equiv C-$)

4. **Addition of halogens: Halogens** add on the double bond to form addition compound known as alkene dihalides. Addition takes place by ionic mechanism, involving electrophilic addition mechanism.

Cl_2, Br_2 and I_2 are nonpolar molecules. The molecule is polarised in the vicinity of electron cloud of carbon to carbon double bond.

$$\text{transition complex} \quad\quad\quad\quad\quad\quad\quad\quad\quad\quad\quad\quad\quad\quad \text{carbocation}$$

Polarised molecule, approaches the ethylene molecule forming a transition complex, which breaks to form a carbocation.

In the final step Br^- gets attached to the +vely charged carbocation to form dibromo compound.

$$\underset{\overset{H}{\underset{H}{}}}{\overset{H}{}}\overset{+}{C}-\underset{Br}{\overset{H}{C}}-H \quad \xrightarrow{Br^-} \quad H-\underset{\overset{H}{}}{\overset{Br}{C}}-\underset{Br}{\overset{H}{C}}-H$$

The relative rate of addition of halogen to olefin, has been observed in the order as given below.

- For halogens $\quad Cl_2 > Br_2 > I_2$
- In the case of alkenes

$CH_2=CHBr < CH_2=CHCOOH < CH_2=CH_2, < CH_3CH=CH_2 < (CH_3)_2 C=CH_2 < CH_3—CH=CHCH_3 < CH_3—CH=C(CH_3)_2$

This order is based on the formation of more stable intermediate carbocations. An alkene, which forms a more stable carbocation reacts more readily. The addition of halogens across the double bond is a *trans*-addition. In the case of alkenes, the fluorinating reagent is PbF_2

$$>C=C< + PbF_2 \longrightarrow \underset{\overset{|}{F}}{-C}-\underset{\overset{|}{F}}{C}-$$

The addition of interhalogen compounds to alkenes is as follows.

$$CH_3—CHCl—CH_2I \xrightarrow{\overset{+}{I}\,\overset{-}{Cl}} CH_3CH=CH_2 \xrightarrow{\overset{+}{Br}/\overset{-}{Cl}} CH_3CHCl\,CH_2Br$$

5. **Addition of halogen acids:** Alkenes react with halogen acids to form alkyl halides.

$$CH_2=CH_2 + HCl \longrightarrow CH_3CH_2Cl$$

Propene reacts with HCl to form isopropyl chloride instead of *n*-propyl chloride

$$CH_3CH=CH_2 + HCl \longrightarrow \underset{\text{Isopropyl chloride}}{CH_3CHClCH_3}$$

The formation of isopropyl chloride is based on Markovnikov's rule, according to this rule, the negative part of the addendum (compound to be added), adds on that doubly bonded carbon which contains lesser number of hydrogen atom in case of an unsymmetrical alkene.

$$CH_3CH=CH_2 \xrightarrow{HX} CH_3CHXCH_3$$

Addition of halogen acid on an unsymmetrical alkene is based on the formation of a more stable intermediate carbocation.

$$CH_3CH=CH_2 \xrightarrow{\overset{+}{H}} \underset{\text{2° carbocation}}{CH_3\overset{+}{C}HCH_3} \xrightarrow{\overset{-}{X}} CH_3CHXCH_3$$

$$\longrightarrow \underset{\text{1° carbocation}}{CH_3CH_2CH_2^+} \qquad \text{(reverts back to alkene)}$$

$$\underset{3°}{\overset{R}{\underset{R}{\overset{|}{R—C}}}+} > \underset{2°}{\overset{R}{\underset{H}{R—\overset{+}{C}}}} > \underset{1°}{\overset{H}{\underset{H}{R—\overset{+}{C}}}} > \underset{\text{methyl}}{\overset{H}{\underset{H}{H—\overset{+}{C}}}}$$

The relative reactivity of halogen acid towards alkenes is in the following order.

$$HI > HBr > HCl$$

The *decreasing order of reactivity of alkenes towards halogen acids* is in the following order:

$(CH_3)_2C=CH_2 > CH_3CH=CH-CH_3 > CH_3CH_2C=CH_2 > CH_3CH=CH_2 > CH_2=CH_2 > CH_2=CHCl$

The addition of halogen acids to a double bond takes place in a *trans*-manner.

$$>C = C< + HX \longrightarrow \overset{\displaystyle H}{\underset{\displaystyle X}{-C-C-}}$$

When addition of HX takes place in presence of a peroxide or a peroxy acid, the addition of HBr to an unsymmetrical alkene, take place in abnormal manner, or opposite to that of Markovnikov's rule (this abnormality is not observed in case of either HCl or HI)

$$CH_3CH=CH_2 + HBr \xrightarrow{\text{peroxide}} CH_3CH_2CH_2Br$$

This abnormal reaction or addition, is termed **peroxide effect or Kharasch reaction or** *anti-Markovnikov's reaction.*

The formation of anti-Markovnikov's product may be explained on the basis of formation of a more stable intermediate free radical from alkene. The attacking free radical in this case is Br•.

Mechanism: It involves the formation of a free radical.

- $((C_6H_5COO)_2 \longrightarrow C_6H_5{}^{•} + 2CO_2$
- $C_6H_5{}^{•} + HBr \longrightarrow C_6H_6 + Br^{•}$

- $CH_3CH=CH_2 \underset{Br^{•}}{\Bigg\langle} \begin{array}{l} \longrightarrow CH_3CHBr-\overset{•}{C}H_2 \\ \quad \text{a 1° free radical} \\ \\ \longrightarrow CH_3\overset{•}{C}H\,CH_2Br \\ \quad \text{a 2° free radical} \end{array}$

A 2° free radical is more stable as compared with a 1° free radical, the final product is formed from the 2° free radical, while the 1° free radical reverts back to form the alkene. The decreasing order of the stability of free radical is in the following order.

$$\underset{3°}{\overset{\displaystyle R}{\underset{\displaystyle R}{R-\overset{|}{\underset{|}{C}}{}^{•}}}} > \underset{2°}{\overset{\displaystyle R}{R-\overset{}{\underset{H}{C}}{}^{•}}} > \underset{1°}{\overset{\displaystyle H}{R-\overset{}{\underset{H}{C}}{}^{•}}} > \underset{\substack{\text{methyl free} \\ \text{radical}}}{\overset{\displaystyle H}{H-\overset{}{\underset{H}{C}}{}^{•}}}$$

The energy required for homolytic fission of halogen acids is as follows.

$H \!:\! Cl \longrightarrow H^{•} + Cl^{•} \qquad$ 107 kcal/mole

$H \!:\! Br \longrightarrow H^{•} + Br^{•} \qquad$ 87 kcal/mole

$H \!:\! I \longrightarrow H^{•} + I^{•} \qquad$ 70 kcal/mole

Hydrochloric acid, HCl does not undergo a homolytic fission, in presence of a peroxide, the energy required is too high, this much energy is not available in the system. HBr and HI undergo such a homolytic fission, in the presence of peroxide.

A Br$^{•}$ liberated from HBr, attacks unsymmetrical alkene and undergoes Kharasch reaction. The I$^{•}$ liberated from HI, immediately combine with another I-atom to form an iodine molecule and therefore iodine atom is not available for a free radical attack on alkene.

$$I\bullet + I\bullet = I_2$$

Addition of a few molecule of an inhibitor like hydroquinone or diphenyl amine, reverses the peroxide effect and the addition of HBr takes place in accordance with Markovnikov's rule.

Effect of addition of oxygen: On adding oxygen in a system, undergoing a free radical reaction, the reaction stops till the added oxygen is consumed by the free radicals.

$$R\bullet + \bullet O\!:\!O\bullet \longrightarrow R-O\!:\!O\bullet$$

6. **Hydration of alkenes:** Olefins may be hydrated to form alcohols.
   - H^+ catalyzed hydration: Alkene is treated with conc. H_2SO_4 to form an alkyl hydrogen sulphate, which upon hydrolysis forms alcohol.

   $$CH_2=CH_2 + H_2SO_4 \longrightarrow C_2H_5HSO_4 \longrightarrow C_2H_5OH$$

   - *Hydroboration*: Alkene react with diborane (B_2H_6) in an ethereal solution in the form of a monomer BH_3, to form an addition compound, which on oxidation with alkaline H_2O_2 gives alcohol.

   $$R-CH=CH_2 + BH_3 \longrightarrow RCH_2CH_2BH_2 \xrightarrow{H_2O_2/OH^+} RCH_2H_2OH$$

   $$BH_3 \text{ adds on as } BH_3 \longrightarrow BH_2^+ + H^-$$

   Therefore the addition across the double bond takes place as per Markovnikov's rule (as the charges on the species is concerned). On the other hand (by ignoring the charges on the species) the addition may be regarded as anti-Markovnikov' rule.

   - *Oxymercuration–demercuration*: In this process, alkene is treated with an aqueous solution of mercuric acetate $(CH_3COO)_2$ Hg, when an addition compound is obtained.

   i. $(CH_3COO)_2Hg \longrightarrow CH_3COO^- + CH_3COOHg^+$

   ii. $CH_3COO^- + H_2O \longrightarrow CH_3COOH + OH^-$

   iii. $CH_3COOHg^+ + RCH=CH_2 \longrightarrow RC^+H-CH_2-HgOCOCH_3$

   iv. $RCH^+CH_2HgOCOCH_3 + OH \longrightarrow RCH(OH)CH_2HgOOCH_3$

   v. $RCH(OH)HgOCOCH_3 + NaBH_4 \longrightarrow RCH(OH)CH_3$

   Mechanism:
   In first step, there is addition of mercuric acetate to form mercurinium ion rather than carbocation. In the second step, H_2O attacks the carbon to form a mere stable transition state, $NaBH_4$ converts C—Hg bond into C—H bond to give alcohol finally.

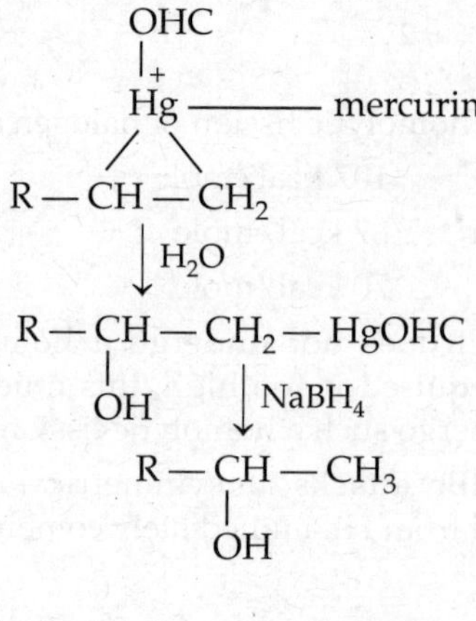

The addition of BH_3, however, does not proceed via carbocation formation. Electrophilic boron attacks electron rich carbon and as transition state begins, the carbon nuclei losing its π-electrons become acidic and begins to pull the nearest hydrogen. This results in the formation of a cyclic transition state having a C—H as well as carbon-boron bond and the partial positive charge on carbon attaches to more carbon as shown below.

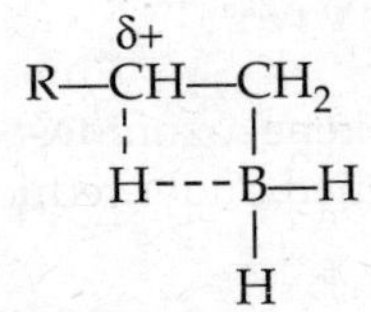

$$
\begin{array}{c}
\delta+ \\
R\!-\!\overset{|}{C}H\!-\!\overset{|}{C}H_2 \\
\overset{|}{H}\text{- - -}\overset{|}{B}\!-\!H \\
\overset{|}{H}
\end{array}
$$

7. **Hydroxylation:** Alkenes are oxidized to alcohols by the process of hydroxylation, which is carried out by the following reagents.
   - *Baeyer's reagent*: A cold dil. solution of $KMnO_4$ oxidize the ethylene to ethylene glycol. The addition is *cis*-addition.

$$
CH_2{=}CH_2 + O + H_2O \longrightarrow \begin{array}{c} CH_2{-}CH_2 \\ |\quad\ \ | \\ HO\quad HO \end{array}
$$

   - *Osmium tetroxide*: OsO_4 adds to the double bond to form an adduct, which upon treatment with $NaHSO_3$ is hydrolyzed to form ethylene glycol. The addition of OsO_4 takes place in a *cis*-manner, to form *cis*-glycol.

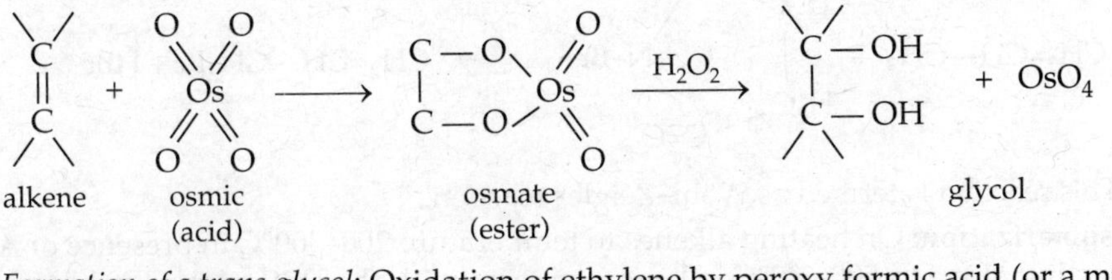

| alkene | osmic (acid) | osmate (ester) | glycol |

Formation of a trans glycol: Oxidation of ethylene by peroxy formic acid (or a mixture of $H_2O_2 + HCOOH$)

$$
CH_2{=}CH_2 \longrightarrow \begin{array}{c} HO \\ |_1 \quad 2 \\ CH_2{-}CH_2 + OsO_4 \\ | \\ OH \end{array}
$$

8. **Addition of HOCl**
$$
CH_2{=}CH_2 + HOCl \longrightarrow \begin{array}{c} CH_2{-}CH_2 \\ |\quad\ \ | \\ HO\quad Cl \end{array}
$$

9. **Ozonide formation: Alkenes** react with O_3 molecule to form ozonide.

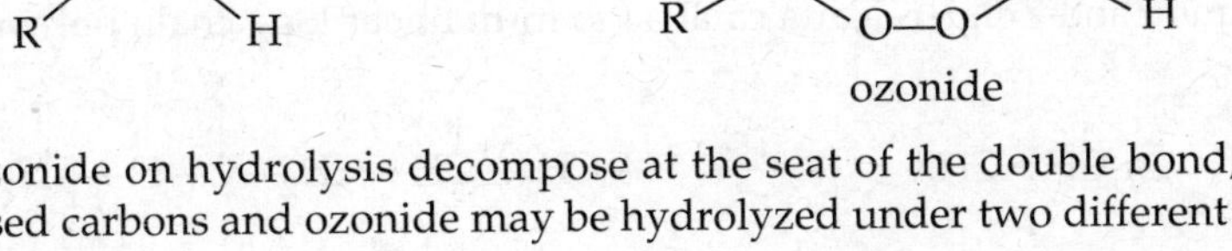

ozonide

Alkene ozonide on hydrolysis decompose at the seat of the double bond, separating the two oxidised carbons and ozonide may be hydrolyzed under two different conditions.

a. **Reductive hydrolysis:** Alkene ozonide is hydrolysed under reducing condition, (where no oxidation may take place).

$$\underset{\overset{|}{O}\text{--------}\overset{|}{O}}{RCH\text{---}O\text{---}CH_2} \xrightarrow{Zn+H_2O \text{ or } CH_3COOH \text{ or } Ni/H_2} RCHO + HCHO$$

aldehyde + aldehyde or
ketone + aldehyde

b. **Oxidative hydrolysis:** Alkene ozonides are hydrolysed in oxidising conditions, whereby the aldehydes formed under reducing conditions are oxidised to acids.

$$\underset{\overset{|}{O}\text{--------}\overset{|}{O}}{RCH\text{---}O\text{---}CH_2} \xrightarrow{\text{Tollen's reagent or alk. } H_2O_2} \text{acid + acid}$$

c. **Reduction with LiAlH$_4$ or NaBH$_4$:** alkene ozonides are reduced **to form a mixture** of alcohols.

10. **Allylic halogenation:** Alkenes containing allyl group in the molecule, may be chlorinated at elevated temperature (or in presence of sunlight) to form chloroalkene.

$$CH_2=CH\text{---}CH_3 + Cl_2 \xrightarrow{500\text{-}600°} CH_2=CH\text{---}CH_2Cl + HCl$$

Similarly allylic bromination may be carried by N-bromo succinimide.

$$CH_2=CH\text{---}CH_3 + \begin{matrix} H_2C \diagup \overset{C=O}{} \\ | \qquad\qquad N\text{--}Br \\ H_2C \diagdown \underset{C=O}{} \end{matrix} \longrightarrow CH_2=CH\text{---}CH_2Br + HBr$$

This reaction is termed as Wohl–Zeigler reaction.

11. **Isomerization:** On heating alkenes to temperature 200–300°C in presence of Al$_2$(SO$_4$)$_3$, or heating alone to 500-600°C, alkene isomerise with the shift of the double bond towards centre.

$$CH_3\,CH_2CH_2CH_2CH=CH_2 \longrightarrow CH_3CH_2CH\text{---}CH=CHCH_3$$

12. **Oxidation:** Alkenes react with oxygen in presence of Ag as catalyst.

$$CH_2=CH_2 + 1/2\,O_2 \xrightarrow{Ag/300°C} \underset{\text{Oxirane}}{\overset{\diagup O \diagdown}{CH_2\text{---}CH_2}}$$

13. **Prileschaiev reaction:** Alkenes are oxidised to form epoxide or oxirane, by treating the alkenes with peroxy acid.

$$CH_2=CH_2^- + \text{peroxy acid} \longrightarrow \overset{\diagup O \diagdown}{CH_2\text{---}CH_2}$$

14. **Polymerisation:** Alkenes polymerize in presence of various reagents, like free radical, cation, anion and Zeigler–Natta catalyst to form linear long chain polymers.

$$n\text{-}CH_2=CHX \longrightarrow \underset{}{\overset{X}{\underset{|}{}}}\text{---}CH_2\text{---}\overset{X}{\underset{|}{CH}}\text{---}CH_2\text{---}\overset{X}{\underset{|}{C}}\text{---}H\text{---}CH_2\text{---}\overset{X}{\underset{|}{CH}}\text{---}$$

Properties

1. Dehydration of RX by a strong base to form alkenes, is not accompanied by rearrangement.
2. Relative ease of the formation of alkene and decreasing order of stability of alkene is as follows.

   $$R_2C=CR_2 > R_2C=CHR > R_2C=CH_2 > RHC=CHR > RCH=CH_2 > CH_2=CH_2$$

3. An alkene, with the greater number of alkyl groups is the preferred product, because it is formed faster than alternate alkenes.
4. **Hydrogenation is an exothermic reaction.**

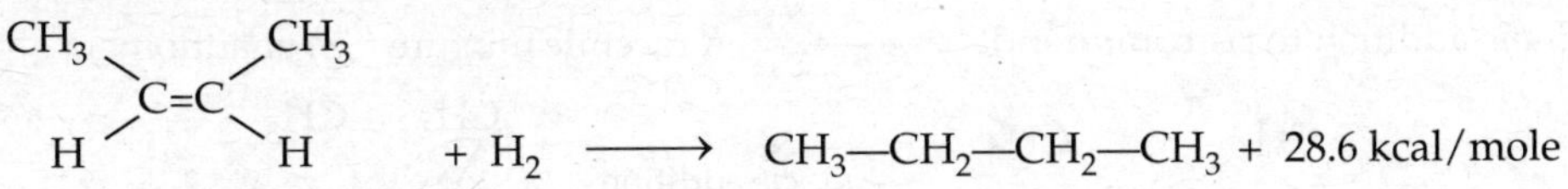

$$+ H_2 \longrightarrow CH_3{-}CH_2{-}CH_2{-}CH_3 + 28.6 \text{ kcal/mole}$$

$-\Delta H = 28.6$ kcal

cis-2-butene

$$+ H_2 \longrightarrow CH_3{-}CH_2{-}CH_2{-}CH_3 + 27.6 \text{ kcal/mole}$$

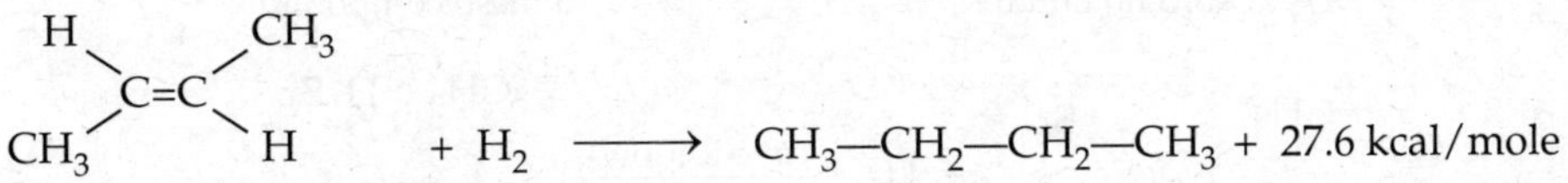

$-\Delta H = 27.6$ kcal

trans-2-butene

(*Trans*-isomer evolves 1 kcal/mole, less energy, on hydrogenation than *cis*-isomer, it means that it contains 1 kcal less energy than the *cis*-isomer, the *trans*-isomer is more stable, than *cis*-isomer).

- Reduction of 2-butyne by Lindlar's catalyst (Pd + $BaSO_4$) gives *cis*-2-butene.
  Lindlar's catalyst (palladium mounted on $BaSO_4$)

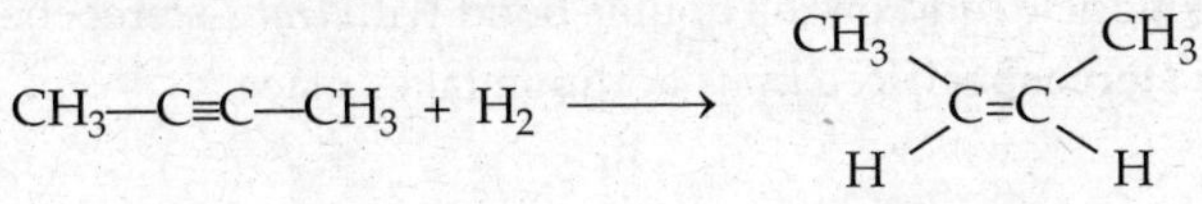

$$CH_3{-}C{\equiv}C{-}CH_3 + H_2 \longrightarrow$$

2-butyne   *cis*-2-butene

- Reduction of 2-butyne by Na + NH_3, or nickel boride gives *trans*-2-butene

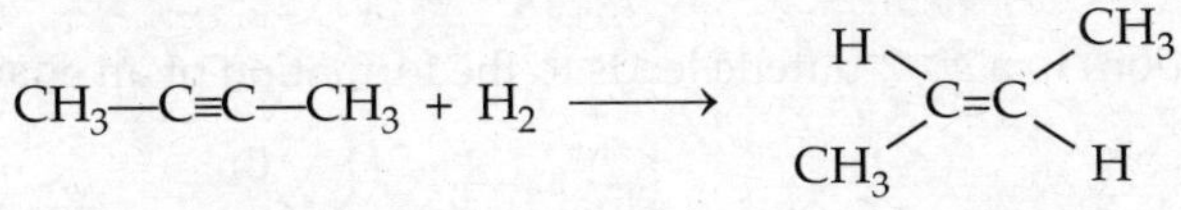

$$CH_3{-}C{\equiv}C{-}CH_3 + H_2 \longrightarrow$$

trans-2-butene

Hydrogen usually adds in a *cis*-manner. When addition of hydrogen to an alkene, leads to the formation of an optically active compound, as in the following cases, *trans*-addition to the *cis*-alkene leads to the formation of a compound.

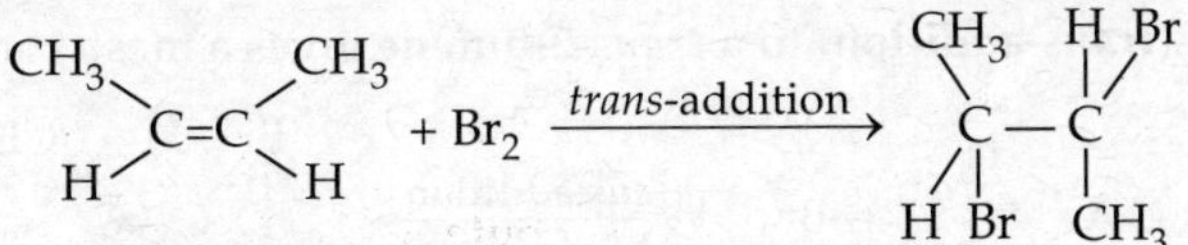

cis-butene   an enantiomer

On the other hand a *trans*-addition to a *trans*-alkene gives an optically active compound, namely an enantiomer.

$$CH_3\!\!\diagdown \!\!{}_{C=C}\!\!\diagup \!\!H \atop H\diagup \quad \diagdown CH_3 \quad + \ Br_2 \xrightarrow{\ \textit{trans-addition}\ } \quad \begin{matrix} Br & & Br \\ \diagdown & & \diagup \\ H-C-C-H \\ \diagup & & \diagdown \\ CH_3 & & CH_3 \end{matrix}$$

trans-butene   a meso compound

cis-addition to *cis*-compound	$\longrightarrow$	a meso compound (RS meso isomer)
cis-addition to *trans*-compound	$\longrightarrow$	a racemic compound (or SS mixture)
trans-addition to *trans*-compound	$\longrightarrow$	a meso compound (RS meso isomer)
trans-addition to *cis*-compound	$\longrightarrow$	A racemic mixture (SS) enantiomer

$$CH_3\!\!\diagdown \!\!{}_{C=C}\!\!\diagup \!\!CH_3 \atop Br\diagup \quad \diagdown Br \quad + \ H_2 \xrightarrow{\ \textit{cis-addition}\ } \quad \begin{matrix} CH_3 & CH_3 \\ | & | \\ C-C \\ \diagup| & \diagup| \\ H\ Br & H\ Br \end{matrix}$$

cis-2,3-bromo butene   a meso compound

$$CH_3\!\!\diagdown \!\!{}_{C=C}\!\!\diagup \!\!Br \atop Br\diagup \quad \diagdown CH_3 \quad + \ H_2 \xrightarrow{\ \textit{cis-addition}\ } \quad \begin{matrix} CH_3 & H\ Br \\ | & |\diagup \\ C-C \\ \diagup| & \diagdown \\ H\ Br & CH_3 \end{matrix}$$

trans-2,3-butene   a meso compound

5. Reactivity of alkenes, towards acids, is as follows (in the decreasing order):
$(CH_3)_2\!-\!C=CH_2 > CH_3\!-\!CH=CH\!-\!CH_3 > CH_3\!-\!CH_2\!-\!CH=CH_2 > CH_3\!-\!CH=CH_2 > CH_2=CH_2 > CH_2=CHCl$
Electron donating group on the olefinic carbon, accelerates the rate of addition reaction.
Halide ion (Cl⁻, Br⁻, I⁻), protic acid, add on the double bond with *trans*-stereochemistry.

6. Addition of **Cl₂ and Br₂ is stereospecific**–a *trans*-addition takes place.

$$>\!C=C\ +\ Br_2 \longrightarrow \begin{matrix} Br \\ | \\ C-C- \\ | \\ Br \end{matrix}$$

Addition of Br_2 (*trans*-addition) to a *cis*-2-butene leads to the formation of an enantiomer.

$$H\!\!\diagdown \!\!{}_{C=C}\!\!\diagup \!\!H \atop CH_3\diagup \quad \diagdown CH_3 \quad + \ Br_2 \xrightarrow{\ \textit{trans-addition}\ } \quad \begin{matrix} H & H & Br \\ \diagdown & | & \diagup \\ Br-C-C \\ \diagup & & \diagdown \\ CH_3 & & CH_3 \end{matrix}$$

cis-2-butene   enantiomer (SS)
a racemic mixture

Addition of Br₂(***trans*-addition) to a *trans*-2-butene gives a meso compound**.

$$CH_3\!\!\diagdown \!\!{}_{C=C}\!\!\diagup \!\!H \atop H\diagup \quad \diagdown CH_3 \quad + \ Br_2 \xrightarrow{\ \textit{trans-addition}\ } \quad \begin{matrix} Br & H & CH_3 \\ \diagdown & |\diagup \\ C-C \\ \diagup & \diagup| \diagdown \\ CH_3 & H & Br \end{matrix}$$

trans-2-butene   a meso compound (R, S)

Addition of I_2 is a reversible reaction proceeds with *trans*-addition.

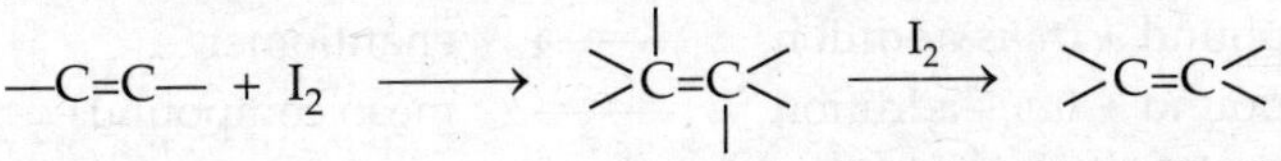

$$-C=C- + I_2 \longrightarrow \underset{|}{>}\overset{|}{C}=C< \xrightarrow{I_2} >C=C<$$

7. Oxymercuration (addition of mercuric acetate $Hg(CH_3COO)_2$ or $Hg(OAc)_2$ of most acyclic and monocyclic olefins is a *trans*-addition, reaction similar to the addition of NO-Cl.

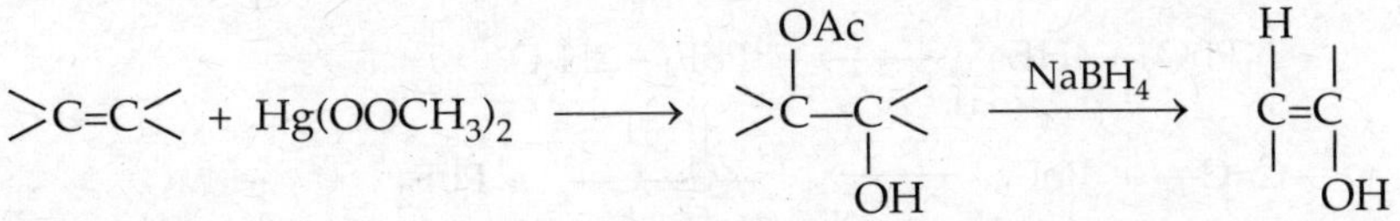

$$>C=C< + Hg(OOCH_3)_2 \longrightarrow \overset{OAc}{\underset{OH}{>C-C<}} \xrightarrow{NaBH_4} \overset{H}{\underset{OH}{C=C}}$$

The reaction is known as oxymercuration-demercuration. It is free from rearrangement. Addition follows Markovnikov's rule. A molecule of H_2O adds in the *trans*-manner

8. Hydroboration of alkenes: Addition of BH_3 to alkenes –H and –OH added in a *cis*-manner follows anti–Markovnikov's rule. It is free from rearrangement.

$$-C=C- + BH_3 \longrightarrow \overset{H\ \ BH_2}{-C-C-} \xrightarrow{H_2O_2/OH} \overset{H\ \ \ \ OH}{-C-C}$$

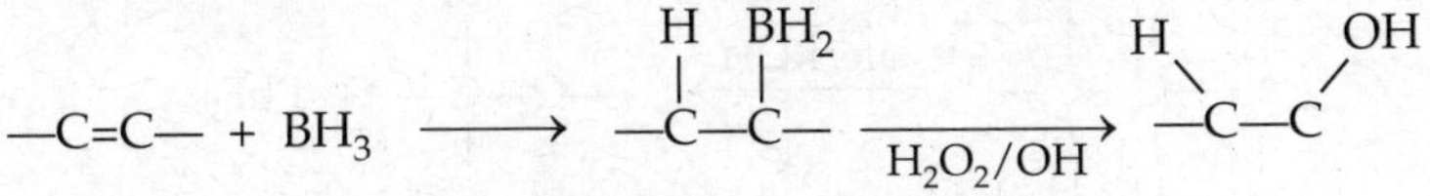

9. **Like olefins, aldehydes also decolourises bromine water.**
10. **Conjugated dienes react more rapidly, than alkenes**; 1,4-product is energetically more favourable than 1,2-product. In dienes 1,4-addition compounds, predominates the equilibrium. It proves that 1,4-addition product is more stable of the two and ionizes more slowly. 1,2-product is not only formed faster than that of 1,4 product, but ionizes rapidly too.
11. (a) Hydroxylation of alkenes by dil. $KMnO_4$ (Baeyer's regent), is a *cis*-addition.

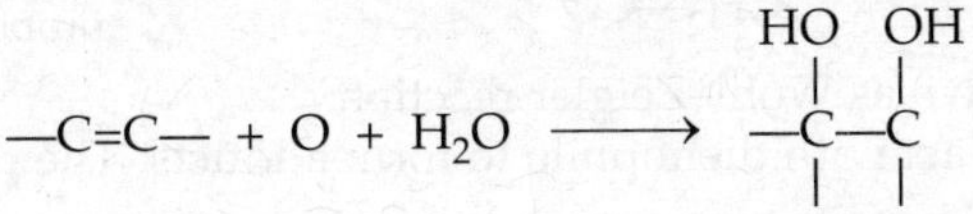

$$-C=C- + O + H_2O \longrightarrow \overset{HO\ \ OH}{-C-C}$$

cis-glycol

(b) Hydroxylation of alkenes by **osmium tetraoxide, (OsO_4)**, followed by treatment by $NaHSO_3$ is a *cis*-addition.

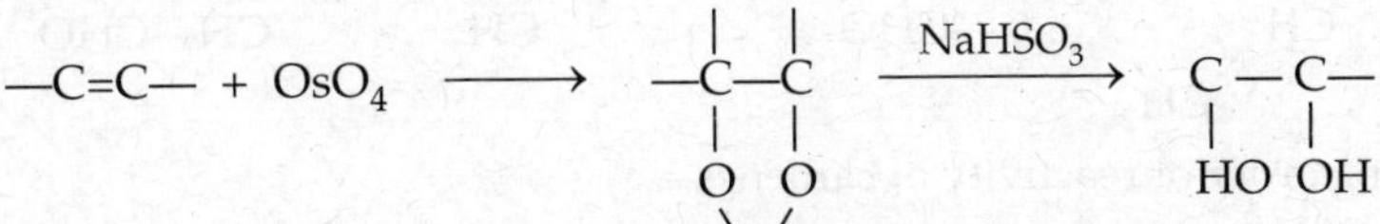

$$-C=C- + OsO_4 \longrightarrow \overset{|\ \ \ |}{-C-C-} \xrightarrow{NaHSO_3} \overset{C-C-}{\underset{HO\ \ OH}{|\ \ \ |}}$$

cis-glycol

(c) **Hydroxylation by peroxy formic acid, HCOOOH** (or a mixture consisting of HCOOH + H_2O_2) is a *trans*-addition.

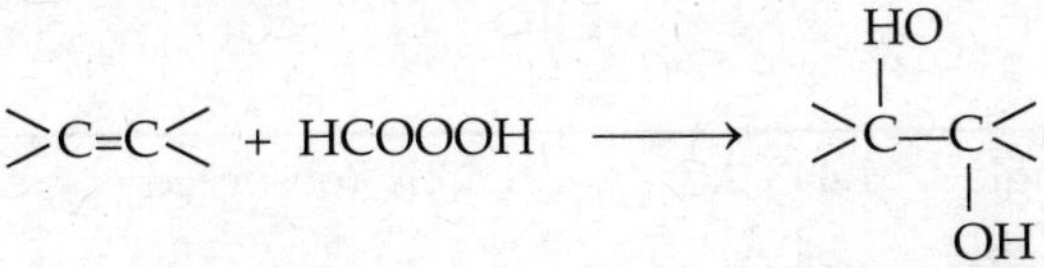

$$>C=C< + HCOOOH \longrightarrow \overset{HO}{\underset{OH}{>C-C<}}$$

trans-glycol

cis-compound + cis-addition $\longrightarrow$ meso compound

cis-compound + $trans$-addition $\longrightarrow$ enantiomer

$trans$-compound + $trans$-addition $\longrightarrow$ meso compound

$trans$-compound + cis-addition $\longrightarrow$ enantiomer

12. **Fluorination of alkenes:** Fluorination of alkenes, may be carried by treating with HF and PbO_2.

$$PbO_2 + 4HF \longrightarrow PbF_4 + 2H_2O$$

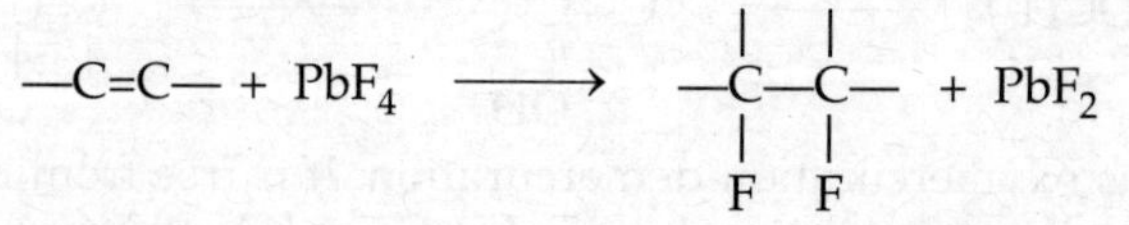

$$-C=C- + PbF_4 \longrightarrow \underset{\overset{|}{F}\ \ \overset{|}{F}}{-\overset{|}{C}-\overset{|}{C}-} + PbF_2$$

13. Dehydrohalogenation of an alkyl halide with alcoholic KOH is a $trans$-elimination or β–elimination.

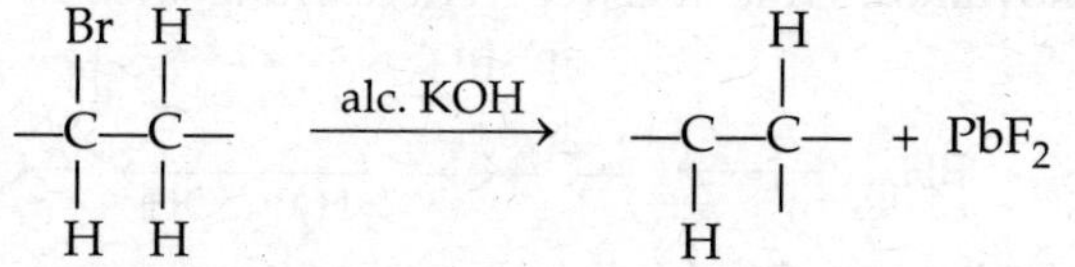

$$\underset{\overset{|}{H}\ \ \overset{|}{H}}{-\overset{\overset{Br}{|}}{C}-\overset{\overset{H}{|}}{C}-} \xrightarrow{\text{alc. KOH}} \underset{\overset{|}{H}}{-\overset{\overset{H}{|}}{C}-\overset{|}{C}-} + PbF_2$$

14. SeO_2–dehydrogenates diaryl ethane to form diaryl ethylene.

$$Ar-CH_2-CH_2-Ar \longrightarrow Ar-CH=CH-Ar$$

15. Allylic brominating by NBS, (N-bromo succinimide), gives an allylic brominated product. It is a specific reagent that is used for brominating alkenes at allylic position.

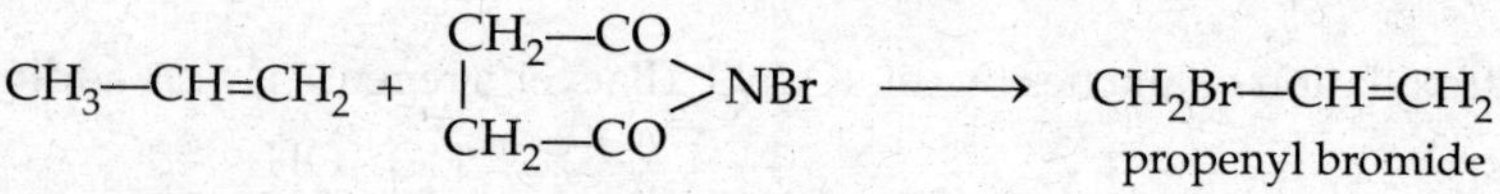

$$CH_3-CH=CH_2 + \underset{CH_2-CO}{\overset{CH_2-CO}{|}}\!\!\!>\!\!NBr \longrightarrow \underset{\text{propenyl bromide}}{CH_2Br-CH=CH_2}$$

The reaction is known as Wohl–Zeigler reaction.

16. Conjugated dienes act with dienophile to form adducts. The reaction is known as Diels–Alder reaction.

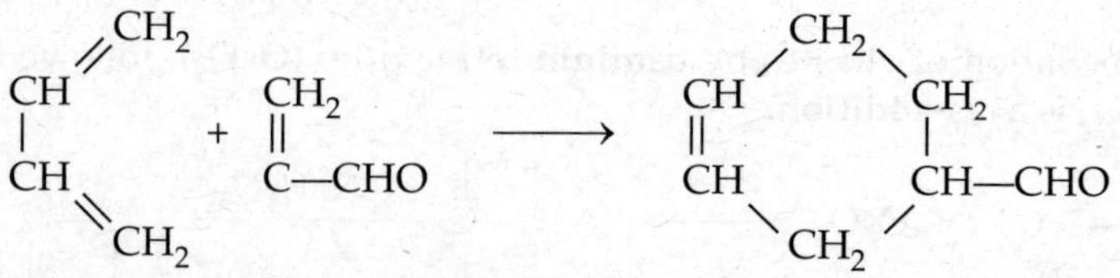

17. Decreasing order of reactivity of carbenes.

$$:CH_2 > :CHCl > :CCl_2 > :CBr_2 > :CF$$

18. The stability of propene has been explained on the basis of hyperconjugation.

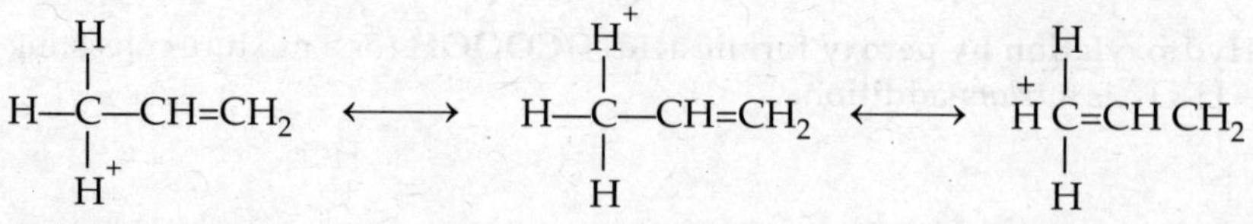

$$\underset{\overset{|}{H}^+}{H-\overset{\overset{H}{|}}{C}-CH=CH_2} \longleftrightarrow \underset{\overset{|}{H}}{H-\overset{\overset{H^+}{|}}{C}-CH=CH_2} \longleftrightarrow \underset{\overset{|}{H}}{H\overset{\overset{H}{+|}}{C}=CH\,CH_2}$$

Bond length–single	1.448 Å	1.45 Å (observed)
Bond length–double	1.34 Å	

REASONING TYPE QUESTIONS (RTQs)

(Group-A)

Complete the following reactions.

1. $CH_2=CH_2 + PdCl_2 + H_2O \longrightarrow$
2. $CH_2=CH_2 + Cl_2^+ + HCl \longrightarrow$
3. $CH_2=CHCl + HCl \longrightarrow$
4. $CCl_3-CH=CH_2 + HCl \longrightarrow$
5. $CH_2=CH-COOH \xrightarrow{+HBr}$
6. $CH_2=CH-CH=CH_2 +$ maleic anhydride $\longrightarrow$
7. $CH_2=CH-CH=CH_2 + Br_2$ (in CH_3COOH) $\longrightarrow$

8.  + hot $KMnO_4 \longrightarrow$

9. $CH_2=CH_2 + CH_2N_2 \longrightarrow$
10. $(C_2H_5)_2CCH=CH_2 + ICl \longrightarrow$
11. $CH_3CH=CHCH_3 +$ reagent $\longrightarrow$
12. $CH_3C\equiv C\ CH_3 +$ reagent $\longrightarrow$

13. 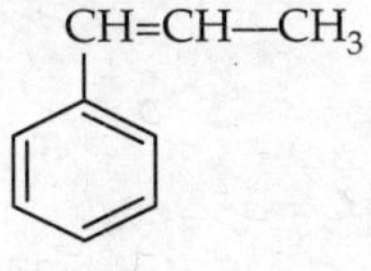 + HBr $\longrightarrow$

14. 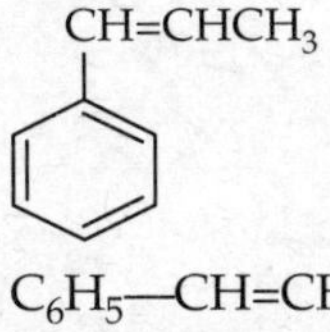 + HBr $\xrightarrow{\text{in the presence of peroxide}}$

15. $C_6H_5-CH=CH-CH_3 +$ cold $H_2SO_4 \longrightarrow$
16. $C_6H_5-CH=CH-CH_3 +$ hot $KMnO_4 \longrightarrow$
17. $CH_3-HC=CH_2 + BH_3/H_2O_2, OH^- \longrightarrow$
18. $CH_3-CH=CH-CH_2-CH_2C\equiv CH + HBr \longrightarrow$
19. $CH_2=CH-C\equiv CH + HBr \longrightarrow$
20. $(CH_3)_3-C-CH=CH_2 + HCl \longrightarrow$

21. 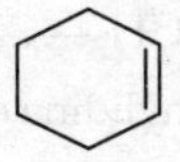 + peroxide $\longrightarrow$

22. $CH\equiv CH + CH_3COOH \longrightarrow$
23. $CH_3-CH=CH_2 + CO + H_2 \longrightarrow$

24. 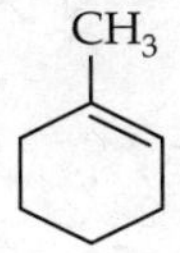 + $Hg(OCOCH_3)_2 \xrightarrow{THF/H_2O}$? $\xrightarrow{NaBH_4}$?

25. $RCH_2-C{\equiv}CH + NaNH_2 \xrightarrow{170°C}$

26. 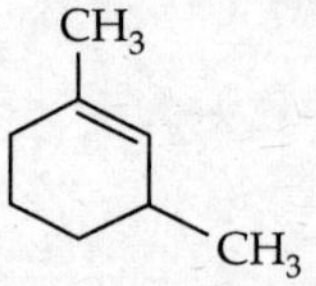 $+ HBr \longrightarrow$

27. $CH_3-CH{=}CH_2 + BrCl \longrightarrow$

28. *cis-* and *trans*-butene $+ I_2 \longrightarrow$?

29. $C_7H_{10} \longrightarrow$

30. $C_9H_{14} \xrightarrow{(1)\ O_3/(2)\ Zn/H_2O}$

31. $C_8H_{14} \xrightarrow{(1)\ O_3,\ (2)\ Zn + H_2O_2}$

32. $CH_2{=}CH-CF_3 + HBr \longrightarrow$

33. $CH_3-\underset{\underset{CH_3}{|}}{\overset{\overset{CH_3}{|}}{C}}-CH{=}CH_2 + HI \longrightarrow$

34. $R_2O + CO \xrightarrow[125–128°C]{BF_3}$

ANSWERS

(Group-A)

1. $CH_2{=}CH_2 + PdCl_2 + H_2O \longrightarrow CH_3CHO + 2HCl$
   This reaction is known as Wacker's process.

2. $CH_2{=}CH_2 + Cl_2 + HCl \longrightarrow$ $\underset{\underset{Cl\quad OH}{|\qquad|}}{CH_2-CH_2}$
   ethylene chlorohydrin

3. $\underset{\text{vinyl chloride}}{CH_2{=}CH-Cl} \longleftrightarrow CH_2-CH{=}Cl^+ \xrightarrow{\text{Addition of HCl; } H^+} CH_3-CH^+-Cl \xrightarrow{Cl^-}$

   $CH_3-CH-Cl_2$
   ethylidine chloride

4. $CCl_3-CH{=}CH_2 \longleftrightarrow CCl_3CH^-{=}CH_2^+ \xrightarrow{HCl} CCl_3-CHCl-CH_3$

5. $CH_2{=}CH-COOH \longleftrightarrow CH_2^+-CH^-COOH \xrightarrow{\text{Addition of HBr; } H^+} CH_2^+-CH_2-COOH$

   $\xrightarrow{Br^-} CH_2Br-CH_2-COOH$

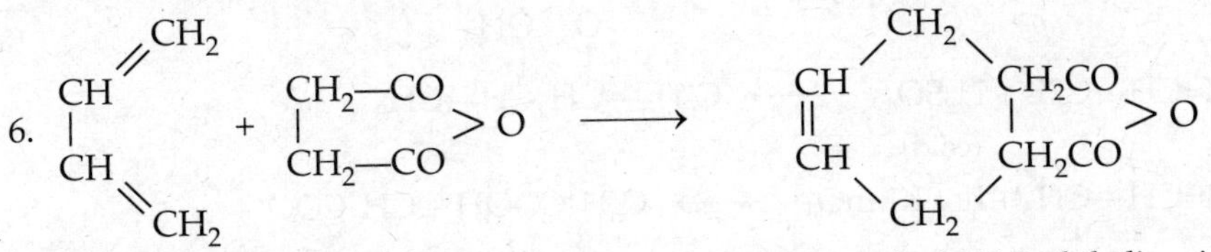

6.
$$\underset{\text{butadiene}}{\begin{matrix} CH_2 \\ \| \\ CH \\ | \\ CH \\ \| \\ CH_2 \end{matrix}} + \underset{\text{maleic anhydride}}{\begin{matrix} CH_2-CO \\ | \\ CH_2-CO \end{matrix}\!\!>\!\!O} \longrightarrow \underset{\substack{\text{1,2,3,6-tetrahydro-1,2-phthalic acid} \\ \text{anhydride}}}{\begin{matrix} CH_2 \\ CH \quad CH_2CO \\ \| \qquad | \\ CH \quad CH_2CO \\ CH_2 \end{matrix}\!\!>\!\!O}$$

The reaction is known as Diels–Alder reaction. An adduct is formed, when a conjugated diene reacts with a dienophile.

7. $\underset{\text{butadiene-1,3}}{CH_2=CH-CH=CH_2} + Br_2 \xrightarrow{CH_3COOH} \underset{\text{1,4-dibromobutene-2}}{CH_2Br-CH=CH-CH_2Br} + \underset{\text{3,4-dibromobutene-1}}{CH_2Br-CHBr-CH=CH_2}$

8.
$$\underset{\text{cyclopentene}}{\begin{matrix} CH_2 \\ / \quad \backslash \\ CH_2 \quad CH_2 \\ | \qquad | \\ CH=CH \end{matrix}} + \text{hot } KMnO_4 \longrightarrow \underset{\text{glutaric acid}}{\begin{matrix} CH_2-COOH \\ / \\ CH_2 \\ \backslash \\ CH_2-COOH \end{matrix}}$$

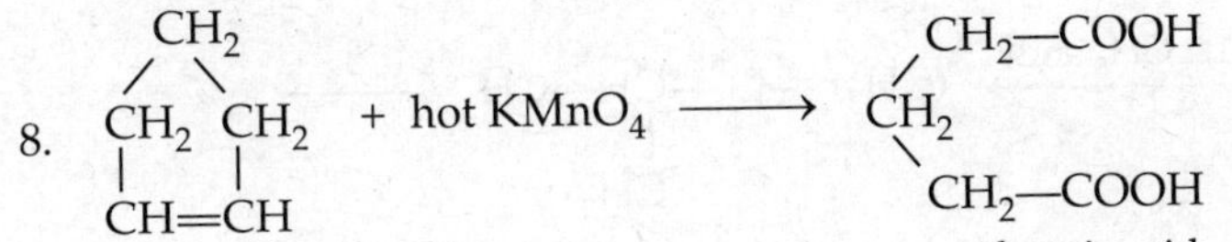

9.
$$\underset{\text{ethylene}}{CH_2=CH_2} + \underset{\text{diazomethane}}{CH_2N_2} \longrightarrow \underset{\text{pyrazoline}}{\begin{matrix} CH_2-CH \\ / \qquad \| \\ CH_2 \qquad N \\ \backslash \quad / \\ N \\ | \\ H \end{matrix}}$$

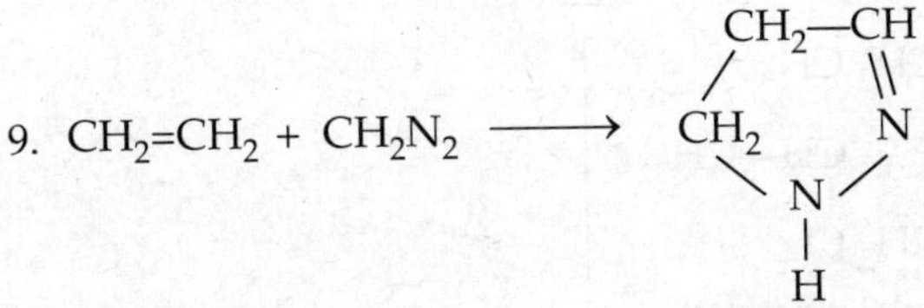

10. $(C_2H_5)_2-CH=CH_2 + I^+Cl^- \longrightarrow (C_2H_5)_2-CH-Cl-CH_2I$

[ICl reacts as $I^+ Cl^-$ because of the electronegativities (I = 2.6 and Cl = 3)]

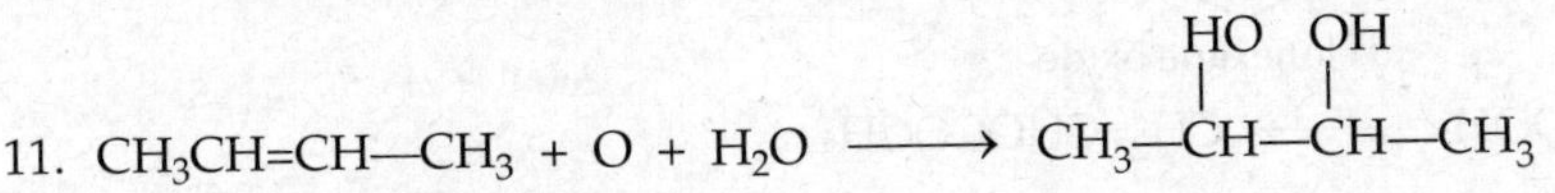

11. $\underset{\text{baeyer's reagent}}{CH_3CH=CH-CH_3} + O + H_2O \longrightarrow \underset{}{\begin{matrix} HO \quad OH \\ | \qquad | \\ CH_3-CH-CH-CH_3 \end{matrix}}$

[*cis*-hydroxylation takes place]

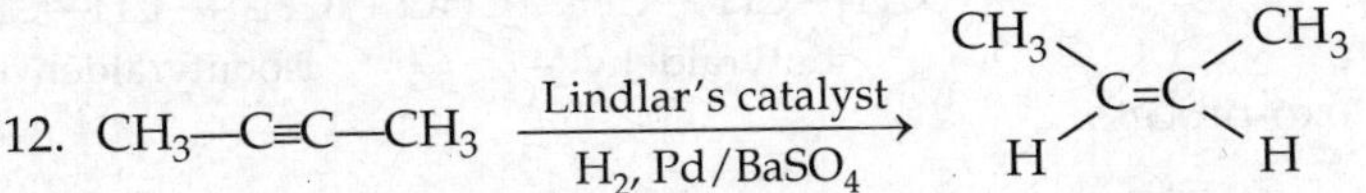

12. $CH_3-C\equiv C-CH_3 \xrightarrow[H_2,\ Pd/BaSO_4]{\text{Lindlar's catalyst}} \underset{\text{cis-butene}}{\begin{matrix} CH_3 \qquad CH_3 \\ \backslash \quad / \\ C=C \\ / \quad \backslash \\ H \qquad H \end{matrix}}$

13. Addition of HBr (in absence of a peroxide)

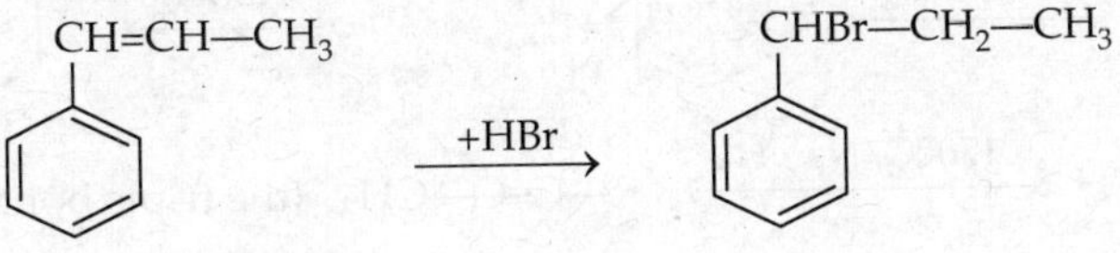

14. Addition in presence of peroxide

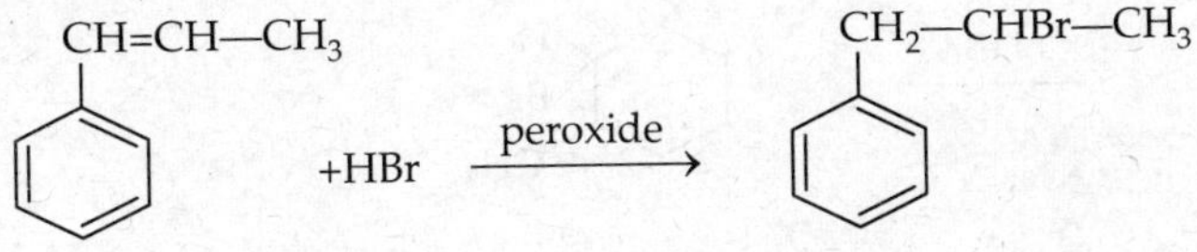

15. $C_6H_5—CH=CH—CH_3 + H_2SO_4 \xrightarrow{} C_6H_5—CH_2\overset{\overset{O}{|}}{C}H—\overset{\overset{OH}{|}}{C}H—CH_3$
    (cold)

16. $C_6H_5—CH=CH—CH_3 + \text{hot } KMnO_4 \xrightarrow{} C_6H_5COOH + CH_3COO$

17. $CH_3—HC=CH_2 + BH_3 \xrightarrow{} CH_3—CH_2—CH_2BH_2 \xrightarrow{H_2O_2} CH_3—CH_2—CH_2—OH$

18. $CH_3—CH=CH—CH_2—C\equiv CH + HBr \xrightarrow{} CH_3—CHBr—CH_2—CH_2—C\equiv CH$

19. $CH_2=CH—C\equiv CH + HBr \xrightarrow{} CH_2=CH—CHBr=CH_2$

20. $(CH_3)_3—C—CH=CH_2 \xrightarrow{\text{addition of HCl}} (CH_3)_3—C—C^+H—CH_2^- \xrightarrow{H^+}$
    3,3-dimethyl-1-butene
    2° carbocation

   $(CH_3)_3—C—C^+H—CH_3 \xrightarrow{\text{1,2 CH}_3 \text{ shift}} (CH_3)_2—\overset{+}{C}—\underset{\underset{CH_3}{|}}{C}H—CH_3 \xrightarrow{}$

   2,3-dimethyl-2-chlorobutane
   3° carbocation

$$\left[CH_3—\overset{\overset{CH_3}{|}}{\underset{\underset{CH_3}{|}}{C}}—\overset{\overset{CH_3}{|}}{\underset{\underset{Cl}{|}}{C}}H—CH_3 \right]$$

2,3-dimethyl-2-chlorobutane

21. 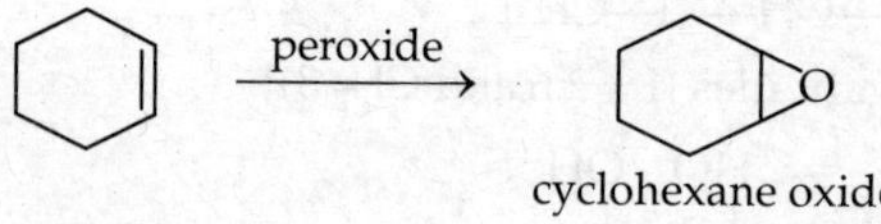 $\xrightarrow{\text{peroxide}}$

   cyclohexane oxide

22. $CH\equiv CH + CH_3COOH \xrightarrow{} CH_2=CHOCOCH_3$

   vinyl acetate

23. $CH_3—CH=CH_2 + CO + H_2 \xrightarrow{CoCO_4/\text{pressure}} CH_3—CH_2—CH_2—CHO + (CH_3)_2—CH—CHO$
    butyraldehyde   isobutyraldehyde

This reaction is termed as oxo-process.

24. 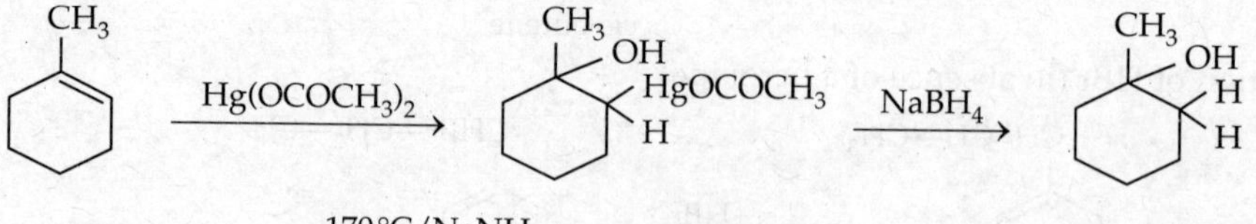 $\xrightarrow{Hg(OCOCH_3)_2}$ $\xrightarrow{NaBH_4}$

25. $R—CH_2—C\equiv CH \xrightarrow{170°C/NaNH_2} R—C\equiv C—CH_3$ (the triple bond shifts in the middle)

26. 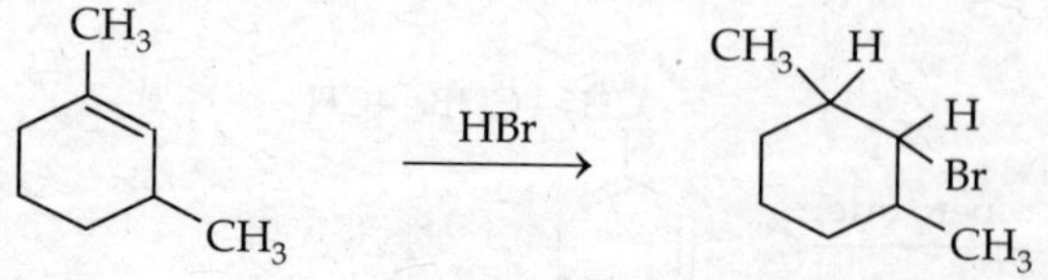 $\xrightarrow{HBr}$

27. CH_3—$CH=CH_2$ + Br^+Cl^- $\longrightarrow$ CH_3—$CHCl$—CH_2Br

28. *cis*- and *trans*-butene + I_2 $\longrightarrow$ interconversion take place

$\qquad\qquad$ *cis*-butene $\longrightarrow$ *trans*-butene and vice versa

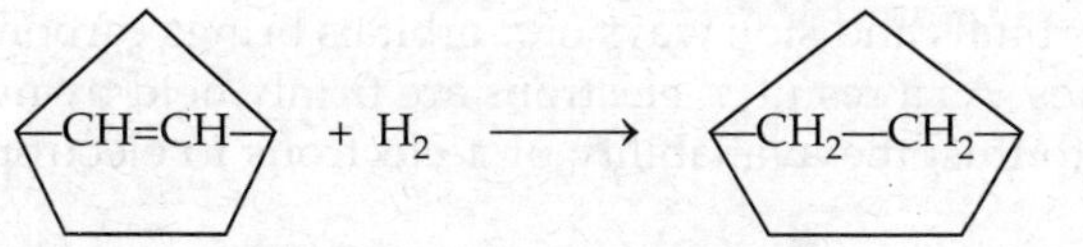

$\langle$CH=CH$\rangle$ + H_2 $\longrightarrow$ $\langle$CH$_2$—CH$_2\rangle$

29. 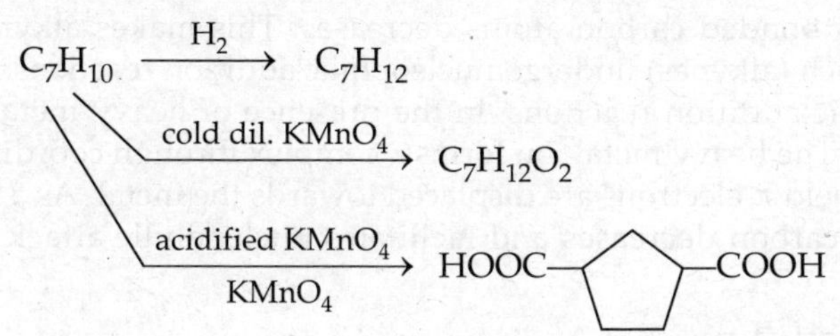 C_7H_{10} $\xrightarrow{H_2}$ C_7H_{12}

$\xrightarrow{\text{cold dil. KMnO}_4}$ $C_7H_{12}O_2$

$\xrightarrow[\text{KMnO}_4]{\text{acidified KMnO}_4}$ HOOC—$\langle\ \rangle$—COOH

30.

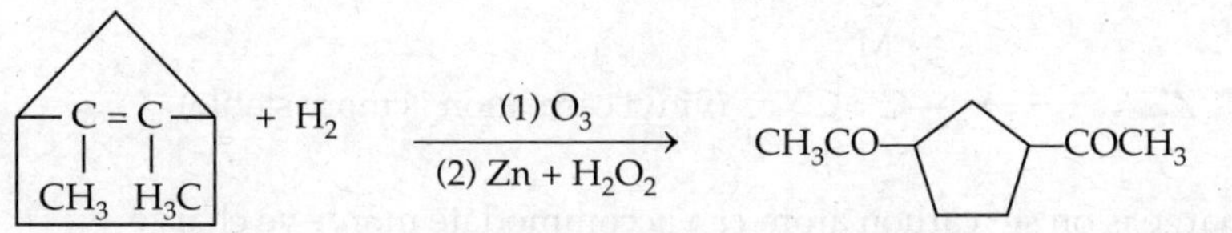

$\begin{array}{c} C=C \\ | \quad | \\ CH_3 \ H_3C \end{array}$ + H_2 $\xrightarrow[(2)\ Zn + H_2O_2]{(1)\ O_3}$ CH_3CO—$\langle\ \rangle$—$COCH_3$

31.

$\xrightarrow[(2)\ Zn + H_2O_2]{(1)\ O_3}$ CH_3—$CO(CH_2)_4$—$COCH_3$

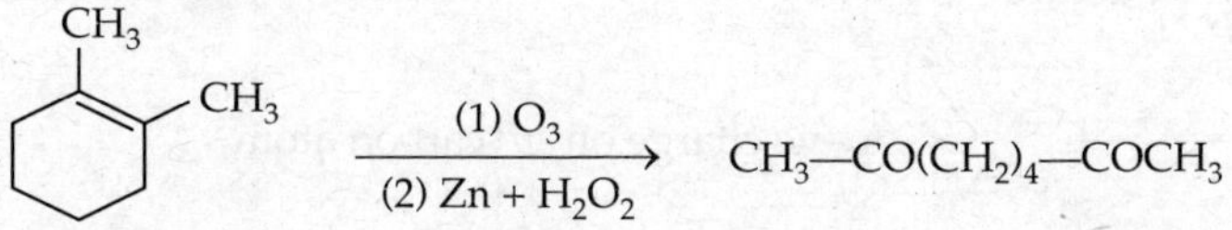

32. $CH_2=CH$—CF_3 $\longrightarrow$ C^+H_2—CH^-—CF_3 $\xrightarrow{HBr}$ CH_2Br—CH_2—CF_3

33. 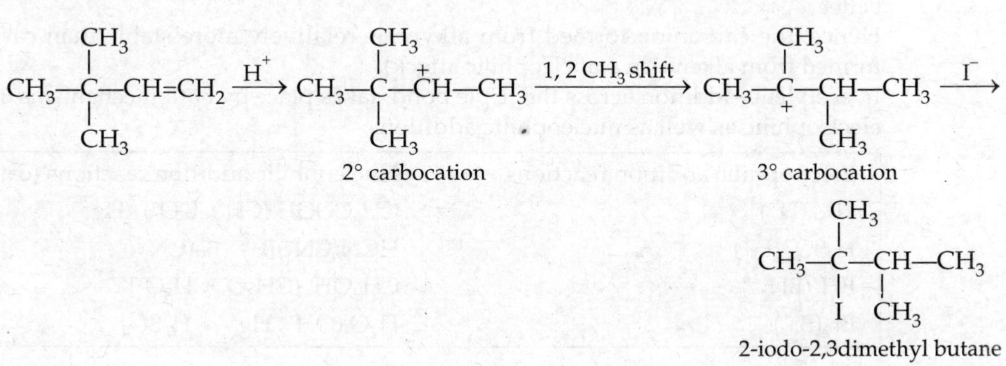 CH_3—$\underset{\underset{CH_3}{|}}{\overset{\overset{CH_3}{|}}{C}}$—$CH=CH_2$ $\xrightarrow{H^+}$ CH_3—$\underset{\underset{CH_3}{|}}{\overset{\overset{CH_3}{|}}{C}}$—$\overset{+}{C}H$—$CH_3$ $\xrightarrow{1,\ 2\ CH_3\ \text{shift}}$ CH_3—$\underset{\underset{CH_3}{|}}{\overset{\overset{CH_3}{|}}{\underset{+}{C}}}$—$CH$—$CH_3$ $\xrightarrow{I^-}$

$\qquad\qquad$ 2° carbocation $\qquad\qquad\qquad\qquad$ 3° carbocation

$\qquad\qquad\qquad\qquad\qquad\qquad\qquad\qquad\qquad\qquad$ CH_3—$\underset{\underset{I}{|}}{\overset{\overset{CH_3}{|}}{C}}$—$\underset{\underset{CH_3}{|}}{CH}$—$CH_3$

$\qquad\qquad\qquad\qquad\qquad\qquad\qquad\qquad\qquad\qquad$ 2-iodo-2,3dimethyl butane

34. R_2O + CO $\xrightarrow{BF_3/125–128\,°C}$ RCOOR

$\quad$ ether $\qquad\qquad\qquad\qquad\qquad\qquad$ ester

ALKYNES

Salient Features

1. Alkynes are less reactive than alkenes, towards electrophilic addition reactions.
   a. The overlapping of sp hybrid orbitals and side ways of π-orbitals brings carbon atoms closer in alkynes than in alkenes. As a result, π-electrons are firmly held by nuclei in alkynes than in alkenes. Due to this, the availability of π-electrons to electrophile is much less than in alkyne.
   b. Due to uniform distribution of cylindrical electron cloud around internuclear axis, electron density of triply bonded carbon atoms decreases. This makes alkyne less reactive than alkenes, which (alkynes) undergo nucleophilic addition reactions.
2. Alkynes undergo nucleophilic addition reactions, in the presence of heavy metal ions, like Pb^{2+}, Hg^{2+}, Ba^{2+} ions etc. The heavy metal ion forms a complex through coordination with π-electrons and loosely held π-electrons are displaced towards the metal. As a result, the electron density around carbon decreases and facilitates nucleophilic attack at the triply bonded carbon atom.

   - $-C{\equiv}C- + M^{2+} \longrightarrow -C{\equiv}C-$
     $\downarrow$
     M^{2+}

   - $-C{\equiv}C- + Z{:}^{-} \longrightarrow -\underset{sp^2}{C^-{=}C^-}$ (vinyl carbanion is more stable)

   - The −ve charge is on sp^2 carbon atom can accommodate more −ve charge
   - Resonance stabilised structure

     $>C{=}C< + Z{:}^{-} \longrightarrow -\underset{sp^3}{C^-}-\overset{Z}{C}<$ A −ve charge on sp^3 carbon atom

     $Z{:}$ (a sp^3 carbon atom is less electronegative, can accommodate less −ve charge)

   - As sp^2 hybridised carbon is more electronegative, it can accommodate the −ve charge better.

     Hence, the carbanion formed from alkyne is relatively more stable than carbanion formed from alkene (in a nucleophilic attack).

     In acetylene–addition across the triple bond, takes place by both mechanisms, namely electrophilic as well as nucleophilic addition.

Electrophilic addition reactions	Nucleophilic addition reactions (catalyst)
HX (H^+)	CH_3COOH (CH_3COO^-) Hg^{2+}
H_2SO_4 (H^+)	$HCN(CN^-)Ba^{2+}(Ba(CN)_2)$
$BH_3(BH_2^+)$	CH_3OH (CH_3O^-)$CH_3O^-K^+$
$Br_2(Br^+)$	H_2O (OH^-) $Hg^{2+} + H_2SO_4$

3. SeO_2 oxidises $CH{\equiv}CH$ to $\underset{\underset{glyoxal}{CHO}}{\overset{CHO}{|}}$

4. The acidic character has been explained on the basis of electro negativity of the carbon atom, which in turn depends upon the state of hybridization of the carbon atom.

		s	*p*
—C—H	sp^3	25%	75%
=C—H	sp^2	33%	67%
≡C—H	sp	50%	50%

- *s*-orbital lies at a lower energy level than *p*-orbital.
- Therefore *s*-electrons are more penetrating towards the nucleus than *p*-electrons.
- *sp*-hybridized orbital with 50% *s*-character is more electronegative than sp^2 hybridized orbital (33.3%), which is in turn is more electronegative than a sp^3 hybridized orbital (25%).
- *sp*-Hybridized orbital's of alkyne, due to their greater *s*-character, have greater attraction for electrons and are more electronegative than sp^2 carbon of alkene and sp^3 carbon of alkanes respectively.
- ≡C—H have greater ionic character than in =C—H alkene or —C—H alkanes.
- This facilitates the release of hydrogen linked to triply bonded carbon atom in alkynes to form H^+.

$$H—C≡C—H \qquad CH_2=CH_2—H \qquad H_3C—H$$
$$K_a = 10^{-26} \qquad K_a = 10^{-36} \qquad K_a = 10^{-42}$$
$$sp \qquad\qquad sp^2 \qquad\qquad sp^3 \text{ hybridization}$$

The more is the *s*-character in the carbon–hydrogen bond, more is the hydrogen acidic in nature.

- —C≡C— is more electron withdrawing than >C=C<.

5. Br_2 and HBr add to the double bond in preference to the triple bond in a compound, containing both a double and a triple bond in an isolated chain.

$$CH_3—CH=CH—CH_2—C≡CH + Br_2 \longrightarrow CH_3—CHBr—CHBr—CH_2—C≡CH$$

6. When a double bond is in conjugation with a triple bond, HBr adds on the triple bond to convert it into more stable conjugated diene.

$$CH_2=CH—C≡C—H + HBr \longrightarrow CH_2=CH—C—Br=CH_2$$

7. On heating with sodamide ($NaNH_2$), the terminal triple bond shifts into middle.

$$R—CH_2—C≡CH \xrightarrow{\ NaNH_2/160°C\ } R—C≡C—CH_3$$

8. Bond lengths in between carbon and hydrogen are in the decreasing order:

—C—H	>	=C—H	>	≡C—H
1.112 Å		1.103 Å		1.06 Å
sp^3–s		sp^2–s		sp–s hybridization

The more the *s* character in the carbon-hydrogen bond, the more is the 'H' acidic.

9. Br_2 (or halogens), add to the double bond in a *trans*-manner. Addition of Br_2 in acetylene leads to the formation of *trans*-dibromo ethylene.

$$H—C≡C—H + HBr \longrightarrow$$

$$\begin{array}{c} H \qquad\quad Br \\ \diagdown \qquad \diagup \\ C=C \\ \diagup \qquad \diagdown \\ Br \qquad\quad H \end{array} \quad \textit{trans}\text{-dibromo ethylene}$$

- R_2BH adds on an alkyne in the *cis*-manner.

$$R—C≡C—R + R_2BH \xrightarrow{CH_3COOH} \underset{\text{cis-dialkyl ethylene}}{\overset{R}{\underset{R_2B}{>}}C=C\overset{R}{\underset{H}{<}}} \xrightarrow{CH_3COOH} \overset{R}{\underset{H}{>}}C=C\overset{R}{\underset{H}{<}}$$

10. For double dehydrohalogenation, $NaNH_2$ is preferred to alcoholic KOH.

$$\underset{\text{acetylene tetrachloride}}{CHCl_2—CHCl_2} \longrightarrow \underset{\text{dichloro vinyl chloride}}{CHCl = CCl_2} \longrightarrow \underset{\text{acetylene}}{H—C≡C—H}$$

- $H—C≡C—H$ does not dissolve in H_2SO_4, because of the more *s*-character in the negatively charged carbon. The less stable is the carbocation, the less likely it is formed.

$$H—C≡C—H + H_2SO_4 \longrightarrow CH_2= C—HSO_4$$

- Lindlar's catalyst (Pd, mounted on $BaSO_4$ poisoned with quinoline) reduces alkyne to alkenes only. It is not reduced further because of deactivation in case of Lindlar's catalyst consisting of Ni/B can be used.

$$H—C≡C—H + H_2 \xrightarrow{\text{Lindlar's catalyst}} CH_2=CH_2 \text{ (–42 kcal/mole, and more exothermic)}$$

- Alkyne are less reactive towards Br_2 than alkenes.

$$H—C≡C—H + Br_2 \longrightarrow \underset{\text{More strained}}{H—\underset{\overset{\diagdown\diagup}{\underset{Br}{+}}}{C}=C—H} \quad CH_2=CH_2 + Br_2 \longrightarrow \underset{\text{Less stable}}{\underset{\overset{\diagdown\diagup}{\underset{Br}{+}}}{CH_2—CH_2}}$$

11. In hydrogenation, alkynes are more reactive than alkenes, in other reactions alkenes are more reactive than alkynes.

12. Copper salts of the alkynes, or copper acetalide undergo oxidative coupling in presence of oxygen.

$$2R—C≡C\,Cu + O_2 \longrightarrow RC≡C—C≡C—R$$

Alkynes are unsaturated hydrocarbons, with molecular composition C_nH_{2n-2}. They contain one pair of carbon atom joined by one σ bond and two π bonds. The molecule is linear in shape and each carbon is *sp* hybridised. The bond distance between the triply bonded carbons is 1.2 Å and the amount of energy required for a homolytic fission is 195 kcal/mole.

On hydrogenation, alkyne are first converted into alkenes and finally into alkanes. Hydrogenation is an exothermic reaction and the amount of energy evolved is = 70 kcal/mole.

Nomenclature: The alkyne is named after the corresponding name of alkane, by substituting the suffix 'ane' by 'yne' in the IUPAC system. Thus, this series is alkyne series. The longest chain, containing the triple bond is chosen as the parent alkyne and the position of the triple bond is indicated by number. The lowest number is given to the triple bond.

Alkynes	Corresponding alkane	IUPAC name
CH≡CH	ethane	ethyne
$CH_3C≡CH$	propane	propyne
$CH_3CH_2C≡CH$	butane	butyne-1
$H_3CC≡CCH_3$	butane	butyne-2

General Methods of Preparation

1. **By dehydrohalogenation of vic-dihalides:** Vic–dichloro alkanes are treated with alcoholic KOH and sodamide ($NaNH_2$) when vic-dichloro alkane are converted to form alkynes.

$$\underset{\text{ethylene chloride}}{\overset{\displaystyle H \quad H}{\underset{\displaystyle Cl \quad Cl}{H-\overset{|}{C}-\overset{|}{C}-H}}} \xrightarrow{\text{alcoholic KOH}} \underset{\text{vinyl chloride}}{\overset{\displaystyle H \quad H}{\underset{\displaystyle Cl}{H-\overset{|}{C}=\overset{|}{C}}}} \xrightarrow{\text{NaNH}_2} \underset{\text{acetylene}}{CH \equiv CH}$$

Vinyl chloride obtained during first stage of the reaction is very unreactive or inert, therefore, it is further treated with a stronger base in the final stage of the reaction.

2. **Dehalogenation of a tetrahalogen derivatives of alkane:**
   1,1,2,2-tetrachloroethane or acetylene tetrachloride is heated with Zn dust in methanolic solution, when acetylene is formed.

$$CHCl_2—CHCl_2 + 2Zn \xrightarrow{\text{Zn/CH}_3\text{OH}/\Delta} CH \equiv CH + 2ZnCl_2$$

3. **By Kolbe's electrolytic synthesis:** By electrolysing an aqueous solution of sodium fumerate, acetylene is obtained.

$$\begin{matrix} CHCOONa \\ \| \\ CHCOONa \end{matrix} \longrightarrow CH \equiv CH + 2CO_2 + 2NaOH + H_2\uparrow$$

4. **From calcium carbide:**

$$CaC_2 \longrightarrow CH \equiv CH + Ca(OH)_2$$

   • Preparation of higher alkynes: Acetylene being acidic in nature, reacts with sodium metal to form a monosodium salt or a disodium salt, which on treatment with alkyl halides gives higher homologues of acetylene.

$$CH \equiv CH + Na \longrightarrow NaC \equiv CH \xrightarrow{+ Na} NaC \equiv CNa \xrightarrow{RX} R—C \equiv C—R \quad \text{(dialkyl)}$$
$$\underset{\text{acetylene}}{} \xrightarrow{RX} RC \equiv CH \quad \text{(mono alkyl acetylene)}$$

Physical properties

Alkynes from C_2 to C_4 are colourless, odourless gases, those containing C_5 to C_{12} are colourless odour less liquid, still higher ones are solids.

   • Acetylene, when prepared from commercial carbide, possess a garlic odour, owing to the presence of PH_3 and H_2S. However, pure acetylene possess a sweet odour.
   • The melting points and boiling points of alkynes increases with increase in molecular weight. The branched chain alkyne have a lower melting point than corresponding linear chain alkynes and boiling points and melting points of alkynes are higher than the corresponding alkenes, due to greater polarity of the bonds in alkyne. Alkynes are sparingly soluble in water but very much soluble in organic solvents.

Chemical properties

Alkynes are unsaturated hydrocarbons, containing one pair of carbon atoms joined one σ-bond and two π-bonds. As the π electrons are held by carbon loosely, the alkynes undergo addition reactions, both nucleophilic as well electrophilic addition reactions.

   • Alkynes are less reactive than alkenes, towards electrophilic addition reactions.
     The bond distance between triply bonded carbon is shorter ($—C \equiv C^-$=1.2 Å) than the double bond in alkenes ($>C=C<$= 1.34 Å), so the electrons are held strongly by the triply bonded carbon atoms, as compared to alkenes. This makes alkynes less reactive towards electrophilic addition reactions.

- In alkynes, four π electrons of the two π-bonds, form an uniform cylindrical cloud around the inter nuclear axis (eg. the triply bonded C to C axis), as a result the electron density about each of the triply bonded carbon decreases. This makes alkyne less reactive about the triple bond but make acetylenic hydrogen more acidic, owing to the pull on carbon hydrogen bond, by each triply bonded deficient carbon atoms, towards itself. As a result hydrogen atoms attached to triply bonded carbon become slightly acidic in nature. This hydrogen is termed as acetylenic hydrogen. Being acidic in nature, alkynes containing acetylenic hydrogen undergo substitution reactions as well as form metal substituted derivative.

Unsaturated Behaviour

1. Like alkenes, alkynes burn with a luminous and sooty flame, decolourise yellow colour of bromine water and decolourise pink colour of dil alk. $KMnO_4$ solution or Baeyer's reagent.
2. Hydrogenation or addition of hydrogen
   i. Alkynes are reduced to alkanes, in presence of H_2/heated Ni as a catalyst. In place of heated Ni, Pt and Pd at room temperature can also be used for hydrogenation. During this process, alkene are formed as intermediate, which cannot be isolated when reduced by hydrogen in presence of a metal catalyst.

$$HC\equiv CH + 2H_2 \longrightarrow CH_3-CH_3$$

   Catalytic reduction of alkynes by hydrogen/metal is termed as **Sabatier Senderens reaction.**

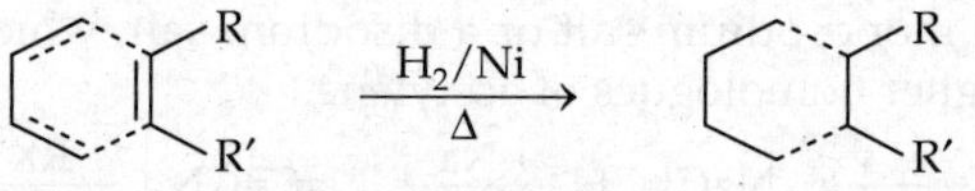

   ii. **Stereospecific reduction of alkynes:** Alkyne can be reduced to alkenes by using some specific reducing agent.
   a. Lindlar's catalyst: Lindlar's catalyst consists of Pd, mounted on barium sulphate, partially poisoned by sulphur or quinoline, dialkyl acetylenes are reduced to form *cis*-alkene.

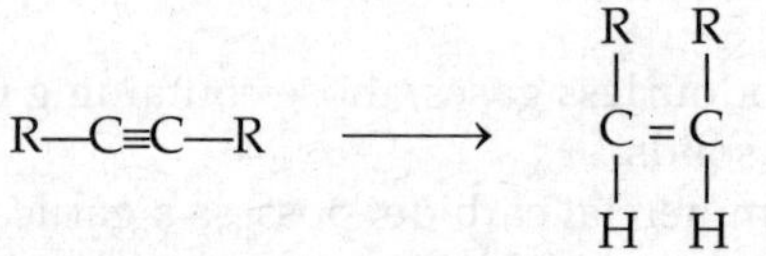

cis-alkene

   b. On reducing dialkyl acetylene either by ROH + Na or Li/Na + NH_3, is reduced to *trans*-dialkyl alkene.

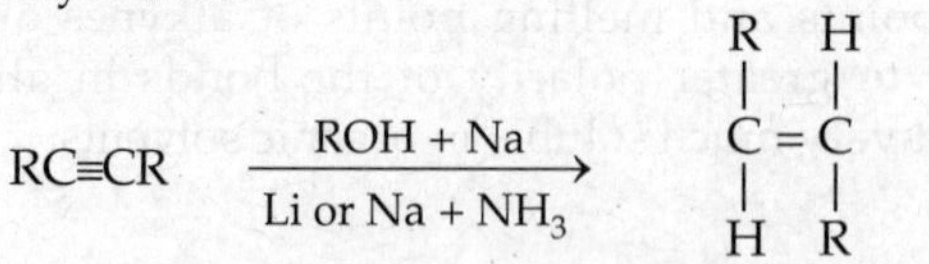

trans-alkene

3. Addition of halogen: (*trans*-alkene)
   Halogens add on across the double bond in a *trans*-manner:

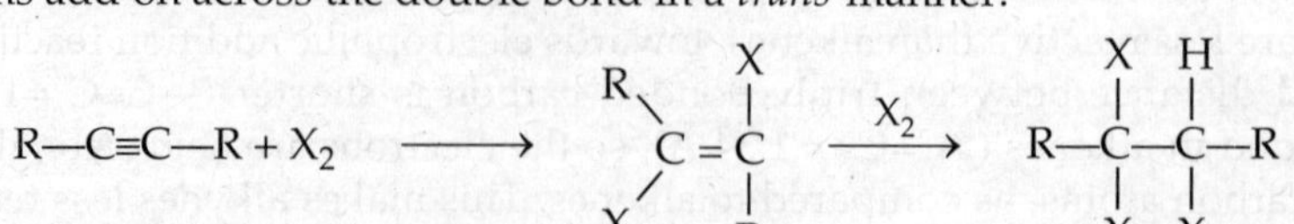

a tetrahalogen derivative

Addition of halogen across the double bond or a triple bond takes place in *trans*-manner Acetylene reacts with Br_2 water to form a dibromo derivative but form a tetrabromide, when treated with liquid bromide. Acetylene di-iodide is obtained when acetylene reacts with an alcoholic solution of iodine.

4. **Addition of halogen acids:**

Halogen acids add across the triple bond to form addition compound, following Markovnikov's rule.

$$HC\equiv CH + HCl \longrightarrow CH_2=CHCl \xrightarrow{+HCl} CH_3CHCl_2$$

Addition of the halogen acid across the triple bond is believed to be an electrophilic addition reaction, and takes place in **trans-manner.**

$$CH_3C\equiv CH \xrightarrow{+H^+ + X^-} CH_3-{}^+C=CH_2 \xrightarrow{+X^-} CH_3CX=CH_2 \xrightarrow{+H^+}$$

$$CH_3C^+X-CH_3 \xrightarrow{X^-} CH_3CX_2CH_3$$

The reactivity of halogen acids is in the following order.

$$HI > BHr > HCl$$

When acetylene reacts with HCl, in presence of $HgSO_4$, as a catalyst acetylene reacts with one molecule only to form vinyl chloride

$$CH\equiv CH + HCl \xrightarrow{HgSO_4/60°C} CH_2=CH\ Cl$$
$$\text{vinyl chloride}$$

Vinyl chloride is used as a starting material for manufacture of polyvinyl chloride, (a polymer) in presence of a suitable catalyst. Peroxides have same effect on addition of HBr.

5. **Addition of sulphuric acid, H_2SO_4:**

H_2SO_4 adds on the alkynes following Markovnikov's rule to form an addition compound, which upon hydrolysis, converts to form either an aldehyde or a ketone.

$$CH\equiv CH + H_2SO_4 \longrightarrow CH_2=CHHSO_4 \xrightarrow{H^+/HSO_4} CH_3CH(HSO_4)_2 \xrightarrow{H^+/H_2O} CH_3CHO$$

Addition of H_2SO_4 across the triple bond follows an electrophilic addition mechanism.

$$H_2SO_4 \longrightarrow H^+ + HSO_4^-$$

This reaction is used to add a molecule of water, across the triple bond.

6. **Addition of diborane (B_2H_6)** Diborane **(reacting as BH_3)** add on across triple bond to form an addition compound.

$$HC\equiv CH + BH_3 \longrightarrow (CH_2=CH)BH_2 \xrightarrow{+CH\equiv CH} (CH_2=CH)_2BH \xrightarrow{+CH\equiv CH} (CH_2=CH)_3B$$
$$\text{acetylene} \qquad\qquad \text{vinyl borane} \qquad\qquad\qquad\qquad\qquad\qquad \text{trivinyl borane}$$

BH_3 is an electron deficient molecule, therefore it acts as an electrophile. It adds on alkyne molecule, coordinating through electrons.

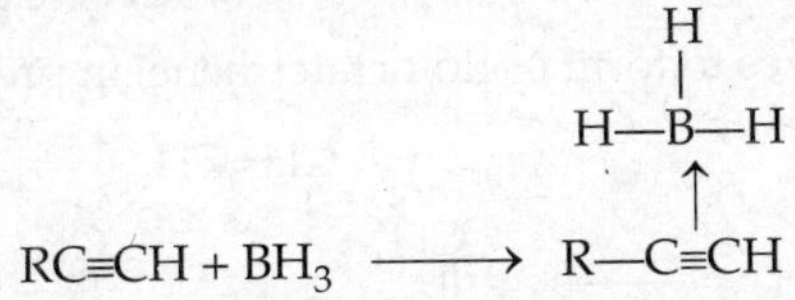

$$\begin{array}{c} H \\ | \\ H-B-H \\ \uparrow \end{array}$$

$$RC\equiv CH + BH_3 \longrightarrow R-C\equiv CH$$

Upon further oxidation with alkaline H_2O_2, the borane-derivatives form the aldehyde or acetone.

$$(CH_2=CH)BH_2 \longrightarrow CH_2=CHOH \longrightarrow CH_3CHO$$
$$(R\text{—}CH=CH)_2BH \longrightarrow RCH=CHOH \longrightarrow RCH_2CHO$$
$$(RCH=C\text{—}R)_3 B \longrightarrow RCH=CR \longrightarrow R\text{—}CH_2COR$$

In case the boro hydrides are treated with acetic acid at 15 °C, boro hydrides decompose back to form alkenes.

$$(RCH=CH)_3 B + 3CH_3COOH \longrightarrow 3RCH=CH_2 + (CH_3COO)_3B$$

Formation of borohydride serve two purpose:

a. It can be used for addition of molecule of water to alkynes to prepare aldehydes or ketones.

b. Hydroboration may also be used for hydration.

Addition of BH_3 across the triple or double bonds takes place as it ionises to

$$BH_3 \longrightarrow BH_2^+ + H^-$$

$$CH_3\text{—}CH=CH_2 + BH_2^+ \longrightarrow RCH^+\text{—}CH_2BH_2 \longrightarrow RCH_2CH_2BH_2$$
$$\longrightarrow RCH_2CH_2OH$$

The ions BH^+_2 and H^- formed, add across the double bond following Markovnikov's rule. By overlooking the charges on the species, one may be tempted to conclude the addition follows anti-Markovnikov's rule. However, the product formed is anti-Markovnikov's product.

Nucleophilic addition reactions: Naturally, alkynes do not undergo nucleophilic addition owing to high electron density on each of the carbon atoms. However, alkynes can be made to undergo nucleophilic addition reactions.

7. **Polymerisation:**

a. Acetylene polymerize, to form benzene on passing the gas through red hot Cu tube kept at 500 °C

$$3C_2H_2 \longrightarrow C_6H_6$$

b. Propyne polymerizes to form mesitylene, or 1,3,5-trimethyl benzene on passing through a red hot tube.

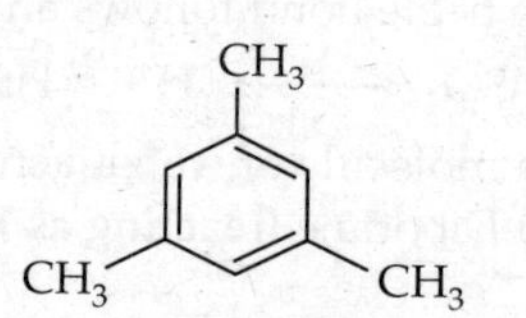

1,3,5-trimethyl benzene

c. Acetylene condenses with itself to form vinyl acetylene in presence of Cu_2Cl_2 and NH_4Cl at ~70°C.

$$CH\equiv CH + CH\equiv CH \xrightarrow{Cu_2Cl_2/NH_4Cl/\sim70°C} CH_2=CH\text{—}C\equiv CH$$

d. Acetylene polymerizes to form cyclo-octatetraene, in presence of nickel cyanide

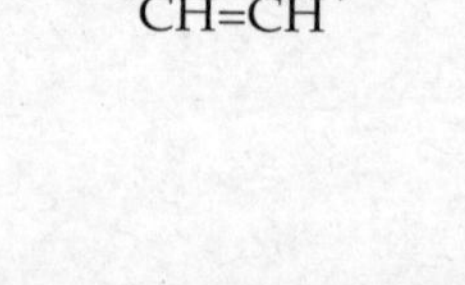

BENZENE AND NAPHTHALENE

Salient Features

1. Benzene is a flat molecule; all carbon and hydrogen atoms are on the same plane. Each carbon is sp^2 hybridised. As the molecule is flat, its thickness should be equal to the diameter of the carbon atom; however the thickness has been found to be more than the diameter of the carbon atom, therefore it can be concluded that there are π-bonds over and below the benzene ring that adds to the thickness.
2. All the six carbon to carbon bond length are equal to 1.39 Å.
3. The six π-electrons, one from each carbon are delocalized and form π-clouds, over and below the benzene ring.
4. All substitutions in benzene ring are electrophilic substitutions. In other words, the attacking reagent is always positively charged.

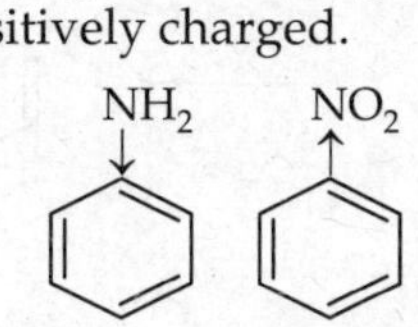

5. There are two types of groups, present in benzene ring.
   i. Electron releasing groups or activating groups.
   ii. Electrons withdrawing groups or deactivating groups.
6. The effect of any group whether activating or deactivating is strongest at o and p-positions.
7. The o- and p-substitutions is faster than m-substitution.
8. If an activating group is competing with a lesser activating group or a deactivating group, the former (the stronger activating group) usually controls the reaction.
9. All other factors being equal, a third group is least likely to enter between two groups, due to steric hindrance, which is further increased by the size of the attacking electrophile.
10. When a m-directing group is in m-position to the o- and p-directing group, the incoming group primarily goes to o-position to the m-directing, group than to p-position.

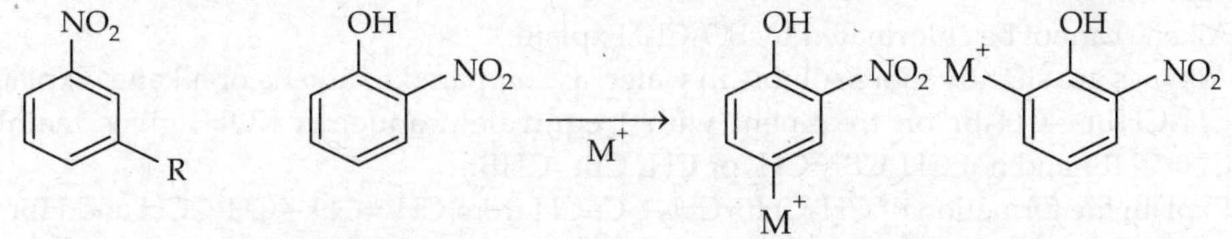

11. The directing influence of o-/p-directing groups in the decreasing order as follows.
    i. —O⁻ > —NR$_2$ > —NH$_2$ > —OH > —OCH$_3$ > —NHCOCH$_3$ > —CH$_3$ > —Cl > —Br > —I.
    ii. —NH$_2$ > —OH > OCH$_3$ > NHCOCH$_3$ > C$_6$H$_5$— > Cl > Br > I > CH$_3$ > m-directing groups.
12. The directive influence of the m-directing groups is in the decreasing order as follows.
    (CH$_3$)$_3$N— or NR$_3$ > —NO$_2$ > —CN > —SO$_3$H > —CHO > —COCH$_3$ > —COOH.

Polynuclear rings–Naphthalene

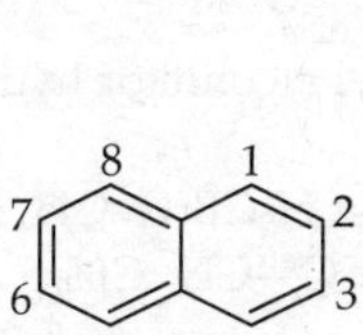

1. In naphthalene, 1,4,5 and 8 are termed α-position and 2,3,6 and 7 as β-positions.
2. When —Cl, Br, CH_3, NHR, or $NHCOCH_3$, are in position number = '1', then homonuclear substitution takes place in position '4', and to a lesser extent in position '2'.
3. When —OH, —CH_3,—NHR, $NHCOCH_3$ groups are in position '2', then a homonuclear substitution takes place in the same ring, i.e. at position '1'.
4. It may be noted that homonuclear substitution usually occurs, when the group is already o- and p-directing.
5. When —NO_2 or —SO_3H are present in positions '1' or '2', heteronuclear substitution takes place in position '5' or '8'. If halogen or —NH_2 group is present in position '2', heteronuclear substitution takes place in position '5' or 7.

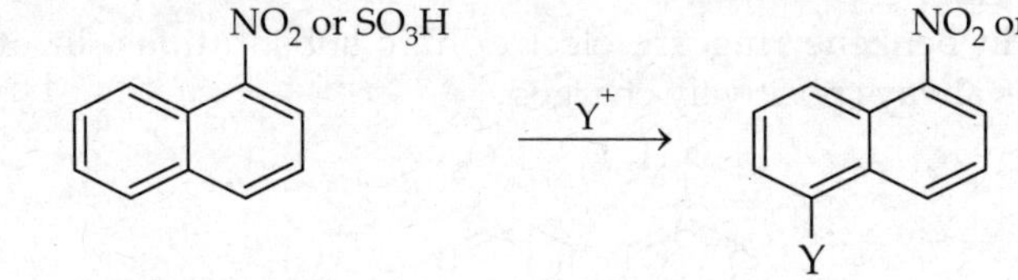

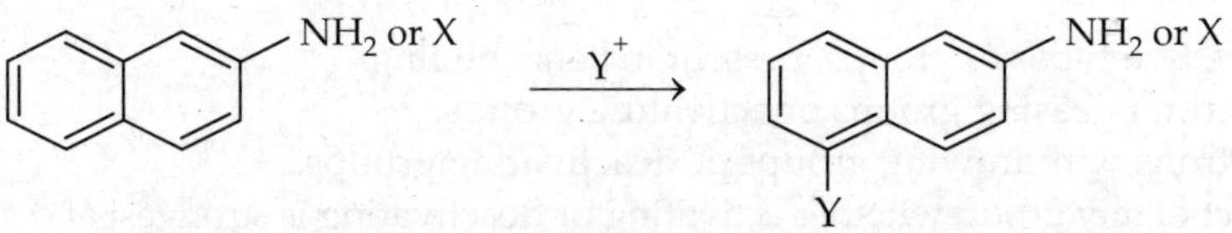

NO_2 or SO_3H NO_2 or SO_3H

$\xrightarrow{Y^+}$

Y

NH_2 or X NH_2 or X

$\xrightarrow{Y^+}$

Y

REASONING TYPE QUESTIONS (RTQs)

(Group-B)

1. How an E structure can be changed to a Z structure?
2. Explain the addition of Cl_2 on $Cl_2C=CCl_2$ in presence of anhydrous $AlCl_3$ and UV radiation.
3. Alkene cannot be chlorinated by SO_2Cl_2. Explain.
4. Alkynes are slightly more soluble in water, as compared to alkene or alkane. Explain.
5. CH_3CHBr—CH_2Br on treatment with 1 equivalent aqueous KOH, gives mainly CH_3 $CH=CHBr$ and not $CH_3CBr=CH_2$ or CH_3 $CBr=CHBr$
6. Explain the formation of $CH_3CHBrCH_2$—$C≡CH$ from $CH_2=CH$—CH_2CCH and HBr.
7. How will you bring about the conversion of 2-methylpropene into isobutyl bromide.
8. What product is formed when propene is treated with SO_2Cl_2.
9. A meso compound is formed on addition of Br_2 to trans-2-butene.
10. Tetraethyl lead [$Pb(C_2H_5)_4$] can inhibit the chlorination of alkanes at 150°C.
11. Account for the acid catalyzed isomerization of *cis*-2-butene to *trans*-2-butene
12. Explain the formation of $CH_3CH_2CH_2$ $CHBr_2$ from butyne-1and HBr in presence of a peroxy acid.
13. Dehydration of cyclohexyl methyl alcohol or hydoxy methyl cyclohexane forms 1-methyl cyclohexene.
14. Explain why Grignard's reagent of $H_2CBrC≡CH$, cannot be prepared.
15. Explain the addition of HBr to CF_3C—$CH=CH_2$.

16. Which of the following electrophile is most strong and why?

    NO_2^+ and NO^+

17. Account for the interconvertibility of *cis*-2-butene and *trans*-2-butene on heating with iodine.

18. The benzyl carbocation $C_6H_5CH_2^+$ is more stable as compared with R_3C^+.

19. Name the product, when Grignard's reagent reacts with $HC\equiv C—CH_3$.

20. Explain why Grignard's reagent, R—MgX or Li do not react with alkanes?

21. Explain the reason for higher selectivity in bromination as compared to chlorination.

22. Explain the product obtained by the ozonolysis followed by the hydrolysis of xylene.

23. Benzene is less reactive than acetanilide.

24. Chlorination of ethyl benzene, in the presence of sunlight gives 91% of $C_6H_5CHClCH_3$ and 9% of $C_6H_5CH_2CH_2Cl$.

25. Explain why at a higher temperature a *p*-substituted product is obtained in larger quantities.

26. Name the compound which would undergo a free radical reaction, with Cl_2 with maximum ease.

27. Name a compound which will give only one mono-substituted derivative?

28. Write the order of sulphonation, with conc. H_2SO_4 in the decreasing order.

ANSWERS

(Group-B)

1. By switching the groups on the same carbon.

2. The four Cl's decrease the electron density or nucleophilicity on the electrons of alkene, hence no addition takes place. On the other hand, $AlCl_3$ increases electrophilicity of Cl^+ which is sufficient to react with it in the presence of sunlight.

$$Cl_2—C{=}CCl_2^- + Cl_2 \xrightarrow{\text{Anh. } AlCl_3 \text{ or sunlight}} Cl_3C—CCl_3$$

3. The use of SO_2Cl_2 as a chlorinating reagent, is more convenient for alkanes which react via free radicals. Peroxide is used an initiator.

$$R—O{:}O—R \longrightarrow 2R—O\bullet$$
$$R—O\bullet + R''H \longrightarrow ROH + R''\bullet$$
$$R''\bullet + SO_2Cl_2 \longrightarrow R''Cl + SO_2Cl\bullet$$
$$SO_2 Cl\bullet \longrightarrow SO_2 + Cl\bullet$$
$$R''\bullet + Cl\bullet \longrightarrow R'' Cl$$

4. Alkynes are more polar and are more soluble in water.

5. Inductive effect of Br increases acidity of the H on the carbon to which Br is attached.

6. Although a $>C{=}C<$ is more reactive than a $—C\equiv C—$ in an electrophilic addition reaction; and being conjugated dienes, the $—C\equiv C—$ reacts with HBr to form the product because the compound formed is more stable.

$$CH_2{=}CH—CH_2C\equiv CH \longrightarrow \begin{cases} CH_3—CHBr_2—C\equiv CH \\ \\ CH_2{=}CH—C\,Br{=}CH_2 \end{cases}$$

7. Addition of HBr in the presence of a peroxide converts 2–methyl propene into isobutyl bromide.

$$CH_2{=}\overset{\overset{\displaystyle CH_3}{|}}{C}{-}CH_3 \xrightarrow{HBr} CH_3{-}\overset{\overset{\displaystyle CH_3}{|}}{HC}{-}CH_2Br$$

8. When propene is treated with SO_2Cl_2, allyl chloride is formed.

$$CH_3{-}CH{=}CH_2 + SO_2Cl_2 \longrightarrow CH_2Cl{-}CH{=}CH_2 + SO_2Cl$$

(mechanism $-SO_2Cl_2 \longrightarrow SO_2Cl{\bullet} + Cl{\bullet} \ ||\ SO_2Cl \longrightarrow SO_2 + Cl{\bullet}$)

9. Addition of Br_2 across the double bond is a two step reaction and proceeds via electrophilic mechanism and forms a *trans*-addition compound.

10. Tetraethyl lead decomposes to form ethyl free radical.

$$Pb(C_2H_5)_4 \longrightarrow Pb + 4\,C_2H_5{\bullet}$$
$$C_2H_5{\bullet} + Cl{\bullet} \longrightarrow C_2H_5Cl$$

11. 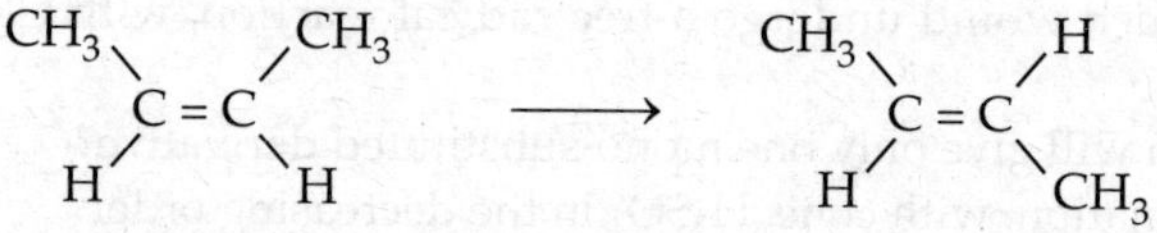
$$\underset{H}{\overset{CH_3}{\diagdown}}C{=}C\underset{H}{\overset{CH_3}{\diagup}} \longrightarrow \underset{H}{\overset{CH_3}{\diagdown}}C{=}C\underset{CH_3}{\overset{H}{\diagup}}$$

12. The reaction is a free radical reaction.

$$R{-}O\!:\!O{-}R \longrightarrow 2RO{\bullet}$$
$$RO{\bullet} + HBr \longrightarrow ROH + Br{\bullet}$$

$$CH_3CH_2\,C{\equiv}CH \xrightarrow{Br{\bullet}} CH_3CH_2C{\bullet}{=}CH{-}Br \xrightarrow{HBr}$$

$$CH_3CH_2CH{=}CHBr_2 \xrightarrow{HBr} CH_3CH_2CHCHBr_2$$

halogen atom destabilises a carbon radical to which it is bonded, hence Br• adds on the carbon.

13.

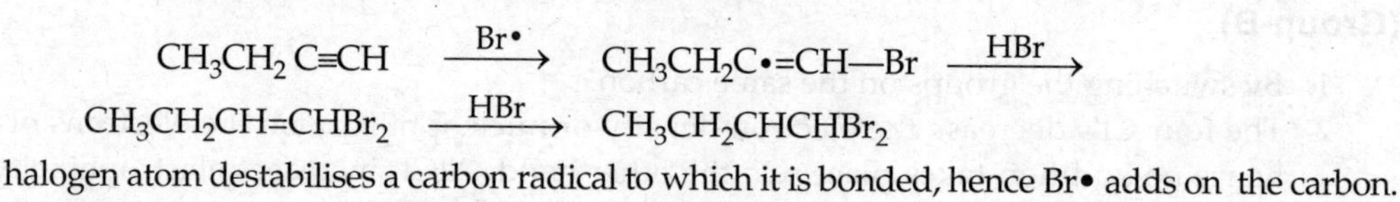

14. This is because the compound contains one acidic hydrogen.

15. On account of strong –I effect of $-CF_3$ group, the addition takes place in the following manner:

$$F_3C{-}CH{=}CH_2 + HBr \longrightarrow F_3C{-}CH_2{-}CH_2Br$$

16. NO_2^+ is the strongest electrophile (N is attached to two oxygen atoms)

17. The addition of I_2 on an alkene molecule can rotate the carbon chain and gives generally a mixture of *cis*- and *trans*-isomers.

18. $C_6H_5CH_2^+$ is more stable because of formation of four resonating structure where as in R_3C^+, the positive charge is on the carbon atoms. Formation of resonating structures decreases the energy of the system, making them more stable.

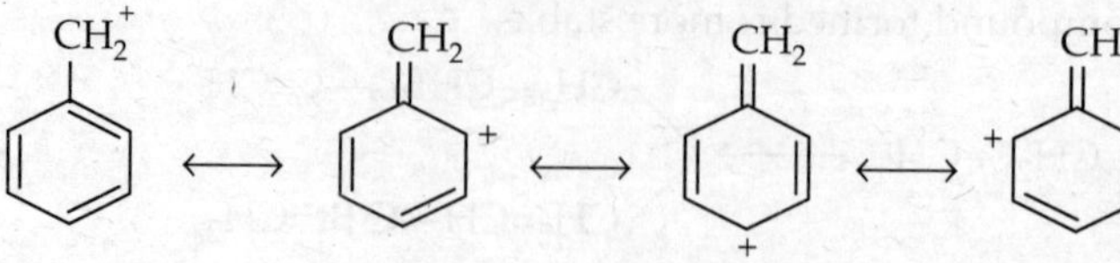

19. $RMgX + HC \equiv C - CH_3 \longrightarrow MgX - C \equiv C - CH_3 + RH.$
20. Alkanes do no have polarity, therefore alkanes do not react with RMgX and Li metal.
21. Bromination is more selective as compared to chlorination, because the intermediate formed is more close to the product. In addition to this, the attack by (Br•) is more endothermic, than the attack by (Cl•)
22. On hydrolysis the products formed in the ratio of:

$$CH_3CO-COCH_3 \quad + \quad CH_3COCHO \quad + \quad (CHO)_2$$

23. Acetanilide is an activated ring, benzene is not.
24. It is a free radical reaction. The intermediate formed in the attack of (Cl•) on the ethyl benzene.

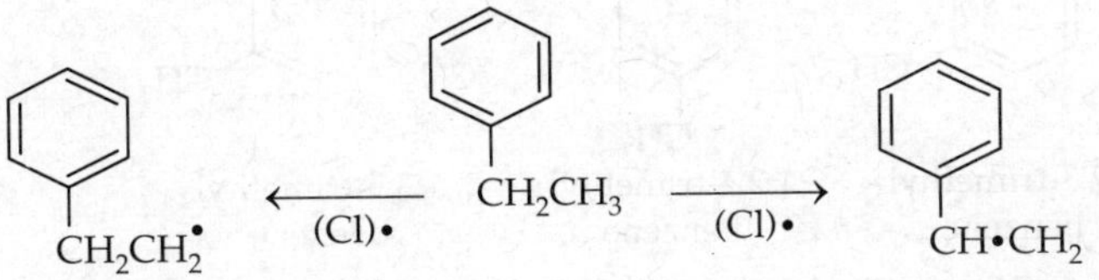

1° free radical
less stable

2° free radical
more stable

25. Thermodynamically, a p-substituted derivative is more stable at high temperature.
26. The compound, $C_6H_5CH(CH_3)_2$ will undergo chlorination very easily.
27. C_6H_5CN will form only one mono-substituted derivative.
28. The substitution of SO_3H, in the benzene ring containing various groups in decreasing order:

$$-OCH_3 > -CH_3 > C_6H_6 > C_6H_5Cl$$

AROMATIC COMPOUNDS

Benzene and its substituted derivatives constitute the class of aromatic compounds. Benzene, the parent hydrocarbon, consists of a monoplanar, cyclic ring of six carbon atoms, with one hydrogen atom attached to each carbon atom. Benzene molecule contains three alternate double bond in the molecule, which are dynamic in nature. It may be represented by any one of the following structures.

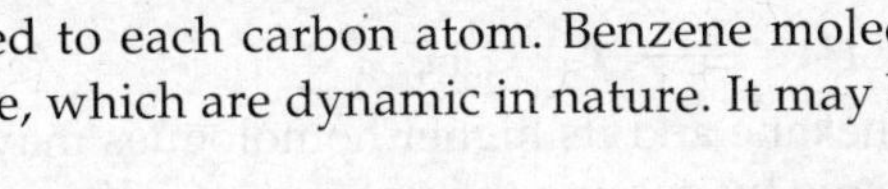

resonance
representation

C—C bond
length 1.397 Å

The molecular composition of the aromatic hydrocarbons is C_nH_{2n-6}. The homologues of benzene do contain a benzene ring, consisting of six carbon atoms, with three alternate double bonds, and the rest of the carbon atoms are joined to the ring, in the form of a side chain. The side chain may be defined as a carbon containing group, joined to the benzene ring through this carbon side chains are oxidised to a COOH group, directly joined to the ring, dil. alk. KMnO$_4$ solution, known as Baeyer's reagent, irrespective of the length of the side chain.

Some of the higher homologues of benzene are:

C_7H_8

C_8H_{10}

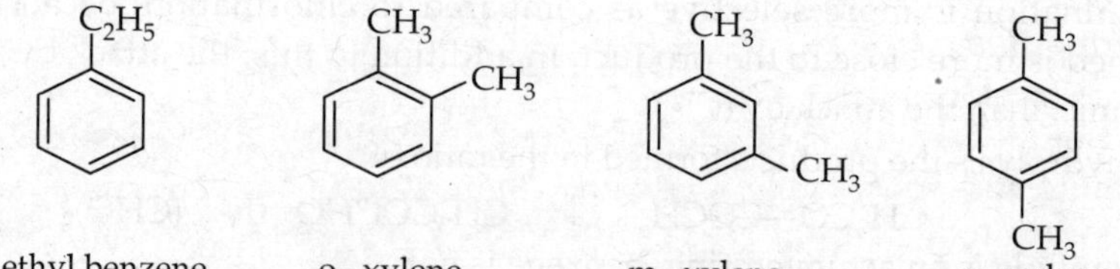

ethyl benzene   o– xylene   m– xylene   p– xylene

C_9H_{12}

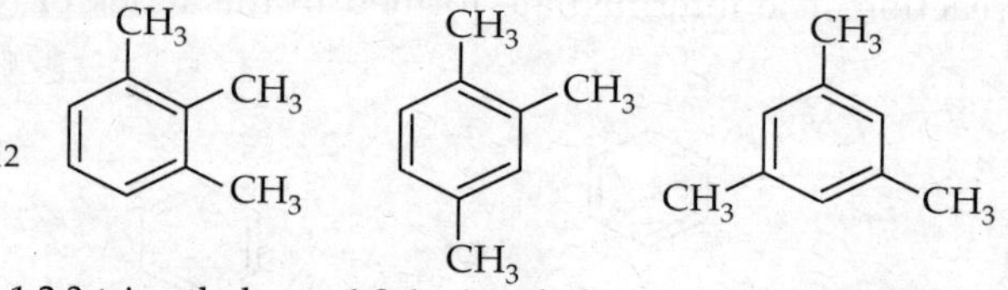

1,2,3-trimethyl
benzene

1,2,4-trimethyl
benzene

1,3,5-trimethyl
benzene

Methods of Preparation

1. **Decarboxylation:**
   On heating sodium salt of benzoic acid with sodalime benzene is formed.

   $$C_6H_5COONa + NaOH \xrightarrow{\text{CaO}} C_6H_6 + Na_2CO_3$$

2. **From phenol**

   $$C_6H_5OH + Zn \longrightarrow C_6H_6 + ZnO$$

3. **From chlorobenzene:** On reduction with Ni–Al catalyst, in presence of NaOH

   $$C_6H_5Cl + 2H \longrightarrow C_6H_6 + 2HCl$$

4. **From benzene diazonium chloride:** By reducing with phosphorous acid

   $$C_6H_5N_2Cl + H_3PO_2 + H_2O \longrightarrow C_6H_6 + N_2 + HCl + H_3PO_3$$

5. **By polymerisation of acetylene:** On passing acetylene gas through a red hot Cu tube at a temperature 500°C.

   $$3CH{\equiv}CH \longrightarrow C_6H_6$$

6. **By Aromatisation:** n-hexane and its higher homologues may be converted into benzene and its higher homologue by passing the vapours of n-hexane or its higher homologues over a catalyst consisting of a mixture of oxide of Mo, V and Cr, where the alkane is first dehydrogenated, followed by cyclisation. If possible isomerization may also take place. Besides, the catalyst used oxides of MoV and Cr, a mixture of CrO_3 and Al_2O_3 or a mixture of Pt and Al_2O_3, kept at 450-550°C (process is termed as platforming) for the conversion of n-hexane or its higher homologue into benzene or its higher homologues.

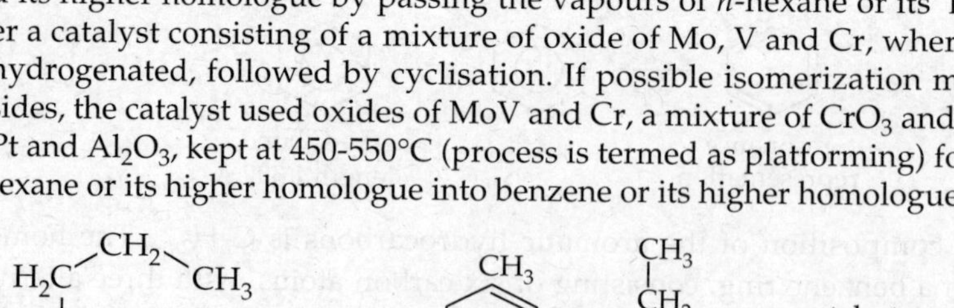

7. **From coaltar:** Coal is a bituminous mixture of higher hydrocarbons mixed with elementary carbon, its black colour is because of subjecting it to destructive distillation it gives first

fraction (boiling up to 170°C, is worked up for the recovery of benzene and other aromatic compound.

Physical properties

Benzene is a colourless liquid having odour like kerosene, immiscible with water (but lighter than water) but mixes with all organic solvents (BP 80°C and FP = 5.5°C, sp. gr. 0.784). It burns with a luminous and sooty flame, a property common to all aromatic compounds.

Benzene is a highly unsaturated compound, it does not show characteristic properties of unsaturated compounds.

a. It **does not decolourise** the pink colour of alk. dil. **KMnO$_4$** solution.
b. It does not decolourise yellow colour of **bromine** solution.
c. It does not form addition compounds with HCl, H$_2$SO$_4$, HOCl, BH$_3$ etc.

Instead of forming addition compound as all olefinic and acetylinic compounds do, benzene forms substituted derivatives, keeping the double bonds intact in the molecule.

Chemical properties

1. **Hydrogenation:** Hydrogenation is an addition reaction. Benzene reacts with hydrogen in presence of heated Ni as a catalyst to form cyclohexane, a cyclic hydrocarbon. A benzene molecule adds on three molecules of hydrogen to give a saturated hydrocarbon.

2. **Reaction with chlorine:** Benzene react with chlorine in presence of sunlight to form an addition compound (C$_6$H$_6$Cl$_6$) known as benzene hexa chloride.

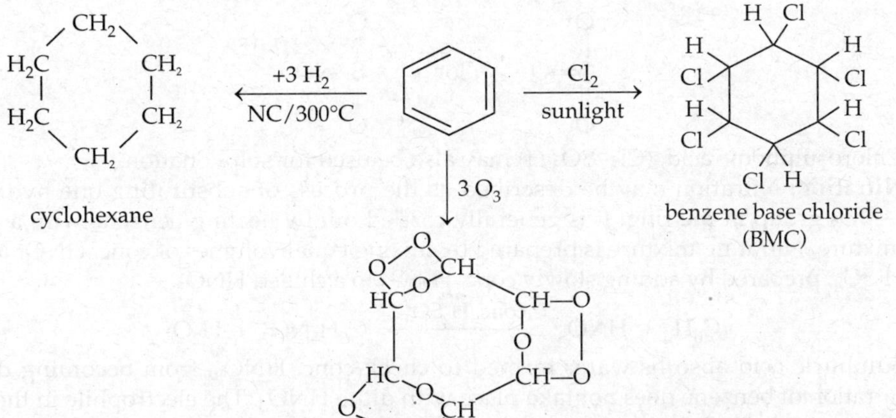

cyclohexane

benzene base chloride (BMC)

benzene triozonide

Benzene hexachloride exist in eight isomeric forms, one of which is known as 'lindane,' used as an insecticide, a 15% solution of lindane is marketed as 'gammaxane'.

3. **Halogenation:** Halogenation of benzene is a substitution reaction. Cl$_2$ and Br$_2$ react with benzene at a low temperature, in the presence of a halogen carrier like Fe or FeCl$_3$ to for chlorobenzene and bromo benzene respectively.

$$C_6H_6 + Cl_2 \xrightarrow{\text{Fe or FeCl}_3} C_6H_5Cl + HCl$$

Bromination is a slower reaction, than chlorination. Iodine reacts with benzene to form iodo benzene. As the reaction is reversible, it is generally carried in presence of some oxidising reagent like HIO$_4$, HNO$_3$ or HgO.

Mechanism: As the reactions in benzene are electrophilic substitution reactions, it is the electrophile, produced from chlorine molecule in presence of a halogen carrier, that attacks the ring.

$$Cl:Cl + FeCl_3 \longrightarrow Cl^+ + FeCl_4^-$$
$$\text{chloronium ion}$$

The Cl^+ ion then attacks the benzene ring to form π-complex, a loosely attached electrophile through π-electrons.

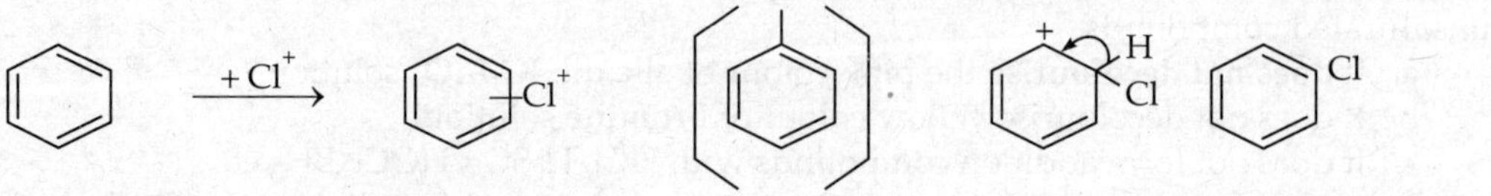

The complex then change to O^- complex through the localisation of π-electrons, creating three alternate double bonds and making one pair of electrons available for electrophile.

$$H^+ + FeCl_4^- \longrightarrow HCl + FeCl_3$$

4. **Sulphonation:** It is a process of substituting one hydrogen by a sulphonic acid group, SO_3H group in the benzene ring. It is carried by heating benzene with conc. H_2SO_4 or with oleum or fuming sulphuric acid. ($H_2SO_4 + SO_3$).

$$C_6H_6 + H_2SO_4 \xrightarrow{\Delta/80°C} C_6H_5SO_3H + H_2O$$

The electrophile in the process of sulphonation is SO_3.

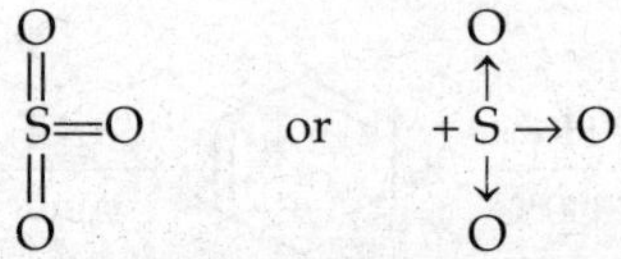

Chlorosulphonic acid, (Cl—SO_3H) may also be used for sulphonation.

5. **Nitration:** Nitration may be described as the process of substituting one hydrogen by —NO_2 group in the ring. It is generally carried out by heating benzene with a nitrating mixture. Nitrating mixture is prepared by mixing equal volumes of conc. HNO_3 and conc. H_2SO_4, prepared by adding slowly conc. H_2SO_4 to a chilled HNO_3.

$$C_6H_6 + HNO_3 \xrightarrow{\text{conc. } H_2SO_4} C_6H_5NO_2 + H_2O$$

Sulphuric acid absorbs water formed to check conc. HNO_3, from becoming dilute, as nitration of benzene does not take place with dlute HNO_3. The electrophile in the process of nitration is $^+NO_2$ or a nitronium ion.

$$H_2SO_4 \longrightarrow H^+ + HSO_4^-$$

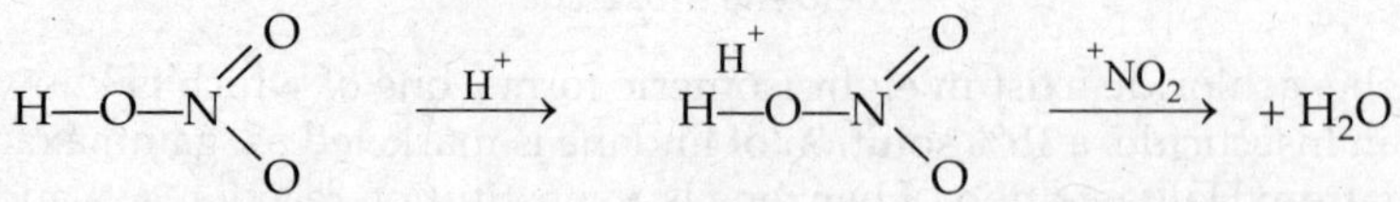

Mechanism: The attacking electrophile is $^+NO_2$ which attacks the ring forming a π complex changing to a σ-complex and finally changing to nitro benzene.

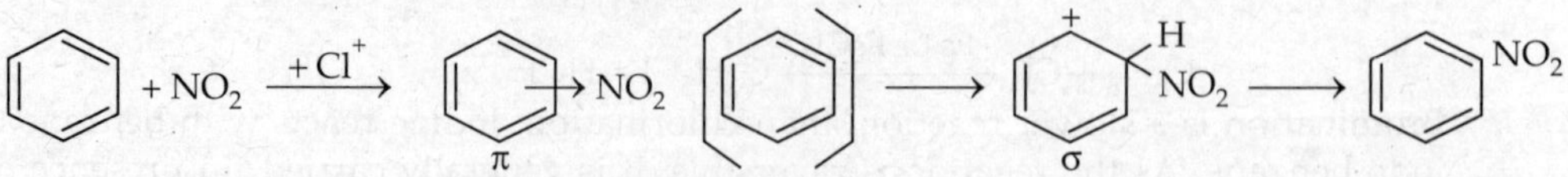

6. **Alkylation:** Benzene is converted to alkyl benzene, when it is treated with alkyl halide in presence of anhydrous $AlCl_3$.

$$RX + AlCl_3 \longrightarrow R^+ + AlCl_4^-$$
$$C_6H_6 + R^+ \longrightarrow C_6H_5—R + H^+$$
$$AlCl_4^- + H^+ \longrightarrow AlCl_3 + HX$$

The reaction is known as Friedel–Craft reaction. Besides alkyl radical, acyl radical (RCOCl) may be introduced in the ring. On treating benzene with alkenes, in the presence of anhydrous $AlCl_3$ alkyl benzene is formed.

7. *Mechanism of Friedel–Craft reaction*: The mechanism of the formation of toluene, (or alkyl benzene) proceeds via formation of a carbocation, which attacks the benzene ring forms initially a complex, changing to a complex and finally forming toluene (or alkyl benzene). The role of anhydrous. $AlCl_3$ is to liberate carbocation or an electrophile form alkyl halide.

$$CH_3X + AlCl_3 \longrightarrow CH_3^+ + AlCl_4^-$$

Alkyl radical so formed, then attacks the ring and through the formation of a complex and then the complex is finally converted to form alkyl benzene.

Similarly with CH_3COCl, $C_6H_5COCH_3$ (acetyl benzene) may be formed. In place of anhydrous $AlCl_3$, $BeCl_2$, BF_3 may also be used.

8. **Side chain oxidation:** Any carbon containing group, when attached to the benzene ring through this carbon, is known as a side chain. Such side chain containing benzene derivatives, on oxidation with dil. alk. $KMnO_4$, are oxidised to benzoic acid (irrespective of length of the side chain) i.e. the side chain is oxidised to a COOH group directly attached to the ring.

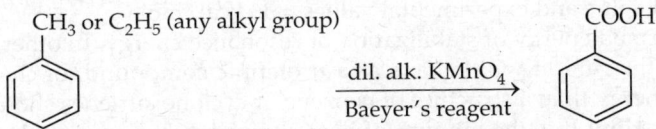

9. **Reaction with O_3:** One molecule of benzene reacts with three molecule of O_3 to form benzene triozonide ($C_6H_6O_9$).

$$C_6H_6O_9 \longrightarrow 3\,CHO—CHO + 3H_2O_2$$
$$\text{Glyoxal}$$

Formation of a triozonide and subsequent formation of three molecules of glyoxal proves the molecule of benzene contains three double bonds in the molecule.

Structure of benzene. It is best explained on the basis of molecular orbital theory.

Each carbon of benzene is sp^2 hybridised.

C_6-$1s^2$, $2s^2$, $2p_x^1$, $2p_y^1$ (in ground state)

One 2s electron takes some energy from the surroundings and jumps to vacant 2p orbital.

C_6-$1s^2$, $2s^1$, $2p_x^1$, $2p_y^1$ and $2p_z^1$ (in exited state)

One $2s^1$ and two $2p$ orbitals, pool up their energy and create three orbitals equivalent in every respect, inclined at an angle of 120°, with each other. This leaves one $2p$ orbital on each carbon atom, perpendicular to sp^2 hybridised carbon. Benzene is a flat molecule, with all the six carbon the same plane, and the angle in between C—C—C and C—C—H is 120° each. The unhybridised $2p$ orbital, one on each carbon, consists of two equal lobes one lying below and the other above the plane of the ring. The unhybridised p-orbital of each carbon can overlap with p-orbital on the adjacent carbon, on either side to satisfy the fourth valency of carbon atom, thus a continuous molecular orbital is formed. As each $2p$-molecular orbital overlaps the $2p$-orbital overlaps of the adjacent carbon, there are formed two continuous clouds, one below and the other above the hexagonal benzene ring. This extensive delocalisation of π electrons gives an intense electronegative charge on the ring, facilitating the formation of bond when an electrophile attacks the ring, resulting in the formation of an initial π complex. In addition to the formation of π clouds over the ring, gives an extra stability to the benzene ring, making all the C to C bonds equal to each other. All C to C bond length are equal to 1.39 Å.

All the reactions in benzene are electrophilic substitution reactions, the attacking reagent is an electrophile. Electrophile attacks the ring, forms a loosely held complex through the π electrons. Between the electrophile and the benzene ring, an initial product is formed. (an attack by a nucleophile would result in repulsion in between clouds and nucleophile)

After the formation of σ-complex, the delocalized bond return back to form three alternate double bonds. This π-complex then makes a pair of electrons available for π-electrophile to shift to newly created negative centre and joins through σ-bond, joined to benzene ring. Removal of a proton from the σ-complex and return of this vacated electron pair back to benzene converts it into a monosubstituted benzene derivative.

Further support for the concept of delocalisation and consequent stabilization of the benzene ring is provided by heat of hydrogenation and heat of combustion .

Heat of hydrogenation: Hydrogenation of alkenes is an exothermic reaction. On hydrogenation, benzene evolves only 49.8 kcal/mole, where as the theoretical value is 85 kcal/mole.

For calculation of the theoretical value:
For one double bond = 28.6 kcal/mole
Therefore for three double bonds, the value = 28.6 × 3 = 85.8 kcal/mole. The difference between the calculated value and experimental values = 36 kcal/mole

This value is termed energy of stabilization or resonance energy. In other words if this much energy is given to the ring, it will behave as a triolefinic compound, or the molecule contains 36 kcal/mole less energy than a structure of benzene or cyclohexatrienes should possess.

Heat of combustion: It is the amount of heat evolved, when one mole of an organic compound is completely oxidised to CO_2 and H_2O.

For benzene	Calculated value	Experimental value
	824.1 cal/mole	789.1 cal/mole

The difference = 35 kcal/mole, know as energy of stabilization or a resonance energy. Benzene, therefore contains 36 kcal/mole less energy, as a result the three double are deactivated. This is why double bonds in benzene do no show olefinic behavior. In case an energy of 36 kcal/mole is provided to the molecule, the three double-bonds will become active and would show all the properties of alkenes. When benzene is reacted with chlorine in the presence of sunlight, the double bonds become active (energy is supplied by sunlight) and it forms an addition compound.

MULTIPLE CHOICE QUESTIONS

(Group-C)

1. Reduction of an alkyne by Lindlar's catalyst gives a
   (a) *cis*-isomer
   (b) *trans*-isomer
   (c) mixture of *cis*- and *trans*-isomers
   (d) none
2. Which of the following bond is most reactive?
   (a) C≡C
   (b) C=C
   (c) —C—C—
   (d) none
3. The chief reaction product in between *n*-butane and bromine at 130°C is,
   (a) $CH_3CH_2CH_2CH_2Br$
   (b) $CH_3CH_2 CHBrCH_3$
   (c) $(CH_3)CH_2—CH_2Br$
   (d) $(CH_3)_3$_C—Br

4. The compounds formed when 2-butene is treated with acidified $KMnO_4$
   - (a) CH_3CHO
   - (b) CH_3COOH
   - (c) $CH_2(OH)—CH_2(OH)$
   - (d) $CH_3CH_2COCH_3$
5. An alkane forms isomers, if the number of carbons are
   - (a) ≥ 1
   - (b) ≥ 2
   - (c) ≥ 3
   - (d) ≥ 4
6. The structure of tertiary butyl carbocation is
   - (a) pyramidal
   - (b) triangular
   - (c) tetrahedral
   - (d) square planar
7. The restricted rotation about the double bond in 2-butene is due to overlapping of
   - (a) two p-orbitals
   - (b) one p- and other sp^2 orbital
   - (c) two sp^2 hybridise orbitals
   - (d) one s and sp^2 orbitals
8. A neutral double bond is formed when
   - (a) two free radicals combine
   - (b) one carbonium and one carbanion combine
   - (c) sp^2 carbon combine together
   - (d) none
9. In the presence of peroxide, HCl and HI do not give anti-Markovnikov's addition to alkene because
   - (a) both are highly ionic
   - (b) one is oxidising and other is reducing
   - (c) one of the step in endothermic
   - (d) in all the steps, one is exothermic
10. During debromination of meso-dibromo butane, the major compounds formed is
    - (a) n-butane
    - (b) 1-butene
    - (c) cis-2-butene
    - (d) $trans$-2-butene
11. Reactivity of the hydrogen atom attached to different carbons in an alkane has the order
    - (a) $3° > 1° > 2°$
    - (b) $1° > 2° > 3°$
    - (c) $3° > 2° > 1°$
    - (d) none
12. Cyclopentene on treatment with alk. $KMnO_4$ gives
    - (a) cyclopentanol
    - (b) $trans$ 1,2-cyclopentan diol
    - (c) cis 1,2-cyclopentan diol
    - (d) a mixture of (a) and (b)
13. When 1-butene is mixed with excess of bromine, the products formed are
    - (a) HBr
    - (b) butylene gas
    - (c) 1,2-dibromobutane
    - (d) a polybromo butane
14. Photochemical chlorination is termed
    - (a) pyrolysis
    - (b) substitution
    - (c) homolysis
    - (d) peroxidation
15. Conjugated double bond is present in
    - (a) propylene
    - (b) isobutylene
    - (c) butylene
    - (d) 1,3-butadiene
16. Active hydrogen is present in
    - (a) ethyne
    - (b) ethane
    - (c) ethene
    - (d) benzene
17. The compound that will not be formed during the reaction of methane with chlorine
    - (a) CH_3Cl
    - (b) $CHCl_3$
    - (c) $CH_3—CH_3$
    - (d) $CH_3CH_2CH_3$
18. $C_6H_5CH=CH—NO_2$ can be reduced to $C_6H_5CH_2CH_2—NH_2$ by
    - (a) alcoholic $NaBH_4$
    - (b) $LiAlH_4$
    - (c) H_2/catalyst
    - (d) all
19. Natural gas is a mixture of
    - (a) $CO + CO_2$
    - (b) $CO + N_2$
    - (c) $CO + H_2 + CH_4$
    - (d) $CH_4 + C_2H_6 + C_3H_8$
20. The alkene which on ozonolysis yields acetone is
    - (a) $CH_2=CH_2$
    - (b) $CH_3—CH=CH_2$
    - (c) $(CH_3)_2C=C(CH_3)_2$
    - (d) $CH_3—CH=CH—CH_3$

21. The reaction of HOCl on propene proceeds via addition of
    (a) H^+ in the first step
    (b) Cl^- in the first step
    (c) OH^- in the first step
    (d) Cl^+ adds first

22. On treating propyne with dil. H_2SO_4 in the presence of $HgSO_4$, the major product formed is
    (a) propanal
    (b) propyl hydrogen sulphate
    (c) acetone
    (d) propanol

23. On ozonolysis, one molecule of a hydrocarbon gives two molecule of formaldehyde. The hydrocarbon is
    (a) but-2-ene
    (b) ethylene
    (c) propylene
    (d) acetylene

24. Ethylene forms ethylene chlorohydrin by the action of
    (a) dry Cl_2 gas
    (b) dry HCl gas
    (c) a solution of Cl_2 in water
    (d) dil. HCl

25. The compound that cannot decolourise alk. $KMnO_4$ is
    (a) acetylene
    (b) ethanol
    (c) ethanal
    (d) ethane

26. Kerosene is a mixture of
    (a) alkanes
    (b) alkenes
    (c) alkynes
    (d) arenes

27. A hydrocarbon of the molecular formula C_6H_{10}, absorbs one molecule of H_2, upon catalytic hydrogenation. Upon ozonolysis, the compound yields OHC—CH_2—CH_2—CH_2—CH_2—CHO, the hydrocarbon is
    (a) cyclohexene
    (b) cyclopentane
    (c) cycloheptene
    (d) cyclobutene

28. Identify Z in the sequence CH_3—CH_2—$\overset{\overset{\displaystyle CH_3}{|}}{C}$=$CH_2$ $\xrightarrow{\text{HBr/peroxide}}$ Y $\xrightarrow{C_2H_5ONa}$ Z
    (a) $(CH_3)_2$—CH—CH_2—OC_2H_5
    (b) CH_3—CH_2—CH—O—C_2H_5
    (c) CH_3—$(CH_2)_3$—O—C_2H_5
    (d) CH_3—$(CH_2)_4$—OCH_3

29. A mixture of ethyl iodide and n-propyl iodide were subjected to Wurtz reaction. A hydrocarbon that will not be formed is
    (a) n-butane
    (b) n-propane
    (c) n-pentane
    (d) n-hexane

30. Order of reactivity of tertiary H, sec. H and primary H, toward elimination is
    (a) tert > sec > prim
    (b) sec > pri > tert
    (c) sec > pri > tert
    (d) pri > sec > tert

31. Addition of Br_2 to 1,3-butadiene gives
    (a) 1,2-addition product
    (b) 1,4-addition product
    (c) both 1,2-addition and 1,4-addition
    (d) none

32. Which of the following will have the smallest heat of hydrogenation per mole?
    (a) 1-butene
    (b) trans-2-butene
    (c) cis-2-butene
    (d) 1,3-butadiene

33. During pyrolysis of alkanes, the C—C bond breaks rather than a C—H bond because
    (a) the C—C bond is reactive in alkanes
    (b) C—H bond is reactive in alkanes
    (c) Bond energy of C—C bond is lower than C—H bond
    (d) energy of C—C bond is very high

34. Which one does not decolourise Br_2 in CCl_4?
    (a) C_2H_2
    (b) C_3H_6
    (c) C_6H_6
    (d) C_2H_4

35. During halogenation of alkanes, the halogen first gives
    (a) X^-
    (b) free radical
    (c) carbonium ion
    (d) X^+

36. The number of conformers in ethane are
    (a) 1
    (b) 2
    (c) 3
    (d) infinite

37. The synthetic gas is
    (a) C_2H_4
    (b) C_2H_2
    (c) NH_3
    (d) $CO + 3H_2O$

38. If the ozonide formed from ethylene is reduced by $LiAlH_4$, the compound formed is
    (a) HCHO   (b) HCOOH   (c) CH_3OH (d) $CH_2(OH)$—$CH_2(OH)$

39. The bond energy of a π-bond in kcal is about
    (a) 36   (b) 50   (c) 74   (d) 140

40. Ozonolysis of the following compound will give

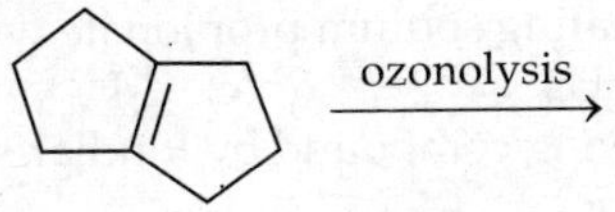

(a)

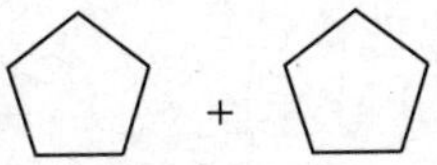

(b) $\begin{array}{c} CO—CH_2—CH_2—CH_2—CO \\ | \qquad\qquad\qquad\qquad | \\ CH_2\text{————}CH_2\text{————}CH_2 \end{array}$

(c)

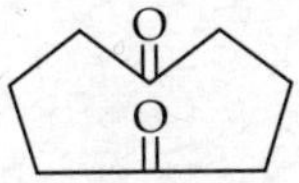

(d) none

41. The dihedral angle between the two methyl groups, in antistaggered form of ethane is
    (a) 120°   (b) 180°   (c) 90°   (d) 60°

42. Stereoisomerism can be exhibited by compounds, possessing
    (a) one chiral center   (b) two chiral centers
    (c) unsymmetrical molecule   (d) all of these

43. Carbocation can
    (a) rearrange to form a more stable carbonium ion
    (b) undergo loss of a proton   (c) combine with a carbanion
    (d) addition to a multiple bond   (e) all

44. Free radicals can undergo/combine
    (a) disproportionation to two species
    (b) arrangement for a more stable free radical
    (c) with other free radical
    (d) all of these

45. Isomers that can be interconverted through rotation about a single bond are termed
    (a) diastereoisomers   (b) conformers
    (c) enantiomers   (d) position isomers

46. In ethane, ethylene and acetylene, the C—H bond energy is
    (a) same in all   (b) greater in ethane
    (c) greatest in ethylene   (d) greatest in acetylene

47. Which of the following can form a metallic derivative?
    (a) alkanes   (b) alkenes   (c) alkynes   (d) arenes

48. Kolbe's synthesis of butanoic acid gives
    (a) butane + NaOH + H_2 (b) iso-butane
    (c) n-haxane   (d) propane

49. Decarboxylation of sodium acetate gives
    (a) methane   (b) ethane   (c) C_3H_8 (d) none

50. Which will react with Na/NH_3?
    (a) CH_3—CH_2—$C\equiv CH$ (b) CH_3—$CC\equiv CH_3$
    (c) CH_2—CH—$CH=CH_2$ (d) CH_3—$CH=CH$—CH_3

51. Aqueous H_2SO_4 reacts with 2-methyl-1-butene, to give predominantly
    (a) isobutyl hydrogen sulphate   (b) 2-methyl-3-butene
    (c) 2-methyl—butene   (d) sec. butyl hydrogen sulphate

52. $CH_3—C\equiv C—CH_3 + NaNH_2 \longrightarrow$ X; X is,
    - (a) $CH_3CH_2 CH_2CH_3$
    - (b) $CH_3—CH_2—C\equiv CH$
    - (c) $(CH_3)_2C=CH_2$
    - (d) $CH_2=C=CH—CH_3$

53. Which will react most readily with bromine?
    - (a) $CH_3CH_2CH_3$
    - (b) $CH_2=CH_2$
    - (c) $CH\equiv CH$
    - (d) $CH_3CH=CH_2$

54. The product obtained on heating sodium propionate with sodalime is
    - (a) CH_4
    - (b) C_2H_6
    - (c) C_2H_4
    - (d) C_2H_2

55. Hydrogenation of the following compound by Lindlar's method gives

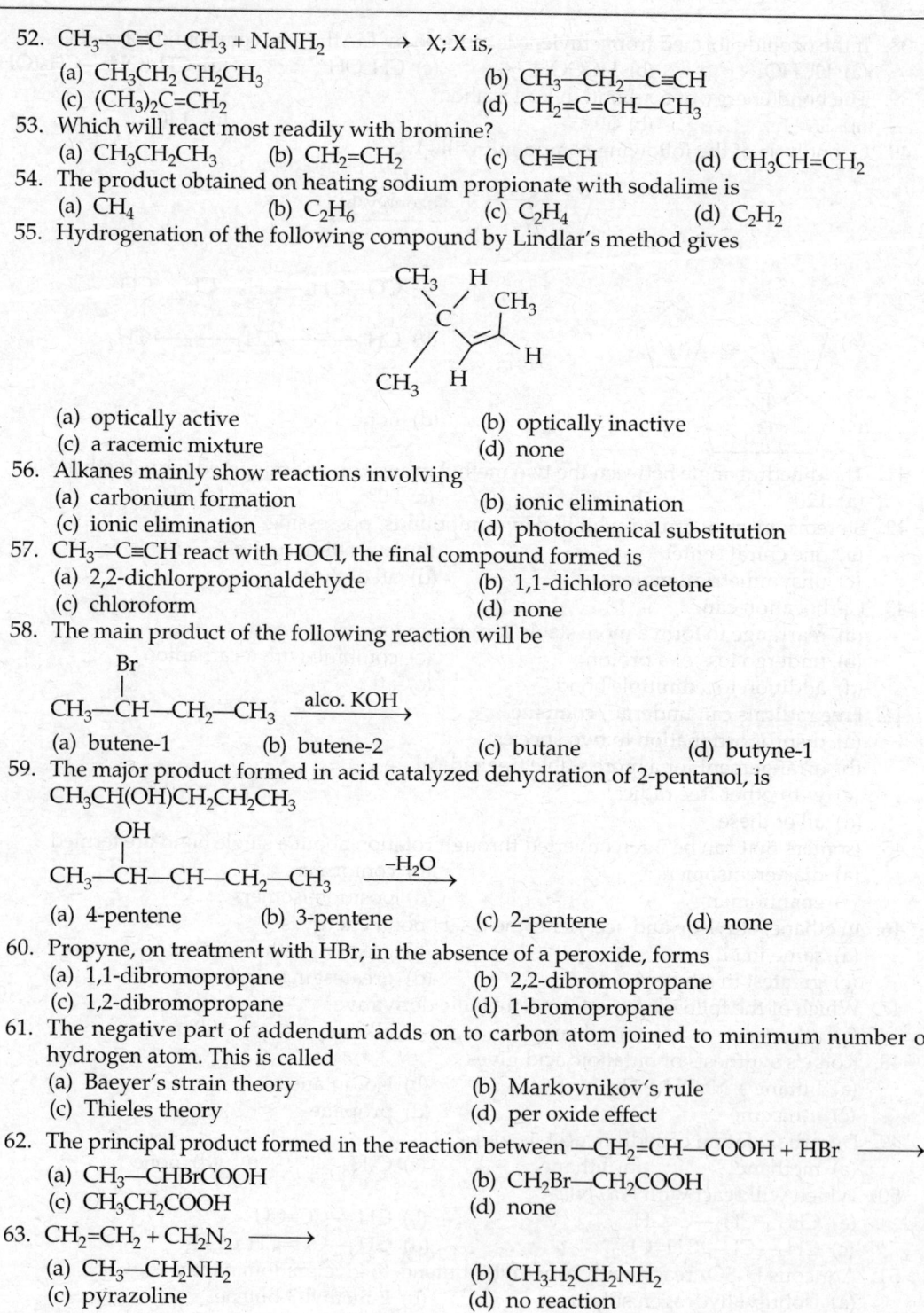

    - (a) optically active
    - (b) optically inactive
    - (c) a racemic mixture
    - (d) none

56. Alkanes mainly show reactions involving
    - (a) carbonium formation
    - (b) ionic elimination
    - (c) ionic elimination
    - (d) photochemical substitution

57. $CH_3—C\equiv CH$ react with HOCl, the final compound formed is
    - (a) 2,2-dichlorpropionaldehyde
    - (b) 1,1-dichloro acetone
    - (c) chloroform
    - (d) none

58. The main product of the following reaction will be

$$CH_3—\overset{\overset{\displaystyle Br}{|}}{CH}—CH_2—CH_3 \xrightarrow{\text{alco. KOH}}$$

    - (a) butene-1
    - (b) butene-2
    - (c) butane
    - (d) butyne-1

59. The major product formed in acid catalyzed dehydration of 2-pentanol, is
    $CH_3CH(OH)CH_2CH_2CH_3$

$$CH_3—\overset{\overset{\displaystyle OH}{|}}{CH}—CH—CH_2—CH_3 \xrightarrow{-H_2O}$$

    - (a) 4-pentene
    - (b) 3-pentene
    - (c) 2-pentene
    - (d) none

60. Propyne, on treatment with HBr, in the absence of a peroxide, forms
    - (a) 1,1-dibromopropane
    - (b) 2,2-dibromopropane
    - (c) 1,2-dibromopropane
    - (d) 1-bromopropane

61. The negative part of addendum adds on to carbon atom joined to minimum number of hydrogen atom. This is called
    - (a) Baeyer's strain theory
    - (b) Markovnikov's rule
    - (c) Thieles theory
    - (d) per oxide effect

62. The principal product formed in the reaction between $—CH_2=CH—COOH + HBr \longrightarrow$
    - (a) $CH_3—CHBrCOOH$
    - (b) $CH_2Br—CH_2COOH$
    - (c) CH_3CH_2COOH
    - (d) none

63. $CH_2=CH_2 + CH_2N_2 \longrightarrow$
    - (a) $CH_3—CH_2NH_2$
    - (b) $CH_3H_2CH_2NH_2$
    - (c) pyrazoline
    - (d) no reaction

64. $(C_2H_5)_2CH=CH_2 + I^+Cl\bullet^- \longrightarrow$
    (a) $(C_2H_5)_2CHCl—CH_2I$
    (b) $(C_2H_5)_2CHI—CH_2Cl$
    (c) both of (a) and (b)
    (d) none

65. $CH_2=CH_2 + BrCF_3 \longrightarrow$ predict the compound formed.

66. $CH_3—CH=CH_2 + Br—Cl \longrightarrow$ predict the compound formed.

67. $CH≡CH + CH_3COOH \longrightarrow$ predict the compound formed.

68. $CH_3—CH=CH—CH_2—C≡CH + HBr \longrightarrow$ predict the compound formed.

69. $CH_2=CH—C≡CH + HBr \longrightarrow$ predict the compound formed.

70. $CH_2=CH—Cl + HCl \longrightarrow$ predict the compound formed.
    (a) $CH_2Cl—CH_2Cl$
    (b) CH_3CHCl_2
    (c) CH_3CH_2Cl
    (d) none

71. $CH_2=CH—CH=CH_2 +$ maleic anhydride $\longrightarrow$
    (a) a linear compound
    (b) cyclic compound
    (c) both (a) and (b)
    (d) none of these

72. Electrolysis of potassium acetate gives
    (a) methane
    (b) ethane
    (c) propane
    (d) mixture of all three

73. As compared to the BP of straight chain isomers, the BP of branched chain are
    (a) lower
    (b) higher
    (c) equal
    (d) b.p. does not depend on branching

74. Buta1,3-diene on treatment with 1 mole of bromine forms
    (a) 1,2 dibromo deri
    (b) 3,4-dibromo addition compound
    (c) 1,4-dibromobutene-2
    (d) none

75. Kolbe's electrolysis of potassium acetate, the formation of ethane can be explained by
    (a) joining of two methyl ions
    (b) two CH_3 free radicals combine to form ethane
    (c) joining of one methyl anion and one methyl cation
    (d) potassium acetate decomposes to form ethane

76. Which will give a monosubstituted derivative on chlorination?
    (a) *n*-pentane
    (b) neopentane
    (c) isopentane
    (d) *n*-butane

77. The type of isomerism exhibited by alkenes is
    (a) chain
    (b) metamerism
    (c) position
    (d) geometrical

78. $RCH=CH_2 \xrightarrow{\text{Na/alc. } NH_3} RCH_2CH_3$, this reaction is an example of
    (a) Clemmensen's reduction
    (b) Wolff-Kishner's reaction
    (c) Birch reaction
    (d) Fischer-Spier reaction

79. Which one has the least octane number
    (a) octane
    (b) cetane
    (c) 2,2,4-trimethylpentane
    (d) *n*-heptane

80. The reaction $C_6H_5N=\overset{\overset{\displaystyle OH}{|}}{C}CH_3 \longrightarrow C_6H_5NHCOCH_3$ is called
    (a) Ellington's reaction
    (b) Glaser's reaction
    (c) Gomberg Beckmann's reaction
    d) Leuckart's reaction

81. A hydrocarbon reacts with HI to give (X), which on reacting with aqueous KOH forms (Y). Oxidation of (Y) gives 3-methyl-2-butanone. The hydrocarbon is

(a) $CH_3-\overset{\overset{\displaystyle CH_3}{|}}{CH}=CH-CH_3$

(b) $CH_2=CH-\overset{\overset{\displaystyle CH_3}{|}}{CH}-CH_3$

(c) $CH_3-CH_2-\overset{\overset{\displaystyle CH_3}{|}}{C}=CH_2$

(d) $CH_3-\overset{\overset{\displaystyle CH_3}{|}}{CH}-\overset{\overset{\displaystyle CH_3}{|}}{CH}-CH_3$

82. Which statement is correct about A and B

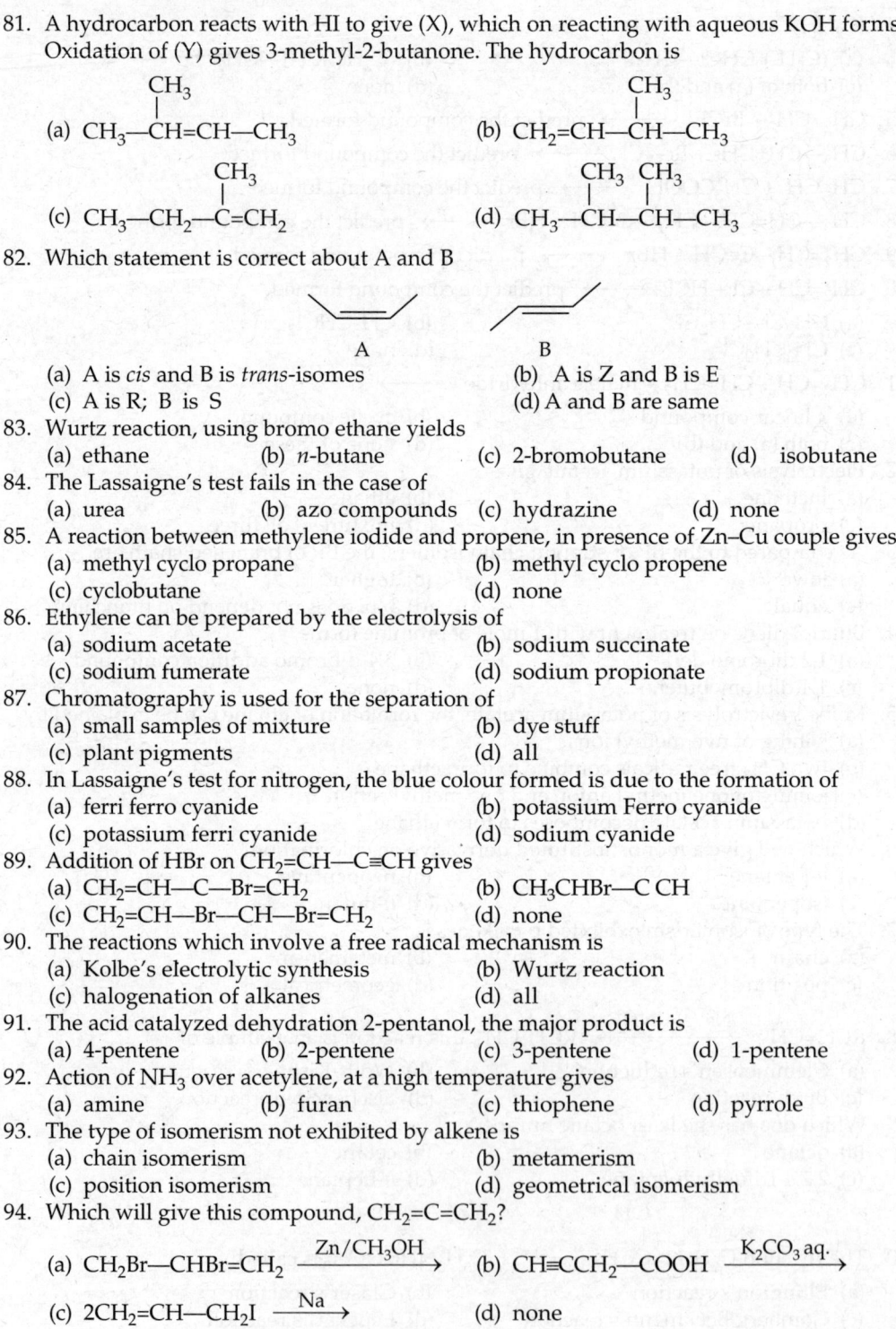

A   B

(a) A is *cis* and B is *trans*-isomes
(b) A is Z and B is E
(c) A is R; B is S
(d) A and B are same

83. Wurtz reaction, using bromo ethane yields
(a) ethane   (b) *n*-butane   (c) 2-bromobutane   (d) isobutane

84. The Lassaigne's test fails in the case of
(a) urea   (b) azo compounds   (c) hydrazine   (d) none

85. A reaction between methylene iodide and propene, in presence of Zn–Cu couple gives
(a) methyl cyclo propane   (b) methyl cyclo propene
(c) cyclobutane   (d) none

86. Ethylene can be prepared by the electrolysis of
(a) sodium acetate   (b) sodium succinate
(c) sodium fumerate   (d) sodium propionate

87. Chromatography is used for the separation of
(a) small samples of mixture   (b) dye stuff
(c) plant pigments   (d) all

88. In Lassaigne's test for nitrogen, the blue colour formed is due to the formation of
(a) ferri ferro cyanide   (b) potassium Ferro cyanide
(c) potassium ferri cyanide   (d) sodium cyanide

89. Addition of HBr on $CH_2=CH-C\equiv CH$ gives
(a) $CH_2=CH-C-Br=CH_2$ (b) $CH_3CHBr-C\ CH$
(c) $CH_2=CH-Br-CH-Br=CH_2$ (d) none

90. The reactions which involve a free radical mechanism is
(a) Kolbe's electrolytic synthesis   (b) Wurtz reaction
(c) halogenation of alkanes   (d) all

91. The acid catalyzed dehydration 2-pentanol, the major product is
(a) 4-pentene   (b) 2-pentene   (c) 3-pentene   (d) 1-pentene

92. Action of NH_3 over acetylene, at a high temperature gives
(a) amine   (b) furan   (c) thiophene   (d) pyrrole

93. The type of isomerism not exhibited by alkene is
(a) chain isomerism   (b) metamerism
(c) position isomerism   (d) geometrical isomerism

94. Which will give this compound, $CH_2=C=CH_2$?

(a) $CH_2Br-CHBr=CH_2 \xrightarrow{Zn/CH_3OH}$

(b) $CH\equiv CCH_2-COOH \xrightarrow{K_2CO_3 \text{ aq.}}$

(c) $2CH_2=CH-CH_2I \xrightarrow{Na}$

(d) none

95. $CH\equiv CH \xrightarrow{\quad} O_3 \xrightarrow{\quad} X \xrightarrow{Zn + CH_3COOH} Y$; What is Y?
    (a) HCHO + HCHO
    (b) COOH—COOH
    (c) CHO—CHO
    (d) none

96. Incomplete combustion of petrol or diesel oil, in an engine, can be tested by testing the fuel gases for the presence or the absence of
    (a) $CO_2 + H_2O$
    (b) CO
    (c) NO
    (d) SO_2

97. The order of relative acidic strength of water, ethyne and propyne is
    (a) H_2O > ethyne > propyne
    (b) propyne > water > ethyne
    (c) ethyne > propyne > water
    (d) none

98. Hydroxylation of the following compound, in presence of Baeyer's reagent gives
    $$CH_3CH=CH—CH_3$$
    (a) *cis*-glycol
    (b) *trans*-glycol
    (c) a mixture of both (a) and (b)
    (d) none

99. One among the following is not an aromatic compound is

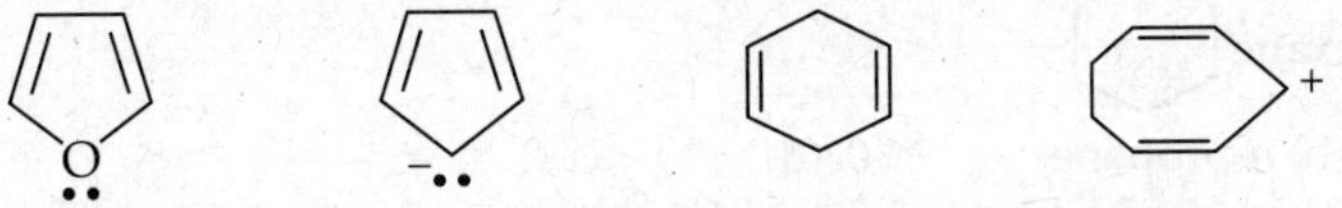

    (a)   (b)   (c)   (d)

100. The compound formed when phenyl acetylene is treated with dil. acid in presence of Hg^{2+}
    (a) acetophenone
    (b) phenyl ethylene
    (c) phenyl methane
    (d) none

101. $CH\equiv CH + R—COOH \longrightarrow$
    (a) CHR=CCOOH
    (b) CH_2=CH—OCOR
    (c) CH_3COOH
    (d) none

102. 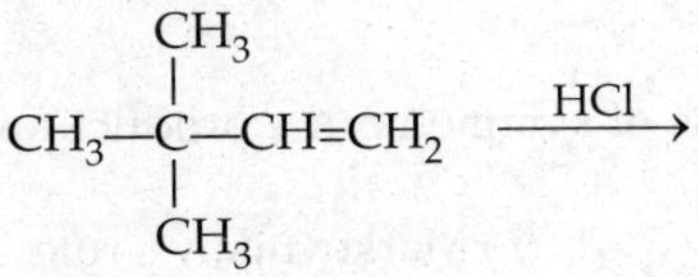
$$CH_3—\underset{\underset{CH_3}{|}}{\overset{\overset{CH_3}{|}}{C}}—CH=CH_2 \xrightarrow{HCl}$$

(a) 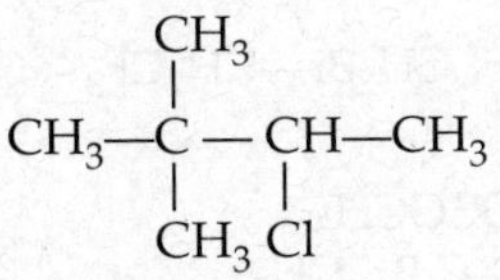 $CH_3—\underset{\underset{CH_3}{|}}{\overset{\overset{CH_3}{|}}{C}}—CH—CH_3$ with Cl
    (b) $(CH_3)CH_2CH_2Cl —CH\,Cl— CH_3$

(c) 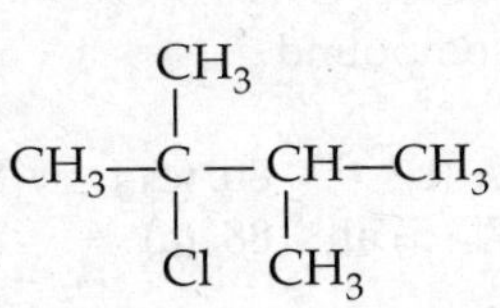 $CH_3—\underset{\underset{Cl}{|}}{\overset{\overset{CH_3}{|}}{C}}—CH—CH_3$ with CH_3
    (d) none

103. Which will have a higher BP?
    (a) *n*-hexane
    (b) *n*-pentane
    (c) 2-methyl butane
    (d) 2,2-dimethyl propane

104. As the molecular weight of an alkane increases; how do the boiling point and melting point change
    (a) increase in boiling point and melting point also increases
    (b) decrease in boiling point and melting point increases
    (c) increase in boiling point but melting point decreases
    (d) boiling point increases; the melting increases sequentially for alkanes containing over 4-carbons

ANSWERS

(Group-C)

1. (a)   2. (a)   3. (b)   4. (b)   5. (d)   6. (b)   7. (a)   8. (c)
9. (c)   10. (d)   11. (c)   12. (c)   13. (c) 1,2-dibromobutane
14. (c) homolysis   15. (d)   16. (a)   17. (d)   18. (c)   19. (d)   20. (c)
21. (d) electrophilic addition HO—Cl ionises as HO^- and Cl^+
22. (c)   23. (b)   24. (c)   25. (d)   26. (a) kerosene contains C_{11} to C_{16} carbons

27. (a) a cyclohexane

28. (d)   29. (b) n-propane   30. (d)
31. (c) conjugated compounds give 1,2-and 1,4-addition compound
32. (d) conjugated dienes are more stable and have a lower heat of hydrogenation
33. (c) The C—C bond involves a $2sp^3$-$2sp^3$ orbitals; the C—H bond involves $2sp^3$–1 s-orbitals in alkanes
34. (c) C_6H_6 35. (b) free radical   36. (d) infinite
37. (d)   38. (c)   39. (b) 50 kcal/mole   40. (c)   41. (b)   42. (d)   43. (e)
44. (d)   45. (b)   46. (d) due to more s-character
47. (c)   48. (a)   49. (a)   50. (b)   51. (a) $CH_3CH_2—C(CH_3)HSO_4^-$ 52. (b)
53. (b) alkenes are most reactive particularly ethene
54. (a)
55. (b) addition occurs at the triple bond, to give a plane of symmetry; the product formed is optically inactive
56. (d)   57. (a)   58. (b)   59. (c)   60. (b)   61. (b) Markovnikov's rule

62. (b)   63. (c)   64. (a) $I^+ Cl^-$ 65. $CH_2Br—CH_2CF_3$

66. $(Br^+—Cl^\bullet)CH_3—CHCl—CH_2Br$ 67. $CH_2=CHOCOCH_3$
68. $CH_3CHBr—CH_2—CH_2C≡CH$ 69. $CH_2=CH—C Br=CH_2$
70. (b) $CH_3—CHCl_2$ 71. (b) cyclic compound
72. (b) ethane   73. (a) lower
74. (c)   75. (b)   76. (b)   77. (d)   78. (c)   79. (c)   80. (c)   81. (b)
82. (a)   83. (b)   84. (b)   85. (c)   86. (b)   87. (d) all   88. (a)   89. (a)
90. (d)   91. (b)   92. (d)
93. (b) metamerism is exhibited by alkenes having same polyvalent groups.
94. (a)   95. (c) $CHO—CHO \longrightarrow CH_2OHCH_2OH$
96. (b) CO   97. (a)   98. (a)   99. (c) is not aromatic
100. (a)   101. (b) $CH_2=CH—OCOR$

102. (c) $CH_3-\underset{\underset{Cl}{|}}{\overset{\overset{CH_3}{|}}{C}}-\overset{\overset{CH_3}{|}}{CH}-CH_3$ 1,2 shift takes place

103. (a) *n*-hexane, on account of van der Walls forces   104. (a)

REASONING TYPE QUESTIONS (RTQs)

(Group-D)

1. Write a short notes on aromaticity.
2. What are antiaromatic compounds? Discuss.
3. Explain whether the substitution in the benzene ring is an electrophilic or a nucleophilic in nature? Explain with reasons.
4. List the activating and deactivating groups? Explain their role, when they are present in the ring.
5. a. As in benzene, majority of reactions are substitution reactions, when carried at normal temperature, but some of the reactions are addition reactions when carried out at high temperature. Explain.
   b. Discuss the compounds formed when benzene reacts with (i) Cl_2 in the presence of sunlight or UV radiation (ii) with H_2, in the presence of heated Ni at 200°C. Also discuss the structure of the products formed, in each case.
6. Name the product, when benzene reacts with (a) O_3 followed by hydrolysis (b) oxidation with air in presence of V_2O_5 at 450°C.
7. Give the general mechanism of electrophilic substitution in benzene ring.
8. Give the mechanism involved in (i) nitration (ii) Sulphonation and (iii) alkylation by methyl chloride in presence of anhydrous $AlCl_3$.
9. How will you distinguish between benzene and cyclohexene?
10. Write a note on Friedel–Craft reaction. What are its drawback? Discuss its mechanism.
11. Explain the role played by the catalyst, $FeBr_3$ in the bromination of benzene.
12. Does benzene undergo Diels–Alder reaction, with maleic anhydride, as benzene also has three double bonds in the molecule.
13. a. What do you understand by the term, "oxidation of the side chain". How it can be used in distinguishing three isomers of xylene? Explain.
    b. Discuss the side chain oxidation of tertiary butyl benzene.
14. How a side chain chlorination may be carried out? Discuss it by taking the example of xylene.
15. Following Huckel's rule, which species is the smallest aromatic compound? Discuss
16. Explain, 1,3,5-cycloheptatrienyl cation is aromatic, but 1,3,5-tricycloheptatriene is not?
17. The halogens are o-and p-directing groups, yet these are deactivating. Explain
18. Name the compound is formed when xylene is treated with ozone, followed by hydrolysis. What conclusions can be drawn from the product, formed.
19. Is it possible that there could be nonbenzenoid, heterocyclic compounds, which will show aromatic character. Discuss.

ANSWERS

(Group-D)

1. Aromaticity may be defined as the properties possessed by some unsaturated cyclic compounds, forming substituted derivatives while keeping the unsaturation intact in the molecule.

Benzene is a highly unsaturated compound, contains three double bonds in the molecule. Majority of its reactions are substitution, whenever it is treated with the reactant under normal conditions. On the other hand, when the reaction is carried out under drastic conditions, it forms addition compounds.

Huckel gave a formula about this self contradictory properties possessed by benzene, known as **Huckel's rule.** According to this rule, if any cyclic unsaturated compound possess the number of electrons as per rule, i.e. $(4n + 2)$ π-electrons. The compounds are expected to behave aromatic in nature.

The requirements are that the compound should be cyclic in nature; all the atoms present must be in the same plane and the number of electrons be $(4n + 2)$ π-electrons, where n is a simple integer $n = 0, 1, 2, 3$ etc.

In benzene the 6 π-electrons are delocalized, all the six C—C bond length are equal. The bond length is equal to 1.39 Å, it is longer than a C=C bond (1.34 Å) but smaller than a C—C single bond (1.54 Å) .

2. Antiaromatic compounds are those compounds, which do not fulfil all the conditions required for the compound to behave as aromatic compounds. Cyclobutadiene-1,3 and cyclooctatetraene-1,3,5,7 are antiaromatic compounds. Both are cyclic in nature, and all the atoms are in the same plane yet these do not behave as aromatic compounds, because the number of π-electron in the rings do not fulfill the third requirement, as per Huckel's rule .

Cyclobutadiene-1,3: It does not behave as an aromatic compound, as the number of π-electrons is only four. However, it can be converted into an aromatic compound, in case it is treated with two equivalent of sodium metal or potassium metal. It is converted into an aromatic ring, as each one of the sodium atom contributes one electron each to the ring, raising the number of π-electrons to six. Thus, disodium salt of cyclobutadiene-1,3, starts behaving as an aromatic compound.

$$
\begin{array}{c}
\text{CH=CH} \\
|\quad\ | \\
\text{CH=CH}
\end{array}
\; + 2K \longrightarrow
\begin{array}{c}
\text{CH=CH} \\
|\,2e\;\; | \\
\text{CH=CH}
\end{array}
\; 2K^{+}
$$

cycolbutadiene-1,3   dipotassium salt of
                                                     cyclobutadiene-1, 3

Cyclo-octatetraene-1,3,5,7 like cyclobutadiene the number of π-electrons in eight only, in the ring. A disodium salt of cyclo octatetraene, would behave as an aromatic ring, as the number of π-electrons is raised to ten, the third condition is also fulfilled.

$$
\begin{array}{c}
\;\;\;\text{CH=CH} \\
/\qquad\quad\backslash \\
\text{CH}\qquad\quad\text{CH} \\
\|\qquad\qquad\| \\
\text{CH}\qquad\quad\text{CH} \\
\backslash\qquad\quad/ \\
\;\;\;\text{CH=CH}
\end{array}
\; + 2K \longrightarrow
\begin{array}{c}
\;\;\;\text{CH=CH} \\
/\qquad\quad\backslash \\
\text{CH}\qquad\quad\text{CH} \\
\|\qquad\qquad\| \\
\text{CH}\qquad\quad\text{CH} \\
\backslash\qquad\quad/ \\
\;\;\;\text{CH=CH}
\end{array}
\; 2K^{+}
$$

cyclooctatetraene-1,3,5,7   dipotassium salt of
                                                             cyclooctatetraene

3 A molecule of benzene, has all the six π-electrons delocalized, and which form a cloud engulfing the ring from top and bottom, like a sandwich, As the group of six π-electrons make the ring sufficient negatively charged, it would repel a negatively charged species (or a nucleophile), but will easily react with a positively species (an electrophile), thus all the reaction of benzene ring are electrophilic in nature.

4. Activating/Deactivating Groups Ortho/meta/para directing groups

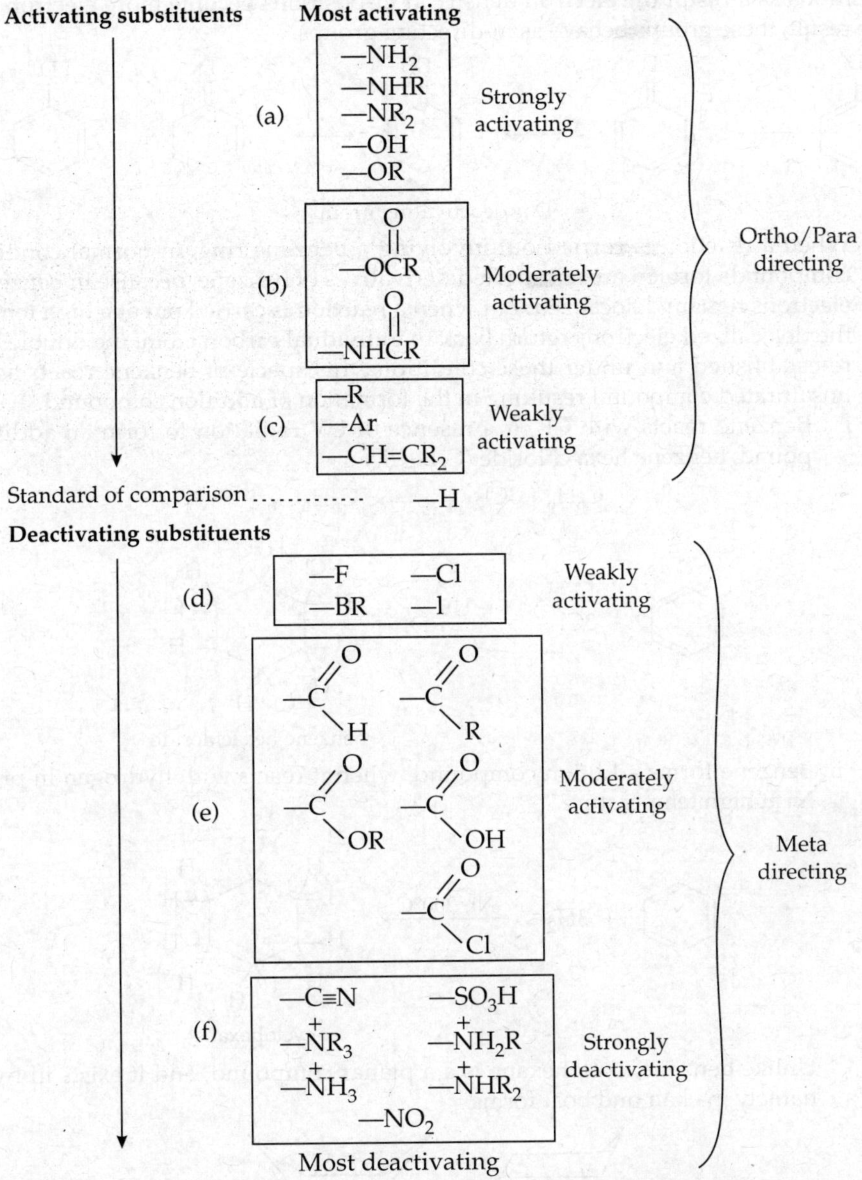

Activating substituents   **Most activating**

(a)
—NH$_2$
—NHR
—NR$_2$
—OH
—OR
Strongly
activating

(b)
$$-OCR\ (\overset{O}{\overset{\|}{})}$$
$$-NHCR\ (\overset{O}{\overset{\|}{})}$$
Moderately
activating

(c)
—R
—Ar
—CH=CR$_2$
Weakly
activating

Ortho/Para
directing

Standard of comparison .... —H

Deactivating substituents

(d)
—F   —Cl
—BR   —1
Weakly
activating

(e)
$$-\overset{O}{\underset{H}{\overset{\|}{C}}}\qquad -\overset{O}{\underset{R}{\overset{\|}{C}}}$$
$$-\overset{O}{\underset{OR}{\overset{\|}{C}}}\qquad -\overset{O}{\underset{OH}{\overset{\|}{C}}}$$
$$-\overset{O}{\underset{Cl}{\overset{\|}{C}}}$$
Moderately
activating

(f)
—C≡N   —SO$_3$H
—NR$_3^+$ —NH$_2$R$^+$
—NH$_3^+$ —NHR$_2^+$
—NO$_2$
Strongly
deactivating

Meta
directing

Most deactivating

Activating groups increase the electron density in o- and p-positions, enabling the electrophile to attack and join at o- and p-positions

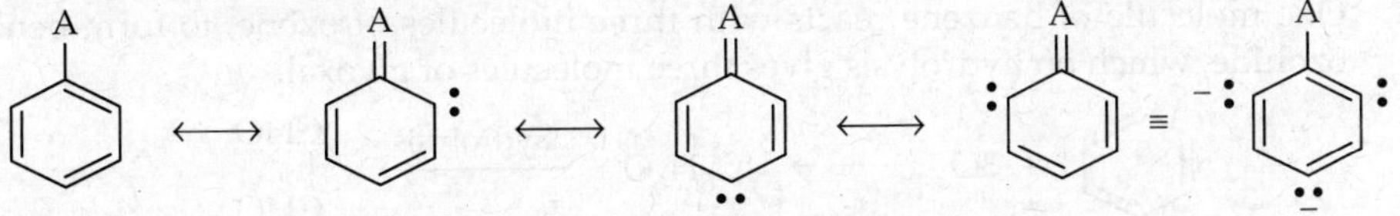

A is the activating group

On the other hand, the deactivating groups, decrease the electron density in o-and-p positions as a result the electron density at m-positions become more electron rich, with the result, these groups behave as m-directing groups.

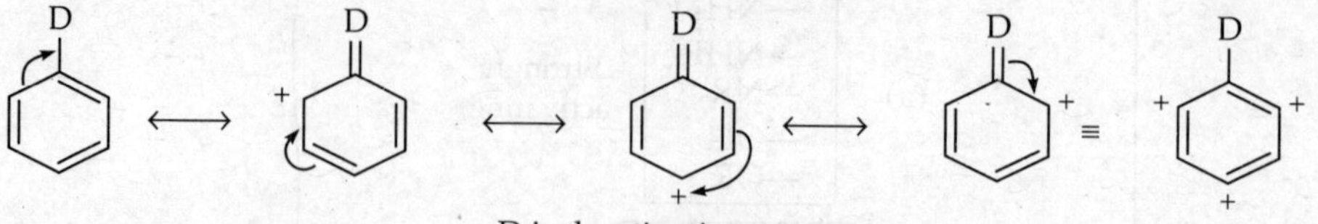

D is deactivating group

5. a. When a reaction is carried out involving a benzene ring, in normal conditions, the compounds formed are substituted derivatives of benzene, because in benzene the six electrons remain delocalized. But when a reaction is carried out at a high temperature, the delocalized electrons return back to individual carbon atom, the double bonds are re-established and under these conditions this specie of benzene reacts as a highly unsaturated compound resulting in the formation of addition compound.

    b. i. Benzene reacts with Cl_2, in presence of UV radiation to form an addition compound, benzene hexa chloride ($C_6H_6Cl_6$)

$$C_6H_6 + 3Cl_2 \longrightarrow C_6H_6Cl_6$$

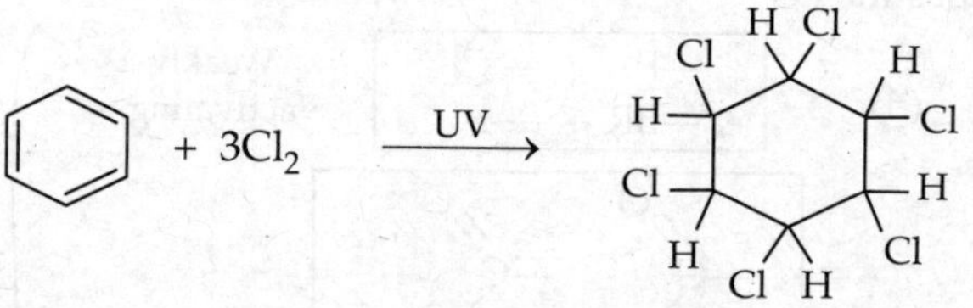

benzene hexachloride

   ii. Benzene forms addition compound, when it reacts with hydrogen in presence of Ni at high temperature.

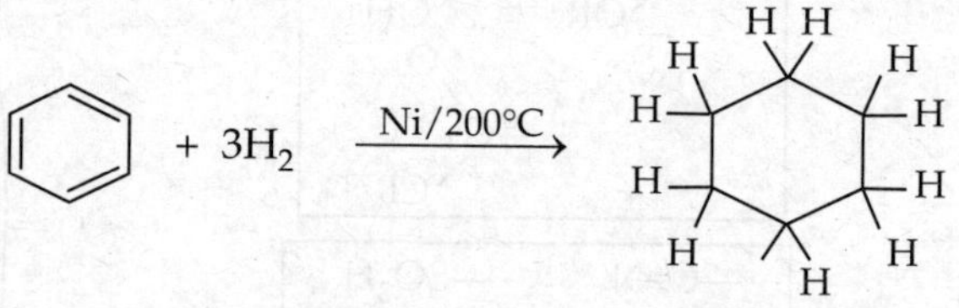

cyclohexane

Unlike benzene, cyclohexane is a a planar compound, and it exists in two forms, namely in chair and boat forms.

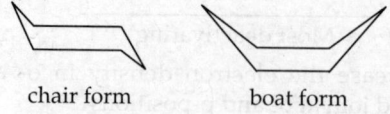

chair form   boat form

6. a. One molecule of benzene reacts with three molecules of ozone, to form benzene tri ozonide, which on hydrolysis gives three molecules of glyoxal.

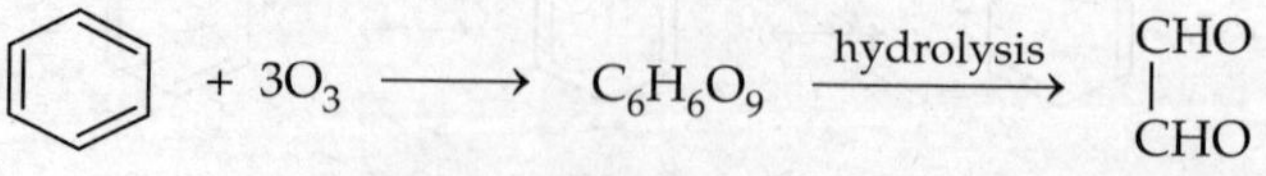

$$\text{benzene} + 3O_3 \longrightarrow C_6H_6O_9 \xrightarrow{\text{hydrolysis}} \begin{array}{c} CHO \\ | \\ CHO \end{array}$$

glyoxal

b. On oxidation, by air in presence of vanadium pentoxide (V_2O_5) as a catalyst, benzene is oxidised to form maleic anhydride.

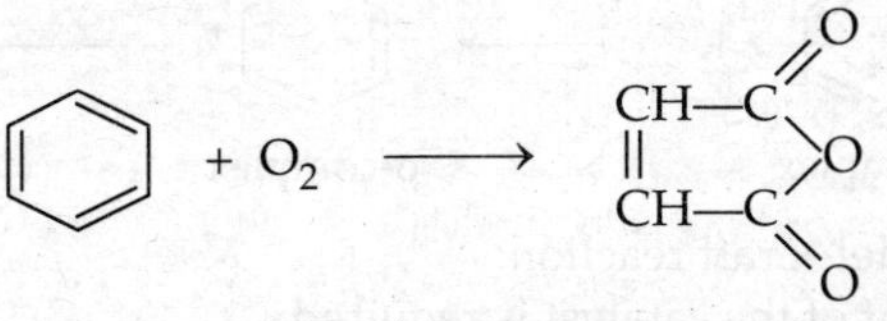

maleic anhydride

7. The reactions of benzene are electrophilic substitution. The electrophile attacks the benzene ring, and forms a lose complex, known a π-complex. This complex, then changes to form a σ-complex, in which a bond is formed between the electrophile and the ring. It then gives up a proton to form the derivative.

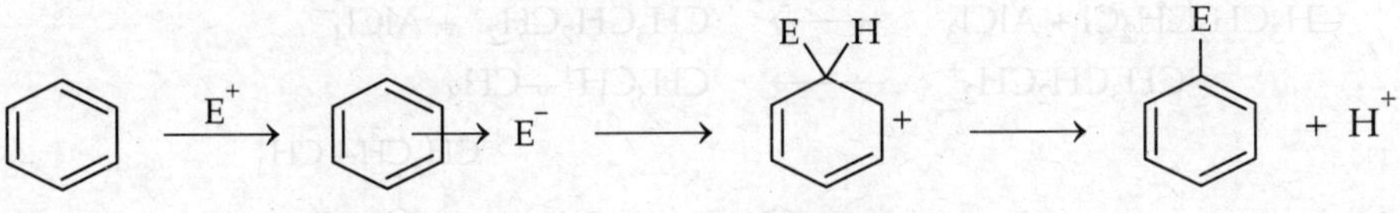

complex   complex
(π-complex)   (σ-complex)

8. i. **Nitration:** Generally nitration is carried by treating the aromatic compound, with a nitrating mixture consisting of HNO_3 to which a chilled conc. H_2SO_4, added slowly. The electrophile generated is nitronium ion $—NO_2^+$.

$$H_2SO_4 \longrightarrow H^+ + HSO_4^-$$

$$HONO_2 + H^+ \longrightarrow H—\overset{H}{\underset{}{\overset{|}{O}}}{}^+NO_2 \longrightarrow H_2O + NO_2^+$$

ii. **Sulphonation:** The electrophile is SO_3. It reacts as

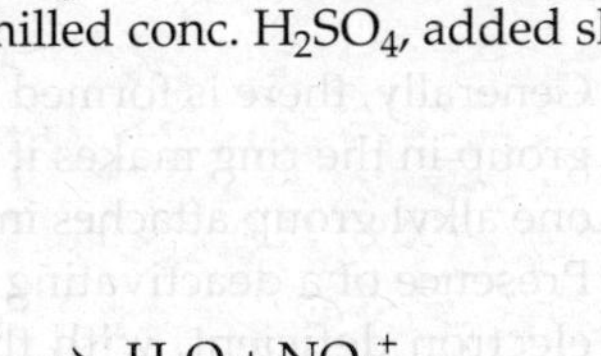

iii. Alkylation with CH_3Cl in presence of anhydrous $AlCl_3$. The role of $AlCl_3$ is to liberate the electrophile, $CH_3^{+\cdot}$

$$CH_3Cl + AlCl_3 \longrightarrow CH_3^+ + AlCl_4^-$$

In all the cases, the electrophile attacks the benzene ring and initially forms a π-complex, which changes to a σ-complex and finally it is converted to form the substituted derivative, after a proton is removed.

9. Cyclohexene will decolourise yellow colour of Br_2 water as well as a pink colour of dil. alk. solution of $KMnO_4$ (Baeyer's reagent) but benzene will not.

10. Friedel–Craft reaction is alkylation and acylation of aromatic rings, in presence of anhydrous $AlCl_3$. The reaction occurs by treating the aromatic compound with alkyl halide or an acyl halide in presence of anhydrous $AlCl_3$. The role of $AlCl_3$ is to liberate the alkyl, R^+ or an acyl carbocation, RCO^+, which then will react with the aromatic ring to form a π-complex, changing to a π-complex and finally forming the derivative.

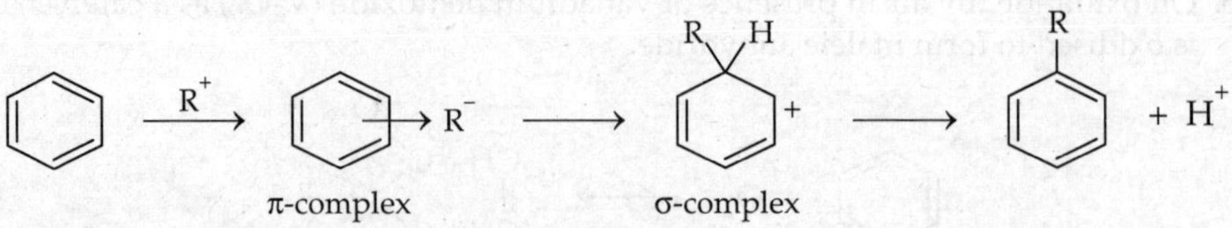

π-complex   σ-complex

Limitations of the Friedel–Craft reaction:
 i. Only one equivalent of the catalyst is required.
 ii. In alkylation, the electrophile formed generally rearranges before it attacks the ring. The alkyl group rearranges to form an alkyl group of higher stability, before it reacts with benzene group.

    For example: In alkylation with *n*-propyl chloride, the compound formed is isopropyl benzene.

$$CH_3CH_2CH_2Cl + AlCl_3 \longrightarrow CH_3CH_2CH_2^+ + AlCl_4^-$$

$$CH_3CH_2CH_2^+ \longrightarrow CH_3CH^+\!-\!CH_3$$

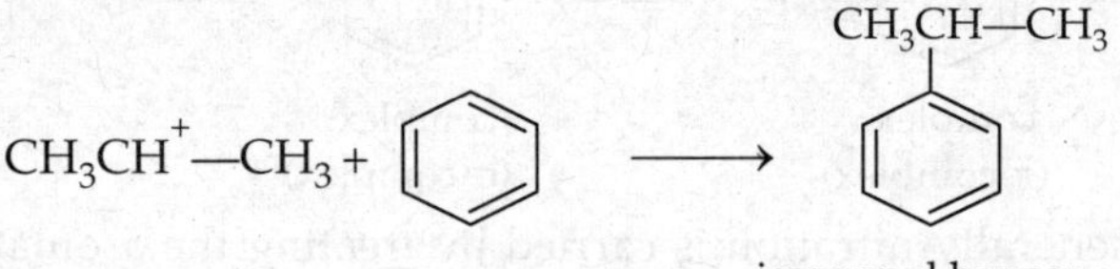

isopropyl benzene

 iii. Generally, there is formed a poly alkyl derivative, because the introduction of an alkyl group in the ring makes it more reactive for electrophilic substitution, thus more than one alkyl group attaches in the ring leading to the formation of a poly alkyl derivative.
 iv. Presence of a deactivating group, or an m-directing group in benzene makes the ring electron deficient, with this result such compounds do not undergo Friedel–Craft reaction.
11. The role played by $FeBr_3$ in bromination of benzene, is that the catalyst, $FeBr_3$ reacts with Br_2 to form a Br^+ ion, an electrophile which will react with benzene to form a bromo derivative. This is an example of F-C reaction.

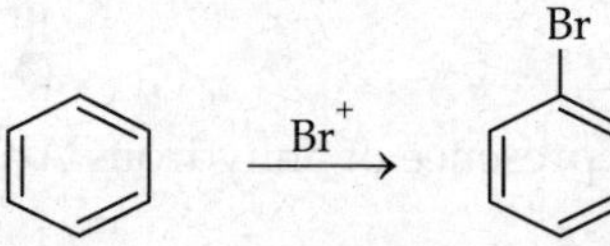

12. Benzene does not undergo Diels–Alder reaction, because benzene molecules does not contain true double bonds in the molecule.
13. i. Any group attached to the benzene ring, containing even one carbon atom, through which the group is attached to the ring is termed a side chain. On oxidation with dil. alk. $KMnO_4$, the whole of the side chain is oxidised to a single carboxylic group, (irrespective of its length) attached to the benzene ring.

    Three isomers of xylene, have their boiling points very close to each other and therefore it is difficult to identify.

    On oxidation, these are converted into the corresponding acids, which on heating form anhydride, at different temperature. o-xylene forms phthalic acid on oxidation, which on heating sublimes, loses a molecule of water to form phthalic anhydride.

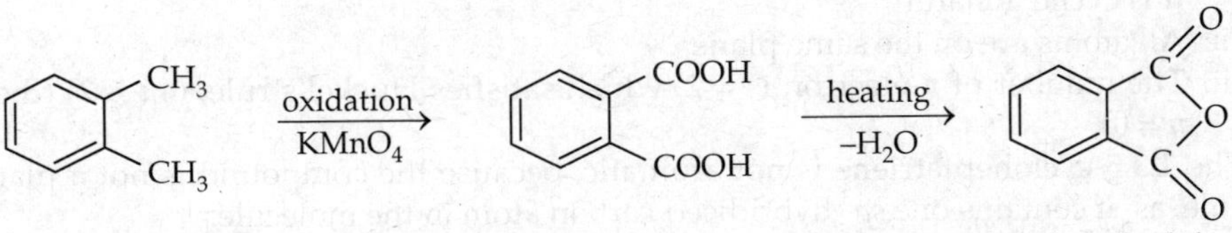

o-xylene   phthalic acid   phthalic anhydide

m-xylene forms, m–dicarboxylic acid on oxidation which on heating remains unchanged.

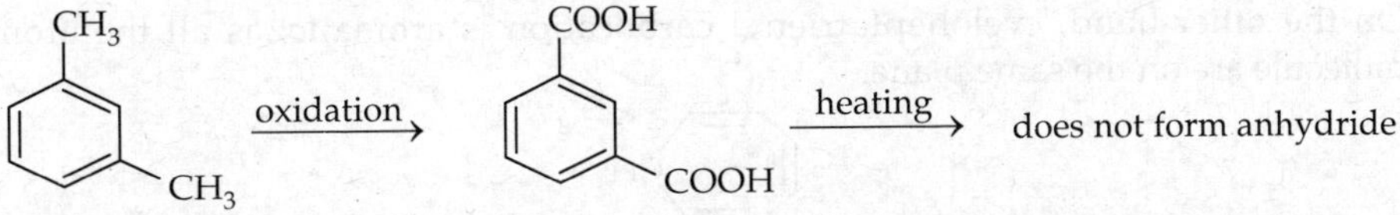

does not form anhydride

p-xylene on oxidation forms terephthalic acid which form an anhydride at a much higher temperature.

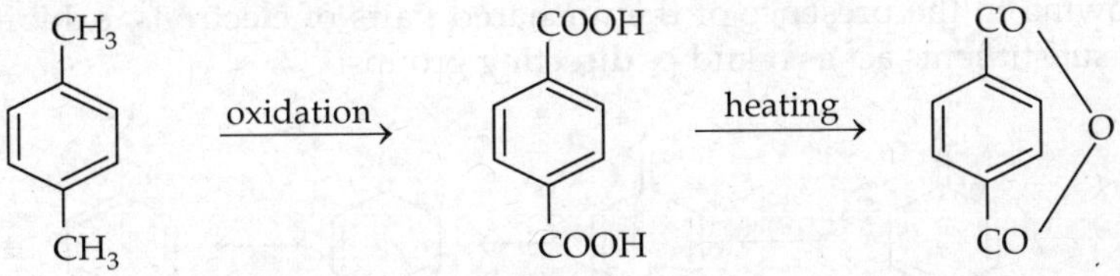

terephthalic acid

ii   Tertiary butyl benzene on oxidation with dil. alk. $KMnO_4$ does not undergo oxidation, as it does not contain the benzylic hydrogen.

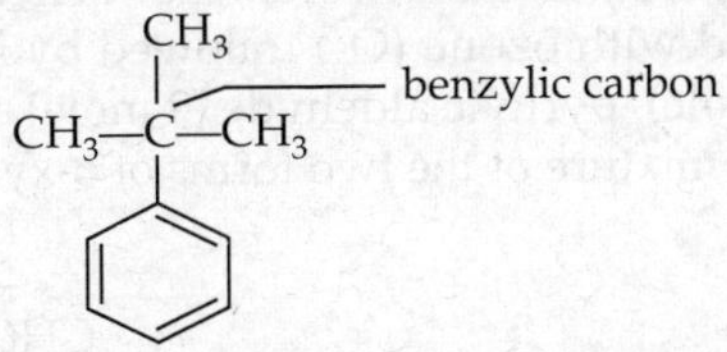

14. The side chain halogenation is a free radical halogenation. It is carried out by treating the compound with Cl_2 or Br_2 either at a high temperature or in presence of UV radiations.

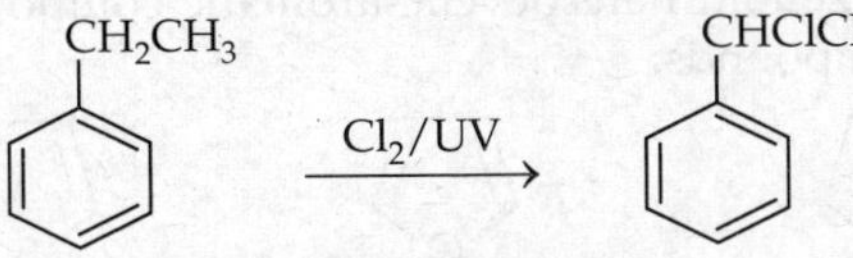

halogenation of xylene

15. The smallest species is cyclo propenyl carbocation.

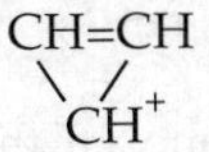

Cyclopropenyl carbocation fulfils all the conditions that are required for a cyclic compound to behave as aromatic compound.

    i. It is cyclic in nature

   ii. All atoms are on the same plane

  iii. The number of π-electron is = 2, which satisfies Huckel's rule, (4n + 2) (the value of n = 0)

16. The 1,3,5-cycloheptatriene is not aromatic, because the compound is not a planar molecule, as it contains one sp^3 hybridised carbon atom in the molecule.

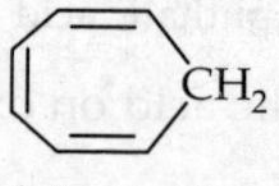

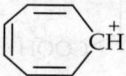

cycloheptatriene

On the other hand, cycloheptatrienyl carbocation is aromatic; as all the atoms in the molecule are on the same plane.

cycloheptatrienyl carbocation

17. The halogens are o-and p-directing groups, yet they deactivate the benzene ring-Halogen atoms, owing to the presence of the unshared pairs of electrons, exhibit resonance. The halogen substituents act as o-and p- directing groups.

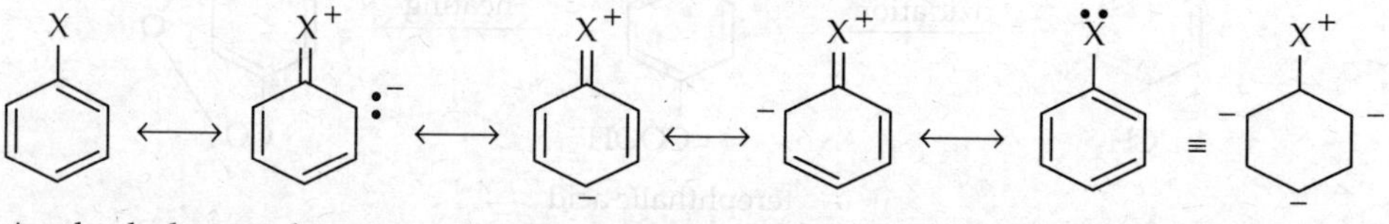

As the halogens because of resonance, act as o-and p-directing groups, its very high electronegativities are attributed to act as deactivator of the ring (the inductive effect is stronger than resonance effect). In other words, the –I effect is stronger than +M effect.

18. When xylenes are treated with ozone (O_3) followed by hydrolysis, the various products formed are diacetyl (1 mole), pyruvic aldehyde (2 mole) and glyoxal (3 mole), indicate the o-xylene is an equivalent mixture of the two forms of o-xylene.

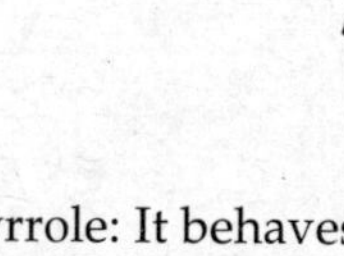

19. The following nonbenzenoid heterocyclic aromatic compounds occur in nature. These behave as aromatic compounds.

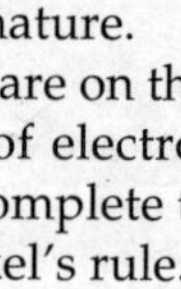

                NH   O   S

      pyrrole 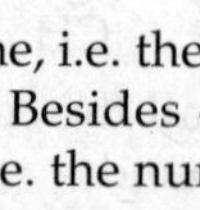   furan 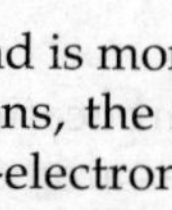   thiophene

Pyrrole: It behaves as an aromatic compound, as it fulfils all the three conditions required to behave as an aromatic compound, i.e.

• It is cyclic in nature.

• All the atoms are on the same plane, i.e. the compound is monoplanar.

• The number of electron is π = 6. Besides 4π electrons, the nitrogen contributes two electrons to complete the sextet, i.e. the number of π-electrons to be = 6, thus fulfilling thus the Huckel's rule.

However, the compound behaves as slightly acidic as lone pair on nitrogen becomes a part of the sextet.

Furan: The compound also behaves as aromatic, as it fulfils all the conditions. (Huckel's rule)

MULTIPLE CHOICE QUESTIONS

(Group-E)

1. The following reaction will proceed through

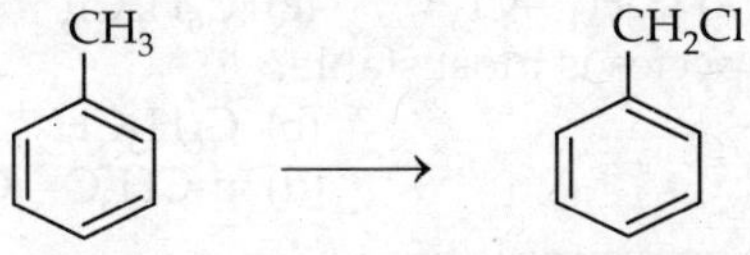

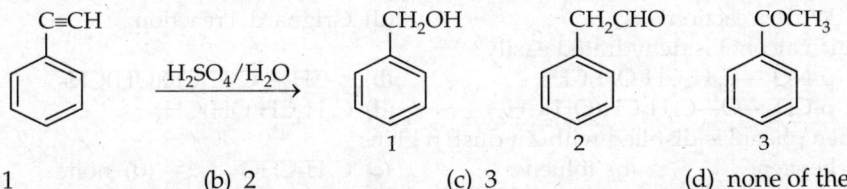

   (a) nucleophilic substitution   (b) electrophilic substitution
   (c) radical substitution   (d) involves more than one process

2. The final product of the reaction is

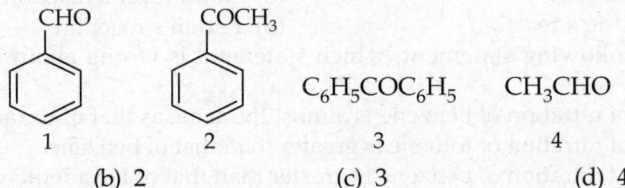

   (a) 1   (b) 2   (c) 3   (d) none of these

3. When phenol is treated with $CHCl_3$ and KOH, followed by acidification, salicyladehyde is formed. Which of the following species are formed as intermediate.

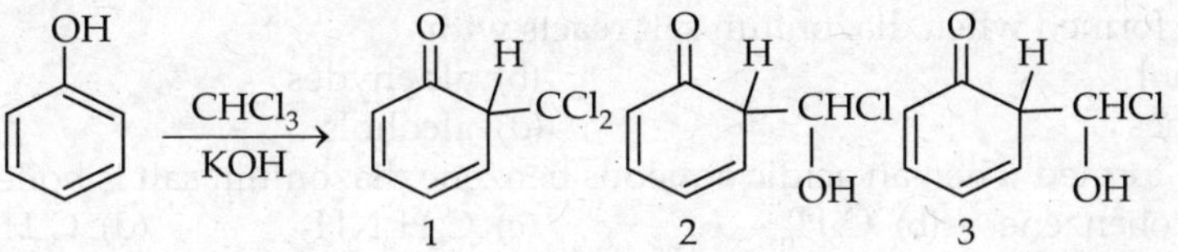

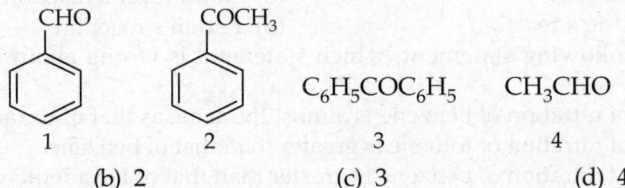

   (a) 1   (b) 2   (c) 3   (d) none

4. Which will form geometrical isomers with NH_2OH?

   $C_6H_5COC_6H_5$ CH_3CHO

   1   2   3   4

   (a) 1   (b) 2   (c) 3   (d) 4

5. Ether (C_6H_5—O—$CH_2C_6H_5$) when treated with HI gives
   (a) C_6H_5—CH_2I (b) $C_6H_5CH_2OH$
   (c) C_6H_5I (d) C_6H_5OH

6. Which of the following will not undergo aldol condensation?
   (a) acetaldehyde   (b) propionaldehye
   (c) benzaldehyde   (d) trideutero acetaldehyde

7. All the m-directing groups in the benzene ring—towards the electrophilic substitution reactions.
   (a) deactivates
   (b) activates
   (c) both activate and deactivate
   (d) none

8. o- and p-directing groups generally
   (a) activates the ring
   (b) deactivates the ring
   (c) neutral group
   (d) none

9. The end product of the reaction

$$C_6H_6 + Cl_2 \xrightarrow{\text{Sunlight}}$$

   (a) C_6H_5Cl
   (b) o-C_6H_4—Cl_2
   (c) $C_6H_6Cl_6$
   (d) p-C_6H_4Cl—CH_2^+

10. Which of the following species is most stable?
    (a) p-NO_2—$C_6H_4CH_2^+$
    (b) $C_6H_5CH_2^+$
    (c) p-Cl—$C_6H_4CH_2^+$
    (d) p-CH_3O—C_6H_4—CH_2^+

11. The following reaction is an example of

$$C_6H_6 + CH_3Cl \xrightarrow{\text{anh. AlCl}_3} C_6H_5CH_3 + HCl$$

    (a) Friedel–Craft reaction
    (b) Kolbe's synthesis
    (c) Wurtz reaction
    (d) Grignard's reaction

12. Which alcohol is dehydrated easily?
    (a) p-NO_2—$C_6H_4CH(OH)CH_3$
    (b) p-CH_3—$C_6H_4CH(OH)CH_3$
    (c) p-CH_3—O—$C_6H_4CH(OH)CH_3$
    (d) $C_6H_5CH(OH)CH_3$

13. When phenol is distilled with Zn dust, it gives
    (a) benzene
    (b) toluene
    (c) C_6H_5CHO
    (d) none

14. Among the following, which is easily sulphonated?
    (a) benzene
    (b) methoxy benzene
    (c) toluene
    (d) chlorobenzene

15. Dyes are formed when diazonium salt reacts with
    (a) phenol
    (b) aldehydes
    (c) ketones
    (d) alcohols

16. Which is formed when an acidic aqueous benzene diazonium salt is boiled with water?
    (a) chlorobenzene
    (b) C_6H_6
    (c) $C_6H_5NH_2$
    (d) C_6H_5OH

17. The reaction is known as $C_6H_5N_2Cl \xrightarrow{\text{Cu}_2\text{Cl}_2 + \text{HCl}} C_6H_5Cl + N_2$

    (a) Etard's reaction
    (b) Sandmeyer's reaction
    (c) Wurtz–Fittig's reaction
    (d) Perkin's reaction

18. Among the following statement, which statement is wrong about nitration of aromatic compounds?
    (a) The rate of nitration of benzene is almost the same as that of hexadeutero benzene.
    (b) The rate of nitration of toluene is greater than that of benzene.
    (c) The rate of nitration of benzene is greater than that of hexadeutero benzene.
    (d) Nitration is an electrophilic substitution reaction.

19.

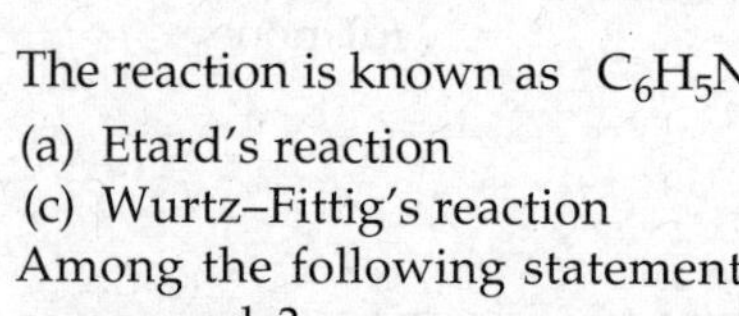

    In the above reaction the X stands for
    (a) NHOH
    (b) NH_2
    (c) OH
    (d) NHCl

20. Aromatic compounds undergo most easily
    (a) nucleophilic substitution reaction
    (b) electrophilic substitution reaction
    (c) nucleophilic addition
    (d) electrophilic addition
21. Benzyl alcohol is obtained from benzaldehyde by
    (a) Fittig's reaction
    (b) Cannizzaro's reaction
    (c) Kolbe's reaction
    (d) Wurtz reaction
22. Carbolic acid is the name used for
    (a) opium
    (b) phenol
    (c) chloroform
    (d) H_2CO_3
23. Between p-nitrophenol and salicyaldehyde, solubility in base is
    (a) almost nil in both cases
    (b) higher in nitrophenol
    (c) higher for salicyaldehyde
    (d) equal in nature
24. Which concept best explains that o-nitrophenol is more volatile than p-nitrophenol?
    (a) resonance
    (b) hydrogen bonding
    (c) hyperconjugation
    (d) steric hindrance
25. When phenol is treated with chloroform and alkali NaOH, the compound formed is salicyaldehyde. If we use pyrene, in place of chloroform, the compound formed is
    (a) salicyaldehyde
    (b) phenolphthalein
    (c) salicylic acid
    (d) cyclohexanol
26. Which of these is strongest acid?

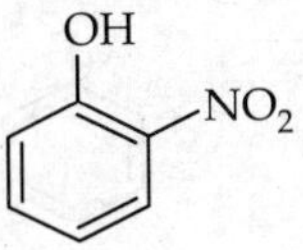

    (a) 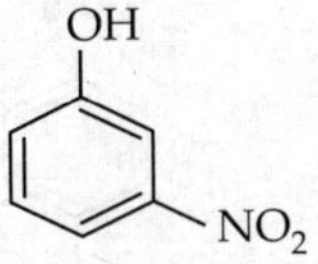   (b) 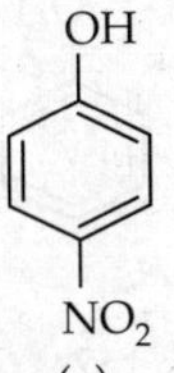   (c)    (d)

27. Phenol reacts with excess of bromine to form
    (a) 2,4,6-tribromophenol
    (b) 2-bromo phenol + HCl
    (c) 2-bromophenol + HBr
    (d) benzyl bromide + HBr
28. The oxidation of toluene to benzaldehyde, in Etards' reaction is by using
    (a) H_2O_2
    (b) Cl_2
    (c) chromium trioxide or by CrO_2Cl_2
    (d) $KMnO_4$
29. When toluene is treated with Cl_2, it forms benzyl chloride
    (a) in presence of sunlight
    (b) in absence of sunlight
    (c) treating toluene with anhydrous $AlCl_3$
    (d) treating toluene with As_2O_3
30. Benzene is converted into toluene by
    (a) Friedel–Craft reaction
    (b) Grignard's reaction
    (c) Wurtz reaction
    (d) Perkin's reaction
31. When ethyl benzoate is hydrolysed by an aqueous alkali, the product formed in the medium are
    (a) $C_6H_5COOH + C_2H_5O^-$
    (b) $C_6H_5COO^- + C_2H_5OH$
    (c) $C_2H_5OH + C_6H_5COOH$
    (d) $C_6H_5COO^- + C_2H_5O^-$
32. The most basic compound among the following is
    (a) benzyl amine
    (b) aniline
    (c) acetanilide
    (d) p-nitro aniline

33. Aniline on heating with conc. H_2SO_4 to a temperature of 200°C
    (a) aniline sulphate
    (b) benzene sulphonic acid
    (c) sulphanilic acid
    (d) none

34. The —NO_2 group in nitrobenzene is
    (a) o-directing   (b) m-directing   (c) p-directing   (d) o- and p-directing

35. What is the function of anhydrous $AlCl_3$ in Friedel–Craft reaction?
    (a) to absorb water
    (b) to absorb HCl
    (c) to produce an electrophile
    (d) to produce a nucleophile

36. Which one is more reactive towards aqueous KOH?
    (a) C_6H_5Cl (b) $C_6H_5CH_2Cl$ (c) C_6H_5Br (d) $C_6H_4Br_2$

37. The strongest acid among the following.
    (a) o-nitrophenol
    (b) p-nitrophenol
    (c) p-chloro phenol
    (d) m-nitrophenol

38. Sulphonation of benzoic acid gives mainly
    (a) o-sulphonic acid
    (b) m-sulphonic acid
    (c) p-sulphonic acid
    (d) o- and p-disulphonic acid

39. The product of oxidation of aniline with $K_2Cr_2O_7$ and conc. H_2SO_4 will be
    (a) p-aminophenol
    (b) p-benzoquinone
    (c) aniline black
    (d) phenyl hydroxylamine

40. Phenol on nitration with dil. HNO_3 at room temperature gives

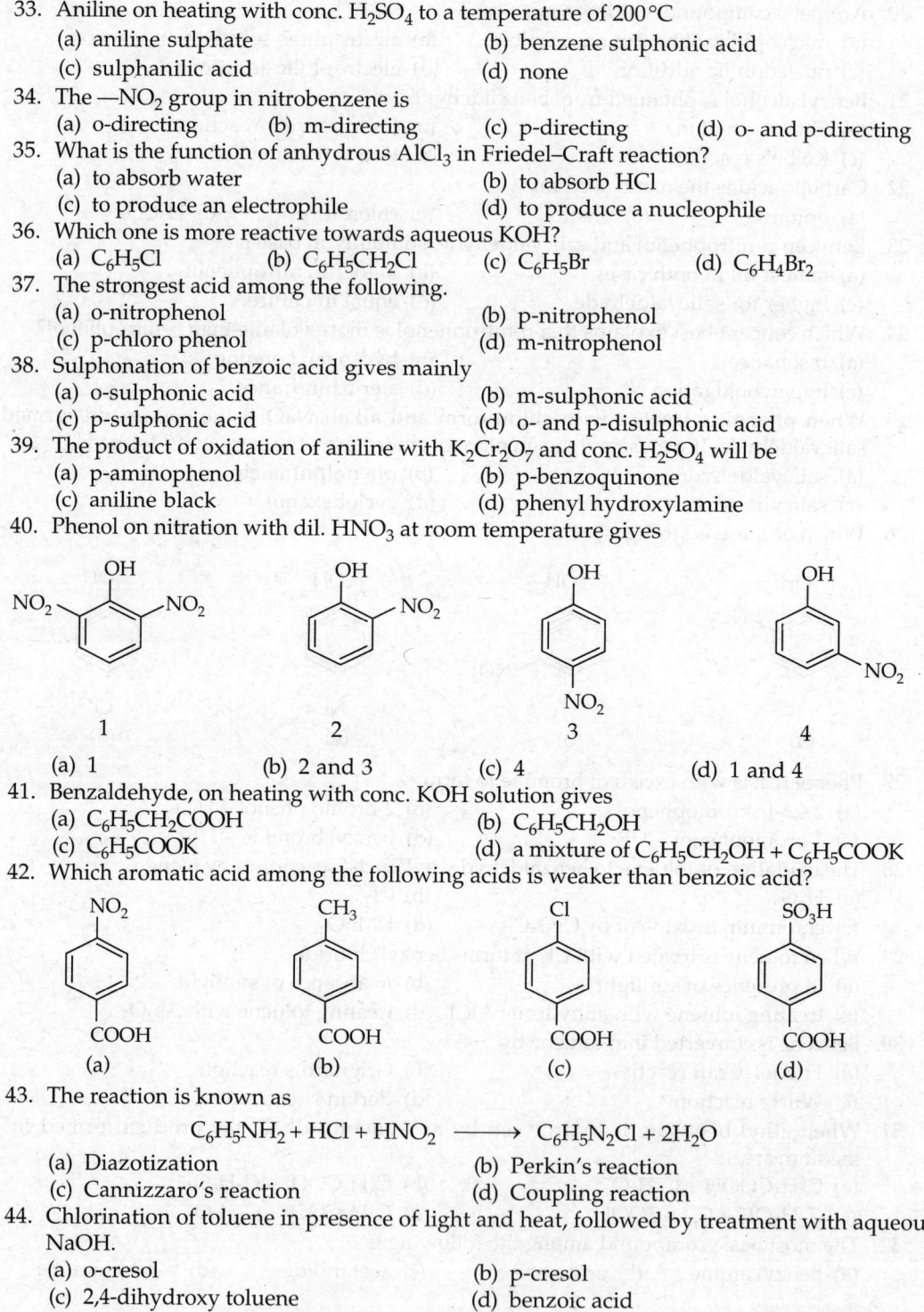

(a) 1   (b) 2 and 3   (c) 4   (d) 1 and 4

41. Benzaldehyde, on heating with conc. KOH solution gives
    (a) $C_6H_5CH_2COOH$
    (b) $C_6H_5CH_2OH$
    (c) C_6H_5COOK
    (d) a mixture of $C_6H_5CH_2OH + C_6H_5COOK$

42. Which aromatic acid among the following acids is weaker than benzoic acid?

43. The reaction is known as

$$C_6H_5NH_2 + HCl + HNO_2 \longrightarrow C_6H_5N_2Cl + 2H_2O$$

    (a) Diazotization
    (b) Perkin's reaction
    (c) Cannizzaro's reaction
    (d) Coupling reaction

44. Chlorination of toluene in presence of light and heat, followed by treatment with aqueous NaOH.
    (a) o-cresol
    (b) p-cresol
    (c) 2,4-dihydroxy toluene
    (d) benzoic acid

45. Increasing order of acid strength among p-methoxy phenol (1) p-methyl phenol (2) and p-nitrophenol (3) is
    (a) $2 > 1 > 3$　　　(b) $3 > 1 > 2$　　　(c) $1 > 2 > 3$　　　(d) $3 > 2 > 1$

46. Choose the correct statement from the ones given below.

　　　　　　　I　　　　　　　　　　II

    (a) II is not an acceptable canonical structure, because carbonium ions are less stable than ammonium ions
    (b) II is not acceptable canonical structure because, it is non aromatic
    (c) II is not acceptable canonical structure, because the nitrogen has 10 valence electrons
    (d) II is an acceptable structure

47. Benzoic acid reacts with $SOCl_2$ to give
    (a) chlorobenzene　　　　　　　　　(b) dichlorobenzene
    (c) benzoyl chloride　　　　　　　　(d) benzyl chloride

48. On boiling phenyl ethyl ether with conc. HBr, the compounds formed are
    (a) $C_6H_5OH + C_2H_5Br$　　　　　(b) bromo benzene + ethanol
    (c) phenol + ethane　　　　　　　　(d) bromo benzene + ethane

49. Benzene reacts with n-propyl chloride in presence of anhydrous $AlCl_3$ to form predominently
    (a) isopropyl benzene　　　　　　　(b) no reaction
    (c) n-propyl benzene　　　　　　　(d) 3-propyl-1-chlorobenzene

50. The correct symbol relating the two Kekule structures of benzene is
    (a) $\leftarrow$　　　(b) $\rightleftharpoons$　　　(c) $\leftrightarrow$　　　(d) $\leftrightarrows$

51. The number of double bonds in BHC are
    (a) 1　　　　　　(b) 2　　　　　　(c) 3　　　　　　(d) zero

52. Which would decolourise cold aqueous $KMnO_4$ solution?
    (a) benzoic acid　　(b) cinnamic acid　　(c) p-toluic acid　　(d) m-toluic acid

53. Which among the following is the strongest acid?
    (a) benzoic acid　　　　　　　　　(b) methoxy benzoic acid
    (c) m-nitro benzoic acid　　　　　　(d) p-nitro benzoic acid

54. Chlorobenzene forms DDT, when it reacts with
    (a) phenol　　(b) naphthalene　　(c) chloral　　(d) CH_3CHO

55. Reimer–Tiemann reaction involves
    (a) a carbonium ion　　　　　　　(b) carbene intermediate
    (c) carbanion intermediate　　　　　(d) a free radical intermediate

56. Toluene can be oxidised to benzoic acid by
    (a) acidic $KMnO_4$　　(b) alk. $K_2Cr_2O_7$　　(c) any of these　　(d) none of these

57. Which statement is wrong about CH_3CHO and C_6H_5CHO?
    (a) both react with hydroxylamine to form oxime
    (b) both react with HCN to form cyanohydrin
    (c) both react with NaOH to form polymer
    (d) both react with hydrazine to form hydrazone

58. Among the following compounds which is most reactive towards nitration?
    (a) benzene　　　　　　　　　　(b) nitrobenzene
    (c) toluene　　　　　　　　　　(d) chlorobenzene

59. When chlorine is passed through a boiling toluene, in presence of sunlight, the reaction take place
    (a) hydrogen is substituted by chlorine in the nucleus
    (b) hydrogen is substituted by chlorine in the side chain
    (c) side chain is oxidised
    (d) nucleus is reduced

60. Benzoyl chloride is formed benzoic acid reacts with
    (a) $SOCl_2$ (b) PCl_5 (c) PCl_3 (d) all of these

61. $C_6H_5CH_2Cl$ can be prepared from toluene by chlorination with
    (a) SO_2Cl_2 (b) $SOCl_2$ (c) S_2Cl_2 (d) $NaOCl$

62. $C_8H_6O_4 \longrightarrow X \xrightarrow{NH_3}$ (benzene ring with CONH₂, CONH₂) , X is
    (a) phthalic anhydride
    (b) phthalic acid
    (c) o-xylene
    (d) benzoic acid

63. Which of the following is used in Friedel–Craft acetylation reaction?
    (a) $(CH_3CO)_2O$ (b) CH_3CH_2Cl (c) CH_3COOCH_3 (d) CH_3Cl

64. In chlorination of benzene, the chlorinating agent is
    (a) Cl_2 (b) Cl^- (c) Cl^+ (d) HCl

65. Phenol is less acidic than
    (a) water
    (b) p-methoxy phenol
    (c) p-nitrophenol
    (d) ethanol

66. When aniline is treated with $NaNO_2$ and dil. HCl at $-5°C$ the compound formed is
    (a) nitro benzene
    (b) benzene diazonium chloride
    (c) benzene
    (d) nitro aniline

67. Schiff's bases are formed when aniline is condensed with
    (a) phenols
    (b) aromatic aldehydes
    (c) aryl chloride
    (d) aliphatic alcohols

68. Tautomerism is not exhibited by which of the following compounds.

        (a)   (b)   (c)   (d)

69. Phenolphthalein is formed on heating phenol with conc. H_2SO_4 and
    (a) salicylic acid   (b) phthalic acid   (c) phenacetin   (d) chlorobenzene

70. Which can be used to distinguish between aniline and benzyl amine?
    (a) Diazotisation followed by coupling with phenol
    (b) Carbylamine reaction
    (c) Reimer–Tiemann reaction
    (d) none

71. Aryl halides are less reactive toward nucleophilic substitution as compared to alkyl halides due to
    (a) formation of less stable carbocation   (b) resonance stabilization
    (c) longer carbon–halogen bond   (d) inductive effect

72. A diazonium salt reacts with phenol to give an azo dye. This reaction is known as
    (a) diazotisation
    (b) condensation
    (c) coupling
    (d) reduction

73. X $\longrightarrow$ benzo tri chloride $\xrightarrow{\text{hydrolysis}}$ Y
    What are X and Y?
    (a) toluene and benzoic acid   (b) toluene and benzyl alcohol
    (c) benzene and benzaldehyde   (d) none

74. Which of the following will show a negative test with phenyl hydrazine?
    (a) glucose   (b) ethanol   (c) acetaldehyde   (d) benzophenone

75. Which of following reaction takes place when a mixture of conc. HNO_3 and conc. H_2SO_4, react with benzene?
    (a) sulphonation   (b) nitration   (c) hydrogenation   (d) dehydration

76. When nitrobenzene is treated with Br_2, in presence of $FeBr_3$, the major product formed is m-nitro bromobenzene. Statements which are related to the m-isomer are
    (a) the relative density on the m-carbon is more than that of o- and p-position
    (b) loss of aromaticity when Br^+ attacks at the o-and p-positions and not at m-positions
    (c) easier loss of a H^+ to regain the aromaticity from the m-position than o-and p-positions
    (d) none

77. The possible number of tri substituted derivatives of benzene is
    (a) 2   (b) 3   (c) 4   (d) 6

78. An aromatic ether cannot be decomposed by HI, even at 525 °K. The compound is
    (a) $C_6H_5OCH_3$ (b) $C_6H_5OC_6H_5$ (c) $C_6H_5OC_3H_7$ (d) tetrahydrofuran

79. Salicylic acid, as compared to benzoic acid
    (a) is more acidic   (b) is less acidic
    (c) with same acidity   (d) none of these

80. 3-chloro-4-methyl-benzene sulphonic acid on steam distillation forms
    (a) m-chloro benzene sulphonic acid   (b) p-methyl benzene sulphonic acid
    (c) toluene   (d) o-chloro toluene

81. p-nitro benzaldehyde reacts with conc. KOH at room temperature, to give
    (a) p-nitro benzamide
    (b) p-nitro benzyl alcohol and p-nitro benzoic acid
    (c) benzaldehyde
    (d) p-nitro toluene

82. Which compound is known as oil of wintergreen?
    (a) phenyl benzoate   (b) phenyl salicylate   (c) phenyl acetate   (d) methyl salicylate

83. When salicylic acid is heated with Zn dust gives
    (a) phenol   (b) salicylaldehyde   (c) benzene   (d) benzoic acid

84. In the following reaction the catalyst used is

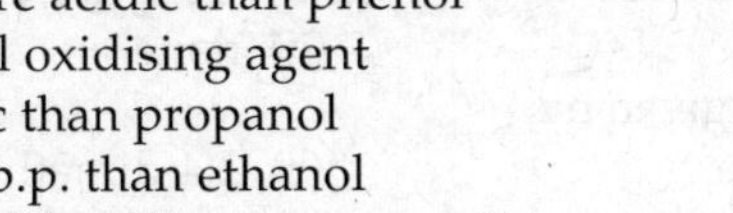

    (a) Cr_2O_3 (b) Al_2O_3 (c) Zn dust   (d) $Cr_2O_3 + Al_2O_3$

85. Which is a correct statement?
    (a) Benzyl alcohol is more acidic than phenol
    (b) Ethanol is a powerful oxidising agent
    (c) Phenol is more acidic than propanol
    (d) Ethane has a higher b.p. than ethanol

86. Which is the most powerful m-directing group?
    (a) —NO_2 (b) —SO_3H (c) —CHO   (d) —COOH

87. Which of the following deactivates the ring?
    (a) —NH_2R (b) —OH   (c) —OR   (d) —COOR

88. A new C to C bond is formed in
    (a) Cannizzaro's reaction
    (b) Friedel–Craft reaction
    (c) Clemmensen's reaction
    (d) None

89. In a reaction, involving a ring substitution of C_6H_5Y, the major compound formed is a m-isomer, the Y group is
    (a) $-NH_2$ (b) $-COOH$ (c) $-CH_3$ (d) $-Cl$

90. Which of the following compounds reacts more slowly than benzene in electrophilic bromination?
    (a) $C_6H_5NO_2$ (b) $C_6H_5NH_2$ (c) C_6H_5OH (d) $C_6H_5CH_3$

91. Which of the following structures correspond to the product expected when excess of benzene reacts with CH_2Cl_2 in presence of anhydrous $AlCl_3$?

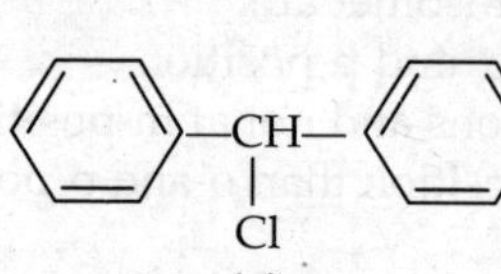

    (a) 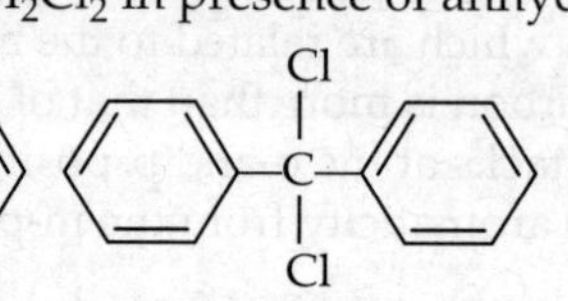   (b) 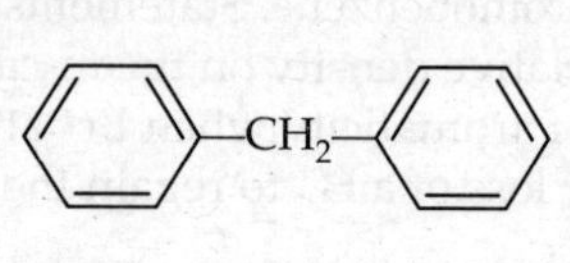   (c) 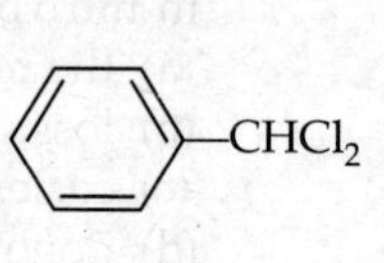   (d)

92. A mixture of aniline and benzene can be separated by
    (a) alcohol   (b) dil. HCl   (c) dil. NaOH   (d) hot water

93. Which reaction can produce R—CO—Ar?
    (a) Ar CO Cl + Ar H   (b) $COCl_2$ + RMgX   (c) R COCl + ArH   (d) RH + CrO_3

94. The reduction of benzyl chloride with H_2/Pd + $BaSO_4$ gives
    (a) C_6H_5CHO (b) $C_6H_5CH_2OH$ (c) C_6H_5COOH (d) $C_6H_5CH_2CN$

95. The major product of the reaction of m-dinitrobenzene with yellow ammonium sulphide is

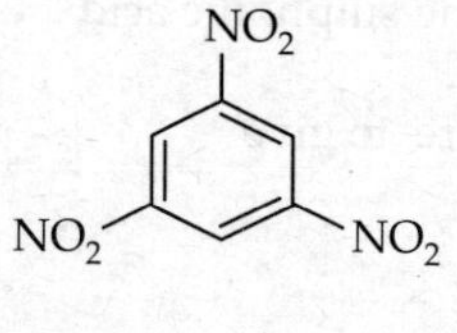

    (a) 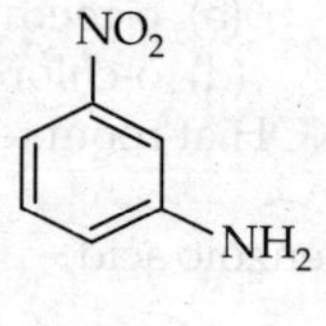   (b) 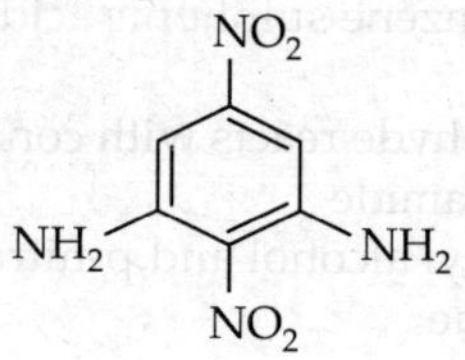   (c) 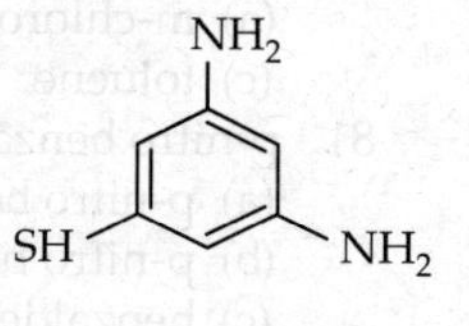   (d)

96. The reaction product of C_6H_5—O—CH_3 + HI $\xrightarrow{\text{heating}}$ is
    (a) $C_6H_5OH + CH_3I$
    (b) $C_6H_5I + CH_3OH$
    (c) $C_6H_5CH_3 + HOI$
    (d) $C_6H_6 + CH_3OI$

97. Arrange in decreasing trend towards S_E reaction

C_6H_5Cl	C_6H_6	$C_6H_5NH_3^+$	$C_6H_5CH_3$
I	II	III	IV

    (a) II > I > III > IV   (b) III > I > II > IV   (c) IV > II > I > III   (d) I > II > III > V

98. The compound formed in a reaction of cinnamic acid with HBr
    (a) C_6H_5CHBr—CH_2—COOH
    (b) $C_6H_5CH_2CHBrCOOH$
    (c) no reaction
    (d) none

99. A +ve carbylamine test is given by
    (a) N, N-dimethyl aniline
    (b) 2,4-dimethyl aniline
    (c) N-methyl-o-methyl aniline
    (d) p-methyl aniline

100. Nitrobenzene can be prepared by using a nitrating mixture consisting of conc. HNO_3 and conc. H_2SO_4. In the nitrating mixture HNO_3 acts as a
    (a) base   (b) acid   (c) reducing agent   (d) a catalyst

101. Which will be attacked by an electrophile most easily?

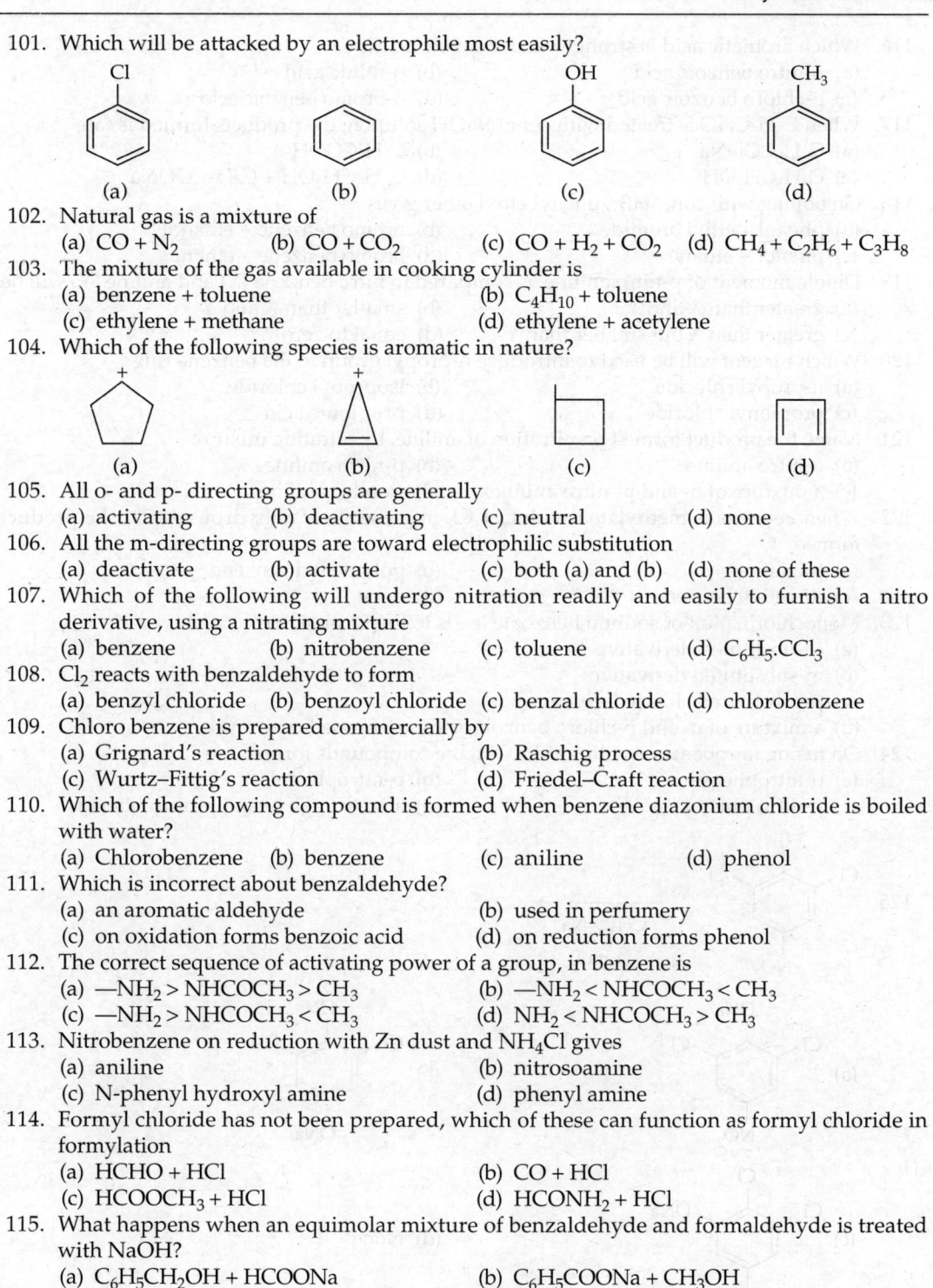

    (a)   (b)   (c)   (d)

102. Natural gas is a mixture of
    (a) $CO + N_2$
    (b) $CO + CO_2$
    (c) $CO + H_2 + CO_2$
    (d) $CH_4 + C_2H_6 + C_3H_8$

103. The mixture of the gas available in cooking cylinder is
    (a) benzene + toluene
    (b) C_4H_{10} + toluene
    (c) ethylene + methane
    (d) ethylene + acetylene

104. Which of the following species is aromatic in nature?

    (a)   (b)   (c)   (d)

105. All o- and p- directing groups are generally
    (a) activating
    (b) deactivating
    (c) neutral
    (d) none

106. All the m-directing groups are toward electrophilic substitution
    (a) deactivate
    (b) activate
    (c) both (a) and (b)
    (d) none of these

107. Which of the following will undergo nitration readily and easily to furnish a nitro derivative, using a nitrating mixture
    (a) benzene
    (b) nitrobenzene
    (c) toluene
    (d) $C_6H_5.CCl_3$

108. Cl_2 reacts with benzaldehyde to form
    (a) benzyl chloride
    (b) benzoyl chloride
    (c) benzal chloride
    (d) chlorobenzene

109. Chloro benzene is prepared commercially by
    (a) Grignard's reaction
    (b) Raschig process
    (c) Wurtz–Fittig's reaction
    (d) Friedel–Craft reaction

110. Which of the following compound is formed when benzene diazonium chloride is boiled with water?
    (a) Chlorobenzene
    (b) benzene
    (c) aniline
    (d) phenol

111. Which is incorrect about benzaldehyde?
    (a) an aromatic aldehyde
    (b) used in perfumery
    (c) on oxidation forms benzoic acid
    (d) on reduction forms phenol

112. The correct sequence of activating power of a group, in benzene is
    (a) $-NH_2 > NHCOCH_3 > CH_3$
    (b) $-NH_2 < NHCOCH_3 < CH_3$
    (c) $-NH_2 > NHCOCH_3 < CH_3$
    (d) $NH_2 < NHCOCH_3 > CH_3$

113. Nitrobenzene on reduction with Zn dust and NH_4Cl gives
    (a) aniline
    (b) nitrosoamine
    (c) N-phenyl hydroxyl amine
    (d) phenyl amine

114. Formyl chloride has not been prepared, which of these can function as formyl chloride in formylation
    (a) $HCHO + HCl$
    (b) $CO + HCl$
    (c) $HCOOCH_3 + HCl$
    (d) $HCONH_2 + HCl$

115. What happens when an equimolar mixture of benzaldehyde and formaldehyde is treated with NaOH?
    (a) $C_6H_5CH_2OH + HCOONa$
    (b) $C_6H_5COONa + CH_3OH$
    (c) $C_6H_5CH_2COONa$
    (d) $CH_3COOH + CH_3ON$

116. Which aromatic acid is stronger than benzoic acid?
    (a) p-nitro benzoic acid
    (b) p-toluic acid
    (c) p-chloro benzoic acid
    (d) p-bromo benzoic acid
117. When C_6H_5CHO is treated with conc. NaOH solution, the products formed is/are
    (a) C_6H_5COONa
    (b) C_6H_5COOH
    (c) $C_6H_5CH_2OH$
    (d) $C_6H_5CH_2OH + C_6H_5COONa$
118. On boiling with conc. HBr, phenyl ethyl ether gives
    (a) phenol + ethyl bromide
    (b) bromo benzene + ethanol
    (c) phenol + ethane
    (d) bromo benzene + ethane
119. Dipole moment of p-nitro aniline as compared to nitro benzene (X) and aniline (Y) will be
    (a) greater than X and Y
    (b) smaller than X and Y
    (c) greater than X but smaller than Y
    (d) equal to zero
120. Which reagent will be used to introduce n-propyl group in the benzene ring?
    (a) n-propyl chloride
    (b) isopropyl chloride
    (c) propionyl chloride
    (d) propionic acid
121. Name the product formed on nitration of aniline, by nitrating mixture
    (a) o-nitro aniline
    (b) p-nitro aniline
    (c) a mixture of o- and p- nitro anilines
    (d) m-nitro aniline
122. When benzene is methylated with CH_3Cl, in presence of anhydrous $AlCl_3$, the products formed
    (a) toluene
    (b) polymethyl benzene
    (c) dimethyl benzene
    (d) none
123. Monochlorination of sodium benzoate leads to the formation of
    (a) o-substituted derivative
    (b) m-substituted derivative
    (c) p-substituted chloro derivative
    (d) a mixture of o- and p-chloro benzoic acids are formed
124. On fusing nitrobenzene with solid KOH, the compounds formed are
    (a) o-nitrophenol
    (b) p-nitrophenol
    (c) p-nitro phenol
    (d) a mixture of o- and p-nitro phenols
125.

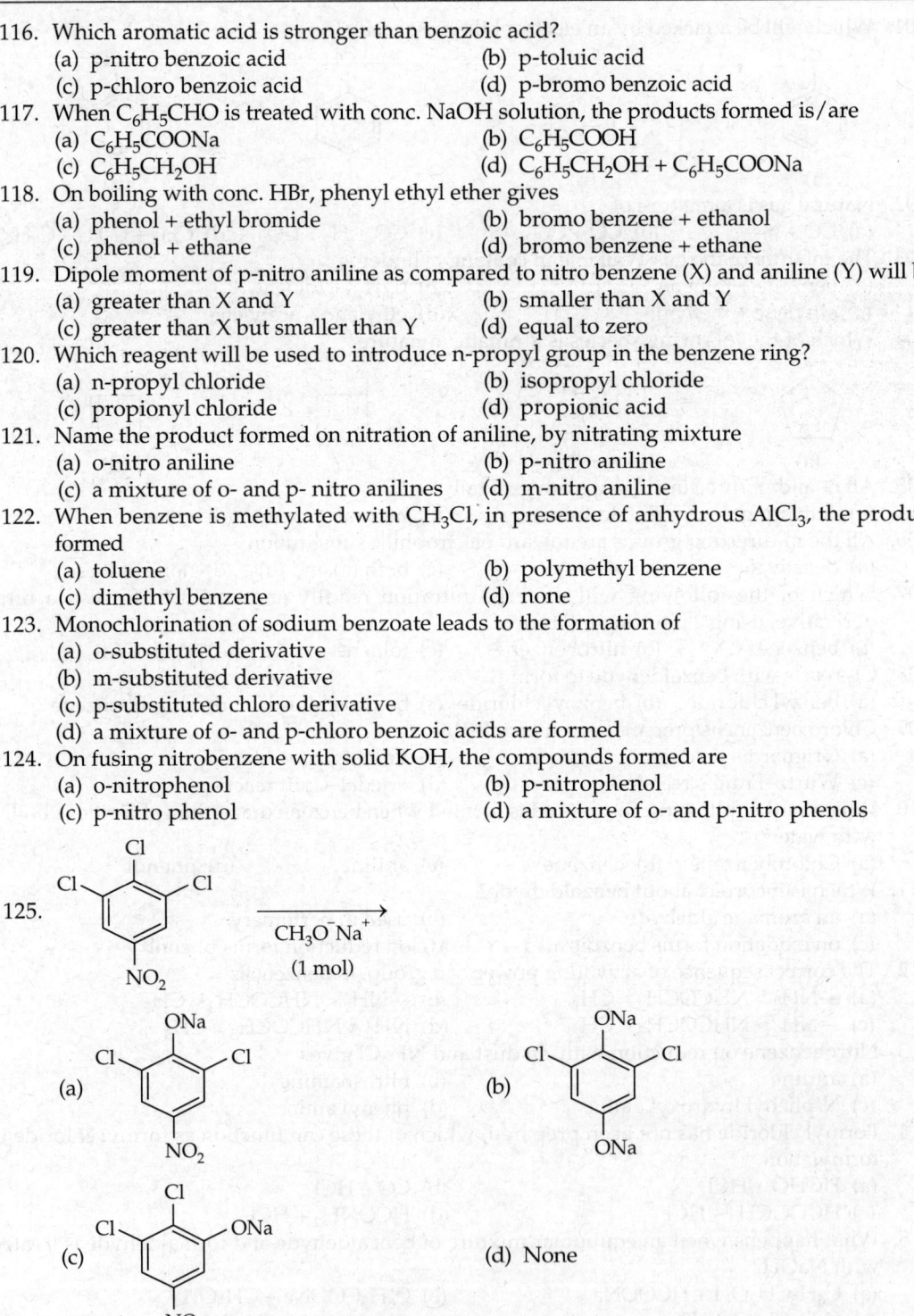

126. The compound formed in the reaction

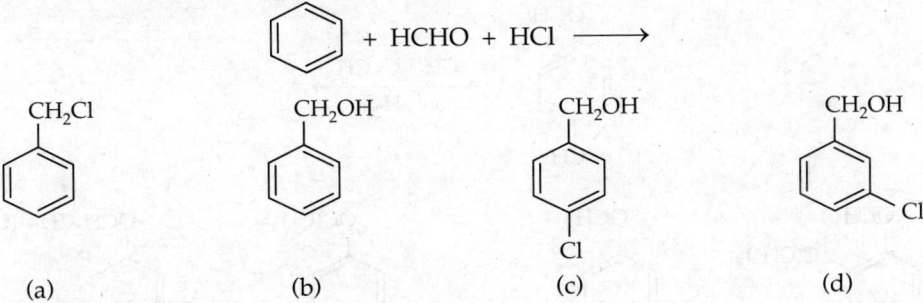

| (a) | (b) | (c) | (d) |

127. The correct order of stability of the ions is

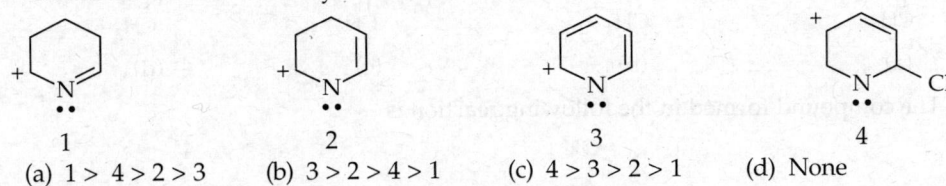

1   2   3   4

(a) $1 > 4 > 2 > 3$ (b) $3 > 2 > 4 > 1$ (c) $4 > 3 > 2 > 1$ (d) None

128. The order of stability of the following compounds

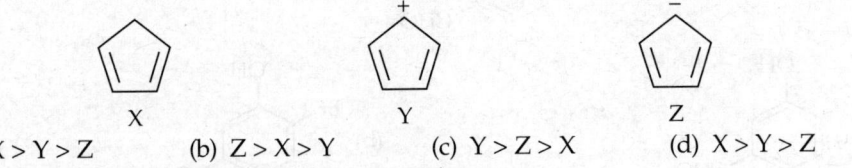

X   Y   Z

(a) $X > Y > Z$ (b) $Z > X > Y$ (c) $Y > Z > X$ (d) $X > Y > Z$

129. Which of the following compound is antiaromatic?

| (a) | (b) | (c) | (d) |

130. The decreasing order of reactivity towards electrophilic substitution is

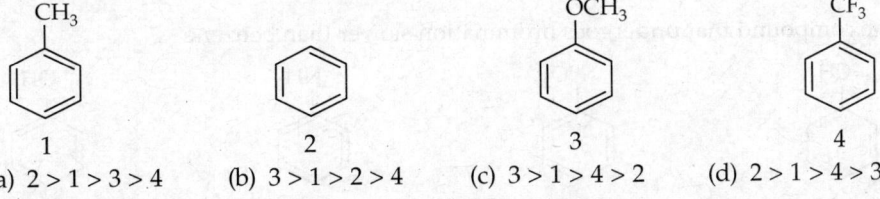

1   2   3   4

(a) $2 > 1 > 3 > 4$ (b) $3 > 1 > 2 > 4$ (c) $3 > 1 > 4 > 2$ (d) $2 > 1 > 4 > 3$

131. The reactivity towards the halonium ion (halogen ion) is

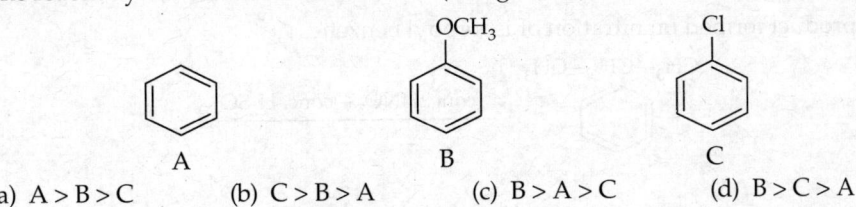

A   B   C

(a) $A > B > C$ (b) $C > B > A$ (c) $B > A > C$ (d) $B > C > A$

132. What are the products formed in the following reaction?

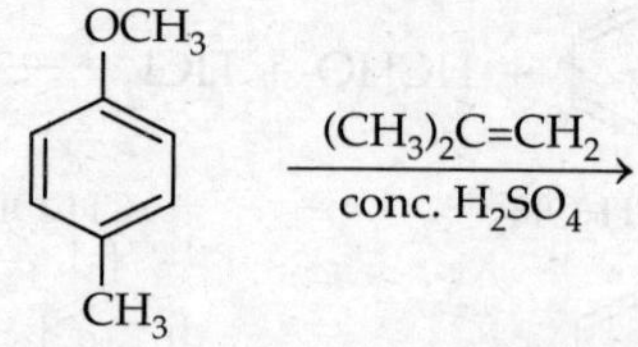

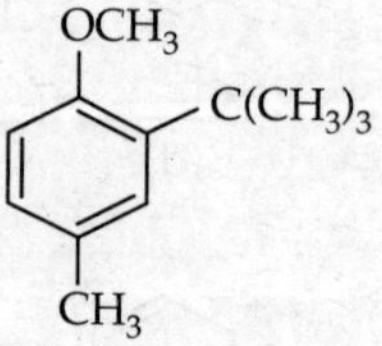

(a)

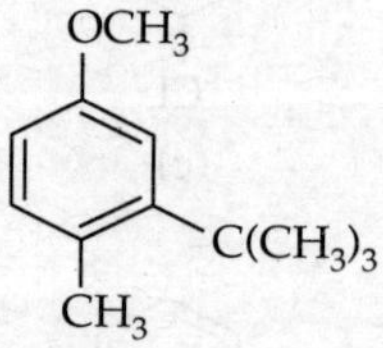

(b)

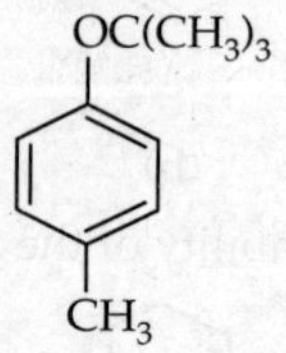

(c)

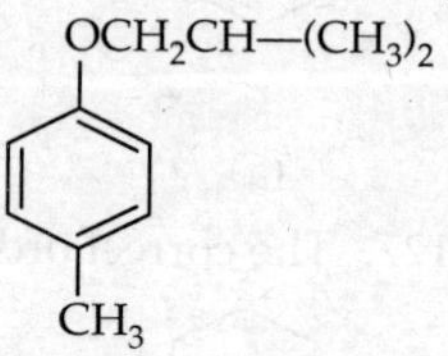

(d)

133. The compound formed in the following reaction is

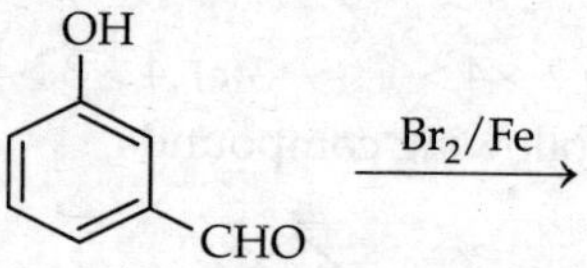

(a)

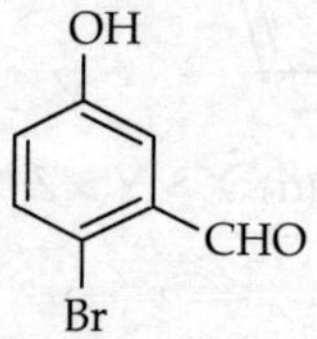

(b)

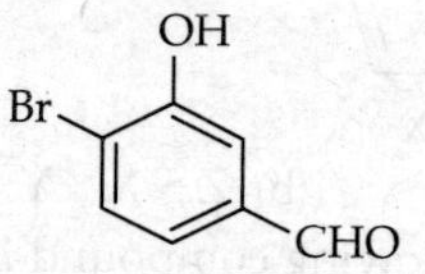

(c)

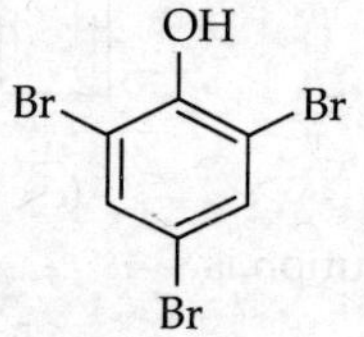

(d) None

134. The compound that undergoes bromination slower than benzene

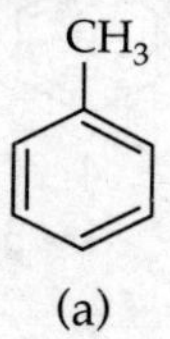

(a)

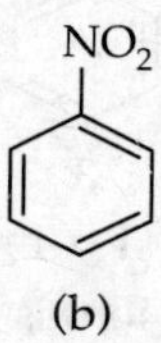

(b)

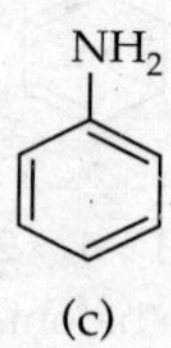

(c)

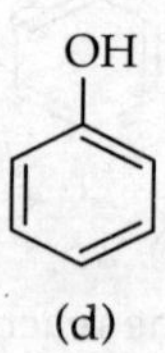

(d)

135. The product formed on nitration of isopropyl benzene

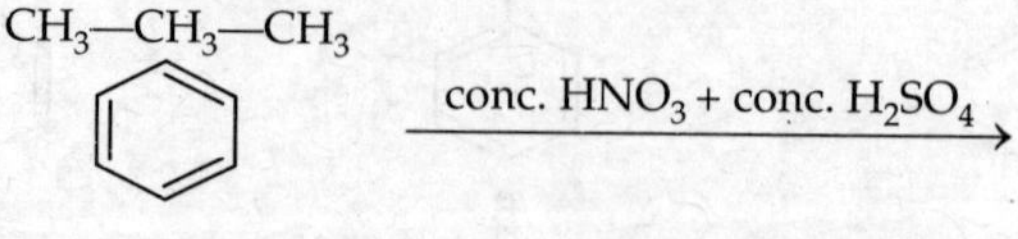

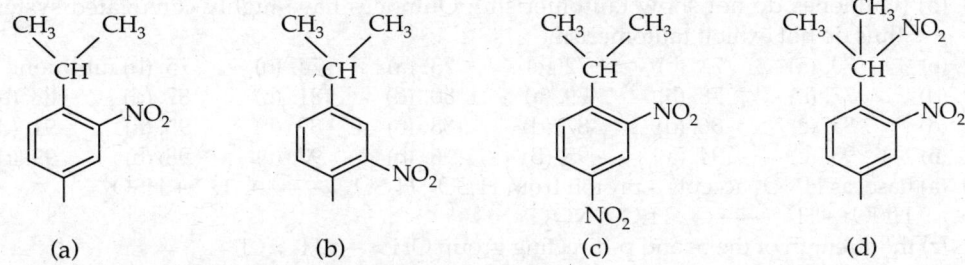

(a)   (b)   (c)   (d)

136. In Friedel–Craft acylation reaction the electrophile is
   (a) R^+ (b) $R-C^+O$ (c) $R-CO$ (d) none

137. Arrange the following compound in the decreasing reactivity for chlorination or bromination.

$N+(CH_3)_3$ $CH_2N^+(CH_3)_3$ $N(CH_3)_2$ CH_3

(i)   (ii)   (iii)   (iv)

   (a) iii > iv > ii > i   (b) iii > iv > i > ii   (c) i > iii > iv > ii   (d) iv > iii > ii > i

138. The F—C reaction of benzene gives mixture of 1 and 1,4-disubstitute derivative because

(a) mono alkylation further activates the ring
(b) the reaction is bimolecular
(c) the t-butyl ion is big, therefore it favours p-substitution
(d) the disubstitution is preferred to monosubstitution

ANSWERS

(Group-E)

1. (c) free radical substitution   2. (c)   3. (c)   4. (a) aldehyde   5. (c)
6. (a) benzaldehyde   7. (a)   8. (a)   9. (c)   10. (a)   11. (a)   12. (c)
13. (a)   14. (b)   15. (a)   16. (d)   17. (b)
18. (a) reactivity of hexadeutero benzene is lesser than benzene (due to isotopic effect)
19. NH_2 20. (b)   21. (b)   22. (b)   23. (b)   24. (b)   25. (a)   26. (c)
27. (a)   28. (b)   29. (a)   30. (a)   31. (b)   32. (a)   33. (c)   34. (b)
35. (c)   36. (b)   37. (b)   38. (b)   39. (c)   40. (b)   41. (d)   42. (b)
43. (a)   44. (d)   45. (a)   46. (c)   47. (c)   48. (a)   49. (a)   50. (d)
51. (d)   52. (b)   53. (d)   54. (c)   55. (b)   56. (a)   57. (c)   58. (c)
59. (b)   60. (d)   61. (a)   62. (b)   63. (a)   64. (c)   65. (c)   66. (b)
67. (b)

68. (b) (Quinones do not show tautomerism). Quinones have highly conjugated system and thus do not exhibit tautomerism

69. (b)   70. (a)   71. (b)   72. (c)   73. (a)   74. (b)   75. (b) nitration
76. (d)   77. (b)   78. (b)   79. (a)   80. (d)   81. (b)   82. (d)   83. (b)
84. (d)   85. (c)   86. (d)   87. (d)   88. (b)   89. (b)   90. (a)   91. (d)
92. (b)   93. (c)   94. (a)   95. (B)   96. (b)   97. (c)   98. (b)   99. (d)

100. (a) base, as HNO_3 accepts a proton from H_2SO_4; $H_2SO_4 \longrightarrow H^+ + HSO_4^-$
$$HNO_3 + H^+ \longrightarrow H_2O + NO_2^+$$

101. (c) the strength of the o- and p-directing group $OH > —CH_3 > Cl$.

102. (d) mostly methane

103. (b) mainly it contains butane + isobutane + propane

104. (b) cyclo propenyl cation is an aromatic species

105. (a)   106. (a)   107. (c)   108. (c)   109. (b)   110. (d)   111. (d)   112. (a)
113. (c)   114. (b)   115. (b)   116. (b)   117. (a)   118. (a)

119. (a)

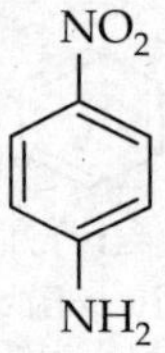

NO$_2$ is an electron withdrawing; —NH$_2$ is electron donating, both are acting in the same direction

120. (c)

121. (d) in presence of an acid, —NH$_2$ group changes to —NH$_3^+$; the group becomes electron deficient, thus it gives a m-nitro aniline

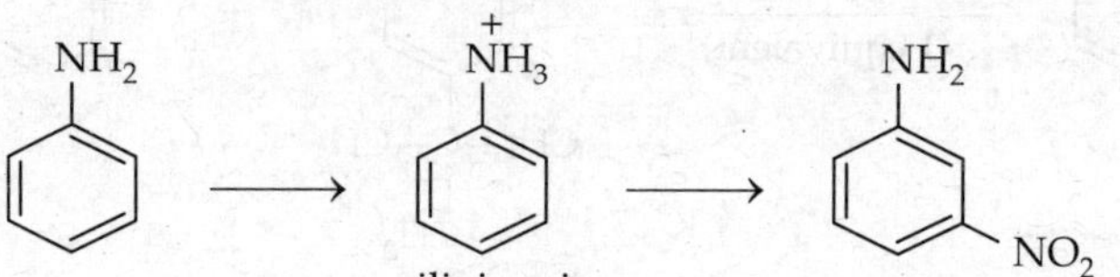

anilinium ion

122. (c)

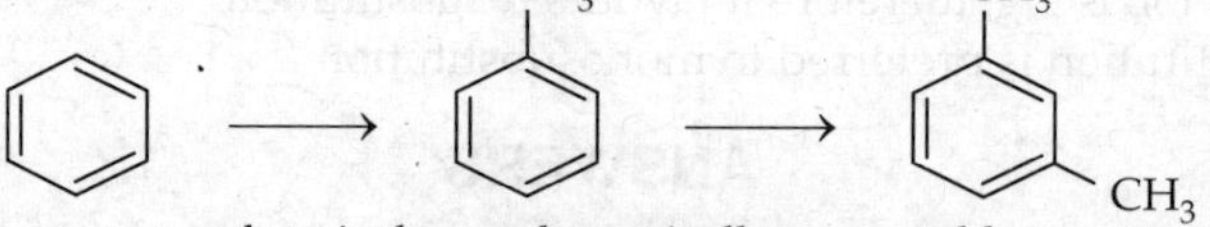

m-xylene is thermodynamically more stable

123. (d) (—COONa) group becomes o- and p- directing)

124. (d) the NO_2 group is a m-directing group, it deactivates the benzene ring. As the attacking group is a nucleophile, it attacks the o- and p-positions, to form o-and p-nitro phenols. It may be mentioned here that a —NO_2 goup is m-directing only in a electrophilic substitution reaction. It is a rare nucleophilic substitution reaction.

nucleophile attacks at electron
deficient positions

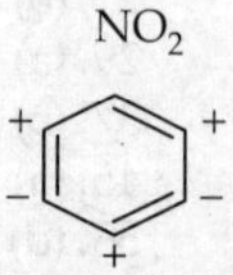

electrophile attacks at electron
rich positions

All the groups, when attached to benzene ring can be divided in two groups.

The o- and p-directing groups activate the ring and as a consequence, the o- and p-positions begin to have higher electron density at o- and p-positions an electrophile then attacks the o- and p-positions to give a mixture of o- and p-substituted derivatives. The m-directing groups deactivate the benzene ring. The o- and p-positions become electron deficient. In an electrophilic attack, the electrophile is compelled to attack the m-positions. (as compared to the o- and p-positions, the m-positions are more –ve)

Nitro group is an –m-directing group, in the electrophilic substitution, the attacking group is –vely charged or in other words OH^- group is a nucleophile.

Whereas, the m-directing groups deactivate the benzene ring, and therefore the attacking electrophile attacks the electron rich positions, the nucleophile attacks the electron deficient positions.

In this case the attacking group is a nucleophile, it attacks the electron deficient positions and converts nitrobenzene to give a mixture of o-nitro and p-nitro phenols.

It is a rare example of nucleophilic substitution reaction.

125. (a)	126. (b)	127. (a)	128. (b)	129. (d)	130. (d)	131. (d)	132. (a)
133. (b)	134. (b)	135. (b)	136. (b)	137. (b)	138. (d)		

ADDITIONAL MCQs ON HYDROCARBONS

(Group-F)

1. Which of the following molecule may be used as an electrophile as well as a nucleophile.
   (a) CH_3NH_2
   (b) CH_3Cl
   (c) CH_3CN
   (d) CH_3OH

2. Which of the having a +I effect?
   (a) OH–
   (b) OCH_3
   (c) –COOH
   (d) –CH_3

3. Pick out a group having a –I effect.
   (a) –$N(CH_3)_3^+$
   (b) –NH_3^+
   (c) –$S(CH_3)_2$
   (d) –F

4. $CH_3CH_2CH_2^+$ is less stable than
   (a) $(CH_3)_3C^+$
   (b) $(CH_3)_2CH_2^+$
   (c) $CH_3CH_2^+$
   (d) CH_3^+

5. The strongest base is
   (a) $(CH_3)_3N$
   (b) $(C_6H_5)_3N$
   (c) $NH_2-\overset{\overset{O}{\|}}{C}-NH_2$
   (d) $(CH_3)_2NH$

6. Arrange the following species in DO of their nucleophilicity:
   (1) OH–
   (2) CH_3O-
   (3) CH_3-
   (4) NH_2-
   (a) $1 > 2 > 3 > 4$
   (b) $3 > 4 > 2 > 1$
   (c) $3 > 2 > 1 > 4$
   (d) $2 > 3 > 1 > 4$

7. Arrange in terms of decreasing order (DO) of stability

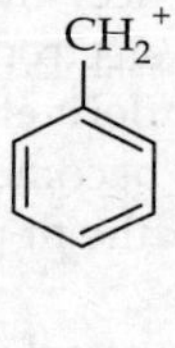

   1   2   3   4

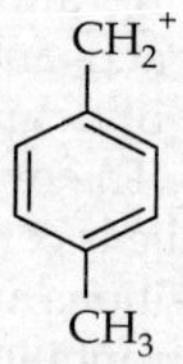

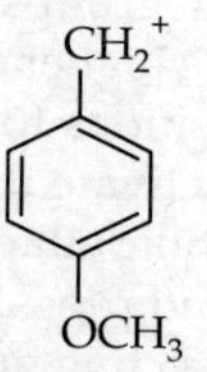

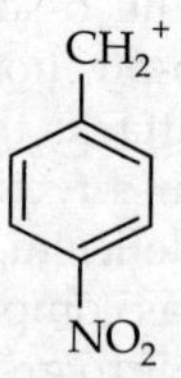

   (a) $4 > 1 > 3 > 2$ (b) $3 > 1 > 2 > 4$ (c) $3 > 4 > 1 > 2$ (d) $1 > 2 > 3 > 4$

8. Discuss the stability in the DO of the free radicals.

   $CH_2=\overset{\bullet}{C}H$ $CH_2=CH—\overset{\bullet}{C}H_2$

   1   2   3   4

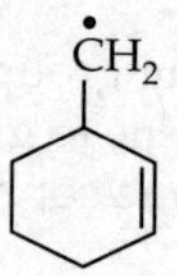

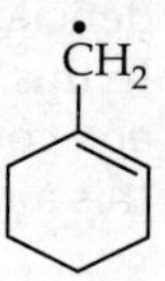

   (a) $1 > 3 > 2 > 4$ (b) $3 > 4 > 2 > 1$ (c) $4 > 3 > 2 > 1$ (d) $1 > 2 > 3 > 4$

9. Decreasing order of the basicity.

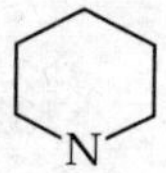

   1   2   3   4

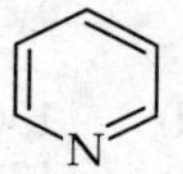

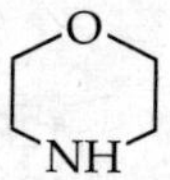

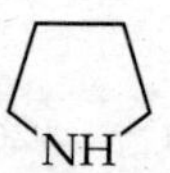

   (a) $1 > 3 > 4 > 2$ (b) $3 > 1 > 2 > 4$ (c) $4 > 3 > 2 > 1$ (d) $1 > 2 > 3 > 4$

10. How many electrons are in the following species.

   (a) 2   (b) 4   (c) 6   (d) 7

11. The C—C single bond length is in the decreasing order as
   (a) $CH_2=CH—CH=CH_2$ (b) $CH_2=CH—C≡CH$
   (c) $CH—≡C—C≡CH$ (d) $CH_3CH_2—C≡CH$

12. Which of the following compounds shows electromeric effect?
   (a) alkanes   (b) alkyl halides
   (c) alkyl amines   (d) aldehydes

13. The correct order of increasing acidity in the following compounds
   (a) $C_6H_5—OH$ (b) C_6H_5COOH (c) $C_6H_{11}—COOH$ (d) C_6H_5CCH
       phenol   benzoic acid   cyclo hexanoic acid   phenyl acetylene

14. Which of the following compounds is not a planar molecule?
   (a) $CH_2=C=CH_2$ (b) $CH_2=C=C=CH_2$ (c) $CH_2=C=C=O$

15. In which pairs, the compound and its resonating structure are correctly represented:
   (a) $HC=O:$ and $H—C≡O:^-$ (b) $—CH_2—N^+≡N:$ and $—CH_2—N=N$
   (c) none   (d) (a) and (b)

16. Which of the following compounds has the longest double bond?
   (a) $CH_2=C=CH_2$ (b) $CH_3CH=CH_2$
   (c) $(CH_3)_2C=CH_2$ (d) $(CH_3)_3C≡C=CH_2$

17. The DO of the length of the of the double bond in the following compounds is

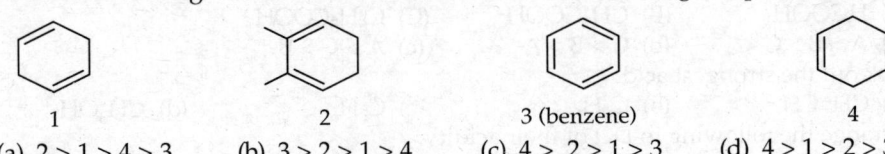

| 1 | 2 | 3 (benzene) | 4 |

(a) $2 > 1 > 4 > 3$ (b) $3 > 2 > 1 > 4$ (c) $4 > 2 > 1 > 3$ (d) $4 > 1 > 2 > 3$

18. The correct order of the C—N bond length in the following is

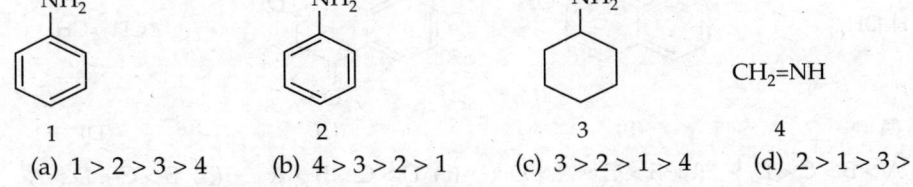

$CH_2=NH$

| 1 | 2 | 3 | 4 |

(a) $1 > 2 > 3 > 4$ (b) $4 > 3 > 2 > 1$ (c) $3 > 2 > 1 > 4$ (d) $2 > 1 > 3 > 4$

19. Which of the following has the longest C=O bond length?

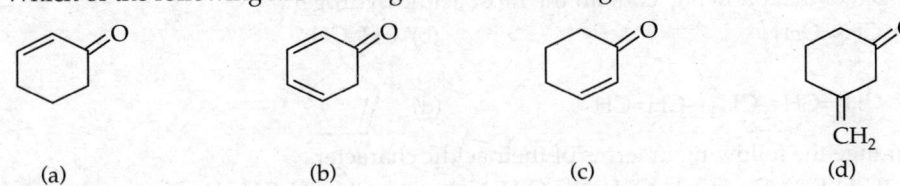

CH_2

| (a) | (b) | (c) | (d) |

20. The correct order of stability of the following anions.

(1) $CH_2=CH$ (2) $C_6H_5CH_2$

(3) $CH_2=CH—CH_2$ (4)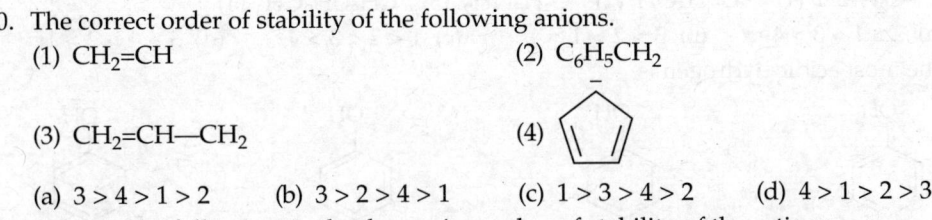

(a) $3 > 4 > 1 > 2$ (b) $3 > 2 > 4 > 1$ (c) $1 > 3 > 4 > 2$ (d) $4 > 1 > 2 > 3$

21. Arrange the following in the decreasing order of stability of the cations.

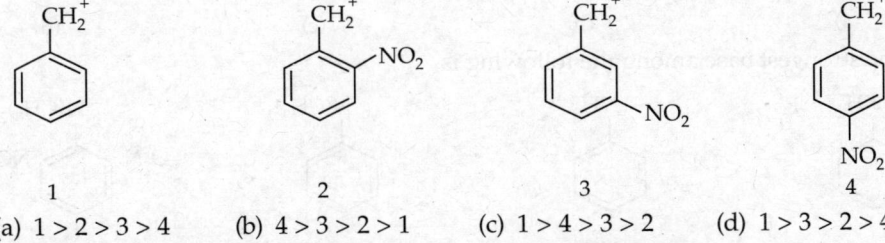

| 1 | 2 | 3 | 4 |

(a) $1 > 2 > 3 > 4$ (b) $4 > 3 > 2 > 1$ (c) $1 > 4 > 3 > 2$ (d) $1 > 3 > 2 > 4$

22. Place these compounds in the decreasing order of their stability.

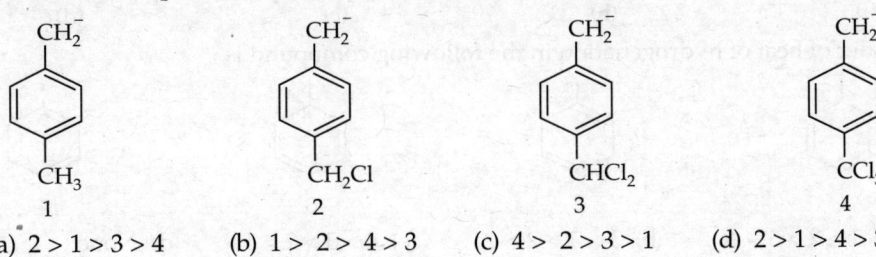

| 1 | 2 | 3 | 4 |

(a) $2 > 1 > 3 > 4$ (b) $1 > 2 > 4 > 3$ (c) $4 > 2 > 3 > 1$ (d) $2 > 1 > 4 > 3$

23. Arrange the following acids in the decreasing order of their acidity.
    (A) HCOOH   (B) CH_3COOH (C) C_6H_5COOH
    (a) A > B > C   (b) C > B > A   (c) A > C > B

24. Pick out the strongest acid.
    (a) $CH \equiv CH$ (b) C_6H_6 (c) C_2H_6 (d) CH_3OH

25. Arrange the following in DO of their acidity.

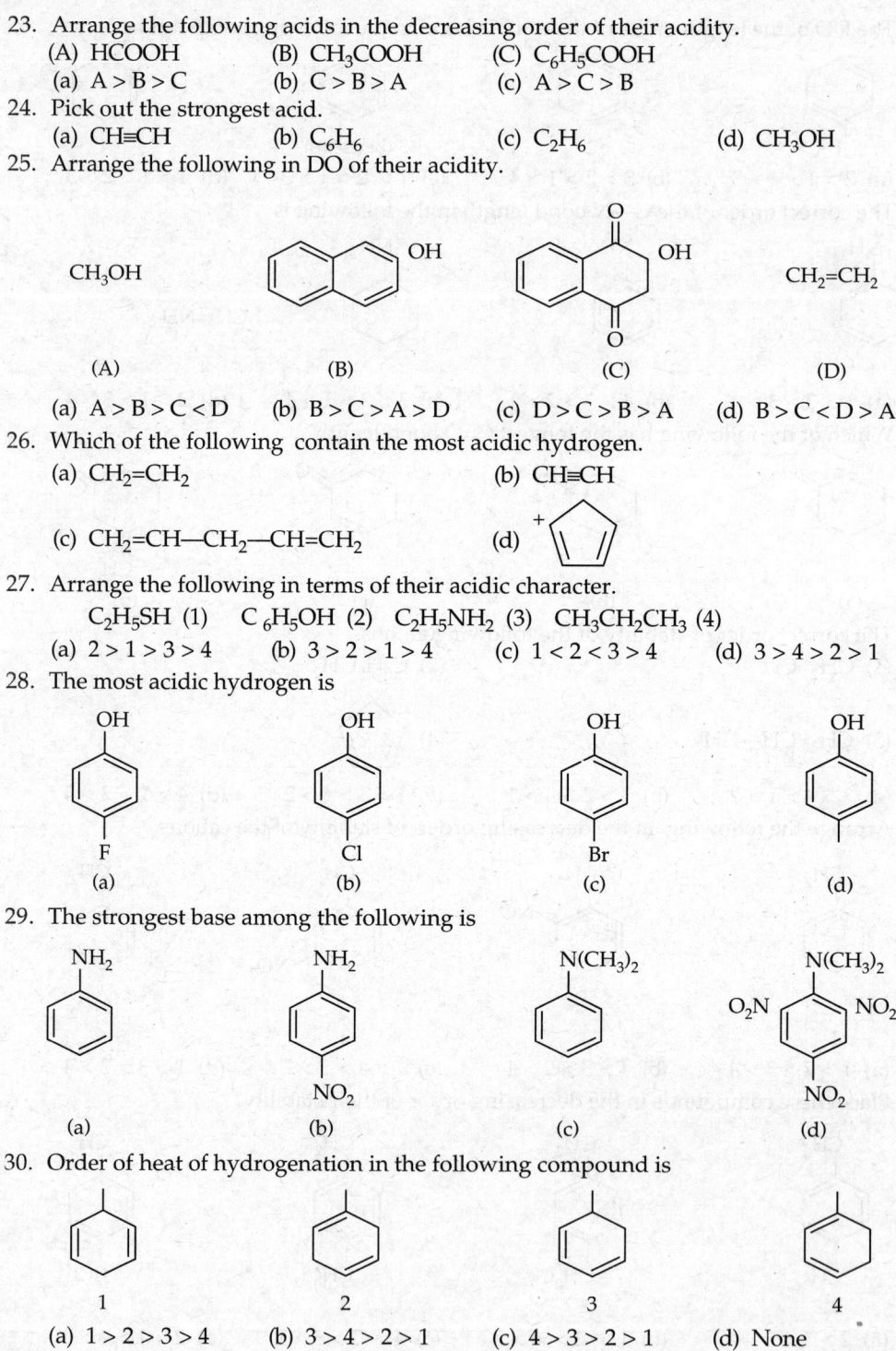

    CH_3OH OH   OH $CH_2=CH_2$

      (A)   (B)   (C)   (D)

    (a) A > B > C < D   (b) B > C > A > D   (c) D > C > B > A   (d) B > C < D > A

26. Which of the following contain the most acidic hydrogen.
    (a) $CH_2=CH_2$ (b) $CH \equiv CH$

    (c) $CH_2=CH-CH_2-CH=CH_2$ (d)

27. Arrange the following in terms of their acidic character.
    C_2H_5SH (1) C_6H_5OH (2) $C_2H_5NH_2$ (3) $CH_3CH_2CH_3$ (4)
    (a) 2 > 1 > 3 > 4   (b) 3 > 2 > 1 > 4   (c) 1 < 2 < 3 > 4   (d) 3 > 4 > 2 > 1

28. The most acidic hydrogen is

    OH   OH   OH   OH

      F   Cl   Br

     (a)   (b)   (c)   (d)

29. The strongest base among the following is

    NH_2 NH_2 $N(CH_3)_2$ $N(CH_3)_2$

                                                       O_2N NO_2

                  NO_2 NO_2

     (a)   (b)   (c)   (d)

30. Order of heat of hydrogenation in the following compound is

     1   2   3   4

    (a) 1 > 2 > 3 > 4   (b) 3 > 4 > 2 > 1   (c) 4 > 3 > 2 > 1   (d) None

31. The correct order of heat of combustion is

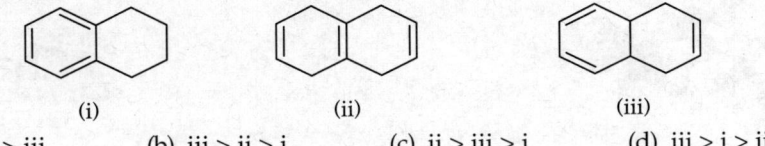

(i)   (ii)   (iii)

(a) i > ii > iii   (b) iii > ii > i   (c) ii > iii > i   (d) iii > i > iii

32. One of the compound which is not aromatic?

(a) 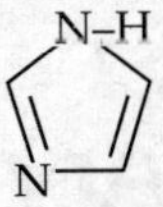   (b) 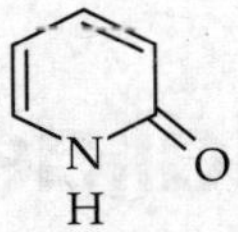   (c) 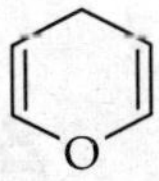   (d)

33. Pick out an anti-aromatic compound from among the following.

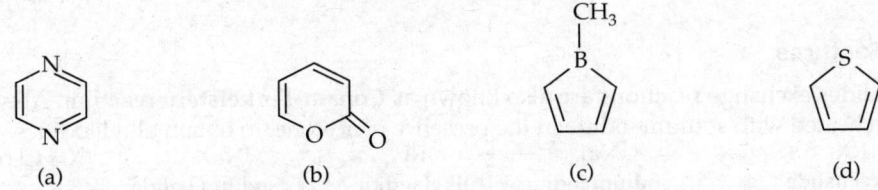

(a)   (b)   (c)   (d)

34. Identify the compound in which all C—C bonds are equal.

(a) 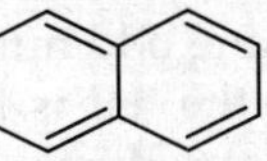   (b)    (c)   (d)

ANSWERS

(Group-F)

1. (c)   2. (d) (CH_3) 3. (a) $—N^+(CH_3)_3$ 4. (a) $(CH_3)_3C^+$

5. (d) $(CH_3)_2NH$ 6. (b) $(CH_3— > NH_2 > OCH_3— > OH—)$

7. (a)   8. (a)   9. (a)   10. (c) 6 electrons   11. (c)   12. (d)

13. (a)   14. (a)   15. (d)   16. (d)   17. (d)   18. (a)   19. (c)

20. (b)   21. (d)   22. (b)   23. (c)   24. (d)   25. (a)   26. (b)

27. (c)   28. (a)   29. (c)   30. (b)   31. (d)   32. (d)   33. (c)

34. (a), (c) (benzene)

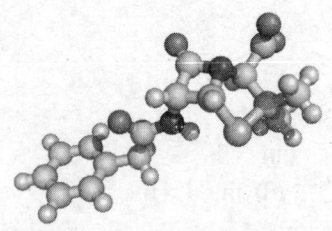

Chapter 2

Alkyl Halides and Aryl Halides

Salient Features

1. Halides exchange reactions are also known as **Conant-Finkelstein reaction**. Alkyl halides are heated with sodium iodide in the presence of acetone, to obtain alkyl iodides.

   RX   +   NaI $\longrightarrow$ RI   +   NaX   (X = Cl or Br)

   alkyl halide   sodium iodide   alkyl iodide   sodium halide

2. Stability of alkyl halides: RF > RCl > RBr > RI

3. Wurtz reaction fails in tertiary alkyl halides; Corey–House synthesis is used for the preparation of branched chain and alkanes containing odd number of carbons.

4. During elimination reaction of alkyl halides, the 'H' is lost from the carbon carrying minimum number of 'H' atoms, to give the major product.

5. The order of deactivation of nuclear halogen, towards nucleophilic substitution:

   —NH_2 > OH > OR > —R

6. CCl_4 resists hydrolysis by water, due to nonavailability of d-orbitals in carbon.

7. CF_2Cl_2 –is known as Freon-12; F_2Cl—C—CF_2Cl – known as Freon-114.

8. $CHCl_2$—$CHCl_2$ is known as Westron; CCl_2=CCl_2 is known as Westrosol.

9. Nitration of chlorobenzene (C_6H_5Cl) is more difficult than benzene because chlorine is more electronegative than phenyl group, it makes the benzene ring more electron deficient so difficult for nitration.

10. The bond length —C—Cl in alkyl chloride 1.77 Å, but in chlorobenzene = 1.69 Å.

    R—CH_2— Cl C_6H_5—Cl

       1.77 Å   1.69 Å

11. Primary alkyl halides can be oxidised to aldehydes in good yield using dimethyl sulphate, $[(CH_3)_2SO_4]$ in presence of trimethyl amine $[(CH_3)_3N]$

    $$RCH_2\,Cl \xrightarrow[\text{(ii) } (CH_3)_3N]{\text{(i) } (CH_3)_2SO_4} RCHO \;+\; (CH_3)_2S$$

    Mechanism:

    $$RCH_2Cl + O_4^-S^+ \xrightarrow{Cl^-} (CH_3)_2 \longrightarrow RCH\!-\!O\!-\!S\,(CH_3)_2 \xrightarrow{H^+} RCHO + (CH_3)_2S$$

12. *Decreasing order of reactivity in S_N1 reaction*:

    CH_2=CH—CH_2Cl > CH_3CH_2Cl > CH_3Cl > CH_2=CH—Cl

    allyl chloride   ethyl chloride   methyl chloride   vinyl chloride

13. *Decreasing order of reactivity in S_N2 reaction*:

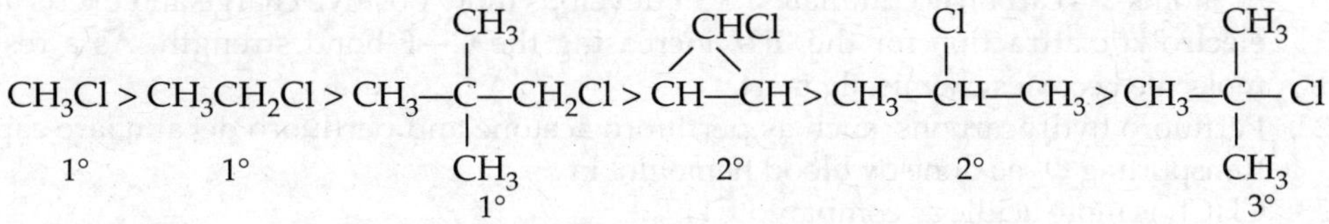

$$CH_3Cl > CH_3CH_2Cl > CH_3-\underset{\underset{1°}{\underset{|}{CH_3}}}{\overset{\overset{CH_3}{|}}{C}}-CH_2Cl > \underset{2°}{CH\overset{\overset{CHCl}{\diagup\diagdown}}{-}CH} > CH_3-\underset{2°}{\overset{\overset{Cl}{|}}{CH}}-CH_3 > CH_3-\underset{\underset{3°}{\underset{|}{CH_3}}}{\overset{\overset{CH_3}{|}}{C}}-Cl$$

  $\qquad$ 1° $\qquad\qquad$ 1° $\qquad\qquad$ 1° $\qquad\qquad$ 2° $\qquad\qquad$ 2° $\qquad\qquad$ 3°

(3° alkyl halides do not undergo S_N2 displacement)

14. The reactivity of tertiary halides: Strong base and medium to low polarity and alcoholic KOH favors elimination.

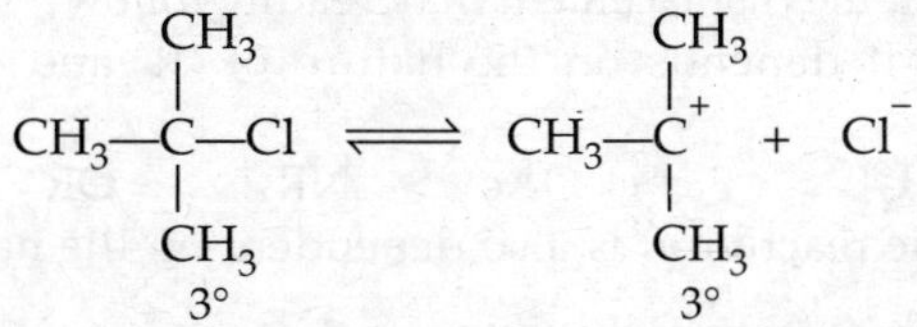

$$CH_3-\underset{\underset{3°}{\underset{|}{CH_3}}}{\overset{\overset{CH_3}{|}}{C}}-Cl \rightleftharpoons CH_3-\underset{\underset{3°}{\underset{|}{CH_3}}}{\overset{\overset{CH_3}{|}}{C}}{}^+ + Cl^-$$

Weak base and medium to high polarity and aqueous KOH favors substitution and forms alcohol

$(NH_2-, CN-C_2H_5O-)$

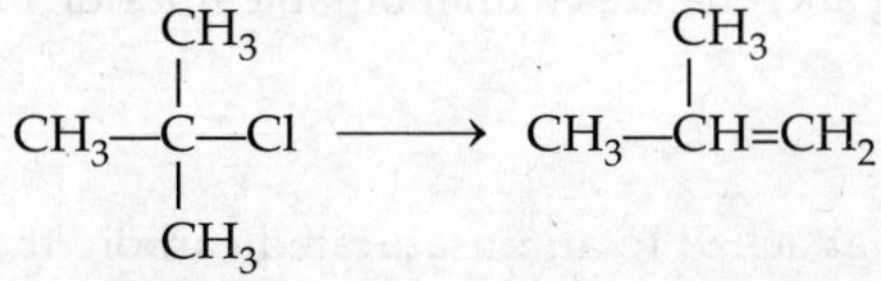

$$CH_3-\underset{\underset{}{\underset{|}{CH_3}}}{\overset{\overset{CH_3}{|}}{C}}-Cl \longrightarrow CH_3-\overset{\overset{CH_3}{|}}{CH}=CH_2$$

15. Stronger the base, stronger the nucleophile, but poor is the leaving group.
Weaker the base, weaker the nucleophile, stronger is the leaving group, e.g. —OH group, H_2O.

16. *Nucleophilicity of a nucleophile* is defined as its ability to react with an electrophilic carbon. *Basicity* is the ability to remove H^+ from an acid.

17. Dry HCl is a stronger acid and better electrophile.

18. 'Iodide ion' is a powerful nucleophile, and a better leaving group.

19. In a reaction between alkyl halide and KCN, alkyl cyanides are formed.

KCN $\longrightarrow$ K$^+$ + C$^-$N:

R^+ + C$^-$N: $\longrightarrow$ RCN (though C$^-$N can join with alkyl radical through nitrogen to form R—NC, through the lone pair; C—N bond is weaker as compared to the C—C bond, which is stronger. Therefore, alkyl cyanide is formed.

On the other hand, a reaction of alkyl halide with AgCN, leads to the formation of alkyl isocyanide, RNC.

AgCN, breaks up to form, silver atom and cyanide free radical.

Ag : CN $\longrightarrow$ Ag + C≡N:

As a lone pair is available only on nitrogen and not on carbon, thus nucleophilic attack takes place from nitrogen side and isocyanides are formed.

R^+ + C≡N$^-$ + Ag $\longrightarrow$ RN≡C: + Ag

20. The dipole moments of methyl halides is as follows:

$\qquad\qquad$ CH_3Cl > CH_3F > CH_3Br > CH_3I
$\qquad\qquad$ $\mu = 1.94D$ $\quad$ 1.82D $\qquad$ 1.79D $\qquad$ 1.64D

21. Perfluoro hydrocarbons are the hydrocarbons in which all the 'H' have been replaced by 'F' atoms. As carbon accumulates 'F', it develops more positive charge and exerts a greater electrolytic attraction for the 'F's, increasing the C—F bond strength. As a result, the molecule becomes chemically inert.

22. Perfluoro hydrocarbons, such as perfluoro acetone and perfluoro decalin, are capable of transporting O_2 as done by blood hemoglobin.

23. $CHCl_3$ is more acidic as compared CHF_3.

24. Bu Li, $(CH_3CH_2CH_2CH_2^- Li^+)$ is a strong base.

25. Halide ion is a weak base and it is easily replaced by stronger bases.

    In general, the stronger the base, the less facile it is, as a leaving group in displacement reactions; but if protonation can occur first, the displacement may readily follow.

    The ease of displacement is not fixed; it depends on the nature of 'R' and solvent conditions. In general, the order is:

    $OTs > OMs > I > Br > Cl > OH_2^+ > F > OAc > NR_3^+ > OR > NR_2$

    The order of nucleophilicity (nucleophilic reactivity) is also dependent on the nature of solvent; in general the order is:

    $C_6H_5S^- > CN^- > I^- > C_2H_5O^- > OH^- > Br^- > C_6H_5O^- > Cl^- > (CH_3)_3N$

    It is the ability to form a bond with the carbon atom, where as basicity is the affinity to protons.

26. Alkyl fluorides may be prepared by heating alkyl halides with inorganic fluoride like AgF, Hg_2F_2.

    $C_2H_5Cl + AgF \longrightarrow C_2H_5F + AgCl$

    This reaction is known as **Swartz reaction.**

27. *Hyperconjugation*: When a —C—H bond is attached to an unsaturated carbon, the sigma electrons of this bond, enters in conjugation with unsaturated system. Such conjugation between electrons of a single and multiple bonds is termed hyperconjugation.

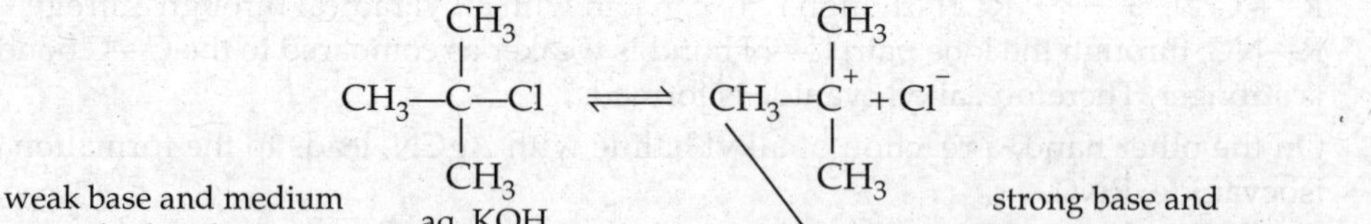

28. Reactions of a tertiary alkyl halides:

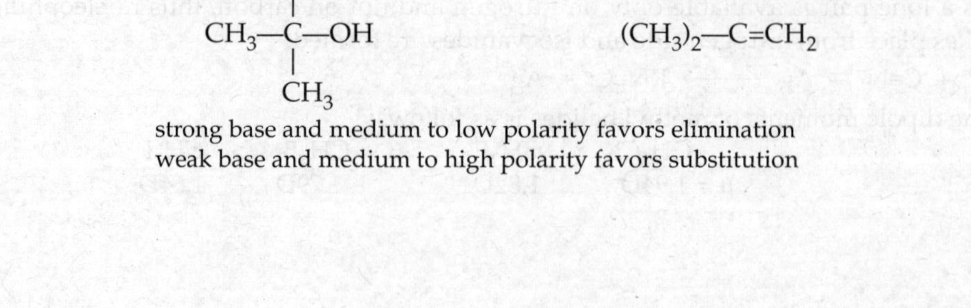

strong base and medium to low polarity favors elimination
weak base and medium to high polarity favors substitution

S_N1 Substitution unimolecular	S_N2 Substitution bimolecular
1. It involves two step mechanism	1. It involves one step mechanism
2. The rate of reaction depends upon concentration of RX only	2. The rate of reaction depends upon the concentration of RX and attacking nucleophile
3. It is a first order reaction	3. It is a second order reaction
4. It involves the formation of a carbocation. In case of an optically active compound, racemisation, takes place	4. It involves the formation of a transition state. In case of an optically active compound, inversion takes place
5. Weak nucleophiles favour S_N1	5. Strong nucleophiles favour S_N2
6. A low concentration favours S_N1	6. A high concentration favours S_N2
7. Polar solvent favours S_N1	7. Nonpolar solvents favour S_N2
8. Order of reactivity of RX: $3° > 2° > 1°$	8. Order of reactivity of RX: $1° > 2° > 3°$

ALKYL HALIDES

Salient Features

When one or more than one hydrogen atom are substituted by halogen atoms, i.e. chlorine, bromine or iodine, mono- or poly-halogen derivatives of alkane are formed.

Alkyl halides, RX: These compounds are mono-halogen derivatives of alkanes, having a molecular composition $C_nH_{2n+1}X$, where R is an alkyl group and X = Cl , Br, or I.

For methane:

	Common Name	IUPAC Name
$ClCH_3$	Methyl chloride	Monochlro methane
$BrCH_3$	Methyl bromide	Monobromo methane
CH_3I	Methyl iodide	Monoiodo methane

For ethane:

C_2H_5Cl	Ethyl chloride	Monochloro ethane
C_2H_5Br	Ethyl bromide	Mthyl iodide
C_2H_5I	Ethyl iodide	Monoiodo ethane

Methods of Preparation

From alcohol: On treating alcohol with phosohorous pentachloride, (PCl_5), phosphorous trichloride, (PCl_3) or thonyl chloride, ($SOCl_2$)

1. $CH_3OH + PCl_5 \longrightarrow CH_3Cl + POCl_3 + HCl$
2. $3CH_3OH + PCl_3 \longrightarrow 3CH_3Cl + H_3PO_3$
3. $CH_3OH + SOCl_2 \longrightarrow CH_3Cl + SO_2 + HCl$

Likewise on treating alcohol with PBr_5, PBr_3 or with $SOBr_2$

1. $C_2H_5OH + PBr_5 \longrightarrow C_2H_5Br + POBr_3 + HBr$
2. $3C_2H_5OH + PBr_3 \longrightarrow 3 C_2H_5Br + H_3PO_3$
3. $C_2H_5OH + SOBr_2 \longrightarrow C_2H_5Br + SO_2 + HBr$

For iodo compounds: PI_3 (PI_5 and SOI_2, do not exist)

$3ROH + PI_3 \longrightarrow 3 RI + H_3PO_3$

On heating alcohol with halogen acids, alkyl halides are formed.

1. *For chlorides*: On heating alcohol with HCl, in presence of $ZnCl_2$

$$ROH + HCl \xrightarrow{ZnCl_2} RCl + H_2O$$

Mechanism:

i. $ZnCl_2 + HCl \longrightarrow ZnCl_3^- + H^+$

ii. $R\!-\!OH + H^+ \longrightarrow R\!-\!\overset{H}{\underset{+}{O}}\!-\!H \longrightarrow R^+ + H_2O$

iii. $R^+ + ZnCl_3^- \longrightarrow RCl + ZnCl_2$

For bromides: On heating alcohol with HBr in presene of conc. H_2SO_4

$ROH + HBr \xrightarrow{\text{conc. } H_2SO_4} RBr + H_2O$

For iodides: On refluxing alcohol with HI (constant boiling at 57% HI)

$ROH + HI \longrightarrow RI + H_2O$

The decreasing order of reactivity of alcohols:

tertiary alcohol > secondary alcohol > primary alcohol

The decreasing order of reactivity of halogen acids:

HI > HBr > HCl

2. *From alkenes*: Alkene react with halogen acids to form alkyl halides

$CH_2{=}CH_2 + HX \longrightarrow CH_3CH_2X$ (HX = HCl, HBr, HI)

$CH_3CH{=}CH_2 + HX \longrightarrow CH_3CHXCH_3$

Mechanism: It is based on the formation of more stable carbocation.

$$HX \longrightarrow H^+ + X^-$$

$$\underset{1°}{CH_3CH{=}CH_2} + H^+ \longrightarrow \underset{2°}{CH_3\overset{+}{C}HCH_3} \longrightarrow \underset{3°}{CH_3CH_2\overset{+}{C}H_2}$$

1° carbocation

As a 2° carbocation is more stable than 1° carbocation, the final product is formed from the 2° carbocation, and the less stable 1° carbocation reverts back to form the starting compound.

Unsymmetrical alkenes react with HBr, in presence of a peroxide (not H_2O_2) to produce a product just opposite to the expected compound, i.e. *n*-alkyl bromide.

$CH_3CH{=}CH_2 + HBr \xrightarrow{\text{peroxide}} CH_3CH_2CH_2Br$

This reaction is termed as *Kharasch reaction*, or peroxide effect or known as anti-Markovnikov 's rule.

Mechanism: It is free radical reaction (the attacking reagent is Br)

$(C_6H_5COO)_2 \longrightarrow 2C_6H_5{\bullet} + 2CO_2$

$C_6H_5 + H:Br \longrightarrow C_6H_6 + Br{\bullet}$

$CH_3CH{=}CH_2 \xrightarrow{Br{\bullet}} CH_3\overset{\bullet}{C}HCH_2Br \xrightarrow{C_6H_6} \underset{2° \text{ free radical}}{CH_3\overset{\bullet}{C}H_2CH_2Br} \longrightarrow \underset{1° \text{ alkyl radical}}{CH_3CHBr\!-\!\overset{\bullet}{C}H_2}$

Since a 2° free radical is more stable than a 1° free radical, the final product is formed from the 2° free radical, (whereas the) 1° less stable radical reverts back to form the starting compound.

3. *Borodine Hunsdiecker reaction*: Alkyl bromide may be prepared by heating silver salt of a fatty acids with Br_2(in CCl_4).

$RCOOAg + Br_2 \longrightarrow RBr + AgBr + CO_2$

Alkyl chloride may be prepared but the yield is poor.

On the other hand, in a reaction of silver salt of a fatty acid with iodine forms an ester. This reaction is known as *Birnbaum–Simonini reaction*:

$2\,RCOOAg + I_2 \longrightarrow RCOOR + CO_2 + 2AgI$

4. *Halogenation of alkane*: Lower alkanes react with halogens in presence of sunlight, the higher ones react on heating to form halogen derivatives of alkanes.

$$RH + X_2 \xrightarrow{\text{sunlight}} RX + HX$$

On prolonged heating, the polyhalogen derivative are formed. In order to increase the yield of alkyl halides, excess of alkane is used.

For the preparation of alkyl iodides, a reaction of alkane is carried with I_2, in presence of an oxidising agent like HNO_3, HIO_4

$$RH + I_2 \longrightarrow RI + HI$$

As the reaction is reversible, the HI formed is oxidised back to form iodine which force the reaction towards completion.

$$HI + O \longrightarrow H_2O + I_2$$

Conant–Finkelstein reaction: Generally alkyl iodides are prepared by heating alkyl chloride or bromides with KI in acetone, when alkyl iodides are formed.

$$RCl \text{ or } RBr + KI \longrightarrow RI + KCl \text{ or } KBr$$

Physical properties

Methyl chloride is a gas, b.p. 24°C, while higher members are pleasant smelling liquids, immiscible and heavier than water (except chloride). All alkyl halides dissolve in organic solvents and burn with a bluish flame.

Alkyl halide are polar compounds, have a dipole moment in the range of 2.05–2.15D. The dipole moments of methyl halide in the decreasing order:

$$\begin{array}{cccc} CH_3Cl & > & CH_3F & > & CH_3Br & > & CH_3I \\ 1.91D & & 1.84D & & 1.79D & & 1.64D \end{array}$$

The higher value of methyl chloride is because of the distance between C—Cl is more than that of between C—F, though the electronegativity of F(4) is more than that of Cl(3.5).

The densities as well as their boiling points are in the decreasing order on account of their decreasing molecular weight.

	RI	>	RBr	>	RCl
Boiling point	102°C		71°C		46.5°C

The reactivity of alkyl halide based on their dissociation energy (heterolytic fission)

	RI	RBr	RCl	RF	
Bond dissociation energy heterolytic fission C : X	212	219	227	256	kcal/mole
Bond dissociation energy (homolytic) fission C : X	51	68	71	107	kcal/mole

As the dissociation energy is lowest for RI, alkyl iodide are most reactive. The decreasing order of reactivity of alkyl halide may also be explained on the basis of polarisability. The C—I bond is easily polarised as compared to C—Cl or C—Br bond, hence the decreasing order.

The reactivity order of the alkyl halide when in R—X, where X is same but R is different

i. $CH_3X > C_2H_5X > C_3H_7X > C_4H_9X$

ii. In RX, where is X same, but R is different

$$\begin{array}{cccc} R_3CX & > & R_2CHX & > & RCH_2X & > & CH_3X \\ 3° & & 2° & & 1° & & \text{methyl halide} \end{array}$$

iii. The boiling point of alkyl halides are higher than those of corresponding alkanes on account of high polarity and dipole–dipole attraction, which are absent in alkanes. Alkanes have relatively weak van der Waal's forces.

Chemical reactions

Alkyl halides are highly polar compounds, their decreasing order of polarity should be in the order of:

$$\begin{array}{ccc} C\!-\!Cl & > & C\!-\!Br & > & C\!-\!I \\ 2.5\text{-}3 & & 2.5\text{-}2.8 & & 2.5\text{-}2.6 \end{array}$$

$$C{-}I \xrightarrow{\text{cation/E}^+} C^+ + I{-}E^+$$

Alkyl halides undergo nucleophilic substitution reactions mainly. Since halide is a weak base, it is substituted by a stronger base. Stronger is the base, less facile it is as a leaving group. The order of displacement of the group is not fixed, it is dependent on the nature of alkyl group and on solvent.

However, the general order is as follows:

$I > Br > Cl > OH_2 > F > OCH_3 > NR_3 > OR > NR_2$

i. Reaction with aq KOH: Alkyl halide react with aqueous KOH to form alcohols

$RX + KOH \longrightarrow ROH$

ii. Reaction with alc. KOH: Dehydrohalogenation takes place

$CH_3CH_2CHBrCH_3$

$\longrightarrow CH_3{-}CH_2CH{:}{=}CH_2$
20% butene-1

$\longrightarrow CH_3CH{=}CHCH_3$
80% butene-2

Saytzeff's rule: In the process of dehydrohalogenation, the major product formed is the one which contain greater number of alkyl groups attached to the doubly bonded carbon atoms.

This type of elimination is termed β-elimination, the 80% product is termed Saytzeff's product and the 20% product is termed Hofmann's product. When an alkene is capable of existing in geometrical isomeric forms, the *trans*-isomer generally predominates over *cis*-form.

iii. Reduction with LiAlH₄ and with NaBH₄

LiAlH₄	NaBH₄
$RCH_2X \longrightarrow RCH_3$	no reduction
$R_2CHX \longrightarrow R_2CH_2$	alkane
$R_3CX \longrightarrow$ alkene + HX	alkane

iv. *Wurtz reaction*: Alkyl halides react with sodium metal in presence of ether to form higher alkanes.

$RX + 2Na + XR \longrightarrow R{-}R + 2NaX$

Alkanes formed contain double the number of carbon or even number of carbon atoms. Only 1° and 2° alkyl halides undergo **Wurtz reaction. It fails with 3° alkyl halides.** For the preparation of an alkane containing odd number of carbon atoms or for the preparation of a branch chain alkane, **Corey–House** synthesis is used.

Corey–House synthesis: Alkyl halide is treated with Li metal preferably in presence of ether, followed by treating the resulting alkyl lithium with cuprous iodide, when dialkyl lithium cuperate or Corey-House reagent is formed.

$RX + 2Li \longrightarrow RLi + LiX$

$2RLi + CuI \longrightarrow R_2CuLi + LiI$

On treating R₂CuLi with an alkyl halide, including a 3° alkyl halide, the desired alkane may be prepared.

$R_2CuLi + R'X \longrightarrow R{-}R' + R\,Cu + LiX$

v. *Reaction with NH₃*: Alkyl halides react with ammonia to give a mixture of primary, secondary, tertiary amines and finally a quarternary ammonium salt.

$RX + NH_3 \longrightarrow RNH_2 \xrightarrow{RX} R_2NH \xrightarrow{RX} R_3N \xrightarrow{RX} R_4NX$

1° 2° 3° tetra alkyl ammonium halide (quarternary ammonium salt)

vi. *Reaction with KCN*:

$$RX + KCN \longrightarrow RCN + KX$$

Alkyl cyanide formed, contains one carbon more than the alkyl halide.

Some reactions of RCN:

  a. On hydrolysis a carboxylic acid is formed

  $$RCN + 2H_2O \longrightarrow RCOOH + NH_3$$

  b. On reduction with Na + alcohol **(Mendius reaction)**

  $$RCN + 4H \longrightarrow RCH_2NH_2$$

  c. On reductive hydrolysis, aldehyde is formed **(Stephen's reaction)**

$$RCN + 2H \xrightarrow{\text{SnCl}_2 + \text{HCl}} RCH=NH \longrightarrow RCHO$$

vii. *Reaction with AgCN*:

$$2\,RX + 2AgCN \longrightarrow RNC + RCN + 2AgX$$

Formation of alkyl isocyanide has been explained on the basis of AgCN, being a covalent compound.

$$AgCN \longrightarrow Ag\bullet \ + \ \bullet C\,N:$$

As the lone pair of electron is available on nitrogen, the nucleophilic attack takes place from nitrogen site, leading to the formation of alkyl isocyanide.

viii. *Reaction with sodium salt of a fatty acid*: Ester is formed.

$$R'COONa + RX \longrightarrow R'COOR + NaX$$

Reaction of a sodium or potassium derivative of an alcohol with alkyl halide leads to the formation of an ether.

$$RX + R'ONa \longrightarrow ROR' + NaX$$

This reaction is termed **Williamsons synthesis.**

xi. *Reaction with AgNO$_2$*

$$RX + AgNO_2 \longrightarrow R{-}N\!\!\begin{array}{c}\diagup\!\!\diagup O \\ \diagdown O\end{array}$$

nitroalkane

$$RX + KNO_2 \longrightarrow R{-}O{-}N{-}=O$$

ARYL HALIDES

When **one or more hydrogen atoms of benzene ring or its side chain**, are **substituted by halogen atoms,** the compounds formed are termed **aryl halides.** The aryl halides are of two types:

  a. **Nuclear substituted aryl halides: One or more** than one **halogen** atoms are **directly attached to the benzene ring.**

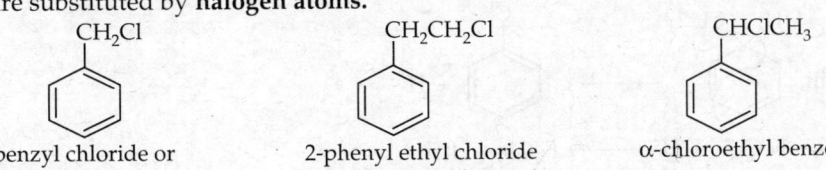

chlorobenzene or   dichlorobenzene or phenylene chlorides
phenyl chloride   (o–, m– and p–positions)

  b. **Side chain substituted aryl halides: One or more than one hydrogen** atoms of the **side chain**, are substituted by **halogen atoms.**

benzyl chloride or   2-phenyl ethyl chloride   α-chloroethyl benzene
chloromethyl benzene   or β-chloroethyl benzene

The side chain substituted halogen derivatives are quite reactive; undergo nucleophilic substitution reactions, like alkyl halides. On the other hand, **nuclear substituted aryl halides are inert,** under ordinary conditions. However, these **undergo nucleophilic substitution** reaction, **under drastic condition.**

Chlorobenzene or Phenyl Chloride, (C_6H_5Cl)

Methods of Preparation

1. **Direct halogenation of benzene:** When benzene is treated with chlorine, in presence of iron or $FeCl_3$, as catalyst, chlorobenzene is formed.

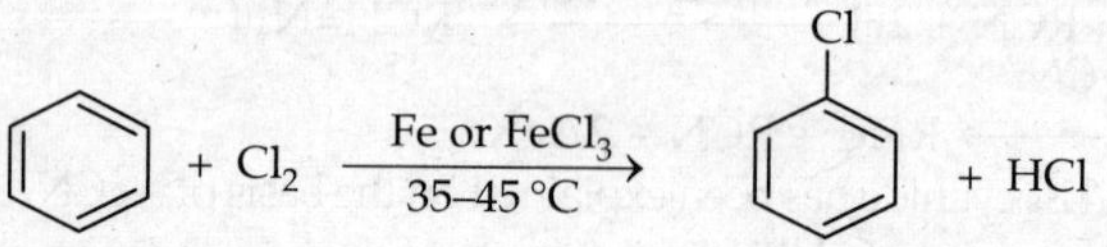

$$\text{benzene} + Cl_2 \xrightarrow[35-45\,°C]{\text{Fe or FeCl}_3} \text{chlorobenzene} + HCl$$

Similarly, bromobenzene is formed, in an interaction between bromine and benzene, in the presence of **iron or $FeBr_3$, as catalyst.** For iodobenzene, a reaction between iodine and benzene is carried out, in the presence of some oxidizing agent like HIO_3, HNO_3 or HgO, as the reaction is reversible, due to the formation of HI. Being a powerful reducing agent, HI reduces iodobenzene back to benzene.

$$C_6H_6 + I_2 \rightleftharpoons C_6H_5I + HI \quad \text{(a reversible reaction)}$$

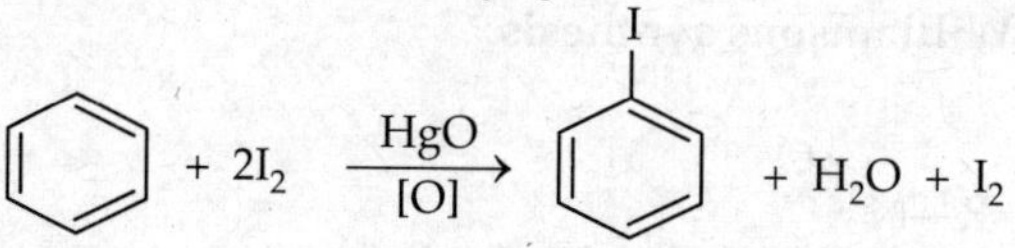

$$\text{benzene} + 2I_2 \xrightarrow[[O]]{HgO} \text{iodobenzene} + H_2O + I_2$$

Mechanism: The role of a catalyst is to release chloronium ion, $(Cl^+$ of in chlorination) or bromonium ion (Br^+ in bromination), which are electrophiles. These electrophiles attack benzene ring to form chloro or bromobenzene.

The formation of chlorobenzene from chlorine and benzene, may be explained as follows:

i. $Cl : Cl + FeCl_3 \longrightarrow Cl^+ + FeCl_4^-$
    chloronium ion

ii.

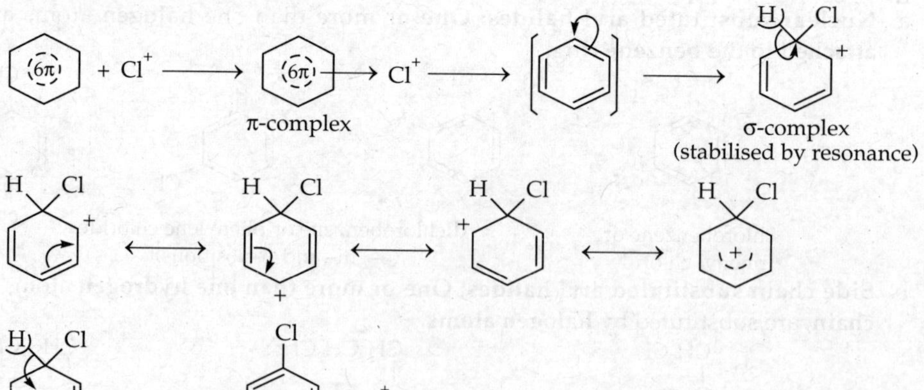

π-complex — σ-complex (stabilised by resonance)

iii.

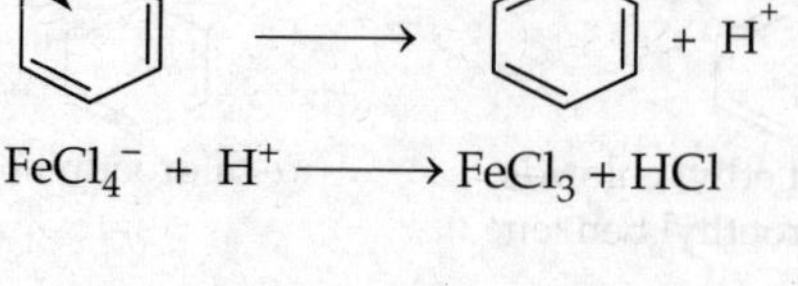

iv. $FeCl_4^- + H^+ \longrightarrow FeCl_3 + HCl$

In chlorination, **(Cl^+) attacks benzene ring** and forms a **π-complex**, involving intense negative charge on benzene ring, on account of delocalised six π-electrons on benzene ring. **π-complex** then undergoes **localization** (i.e. the return of π-electrons to individual carbon atoms, and forming alternate double bonds) and **make available a pair of electrons on one carbon of the ring**, to which **Cl^+ is transferred and Cl^+ joins to form a σ-complex.**
σ-complex is stabilized by resonance. In the final stage, a proton is lost from **σ-complex**, the pair of electrons, thus vacated, return back to benzene ring, which undergoes delocalisation again, forming chlorobenzene.

2. **From dizonium chloride, ($C_6H_5N_2Cl$):** On treating benzene diazonium chloride, ($C_6H_5N_2Cl$) with **cuprous chloride, dissolved in conc. HCl, chlorobenzene is formed.**

$$C_6H_5N_2Cl \xrightarrow[\Delta]{CuCl/HCl} C_6H_5Cl + N_2$$
$$\text{chlorobenzene}$$

Likewise **bromobenzene** is obtained on **treating benzene diazonium chloride** with **cuprous bromide, dissolved in conc. HBr.**

$$C_6H_5N_2Br \xrightarrow[\Delta]{CuBr/HBr} C_6H_5Br + N_2$$
$$\text{bromobenzene}$$

For iodobenzene, benzene diazonium chloride is treated with cuprous iodide and aqueous KI solution.

$$C_6H_5N_2Cl \xrightarrow[\Delta]{CuI/KI} C_6H_5I + N_2$$
$$\text{iodobenzene}$$

Similarly **phenyl cyanide**, may be prepared on treating $C_6H_5N_2Cl$, with **cuprous cyanide and KCN.**

$$C_6H_5N_2Cl \xrightarrow{CuCN/KCN} C_6H_5CN + N_2$$
$$\text{phenyl cyanide}$$

This reaction is termed as **Sandmeyer's reaction.**

Mechanism: A possible mechanism, as suggested is as follows. This involves a **free radical mechanism.**

i. $C_6H_5N_2Cl \rightleftharpoons C_6H_5N_2^+ + Cl^-$

ii. $CuCl + Cl^- \rightleftharpoons CuCl_2^-$

iii. $C_6H_5N_2^+ + Cl-CuCl^- \xrightarrow[\text{fast}]{\text{slow}} C_6H_5^{\bullet} + N_2 + Cl-Cu-Cl$
$$\text{phenyl free radical}$$

iv. $C_6H_5^{\bullet} + CuCl_2 \longrightarrow C_6H_5Cl + CuCl$

This mechansim may also be written as follows:

i. $Cu^+ \longrightarrow Cu^{2+} + e$

ii. $C_6H_5N_2 + e \longrightarrow C_6H_5 + N_2$
$$\text{phenyl free radical}$$

iii. $C_6H_5^{\bullet} + Cu^{2+} + Cl^- \longrightarrow C_6H_5Cl + Cu^+$

The preparation of **phenyl halides** may also be carried, from **benzene diazonium chloride**, by **heating C_6H_5Cl with copper powder and halogen acids.**

$$C_6H_5N_2Cl \begin{cases} \xrightarrow{Cu/HCl} C_6H_5Cl + N_2 \\ \xrightarrow{Cu/HBr} C_6H_5Br + N_2 \\ \xrightarrow{Cu/KI} C_6H_5I + N_2 \\ \xrightarrow{Cu/KCN} C_6H_5CN + N_2 \end{cases}$$

This modification of **Sandmeyer's reaction is known as Gattermann's modification** of Sandmeyer's reaction.

Phenyl cyanide, on reduction with sodium and alcohol, gives benzyl amine ($C_6H_5CH_2$ NH_2) and on hydrolysis gives benzoic acid (C_6H_5COOH)

$$C_6H_5CN \begin{array}{c} \xrightarrow{\text{+4H/Na/C}_2\text{H}_5\text{OH}} \quad C_6H_5CH_2NH_2 \\ \xrightarrow[\text{hydrolysis}]{H^+/H_2O} \quad \begin{array}{l} \text{benzyl amine} \\ C_6H_5COOH \\ \text{benzoic acid} \end{array} \end{array}$$

Similarly, phenyl cyanide on reductive hydrolysis, gives benzaldehyde (C_6H_5CHO). This reaction is known as **Stephen's reaction.**

$$C_6H_5CN \xrightarrow[2H]{SnCl_2/HCl} \underset{\text{benzylidimine}}{C_6H_5CH{=}NH} \xrightarrow{H+/H_2O} \underset{\text{benzaldehyde}}{C_6H_5CHO} + NH_3$$

Fluorobenzene (C_6H_5F) may also be prepared from **C_6H_5Cl,** by subjecting it to **Balz–Schiemann reaction.**

Balz–Schiemann reaction: $C_6H_5N_5Cl$ is treated with **tetrafluoro boric acid (HBF$_4$)**, when a **precipitate of benezene diazonium tetrafluoroborate ($C_6H_5N_2BF_4$)** formed. This on heating gives **fluorobenzene (C_6H_5F).**

$$C_6H_5N_2Cl \xrightarrow{HBF_4} C_6H_5N_2BF_4 \xrightarrow{heat} C_6H_5F + N_2 + BF_3$$

Fluorobenzene (C_6H_5F) may also be prepared from benzene. Benzene is treated with **thallium trifluoroacetate, [Tl(CF$_3$COO)$_3$]**, when benzene is converted into **phenyl trifluoroacetate thallium [$C_6H_5Tl(CF_3COO)_2$].** This on heating, in presence of KF and BF$_3$, gives fluorobenzene.

$$C_6H_6 + Tl(CF_3COO)_3 \longrightarrow C_6H_6Tl(CF_3COO)_2 + CF_3COOH$$
$$\underset{KF + BF_3 |}{} \text{heating}$$
$$C_6H_5F$$

Note: Fluorobenzene cannot be prepared from fluorine and benzene, because fluorine is too reactive, **both addition and polysubstitution occurs in benzene. Fluorobenzene is used as a powerful insecticide.**

3. **From phenol:** Phenol is treated with **phosphorous pentachloride (PCl$_5$)** or **phosphorous trichloride (PCl$_3$)** or **thionyl chloride (SOCl$_2$),** chlorobenzene is formed.

   i. $C_6H_5OH + PCl_5 \longrightarrow C_6H_5Cl + POCl_3 + HCl$
   phosphorous oxychloride

   The yield of chlorobenzene is very poor, as the main product formed is triphenyl phosphate, (C_6H_5O)$_3$PO or (C_6H_5)$_3$PO$_4$. POCl$_3$, reacts with phenol to form triphenyl phosphate.
   $POCl_3 + 3C_6H_5OH \longrightarrow (C_6H_5)_3PO_4 + 3HCl$

   ii. $3C_6H_5OH + POCl_3 \longrightarrow 3C_6H_5Cl + H_3PO_3$
   phosphorous oxychloride

   iii. $C_6H_5OH + SOCl_2 \longrightarrow C_6H_5Cl + SO_2 + HCl$
   thionyl chloride

Bromobenzene (C_6H_5Br) may be formed from phenol, on reaction with either phosphorous pentabromide (PBr$_5$) or phosphorous tribromide (PBr$_3$) or thionyl bromide SOBr$_2$.

Iodobenzene (C_6H_5I) may be prepared on treating phenol with phosphorous triodide (PI$_3$) (phosphorous pentaiodide (PI$_5$) or thionyl iodide (SOI$_2$) do not exist).

$3C_6H_5OH + PI_3 \longrightarrow 3C_6H_5I + H_3PO_3$
phosphorous acid

4. **By Hunsdiecker's reaction:** Silver benzoate, C_6H_5COOAg, is warmed with **bromine, (Br, in CCl_4), or chlorine, (Cl_2 in CCl_4),** for the preparation of **bromobenzene or chlorobenzene**, respectively.

i. $C_6H_5COOAg + Br_2 \xrightarrow{\sim -70\,°C} C_6H_5Br + CO_2 + AgBr$
   silver benzoate   bromo-
                             benzene

ii. $C_6H_5COOAg + Cl_2 \xrightarrow{\sim -70\,°C} C_6H_5Cl + CO_2 + AgCl$
                            chloro-
                            benzene

The **yield of chlorobenzene is very poor**, on account of formation of **various side products**, hence the process is not useful for the preparation of chlorobenzene. **Iodobenzene cannot be prepared by** this method **as a reaction between iodine and silver benzoate** gives an **ester, phenyl benzoate,** as the final product. This reaction is termed **Birnbaum–Simonini reaction.**

$2C_6H_5COOAg + I_2 \longrightarrow C_6H_5COOC_6H_5 + 2AgI + CO_2$
                             phenyl benzoate

5. **By Raschig's method:** In this method, **benzene is reacted with hydrochloric acid (HCl) in presence of oxygen, using $CuCl_2$, as catalyst, at 135 °C, chlorobenzene** is formed.

$C_6H_6 + HCl \xrightarrow[135\,°C]{CuCl_2/O_2} C_6H_5Cl + H_2O$

Physical properties

Chlorobenzene, (C_6H_5Cl), (b.p. = 132°C), bromobenzene, C_6H_5Br (b.p. = 156°C), and iodobenzene (C_6H_5I), (b.p. = 189°C) are colourless liquids, immiscible in water, but miscible in organic solvents.

The **dipole moments of aryl halides is less than that of alkyl halides.**

                ϕX   R X
                1.7 D   2.2 D
            aryl halide   alkyl halide

Chemical properties

As compared to alkyl halides, the aryl halides are inert and undergo nucleophilic reaction less readily.

The **halogen** attached to **benzene ring** is **more firmly attached**, hence they are **inactive.** Aryl halides undergo nucleophilic substitution reactions with great difficulty and often under drastic conditions.

The **inertness and nonreactivity** of **halide group** in **aryl halides** may be explained on the basis of **reasonance effect.** Though **halide group** is **deactivating and should deactivate the phenyl ring,** yet owing to resonance effect, **halogen enters into resonance with phenyl ring,** which is **opposite to —I effect of —Cl group, (resonance effect)** being more powerful than inductive effect, it makes **halide group, acting as o– and p– directing** (when attached alone in the benzene ring).

As **C—Cl bond in phenyl chloride**, acquires a **partial double bond character,** on account of resonance, therefore the cleavage of **C—Cl bond becomes difficult and requires more energy.** Thus, aryl halides undergo various substitution reactions at a much higher temperature and even under drastic conditions.

The bond **length between chlorine and carbon, in aryl halides,** has been found to be **1.69Å,** as compared to **C—Cl bond length in alkyl halides which is 1.77Å.** Moreover, in **aryl halides the —Cl is attached to a** sp^2 **hybridised carbon,** while in **alkyl halides, the halogen is attached to** sp^3 hybridised carbon. Since **electronegativity of** sp^2 **hybridised carbon is more than** sp^3 hybridised carbon, as a result **the electron pair holding carbon and chlorine (i.e. C—Cl), is firmly held by** sp^2 **hybridised carbon than** sp^3 **hybridised carbon.** This **electron pair is not easily displaced in aryl halide (or the C—Cl bond does not become polar easily)** and therefore they are less reactive towards nucleophilic substitution reactions.

i. *Reaction with sodium hydroxide (NaOH) or substitution of —Cl by —OH group:* **Chlorobenzene reacts with 6–8% caustic soda at ~300°C and under 200 atmospheric pressure, in presence of copper salt, whereby —Cl in chlorobenzene is replaced by —OH group, i.e. chlorobenzene is converted into phenol.**

$$C_6H_5Cl + NaOH \xrightarrow[\substack{\text{200 atm. pressure} \\ + \text{ Cu salt}}]{\text{~300°C}} C_6H_5OH + NaCl$$

The reaction is known as **Dow's process** and is used in the **preparation of phenol.**

ii. *Reaction with cuprous cyanide, CuCN, i.e. substitution of —Cl group by —CN group:* **Chlorobenzene reacts with cuprous cyanide,** in presence of pyridine at 200°C to form **benzonitrile or phenyl cyanide (C_6H_5CN).**

$$C_6H_5Cl + CuCN \xrightarrow[\text{200°C}]{\text{pyridine}} C_6H_5\text{—CN} + CuCl$$
$$\text{phenyl}$$
$$\text{cyanide}$$

iii. *Reaction with ammonia (NH_3) i.e. replacement of —Cl group by —NH_2 group:* **Chloro-benzene reacts with ammonia at 200°C and 60 atm. pressure,** in the presence of cuprous oxide as catalyst, to form **aniline.**

$$2C_6H_5Cl + 2NH_3 + Cu_2O \xrightarrow[\text{60 atm. pressure}]{\text{200°C}} C_6H_5NH_2 + Cu_2Cl_2 + H_2O$$
$$\text{aniline}$$

iv. *Reaction with sodamide ($NaNH_2$) in liquid ammonia:* When **chlorobenzene** is treated with **sodamide in liquid ammonia at 75°C** an **amino group enters the ring, adjacent to the position held by chloride in the benzene ring, through the formation of benzyne,** as an intermediate.

v. *Reaction with sodium methoxide (CH_3ONa), i.e. substitution of —Cl by a methoxy group (—OCH_3):* **Chlorobenzene reacts with sodium methoxide,** in the presence of **copper salt, as catalyst, at 200°C,** to form **methoxy benzene or anisole.**

$$C_6H_5Cl + CH_3ONa \xrightarrow[\text{copper salt}]{\text{200°C}} C_6H_5OCH_3 + NaCl$$
$$\text{anisole}$$

vi. *Fittig's reaction, formation of diphenyl*: **Chlorobenzene reacts** with **sodium metal, in the presence of dry ether, to form diphenyl or biphenyl.**

$$C_6H_5—Cl + 2Na + Cl—C_6H_5 \xrightarrow{\text{dry ether}} \underset{\text{diphenyl or biphenyl}}{C_6H_5—C_6H_5} + 2NaCl$$

6. **Wurtz-Fittig's reaction, formation of alkyl benzene: Alkyl halides react with chlorobenzene, (or any aryl halide), in the presence of sodium metal**, taken in dry ether, to form **alkyl benzenes.**

$$C_6H_5—Cl + 2Na + \underset{\text{alkyl halide}}{Cl—R} \xrightarrow{\text{dry ether}} \underset{\text{alkyl benzene}}{C_6H_5—R} + 2NaCl$$

7. **Ullmann reaction: Iodobenzene reacts with copper,** in the presence of dry ether, to form **diphenyl.**

$$C_6H_5I + Cu + IC_6H_5 \xrightarrow{\text{dry ether}} \underset{\text{diphenyl}}{C_6H_5—C_6H_5} + CuI_2$$

8. **Reduction: Chlorobenzene** is reduced to **benzene, by a reducing agent,** consisting of **Ni—Al alloy,** in the presence of **caustic soda.**

$$C_6H_5Cl + 2H \xrightarrow{\text{Ni-Al/NaOH}} C_6H_6 + HCl$$

9. **Reaction with magnesium metal, formation of aryl magnesium halide, (C_6H_5MgX):**

i. $C_6H_5Cl + Mg \xrightarrow{\text{tetrahydrofuran (THF)}} C_6H_5MgCl$

ii. $C_6H_5Br + Mg \xrightarrow{C_2H_5OC_2H_5} C_6H_5MgBr$

iii. $C_6H_5I + Mg \xrightarrow{C_2H_5OC_2H_5} C_6H_5MgI$

10. **Electrophilic substitution:**

i. **Chlorination: Chlorobenzene undergoes electrophilic substitution,** when treated with **Cl_2 in presence of $FeCl_3$** as catalyst to give a mixture of **o– and p– dichlorobenzenes.**

o– p–
dichlorobenzenes

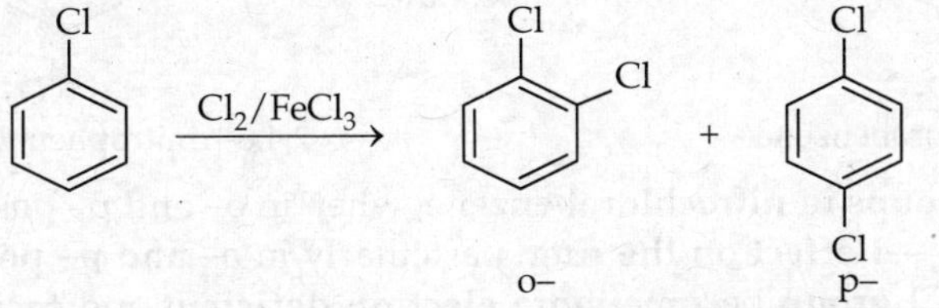

ii. **Nitration: Nitration of chlorobenzene,** with a **nitrating mixture, (conc. HNO_3 and conc. H_2SO_4),** gives a mixture of **o– and p– nitrochlorobenzenes.**

o– p–
chloronitrobenzene

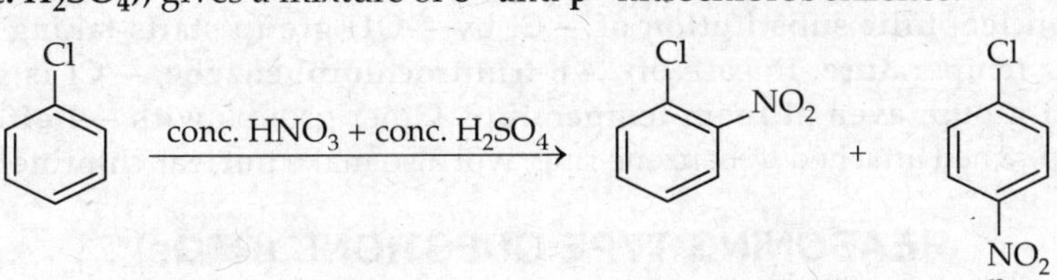

On the other hand, **nitration of chlorobenzene, with a mixture of fuming HNO_3 and fuming H_2SO_4,** gives **2,4,6-trinitrochlorobenzene.**

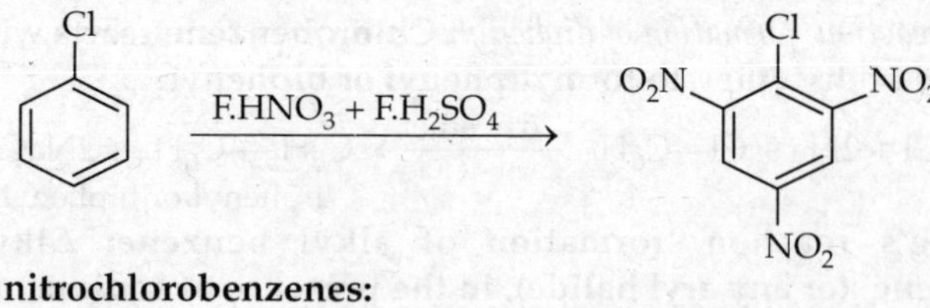

iii. Reactivity of nitrochlorobenzenes:

Chlorine in chlorobenzene, containing one or more **nitro group** in the benzene ring, is much more reactive than that in chlorobenzene, and it readily **undergoes nucleophilic substitution rections.** On boiling with **NaOH, —Cl group is substituted** by a **—OH** group, **o– and p–** nitrochlorobenzene, on boiling with NaOH, are converted into corresponding nitrophenols.

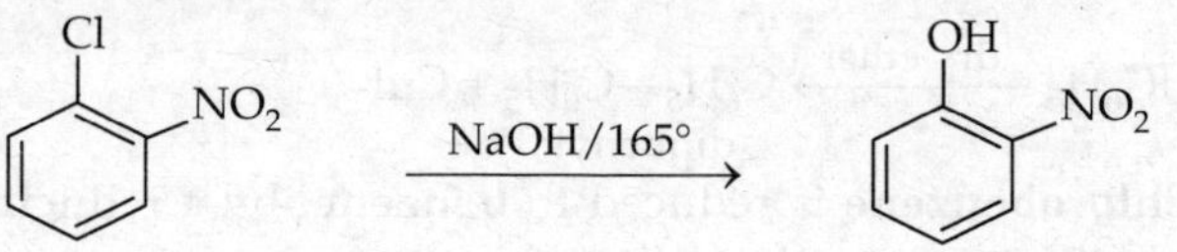

o-nitrochlorobenzene — o-nitrophenol

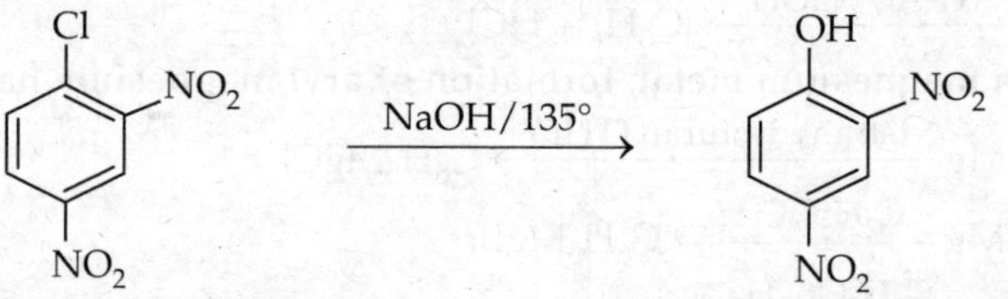

2,4-di nitrochlorobenzene — 2,4-di nitrophenol

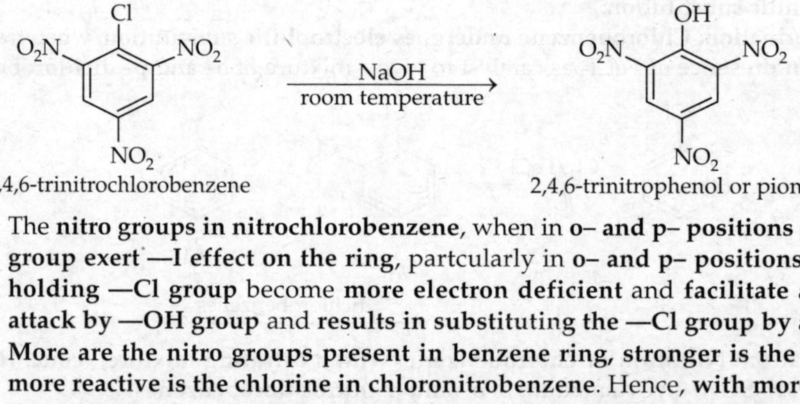

2,4,6-trinitrochlorobenzene — 2,4,6-trinitrophenol or pionic acid

The **nitro groups in nitrochlorobenzene,** when in **o– and p– positions change** to **—Cl group exert —I effect on the ring,** partcularly in **o– and p– positions.** Thus, **carbon holding —Cl group** become **more electron deficient** and **facilitate a nucleophilic attack by —OH group** and **results in substituting the —Cl group by a —OH group. More are the nitro groups present in benzene ring, stronger is the —I effect and more reactive is the chlorine in chloronitrobenzene.** Hence, **with more nitro groups, the nucleophilic substitution of —Cl by —OH group starts taking** place at much **lower temperature.** In case of **2,4,6-trinitrochlorobenzene, —Cl is substituted by —OH group, even at room temperature.** Other groups **with —I effect,** like **—CN group,** when attached to benzene ring, will also make **nuclear chlorine more reactive.**

REASONING TYPE QUESTIONS (RTQs)

(Group-A)

1. The electronegativity of halogens is in the decreasing order of F > Cl > Br > I, yet the reactivity of alkyl halides is in the reverse order. Explain.
2. Allyl chloride is very reactive, but vinyl chloride is not. Explain.

3. Alkyl halides are highly reactive compounds but aryl chlorides are not reactive at all. Explain the reason of the reactivity of ethyl chloride and the nonreactivity of phenyl chloride.

4. The dipole moments of monochlorobenzene and vinyl chloride are lower than those of alkyl chlorides, i.e. methyl chloride. Explain.

5. Arrange the four halobenzenes and the three dichlorobenzene in the order of their decreasing dipole moment.

6. Give the decreasing order of the bond lengths of carbon–halogens in halobenzenes.

7. Chlorobenzene is nonreactive, yet the halogen atom can be substituted by a very strong base like an aminogroup. Suggest a mechanism for this. Also give a suitable explanation, why 2, 6 –dimethyl bromobenzene does not react with a strong base like $NaNH_2$.

8. Discuss the mechanism of nucleophilic substitution reactions.

9. What are elimination reactions? What is its type? Discuss.

10. Discuss the direction of elimination reactions in the case of alkyl halides and Saytzeff's rule.

11. What are products formed, when chloroform reacts with (a) air in the presence of sunlight, (b) aniline and alcoholic KOH on warming.

12. Discuss the precautions that must be taken in storing chloroform for its use as an anesthetic.

13. Write a short note on Kharasch effect or peroxide effect or anti-Markovnikov's rule, taking the example of addition of HBr to propene in presence of a peroxide.

14. Write a mechanism that explains the course of the following reaction:

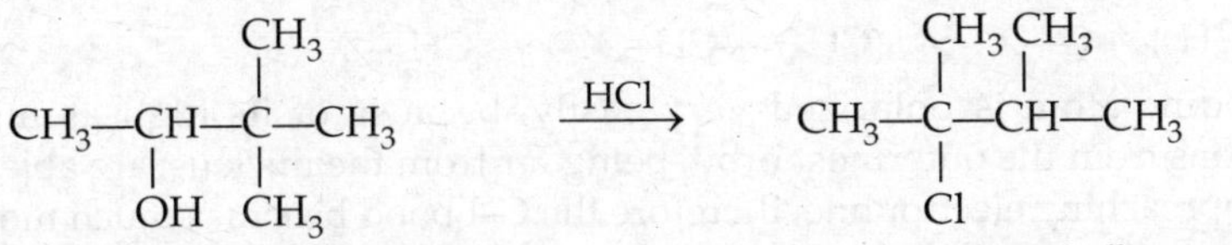

$$CH_3-\underset{\underset{OH}{|}}{CH}-\underset{\underset{CH_3}{|}}{\overset{\overset{CH_3}{|}}{C}}-CH_3 \quad \xrightarrow{HCl} \quad CH_3-\underset{\underset{Cl}{|}}{\overset{\overset{CH_3}{|}}{C}}-\overset{\overset{CH_3}{|}}{CH}-CH_3$$

2 hydroxy-3, 3-dimethylbutane   2-chloro-2, 3-dimethylbutane

15. In the preparation of n-propyl benzene, by treating benzene with *n*-propyl chloride, $n\text{-}CH_3-CH_2-CH_2Cl$, by Friedel-Craft reaction, the compound formed is isopropyl benzene and not *n*-propyl benzene. Suggest a mechanism.

16. Explain the reactivity order of the following compounds.
    Tertiary butyl bromide   >   sec. butyl bromide   >   *n*-butyl bromide.

17. Explain the reaction of ethyl bromide with aq. KOH and alc. KOH.

18. Discuss Wurtz reaction and discuss its limitation?

19. A reaction of vinyl chloride with HBr.

20. Write a short note on Finkelstein reaction.

21. Explain the formation of a mixture of 2-bromo and 3-bromo-pentane, when 3-pentanol reacts with HBr.

22. Explain the formation of 2-methyl butene from 2, 2-dimethyl bromo butane or neo pentyl.

23. What effect should the resonance in vinyl group have on its dipole moment value?

24. The C—Cl bond length is shorter in vinyl chloride as compared to that of allyl chloride.

25. Why are flouro hydrocarbons are chemically inert?

26. Why is hydrogen chloride in gaseous state or in an inert solvent is used to convert alkenes into alkyl halides?

27. A reaction takes place between $Cl_2-C=C-Cl_2$ and Cl_2 only in presence of $AlCl_3$ or in presence of UV radiations, explain.

28. $CH_2=CH-CH_2Cl$ is more reactive than $CH_3CH_2CH_2Cl$.

29. Why it is difficult to replace chlorine in chloro benzene? Explain.

30. Why iodination in benzene is reversible? Explain.

31. Meso-2, 3-dibromobutane, on treatment with NaI gives *trans*-2-butene.
32. A reaction of NaOI with ethanol gives iodoform, but not with NaI.
33. The value of μ for CH_3Cl (μ = 1.91D) is higher than CH_3F (μ = 1.82D).
34. Phosgene is destroyed by treatment with which reagent.
35. Explain the formation of C_6H_5Br, on treating $C_6H_5SO_3H$ with bromine.
36. Benzene reacts with $CHCl_3$ in presence of anhydrous $AlCl_3$.
37. The reduction of $C_6H_5CH=CH—CH=CHCH_3$ with H_2 one equivalent, in presence of Pt as catalyst gives $C_6H_5CH=CH—CH_2CH_2CH_3$.
38. The compound formed when neo pentyl bromide is reacted with tertiary butoxide.
39. There is no reaction between C_6H_5Cl and Mg in presence of diethyl ether, explain.
40. Comment on decreasing order of formation of RMgX with alkyl halides, which is in the order RI > RBr > RCl > RF.
41. Which of the isomers of C_4H_9Br, wil give only one alkene, on dehydrohalogenation.

ANSWERS

1. The reactivity of the alkyl halides (R-X), is in the given order is because of the following reasons:

    a. Among the primary halides 'X' is same but the 'R' differs. The decreasing order of reactivity is

        i. $CH_3—X$ > $CH_3CH_2—X$ > $CH_3—CH_2—CH_2—X$ etc.

        ii. $(CH_3)_3—C—X$ > $(CH_3)_2—CH—X$ > $CH_3—X$

    b. An iodine atom is polarized very easily, because of its high atomic number; the electrons from the outermost orbit, being far from the nucleus, are able to tilt, towards the approaching electrophile, therefore the C–I bond becomes much more polar. Hence the reactivity order.

2. **Vinyl chloride, $CH_2=CH—Cl$—** Chlorine, in vinyl chloride, is inert, because the C—Cl bond in vinyl chloride is stronger than the C—Cl bond in alkyl halide. This is because a *p*-orbital on chlorine, interacts with the I-orbital on the adjacent carbon atom. This new delocalized orbital formed, gives a partial double bond character, to the C—Cl bond in vinyl chloride, as a result, the carbon–chlorine bond assumes partial double bond character, becomes firmly bound and cannot be replaced by another nucleophile easily.

$$\underset{\overset{|}{H}}{\overset{\overset{|}{H}}{C}} == \underset{\overset{|}{H}}{\overset{}{C}} \overset{\frown}{—} \overset{..}{\underset{..}{Cl}} \longrightarrow :\partial^- \underset{\overset{|}{H}}{\overset{\overset{|}{H}}{C}} — \underset{\overset{|}{H}}{\overset{}{C}} = Cl\ \partial^+$$

vinyl chloride

Allyl chloride, $CH_2=CH—CH_2Cl$: The chlorine in allyl chloride is very reactive. Allyl chloride undergoes nucleophilic substitution reactions very easily. Unlike vinyl chloride, $CH_2=CH—Cl$, in allyl chloride, the chlorine atom is attached to the sp^3 hybridized carbon atom and it is separated from the *p*-MO, by a saturated sp^3 hybridized carbon atom; the *p*-orbital, therefore cannot interact.

$$CH_2=CH—\overset{\overset{\displaystyle sp^3\ \text{hybridized carbon}}{\nearrow}}{CH_2}—Cl$$

Allyl chloride

3. **Ethyl chloride** is reactive towards nucleophilic reagents and gets substituted by nucleophiles to form the various derivatives.

The orbitals of carbon atom in C—Cl bond in alkyl chloride are sp^3 hybridized. Therefore, the carbon–chlorine bond in alkyl halide has **less 's' character**, therefore, the C—Cl bond is **weak** and is therefore easily broken and undergoes nucleophilic substitution easily.

On the other hand, the chlorine in chlorobanzene is not reactive at all, under normal conditions. The orbitals of carbon atom in C—Cl bond of chlorobenzene are sp^2 hybridized, therefore the C—Cl bond has **more 's' character** and the bond is **stronger, than the one in alkyl halide**.

C sp^2—Cl p	C sp^3 —Cl p
in chlorobenzene	in alkyl chloride
more 's' character	less 's' character
(a strong bond)	(a weak bond)

In chlorobenzene, there are two forces working opposite to each other. The inductive force due to C—Cl group is acting as electron withdrawing from the ring, acting as a deactivating group. On the other, the lone pair on chlorine atom, chlorobenzene, exist as a resonance stabilized hybrid structure having a partial double bond between carbon and chlorine.

The structures, II, III, IV have a partial double bond character, as is proved by the C—Cl bond length in chlorobenzene. Hence, elthye chloride is more reactive towards nucleophilic substitution.

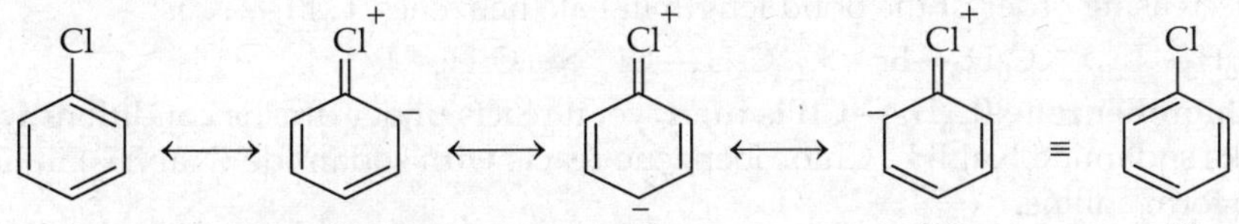

I	II	III	IV	V

C_6H_5—Cl
1.69 Å
a short and strong bond

CH_3—CH_2—Cl
1.78 Å
a weak bond

4. The dipole moment of the monochlorobenzene is $\mu = 1.6$ D

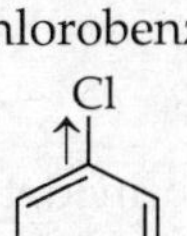

sp^2 hybridized carbon atom

The dipole moment of vinyl chloride $CH_2=CH$—Cl is $\mu =1.4$ D

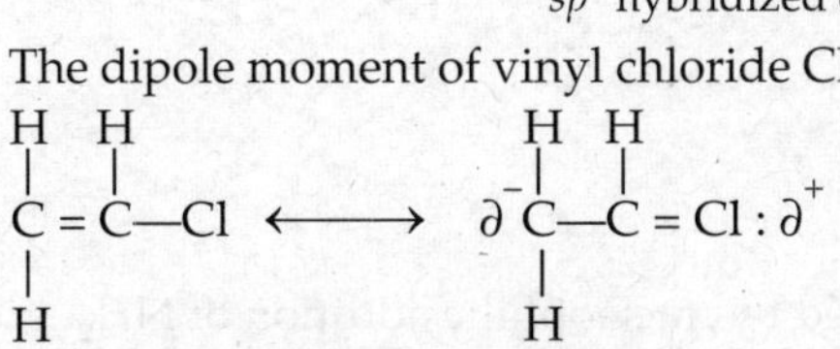

sp^2 hybridized carbon atom

The dipole moment of methyl chloride, CH_3—Cl is $\mu = 1.92$ D

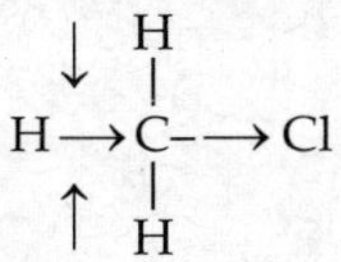

sp^3 hybridized carbon atom

5. **The dipole moments** of chlorobenzene (or aryl halide) and vinyl chloride (or vinyl halides) is lower than alkyl halide due to greater attracting power of sp^2 hybridized carbon atom as compared sp^3 hybridized carbon atom.

   1. Dipole moments of

$$C_6H_5-F\,; \quad \mu = 1.5\ D$$
$$C_6H_5-Cl\,; \quad \mu = 1.69\ D$$
$$C_6H_5-Br\,; \quad \mu = 1.86\ D$$
$$C_6H_5-I\,; \quad \mu = 1.4\ D$$

Dipole moments of o-dichlorobenzene, m-dichlorobenzene and p-dichlorobenzene in the decreasing order:

$$OC_6H_4Cl_2 \quad > \quad mC_6H_4Cl_2 \quad > \quad pC_6H_4Cl_2$$
$$2.30\ D \qquad\qquad 1.48\ D \qquad\qquad zero$$

6. **Carbon–halogen bond length** in alkyl halides as compared to C—X in halobenzene, (C_6H_5X) is given below:

alkyl halides	bond length	aryl halides	bond length	Angstrom units ($1\ \text{Å}=10^{-8}$ cm)
C-F	1.42Å	C_6H_5F	1.6Å	
C-Cl	1.77Å	C_6H_5Cl	1.69Å	
C-Br	1.91Å	C_6H_5Br	1.86Å	
C-I	2.12Å	C_6H_5-I	1.4Å	

Decreasing order of the bond length in halo benzenes, C_6H_5—X, is:

$$C_6H_5{-}I \;>\; C_6H_5{-}Br \;>\; C_6H_5{-}Cl \;>\; C_6H_5{-}F$$

7. **Chlorobenzene (C_6H_5 —Cl) is inert,** yet it reacts under drastic conditions with strong base like sodamide $NaNH_2$. Chlorobenzene reacts with sodamide ($NaNH_2$) in liquid ammonia to form aniline.

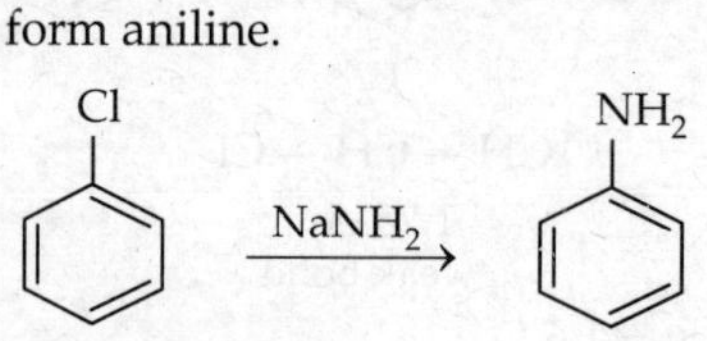

chlorobenzene   aniline

Mechanism:

Step 1 (Elimination step): A benzyne is formed, through removal of H^+ by NH_2^- and loss of a proton

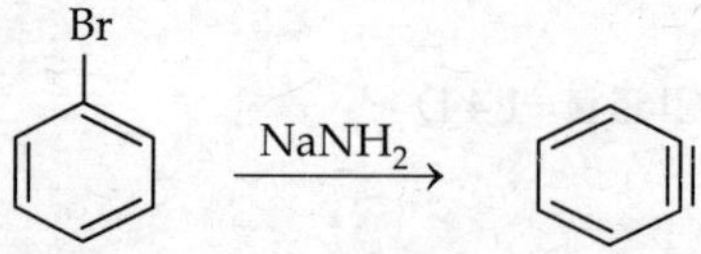

bromobenzene   benzyne

Step 2 (Addition step): A carbanion is formed by nucleophilic addition of NH_2–group to benzyne. The —NH_2 group adds to ortho position to the position held by the leaving group, i.e. —Cl. The carbanion abstracts a proton from the solvent (NH_3) to give the substitution product.

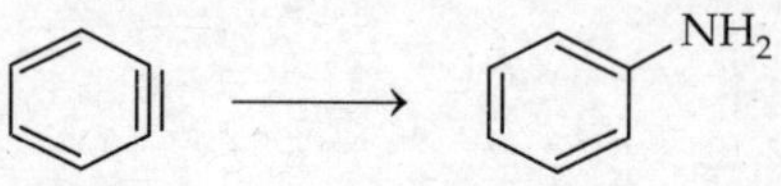

benzyne   aniline

2,6-dimethyl bromobenzene, does not react with $NaNH_2$, because of the steric hindrance. On account of the presence of the two methyl groups at 2 and 6 positions, NH_2 group is not able to approach and attack the —Br atom.

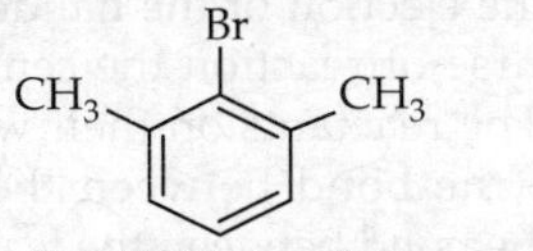

CH₃ (structure) CH₃

2, 6-dimethyl bromobenzene

8. **Nucleophilic substitution reactions:**

   The C—X bond in alkyl halides is polar, because of high electronegativity of the halogen atom, as compared to carbon atom. On account of high polarity of the C—X bond, the carbon becomes electron deficient, and bears a slight positive charge. A nucleophile, therefore, is able to attack and replace the halide ion. These reactions are termed nucleophilic substitution reactions, and represented as 'S_N' reactions ('S' stands for substitution and 'N' stands for nucleophilic substitution).

$$\overset{\delta +}{R}—X + : \overset{-}{Nu} \longrightarrow R—Nu + \overset{\delta -}{\overset{-}{X}}$$

nucelophilic   leaving group

There are two types of nucleophilic substitution reactions:

1. *S_N1 reactions or unimolecular nucleophilic substitution reactions:*
   - The reaction takes place in two steps.
   - The rate of a nucleophilic substitution depends upon the concentration of the substrate only.
   - The reaction is of first order and is represented as S_N1.
   - The reaction is favoured by polar solvents.
   - The reaction can take place with weak bases.
     i. First step: The alkyl halide forms a carbocation by the removal of the leaving group, by solvation. A carbocation is a triangular planar species; all the three groups lie on the same plane, inclined at an angle of 120°, with each other, and the central carbon is in the sp^2 hybridized state.

     ii. Second step: The nucleophile attacks it from either side, from the top as well as from the bottom, to form the nucleophile substituted product.

        In case the starting compound is an optically active compound it contains an asymmetric carbon atom, then the nucleophilic substituted derivative would be a racemic mixture and would be optically inactive (it can be resolved). In other words, racemisation takes place.

     iii. The order of reactivity of the different classes of alkyl halides, towards S_N1 reactions, is as follows:

        $S_N1 — R_3C—X \quad > \quad R_2CH—X \quad > \quad R—CH_2—X$

        $3° \quad\quad > \quad\quad 2° \quad\quad > \quad\quad 1°$

        (depends on stability of carbocation formed from the alkyl halides)

2. *S_N2 reactions or bimolecular nucleophilic substitution reactions:*
   - The reaction takes place in one step only.
   - The rate of the reaction depends upon the concentration of both the substrate as well as that of the attacking nucleophilic reagent.

- The reaction is bimolecular.
- The reaction is favoured by nonpolar solvents.
- Reaction requires strong bases.

In S_N2 reactions, the attack by a nucleophile and the ejection of the halide ion, takes place simultaneously. At one stage during the course of reaction the central carbon atom is attached to five groups or atoms or both. The reactants on their way to form products, pass through a transition state, where the bond between the attacking nucleophile and the substrate is being formed and the bond between the leaving group and the substrate is being broken. Since in S_N2 reactions, the attack of the nucleophile takes from the side opposite to that of the leaving group, namely: X^-. Therefore an inversion takes place in these reactions.

$$X—\overset{\overset{\displaystyle H}{|}}{\underset{\underset{\displaystyle H}{|}}{CH}}—Cl \longrightarrow X—\overset{\overset{\displaystyle H}{|}}{\underset{\underset{\displaystyle H}{|}}{CH}}—Cl \longrightarrow X—\overset{\overset{\displaystyle H}{|}}{\underset{\underset{\displaystyle H}{|}}{C}}—H + Cl$$

- The order of reactivity of the alkyl halides towards S_N2 reactions is as follows:

$$\begin{array}{ccccccc} CH_3—X & > & RCH_2—X & & R_2CH—X & > & R_3C—X \\ 1° & & & > & 2° & > & 3° \\ \text{primary} & & & & \text{secondary} & & \text{tertiary} \end{array}$$

9. **Elimination reactions:** In most of the elimination reactions, two atoms or groups are removed from the adjacent carbon atoms to form a carbon to carbon (C=C) double bond (a π-bond). In the dehydrohalogenation of alkyl halides, a halogen atom and a hydrogen atom are removed from the adjacent carbon atoms to form a double bond between the two carbon atoms. It is called as β-elimination. The reagent used in dehydrohalogenation of alkyl halides, is hot and concentrated strong base, an alcoholic solution of KOH or NaOH.

$$\underset{\text{alkyl halide}}{R—\overset{\overset{\displaystyle H}{|}}{CH}—\overset{\overset{\displaystyle X}{|}}{CH_2}} + KOH \xrightarrow[\Delta]{\text{alcohol}} \underset{\text{alkene}}{R—CH=CH_2} + KX + H_2O$$

There are two types of the elimination reactions:

a. E2 (Elimination reactions):

- It is one step elimination.
- The rate of the reaction depends upon the concentrations of both alkyl halide and the nucleophile

   Rate $= k$ (RX) (Nu:$^-$) 2nd order
- The reaction is therefore a bimolecular reaction, or a second order reaction.
- The halide ion is removed from the α-carbon and the hydrogen atom is removed from the β-carbon simultaneously.
- A strong base and a nonpolar solvent favours E2 elimination reactions.

$$R—\overset{\overset{\displaystyle H}{|}}{CH}—CH_2—X \longrightarrow R—CH=CH_2 + H_2O + X:$$

The order of elimination can be given as:

$$\begin{array}{ccccc} CH_3—X & > & CH_3—CH_2—X & > & R_3C—X \\ 1° & > & 2° & > & 3° \text{ alkyl halides} \end{array}$$

   b. E1 (Elimination reactions):
   - It is a two step elimination.
   - The rate of the reaction depends upon concentration of the alkyl halide.
     Rate = k (R—X)   (first order)
   - The reaction is a unimolecular reaction or a first order reaction.

     Step 1. Alkyl halide slowly ionises to form carbocation.

     Step 2. A proton is then abstracted by the nucleophile, from the carbocation to form alkene.
   c. A polar solvent and a weak base favours the reaction.
$$R_3C—X \; > \; R_2—CH—X \; > \; RCH_2—X$$
$$3° \quad > \quad 2° \quad > \quad 1°$$

10. **The dehydrohalogenation of an alkyl halide, can yield more than one alkene, then according to Saytzeff's rule**, the main product is the highly substituted alkene. In other words, alkene formed in larger quantity (or the major product) is the one that has more alkyl groups, attached to the doubly bonded carbon atoms, for two alkenes are possible, when 2-bromobutane is heated with alcoholic KOH.

$$CH_3CHBrCH_2CH_3 \xrightarrow{\text{alc. KOH}} CH_3CH=CH—CH_3 + CH_3—CH_2CH=CH_2$$
$$\text{2-butene (80\%)} \qquad\qquad \text{1-butene (20\%)}$$

11. Chloroform ($CHCl_3$) reacts with oxygen (or air), in the presence of sunlight to form carbonyl chloride or phosgene ($COCl_2$) a very poisonous substance. As phosgene is soluble in chloroform, it remains dissolved in chloroform.

    Chloroform is used as an anesthetic as it contains phosgene and can harm the patient. Therefore, chloroform must be free from phosgene or it should be destroyed before using chloroform.

$$2CHCl_3 + O_2 \xrightarrow{\text{sunlight}} 2COCl_2 + 2\,HCl$$
$$\text{chloroform} \qquad\qquad \text{phosgene}$$

Test for the presence of phosgene ($COCl_2$) in chloroform: Formation of a white precipitate with a $AgNO_3$ solution with chloroform indicates the presence of $COCl_2$ in chloroform.
$$COCl_2 + 2AgNO_3 + H_2O \longrightarrow AgCl + 2HNO_3 + CO_2$$
$$\text{white}$$
$$\text{precipitate}$$

   b. On warming aniline ($C_6H_5NH_2$) with $CHCl_3$, in the presence of alcoholic KOH, it is converted into phenyl cyanide, which has a very sharp and untolerable smell.
$$C_6H_5NH_2 + CHCl_3 + 3KOH \longrightarrow C_6H_5NC + 3KCl + 3H_2O$$
   This reaction is known as a *carbylamine test* or an *isocyanide test*. This serves as test for an amino group (–NH$_2$ group) as well as a test for chloroform.

12. Precautions:
   - A brown coloured bottled is invariably used to store, so as to minimize the exposure to light.
   - The bottle is filled up to the brim, in order to expel air from the bottle.
   - Some ethyl alcohol is added to destroy phosgene, in case it is formed. Ethyl alcohol reacts with $COCl_2$ to form an insoluble ethyl carbonate, which can be filtered off.
$$2\,C_2H_5—OH + COCl_2 \longrightarrow (C_2H_5)_2CO_3 + 2HCl$$
$$\text{ethyl alcohol} \quad \text{phosgene} \qquad \text{ethyl carbonate}$$
$$\text{insoluble}$$

13. **Peroxide effect or anti-Markovnikov's rule or Kharasch reaction:** The addition of HBr to a double bond in presence of a peroxide is a free radical reaction.

Following steps are involved:

Step 1. Peroxide dissociates to give alkoxy free radical.
$$R-O-O-R \longrightarrow 2\,R-O^\bullet + energy$$

Step 2. Free radical attacks HBr to form a bromine free radical.
$$R-O^\bullet + H:Br \longrightarrow R-OH + Br^\bullet$$

Step 3. $CH_3-CH{=}CH_2 + Br^\bullet \longrightarrow \overset{\bullet}{C}H_3-\overset{\bullet}{C}H-CH_2Br^\bullet \longrightarrow CH_3-CHBr-CH_2^\bullet$

　　　　　　　　　　　　　　　　　　2° free radical　　　　　　　　1° free radical

Since a 2° free radical is more stable than a 1° free radical, the further reaction thus takes place with the 2° free radical. The other free radical reverts back to the starting material.

$$CH_3o\ \overset{\bullet}{C}Ho\ CH_2Br + H:Br \longrightarrow CH_3o\ CH_2o\ CH_2Br + Br^\bullet$$

　　　　　　　　　　　　　　　　　　　　　　　　　　n-propyl bromide

HCl and HI do not give anti-Markovnikov's product, in the presence of peroxide. This is because
- H—Cl bond requires 103 kcal of heat energy, to break the bond between H and Cl, since this much amount of energy is not available in the system, therefore HCl does not give **anti-Markovnikov's product.**
- H—I bond requires 87 kcal of heat energy to break the bond between H and I, as this much amount of energy is available, therefore the H—I bond breaks to give H• and I•. But as soon I• is liberated, it combines immediately with another I atom to form an iodine molecule, rather than attacking the double bond in alkene.

$$I\bullet\ +\ I\bullet \longrightarrow I_2$$

　　　　　　　　　　　　Iodine
　　　　　　　　　　　　molecule

(The energy for the reaction is provided by the decomposition of peroxide)

14. In the formation of the product, following steps take place.

   i. Formation of a carbocation:

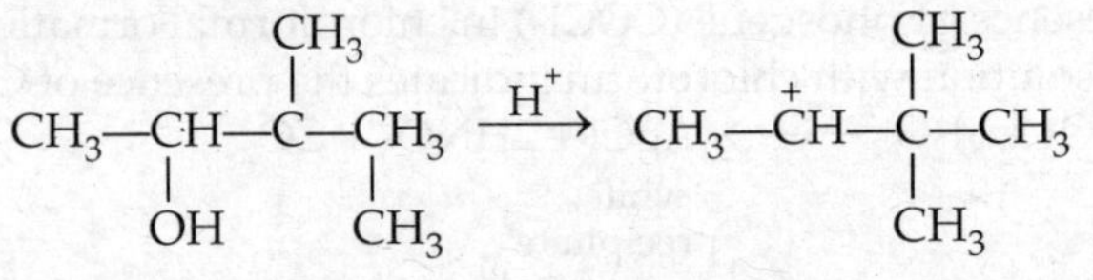

   ii. Rearrangement of a lower stability of carbocation to form a carbocation of higher stability:

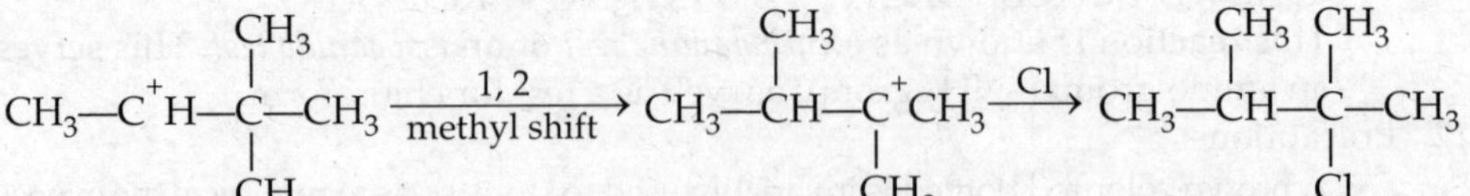

   2° carbocation　　　　　　　3° carbocation　　　　2,3-dimethyl 2-carbocation

15. It is an example of Friedel–Craft reaction.

   $CH_3CH_2CH_2Cl$ reacts with anhydrous $AlCl_3$ to liberate a propyl carbocation.

   $$CH_3CH_2CH_2Cl + AlCl_3 \longrightarrow CH_3CH_2CH_2^+ + AlCl_4^-$$

   This 1° carbocation then rearranges to form a more stable 2° carbocation by shifting a hydride ion (H^-), which reacts with benzene to form isopropyl benzene.

$$\underset{\underset{H}{|}}{CH_3CHCH_3} \xrightarrow{\text{1,2-H}^- \text{shift}} \underset{\text{2° carbocation}}{CH_3\overset{+}{CH}CH_3}$$

isopropyl

16. Alkyl halides, undergo various changes, it first forms an alkyl carbocation
    The stability of the carbocation is as follows:
    3° carbocation > 2° carbocation > 1° carbocation > methyl carbocation
    Alkyl groups are electrons repelling or electron donating, with the result, that the partial positive charge on the central carbon atom is partially compensated. Since a 3° carbocation contains 3° alkyl groups, it is more stable than a 2° carbocation, which contains two alkyl groups which is more stable than 1° carbocation, which contains only one alkyl group in the species.
    Hence the order of reactivity of the various butyl chlorides is in the order as given.

17. The reaction of ethyl bromide with aq. KOH gives an alcohol. In this reaction, the OH is a nucleophile, therefore halide is substituted by the OH group.
    $$KOH \rightleftharpoons K^+ + OH^-$$
    $$C_2H_5Br + OH^- \longrightarrow C_2H_5OH + KBr$$
    On the other hand, a reaction of ethyl bromide with alc. KOH forms ethylene
    $$C_2H_5Br + KOH \longrightarrow C_2H_4 + KBr + H_2O$$

18. **Wurtz reaction: 1° and 2° alkyl halides** react with sodium metal in presence of dry ether to form a higher alkane, containing double the number of carbon atom as present in the parent alkyl halide. The molecule formed has a symmetrical structure.
    $$\underset{\text{1° alkyl halide}}{RCH_2X} + 2Na + XCH_2R \xrightarrow{\text{ether}} RCH_2-CH_2R + 2NaX$$
    $$\underset{\text{2° alkyl halide}}{R_2CHX} + 2Na + XCHR_2 \longrightarrow R_2CH-CHR_2 + 2NaX$$
    A reaction of a 3° alkyl halide with the sodium metal may lead to the formation of alkene as a major product.

19. A reaction of vinyl chloride with HBr
    $$CH_2=CHCl + HBr \quad CH_2=CH-Cl \rightleftharpoons CH_2^- -\overset{+}{CH}=Cl \xrightarrow{HBr} CH_3-CH(Cl)Br$$

20. **Finklestein reaction:** Halide exchange reactions are also known as **Conant–Finklestein reaction**. Alkyl halides are heated with sodium iodide in the presence of acetone, when the alkyl iodides are obtained.
    $$\underset{\text{alkyl halide}}{RX} + \underset{\text{sodium iodide}}{NaI} \xrightarrow{\text{acetone}} \underset{\text{alkyl iodide}}{RI} + \underset{\text{sodium halide}}{NaX} \quad (X = Cl \text{ or } Br)$$

21. A hydride shift takes place:
    $$\underset{\underset{OH}{|}}{CH_3-CHCHCH_2CH_3} \xrightarrow[H_2O]{HBr} \underset{\underset{\underset{\text{3-bromopentane}}{Br}}{|}}{CH_3CH_2CHCH_2CH_3} \longrightarrow \underset{\underset{\underset{\text{2-bromopentane}}{Br}}{|}}{CH_3\overset{+}{CH}CH_2CH_2CH_3}$$

22. A methide shift (CH_3^-) takes place

$$CH_3C\overset{\displaystyle CH_3}{\underset{\displaystyle CH_3}{|}}CH_2-X^- \longrightarrow CH_3\overset{\displaystyle CH_3}{\underset{\displaystyle CH_3}{C}}-CH_2^+ \longrightarrow CH_3\overset{\displaystyle CH_3}{C^+}CH_2CH_3$$

        neo pentyl   1° carbocation   3° carbocation

23. The dipole moment value decreases.
24. On account of resonance in vinyl chloride.
25. Perflouro compounds are inert on account of back bonding of flourine with carbon.
26. In aqueous solution, Cl_2 reacts as HOCl
    $$Cl_2 + 2H_2O \longrightarrow HCl + HOCl$$
    Therefore addition of chlorine is done in an inert solvent.
27. Addition of Cl_2 to $Cl_2C=CCl_2$ takes place either in presence of high energy or in presence of $AlCl_3$ as aluminum binds the double bond and make the compound reactive.
28. The $CH_2=CH-CH_2Cl$ undergoes S_N1 reation to form allyl carbocation which is more stable than propyl carbocation on account of resonance.
29. In chlorobenzene owing to resonance, develops a partial double bond character between C and Cl which becomes stronger and diffcult to break.

30. The reaction is reversible owing to the formation of HI, in a reaction between benzene and I_2. HI is a powerful reducing agent.
    $$C_6H_6 + I_2 \rightleftharpoons C_6H_5I + HI$$

31. $$CH_3-\overset{\displaystyle }{\underset{\displaystyle Br}{CH}}-\overset{\displaystyle }{\underset{\displaystyle Br}{CH}}-CH_3 + NaI \longrightarrow \overset{CH_3}{\underset{I}{\diagup}}CH-CH\overset{I}{\underset{CH_3}{\diagdown}} \xrightarrow{I_2} \overset{CH_3}{\underset{H}{\diagup}}C=C\overset{H}{\underset{CH_3}{\diagdown}}$$

32. $CH_3CH_2OH + NaOI \longrightarrow CH_3CHO + NaI$
    $CH_3CHO + 3NaOI \longrightarrow CI_3CHO + 3NaI$
    $CI_3CHO + NaOH \longrightarrow CHI_3 + HCOONa.$
                          Iodoform
33. $\mu = d \times$ charge $\longrightarrow d$ is larger in the case of CH_3Cl.
34. Phosgene is destroyed by addition of ethanol. Reaction is as follows:
    $$COCl_2 + 2C_2H_5OH \longrightarrow (C_2H_5)_2CO_3 + 2HCl$$
35. $C_6H_5SO_3H + Br_2 \xrightarrow{FeBr_2} C_6H_5Br + H_2SO_3$
36. $3C_6H_6 + CHCl_3 \xrightarrow{AlCl_3} (C_6H_5)_3CH + 3HCl$
                   triphenyl methane
37. The reduction of $C_6H_5CH=CH-CH=CH-CH_3$ to form $C_6H_5CH=CH-CH_2-CH_2-CH_3$ because the double bond of the CH=CH, attached directly to the ring enters into conjugation with the benzene ring, there by making it less reactive.

38. 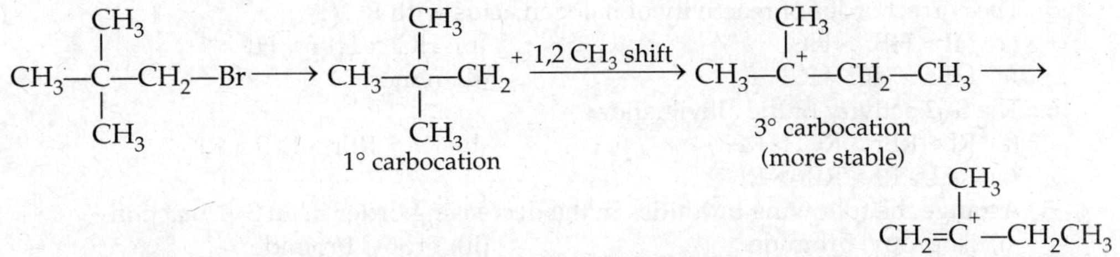
$$CH_3-\underset{\underset{CH_3}{|}}{\overset{\overset{CH_3}{|}}{C}}-CH_2-Br \longrightarrow CH_3-\underset{\underset{CH_3}{|}}{\overset{\overset{CH_3}{|}}{C}}-CH_2^+ \xrightarrow{+\,1,2\ CH_3\ shift} CH_3-\overset{\overset{CH_3}{|}}{\overset{|}{C}}{}^+-CH_2-CH_3 \longrightarrow$$

1° carbocation

3° carbocation
(more stable)

$$CH_2=\overset{\overset{CH_3}{|}}{C}{}^+-CH_2CH_3$$

39. No reaction takes place between C_6H_5Cl and Mg, in presence of ether, because dielectric constant of ether is very low, it does not help ionise C_6H_5Cl.

40. It is based on the dissociation energy for $HI = 70\ kcal/mol$

$$HBr = 80\ kcal/mol$$
$$HCl = 110\ kcal/mol$$

41. Tert butyl bromide will give only one alkene.

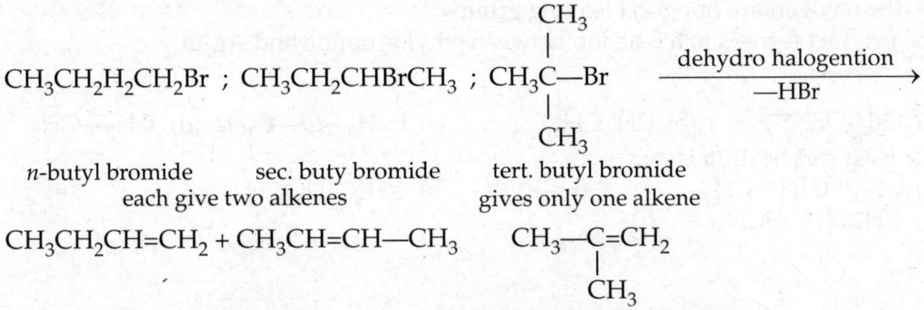

$$CH_3CH_2H_2CH_2Br \ ; \ CH_3CH_2CHBrCH_3 \ ; \ CH_3\underset{\underset{CH_3}{|}}{\overset{\overset{CH_3}{|}}{C}}-Br \xrightarrow[-HBr]{dehydro\ halogention}$$

n-butyl bromide   sec. buty bromide   tert. butyl bromide
each give two alkenes   gives only one alkene

$$CH_3CH_2CH=CH_2 + CH_3CH=CH-CH_3 \qquad CH_3-\underset{\underset{CH_3}{|}}{C}=CH_2$$

MULTIPLE CHOICE QUESTIONS

(Group-B)

1. The compound formed in the following reaction.
   $$2Br-CH_2-CH_2-Cl + Mg \longrightarrow$$

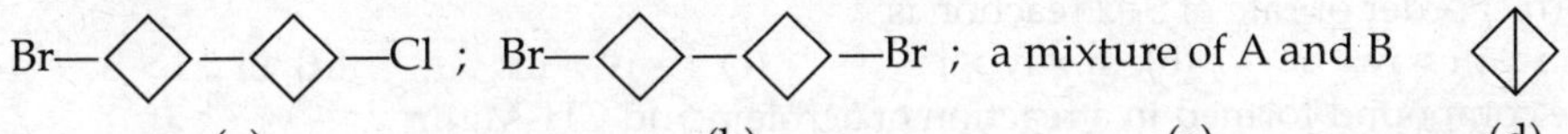

   (a) Br—⬡—⬡—Cl ;   (b) Br—⬡—⬡—Br ;   (c) a mixture of A and B   (d) ◁▷

2. 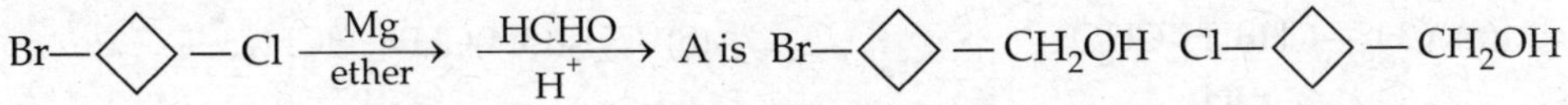 Br—⬡—Cl $\xrightarrow[ether]{Mg}$ $\xrightarrow[H^+]{HCHO}$ A is   (a) Br—⬡—CH_2OH (b) Cl—⬡—CH_2OH

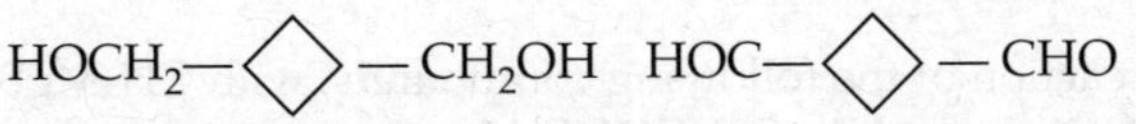

   (c) $HOCH_2$—⬡—CH_2OH (d) HOC—⬡—CHO

3. The decreasing order of the following species on their nucleophilicity is
   (a) $OH > CH_3O^- > CH_3^- > NH_2^-$
   (b) $CH_3^- > NH_2^- > OH^- > F^-$
   (c) $OH^- > CH_3COO^- > OCH_3^- > C_6H_5O^-$

4. The order of leaving group
   (a) $I^- > Br^- > Cl^- > F^-$ (b) $F^- > Cl^- > Br^- > I^-$
   (c) $Br^- > I^- > Cl^- > F^-$ (d) none

5. The correct order of reactivity of halogen acids with ROH
   (a) HI > HBr > HCl
   (b) HCl > HBr > Hl
   (c) HBr > HI > HCl
   (d) none

6. The S_N2 activity of the alkyl halides
   (a) RI > RBr > RCl > RF
   (b) RF > RBr > RCl > RI
   (c) RCl > RI > RBr > RF

7. Arrange the following bromides in the decreasing order in an S_N1 reaction
   (i) isopropyl bromide
   (ii) propyl bromide
   (iii) tertiary butyl bromide
   (iv) methyl bromide
   (a) iii > i > ii > iv   (b) iv > ii > i > iii   (c) i > ii > iv > iii   (d) i > ii > iii > iv

8. The vinyl halides are unreactive towards nucleophilic reaction
   (a) C—Cl bond is strong
   (b) halogen is attached to sp^2 carbon
   (c) because a double bond character develops in C—Cl bond
   (d) the halogen are not good leaving groups

9. The product formed in a reaction between ethyl bromide and Ag_2O.

   (a) C_2H_5OH (b) $CH_3 CHO$ (c) $C_2H_5—O—C_2H_5$ (d) $\underset{CH_2—CH_2}{\overset{O}{\triangle}}$

10. The least carbocation is
    (a) $CH_3—C^+H—CH_3$
    (b) $CH_3C^+HOCH_3$
    (c) $CH_3CH^+COCH_3$

11. $(CH_3)_2—\underset{\underset{i}{|}}{\overset{\overset{Cl}{|}}{C}}—CH_2CH_3 \xrightarrow{\text{Hoffman's elimination}} \underset{ii}{(CH_3)_2C=CHCH_3} \xrightarrow{HBr}$

    $(CH_3)_2CH—\underset{\underset{iii}{|}}{\overset{\overset{Br}{|}}{CH}}—CH_3(C) \downarrow HBr/uv\ \underset{iv}{(CH_3)_2CH—CHBr—CH_3}$

    Correct order of rate of S_N2 reaction is
    (a) i > iii > iv   (b) iii > iv > i   (c) i > iv > iii   (d) iii > i > iv

12. The compound formed in a reaction of acrolein and CH_3MgBr
    $CH_2=CH—CHO + CH_3MgBr \longrightarrow$
    (a) $CH_3—CH_2CH_2CHO$
    (b) $CH_2(OH)CH=CHCH_3$
    (c) $CH_2=CH\overset{\overset{OH}{|}}{C}—CH_3$

13. The decreasing order of rate of reaction of the following compounds, with CH_3MgBr
    (i) CH_3COCl (ii) CH_3CHO (iii) $CH_3COOC_2H_5$
    (a) A > B > C   (b) C > A > B   (c) C > B > A

14. Identify the product

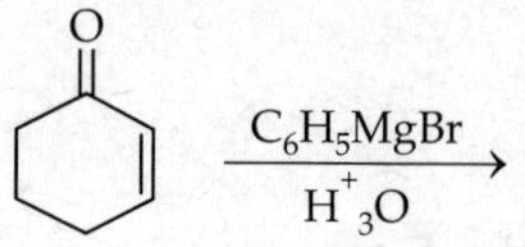

    $\xrightarrow[H^+_3O]{C_6H_5MgBr}$

(a) 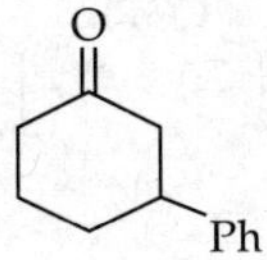 [cyclohexanone with Ph substituent]

(b) 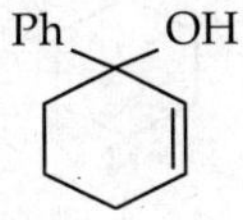 [Ph, OH on cyclohexene]

(c) 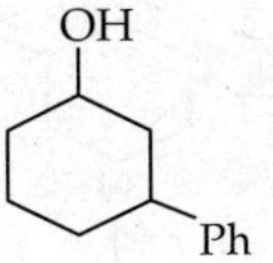 [OH on cyclohexane with Ph]

(d) 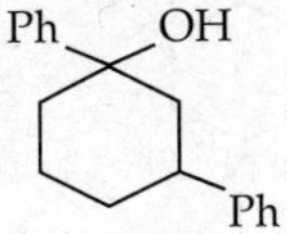 [Ph, OH on cyclohexane with Ph]

15. The reaction $CH_2=CH-\overset{O}{\underset{||}{C}}-CH_3$ + $(CH_3)_2CuLi \longrightarrow$

(a) $CH_3CH_2CH_2COCH_3$

(b) $CH_3CH_2\overset{OH}{\underset{|}{C}}(CH_3)_2$

(c) $CH_2=CH_2CH_2\overset{O}{\underset{||}{C}}CH_3$

(d) 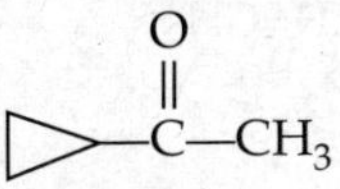 [cyclopropyl]$-\overset{O}{\underset{||}{C}}-CH_3$

16. The product formed in following reaction:

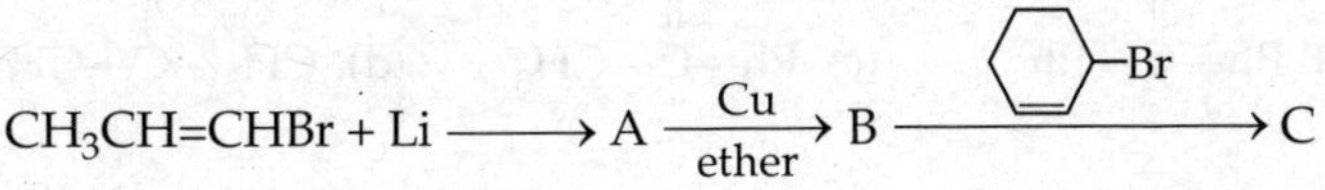

$CH_3CH=CHBr + Li \longrightarrow A \xrightarrow[\text{ether}]{Cu} B \xrightarrow{\text{[cyclohexenyl]}-Br} C$

(a) $(CH_3-CH=CH_2)_2\,Cu\,Li$

(b) 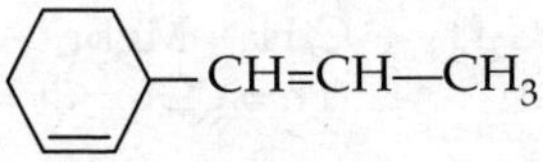 [cyclohexenyl]$-CH=CH-CH_3$

(c) 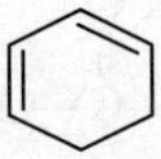 [cyclohexadiene]

(d) none

17. Predict the compound formed in the reaction.

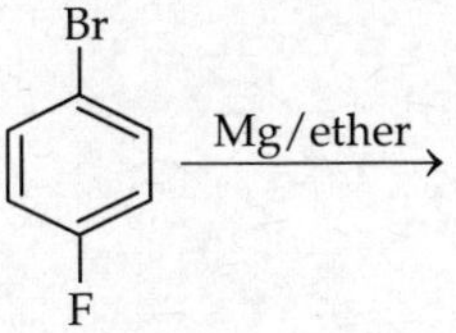

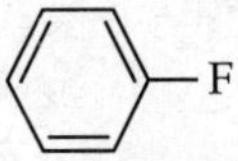

[p-bromofluorobenzene: Br top, F bottom] $\xrightarrow{\text{Mg/ether}}$

[benzene]$-F$

(a)

$F-$[benzene]$-MgBr$

(b)

[benzene]$-Br$

(c)

$Br-$[benzene]$-MgBr$

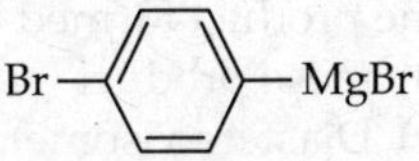

(d)

18. 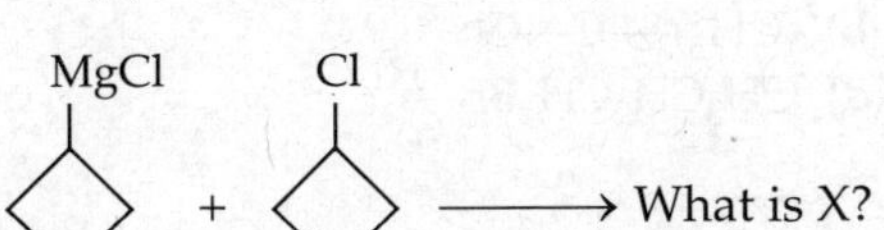 MgCl + 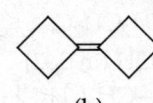 Cl $\longrightarrow$ What is X?

(a)

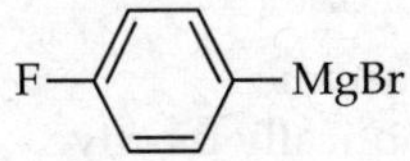

(b)

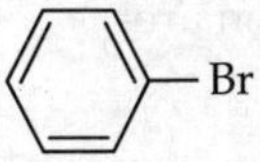

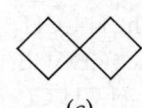

(c)

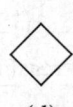

(d)

19.

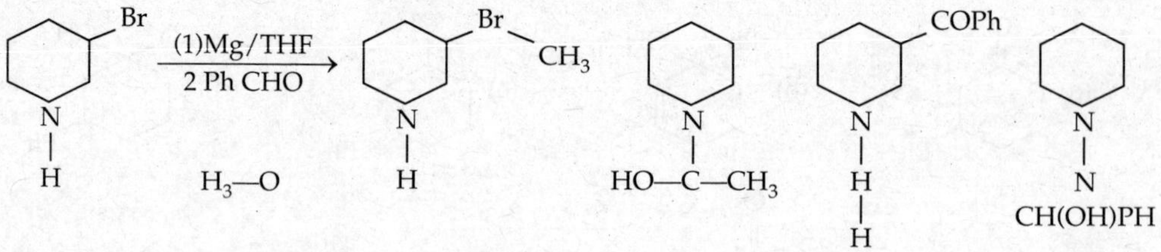

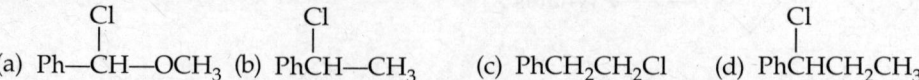

(1)Mg/THF
2 Ph CHO

H₃—O

(a)   (b)   (c)   (d)

20. $CH_2=CHCH_2\ Br \xrightarrow[\text{(2-mole)}]{Mg/THF}$ z. What is z?
   (2-mole)

   (a) $CH_2=CHCH_2—CH=CH_2$ (b) $CH_2=CH—CH_2—CH_2—CH=CH_2$
   (c) $CH_2=CH\ CH=CH_2$ (d) $CH_3—CH=CH—CH=CHCH_2$

21. $PhMgBr + CH_3—CN$

   (a) $CH_3\underset{Ph}{\overset{Ph}{\underset{|}{\overset{|}{C}}}}—OH$ (b) $Ph—\overset{O}{\overset{||}{C}}—Ph$ (c) $Ph—\overset{O}{\overset{||}{C}}—CH_3$ (d) $CH_3—\overset{O}{\overset{||}{C}}—CH_3$

22. $Cl—CO—OC_2H_5 + C_6H_5—MgBr \longrightarrow$
      (in excess) $(Ph = C_6H_5;\ Et = C_2H_5\ Me = CH_3\ groups)$

   (a) $Ph\underset{PH}{\overset{OH}{\underset{|}{\overset{|}{C}}}}—Ph$ (b) $Ph—\overset{O}{\overset{||}{C}}—Cl$ (c) $Ph—\overset{O}{\overset{||}{C}}—Ph$ (d) $Ph—\overset{O}{\overset{||}{C}}—OEt$

23. The reaction product:

   $C_6H_5\ MgBr + CH_3—CH_2—CHO \xrightarrow{H_2O} CH_3—CH_2—\underset{PH}{\overset{C_6H_5}{\underset{|}{\overset{|}{C}}}}—OH$

   The product formed in reaction are-
   (a) meso product   (b) racemic
   (c) Diastereo isomer   (d) optically inactive

24. Which of the following will react faster with methanol?
   (a) $(CH_3)_3C—Br$ (b) $(CH_3)_3—C—Cl$
   (c) $(CH_3)_2CH—Br$ (d) $CH_3CH_2CH_2Br$

25. Which will react faster with aq NaOH

   (a) $Ph—\overset{Cl}{\overset{|}{C}}H—OCH_3$ (b) $Ph\overset{Cl}{\overset{|}{C}}H—CH_3$ (c) $PhCH_2CH_2Cl$ (d) $Ph\overset{Cl}{\overset{|}{C}}HCH_2CH_3$

26. The decreasing order of reactivity with aqueous NaOH (in S_N2)
   (i) C_2H_5Cl (ii) $C_2H_5CH_2Cl$ (iii) $HO—CH_2CH_2Cl$ (iv) $(CH_3)_2CH—Cl$
   (a) iii > ii > iv > i   (b) i > ii > iii > iv
   (c) iv > iii > ii > i   (d) none

27. Which compound will be formed in following reaction?

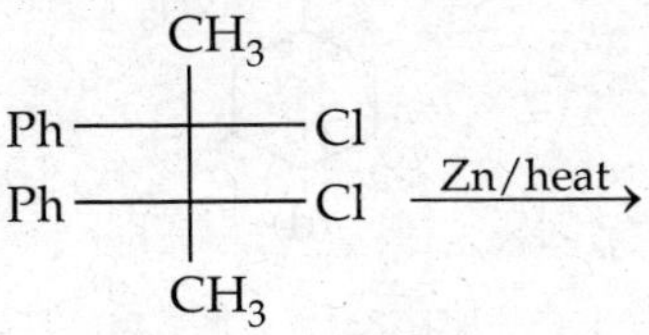

$$\underset{\substack{Ph \\ Ph}}{\overset{CH_3}{\underset{CH_3}{\bigg|}}}\overset{Cl}{\underset{Cl}{\bigg|}} \xrightarrow{Zn/heat}$$

(a) 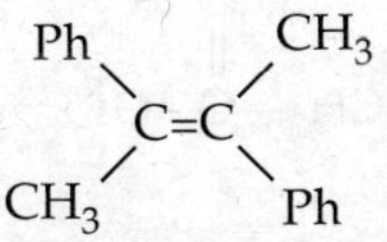 $\underset{CH_3}{\overset{Ph}{\diagup}}C=C\underset{Ph}{\overset{CH_3}{\diagup}}$ (b) $\underset{CH_3}{\overset{Ph}{\diagup}}C=C\underset{CH_3}{\overset{Ph}{\diagup}}$ (c) $\underset{Ph}{\overset{Ph}{\diagup}}C=C\underset{CH_3}{\overset{CH_3}{\diagup}}$ (d) none

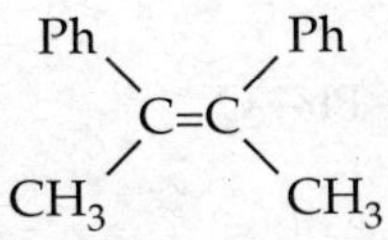

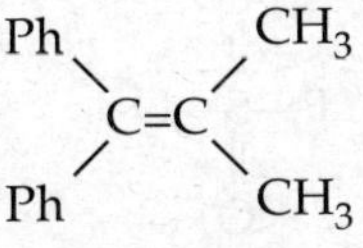

28. Which compound will be formed in the following reaction?

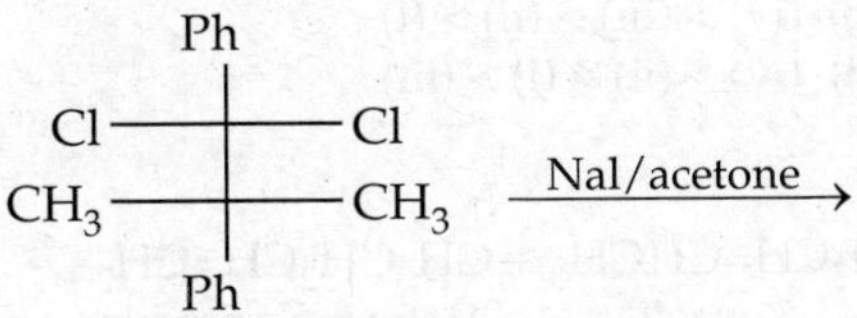

$$\underset{\substack{CH_3 \\ Ph}}{\overset{Ph}{\underset{}{\bigg|}}}\overset{Cl}{\underset{CH_3}{\bigg|}} \xrightarrow{NaI/acetone}$$

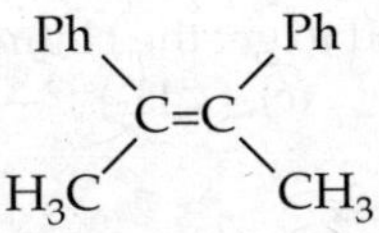

(a) 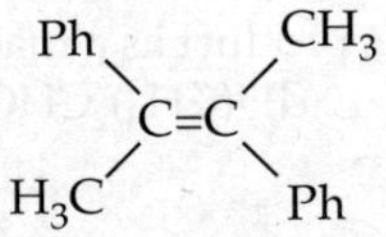 $\underset{H_3C}{\overset{Ph}{\diagup}}C=C\underset{Ph}{\overset{CH_3}{\diagup}}$ (b) $\underset{H_3C}{\overset{Ph}{\diagup}}C=C\underset{CH_3}{\overset{Ph}{\diagup}}$ (c) $\underset{Ph}{\overset{Ph}{\diagup}}C=C\underset{CH_3}{\overset{CH_3}{\diagup}}$ (d) none

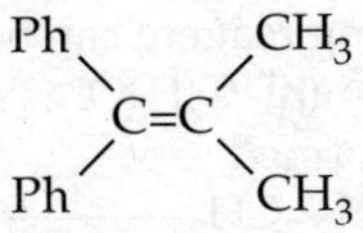

29. The major product in the following reaction

$\xrightarrow{H_2O}$

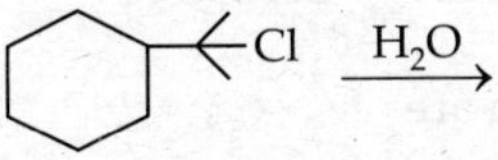

(a) 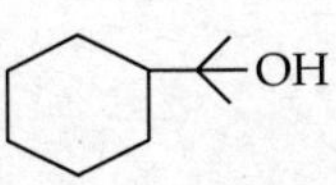   (b)

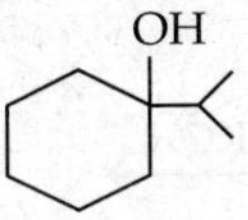

(c) a ring expansion takes place

30. The reaction product

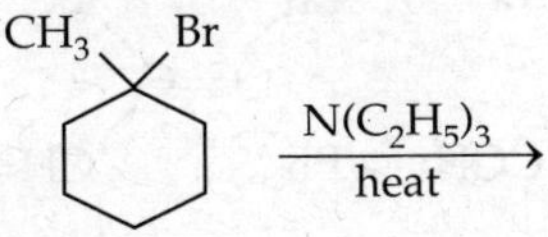

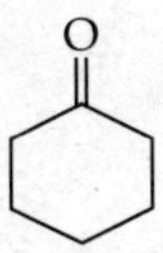

$\xrightarrow[\text{heat}]{N(C_2H_5)_3}$

(a) 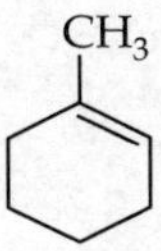   (b) 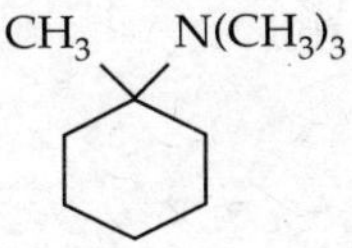   (c) 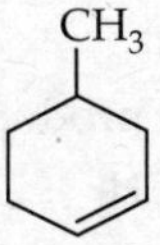   (d)

31. 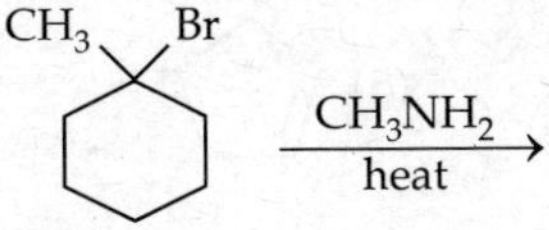 $\xrightarrow[\text{heat}]{CH_3NH_2}$

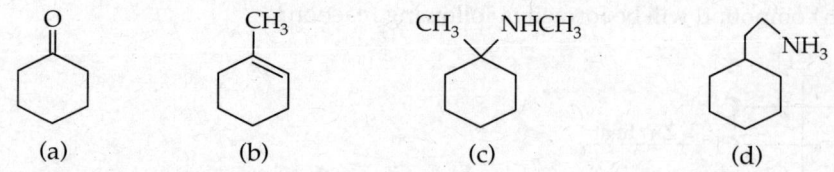

(a) (b) (c) (d)

32. The decreasing order of the leaving groups

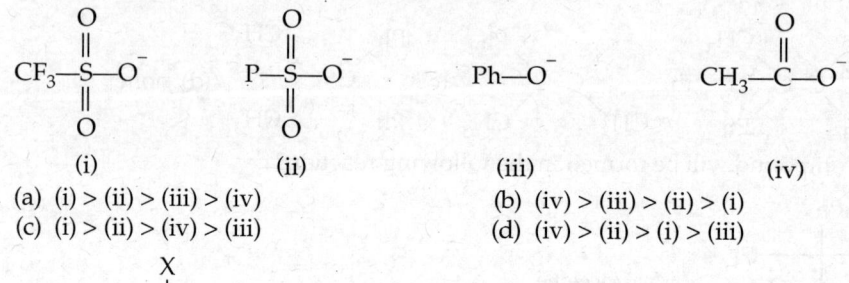

(i) (ii) (iii) (iv)

(a) (i) > (ii) > (iii) > (iv) (b) (iv) > (iii) > (ii) > (i)
(c) (i) > (ii) > (iv) > (iii) (d) (iv) > (ii) > (i) > (iii)

33.
$$CH_3-CH_2-\overset{\overset{\displaystyle X}{|}}{CH}-CH_3 \xrightarrow{\text{E2 elimiation}} CH_3CH=CHCH_3 + CH_3CH_2CH=CH_2$$

Saytzeff'   Hofmann's product

Which of the following reagent can be used to get the Hofmann's product as a major product
(a) CH_3O^- (b) $C_2H_5O^-$ (c) $(CH_3)_3C-O^-$ (d) $(CH_3)_2CHO^-$

34. The major product formed
$$(CH_3)_2C=CH\ Li + CH_2=CH_2 \xrightarrow[CO_2/H_3O^+]{}$$

(a) $(CH_3)_2-CH-COOH$ (b) $(CH_3)_2-CH-CH_2-CH_2COOH$
(c) $CH_2=CH-COOH$ (d) all correct

35. The alkenes is expected to be formed in the following reaction are

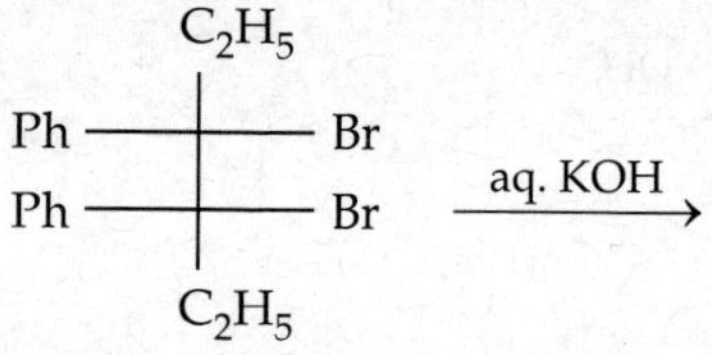

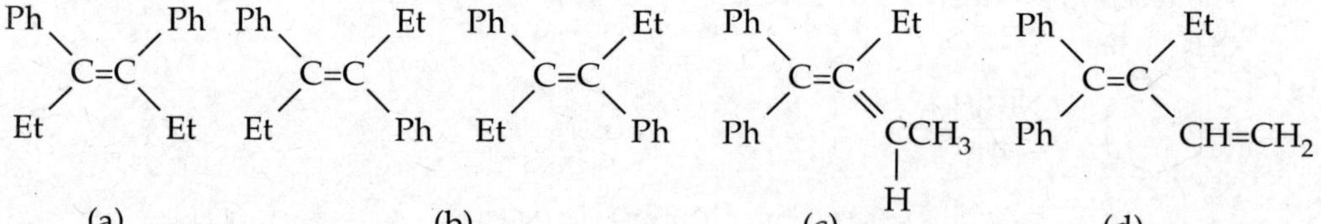

$$\begin{array}{c} C_2H_5 \\ Ph - \!\!\!\! \begin{array}{|c} \\ \\ \end{array} \!\!\!\! - Br \\ Ph - \!\!\!\! \begin{array}{|c} \\ \\ \end{array} \!\!\!\! - Br \\ C_2H_5 \end{array} \xrightarrow{\text{aq. KOH}}$$

(a) Ph, Et / C=C / Et, Ph
(b) Ph, Et / C=C / Et, Ph
(c) Ph, Et / C=C / Ph, CCH₃ H
(d) Ph, Et / C=C / Ph, CH=CH₂

(a) (b) (c) (d)

36. The major product formed

$$\overset{CH_3}{\underset{N^+Me_3}{\diagup}} \xrightarrow[\text{heat}]{OH^-}$$

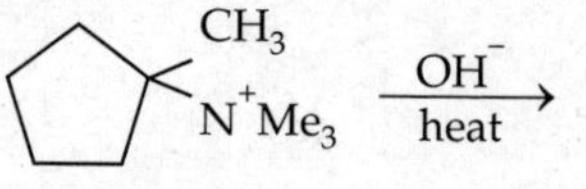

(a) 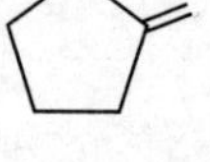 (b) 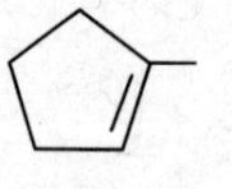 (c) 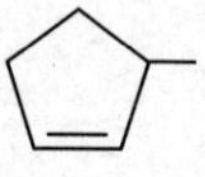 (d)

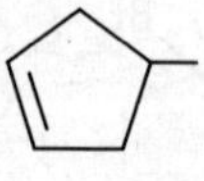

37. $CH_2=CHCH_2—Cl + CH_3COOH \longrightarrow$
    (Note: $Ph=C_6H_5$ or phenyl group; $Me=CH_3$ or methyl group)
    (a) $CH_2=CHCH_2—OCOCH_3$
    (b) [structure]$—CH_2 O COCH_3$
    (c) [structure with CH_3 and $O COCH_3$]
    (d) none

38. Which of the following is the most reactive halide to form Grignard's reagent?
    (a) [cyclopentane with F]
    (b) [cyclopentane with Br]
    (c) [cyclopentane with Cl]
    (d) [cyclopentane with I]

39. Arrange the reactivity of various bromine atoms in decreasing order. In the compound
    [structure with (1) Br, (2) Br, (3) Br, (4) Br, O, O]
    (a) $1 > 2 > 3 > 4$ (b) $4 > 3 > 2 > 1$ (c) $4 > 3 > 2 > 1$ (d) $3 > 1 > 4 > 2$

40. Arrange the following compounds, in decreasing order, in terms of their reactivity towards Grignard's reagent.
    (i) Methyl benzoate (ii) benzaldehyde (iii) benzoyl chloride (iv) acetophenone
    (a) ii > iii > i > iv   (b) i > ii > iii > iv   (c) iii > ii > iv > i   (d) ii > iv > i > iii

41. Arrange in decreasing order of reactivity towards original reagent.
    (i) CH_3CHO (ii) CH_3COCH_3 (iii) $Cl—CO—Cl$ (iv) C_6H_5CHO
    (a) iii > i > iv > ii   (b) i > ii > iii > iv   (c) iii > iv > i > ii   (d) iv > iii > ii > i

42. For following compounds which will react fast with aqueous KOH
    (a) $Me_3—C—Br$ (b) $Me_3—C—Cl$ (c) $Me_3—CCH_2Cl$ (d) $CH_3CH_2CH_2$-Br

43. The correct order of the reactivity of the chloride toward aqueous KOH
    (i) CH_3CH_2Cl (ii) $CH_3CH_2CH_2Cl$
    (iii) $(HO)—CH_2CH_2Cl$ (iv) $(CH_3)_2CHCl$
    (a) (i) > (ii) > (iii) > (iv)   (b) (iv) > (iii) > (ii) > (i)
    (c) (iii) > (iv) > (ii) > (i)   (d) (iv) > (iii) > (i) > (ii)

44. The decreasing order of the reactivity of the following halides in S_N2 reaction
    CH_3COCl CH_3CH_2Cl $CH_3—O—CH_2Cl$ $C_6H_5CH_2Cl$
       M   N   O   P
    (a) $M > N > O > P$ (b) $P > O > M > N$ (c) $O > P > N > M$ (d) $P > O > N > M$

45. Decreasing order of reactivity of the halides in S_N1 reaction is:
    [benzene—Br]   [benzene—CH_2Br]   [benzene—CH_2CH_2Br]   [benzene—$CH—CH_3$ with Br]
       M   N   O   P
    (a) $P > N > O > M$ (b) $M > N > O > P$ (c) $P > O > M > N$ (d) $M > O > N > M$

46. The reactivity of C_6H_5MgBr with the following compound is
    (M) C_6H_5OH (N) CH_3CHO (O) $R—C=N$ (P) CH_3COCH
    (a) $P > N > O > M$ (b) $M > O > N > P$ (c) $N > O > M > P$ (d) $P > N > M > O$

47. Reactivity of the following halide in S_N2 in the following order (decreasing).
    (M) CH_3Cl (N) C_2H_5Cl (O) $CH_3CH_2CH_2Cl$ (P) $(CH_3)_2CHCl$

(a) M > N > P > O (b) M > N > O > P
(c) P > O > M > N (d) O > P > M > N

48. Arrange the following halides in decreasing order in S_N2 reaction
   (i) CH_3CH_2Br (ii) $CH_2=CH CH_2Br$
   (iii) CH_3Br (iv) $CH=CH Br$
   (a) (ii) > (iii) > (i) > (iv) (b) (ii) > (iii) > (iv) > (i)
   (c) (i) > (ii) > (iii) > (iv) (d) (iv) > (iii) > (ii) > (i)

49. Arrange the following halides in decreasing order in S_N1

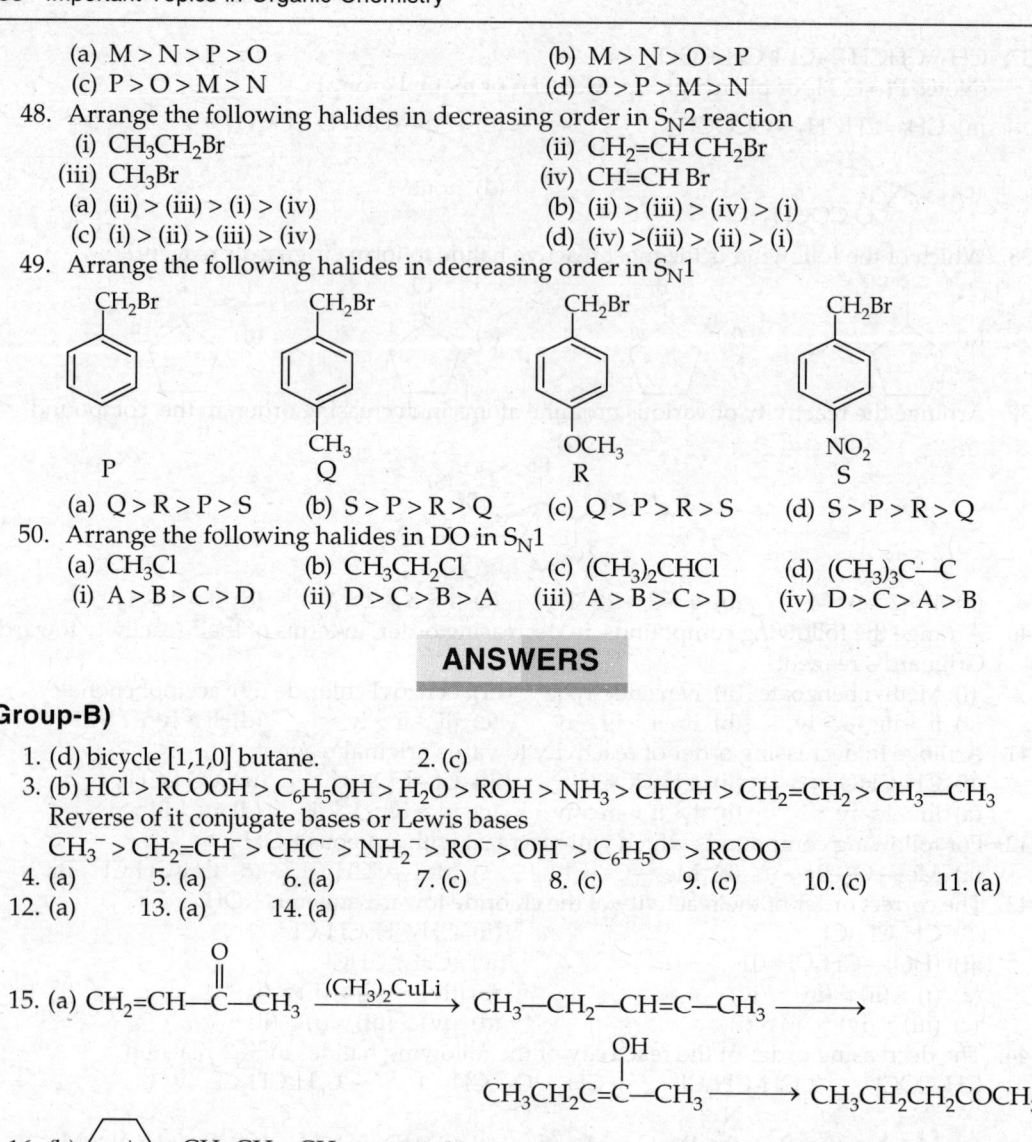

P Q R S

   (a) Q > R > P > S (b) S > P > R > Q (c) Q > P > R > S (d) S > P > R > Q

50. Arrange the following halides in DO in S_N1
   (a) CH_3Cl (b) CH_3CH_2Cl (c) $(CH_3)_2CHCl$ (d) $(CH_3)_3C$—C
   (i) A > B > C > D (ii) D > C > B > A (iii) A > B > C > D (iv) D > C > A > B

ANSWERS

(Group-B)

1. (d) bicycle [1,1,0] butane. 2. (c)
3. (b) $HCl > RCOOH > C_6H_5OH > H_2O > ROH > NH_3 > CHCH > CH_2=CH_2 > CH_3$—$CH_3$
   Reverse of it conjugate bases or Lewis bases
   $CH_3^- > CH_2=CH^- > CHC^- > NH_2^- > RO^- > OH^- > C_6H_5O^- > RCOO^-$
4. (a) 5. (a) 6. (a) 7. (c) 8. (c) 9. (c) 10. (c) 11. (a)
12. (a) 13. (a) 14. (a)

15. (a) $CH_2=CH$—$\overset{\overset{O}{\|}}{C}$—$CH_3$ $\xrightarrow{(CH_3)_2CuLi}$ CH_3—CH_2—$CH=C$—CH_3 $\longrightarrow$

$CH_3CH_2C=\overset{\overset{OH}{|}}{C}$—$CH_3$ $\longrightarrow$ $CH_3CH_2CH_2COCH_3$

16. (b) ⬡—$CH=CH$—CH_3 17. (b) 18. (c) 19. (b) 20. (b) 21. (c)
22. (a) 23. (b) 24. (a) 25. (a) 26. (a) 27. (a) 28. (a) 29. (a)
30. (a) 31. (b) 32. (c) 33. (d) 34. (a) 35. (b) 36. (a) 37. (b)
38. (d) 39. (d) 40. (a) 41. (a) 42. (c) 43. (c) 44. (a) 45. (a)
46. (b) 47. (a) 48. (a) 49. (a) 50. (b)

MULTIPLE CHOICE QUESTIONS

(Group-C)

1. For a given alkyl group, the boiling points are in the order of
   (a) RI > RBr > RCl (b) RI > RCl > RBr (c) RBr < RI < RCl (d) RCl < RBr > RI

2. Chloroform, when treated with benzene in presence of anhydrous $AlCl_3$, the product formed is
   (a) chlorobenzene
   (b) toluene
   (c) a mixture of o- and p- chloro toluen
   (d) triphenyl methane

3. 1-chlorobutane, on reaction with alc. KOH gives
   (a) 1-butene
   (b) 1-butanol
   (c) 2-butene
   (d) 2-butanol

4. Chlorine is most reactive in which compond towards NaOH is
   (a) $CHCl_3$
   (b) $CH_2=CHCl$
   (c) C_6H_5Cl
   (d) $C_6H_5CH_2Cl$

5. Which is most reactive in S_N2 reactions?
   (a) CH_3I
   (b) C_2H_5I
   (c) C_3H_7I
   (d) C_4H_9I

6. CHI_3 can be obtained from
   (a) methanol
   (b) methanal
   (c) propanol-1
   (d) propanol-2

7. Alc. KOH is used for
   (a) dehalogenation
   (b) dehydrohalogenation
   (c) dehydration
   (d) dehydrogenation

8. On heating chloroform, in presence of alc. KOH, with a primary amine it forms a iso-cyanide, which has a very foul odour, the compound formed in carbyl amine test is
   (a) CH_3CN
   (b) CH_3NC
   (c) $CH_3—N=C=O$
   (d) none

9. In the reaction of phenol with chloroform and aqueous KOH on warming the electrophile attacking ring is:
   (a) $CHCl_3$
   (b) $CHCl_2$
   (c) $:CCl_2$
   (d) $COCl_2$

10. Isobutyl magnesium bromide with dry ether and absolute alcohol gives:
    (a) $CH_3CH—CH_2—OH + CH_3CH_2MgBr$
        $\quad\quad |$
        $\quad\quad CH_3$
    (b) $CH_3CH—CH_2—OH + CH_3CH_2MgBr$
        $\quad\quad |$
        $\quad\quad CH_3$
    (c) $CH_3—CH—CH_3 + CH_2=CH_2$
        $\quad\quad\quad |$
        $\quad\quad\quad CH_3$
    (d) none

11. For the reaction:

    $$\underset{\underset{\displaystyle CH_3}{|}}{CH_3CH}—CH_2CH_3 \xrightarrow[100\,°C]{H_2SO_4} \begin{cases} CH_3—CH=CH—CH_3 \text{ A} \\ CH_2=CH—CH—CH_3 \text{ B} \end{cases}$$

    (a) A dominates
    (b) B dominates
    (c) both are formed in equal amounts
    (d) ratio of the product depends on the halogen X

12. Vicinal and gem dihalides can be distinguished by
    (a) aqueous KOH
    (b) alc. KOH
    (c) Zn dust
    (d) none

13. Iodoform gives a yellow precipitate with $AgNO_3$, but chloroform does not, because,
    (a) iodoform is ionic in nature
    (b) chloroform is co-valent
    (c) a C—I bond is weak in CHI_3 and C—Cl bond is strong in $CHCl_3$
    (d) none

14. Which compound is used in war as an insecticide and a war gas
    (a) $CHCl_2CCl_3$ (b) $C(NO_2)Cl_3$ (c) $CHCl_2CH_2(OH)$ (d) none

15. Freon is
    (a) CCl_2F_2 (b) $CHCl_3$ (c) CH_3F (d) CHF_3

16. The reaction condition leading to the best yield of C_2H_5Cl are

    (a) C_2H_6 (taken in excess) $+ Cl_2 \xrightarrow{UV}$ (b) $C_2H_6 + Cl_2 \xrightarrow{dark/room\ temp.}$

    (c) $C_2H_6 + Cl_2$ (In excess) $\xrightarrow{UV}$ (d) $C_2H_6 + Cl_2 \xrightarrow{UV}$

17. Ethylidine chloride, on reaction with aqueous KOH, gives
    (a) CH_3CHO (b) CH_2OH (c) HCHO   (d) CHO
                             | |
                          CH_2OH CHO

18. 2-Bromopentane is heated with potassium ethoxide in ethanol. The major product is
    (a) *trans*-pent-2-ene   (b) 2-ethoxypentane
    (c) pent-1-ene   (d) *cis*-pent-2-ene

19. $CH_2=CH—Cl$ reacts with HCl to form
    (a) $CH_2Cl—CH_2Cl$ (b) CH_3CHCl_2
    (c) $CH_2=CHCl.HCl$ (d) none

20. $2CHCl_3 + O_2 \xrightarrow{X} 2COCl_2 + 2HCl$; X stands for
    (a) an oxidant   (b) a reducing agent   (c) light and air   (d) nothing

21. In S_N1 reactions, the first step is the formation of
    (a) free radical   (b) carbanion   (c) carbocation   (d) final product

22. Which will produce a tertiary butyl alcohol?
    (a) $CH_3MgBr + CH_3COCH_3$ (b) $C_2H_5MgBr + CH_3COCH_3$
    (c) $CH_3MgBr + (C_2H_5)_2CHBr$ (d) $CH_3MgBr + CH_3COOH$

23. The following reaction described as

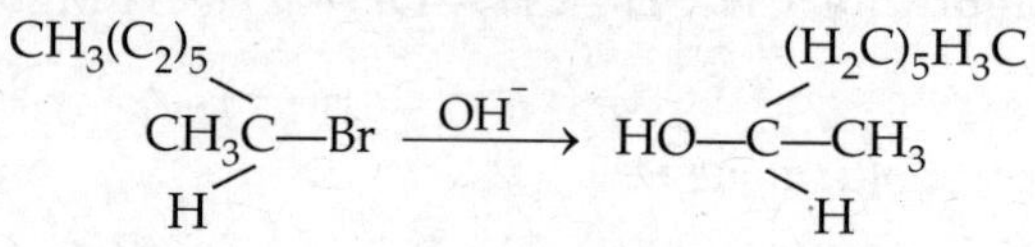

    $CH_3(C_2)_5$ $(H_2C)_5H_3C$

    $\underset{H}{CH_3C}—Br \xrightarrow{OH^-} HO—\underset{H}{C}—CH_3$

    (a) S_E1 (b) S_N1 (c) S_N2 (d) S_E2

24. $(CH_3)_2—C—HCl \xrightarrow{OH^-} (CH_3)_2CHOH$ is a result of
    (a) reduction   (b) oxidation
    (c) neutralization   (d) nucleophilic substitution

25. Correct order of reactivity of halide is
    (a) vinyl chloride > allyl chloride > propyl chloride
    (b) propyl chloride > vinyl chloride > allyl chloride
    (c) allyl > propyl > vinyl
    (d) none

26. The reaction $CH_3—Br + OH^- \longrightarrow CH_3OH + Br$ is
    (a) S_E1 (b) S_N2 (c) S_E2 (d) S_N2

27. Dehydrohalogenation of alkyl halide results in formation of
    (a) a single bond   (b) a double bond   (c) a triple bond   (d) fragmentation

28. Grignard's reagent reacts with HCN to produce
    (a) aldehyde   (b) ketone   (c) both a and b   (d) none

29. Which alkyl halide is hydrolysed preferentially by S_N1 mechanism?
    (a) CH_3Cl (b) C_2H_5Cl (c) $CH_3CH_2CH_2Cl$ (d) $(CH_3)_3C—Cl$

30. 1, 3-dibromopropane will react with metallic zinc to produce
    (a) cyclopropane derivatire   (b) propene
    (c) propyne   (d) hexane

31. Ethylidine chloride can be prepared by adding HCl to
    (a) ethane   (b) ethene   (c) ethyne   (d) ethylene glycol

32. S_N1 reaction is favoured by
    (a) nonpolar solvents
    (b) more alkyl groups attached to the carbon carrying the halogen atom
    (c) small groups attached to the carbon carrying halogen

33. In elimination reactions, i.e. in the formation of alkenes, the reactivity of alkyl halides shows the following order

         C_2H_5Cl $CH_3CH_2CH_2Cl$ $(CH_3)_3C\ Cl$
           (S)   (P)   (T)
    (a) $T > S > P$ (b) $P > S > T$ (c) $S > P > T$ (d) none

34. The reagent used in the conversion of 1-butanol to 1-bromobutane is
    (a) $CHBr_3$ (b) Br_2 (c) CH_3Br (d) $P + Br_2$

35. Methyl magnesium iodide on treatment with D_2O, furnishes a hydrocarbon along with MgODI, the hydrocarbon is
    (a) CH_3D (b) CH_3CH_2D (c) CH_4 (d) one

36. Which of the following will yield ethylene, upon treatment with Zn
    (a) $CH_2Br—CH_2Br$ (b) $CHBr=CHBr$ (c) $CHBr_2—CHBr_2$ (d) none

37. Identify 'Z' in the series

    $$CH_3—CH_2—CH_2OH \xrightarrow[160–180\,°C]{conc.\ H_2SO_4} X \longrightarrow Y \xrightarrow[(2)\ NaNH_2]{Br_2\ (1)\ alc.\ KOH} Z$$

          X   Y   Z
    (a) $CH_3CH=CH_2$ (b) $CH_3CHBr—CH_2Br$
    (c) $CH_3C≡CH$ (d) none

38. What method is used to convert an alkyl halide to an alcohol?
    (a) addition   (b) substitution
    (c) dehydration   (d) rearrangement

39. Aryl halides are less reactive towards nucleophiles than alkyl halides, due to
    (a) resonance   (b) stability of carbocation
    (c) high boiling point   (d) none of these

40. Pick up the correct statement about alkyl halides
    (a) They show H-bonding   (b) They are miscible with water
    (c) They are soluble in organic solvents   (d) They do not contain any polar bond

41. Grignard's reagent adds on multiple bonds
    (a) $>C=O$ (b) $>C=S$ (c) $C\ N$ (d) all

42. S_N2 reactivity order for halides   RF (1),   RCl (2),   RBr (3),   RI (4)
    (a) $1 > 2 > 3 > 4$ (b) $4 > 3 > 2 > 1$ (c) $3 > 4 > 2 > 1$ (d) $2 > 3 > 4 > 1$

43. The reaction is described as

(a) S_E1 (b) S_N1 (c) S_N0 (d) S_N2

44. Ethylidine chloride and ethylene chloride, are isomeric compounds mark the statement that is not applicable to both of them.
    (a) react with alc. KOH
    (b) react with alc. KOH and give the same product
    (c) are di halide
    (d) are positive to Beilstein's test
45. A reaction between ethyl bromide and sodium metal in presence of dry ether, to form butane, is called
    (a) Friedel-Craft reaction
    (b) Wurtz reaction
    (c) Cannizzaro's reaction
    (d) Williamson's synthesis
46. Grignard's reagent is prepared by reacting magnesium with
    (a) methyl amine   (b) diethyl ether   (c) ethyl iodide   (d) methyl alcohol
47. In the reaction $CH_3—CH\equiv C—Na^+ + Cl\,CH(CH_3)_2$, the products formed is
    (a) 4-methyl-2-pentyne
    (b) propyne
    (c) propyne + propene
    (d) none
48. On treating a silver salt of a fatty acid with Br_2 (or Cl_2 in CCl_4), it gives an alkyl bromide (or chloride). The reaction is known as
    (a) Hofmann's reaction
    (b) Borodine-Hunsdiecker reaction
    (c) Perkin's reaction
    (d) Birnbaum reaction
49. The common method for the preponation of alkyl iodide is by
    (a) Perkin's reaction
    (b) Sandmeyer's reaction
    (c) Finklestein reaction
    (d) reaction with iodine
50. Nonsticky frying pans are coated with
    (a) styrene
    (b) tetra fluoro ethylene (teflon)
    (c) tetra chloro ethylene
    (d) none
51. The reactivity of an alkyl halide depends upon
    (a) nature of alkyl group
    (b) nature of halogen atom
    (c) nature of alkyl group and halogen atom
    (d) none
52. Ethylidine chloride can be prepared by the reaction of HCl on
    (a) ethylene   (b) acetylene   (c) ethane   (d) ethanol
53. Freon is used for
    (a) refrigeration   (b) antiseptic   (c) anesthetic   (d) insecticide
54. $C_3H_7Br \xrightarrow[CH_3OH]{Mg\ ether}$
    (a) propyl alcohol
    (b) propylene peroxide
    (c) n-propyl bromide
    (d) isopropyl bromide
55. On pouring a solution of $AgNO_3$ in CCl_4,
    (a) a pale yellow precipitate is formed
    (b) curdy white ppt is formed
    (c) no ppt is formed
    (d) none
56. A sample of $CHCl_3$, before being used, is tested with/after
    (a) $AgNO_3$
    (b) boiling with KOH soln is tested again with $AgNO_3$
    (c) Fehling solution
    (d) Tollen's reagent
57. For a given alkyl group, density, melting point/boiling point are in the order
    (a) RI < RBr < RCl   (b) RI < RCl < RBr   (c) RBr < RCl < RI   (d) RCl < RBr < RI

58. CH_2=CH—Cl reacts with HCl to form
    (a) CH_2Cl—CH_2Cl (b) CH_3CHCl_2 (c) CH_2=CHCl.HCl   (d) none
59. 2-bromopentane is heated with potassium ethoxide, the compound formed is
    (a) *trans*-2-pentene   (b) 2-ethoxypentene   (c) pent-1-ene   (d) *cis*-2-pentene
60. RCl + NaI $\longrightarrow$ RI + NaCl; the reaction is known as
    (a) Wurtz reaction   (b) Fittig's reaction
    (c) Finklestein reaction   (d) Frankland reaction
61. Which is the most stable halide?
    (a) C_6H_5Cl (b) $(C_6H_5)_2CHCl$ (c) $C_6H_5CH_2Cl$ (d) $(C_6H_5)_3CCl$
62. Formation of an alkane by action of Zn on an alkyl halide is known as
    (a) Wurtz reaction   (b) Kolbe's reaction
    (c) Cannizzaro's reaction   (d) Frankland reaction
63. 1,3-dibromo propane reacts with metallic Zn, to form
    (a) propene   (b) cyclopropane   (c) propane   (d) hexane
64. Dehydrohalogenation of haloalkanes produces
    (a) a single bond   (b) a double bond   (c) a triple bond   (d) fragmentation
65. When sp^3-sp^3 bond changes to a sp^2-sp^2 bond, the total number of bonds
    (a) remain the same   (b) decreases by one
    (c) decreases by two   (d) or decreases by more
66. The S_N1 reactivity of ethyl chloride is
    (a) more or less equal to that of benzyl chloride
    (b) less than that of benzyl chloride
    (c) more or less equal to that of chlorobenzene
    (d) less than that of chlorobenzene
67. $(CH_3)_3C$—MgCl— + D_2O $\longrightarrow$
    (a) $(CH_3)_3C$—D   (b) $(CH_3)_3C$—OD
    (c) $(CD_3)_3CD$ (d) $(CD_3)_3C$—OD
68. Isobutyl chloride and *n*-butyl chloride are
    (a) position isomers   (b) chain isomers
    (c) functional isomers   (d) metamers
69. When a sp^2-sp^2 bond gets converted into sp-sp bond the total number of bonds
    (a) decreases by one   (b) total number of bonds remain same
    (c) increases by one   (d) none
70. When ethane is converted to acetylene the total number of bonds?
    (a) increases by one   (b) decreases by one
    (c) decreases by two or more   (d) there is no increase or a increase
71. Which group is displaced by a halogen atom?
    (a) —OH   (b) —CHO
    (c) —NO_2 (d) >C=O
72. The compound that will not form iodoform, on treatment with I_2 and alkali
    (a) acetone   (b) ethanol
    (c) diethyl ketone   (d) isopropyl alcohol
73. In the preparation of chloroform, from ethanol and bleaching powder, the later provides
    (a) Cl_2 (b) $Ca(OH)_2$
    (c) both Cl_2 and $Ca(OH)_2$ (d) none
74. Which does not give iodoform reaction?
    (a) $C_6H_5COCHH_3$ (b) CH_3OH
    (c) CH_3CH_2OH (d) CH_3CHO

75. In Wurtz reaction of alkyl halides with sodium, the reactivity order is
    (a) RI > RBr > RCl
    (b) RCl > RBr > RI
    (c) RBr > RI > RCl
    (d) none

76. In elimination reaction (the formation of alkenes), the order of reactivity of alkyl halides is
    (a) I > Br > Cl
    (b) Cl > Br > I
    (c) Br > Cl > I
    (d) none

77. Ethyl bromide and isopropyl chloride can be distinguished by
    (a) alc. $AgNO_3$
    (b) comparing their colours
    (c) burning the compounds on spatula
    (d) aqueous KOH

78. The alkyl group of Grignard's reagent acts as
    (a) free radical
    (b) carbocation
    (c) carbanion
    (d) none

79. In the reaction, $C_2H_5OH + HX \xrightarrow{ZnCl_2} C_2H_5X + H_2O$; the reactivity of hydrogen halides is
    (a) HI > HCl > HBr
    (b) HI > HBr > HCl
    (c) HCl > HB > HI
    (d) HBr > HI > HCl

80. Grignard's reagent undergoes
    (a) nucleophilic substitution
    (b) nucleophilic addition
    (c) both (a) and (b)
    (d) none

81. The Mg—Br bond in Grignard's reagent is
    (a) ionic
    (b) covalent
    (c) nonpolar
    (d) none

82. Which of the following will not form Grignard's reagent?
    (a) CH_3F
    (b) CH_3Cl
    (c) CH_3Br
    (d) CH_3I

83. When vinyl chloride is passed through alco. KOH solution it
    (a) dissolves
    (b) forms vinyl alcohol
    (c) forms acetylene
    (d) has no action

84. A camphor substitute is
    (a) $CHCl_3$
    (b) CCl_2F_2
    (c) CF_3CHCl_2
    (d) C_2Cl_6

85. 1,3-dibromopropane + metallic Zn $\longrightarrow$
    (a) propene
    (b) cyclopropane
    (c) hexane
    (d) propane

86. Only two monochloro derivatives are possible from
    (a) n-pentane
    (b) 2, 4-dimethylpentane
    (c) benzene
    (d) 2-methyl propane

87. Under IUPAC rules which is the principal functional group?
    (a) —CN
    (b) —SO_3H
    (c) —COOH
    (d) —CHO

88. Which alkyl halides will react most easily by nucleophilic substitution?
    (a) C_2H_5Cl
    (b) C_2H_5Br
    (c) C_2H_5I
    (d) C_2H_5F

89. Which of the following compound would react most rapidly in an S_N2 reaction?
    (a) C_2H_5I
    (b) $CH_2=CHI$
    (c) $(CH_3)_2CHI$
    (d) $(CH_3)_3CI$

90. Which of the following factors influence whether a reaction will proceed by an S_N1, S_N2, E1 or E2 mechanism?
    (a) structure of the alkyl halide
    (b) solvent
    (c) nature of the nucleophile
    (d) concentration of the reagents
    (e) all of these

ANSWERS

1. (a)   2. (d)   3. (a)   4. (b)   5. (d)   6. (d)   7. (b)   8. (b)
9. (c)   10. (b)   11. (b)   12. (a)   13. (c)   14. (b)   15. (a)   16. (a)
17. (a)
18. (a) β – elimination gives a *trans*-product   19. (b)
20. (c) ($COCl_2$ is also named as phosgene; being a deadly gas it is destroyed, before using $CHCl_3$)
21. (c)   22. (a)   23. (d)   24. (d)   25. (c)
26. (c) S_N2 ; the order is $3° < 2° < 1° < CH_3$
27. (b)
28. (a) aldehyde, $HCN + R MgX \longrightarrow R—CH=N Mg Br \longrightarrow RCHO$
29. (d)   30. (a)   31. (c)   32. (b)   33. (a)   34. (d)   35. (a)   36. (a)
37. (c) $X=CH_3CH=CH_2$; $Y=CH_3CHBr—CH_2Br$; $Z=CH_3C\equiv CH$
38. (b)
39. (a) because of resonance
40. (c)   41. (d)
42. (b) on account of longer bond
43. (d) S_N2
44. (b)   45. (b)   46. (c)   47. (c)   48. (b)   49. (c)   50. (b)   51. (c)
52. (b)   53. (a)   54. (b)   55. (d)   56. (b)
57. (b) iodide > bromide > chloride > fluoride
58. (b)
59. (a) elimination gives *trans*-derivative
60. (c)
61. (d) tertiary halide is most stable
62. (d)   63. (b)   64. (b)   65. (b) by one
66. (b) S_N1 order : benzyl > allyl > 3 > 2 > 1 > phenyl halide
67. (a) $(CH_3)_3C—D$
68. (b) chain
69. (a) decrease by one
70. (c) decreases by two
71. (a)   72. (c)   73. (c)   74. (b)   75. (a)   76. (a)
77. (a) a yellow precipitate in case of RBr and a white precipitate in case of RCl
78. (c) $R^- + MgX^+$ 79. (b)   80. (b)
81. (a) polar or ionic bond
82. (a) RMgF are unstable compounds
83. (c)   84. (d)   85. (b)   86. (d)   87. (b)   88. (c)   89. (a)   90. (e) all

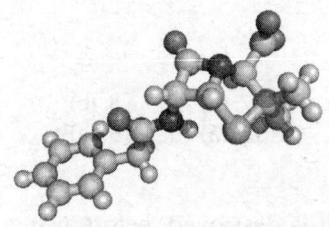

Chapter 3

Alcohols, Phenols and Ethers

Salient Features

1. Alcohols cannot be dried over $CaCl_2$, as they form solid derivatives with $CaCl_2$, such as $CaCl_2.4CH_3OH$, with methanol and $CaCl_2.4C_2H_5OH$ with ethanol. These solid derivatives are known as *alcoholates*.

2. Alcohols may be dried over burnt lime, CaO.

3. Power alcohol is a mixture of 80% petrol and 20% of ethanol with cosolvent benzene. This mixture is known as *'power alcohol'*.

4. Phenols do not liberate CO_2, when treated with $NaHCO_3$, carbonic acid, (pKa = 10) is stronger than phenol (pKa = 6), but weaker than carboxylic acids (pKa = 4.5).

$$\underset{\text{weak acid}}{C_6H_5OH} + \underset{\text{weak base}}{NaHCO_3} \longrightarrow \underset{\text{strong base}}{C_6H_5ONa} + \underset{\text{strong acid}}{H_2CO_3}$$

5. **Dunston's test for glycerol:** On mixing borax with phenolphthalein, a pink colour develops. On adding glycerol, the pink colour disappears, which again appears on heating.

6. **Ceric ammonium test for alcohols:** A freshly prepared solution of ceric ammonium nitrate (yellow colour) turns red on adding alcohols.

7. Acetoacetic ester $CH_3CO\ CH_2COOC_2H_5$ exists in dynamic equilibrium with its tautomeric form, i.e. enol form $CH_3C(OH)\ CH\ COOC_2H_5$.

$$CH_3COCH_2COOC_2H_5 \rightleftharpoons CH_3\overset{\overset{\displaystyle OH}{|}}{C}=CH-COOC_2H_5$$

Tautomers always exist in dynamic equilibrium.

8. The presence of the enol form may be tested with $FeCl_3$, which forms a cherry red colour with enol form.

9. Absolute alcohol or 100% alcohol: When rectified spirit is distilled to concentrate, as soon as the concentration reaches 95.87% alcohol and 4.5% of water, it distils unchanged, at a temperature of 78.15°C, beyond which it cannot be concentrated further, by simple distillation.

To concentrate it further, it is subjected to azeotropic distillation. In this process, it is mixed with sufficient quantity of benzene and distilled again, when the following fractions are obtained.

Mixture	b.p.	Alcohol	Water	Benzene
First fraction-ternary mixture	64.8 °C	18.4%	7.4%	74.1%
Second fraction-binary mixture	68.2 °C	32.4%	nil	67.6%

When all benzene has been distilled over, pure alcohol (100%) is left, which distils at a temperature of 78.3 °C.

10. Alcohols have extensive hydrogen bonding due to which their volatility is low. As a result the boiling points of alcohols are much higher than can be predicted from their molecular weight.

 Alcohols are highly miscible with water but their solubility goes on decreasing with rise in molecular weight.

11. Reduction of RCHO, RCOR and RCOOR, by sodium metal and alcohol is known as Bouveault–Blanc reduction.

12. In phenols, groups with –I effect increases the acidity in o- and p- positions, and the groups with +I effect decrease the acidity in o- and p- positions, but increase the acidity in m-position.

13. Proof spirit: A mixture of alcohol (57.1 g) and water (49.3 g)

14. Oxygen in ether is sp³ hybridised.

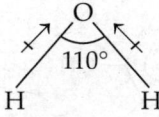

15. μ for ether is 1.18D.

16. Ethers on prolonged contact with air or oxygen in the presence of light forms peroxides.

$$CH_3CH_2—O—CH_2CH_3 \xrightarrow{O_2 + light} \overset{\overset{\textstyle O—OH}{|}}{CH_3CH}—O—CH_2CH_3$$
1-ethoxy-ethyl hydrogen peroxide

17. Presence of peroxide is detected by shaking it with $FeSO_4$ (freshly prepared), followed by addition of KCNS, (potassium. sulphocyanide), formation of a red colour indicates the presence of peroxide.

18. In decomposing unsymmetrical ether with HI, alkyl iodide is always formed from smaller alkyl group, because the I⁻ being lager in size, approaches smaller alkyl group to avoid steric hindrance.

19. Ethylene epoxide is a three membered ring and is extremely reactive towards nucleophilic displacement reaction, due to ring strain.

20. Cl⁻ is a weak nucleophilic reagent, Br⁻ and I⁻ are good nucleophiles.

21. Electron releasing groups increase the nucleophilic strength.

22. Electron withdrawing groups, decrease then nucleophilic strength.

$$CH_3—O^- > H—O^- > CH_3 \overset{\overset{\textstyle O}{||}}{→C}←O^-$$

23. Nucleophilicity is in the order of
 I⁻ > C_2H_5OH > Cl⁻

$$CH_2=CH_2 + \xrightarrow{HI} \begin{cases} C_2H_5^+ \xrightarrow{I^-} C_2H_5I \\ \\ C_2H_5^+ \xrightarrow{C_2H_5OH} C_2H_5\overset{+}{\underset{H}{O}}-C_2H_5 \longrightarrow C_2H_5-O-C_2H_5 + H^+ \end{cases}$$

24. Dialkyl cuperate is less reactive than RMgX.
25. Ethers are readily split by Lewis acids.
    $$R-O-R + BCl_3 \longrightarrow RCl + RBCl_2$$
26. Ethers although are polar compounds yet they form coordination complex.
27. Phenols are weakly acidic in nature, they turn blue litmus red, dissolve in alkali but do not liberate CO_2 from $NaHCO_3$.
28. Phenols give a characteristic colour with $FeCl_3$ solution, which serve as a test for phenolic group. C_6H_5OH gives a violet-blue colour with $FeCl_3$.
29. When a hydroxyl group (—OH group) is attached to a side chain as in C_6H_5—CH_2—OH, (benzyl alcohol) such compounds behave as alcohols and do not form any colour with $FeCl_3$.
30. When electron withdrawing groups, such as —NO_2, —Cl, are attached in the benzene ring along with —OH group, they make phenols acidic (stronger), from all the positions.
    In nitro phenols, decreasing order of their acidic character is as follows.

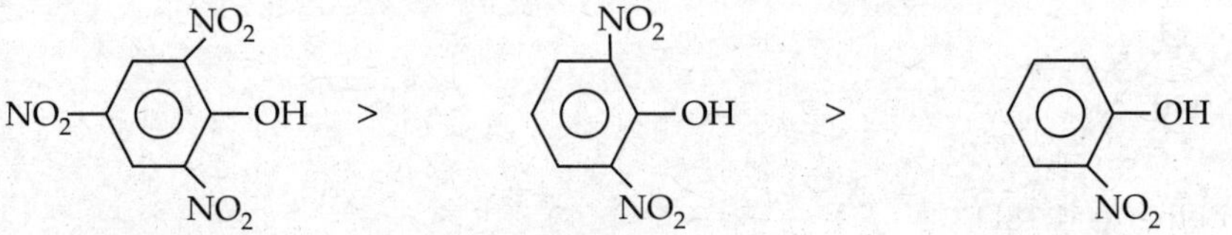

2,4,6-tri nitrophenol > 2,4-dinitro phenol > 4-nitro phenol

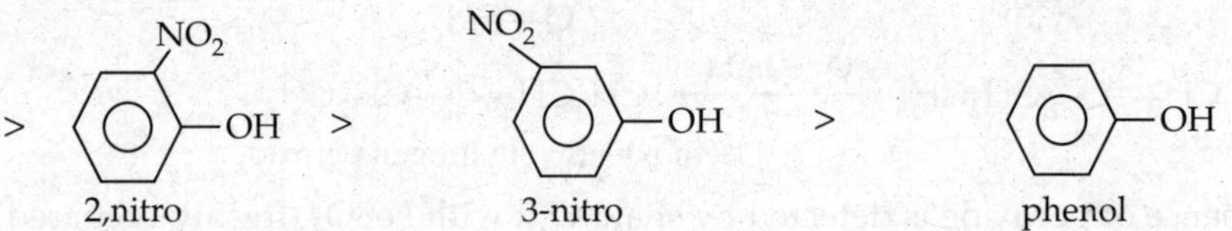

> 2,nitro phenol > 3-nitro phenol > phenol

In chlorophenols, the decreasing order of their acidic character is as follows:

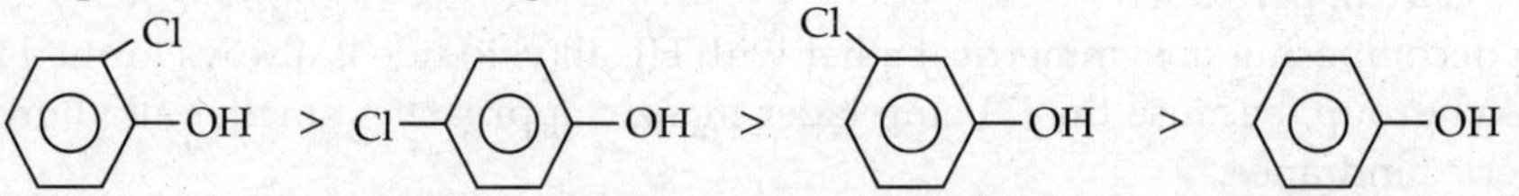

o-chlorophenol > p-chlorophenol > m-chlorophenol > phenol

31. Electron donating groups, such as —CH_3 make phenol weaker from all the positions.

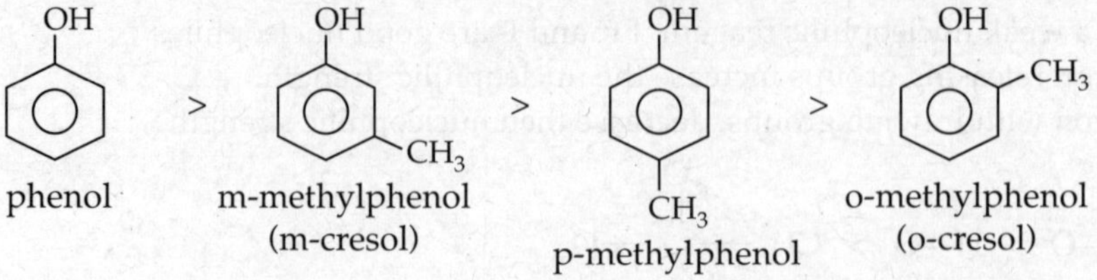

phenol > m-methylphenol (m-cresol) > p-methylphenol (p-cresol) > o-methylphenol (o-cresol)

32. Alkoxy group, such as —OCH_3, acts as electron donating from both o- and p- positions (making phenol weaker), but act as electron withdrawing from m-position, (making phenol stronger).

$$\underset{\text{m-methoxyphenol}}{\text{OCH}_3\!\!-\!\!\bigcirc\!\!-\!\!\text{OH}} \quad > \quad \underset{\text{phenol}}{\bigcirc\!\!-\!\!\text{OH}} \quad > \quad \underset{\text{o-methoxyphenol}}{\overset{\text{OCH}_3}{\bigcirc}\!\!-\!\!\text{OH}} \quad > \quad \underset{\text{p-methoxyphenol}}{\text{CHO}_3\!\!-\!\!\bigcirc\!\!-\!\!\text{OH}}$$

ALCOHOLS

Alkanols

Hydroxy derivatives of alkanes are known as alcohols or alkanols. These may be mono-hydroxy derivatives (alcohols containing only one —OH group, attached to alkyl group, **ROH**), these may be dihydroxy derivatives (alcohols containing two —OH groups, in the molecule, $C_nH_{2n}(OH)_2$) or these may be trihydroxy derivatives (alcohols containing three —OH groups in the molecule, $C_nH_{2n-3}(OH)_3$, such as glycerol).

Monohydroxy alcohols, ROH: When one hydrogen atom in alkanes, is substituted by a single —OH group, alcohols with the composition, ROH, are obtained, where R is an alkyl group ($R = C_nH_{2n} + 1$).

Example:

alcohol	common name	IUPAC names
CH_3OH	methyl alcohol	methanol
C_2H_5OH	ethyl alcohol	ethanol
C_3H_7OH	propyl alcohol	propanol
isomers of propyl alcohol		
$CH_3CH_2CH_2OH$	n-propyl alcohol	propanol-1
$CH_3CH(OH)CH_3$	isopropyl alcohol	propanol-2
C_4H_9OH	butyl alcohol	butanol
isomers of butyl alcohol		
$CH_3CH_2CH_2CH_2OH$	n-butyl alcohol	butanol-1
$CH_3CH_2CH(OH)CH_3$	secondary butyl alcohol	butanol-2
$(CH_3)_2CH_2CH_2OH$	isobutyl alcohol	2-methyl-1-propanol
$(CH_3)_3C—OH$	tertiary butyl alcohol	2-methyl-2-propanol

Dihydric alcohols: These contains two —OH groups, present in the molecule. There are three types of possible dihydroxy derivatives.

1. When both —OH groups are attached on the same carbon atom, like $> \overset{\overset{\displaystyle OH}{|}}{C}\!\!-\!\!OH$. Such dihydroxy derivatives are highly unstable. These lose a molecule of water, and are converted into corresponding carbonyl compounds.

$$>C\!\!\Big\langle\!\!\begin{array}{l}OH\\OH\end{array} \longrightarrow >C=O + H_2O$$

2. When the two —OH groups are attached on the adjacent carbon atoms, like $\overset{\overset{\displaystyle HO\ \ \ OH}{|\ \ \ \ \ |}}{-C\!-\!C-}$, such dihydroxy derivatives are very stable, and are termed as glycols.

For example: $\underset{\text{ethylene glycol}}{\overset{\displaystyle CH_2\!-\!CH_2}{\underset{\displaystyle |\quad\ |}{\underset{\displaystyle OH\ \ OH}{}}}}$

3. When the two —OH groups are attached on the distant carbon atoms, such dihydroxy derivatives do not have any special name or any special property. Both the —OH groups react independent of each other.

$$\begin{array}{c}\quad\;\; OH \quad\;\; OH \\ \quad\;\; | \qquad | \\ -C-C-C- \\ \quad\;\; | \qquad | \quad | \end{array}$$

Trihydric alcohols: These contains three —OH groups in the molecule. The first member is glycerol, that contains three —OH groups, one OH attached to the each of the three carbon atom, that are present in the molecule.

$$\begin{array}{c} OH \;\; OH \;\; OH \\ |\qquad |\qquad | \\ H-C-C-C-H \\ |\qquad |\qquad | \\ H \quad\; H \quad\; H \end{array} \qquad \text{glycerol or propane-1,2,3-triol}$$

Monohydric alcohols, ROH (R = alkyl radical, R = C_n H_{2n} + 1): Monohydric alcohols may be divided into three sub-classes. These are

1. **Primary alcohol or 1°alcohol, RCH_2OH:** The carbon containing —OH group is attached to minimum two hydrogen atoms (there can be three hydrogen atoms, attached to carbon containing —OH group, as in the case of CH_3OH).
2. **Secondary alcohol or 2° alcohol, R_2CHOH:** The carbon containing —OH group is attached to only one hydrogen atom.
3. **Tertiary alchol or 3°alcohol, R_3C—OH:** The carbon containing —OH group is not attached to any hydrogen atom.

Methods of Preparation

1. **By the action of aqueous alkali on alkyl halides, RX:**
   $$RX + NaOH \xrightarrow{\quad} ROH + NaX$$
   aqueous

   In place of aqueous alkali, moist Ag_2O may also be used.
   $$Ag_2O + H_2O \longrightarrow 2AgOH$$
   $$RX + AgOH \longrightarrow ROH + AgX$$

2. **By the reduction of aldehydes or ketones, by sodium and alcohol (C_2H_5OH + Na) or lithium aluminium hydride ($LiAlH_4$):**
   $$C_2H_5OH + Na \longrightarrow C_2H_5ONa + H$$
   sodium ethoxide
   $$\underset{\text{aldehyde}}{RCHO} + 2H \longrightarrow \underset{\text{1° alcohol}}{RCH_2OH}$$
   $$\underset{\text{ketone}}{RCOR} + 2H \longrightarrow \underset{\text{2° alcohol}}{RCH(OH)R}$$

   Reduction of aldehydes and ketones, by sodium and alchol is known as **Bouveault–Blanc reaction or reduction.**
   $$\underset{\text{aldehyde}}{RCHO} \xrightarrow{\;LiAlH_4\;} \underset{\text{1° alcohol}}{RCH_2OH}$$
   $$\underset{\text{ketone}}{RCOR} \xrightarrow{\;LiAlH_4\;} \underset{\text{2° alcohol}}{RCH(OH)R}$$

3. **By action of nitrous acid, HNO_2, on primary or 1°amine–**
   $$KNO_2 + H_2SO_4 \longrightarrow HNO_2 + KHSO_4$$
   $$RNH_2 + HNO_2 \longrightarrow ROH + N_2 + H_2O$$

4. **By hydration of alkenes:** Alkenes add on a molecule of water, in the presence of suitable reagent, and are converted into corresponding alcohols.

The various methods are:

a. By H$^+$ catalysed hydration: Alkene is treated with conc. **H$_2$SO$_4$** to form alkyl hydrogen sulphate, which upon hydrolysis, gets converted into alcohol.

Addition of **H$_2$SO$_4$** takes place by **Markovnikov's rule**.

$$H_2SO_4 \rightleftharpoons H + HSO_4^-$$

1. $CH_2{=}CH_2 \xrightarrow{H^+} CH_3CH_2^+ \xrightarrow{HSO_4^-} CH_3CH_2HSO_4 \xrightarrow{HOH} CH_3CH_2OH + H_2SO_4$

    ethene   ethyl   ethyl hydrogen   ethyl

              carbocation   sulphate   alcohol

2. $RCH{=}CH_2 \xrightarrow{H^+} RCH^+{-}CH_3 \xrightarrow{HSO_4^-} \underset{\underset{HSO_4}{|}}{RCHCH_3} \xrightarrow{HOH} \underset{\underset{OH}{|}}{RCH_2CH_3} + H_2SO_4$

   alkene   alkyl   alkyl hydrogen   alcohol

              carbocation   sulphate

b. By hydroboration: In this process, alkene is treated with borane hydride, **BH$_3$**, to form **borohydride addition compound**, which upon hydrolysis, with **alkaline H$_2$O$_2$** gives alcohol.

1. $CH_2{=}CH_2 \xrightarrow{BH_3} \underset{\pi\text{-complex}}{CH_2{\stackrel{\uparrow}{=}}CH_2} \xrightarrow{H^-/BH_2^+} \underset{\substack{\text{borohydride} \\ \text{addition compound}}}{CH_3{-}CH_2BH_2} \xrightarrow[H_2O_2]{\text{alkaline}} CH_3CH_2OH + BH_3$

   ethene   ethanol

(where H—B—H with H above)

2. $RCH{=}CH_2 \xrightarrow{BH_3} \underset{\pi\text{-complex}}{RCH{\stackrel{\uparrow}{=}}CH_2} \xrightarrow{H^-/BH_2^+} \underset{\text{borohydride}}{RCH_2CH_2BH_2} \xrightarrow[H_2O_2]{\text{alkaline}} \underset{1° \text{ alcohol}}{RCH_2CH_2OH} + BH_3$

   alkene

(where H—B—H with H above)

In π-complex, **BH$_3$** then ionizes as **hydride ion, H$^-$**, and **borohydride cation, BH$_2^+$**, and these two ions add on the alkene molecule, following Markovnikov's rule. The **alcohol**, formed upon alkaline hydrolysis, is an **anti-Markovnikov's** product.

c. By oxymercuration–demercuration process: In this process, alkene is treated with an aqueous solution of mercuric acetate, [Hg(OCOCH$_3$)$_2$] and the addition product formed is hydrolysed with sodium borohydride, NaBH$_4$. Formation of alcohol follows Markovnikov's rule.

i. $Hg(OCOCH_3)_2 \longrightarrow CH_3COO^- + HgOCOCH_3^+$

ii. $H_2O \longrightarrow H^+ + OH^-$

iii. $CH_3COO^- + H^+ \longrightarrow CH_3COOH$ (does not dissociated further)

iv. $CH_2{=}CH_2 \xrightarrow{HgOCOCH_3^+} \underset{\underset{HgOCOCH_3}{|}}{CH_2{-}CH_2^+} \xrightarrow{OH^-} \underset{\underset{HgOCOCH_3}{|}}{CH_2{-}CH_2OH} \xrightarrow{NaBH_4} CH_3{-}CH_2OH$

    ethane   hydrolysis

                            carbocation

v. $RCH{=}CH_2 \xrightarrow[OH^-]{HgOCOCH_3^+} \underset{\underset{OH \quad HgOCOCH_3}{| \quad\quad |}}{RCH{-}CH_2^+} \xrightarrow{NaBH_4} \underset{\underset{OH}{|}}{RCHCH_3}$

Industrial alcohol: Methanol (CH$_3$OH) and ethanol (CH$_3$CH$_2$OH) are industrial alcohols. These are manufactured by some special methods.

Manufacture of Methanol (CH_3OH):

1. From water gas—Steam is passed over red hot coke, when water gas is formed.

$$C + H_2O \longrightarrow CO + H_2$$
$$\text{water gas}$$

Water gas is then mixed with its half volume of hydrogen and the mixture, **($CO + 2H_2$)**, is passed over a mixture of catalyst, consisting of oxides of Zr, Cr, V, when methanol is formed along with small amount of ethanol, propanol and isobutanol.

$$CO + 2H_2 \longrightarrow CH_3OH + (C_2H_5OH + C_3H_7OH + (CH_3)_2CHCH_2OH)$$
$$\text{formed in small amount}$$

2. By partial oxidation of methane, CH_4: Methane is mixed with oxygen, in the ratio of 9 : 1 by volume and the mixture is compressed to ~ 100 atmospheric pressure and passed through a hot copper tube, kept at a temperature of 250 °C, when about 25% of methane is converted into methanol.

$$\underset{\text{9 vol}}{CH_4} + \underset{\text{1 vol}}{O_2} \xrightarrow[\text{100 atm/250 °C}]{\text{copper tube}} CH_3OH$$

3. From pyroligenous acid: When greenwood, with ~25% moisture is subjected to destructive distillation (heating out of contact with air), the following fraction are obtained.

  a. **Wood gas**-consists of mainly **hydrocarbons, CO_2, CO** etc. used as fuel.
  b. **Ammoniacal liquor:** It is condensed along with pyroligenous acid. Where as pyroligenous acid forms lower layer, ammoniacal liquor forms upper layers, which is separated and used for the preparation of ammonium compounds.
  c. **Pyroligenous acid:** It forms the lower layer. It is separated and is subjected to further treatment.
  d. **Pitch:** Mainly consists of carbon.

Further treatment of pyroligneous acid layer: This fraction contains acetic acid (10%), acetone (0.5%) and methanol (2-3%), besides water. Where as acetic acid is separated by passing the vapours of pyroligneous acid through hot milk of lime, which reacts with acetic acid to form insoluble calcium acetate, other uncondensed vapours, consisting of acetone and methanol distilled over, condensed and collected. Methanol (b.p. 65°C) and acetone (b.p. 56°C), are separated by fractional distillation.

$$2CH_3COOH + Ca(OH)_2 \longrightarrow (CH_3COO)_2Ca + H_2O$$
$$\text{solid}$$

On treating calcium acetate [$(CH_3COO)_2Ca$] with sulphuric acid, acetic acid is recovered. Finally it is purified by distillation (b.p. 118 °C).

$$(CH_3COO)_2Ca + H_2SO_4 \longrightarrow 2CH_3COOH + CaSO_4$$

Ethyl alcohol: It is manufactured by the process of fermentation.

Fermentation may be defined as breakdown of complex organic molecule into simpler ones, by the activities of enzymes, secreted by living organisms.

The starting material for the manufacture of ethyl alcohol, may be starch, cane sugar or glucose.

Manufacture of ethyl alcohol:

  a. **From starch, $(C_6H_{12}O_6)n$:** The raw material taken for fermentation, may be cereals like maize (65%), wheat (60–65%), rice (75–80%) or vegetable like potato (15–20%). The raw material is heated under pressure with steam, when starch granules burst and starch is exposed, obtained as slurry. Starch is then treated with germinated barley, when diastase or amylase, produced during germination of barley, hydrolyses starch into maltose.

$$\underset{\text{starch}}{(C_6H_{12}O_6)n} + H_2O \xrightarrow{\text{diastase or amylase}} \underset{\text{maltose}}{C_{12}H_{22}O_{11}}$$

Maltose is then treated with yeast, which contains an enzyme, maltase, which hydrolyses maltose into glucose.

$$C_{12}H_{22}O_{11} + H_2O \xrightarrow{\textbf{maltase}} 2C_6H_{12}O_6$$
$$\text{maltose} \qquad\qquad\qquad \text{glucose}$$

Zymase, another enzyme present in yeast, then decomposes glucose into ethyl alcohol and carbon dioxide.

$$C_6H_{12}O_6 \xrightarrow{\text{zymase}} 2C_2H_5OH + 2CO_2$$
$$\text{glucose} \qquad\qquad \text{ethyl alcohol}$$

From cane sugar or sucrose, $C_{12}H_{22}O_{11}$: Uncrystallisable cane juice (~40% sucrose) is known as molasses. It is used as a raw material for manufacture of ethyl alohol. The **diluted solution of molasses** is treated with yeast, when sucrose in hydrolysed to form glucose and fructose, by **invertase**, an enzyme present in yeast.

$$C_{12}H_{22}O_{11} + H_2O \xrightarrow{\text{invertase}} C_6H_{12}O_6 + C_2H_{12}O_6$$
$$\text{sucrose} \qquad\qquad\qquad \text{glucose} \quad \text{fructose}$$

Zymase present in yeast, then decomposes glucose into ethyl alcohol and carbon dioxide.

$$C_6H_{12}O_6 \xrightarrow{\text{zymase}} 2C_2H_5OH + 2CO_2$$

From glucose, $C_6H_{12}O_6$: Glucose is treated with yeast, when zymase, present in yeast, decomposes into ethyl alcohol and CO_2.

Fructose is also fermented, like glucose, to give ethyl alcohol, but the process is slow.

Ethyl alcohol (or fermented wort), formed by fermentation is not allowed to go beyond 8–10%, as the bacteria will be killed. Dilute alcohol, so obtained, is unfit for any industrial use.

In order to make it fit for industrial use, it is distilled in Coffey's still, when in a single operation, **alcohol, with 85–95%, known as rectified spirit**, is obtained, which is fit practically for all industrial uses. However, rectified alcohol does not mix with petrol, as it is immiscible with it, for which a 100% alcohol is required.

A **100% ethyl alcohol** is also known as **absolute alcohol or power alcohol**, is mixed with petrol in 10-20% by volume and is used as fuel for piston engines (motor cars etc).

Rectified spirit can not be converted into a 100% alcohol, by simple distillation as when the **concentration of alcohol with water reaches to 95.87%, it distils at a constant temperature of 78.15°C.**

Rectified spirit can be converted into absolute alcohol by **azeotropic distillation** or **by chemical methods. Azeotrope** may be defined as a **mixture of liquids** that **distils unchanged.**

Azeotropic distillation of rectified spirit: Rectified spirit is mixed with sufficient quantity of benzene and the mixture is distilled, when temperature reaches to **64.8°C**, an **azeotrop distils over.**

temperature	alcohol	water	benzene
at 64.8°C	18.4%	7.4%	74.2%

When all water has been distilled off, another azeotropic fraction distils over at **68.2°C**.

temperature	alcohol	water	benzene
at 68.2°C	32.4%	nil	67.6%

When all benzene has been distilled over, only 100% alcohol or absolute alcohol is left, which distils at a temperature of **78.3°C.**

Rectified alcohol is mixed with 9% methanol (to make it unfit for drinking) and 1% pyridine (to make it repulsive in its odour), known as **methylated spirit**, is marketed and is used as a fuel or as a solvent for varnishes etc.

Properties: Alcohols are colourless, sharp smelling liquids, miscible with water and other organic solvents.

Methanol is oxidized in human blood system to form formaldehyde **(HCHO)** which causes blood coagulation that can prove fatal. That is why it is mixed with rectified spirit, to make it unfit for drinking. **Ethyl alcohol,** when consumed in small quantity helps in digestion. However, when consumed in larger quantity, produces mad like behaviour and damages liver. Alcohols are slightly acidic in nature, but insufficient to turn blue litmus red. The **Ka value for alcohols is 1×10^{-18}**.

Chemical properties

1. **Reaction with sodium metal:** Alcohols react with sodium metal to form sodium alkoxides.

$$CH_3OH + Na \longrightarrow CH_3O^- Na^+ + \frac{1}{2}H_2$$
methanol   sodium methoxide

$$C_2H_5OH + Na \longrightarrow C_2H_5O^- Na^+ + \frac{1}{2}H_2$$
ethanol   sodium ethoxide

$$ROH + Na \longrightarrow RO^-Na^+ + \frac{1}{2}H_2$$
alcohol   sodium alkoxide

In a reaction of alcohols, with sodium metal, the **—O—H bond** is broken, which breaks faster in 1° alcohol and the decreasing order of reactivity in alcohols is as follows:

Primary or 1° alcohol > secondary or 2° alcohol > tertiary or 3° alochol
$$RCH_2OH > R_2CHOH > R_3COH$$

2. **Reaction with phosphorous pentachloride (PCl_5), phosphorous trichloride (PCl_3) and thionyl chloride, $SOCl_2$:** Alcohols are converted into corresponding **alkyl chlorides (RCl)** when react with these reagents. In alcohols, the —OH group is substituted by —Cl group.

$$CH_3OH + PCl_5 \longrightarrow CH_3Cl + HCl + POCl_3$$
methanol   methyl chloride   phosphorous
          oxychloride

$$3CH_3OH + PCl_3 \longrightarrow 3CH_3Cl + H_3PO_3$$
          phosphorous
          acid

$$CH_3OH + SOCl_2 \longrightarrow CH_3Cl + SO_2 + HCl$$

Similarly, alcohols react with **phosphorous pentabromide PBr_5, phosphorous tribromide, PBr_3 and thionyl bromide $SOBr_2$,** to form corresponding **alkyl bromide, RBr.** The —OH group in alcohols is substituted by —Br group.

In the same way, alcohols react with **phosphorous triiodide, PI_3,** (PI_5 and SOI_2 do not exist), to form **alkyl iodide, RI.**

$$3ROH + PI_3 \longrightarrow 3RI + H_3PO_3$$
          alkyl
          iodide

In general

$$ROH + PX_5 \longrightarrow RX + HX + PO_3$$
$$3ROH + PX_3 \longrightarrow 3RX + H_3PO_3$$
$$ROH + SOX_2 \longrightarrow RX + SO_2 + HX \ (X : Cl, Br)$$

3. **Reaction with halogen acids:** Alcohols react with halogen acids, to form corresponding alkyl halides.

$$ROH + HCl \xrightarrow{ZnCl_2, \ \Delta} RCl + H_2O$$
          alkyl
          chloride

$$ROH + HBr \xrightarrow[\Delta]{\text{conc. } H_2SO_4} RBr + H_2O$$

alkyl
bromide

$$ROH + HI \xrightarrow[\Delta]{\text{red P}} RI + H_2O$$

alkyl
iodide

In reaction of alcohols, with halogen acids, it is the $-\overset{|}{\underset{|}{C}}-O$ bond that breaks, to form alkyl

halides. Since —C—O bond breaks faster in tertiary alcohols, the decreasing rate of reaction of alcohols with halogen acids is as follows:

R_3—C—OH > R_2CH—OH > RCH_2OH

3° alcohol > 2° alcohol > 1° alcohol

In the case of halogen acids

HI > HBr > HCl

Mechanism:

$$HX \rightleftharpoons H^+ + X^-$$

$$R—O—H + H^+ \longrightarrow R—\overset{+}{\underset{\underset{H}{|}}{O}}—H \longrightarrow R^+ + H_2O$$

$$R^+ + X^- \longrightarrow RX$$

The decreasing order of reactivity in the case of halogen acids is, because a lower amount of energy is involved in dissociation of HI into its ions, by a heterolytic fission, as compared to either HBr or HCl.

halogen acid	heterolytic fission	amount of energy required
H : I $\longrightarrow$	$H^+ + I^-$	315 kcal/mol
H : Br $\longrightarrow$	$H^+ + Br^-$	324 kcal/mol
H : Cl $\longrightarrow$	$H^+ + Cl^-$	335 kcal/mol

In the case of alcohols, involving the —O—H group, it is the —O—H bond that is under greater strain in 1° or a primary alcohol, as compared to either 2° or a 3° alcohol. So when a —O—H bond is attacked by sodium metal, it breaks faster in 1° alcohol. **The order is**

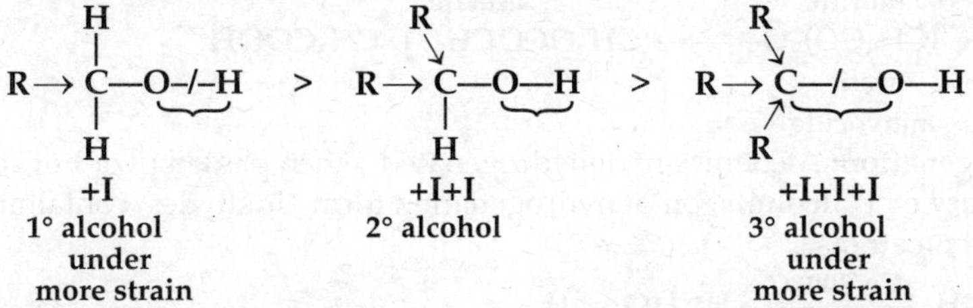

Similarly, when a —C—O bond is attacked, it breaks faster in 3° alcohol, as compared to either 2° or 1°alcohol. **The order is**

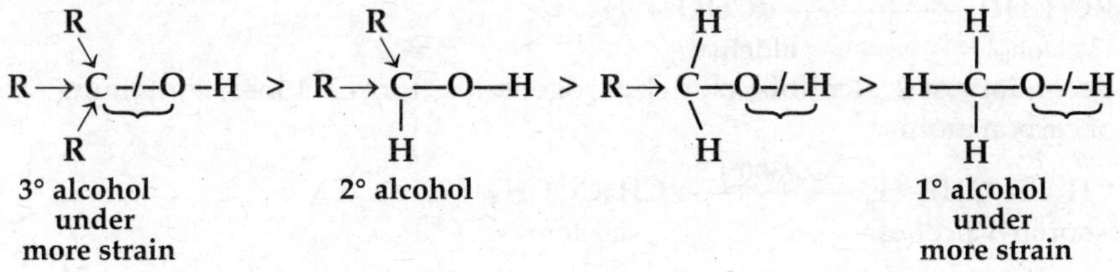

4. **Reaction with organic acids: Ester formation**

   Alcohols react with organic acids, in the presence of little conc. H_2SO_4 to forms esters.

   $$RCOOH + R'OH \xrightarrow{\text{conc. } H_2SO_4} RCOOR' + H_2O$$

   organic acid   ester

   The reaction may be termed *esterification*. When HCl is used as a catalyst, for esterification, the reaction is termed *Fischer–Spears esterification*.

Mechanism:

i. $H_2SO_4 \longrightarrow H^+ + HSO_4^-$

ii.
$$R-\overset{\overset{\displaystyle O}{\|}}{C}-O-H + H^+ \longrightarrow R-\overset{\overset{\displaystyle O}{\|}}{C}-\underset{\underset{\displaystyle H}{|}}{O^+}-H \longrightarrow R-\overset{+}{C}=O + H_2O$$

                                                               acyl carbocation

iii.
$$R-\overset{+}{C}=O + R'OH \longrightarrow R-\overset{\overset{\displaystyle O}{\|}}{C}-\underset{\underset{\displaystyle H}{|}}{O^+}-R' \longrightarrow R-\overset{\overset{\displaystyle O}{\|}}{C}-O-R' + H^+$$

                                   protonated ester   ester

Transesterification: The replacement of an alkoxy group ($-\overset{..}{O}R'$), in an ester by another alkoxy group ($-OR''$), is known as transesterification. In this process, the ester is refluxed with the substituted alcohol, taken in excess, in the presence of a little conc. H_2SO_4.

$$RCOOR' + R''OH \underset{}{\overset{H^+}{\rightleftharpoons}} RCOOR'' + R'OH$$

Distilling off one of the product, pushes the equilibrium, towards formation of more transerified ester.

5. **Reaction with acetyl chloride, CH_3COCl, (or acyl chloride, $RCOCl$) and acetic anhydride, $(CH_3CO)_2O$:** Alcohols react with acetyl chloride or acetic anhydride, to form esters, in presence of either a little conc. H_2SO_4 or pyridine.

   This process is known as acetylation or acylation.

   $$CH_3OH + CH_3COCl \xrightarrow{\text{conc. } H_2SO_4} CH_3OCOCH_3 + HCl$$

                    acetyl   methyl

                    chloride   acetate

   $$CH_3OH + (CH_3CO)_2O \longrightarrow CH_3OCOCH_3 + CH_3COOH$$

                    acetic anhydride

6. **Dehydrogenation:** Alcohols are dehydrogenated, when passed over **hot copper at 300 °C.**

   a. **Primary or 1° alcohols,** on dehydrogenation form aldehydes, containing same number of carbon atoms.

   $$CH_3OH \xrightarrow{Cu/300\,°C} HCHO + H_2$$

                                 formaldehyde

   $$RCH_2OH \xrightarrow{Cu/300\,°C} RCHO + H_2$$

          1° alcohol   aldehyde

   b. **Secondary or 2° alcohols,** on dehydrogenation, form ketones, containing same number of carbon atoms.

   $$CH_3CH(OH)CH_3 \xrightarrow{Cu/300\,°C} CH_3COCH_3 + H_2$$

          isopropyl alcohol   acetone

$$RCH(OH)R \xrightarrow{Cu/300°C} RCOR + H_2$$
secondary alcohol   ketone

  c. **Tertiary or 3°alcohols,** on dehydrogenation, form alkene and water.

$$CH_3-\underset{\underset{CH_3}{|}}{\overset{\overset{CH_3}{|}}{C}}-OH \xrightarrow{Cu/300°C} \underset{\underset{CH_3}{|}}{\overset{\overset{CH_3}{|}}{C}}=CH_2 + H_2O$$

tertiary butyl alcohol   iso butylene
3° alcohol   alkene

The products formed, on dehydrogenation of alcohols, may be identified and this can be used for distinguishing alcohols.

7. **Oxidation:** Alcohols are easily oxidized by various agents.
   a. **Oxidation with acidified $K_2Cr_2O_7$ or acidified $KMnO_4^-$**
   i. **Primary alcohols or 1° alcohols** are first oxidized to **aldehydes** and then to **carboxylic acid, each containing same number of carbon atoms.**

$$RCH_2OH \xrightarrow{oxidation} RCHO \xrightarrow{oxidation} RCOOH$$
1° alcohol   aldehyde   carboxylic acid

   ii. **Secondary alcohols or 2° alcohols** are first oxidized to ketones, containing same number of carbon atoms and then to carboxylic acid, containing lesser number of carbon atoms.

$$RCHOHR \xrightarrow{oxidation} RCOR \xrightarrow{oxidation} R'COOH + CO_2 + H_2O$$
2° alcohol   ketone with same number   carboxylic acid with lesser number
           of carbon atoms   of carbon atoms

   iii. **Tertiary alcohols or 3° alcohols** are first oxidized to ketones, with lesser number of carbon atoms and finally to a mixture of carboxylic acids, each containing even lesser number of carbon atoms.

$$R'-\underset{\underset{R'''}{|}}{\overset{\overset{R''}{|}}{C}}-O-H \longrightarrow R'-COR'' \longrightarrow R'COOH + R''COOH + R'''COOH$$

3° alcohol   ketone with   a mixture of carboxylic acid with
          lesser number carbon atoms   still lesser number carbon atoms

   b. **Oxidation with Sorret–Collins reagent and PCC (pyridinium chlorochromate, $C_5H_5N.$ HCl·CrO_3) –**
   i. **Sorret–Collins reagent:** It consists of **chromium oxide (CrO_3)** and **pyridine, (C_5H_5N)** dissolved in **dichloromethane, (CH_2Cl_2)** as a solvent. **Sorret–Collins** reagent **oxidizes a primary alcohol** to the corresponding **aldehyde** only. The oxidation stops at aldehyde stage, which is not oxidized further.
   ii. **PCC**–The reagent consists of **chromum oxide, HCl and pyridine,** dissolved in **dichloromethane, CH_2Cl_2.** The reagent oxidizes only a primary alcohol to an aldehyde. The aldehyde is not oxidized further.

$$RCH_2OH \xrightarrow{Sorret\text{-}Collins \ reagent \ or \ PCC} RCHO + H_2O \ (no \ further \ oxidation)$$

   c. **Jones' reagent**–It is a solution of chromic acid in aqueous acetone. Jones reagent is a sufficiently mild oxidizing reagent, it oxidizes a primary alcohol to aldehyde and a secondary alcohol to a ketone (without oxidizing or rearranging a double bond, when present together).

$$RCH_2OH \longrightarrow RCHO \ no \ further \ oxidation$$
1° alcohol   aldehyde

$$\underset{\text{2° alcohol}}{RCH(OH)R} \longrightarrow \underset{\text{ketone}}{RCOR} \text{ no further oxidation}$$

Dihydric Alcohols

Dihydric alcohols contain **two hydroxyl groups,** in the molecule. These dihydric alcohols may be divided into **three categories.**

1. Those alcohols which contain **two —OH groups, attached to the same carbon atom,** like, **RCH(OH)$_2$.** Such **gem dihydroxy compounds** are quite **unstable. These lose a molecule of water** and are converted into other class of compounds.

$$\overset{\diagdown}{\underset{\diagup}{C}} \overset{OH}{\underset{OH}{\diagup}} \longrightarrow \overset{\diagdown}{\underset{\diagup}{C}} = O + H_2O$$

2. The other class of dihydroxy alcohols, consists of compounds **containing two —OH group,** attached to **two adjacent carbon atoms.**
   For example:

$$\begin{array}{cc} H & H \\ | & | \\ H-C-C-H \\ | & | \\ OH & OH \end{array} \qquad \text{ethylene glycol}$$

   Such dihydroxy alcohols are termed **glycols.** These are **quite stable** and do not lose a water molecule, under ordinary conditions.

3. The third category of dihydric alcohols are those which contain **two —OH groups, attached to two distant carbon atoms in the molecule.**

$$\underset{\underset{OH}{|}}{CH_3-CH-CH_2-CH_2-CH_2-OH} \qquad \text{pentane-1, 4-diol}$$

Methods of Preparation

1. **By hydroxylation of ethylene:**
   a. **Ethylene** may be **hydroxylated by dil. alk. KMnO$_4$ solution (Baeyer's reagent)**
   $$2KMnO_4 + 2H_2O \longrightarrow 2KOH + 2MnO_2 + 3O$$
   $$CH_2=CH_2 + O + H_2O \longrightarrow CH_2(OH)-CH_2(OH)$$
   The addition of two —OH groups, across the double bond, takes place in a *cis*-manner.

   b. **By osmium tetraoxide (OsO$_4$):** Ethylene forms a **complex with OsO$_4$,** which upon **hydrolysis with NaHSO$_3$,** gives *cis*-ethylene glycol.

$$CH_2=CH_2 + OsO_4 \longrightarrow \begin{array}{c} CH_2-CH_2 \\ | \quad\quad | \\ O \quad\quad O \\ \diagdown \;\; \diagup \\ Os \\ \diagup \;\; \diagdown \\ O \quad\quad O \end{array} \xrightarrow{\;NaHSO_3\;} CH_2(OH)-CH_2(OH)$$

   The two —OH groups, add across the double bond, in a *cis*-manner.

   c. **By using peroxy formic acid (HCO·O·OH):** Peroxy formic acid or formic acid, **HCOOH, containing hydrogen per oxide (H$_2$O$_2$)** hydroxylates ethylene, to give *trans*-ethylene glycol.

$$CH_2=CH_2 + HCO\cdot O\cdot O\cdot H + H_2O \longrightarrow \begin{array}{c} OH \\ | \\ CH_2-CH_2 \\ | \\ OH \end{array}$$

The **addition of two —OH groups, across the double bond, takes place in a** *trans-*
manner.

2. **From ethylene oxide, CH_2—CH_2:** with O below — **Ethylene oxide or oxirane,** may be prepared from
ethylene, by treating **ethylene with oxygen,** in the presence of **silver as a catalyst, at a
temperature 250 °C,** which on hydrolysis with dil. acid, gives ethylene glycol.

$$CH_2=CH_2 + 1/2O_2 \xrightarrow{Ag/250\,°C} CH_2\text{—}CH_2 \text{ (with O)} \xrightarrow{H^+/H_2O} \underset{\underset{OH\ \ OH}{|\ \ \ |}}{CH_2\text{—}CH_2}$$

3. **Ethylene glycol** may also be prepared from ethylene by first treating it with **hypochlorous
acid (HOCl)** followed by subjecting the compound formed, namely **ethylene chlorhydrin,
$(CH_2(OH)$—$CH_2Cl)$ to alkaline hydrolysis.**

$$CH_2=CH_2 + HOCl \longrightarrow \underset{\underset{OH\ \ Cl}{|\ \ \ |}}{CH_2\text{—}CH_2} \xrightarrow{NaOH} \underset{\underset{OH\ \ OH}{|\ \ \ |}}{CH_2\text{—}CH_2}$$

4. **Ethylene glycol** may also be prepared from ethylene, first by reacting **ethylene with bromine, to get ethylene bromide,** which is then treated with **sodium acetate, CH_3COONa,
to prepare ethylene diacetate. Ethylene diacetate,** on **hydrolysis** with dil. acid gives
ethylene glycol.

$$CH_2=CH_2 + Br_2 \longrightarrow \underset{\underset{Br\ \ Br}{|\ \ \ |}}{CH_2\text{—}CH_2} \xrightarrow{CH_3COONa} \underset{\underset{CH_3COO\ \ OCOCH_3}{|\ \ \ \ \ \ \ \ \ \ |}}{CH_2\text{—}CH_2} \xrightarrow{H^+/N_2O} \underset{\underset{OH\ \ OH}{|\ \ \ |}}{CH_2\text{—}CH_2}$$

$$\qquad\qquad\qquad\quad \text{ethylene bromide} \qquad\qquad \text{ethylene diacetate} \qquad\qquad \text{ethylene glycol}$$

A direct treatment of ethylene bromide with alkali, converts ethylene bromide to ethylene
oxide.

Physical properties

It is a colourless, syrupy liquid, **b.p. = 190 °C,** miscible with water, sweet in taste, but
poisonous in nature. Therefore, it is not **consumed orally. A 40% aqueous solution of ethylene
glycol has a freezing point = – 40 °C,** therefore it is used for **cooling motor engines** in countries,
like Russia, where during winter season, the temperature touches as low as – 40 °C.

Chemical properties

It contains two hydroxyl groups in the molecule, and therefore it behaves as an alcohol.
However, it is less reactive than a monohydric alcohol.

1. **Reaction with sodium metal:** Ethylene reacts with sodium metal, to form both mono and
disodium derivatives.

$$\underset{\underset{CH_2\text{—}OH}{|}}{CH_2\text{—}OH} \xrightarrow[-H^+]{Na} \underset{\underset{CH_2\text{—}OH}{|}}{CH_2\text{—}Na} \xrightarrow[-H^+]{Na} \underset{\underset{CH_2\text{—}O^-Na^+}{|}}{CH_2\text{—}O^-Na^+}$$

$$\qquad\quad \text{glycol} \qquad\quad \text{monosodium glycollate} \qquad \text{disodium glycollate}$$

2. **Reaction with HI:** Ethylene glycol reacts with small amount of HI and forms ethylene on
heating.

$$\underset{\underset{CH_2}{|}}{CH_2}\boxed{\begin{matrix}\text{—OH + H}\\\text{—OH + H}\end{matrix}}\begin{matrix}I\\I\end{matrix} \xrightarrow{\Delta} \underset{\underset{CH_2[I]}{|}}{CH_2[I]} \xrightarrow{\Delta} \underset{\underset{CH_2}{||}}{CH_2} + I_2$$

$$\qquad\qquad\qquad\qquad\qquad \text{ethylene iodide}$$

On the other hand, on **heating ethylene glycol with a large excess of HI,** the final product is **ethyl iodide, CH$_3$CH$_2$I.**

$$\begin{array}{c} CH_2 \\ \parallel \\ CH_2 \end{array} + \; HI \longrightarrow \begin{array}{c} CH_3 \\ \mid \\ CH_2I \end{array}$$

3. **Reaction with conc. H$_2$SO$_4$ or anhydrous ZnCl$_2$:** Ethylene glycol, on treatment with **conc. H$_2$SO$_4$ or anhydrous ZnCl$_2$,** loses a molecule of water and rearranges to form **acetaldehyde, CH$_3$CHO.**

$$\begin{array}{c} CH_2-OH \\ \mid \\ CH_2-OH \end{array} \xrightarrow{\text{conc. H}_2\text{SO}_4 \text{ or ZnCl}_2} CH_3CHO + H_2O$$

4. **Oxidation:** On oxidation with **acidified K$_2$CrO$_4$,** ethylene glycol forms a number of intermediate oxidation products, forming finally **oxalic acid.**

$$\begin{array}{ccccccc} CH_2-OH & \longrightarrow & CH_2-OH & \longrightarrow & CHO & \longrightarrow & COOH & \longrightarrow & COOH \\ \mid & & \mid & & \mid & & \mid & & \mid \\ CH_2-OH & & CHO & & CHO & & CHO & & COOH \\ & & \text{glycolic} & & \text{glyoxal} & & \text{glyoxalic} & & \text{oxalic acid} \\ & & \text{aldehyde} & & & & \text{acid} \end{array}$$

5. **Reaction with periodic acid (HIO$_4$) or periodic acid oxidation:** HIO$_4$ **oxidises** ethylene glycol, to form two molecules of **formaldehyde, HCHO. A carbon to carbon bond, holding —OH groups, on each adjacent carbons, breaks up to give oxidized compounds.**

$$\begin{array}{c} CH_2-OH \\ \mid \\ CH_2-OH \end{array} \xrightarrow{\text{HIO}_4} \underset{\text{formaldehyde}}{HCHO + HCHO}$$

This may be used as a test, to detect the presence of two —OH groups on adjacent carbons.

Trihydric Alcohols

$$\begin{array}{l} CH_2-OH \\ \mid \\ CH-OH \qquad \text{Glycerol or glycerine or 1,2,3-hydroxy propane or propane–1,2,3–triol} \\ \mid \\ CH_2-OH \end{array}$$

Glycerol is present in nature, in combined form as oils or fats. When seeds are squeezed, oils (liquid glycerides), or fats, (solid glycerides) are obtained. **Oils or fats, are esters of glycerol with higher fatty acids, like stearic acid, (C$_{17}$H$_{35}$COOH), oleic acid, (C$_{17}$H$_{33}$COOH), and palmitic acid, (C$_{15}$H$_{31}$COOH).**

These may be formulated as—
$$\begin{array}{ll} CH_2-OCOR & \text{where} \quad R = C_{17}H_{35} \\ \mid \\ CH-OCOR & \qquad\qquad R = C_{17}H_{33} \\ \mid \\ CH_2-OCOR & \qquad\qquad R = C_{15}H_{31} \end{array}$$

All 'R' may be same or different in the same molecule.

For example:
$$\begin{array}{l} CH_2-OCOC_{17}H_{35} \\ \mid \\ CH-OCOC_{17}H_{35} \qquad \text{glyceryl tristearate} \\ \mid \\ CH_2-OCOC_{17}H_{35} \end{array}$$

Manufacture:

1. By hydrolysis of oils or fats:

a. **Alkaline hydrolysis or saponification:** When oils or fats are boiled with caustic alkali, they are hydrolysed and form **insoluble sodium salt of fatty acids,** as precipitate, and glycerol. **Sodium salt of fatty acids is used as soap.**

$$
\begin{array}{l}
CH_2\text{—O}\,|\,COR \\
\quad\; H\,|\,ONa \\
CH\text{—O–}|\,COR \\
\quad\; H\,|\,ONa \\
CH_2\text{—O}\,|\,COR \\
\quad\; H\,|\,ONa
\end{array}
\xrightarrow{\text{alkaline hydrolysis}}
\begin{array}{l}
CH_2\text{—OH} \\
| \\
CH\text{—OH} \\
| \\
CH_2\text{—OH} \\
\text{glycerol}
\end{array}
+ \;\; 3RCOO^- Na^+
$$

Soap or
sodium salt of
higher fatty acids

After filtering the precipitated **insoluble soap,** the filtrate, known as **mother liquor,** is purified and processed further to recover glycerine from it.

b. **Acid hydrolysis:** Oils or fats are boiled with **dilute H_2SO_4, under pressure,** when oils and fats are hydrolysed to give glycerol and solid fatty acids. Solid fatty acids are filtered off and used for the manufacture of candles. Glycerol is recovered from the filtrate, after purification.

$$
\begin{array}{l}
CH_2\text{—O}\,|\,COR \\
\quad\; H\,|\,OH \\
CH\text{—O}\,|\,COR \\
\quad\; H\,|\,OH \\
CH_2\text{—O}\,|\,COR \\
\quad\; H\,|\,OH
\end{array}
\xrightarrow{H^+/H_2O}
\begin{array}{l}
CH_2\text{—OH} \\
| \\
CH\text{—OH} \\
| \\
CH_2\text{—OH} \\
\text{glycerol}
\end{array}
+ \;\; 3RCOOH
$$

fatty acids

c. **Synthesis of glycerine**

  i. $CaO + 3C \longrightarrow CaC_2 + CO$

 ii. $CaC_2 + 2H_2O \longrightarrow Ca(OH)_2 + CH{\equiv}CH$
    calcium carbide $\qquad\qquad$ acetylene

iii. $CH{\equiv}CH + H_2O \xrightarrow{Hg^{2+}} CH_3CHO\text{—(oxidation)} \longrightarrow CH_3COOH$
    $\qquad\qquad\qquad\qquad$ acetaldehyde $\qquad\qquad\qquad$ acetic acid

 iv. $2CH_3COOH + CaO \xrightarrow{\text{dry distillation}} (CH_3COO)_2Ca + H_2O$
    $\qquad\qquad\qquad\qquad\qquad$ calcium acetate

  v. $(CH_3COO)_2Ca \xrightarrow{Hg^{2+}} CH_3COCH_3 + CaCO_3$
    $\qquad\qquad\qquad\qquad$ acetone

 vi. $CH_3COCH_3 + 2H \xrightarrow{\text{reduction}} CH_3CH(OH)CH_3 \xrightarrow[-H_2O]{\text{conc. } H_2SO_4} CH_3\text{—CH=CH}_2$
    $\qquad\qquad\qquad\qquad\qquad$ isopropyl alcohol $\qquad\qquad\qquad$ propene

Glycerol is manufactured from propene, which is the modern method, as this method gives pure glycerine. In industry, propene is obtained from petroleum gas.

Manufacture:

$$
\begin{array}{l}
CH_3 \\
| \\
CH \\
\| \\
CH_2
\end{array}
\xrightarrow{Cl_2/500^\circ C}
\begin{array}{l}
CH_2Cl \\
| \\
CH \\
\| \\
CH_2
\end{array}
\xrightarrow{\text{aq. NaHCO}_3}
\begin{array}{l}
CH_2OH \\
| \\
CH \\
\| \\
CH_2
\end{array}
\xrightarrow{HOCl}
\begin{array}{l}
CH_2OH \\
| \\
CHCl \\
\| \\
CH_2OH
\end{array}
\xrightarrow{\text{aq. NaHCO}_3}
\begin{array}{l}
CH_2\text{—OH} \\
| \\
CH\text{—OH} \\
\| \\
CH_2\text{—OH}
\end{array}
$$

propene $\qquad\qquad$ allyl chloride $\qquad\qquad$ allyl alcohol

Physical properties

Glycerine is a colourless, odourless, syrupy liquid, **b.p. = 290°C (decomposition)**, it is sweet in taste. It is not poisonous to human system; therefore it is used for internal consumption.

It is miscible with water and other organic solvents; but it is immiscible with diethyl ether.

Chemical properties

1. As the molecule contains three —OH groups, it reacts with sodium metal, to form mono-, di- and ultimately a tri sodio-derivatives.

$$
\begin{array}{ccccc}
\text{CH}_2\text{—OH} & & \text{CH}_2\text{O}^-\text{Na}^+ & & \text{CH}_2\text{—O}^-\text{Na}^+ & & \text{CH}_2\text{—O}^-\text{Na}^+ \\
| & \xrightarrow[-\text{H}^+]{\text{Na}} & | & \xrightarrow[-\text{H}^+]{\text{Na}} & | & \xrightarrow[-\text{H}^+]{\text{Na}} & | \\
\text{CH—OH} & & \text{CH—OH} & & \text{CH—OH} & & \text{CH—O}^-\text{Na}^+ \\
| & & | & & | & & | \\
\text{CH}_2\text{—OH} & & \text{CH}_2\text{—OH} & & \text{CH}_2\text{—O}^-\text{Na}^+ & & \text{CH}_2\text{—O}^-\text{Na}^+ \\
\text{glycerine} & & \text{mono sodio deri-} & & \text{di sodio deri-} & & \text{tri sodio derivative}
\end{array}
$$

2. **Reaction with HI (small amount):** When heated with small amount of HI, glycerol reacts with **HI, to form allyl iodide.**

$$
\begin{array}{ccc}
\text{CH}_2\text{—}\boxed{\text{OH + H}}\text{I} & & \text{CH}_2\text{[I]} & & \text{CH}_2 \\
| & & | & & \| \\
\text{CH—}|\text{OH + H}|\text{I} & \xrightarrow{-3\text{H}_2\text{O}} & \text{CH [I]} & \xrightarrow{-\text{I}_2} & \text{CH} \\
| & & | & & | \\
\text{CH}_2\text{—}\boxed{\text{OH + H}}\text{I} & & \text{CH}_2\text{I} & & \text{CH}_2\text{I} \\
& & \text{glyceryl tri iodide} & & \text{allyl iodide}
\end{array}
$$

However, in the presence a large excess of HI, glycerine forms isopropyl iodide, as a final product.

$$
\begin{array}{ccccccc}
\text{CH}_2 & & \text{CH}_3 & & \text{CH}_3 & & \text{CH}_3 \\
\| & \xrightarrow{+\text{HI}} & | & \xrightarrow{-\text{I}_2} & | & \xrightarrow{+\text{HI}} & | \\
\text{CH} & & \text{CH} \,|\text{I}| & & \text{CH} & & \text{CHI} \\
| & & | & & \| & & | \\
\text{CH}_2\text{I} & & \text{CH}_2\,|\text{I}| & & \text{CH}_2 & & \text{CH}_3 \\
\text{allyl iodide} & & \text{1,2-di iodo propane} & & \text{propene} & & \text{isopropyl iodide}
\end{array}
$$

3. **Reaction with oxalic acid**

   a. **At 110°C:** on heating **oxalic acid with glycerine** to a **temperature of 110°C, formic acid** is formed.

$$
\begin{array}{ccccc}
\text{CH}_2\text{—OH + HOOC} & & \text{CH}_2\text{—OCO—}\boxed{\text{COO}}\text{H} & \text{CH}_2\text{—O}|\text{COH} \\
| & | & \xrightarrow[110°C]{-\text{H}_2\text{O}} & | & | \\
\text{CH—OH} & \text{HOOC} & & \text{CH—OH} & \text{CH—OH} \\
| & & & | & | \\
\text{CH}_2\text{—OH} & & & \text{CH}_2\text{—OH} & \text{CH}_2\text{—OH} \\
& \text{oxalic acid} & & \text{glyceryl mono-oxalate} & \text{glyceryl mono formate}
\end{array}
$$

with $\xrightarrow{-\text{CO}_2}$ and $\xrightarrow[\text{hydrolysis}]{\text{H}^+/\text{H}_2\text{O}}$ leading to:

$$
\begin{array}{c}
\text{H}|\text{OH} \\
\text{CH}_2\text{—OH} \\
| \\
\text{CH—OH + HCOOH} \\
| \qquad \text{formic acid} \\
\text{CH}_2\text{—OH}
\end{array}
$$

b. At 260 °C: on heating **oxalic acid with glycerol, to a temperature of 260 °C, allyl alcohol** is obtained.

$$
\begin{array}{ccc}
\underset{\text{oxalic acid}}{
\begin{array}{l}
CH_2-O\boxed{H}\;+\;HO\boxed{OC}\\
|\\
CH-O\boxed{H\quad HO}\;OC\\
|\\
CH_2-OH
\end{array}}
&\xrightarrow[260\,°C]{-2H_2O}&
\underset{\text{glyceryldioxalate}}{
\begin{array}{l}
CH_2-\boxed{OCO}\\
|\\
CH-\;\boxed{OCO}\\
|\\
CH_2-OH
\end{array}}
&\xrightarrow{-CO_2}&
\underset{\text{allyl alcohol}}{
\begin{array}{l}
CH_2\\
\parallel\\
CH\\
|\\
CH_2-OH
\end{array}}
\end{array}
$$

4. **Reaction with nitrating mixture:** Nitrating mixture consists of a solution, containing **an equal volume of conc. HNO_3 and conc. H_2SO_4,** prepared by adding **conc. H_2SO_4, drop by drop, to a chilled conc. HNO_3** with constant shaking.

Glycerol reacts with nitric acid, to form glyceryl trinitrate, which is a yellow coloured oil. It is highly explosive in nature and is used in making bombs and other explosives.

$$
\begin{array}{l}
CH_2-\boxed{O}\boxed{H\;+\;HO}\,NO_2\\
|\\
CH-\;\boxed{O}\boxed{H\;+\;HO}\,NO_2\\
|\\
CH_2-\boxed{O}\boxed{H\;+\;HO}\,NO_2
\end{array}
\xrightarrow{\text{conc. } H_2SO_4}
\begin{array}{l}
CH_2-ONO_2\\
|\\
CH-ONO_2\quad+\;3H_2O\\
|\\
CH_2-ONO_2
\end{array}
$$

glyceryl trinitrate
(wrongly named as trinitro glycerine)

Trinitroglycerine or TNG was known as an unpredictable and uncontrollable explosive. It was, **Alfred Nobel in 1867,** who discovered its controlled use as an explosive, in the form of **TNG, absorbed in keisulguhr** earth, known as **dynamite. Dynamite would explode only when detonated.**

It was after **Alfred Nobel that Nobel price has been instituted for a noble cause.**

5. **Reaction with potassium bisulphate, $KHSO_4$ or conc. H_2SO_4 or dehydration of glycerol:**

$$
\begin{array}{l}
CH_2-OH\\
|\\
CH-OH\\
\parallel\\
CH_2-OH
\end{array}
\xrightarrow[-H_2O]{KHSO_4 \text{ or conc. } H_2SO_4\,/\,\Delta}
\begin{array}{l}
CH_2\\
\parallel\\
CH\\
|\\
CHO
\end{array}
$$

acrolein or acrylic aldehyde
a sharp smelling liquid

As acrolein in also formed, when oils or fats are heated with $KHSO_4$ or conc. H_2SO_4, glycerol is distinguished from oils or fats by Dunston's test, which is given by glycerol but not by oils or fats.

Dunston's test for glycerol: A drop of phenolphthalein is added to an aqueous solution of **borax or sodium tetra borate, $Na_2B_4O_7$,** when a **pink colour develops.** On adding glycerol to this pink coloured solution, the **pink colour disappears,** which **appears again on warming and disappears and disappears again on cooling.**

6. **Oxidation:**

$$
\begin{array}{ccccccc}
\text{CHO} & & \text{COOH} & & \text{COOH} & & \text{COOH} \\
| & & | & & | & & | \\
\text{CH—OH} & \xrightarrow{[O]} & \text{CH—OH} & \xrightarrow{[O]} & \text{CH—OH} & \xrightarrow{[O]} & \text{CO} \\
| & & | & & | & & | \\
\text{CH}_2\text{—OH} & & \text{CH}_2\text{—OH} & & \text{COOH} & & \text{COOH}
\end{array}
$$
glyceric aldehyde   glyceric acid   tartronic acid   meso oxalic acid

$$
\begin{array}{c}
\text{CH}_2\text{—OH} \\
| \\
\text{CH—OH} \\
| \\
\text{CH}_2\text{—OH}
\end{array}
\begin{array}{c}
\nearrow \!\!{}^{[O]} \\
\\
\searrow \!\!{}_{[O]}
\end{array}
$$

$$
\begin{array}{ccccc}
\text{CH}_2\text{—OH} & & \text{COOH} & & \text{COOH} \\
| & & | & & | \\
\text{CO} & \xrightarrow{[O]} & \text{CO} & \xrightarrow{[O]} & \text{COOH} \\
| & & | & & \\
\text{CH}_2\text{—OH} & & \text{COOH} & &
\end{array}
$$
     dihydroxy acetone   oxalic acid

The different oxidation products are obtained by oxidation of glycerol by different oxidizing agents.

   i. conc. $HNO_3 \longrightarrow$ glyceric acid

   ii. dil. $HNO_3 \longrightarrow$ glyceric acid and tartronic acid

   iii. acidified $KMnO_4 \longrightarrow$ oxalic acid

   iv. bismuth nitrate, $Bi(NO_3)_3 \longrightarrow$ meso oxalic acid

   v. Fenton's reagent $\longrightarrow$ glyceraldehyde and dihydroxy acetone

     (aqueous $FeSO_4$ + a little H_2O_2)   (this mixture is known as glycerose)

$$
\begin{array}{ccccccc}
\text{CH}_2\text{—OH} & & \text{CH}_2\text{—OH} & & \text{H} & & \text{CH}_2\text{—OH} \\
| & & | & & | & & | \\
\text{CH—OH} & \longrightarrow & \text{C=O} & + & \text{C=O} & \xrightarrow{\text{alkali}} & \text{C=O} \\
| & & | & & | & & | \\
\text{CH}_2\text{—OH} & & \text{CH—OH} & & \text{CH—OH} & & \text{CH—OH} \\
& & & & | & & | \\
& & & & \text{CH}_2\text{—OH} & & \text{CH—OH} \\
& & & & & & | \\
& & & & & & \text{CH—OH} \\
& & & & & & | \\
& & & & & & \text{CH}_2\text{—OH}
\end{array}
$$
            dihydroxy acetone   glyceric aldehyde   inactive fructose

7. **Reaction with periodic acid, HIO_4 or periodic acid oxidation:** Glycerol, on warming with **HIO_4 or lead tetra acetate, $Pb(CH_3COO)_4$,** undergoes decomposition and oxidation, to form formaldehyde and formic acid.

$$
\begin{array}{ccc}
\text{CH}_2\text{—OH} & & \text{HCHO}^+ \\
| & & \\
\text{CH—OH} & \xrightarrow{HIO_4 \text{ or } Pb(CH_3COO)_4} & \text{HCOOH}^+ \\
| & & \\
\text{CH}_2\text{—OH} & & \text{HCHO}
\end{array}
$$

Periodic acid oxidation is used to determine the **glycolic structure,** i.e. the **presence of —OH groups on vicinal carbons.** As a result of this oxidation, a **primary alcoholic group, CH_2—OH group,** is oxidized to **formaldehyde** molecule and a **secondary alcoholic group, —CHOH group,** is oxidized to **formic acid.**

Phenols

Phenols are the hydroxy compounds, containing —OH groups, attached to the benzene ring. When one or more hydrogen, in benzene ring are substituted by —OH groups, phenols are formed. These can be mono-, di- or poly hydroxy phenols.

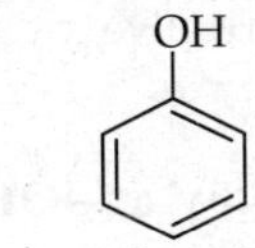

phenol 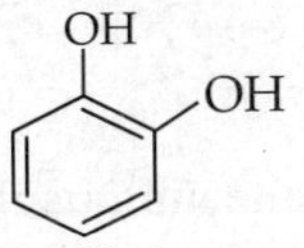 catechol 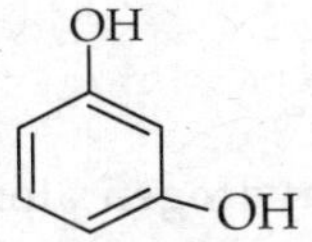 resorcinol 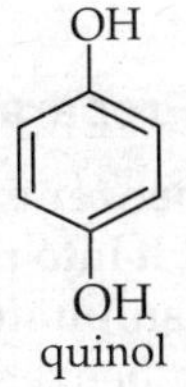 quinol

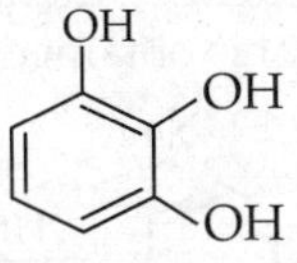

pyrogallol 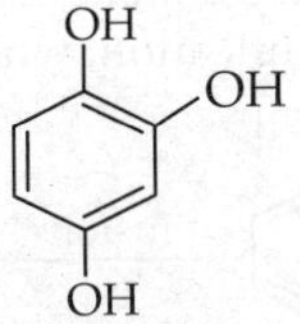 hydroxyquinol 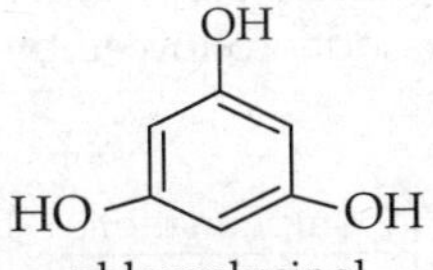 phloroglucinol

Unlike alcohols, **phenols are slightly acidic in nature**, their **Ka value** $= 1 \times 10^{-10}$. **These are sufficiently acidic in nature. These turn blue litmus red, react with alkali to form salt and water, but do not liberate CO_2 from $NaHCO_3$, like mineral acids.**

Ka values:	RCOOH	>	C_6H_5OH	>	C_6H_5SH	>	ROH
	carboxylic acid		phenols		thioalcohols		alcohols
	1×10^{-5}		1×10^{-10}		1×10^{-11}		1×10^{-18}

The acidic character of phenols may be explained, on the basis of resonance. Phenol forms the following resonating structures:

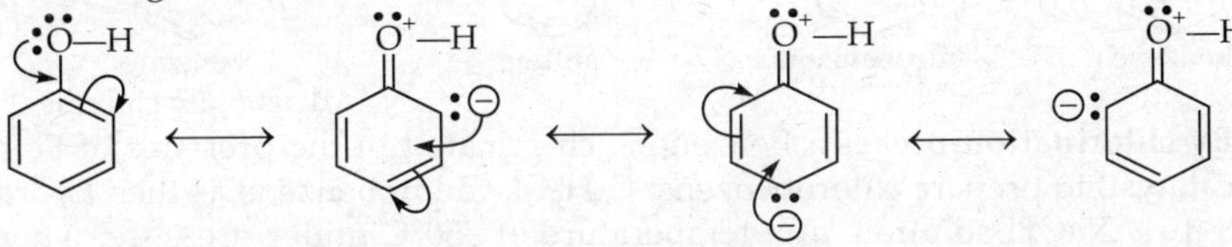

As the **resonating structures contain two charges in the same molecule, the system becomes thermodynamically unstable, so it releases the proton from —OH group, to form phenoxide ion, in order to stablize the species. Phenoxide ion contains only one charge.** Therefore, the driving force in the ionization of phenol molecule is to form phenoxide ion and to release a proton, so as to **stabilize.**

Phenoxide ion is stabilized by forming a number of resonating structures.

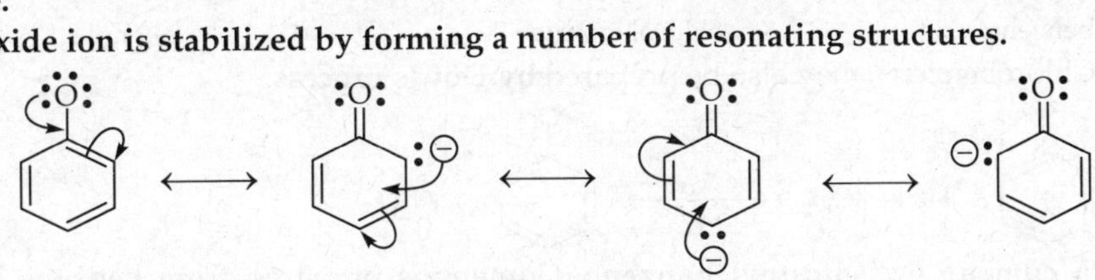

Owing to the formation of resonating structures, a phenol molecule develops a negative charge on ortho- and para- positions. In further substitution, by an electrophile, the entering group occupies and substitutes ortho- and para- positions and forms a mixtue of o- and p-disubstituted phenol. The **—OH group acts as a o- and p-directing group.**

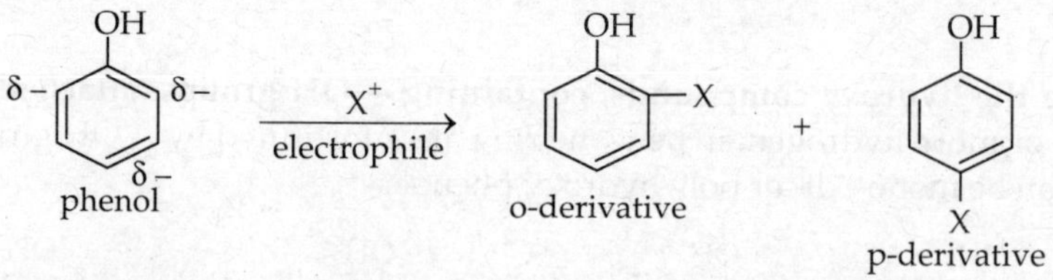

phenol → o-derivative + p-derivative

Methods of Preparation

1. **From benzene: One hydrogen atom in benzene is substituted by a —OH group, to convert it into phenol**

   a. **By sulphonation process:** Benzene is heated with conc. sulphuric acid to convert it into benzene sulphonic acid. Benzene sulphonic acid is then fused with excess of caustic soda, followed by acidification, when phenol is obtained.

   benzene → benzene sulphonic acid → sodium phenoxide → phenol

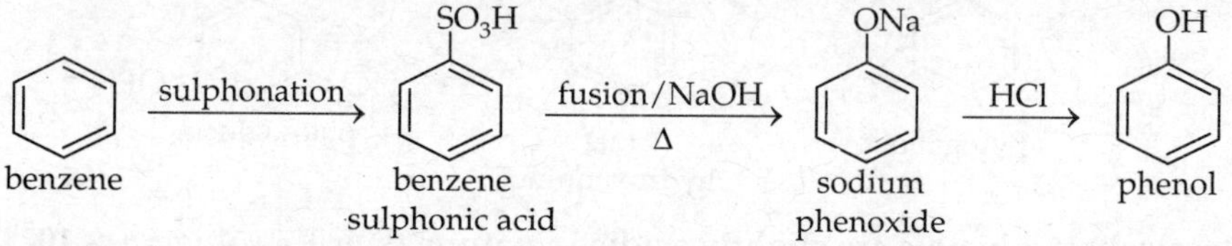

   b. **By nitration process:** Benzene is first nitrated, by heating it to 80 °C with a nitrating mixture (equal volume of conc. H_2SO_4 and conc. HNO_3), to convert benzene into nitrobenzene. Nitrobenzene is then reduced with metal and acid, to convert it into aniline. Aniline is then diazotised (treated with HNO_2 and HCl at O°C), to form **benzene diazonium chloride, which upon hydrolysis gives phenol.**

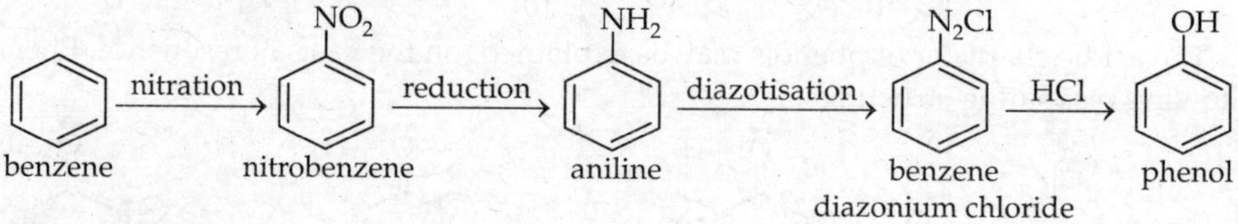

   benzene → nitrobenzene → aniline → benzene diazonium chloride → phenol

   c. **By chlorination process:** Benzene is chlorinated in the presence of **Fe or $FeCl_3$ as catalyst,** to prepare chlorobenzene, C_6H_5Cl. Chlorobenzene, is then hydrolysed with 8–10% NaOH solution, at a temperature of 350°C under pressure, when phenol is obtained (after acidification with acid).

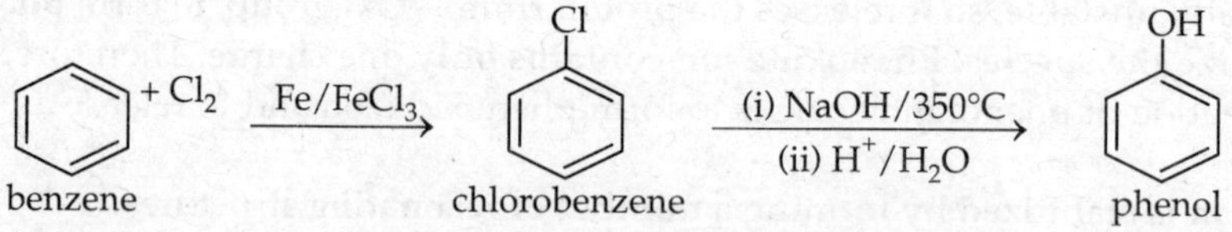

   benzene → chlorobenzene → phenol

   Chlorobenzene, may also be prepared **by Dow's process.**

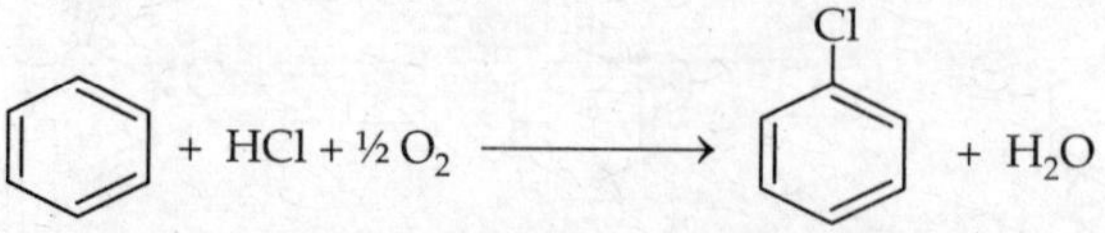

2. **From cumene or isopropyl benzene:** Cumene is prepared from benzene by treating benzene with either *n*-propyl chloride or isopropyl chloride, in the presence of anhydrous $AlCl_3$, i.e. by **Friedel–Craft** reaction.

   i. $CH_3CH_2CH_2Cl + AlCl_3 \longrightarrow CH_3CH_2CH_2^+ + AlCl_4^-$
   n-propyl chloride

ii. $CH_3CH_2CH_2^+$ $\xrightarrow{\text{rearrangement}}$ $CH_3–CH^+–CH_3$
    1° carbocation   2° carbocation

In the case of isopropyl chloride, it forms isopropyl carbocation with anhydrous $AlCl_3$.

$$CH_3CHClCH_3 + AlCl_3 \longrightarrow CH_3–CH^+–CH_3 + AlCl_4^-$$
                                               2° carbocation

1° carbocation rearranges to **2° carbocation,** as it is more stable than a 1° carbocation. This reacts with benzene to form isopropyl benzene, common names cumene.

iii. ⬡ $+ CH_3CH^+CH_3 \longrightarrow$ ⬡ (with CH_3CHCH_3 group) $+ H^+$
                                                 cumene

iv. $AlCl_4^- + H^+ \longrightarrow AlCl_3 + HCl$

Cumene, is then treated with oxygen to form cumene peroxide, which on hydrolysis forms phenol and acetone.

cumene $\xrightarrow{O_2}$ cumene peroxide $\xrightarrow{H_2O}$ phenol $+ CH_3COCH_3$ acetone

3. **From middle oil fraction (boiling range –170° to 230°C):** This fraction from petroleum, contains phenolic compounds, besides naphthalene, etc. It is treated with aqueous caustic alkali, which reacts with phenolic compounds, to form sodium salts, which being soluble in water, dissolve and form a separate layer. This aqueous alkaline layer is separated and alkaline solution is acidified, when phenols are liberated. On further distillation of phenolic compounds mixture, **phenol is obtained, boiling at 182°C.**

Physical properties

Phenol also known as **carbolic acid,** is a colourless crystalline compound, turns pink, owing to atmospheric oxidation, **mp = 44°C, b.p. = 182°C,** slimy in nature and caustic on touch. It is slightly soluble in water, more so in alcohol and other organic solvents.

It is slightly acidic in nature, (Ka value = 1×10^{-10}), turns blue litmus red and gives a red colour with $FeCl_3$. Formation of a violet colour with $FeCl_3$ is a characteristic test of all phenolic compounds.

Chemical properties

1. **Reaction with sodium metal:** Phenol reacts with sodium metal to form sodium salt and liberates hydrogen.

$$C_6H_5–OH + Na \longrightarrow C_6H_5–O^- Na^+ + \frac{1}{2} H_2 \uparrow$$
                                sodium phenate
                         or sodium phenoxide

2. **Reaction with caustic alkali:** Phenol reacts with caustic soda, to form sodium phenoxide and water.

$$C_6H_5–OH + NaOH \longrightarrow C_6H_5–ONa + H_2O$$

However it does not liberate CO_2 when treated with $NaHCO_3$.

$$C_6H_5-OH + NaHCO_3 \rightleftharpoons C_6H_5-ONa + H_2CO_3$$

pKa = 10   weak alkali   strong base pKa = 6 (strong acid)

The products formed are stronger base and stronger acid, which react back to form starting material; hence CO_2 is not liberated from $NaHCO_3$.

3. **Reaction with PCl_5, PCl_3 or $SOCl_2$:**

a. $C_6H_5-OH + PCl_5 \longrightarrow C_6H_5Cl + POCl_3 + HCl$

        phosphorous   chlorobenzene phosphorous

        pentachloride   oxychloride

However, the main product in the above reaction is **triphenyl phosphate, $(C_6H_5-O)_3$ PO or $(C_6H_5)_3 PO_4$.**

$$C_6H_5-OH + PCl_5 \longrightarrow C_6H_5Cl + POCl_3$$
$$3C_6H_5-OH + POCl_3 \longrightarrow (C_6H_5-O)_3 PO + 3HCl$$

b. $3C_6H_5-OH + PCl_3 \longrightarrow 3C_6H_5Cl + H_3PO_3$

        phosphorous   phosphorous acid

        trichloride

c. $C_6H_5-OH + SOCl_2 \longrightarrow C_6H_5Cl + SO_2 + HCl$

        thionyl chloride

4. **Reaction with acetyl chloride, CH_3COCl or acetic anhydride, $(CH_3CO)_2O$:**

Phenol reacts with acetyl chloride or acetic anhydride, in the presence of a little conc. H_2SO_4 or pyridine, to form acetyl derivative.

$$C_6H_5-OH + CH_3COCl/(CH_3CO)_2O \longrightarrow C_6H_5OCOCH_3 + HCl$$

                                           phenyl acetate

Fries migration: Phenyl acetate, on heating with anhydrous $AlCl_3$, rearrange to from a mixture of o- and p-acetyl phenols.

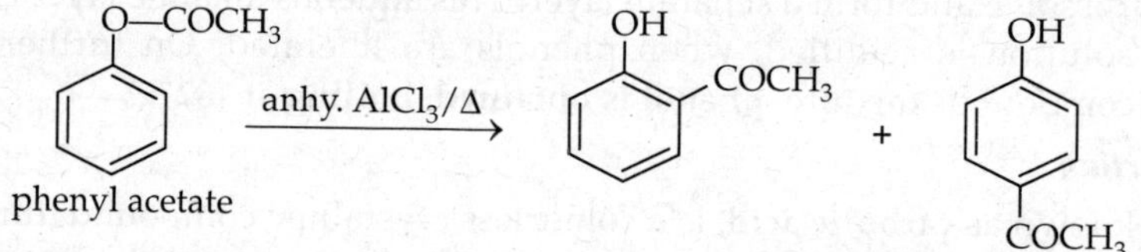

phenyl acetate   anhy. $AlCl_3/\Delta$ o-   +   p-
hydroxy acetophenone

5. **Reaction with zinc dust:** On heating phenol with zinc dust, it is converted into benzene.

$$C_6H_5-OH + Zn \xrightarrow{\Delta} C_6H_6 + ZnO$$

6. **Reaction with bromine:** Phenol reacts with bromine, both aqueous or alcoholic, to form a tribromoderivative.

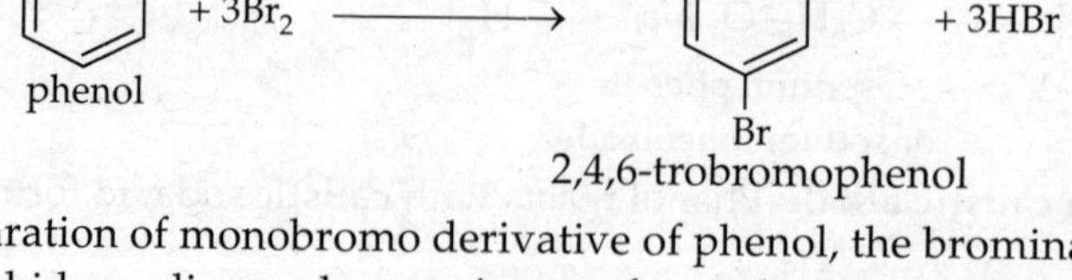

phenol   + $3Br_2$ + 3HBr

2,4,6-trobromophenol

For the preparation of monobromo derivative of phenol, the bromination is carried out in carbon disulphide medium, when a mixture of o- and p-bromo phenols is obtained.

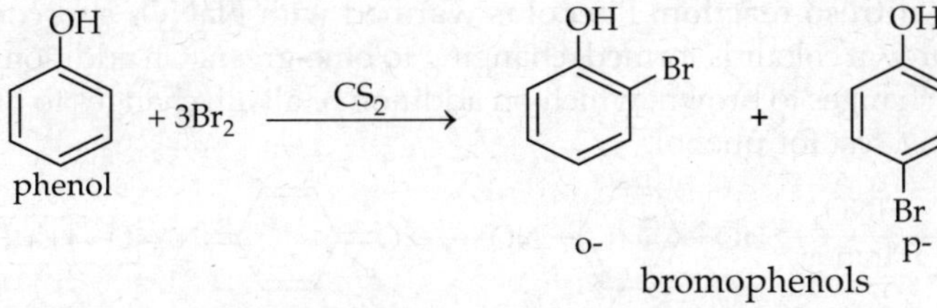

phenol + 3Br$_2$ $\xrightarrow{CS_2}$ o-bromophenols + p- bromophenols

Alternatively, phenol is first acetylated, and then treated with bromine, when it forms a mixture of o- and p- bromo phenyl acetates. These on further hydrolysis are converted into o- and p-bromo phenols.

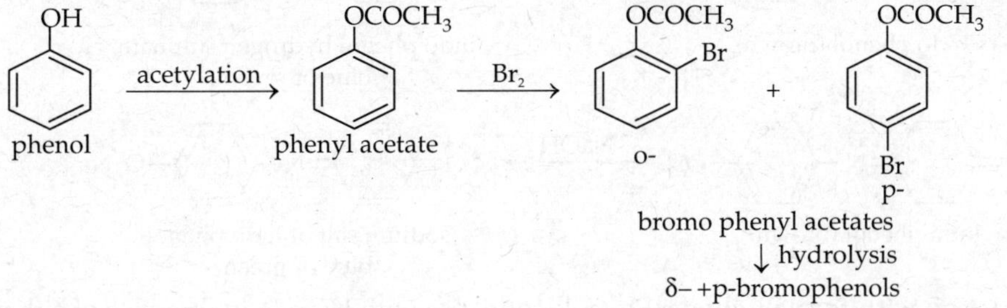

phenol $\xrightarrow{acetylation}$ phenyl acetate $\xrightarrow{Br_2}$ o- + p- bromo phenyl acetates

$\downarrow$ hydrolysis

$\delta-$ +p-bromophenols

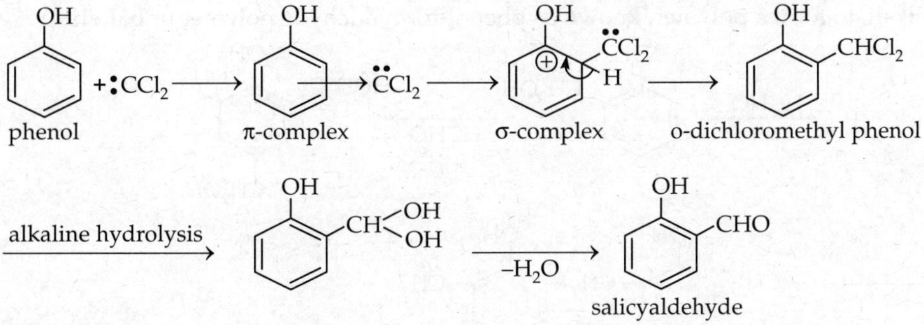

phenol +$:CCl_2$ $\longrightarrow$ π-complex $\rightarrow \ddot{C}Cl_2 \longrightarrow$ σ-complex $\longrightarrow$ o-dichloromethyl phenol

alkaline hydrolysis $\xrightarrow{}$ $\xrightarrow{-H_2O}$ salicyaldehyde

Some p-hydroxy benzaldehyde is also formed (only the o-derivative is known as salicyaldehyde).

7. **Kolbe's–Schmidt reaction:** On heating **sodium phenate, C$_6$H$_5$ONa,** with **CO$_2$,** at **120–140°C,** under pressure **salicylic acid** is formed.

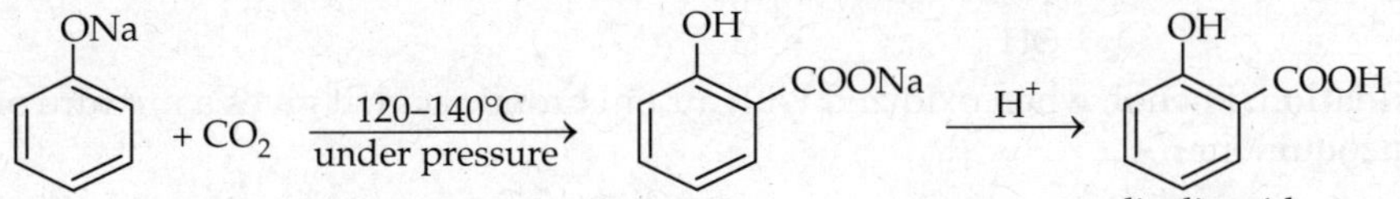

+ CO$_2$ $\xrightarrow[\text{under pressure}]{120-140°C}$ COONa $\xrightarrow{H^+}$ salicylic acid

Some p-hydroxy benzoic acid is also formed.

Mechanism:

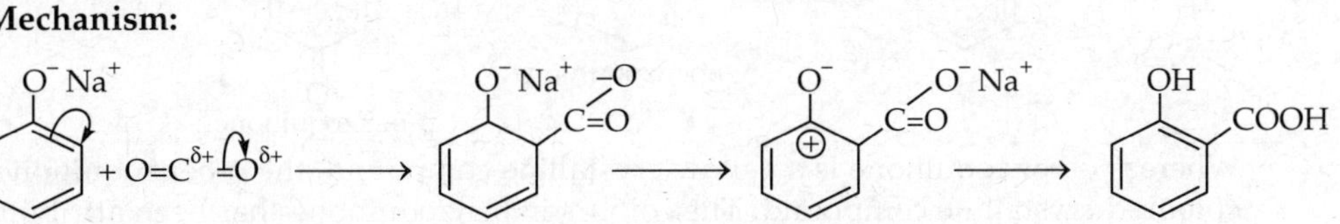

+ O=C$^{\delta+}$=O$^{\delta+}$ $\longrightarrow$ $\longrightarrow$ $\longrightarrow$ COOH

8. **Libermann's nitroso reaction:** Phenol is warmed with $NaNO_2$ and conc. H_2SO_4, when a momentary brown colour is formed, changing to blue-green. On addition of water, this blue-green colour changes to brown, which on addition of alkali, changes to blue again finally. This serves as a test for phenol.

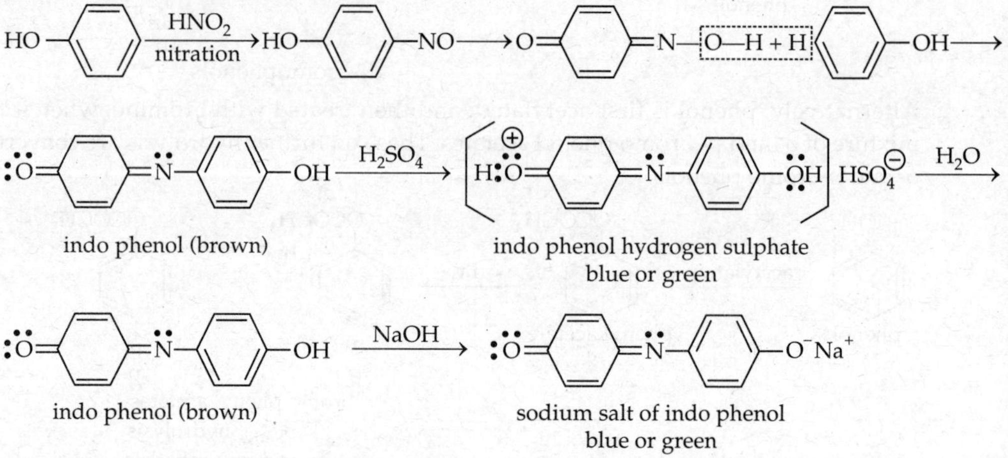

indo phenol (brown)   indo phenol hydrogen sulphate
                                      blue or green

indo phenol (brown)   sodium salt of indo phenol
                                      blue or green

9. **Reaction with formaldehyde:** Phenols condense with HCHO, in presence of either acid or alkali, to form a polymer, known as phenol-formaldehyde polymer or **bakelite**.

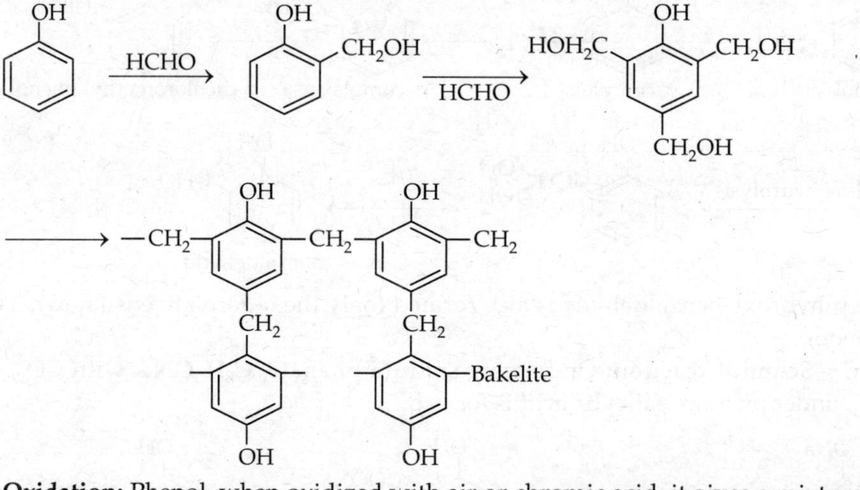

—Bakelite

10. **Oxidation:** Phenol, when oxidized with air or chromic acid, it gives a mixture of o- and p-benzoquinones.

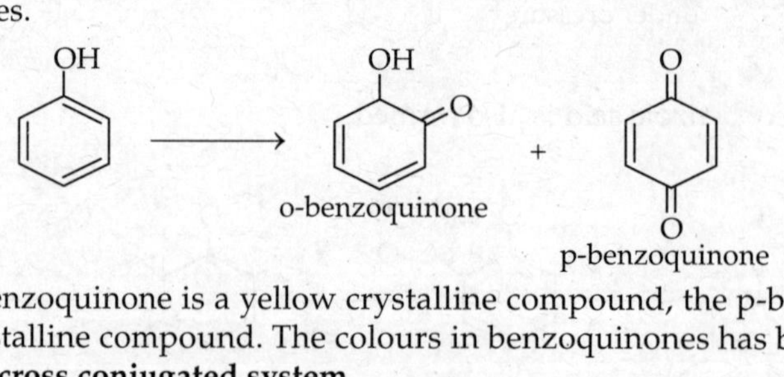

o-benzoquinone   +   p-benzoquinone

Whereas o-benzoquinone is a yellow crystalline compound, the p-benzoquinone is a red coloured crystalline compound. The colours in benzoquinones has been attributed to the presence of a **cross conjugated system.**

It has been inferred that **any compound, it could be represented by a cross conjugated**

structure =⟨   ⟩= **, it will be coloured.**

11. **Reaction with phthalic anhydride:** Phenol, condenses with **phthalic anhydride, in the** presence of conc. H_2SO_4, to form **phenolphthalein,** a colourless crystalline compound. It **gives a pink colour with alkali,** and therefore, used as an **indicator, in the titration in between acid and strong alkali.**

phenolphthalein   pink colour

REASONING TYPE QUESTIONS (RTQs)

(Group-A)

1. What are primary, secondary and tertiary alcohols? Discuss the same by taking the very first alkanol, which exist in the three forms, as mentioned.
2. What are the various methods, which can be used to distinguish the primary, secondary and tertiary alcohols from one another?
3. How will you distinguish between (i) ethyl alcohol and allyl alcohol (ii) propyl alcohol and isopropyl alcohol (iii) methanol and ethanol? Discuss with the chemistry involved.
4. What are the requirements for a compound to undergo 'haloform' reaction? Discuss.
5. Ethyl alcohol is not harmful to the human system, but methanol is? Explain.
6. Glycerol is used in the various medicines for internal use but glycol is not, and actually is poisonous. Explain.
7. Discuss the mechanism of acid catalyzed dehydration of ethyl alcohol.
8. What is the difference between ordinary alcohol, absolute alcohol, denatured alcohol, and proof spirit? Explain.
9. a. What products are formed in a reaction between glycerol and oxalic acids, when heated to 230°C and 110°C? Write equations to support your answer.
   b. How does glycerol reacts with potassium hydrogen sulphate, $KHSO_4$, or conc. H_2SO_4, on heating? Write the equation.
10. Write a note on periodic acid oxidation, taking the case of ethylene glycol.
11. How does glycerol react with HI, when taken in small quantity and when taken in excess? How these compounds can be prepared starting from the corresponding alkene? Discuss.
12. Discuss the chemistry involved, in the breath-test performed on the spot, on drivers, suspected to be driving drunk.
13. Discuss the pinacol and pinacolone rearrangement.
14. Phenol is much more acidic in nature as compared to alcohols. Explain.
15. Picric acid is a strong acid as carboxylic acid. Explain.
16. How phenol may be prepared from cumene hydroperoxide? How phenol can be tested by colour reactions?

17. How does phenol reacts with (i) Zinc dust on heating, (ii) nitrous acid, (iii) phthalic anhydride/conc. H_2SO_4, (iv) Bromine water and (v) with chloroform and aqueous NaOH, followed by acidification with dilute acid.

18. Write a note on Kolbe's reaction, and Fries migration.

19. The alcohols have a C—O bond and a O—H bond, discuss how they break when treated with the various reagents.

20. Discuss the cleavage of ethers by HI, with respect to methyl ethyl ether.

ANSWERS

(Group-A)

1. The first alkanol that exist in the three isomeric forms is butanol, C_4H_9—OH.

n-Butyl alcohol	isobutyl alcohol	tertiary butyl alcohol
	OH	CH_3
CH_3—CH_2—CH_2—OH	CH_3—CH_2—CH—OH_3	CH_3—C—OH
primary alcoholic group	secondary alcoholic group	CH_3
		tertiary alcoholic group

2. The methods used are

   i. **Luca's reagent test:** Luca's reagent consists of a solution of $ZnCl_2$, in conc. HCl.

   $$R—OH + HCl \xrightarrow{ZnCl_2} R—Cl + H_2O$$
   alcohol   alkyl chloride

   The three alcohols react with Lucas reagent at different rates.
   a. tertiary alcohols react with Lucas reagent very fast,
   b. secondary react rather slowly,
   c. primary alcohols react very very slowly.

Method – The alcohol is mixed with Lucas reagent, at room temperature. The alkyl chloride formed is insoluble in the medium. It causes the solution to become cloudy, before it appears as a separate distinct layer.
   • With tertiary alcohols, the cloudiness appears immediately,
   • With secondary alcohols, cloudiness appears, in 4-5 minutes,
   • With primary alcohols, the solution remains clear, even after 15 minutes.

   ii. **Action on hot reduced copper:** The vapours of the given alcohol, are passed over hot copper, kept at 300°C, and the product formed is identified.
   a. Primary alcohol undergoes dehydrogenation to give aldehyde

   $$R—CH_2OH \longrightarrow RCHO + H_2$$
   primary alcohol   aldehyde

   b. Secondary alcohol also dehydrogenates to give a ketone

   $$\begin{array}{c} R \\ | \\ R—CHOH \end{array} \longrightarrow R—CO—R + H_2$$
   secondary alcohol   ketone

   c. Tertiary alcohols lose a molecule of water to give an alkene.

   $$\begin{array}{c} R \\ \backslash \\ R—CHOH \\ / \\ R \end{array} \longrightarrow \begin{array}{c} R \\ \backslash \\ R—C=CH_2 + H_2O \end{array}$$
   tertiary alcohol   alkene   water

A positive test for aldehyde, ketone and a test for a double bond may serve as a test for primary, secondary and tertiary alcohol.

iii. **Oxidation test:** The mode of oxidation of the three types of alcohols, is characteristic of each alcohol.

   a. A **primary alcohol** on oxidation gives an aldehyde, which oxidises further to form an acid, containing the same number of carbon atoms, as present in the original alcohol.

$$RCH_2\text{—}OH \xrightarrow{\text{oxidation}} RCHO \xrightarrow{\text{oxidation}} RCOOH$$

primary alcohol $\qquad\qquad$ aldehyde $\qquad\qquad$ carboxylic acid
1° alcohol

Number of carbon atoms remains same, all through out.

   b. **A secondary alcohol** on oxidation first gives a ketone, containing the same number of carbon atoms. However, on further oxidation, with more vigorous oxidation, ketone formed, is oxidised to give a carboxylic acid, containing lesser no of carbon atoms.

$$\overset{\displaystyle R}{\underset{\displaystyle }{R\text{—}CH\text{—}OH}} \xrightarrow{(O)} RCOR \xrightarrow{(O)} RCOOH + CO_2 \text{ etc.}$$

secondary alcohol $\qquad$ a ketone $\qquad\qquad$ a carboxylic acid
2° alcohol $\qquad\qquad$ containing same number $\qquad$ containing less number
$\qquad\qquad\qquad$ of carbon atoms $\qquad\qquad$ of carbon atoms

   c. A **tertiary alcohol** on oxidation gives a ketone containing less number of carbon atoms. On further oxidation, ketone formed, gives a carboxylic acid, containing still lesser number of carbon atoms.

$$\overset{\displaystyle R}{\underset{\displaystyle R}{R\text{—}C\text{—}OH}} \xrightarrow{(O)} RCOR \xrightarrow{(O)} RCOOH + CO_2 \text{ etc.}$$

tertiary alcohol $\qquad$ a ketone $\qquad\qquad$ a carboxylic acid
3° alcohol $\qquad\qquad$ containing less number $\qquad$ containing still lesser
$\qquad\qquad\qquad$ carbon atoms $\qquad\qquad$ of carbon atoms

iv. **Victor Meyer's Test:** The alcohol is first converted into an iodide, on treatment with iodine and phosphorous, which in turn is then treated with $AgNO_2$ to convert the iodide formed, into a nitro alkane. The nitro alkanes so formed, are then treated with an alkaline solution of sodium nitrite $NaNO_2$, then acidified with HCl and finally made alkaline with NaOH solution.

$$R\text{—}OH \xrightarrow{P + I_2} RI \xrightarrow{AgNO_2} R\text{—}NO_2$$

primary $\qquad\qquad$ secondary $\qquad\qquad$ tertiary
alcohol $\qquad\qquad$ alcohol $\qquad\qquad$ alcohol

$RCH_2\text{—}NO_2 \qquad\qquad R_2\text{—}CH\text{—}NO_2 \qquad\qquad R_3\text{—}C\text{—}NO_2$

$\downarrow O{=}N\text{—}OH \qquad\qquad \downarrow HO\text{—}N{=}O \qquad\qquad \downarrow HO\text{—}N{=}O$

$R\text{—}C\text{—}NO_2 \qquad\qquad R_2\text{—}C\text{—}NO_2 \qquad\qquad$ NO reaction

$\;\;\|$ $\qquad\qquad\qquad\quad\;\; |$
NOH $\qquad\qquad\qquad\qquad$ NO

(nitrolic acid $\qquad\qquad$ (pseudo nitrol
gives blood red colour $\qquad$ gives blue colour
with alkali) $\qquad\qquad\qquad$ with alkali)

3. **Distinction:**
   a. Between ethyl alcohol, and allyl alcohol
   C_2H_5—OH gives a yellow precipitate of iodoform with iodine and NaOH solution
   $$C_2H_5—OH + 3I_2 + 4\,NaOH \longrightarrow CHI_3 + 3NaI + HCOONa + 3H_2O$$
   $$\text{Iodoform}$$
   Allyl alcohol decolourises yellow colour of Br_2 water.
   $$CH_2=CH—CH_2—OH + Br_2 \longrightarrow CH_2Br—CHBr—CH_2OH$$
   b. *n*-propyl alcohol and isopropyl alcohol
   Isopropyl alcohol gives a yellow precipitate of CHI_3 with iodine and NaOH.
   $$CH_3—CH(OH)—CH_3 + I_2 \longrightarrow CH_3—CO—CH_3 + 2HI$$
   $$CH_3—CO—CH_3 + 3I_2 + 4NaOH \longrightarrow CHI_3 + 3NaI + CH_3COONa + 3H_2O$$
   n-propyl alcohol does not give iodoform test.
   c. Methanol and ethanol
   C_2H_5—OH gives forms iodoform with iodine and NaOH.
   CH_3—OH does not react with iodine and NaOH.
4. **Haloform reaction:**
   Any compound, containing either a methyl keto group, CH_3—CO—, or a CH_3—CH(OH)— group in the molecule, respond to the haloform test. The compounds containing a CH_3—CH(OH)— are oxidised to a CH_3—CO— group, on treatment with either bleaching powder or with iodine and caustic alkali.
   i. $CH_3—CH(OH)—R + I_2 \longrightarrow CH_3—CO—R + 2HI$
   ii. $CH_3—CO—R + I_2 + 4NaOH \longrightarrow CHI_3 + 3NaI + RCOONa + H_2O.$
5. **Ethyl alcohol** is known to be helpful, in the digestion process, when consumed in small quantity. However, when consumed in large quantity, it produces a mad like behavior.
   On the other hand, methanol is oxidised to formaldehyde in the human body, which is very harmful, being a powerful reducing compound. This may cause a blindness or even may cause death.
   $$CH_3—OH \xrightarrow{\text{(O)}} HCHO$$
   $$\text{formaldehye}$$
   Methanol is therefore, used to make ethanol unfit for drinking, which is marketed as denatured alcohol or as methylated spirit, to be used as a fuel and a source for heat and light.
6. The oils and fats are esters of glycerol and fatty acids. When oils and fats are hydrolysed by alkalis, these break up into sodium salt of fatty acids and glycerol. As these oils and fats are also consumed by the living beings, these oils and fats are also hydrolyzed to fatty acids and glycerol, in the bodies of the living beings. Therefore, glycerol is not harmful.
   On the other hand, glycol is oxidised to oxalic acid, when consumed. As oxalic acid is poisonous in nature, it is therefore harmful to human body.

$$
\begin{array}{l}
CH_2—O—COR \\
| \\
\bullet \;\; CH—O—COR \xrightarrow[\text{acid}]{NaOH /} \\
| \\
CH_2—O—COR \\
\text{oil or fats} \qquad \text{hydrolysis}
\end{array}
\qquad
\begin{array}{l}
CH_2—OH \\
| \\
CH—OH \quad + \;\; 3\,RCOONa \\
| \\
CH_2—OH \\
\text{glycerol} \qquad \text{sodium salt of fatty acid}
\end{array}
$$

$$
\begin{array}{l}
CH_2—OH \\
| \\
\bullet \;\; CH_2—OH \xrightarrow{\text{(O)}} \\
\text{glycol}
\end{array}
\qquad
\begin{array}{l}
COOH \\
| \\
COOH \\
\text{oxalic acid}
\end{array}
$$

7. **Acid catalysed dehydration of ethyl alcohol**

   Alcohols may be dehydrated to corresponding olefins. The order of ease of dehydration of alcohol is

   tertiary alcohol > secondary alcohol > primary alcohol

   Alcohols may be dehydrated by simple heating (400–700°C) or by passing the vapours of the alcohols over heated alumina, Al_2O_3 (350°C).

   Ethyl alcohol reacts with conc. H_2SO_4, at room temperature to form ethyl hydrogen sulphate.

   a. When heated at 170°C, ethyl hydrogen sulphate, loses a molecule of H_2SO_4 to form ethene, in the presence of an excess of conc. H_2SO_4.

   $$H_2SO_4 \longrightarrow H^+ + SO_4^-$$

   $$CH_3CH_2{-}\overset{\cdot\cdot}{O}{-}H \xrightarrow{\;H^+\;} CH_3{-}CH_2{-}\overset{\overset{\textstyle H}{|}}{O}{-}H \longrightarrow CH_3{-}\overset{+}{C}H_2{-}H_2O$$
   ethyl carbocation

   $$CH_3{-}\overset{+}{C}H_2 + HSO_4^- \longrightarrow CH_3{-}CH_2{-}HSO_4$$
   ethyl hydrogen sulphate

   $$H{-}\underset{\underset{\textstyle H}{|}}{\overset{\overset{\textstyle H}{|}}{C}}{-}\underset{\underset{\textstyle HSO_4}{|}}{\overset{\overset{\textstyle H}{|}}{C}}{-}H \longrightarrow CH_2{=}CH_2 + H_2SO_4$$

   b. When heated to a temperature of 145°C, in the presence of an excess of ethyl alcohol, diethyl ether is formed.

   $$H{-}\underset{\underset{\textstyle H}{|}}{\overset{\overset{\textstyle H}{|}}{C}}{-}\underset{\underset{\textstyle H}{|}}{\overset{\overset{\textstyle H}{|}}{C}}({-}HSO_4 + H){-}O{-}C_2H_5 \xrightarrow{\;145°C\;} CH_3{-}CH_2{-}O{-}CH_2{-}CH_3 + H_4SO_4$$
   diethyl ether

8. i. *Rectified alcohol or spirit:* It is about 93-95% alcohol.

   ii. *Absolute alcohol:* It is a 100% alcohol. It is easily miscible with petrol, therefore it is mixed with petrol and it is known as power alcohol.

   iii. *Methylated spirit* or *denatured alcohol:* Rectified alcohol is mixed with ~9% methanol to make it unfit for drinking purposes or to be used for beverage. About 1% pyridine is also added to make it repulsive.

   iv. *Proof spirit:* It is a mixture of alcohol (57.1%) by volume and rest being water.

9. **Glycerol:** It reacts with hydrogen iodide (HI) taken in a small amount, on heating to form allyl iodide $CH_2{=}CH{-}CH_2I$.

   $$\begin{array}{l} CH_2{-}OH + HI \\ | \\ CH{-}OH + HI \\ | \\ CH_2{-}OH + HI \end{array} \longrightarrow \begin{array}{l} CH_2(I) \\ | \\ CH(I) \\ | \\ CH_2I \end{array} \longrightarrow \begin{array}{l} CH_2 \\ \| \\ CH \\ | \\ CH_2I \end{array} + I_2$$

   glycerol   glyceryl tri iodide   allyl iodine   iodine
   unstable

a. Glycerol, when heated, with a large excess of HI, is finally converted into isopropyl iodide.

$$\begin{array}{l} CH_2-OH + HI \\ | \\ CH-OH + HI \\ | \\ CH_2-OH + HI \end{array} \longrightarrow \begin{array}{l} CH_2I \\ | \\ CHI \\ | \\ CH_2I \end{array} \longrightarrow \begin{array}{l} CH_2 \\ \| \\ CH \\ | \\ CH_2I \end{array} \xrightarrow{HI} \begin{array}{l} CH_3 \\ | \\ CHI \\ | \\ CH_2I \end{array} \xrightarrow{-I_2} \begin{array}{l} CH_3 \\ | \\ CH \\ | \\ CH_2 \end{array} \xrightarrow{HI} \begin{array}{l} CH_3 \\ | \\ CHI \\ | \\ CH_3 \end{array}$$

tri iodide   isopropyl iodide

b. Glycerol, on heating with $KHSO_4$ or conc. H_2SO_4 is dehydrated, loses two molecule of water, and rearranges to form acrolein or acryldehyde.

$$\begin{array}{l} CH_2-OH \\ | \\ CH-OH \\ | \\ CH_2-OH \end{array} \xrightarrow[\text{conc. } H_2SO_4/\Delta]{KHSO_4} \begin{array}{l} CH_2 \\ \| \\ CH \\ | \\ CHO \end{array} + 2H_2O$$

glycerol   acrolein

10. Periodic oxidation of glycol

Glycol is a vic- diol, which contains two hydroxyl groups on the adjacent carbon atoms. All such di -ols, are oxidised by, periodic acid, HIO_4, or lead tetra acetate, $Pb\,(CH_3COO)_4$. The bond, between the two adjacent carbon atoms, bearing one —OH group each, is cleaved.

Glycol forms formaldehyde, on oxidation with HIO_4.

$$\begin{array}{l} CH_2-OH \\ | \\ CH_2-OH \end{array} \xrightarrow[Pb(CH_3COO)_4]{HIO_4} 2HCHO$$

glycol   formaldehyde

$$\begin{array}{l} CH_2-OH \\ | \\ CH-OH \\ | \\ CH_2-OH \end{array} \xrightarrow{HIO_4} 2HCHO + HCOOH$$

glycol   formaldehyde   formic acid

11. Glycerol reacts with oxalic acid, at 110°C to give formic acid.

$$\begin{array}{l} CH_2-OH \\ | \\ CH-OH \\ | \\ CH_2-OH \end{array} + \begin{array}{l} HOOC \\ | \\ COOH \end{array} \longrightarrow \begin{array}{l} CH_2-O-COH \\ | \\ CH-OH \\ | \\ CH_2OH \end{array} + CO_2 + H_2O$$

glycerol   oxalic acid   glyceryl monoformate

Glyceryl monoformat is then hydrolyzed to form formic acid.

$$\begin{array}{l} \quad\quad H|OH \\ CH_2-O|COH \\ | \\ CH-OH \\ | \\ CH_2-OH \end{array} \xrightarrow{\text{hydrolysis}} HCOOH + \begin{array}{l} CH_2-OH \\ | \\ CHOH \\ | \\ CH_2OH \end{array}$$

glyceryl monoformate   formic acid   glyceryl

a. Glycerol reacts with oxalic acid at 260°C, to form allyl alcohol.

$$\begin{array}{cccc}
CH_2{-}OH & HOOC & & \\
| & | & \xrightarrow{260°C} & \\
CH{-}OH & HOOC & & \\
| & & & \\
CH_2{-}OH & & & \\
\text{glycerol} & \text{oxalic acid} & & \\
\end{array}
\quad
\begin{array}{c}
CH_2{-}OOC \\
| \\
CH{-}OOC \\
| \\
CH_2OH \\
\text{glyceryl dioxalin} \\
\end{array}
\xrightarrow{-2CO_2}
\begin{array}{c}
CH_2 \\
\| \\
CH \\
| \\
CH_2OH \\
\text{allyl alcohol} \\
\end{array}$$

12. Very often, drivers of the vehicles have been found driving under the influence of alcohol. They are tested for the presence of alcohol in the breath.

    Such persons are required to blow into the tube containing potassium dichromate and dilute sulphuric acid.

    In case the contents turn green on blowing the breath in the tube, this confirms that the driver has consumed alcohol.

    $K_2Cr_2O_7 + 4H_2SO_4 + C_2H_5{-}OH \longrightarrow Cr_2(SO_4)_3 + K_2SO_4 + CH_3CHO + H_2O$

    Green colour is due to the formation of chromium sulphate. The reaction takes place, only when alcohol is present.

13. **Pinacol–Pinnacolone rearrangement**

    a. Pinacol is formed when acetone is reduced with magnesium metal and HCl.

$$\begin{array}{c}
CH_3 \\
\diagdown \\
\quad\quad C{=}O + H_2 \\
\diagup \\
CH_3 \\
\end{array}
\begin{array}{c}
CH_3 \\
\diagup \\
+O{=}C \\
\diagdown \\
CH_3 \\
\end{array}
\longrightarrow
\begin{array}{c}
CH_3\ CH_3 \\
| \quad\ | \\
CH_3{-}C{-}C{-}CH_3 \\
| \quad\ | \\
HO\ OH \\
\text{pinacol} \\
\end{array}
\xrightarrow{HCl}
\begin{array}{c}
CH_3 \\
| \\
CH_3{-}C{-}COCH_3 \\
| \\
CH_3 \\
\text{pincolone} \\
\end{array}$$

    b. Pinacol, on treatment with mineral acids (HCl), undergoes dehydration and rearranges to form a ketone, namely pinacolone.

$$\begin{array}{c}
CH_3\ CH_3 \\
| \quad\ | \\
CH_3{-}C{-}C{-}CH_3 \\
| \quad\ | \\
OH\ OH \\
\text{pinacol} \\
\end{array}
\xrightarrow{H^+}
\begin{array}{c}
CH_3\ CH_3 \\
| \quad\ | \\
CH_3{-}C{-}C{-}CH_3 \\
| \quad\ | \\
OH\ \overset{\oplus}{O}H_2 \\
\end{array}
\longrightarrow$$

$$\begin{array}{c}
CH_3 \\
| \\
H_2O + CH_3{-}C{-}C{-}CH_3 \\
\| \quad | \\
O\ \ CH_3 \\
\text{pinacolone} \\
\end{array}
\longrightarrow
\begin{array}{c}
CH_3\ CH_3 \\
| \quad\ | \\
CH_3{-}C{-}C{-}CH_3 \\
| \quad \oplus \\
OH \\
\end{array}$$

14. Phenol is acidic in nature, because of formation of a stable phenoxide ion in the aqueous solution. The phenoxide ion is stable due to resonance. The various resonating structures may be represented as follows. The driving force is the formation of a resonance stabilized phenoxide ion. The negative charge is spread throughout the benzene ring and is effectively dispersed.

    The acidity of the various substances may be as listed as follows:

    $RCOOH > C_6H_5{-}OH > H_2O > R{-}OH > CH{\equiv}CH > CH_2{=}CH_2 > CH_3{-}CH_3$

    $Ka = 1 \times 10^{-5}\ \ 1 \times 10^{-10}\ \ 1 \times 10^{-14}\ \ 1 \times 10^{-18}\ \ 1 \times 10^{-22}\ \ 1 \times 10^{-36}\ \ 1 \times 10^{-42}$

15. **Picric acid:** 2, 4, 6–Trinitro phenol is also known as picric acid. Due to the effect of the three electrons withdrawing $-NO_2$ groups, in the molecule, the acidity is greatly increased, making picric acid as acidic as mineral acids. Like mineral acids, picric acid liberates CO_2, from the carbonates.

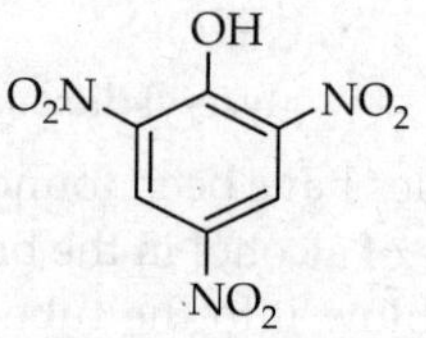

picric acid - pKa value = 0.38

16. **Cumene**–isopropyl benzene, C_6H_5—$CH(CH_3)_2$
    It is prepared by Friedel–Craft reaction of propene with benzene.

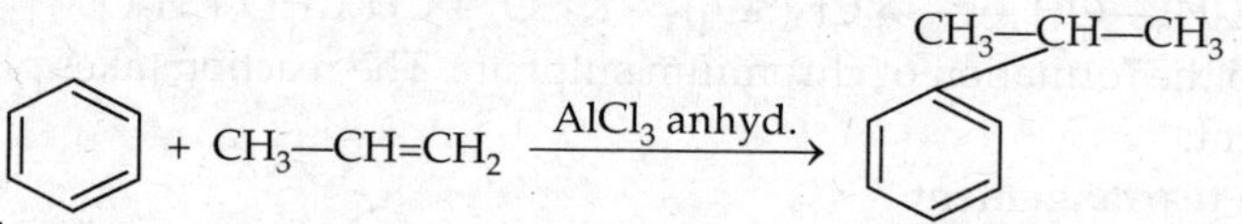

benzene   propene   cumene

Acetone is manufactured on a large scale from cumene. Oxygen is bubbled through an emulsion of cumene, containing a metal catalyst at 130 °C, when cumene is converted into cumene hydro peroxide. On treating, cumene hydro peroxide with 10% H_2SO_4, it splits to form phenol and acetone.

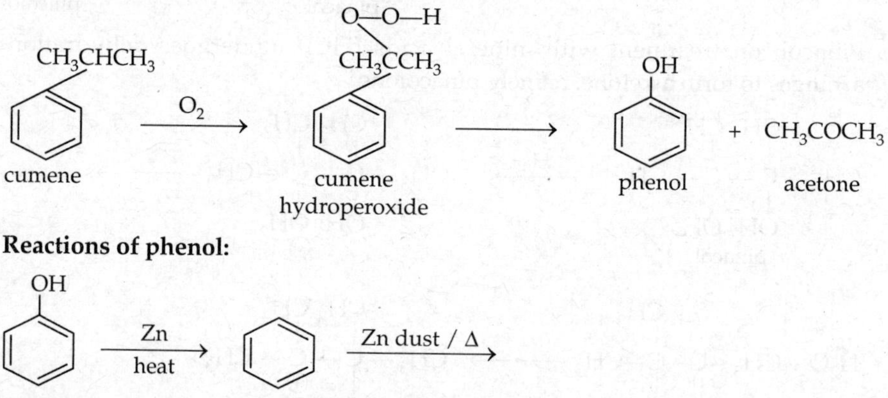

cumene   cumene   phenol   acetone
         hydroperoxide

17. **Reactions of phenol:**

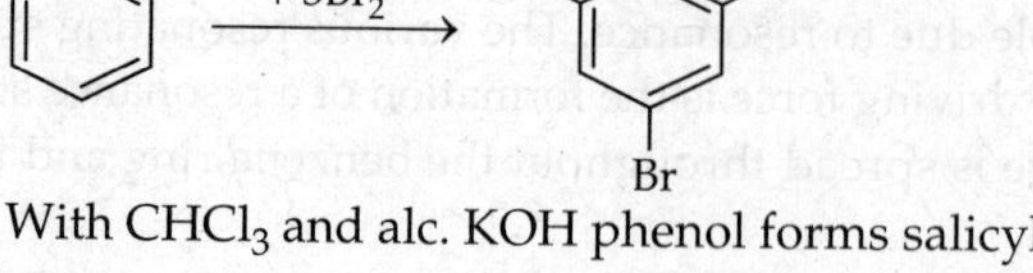

With Br_2 water, it forms 2,4,6-tri bromophenol

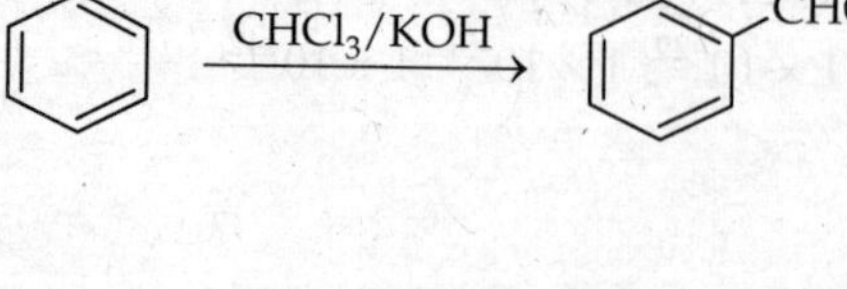

With $CHCl_3$ and alc. KOH phenol forms salicylaldehyde

18. **Kolbe's reaction:**
On heating sodium salt of phenol or sodium phenoxide, C_6H_5ONa, with carbon dioxide, CO_2, under pressure, salicylic acid is formed.

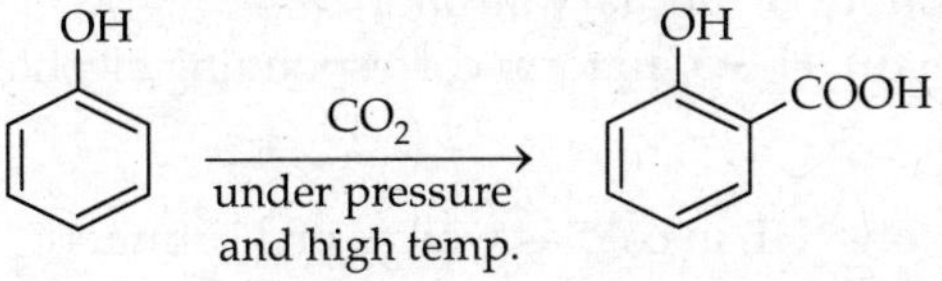

19. In alcohols, two bonds are involved, which undergo a change.
A C—O—H may break in two ways

—C | —O—H   —C—O | —H

The shared pair of electrons in C—O—H, are shifted towards the oxygen atom, due to the higher electronegativity of the oxygen atom, as compared to carbon and hydrogen. It may be represented as follows.

$$\overset{\delta+}{—C} \longrightarrow \overset{\delta-}{O} \longleftarrow \overset{\delta+}{H}$$

The greater the polarity of the —C—O bond, the smaller is the polarity of the —O—H bond. The strength of the bond decreases, as the polarity increases.
Since alkyl groups are electron donating, the larger the number of alkyl groups, attached to the carbon atom of —C—O—H group, the greater will be the polarity of the —C—O bond. Therefore, the polarity of the —C—O bond will be **greatest in the tertiary alcohols** and will be **least in the case** of **primary alcohols.**

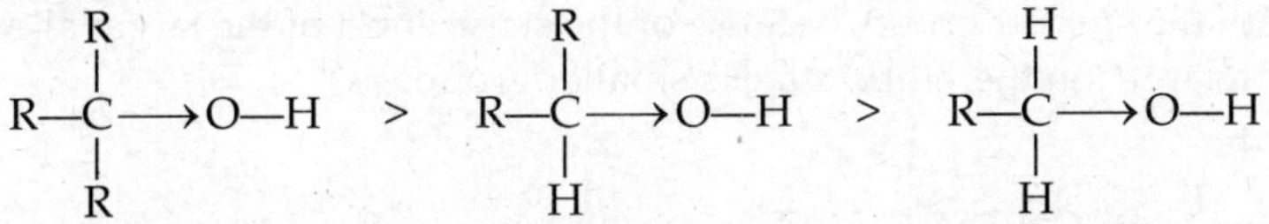

3° alcohol   2° alcohol   1° alcohol

Therefore the tendency to break as —C— | O—H bond, is greatest in the case of tertiary alcohols and least in the case of primary alcohols.
Consequently, the **polarity of the —O—H bond** will be smallest in the case of **tertiary alcohols** and **greater in the case of primary alcohols.**

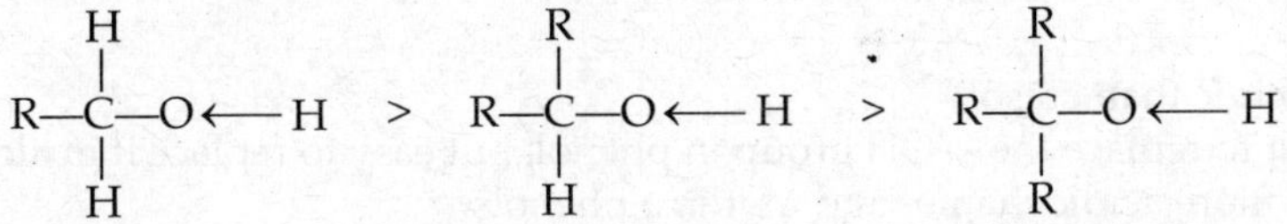

1° alcohol   2° alcohol   3° alcohol

Therefore, the tendency to break alcohols to break as —C—O | —H bond will be the greatest in the case of primary alcohols and will be least in the case of tertiary alcohols.

i. A C | —OH bond undergoes a change, in the reactions with HCl, PCl_5, H_2SO_4, etc. the —OH group is substituted by another nucleophile.
The reactivity of an alcohol with a given halogen acid is in the following order:

A tertiary alcohol > secondary alcohol > primary alcohol

Similarly, the reactivity of the alcohol towards halogen acid is the following order:

HI > HBr > HCl

ii. A C—O |—H bond breaks up, when treated with the reagents like sodium metal, in ester formation, in acylation, etc. the decreasing order of the reactivity of the alcohols is as follows.

primary alcohol   >   secondary alcohol   >   tertiary alcohol

20. Ethers are hydrolyzed when heated with dil. H_2SO_4 to form corresponding alcohols.

$$R—O—R + H_2O \xrightarrow{H_2SO_4} 2R—O—H$$

Similarly ethers are readily attacked by conc. HI, in cold as well as on heating.

$$CH_3—O—C_2H_5 + HI \xrightarrow{\text{in cold}} CH_3I \quad + \quad C_2H_5OH$$
methyl ethyl ether   methyl iodide   ethyl alcohol

On heating ethers with conc. HI, alkyl iodides are formed.

$$CH_3—O—C_2H_5 + 2HI \xrightarrow{\text{heating}} CH_3I \quad + \quad C_2H_5I \quad + \quad H_2O$$
methyl ethyl ether   methyl iodide   ethyl alcohol

$$HI \longrightarrow H^+ + I^-$$

H^+ then attacks oxygen of the ether

$$R—O—R + H^+I^- \longrightarrow [R—\overset{\overset{H}{|}}{O}{}^+—R]\ I^-$$
oxonium salt

Cleavage of oxonium salt takes place to form alkyl iodide and one molecule of alcohol. In the case of mixed ethers, or unsymmetrical ethers the I^- attacks the smaller alkyl group, because of steric hindrance of the large alkyl group. In the case of unsymmetrical ethers, the halide goes with the smaller group, because of the steric effect of the larger alkyl group. Iodide ion itself is a large ion therefore, attacks smaller group.

$$CH_3—\overset{\overset{H}{|}}{O}{}^+—CH_5 \xrightarrow{I^-} CH_3I + C_2H_5OH$$

The reactivity of the alcohol towards halogen acid is in the following order:

HI > HBr > HCl

REASONING TYPE QUESTIONS (RTQs)

(Group-B)

1. Phenol is more acidic than cresol.
2. It is more difficult to replace the —OH group in phenol, but easy to replace it in alcohols.
3. p-nitro phenol is more acidic than o- and m- nitro phenols.
4. Ethers can be cleaved, with HI but not with HCl.
5. A tertiary alcohol cannot be prepared by Williamsons' synthesis.
6. Most ethers are inert towards bases, but ether 2,4-dinitro anisole is easily cleaved to yield methanol and 2,4-dinitro phenol.
7. Inversion takes place, when NaI reacts with 2-bromo butane, in the presence of acetone.
8. 2,4,6-trinitro chlorophenol is easily hydrolyzed, where as chlorobenzene needs drastic conditions for hydrolysis.
9. The reduction of $CH_3CH=CH—CHO$ to $CH_3CH=CH—CH_2OH$ can be brought about by a specific reducing agent, name the reducing agent.
10. Ether must be distilled over $FeSO_4$, before use, because on standing, it forms a compound, which explodes on heating. Name the compound.

11. Phenol can be brominated but cannot be chlorinated.
12. The C—O bond is much shorter in phenols than in alcohols.
13. The keto form of phenol is much less stable than the enol form.
14. Alcoholic solution of KOH is a stronger base than an aqueous solution.
15. RX loses HX easily in presence of bases, but alcohols do not undergo base induced dehydration, in the same way as RX.
16. Phloroglucinol reacts with hydroxyl amine.
17. The boiling points of phenols are much higher than the corresponding hydrocarbons of same molecular wt.
18. Lower alcohols are water miscible, but the higher ones are not.
19. Phenols are water soluble but arenes are not.
20. 2,4-dinitrochloro benzene resists further substitution (nitration) but 2,4-dinitro benzene can be further nitrated to give picric acid.
21. The cleavage of a C—O bond, some acid when added, facilitate the cleavage.
22. No salt formation takes place when phenol is treated with $NaHCO_3$.
23. Ethyl alcohol reacts with HCl but not with HCN.
24. o-nitro phenol is steam volatile but p-nitro phenol is not.
25. Dehydration of $C_6H_{11}CH_2OH$ gives 1-methyl cyclohaxene.
26. Write the IUPAC name for the keto form of phenol.
27. Oxidation of glycerine by Fenton's reagent gives two compounds, one of which is used in carbohydrate chemistry, as a standard for assigning configuration.
28. Explain a reaction between $CH_3CH(OH)$—CH=CH—$COCH_3$ and CH_3MgBr, gives the starting material.
29. Oils and fats are identified by heating it with a compound, when a pungent smelling compound is formed. What it is ?
30. Water solubility of ethers in general is comparable with isomeric alcohols.
31. CH_2=CH—CH_2Cl is highly reactive towards nucleophilic substitution, CH_2=CH Cl is not.
32. Arrange the following compounds in the decreasing order of the bond angles:
    (a) NH_3 (b) CH_4 (c) R—O—R
    (d) H_2O (d) H_2S
33. Why ether is used for the preparation of Grignard's reagent?
34. Ethers can't be cleaved by HCl, though they can be cleaved by hot HI and hot HBr.
35. On decomposing an unsymmetrical ether molecule, iodine always goes to smaller alkyl group.
36. The b.p. of ethers and alkanes of comparable molecular wt. are almost same.
37. The solubility of alcohols decreases with rise in molecular weight.
38. Reaction of potassium metal with alcohols is in the order of: $1° > 2° > 3°$.
39. The boiling points of alcohols is in the order of: $1° > 2° > 3°$.
40. Esterification of an acid with a tertiary alcohol is more difficult as compared with a primary alcohol.

ANSWERS

(Group-B)

1. As methyl group is electrons releasing, therefore, it makes cresol much weaker.

2. In phenol, the —OH group, develops some double bond character, on account of resonance.

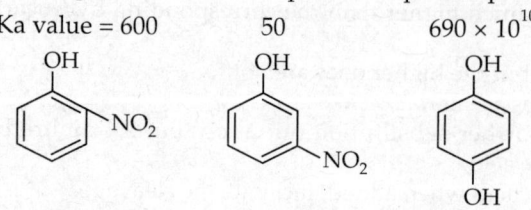

3. o-nitro phenol   m-nitro phenol   p-nitro phenol

    Ka value = 600   50 690×10^{10}

4. The dissociation energy of HI is = 70 kcal/mole is less than that of HBr (87 kcal/mole) or HCl (110 kcal/mole).

5. Owing to elimination, leading to formation of an alkene.

6. Nucleophilic attack by-OH-on the C—O—C, leads to the cleavage of C—O—C group. On account of the presence of two nitro groups in the ring, the ring becomes electron deficient, this facilitate the attack. In other words, by an attack of-OH-on C—Cl , is facilitated by an electron deficient ring.

7. As the reaction takes place by S_N2 mechanism, the attack takes place from back side

8. 2,4,6-trinitro chlorobenzene is easily hydrolyzed, by the attack of —OH group, because the ring is very much electron deficient on account of presence of three NO_2 groups.

9. $NaBH_4$ reduces the —CHO to —CH_2OH group, but it does not reduce a double bond.

10. Ether forms a peroxide, which explode on heating. On distilling ether containing peroxide in the presence of $FeSO_4$, the peroxide is destroyed.

$$Fe_2^+ \longrightarrow Fe_3^+$$

$$CH_3CH_2O—CH_2CH_3 \xrightarrow[\text{from air}]{O_2} CH_3CH_2—O—\underset{\underset{O—OH}{|}}{CH}—CH_3$$

11. Phenol is susceptible to oxidation by chlorine

12. As phenol ionises to form phenoxide, the phenoxide ion is stabilized by resonance. Owing to resonance, the C—O bond develops a partial double bond character.

C—O bond in phenol	C—O in alcohols
carbon is sp^2 hybridised has more s–character	carbon is sp^3 hybridised has less s–character

In addition the delocalization of π-electrons from O to the ring, by extended bonding, provides some double bond character, making it shorter.

13. A keto form of phenol is much less stable than enol form, because the keto form can be attained or obtained by the loss of benzene ring.

14. $$KOH \rightleftharpoons K^+ + OH^- \text{ ionises in water}$$

 In alcoholic solution $$ROH + KOH \rightleftharpoons RO^- K^+ + H_2O$$

 $$ROK + OH^- \rightleftharpoons RO^- + KOH$$

 RO^- is a stronger base compared to OH^-. In other words $RO^- K^+$ is stronger base than KOH

15. The weaker the base, better the leaving group. In alcohols the leaving group would be $-OH^-$ group, which is a strong base too, for the reaction to proceed.

16. The keto form of phloroglucinol has a considerable stability, because of the large amount of resonance energy due to three C=O groups. It is the keto form that reacts with NH_2OH.

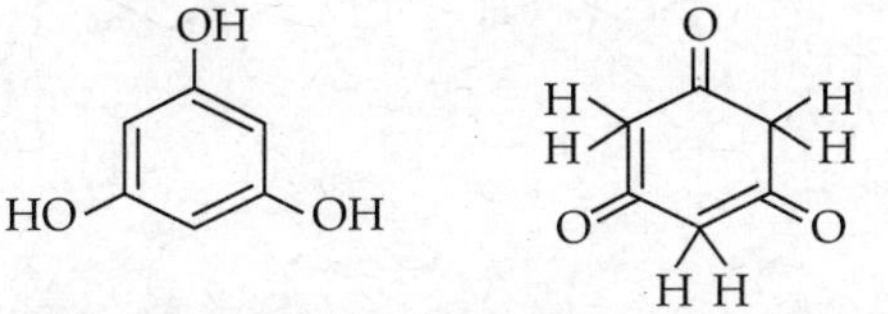

17. Owing to intramolecular H-bonding.

18. Higher alcohols have a longer hydrophobic carbon chain. The water solubility of alcohols follow the given pattern, as 3° alcohols have a shorter carbon chain and the shape of the molecule is almost the rounded one, in 2° and 1° alcohols the chain and the solubility goes on decreasing (ethers can also form H-bonding).

19. Phenols are water soluble, on account of H-bonding, involving —OH and H_2O.

20. The two —NO_2 groups and —Cl group in 2,4-dinitrochloro benzene deactivate the ring, towards further substitution.

 In case of 2,4-dinitrophenol, the presence of —OH group activate it for further substitution.

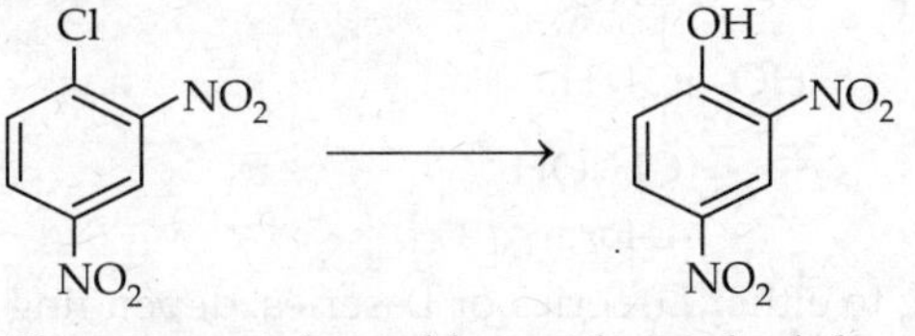

21. The proton released by acid attacked the —C—O bond, with the result that oxygen atom become electron deficient, and acquires a +ve charge. This helps the decomposition of a C—O bond.

22. As both are weak, so no salt formation takes place.

23. Ethanol is weakly acidic in nature therefore no salt formation takes place.

24. o-nitro phenol has intramolecular hydrogen bonding, and therefore it has a minimal attraction for water but p-nitro phenol has intermolecular hydrogen bonding. Whereas the o-nitro phenol in spite of intramolecular bonding remains an individual molecule, is easily distilled off in the presence of steam, p-nitro phenol, on account of intermolecular bonding, becomes a mega molecule, which the vapours simply cannot lift. This is why o-nitro phenol is steam volatile, but p-nitro phenol is not. A compound which can be steam distilled must have an appreciable vapour pressure, at the boiling point of water (attraction to water, greatly lowers the vapour pressure preventing steam distillation).

25. *Dehydration of 1-methylhexane gives 1-methyl cyclohexene*

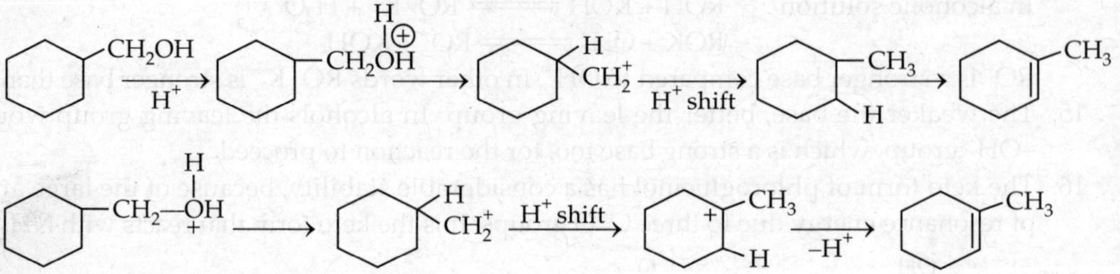

26. 1-keto-cylohexyl-2, 4-diene

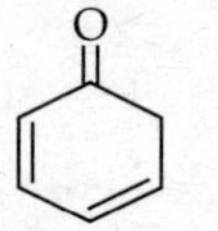

27. Glyceric aldehyde is one of the products formed, that is used, as a reference standard for configuration.

$$\begin{array}{l} CH_2OH \\ | \\ CH_2OH \\ | \\ CH_2OH \end{array} \xrightarrow[FeSO_4\ (H_2O_2)]{\text{Fenton's reagent}} \begin{array}{l} CH_2OH \\ | \\ C{=}O \\ | \\ CH_2OH \end{array} + \begin{array}{l} CHO \\ | \\ CHOH \\ | \\ CH_2OH \end{array}$$

glycerine or glycerol — di hydroxy acetone — glyceraldehyde

The glyceric aldehyde can be represented in two forms, the structure that contains —OH group can be written to the right hand side; it has been assigned a D-configuration, the other one is assigned as L-configuration.

$$\begin{array}{l} CHO \\ | \\ CHOH \\ | \\ CH_2OH \end{array} \qquad \begin{array}{c} CHO \\ | \\ H{-}C{-}OH \\ | \\ CH_2OH \end{array} \qquad \begin{array}{c} CHO \\ | \\ HO{-}C{-}H \\ | \\ CH_2OH \end{array}$$

glyceraldehyde — D-form — L-form

All carbohydrates are classified belonging to either D–series or L–series, depending upon, whether the —OH group on the second last carbon is on the right hand side (they belong to D series) or to the left hand side (they belong to L–series). However, this is total arbitrary. The D and L-are not the substitute for 'd' or dextro and 'l'-or laevo sugars respectively. A sugar may belong to a "D" series may be laevo rotatory. For ex. Fructose is laevo rotator, yet it belongs the D-series.

28. $CH_3CH(OH)CH_2CH_2COCH_3 + CH_3MgBr \longrightarrow CH_3{-}\underset{}{\overset{OMgBr}{\overset{|}{C}}}HCH_2CH_2COCH_3 + CH_4$

$$\xrightarrow{H^+} CH_3CH(OH)CH_2CH_2COCH_3$$

29. On heating oils and fats with $KHSO_4$ (or with conc. H_2SO_4), a pungent smelling substance, namely acrolein is formed.

$$\begin{array}{l} CH_2OH \\ | \\ CHOH \\ | \\ CH_2OH \end{array} \xrightarrow[\text{conc. } H_2SO_4]{KHSO_4} CH_2{=}CH{-}CHO$$

glycerol — acrolein or acrylaldehyde

30. Ethers are capable of forming H-bonding, just like alcohols.

31. Allyl alcohol reacts by S_N1 mechanism, and is stabilized by resonance

$CH_2=CH—CH_2Cl \longrightarrow CH_2=CH—CH_2^+ + Cl^-$

$CH_2=CH—CH2^+ \longrightarrow CH2^+—CH=CH_2$

In $CH_2=CHCl$, on account of resonance the Cl acquires a +ve charge

$CH_2=CH—\underset{\cdot\cdot}{\overset{\cdot\cdot}{\ddot{C}l}} \longrightarrow CH_2^{\;-}—CH=Cl^+$

32.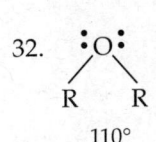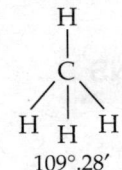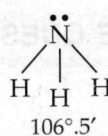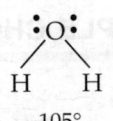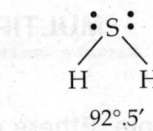

33. Ether is used in the preparation of Grignard's reagent, because two molecules of ether co-ordinate with the Mg atom, which is soluble in ether. Ethers form co-ordination complexes.

$$R — \underset{\cdot\cdot}{O} — R$$
$$\downarrow$$
$$R — Mg — X$$
$$\uparrow$$
$$R — \underset{\cdot\cdot}{O} — R$$

An ethereal solution is used.

34. Cl^- is a weak nucleophile as compared to the Br^- and I^-, therefore HCl does not decompose the ether molecule.

35. The I^-, being larger in size approaches the smaller alkyl group to avoid steric hindrance.

36. In alkanes and in ethers there is no hydrogen bonding. The alkanes contain very weak van der Waals forces only, which do not effect the boiling points of the compounds. In ethers, like alcohols, there are no attractive forces of any type. Therefore, their boiling points are almost same.

37. As the molecular weight increases, the solubility decrease because the long chain of alkyl group which being non-polar, have less tendency to form H–bonding.

38. The **O—H bond in alcohols, is under strain in 1° alcohols**, because of a large –I effect of oxygen. On the other hand **the C—O bond is under strain, in 3° alcohols** because of a large +I effect of the alkyl groups.

Therefore, in **1° alcohols the O—H break faster** and in the **3° alcohols** it is the **C—O bond breaks faster.**

39. In 1° alcohols beside H-bonding, there are van der Waals forces operating, depending upon the length of the alkyl chain. In the 3° alcohols, since the molecule is in the form of coiled form, not only the H-bonding is not that effective but the van der Waal forces are also quite weak. As it takes some heat energy to break the H-bonding, the 1° alcohol will require more heat. For a 2° alcohols the boiling point is between the two. Therefore, the order of boiling points will be in the order of 1° > 2° > 3°.

40. It is on account of the steric hindrance. The proton cannot reach easily the O—H group of a tertiary alcohol, to protonate it, to initiate the reaction. On the other hand, a primary alcohol will react easily with an acid to form an ester.

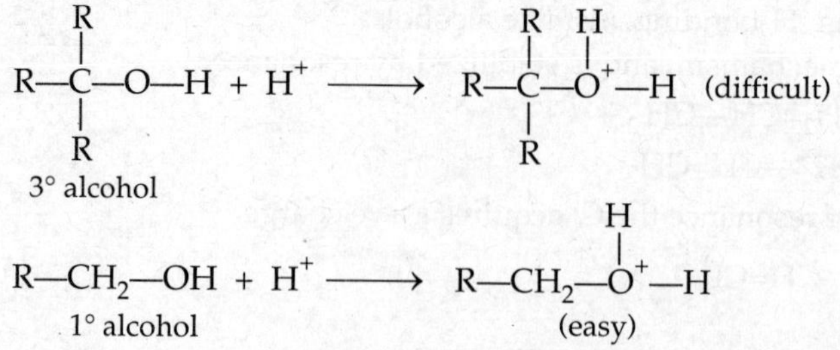

$$R-\underset{\underset{R}{|}}{\overset{\overset{R}{|}}{C}}-O-H + H^+ \longrightarrow R-\underset{\underset{R}{|}}{\overset{\overset{R}{|}}{C}}-\overset{\overset{H}{|}}{O^+}-H \quad \text{(difficult)}$$

3° alcohol

$$R-CH_2-OH + H^+ \longrightarrow R-CH_2-\overset{\overset{H}{|}}{O^+}-H$$

1° alcohol   (easy)

MULTIPLE CHOICE QUESTIONS

(Group-C)

Questions on Alcohols, Ethers and Phenols

1. Predict the compounds formed

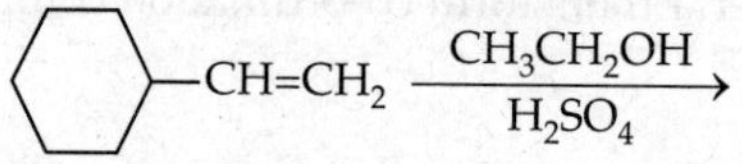

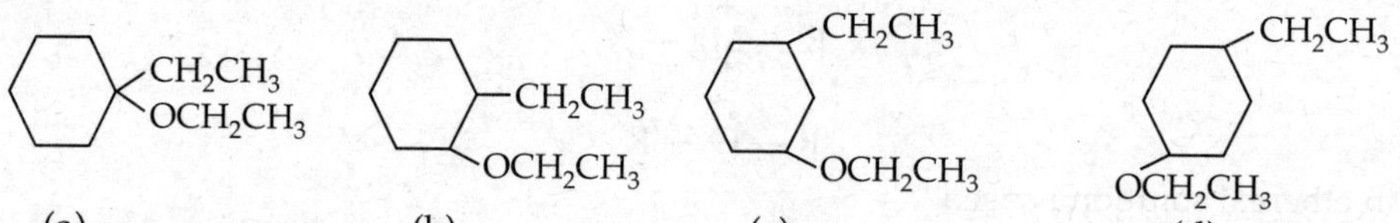

   (a)   (b)   (c)   (d)

2. The following reaction given

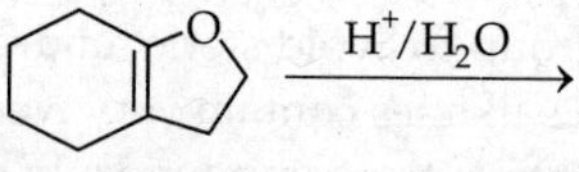

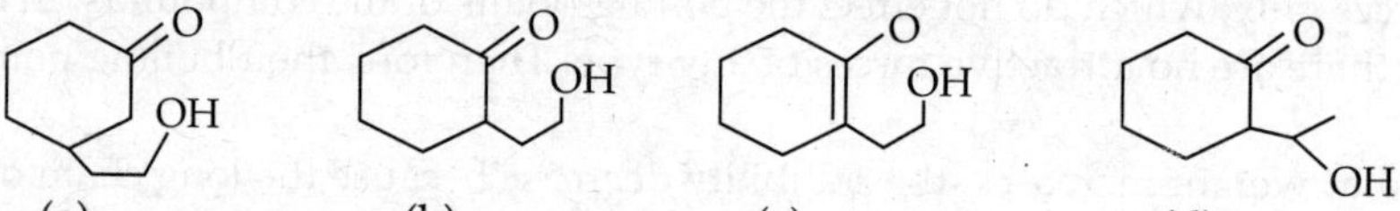

   (a)   (b)   (c)   (d)

3. For the given reaction predict the formation of product

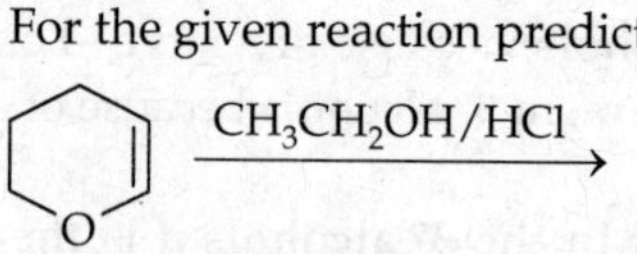

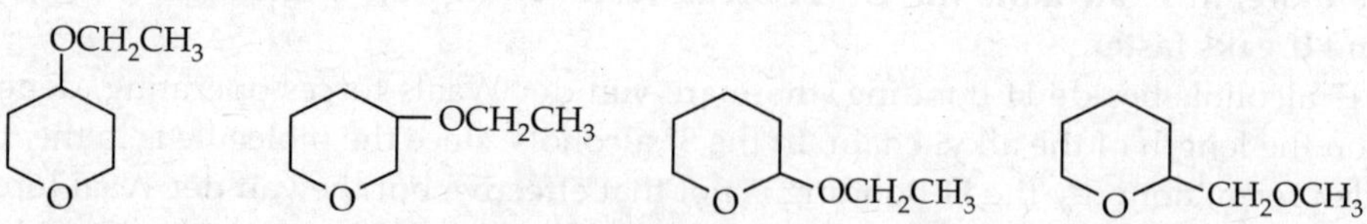

   (a)   (b)   (c)   (d)

4. The decreasing order of solubility (in water)

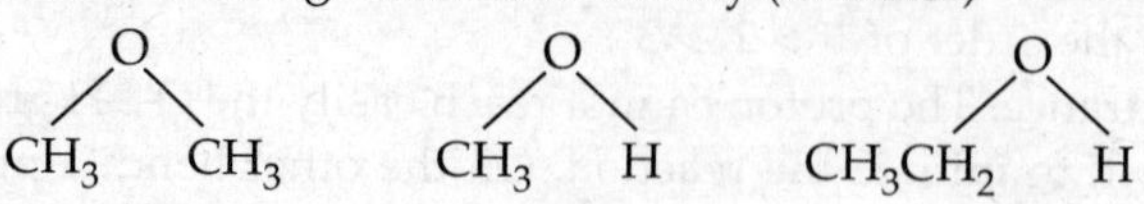

   (i)   (ii)   (iii)

   (a) i > ii > iii   (b) iii > ii > i   (c) ii > iii > i   (d) ii > i > iii

5. In the following reaction

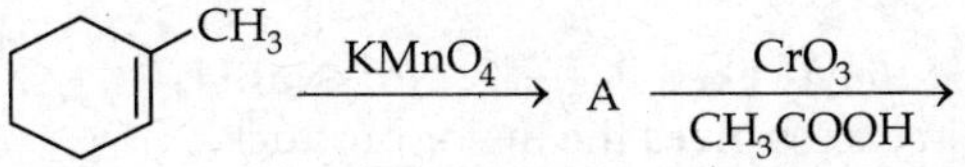

$\xrightarrow{\text{KMnO}_4}$ A $\xrightarrow[\text{CH}_3\text{COOH}]{\text{CrO}_3}$ B, what are A and B?

6. $Ph-COCH_3 \xrightarrow{\text{Mg-Hg/H}_2\text{O}} \xrightarrow{\text{Conc.H}_2\text{SO}_4} \xrightarrow{\text{KMnO}_4/\text{H}^+}$

(a) $CH_3CO-COCH_3$

(b) $CH_3-\underset{\underset{HO}{|}}{\overset{\overset{Ph}{|}}{C}}-\underset{\underset{OH}{|}}{\overset{\overset{Ph}{|}}{C}}-CH_3$

(c) $Ph-CO-CO-Ph$

(d) $CH_3-\underset{\underset{Ph}{|}}{\overset{\overset{Ph}{|}}{C}}-CO-CH_3$

7. Dehydration of the following compounds is in the order

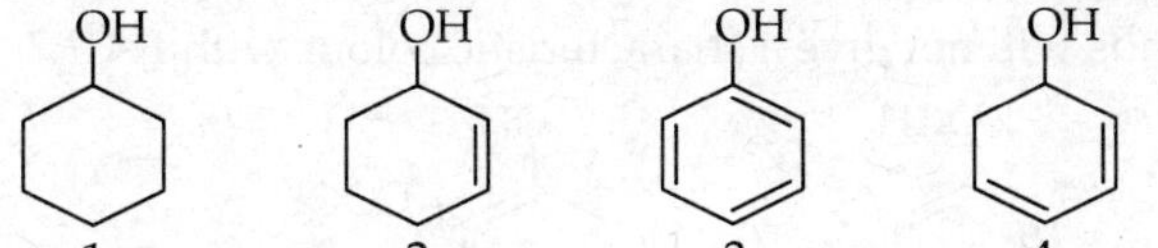

   1   2   3   4

(a) $1 > 2 > 3 > 4$
(c) $4 > 2 > 1 > 3$

(b) $4 > 3 > 2 > 1$
(d) $4 > 3 > 1 > 2$

8. $Ph-\underset{\underset{H}{|}}{\overset{\overset{HO}{|}}{C}}-\underset{\underset{H}{|}}{\overset{\overset{OH}{|}}{C}}-CH_3 \xrightarrow{\text{H}^+/\text{H}_2\text{O}}$

(a) $Ph-\underset{\underset{H}{|}}{\overset{\overset{OH}{|}}{C}}-COCH_3$

(b) $C_6H_5COCH_2CH_3$

(c) $C_6H_5CH_2COCH_3$

(d) $Ph-\underset{\underset{CH_3}{|}}{CH}-CHO$

9. $\underset{CH_3}{\overset{CH_3}{>}}C \underset{\diagdown O \diagup}{-} CH_2 \xrightarrow{\text{CH}_3\text{OH/H}^+}$

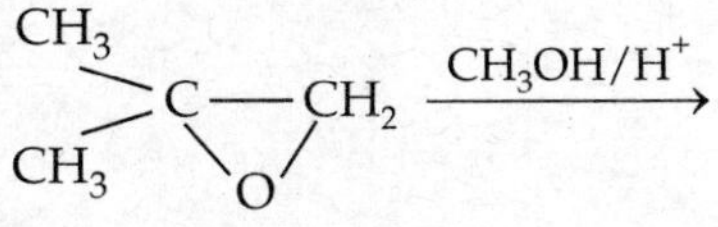

(a) $(CH_3)_2-\underset{\underset{CH_3}{|}}{C}-CH_2OH$

(b) $(CH_3)_2\underset{\underset{CH_3}{|}}{C}-CH_2OH$

(c) $(CH_3)_3-\underset{\underset{OH}{|}}{C}-OCH_3$

(d) $(CH_3)-\underset{\underset{CH_3}{|}}{CH}-CH_2-OCH_3$

10. $(CH_3)_3C-O-CH_2CH_3 + HI \text{ (mole)} \longrightarrow$
    (a) $(CH_3)_3COH + C_2H_5I$
    (c) $(CH_3)_3C-I + C_2H_5I$

    (b) $(CH_3)_3CI + C_2H_5OH$
    (d) $(CH_3)_3COH + C_2H_5OH$

11. The reduction of the alcohol may be carried by
$$RCH_2OH \longrightarrow RCH_3$$
   (a) $LiAlH_4$ (b) H_2/Ni (c) I_2 + red P   (d) $NaBH_4$

12. Reduction of the given compound by HI in excess gives the major product

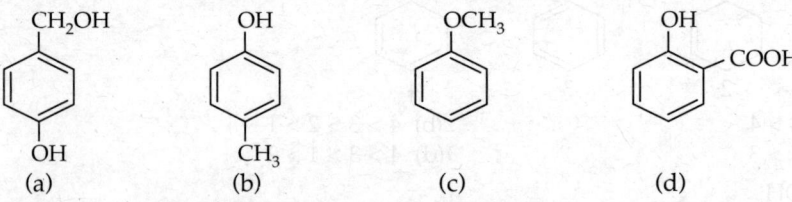

      (a)   (b)   (c)

13. Which of the following compounds will not give a characteristic colour with $FeCl_3$?

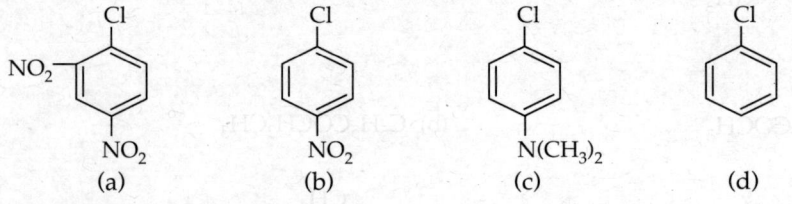

      (a)   (b)   (c)   (d)

14. Which one will be hydrolyzed most rapidly?

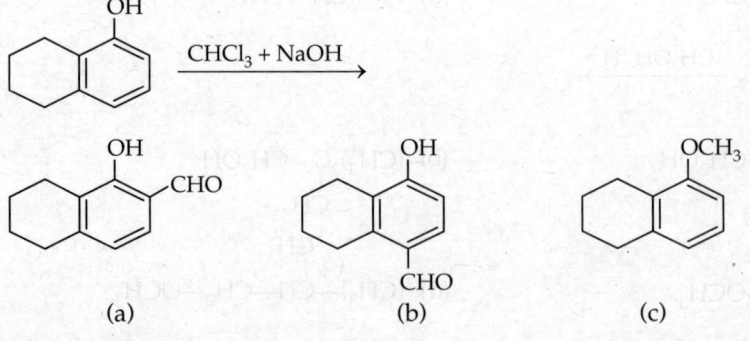

      (a)   (b)   (c)   (d)

15. The compound formed in the reaction

      $\xrightarrow{\text{CHCl}_3 + \text{NaOH}}$

      (a)   (b)   (c)

16. Which will not give Victor Meyer test?
   (a) CH_3CH_2OH (b) $(CH_3)_3C—OH$
   (c) $C_2H_5—I$ (d) $(CH_3)_2CHNO_2$

17. Which of the following will not respond to a Haloform test?

    (a) C_2H_5OH

    (b) $CH_3\overset{\overset{\displaystyle OH}{|}}{CH}—C_6H_5$

    (c) $C_2H_5—O—C_2H_5$

    (d)

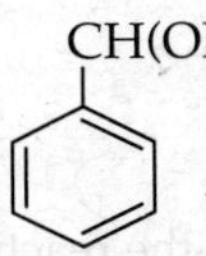

18. The compound formed, when the given ether is treated with HI.

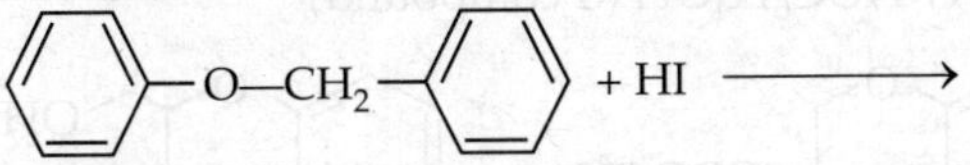

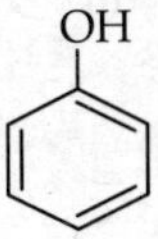

    + HI ⟶

    (a) OH

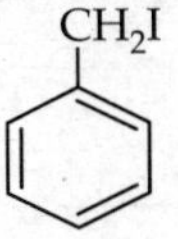

    (b) CH_2I

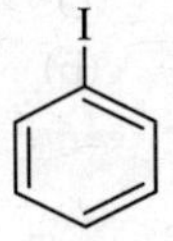

    (c) I

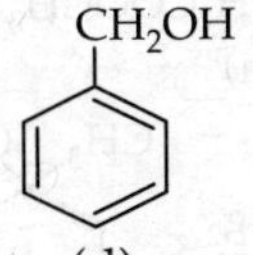

    (d) CH_2OH

19. Predict the reaction product.

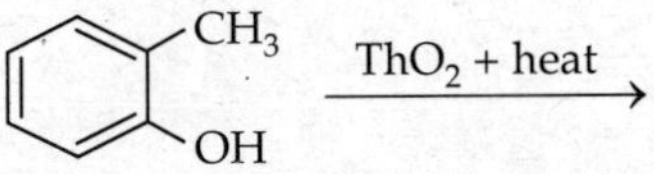

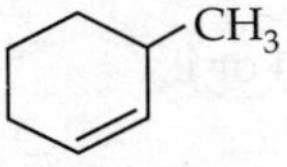

    $\xrightarrow{ThO_2 + heat}$

    (a) CH_3

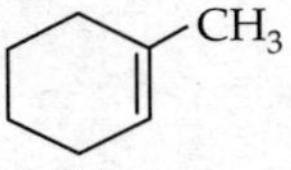

    (b) CH_3

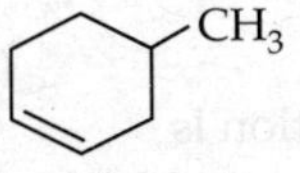

    (c) CH_3

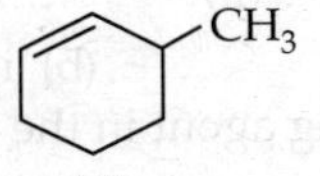

    (d) CH_3

20. The major compound formed

    $(CH_3)_3—C—CH(CH_3)\ OH \xrightarrow{conc.\ H_2SO_4}$

    (a) $(CH_3)_3—C=CH_2$

    (b) $(CH_3)_2C=C—(CH_3)_2$

    (c) $CH_2=CH—CH—(CH3)_2$

    (d) none

21.

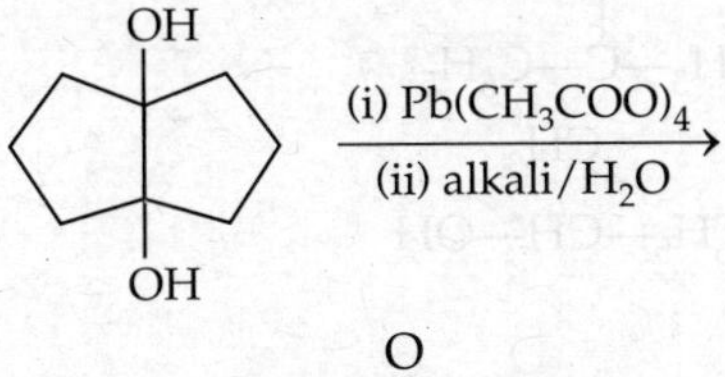

    $\xrightarrow[\text{(ii) alkali/H}_2\text{O}]{\text{(i) Pb(CH}_3\text{COO)}_4}$

    (a)

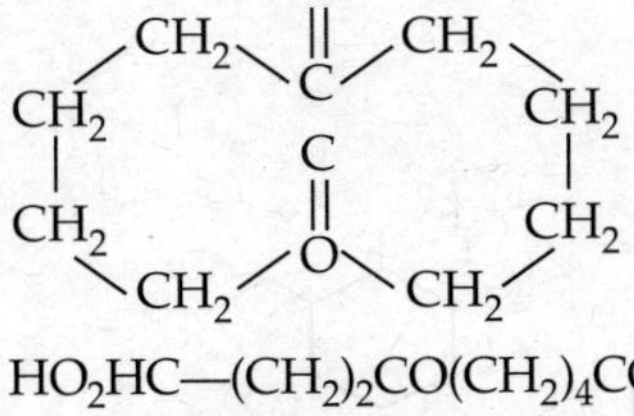

    (b) $HOOC—(CH_2)_3CO(CH_2)_4COOH$

    (c) $HO_2HC—(CH_2)_2CO(CH_2)_4COOH$

    (d) $HO_2HC—(CH_2)_3CO(CH_2)_4CH_2OH$

22. The major products formed

    $CH_3CH_2CH_2CHOH + conc.\ H_3PO_4 \longrightarrow$

    (a) $CH_3H_2CH=CH_2$

    (b) $cis\text{-}CH_3CH=CH—CH_3$

    (c) $trans\text{-}CH_3CH=CH—CH_3$

    (d) none

23. On passing the vapours of a compound X, an alkene is formed, the compound is
    (a) $(CH_3)_2CHOH$
    (b) C_2H_5OH
    (c) $Ph\ CH(CH)CH_2OH$
    (d) $(CH_3)_3COH$

24. The product of the reaction

    $CH_3CH_2Br + Ag_2O \xrightarrow{\text{heat}}$

    (a) $CH_2=CH_2$
    (b) $CH_3CH_2CH_2CH_3$
    (c) $C_2H_5—O—C_2H_5$
    (d) C_2H_5OH

25. The compound formed in the reaction is

    $CH_2=CH—COOCH_3 + Br_2 \longrightarrow A\ ;\ A + HOC_6H_4OH\ \text{(o-compound)} \xrightarrow{K_2CO_3/\text{acetone}} B$

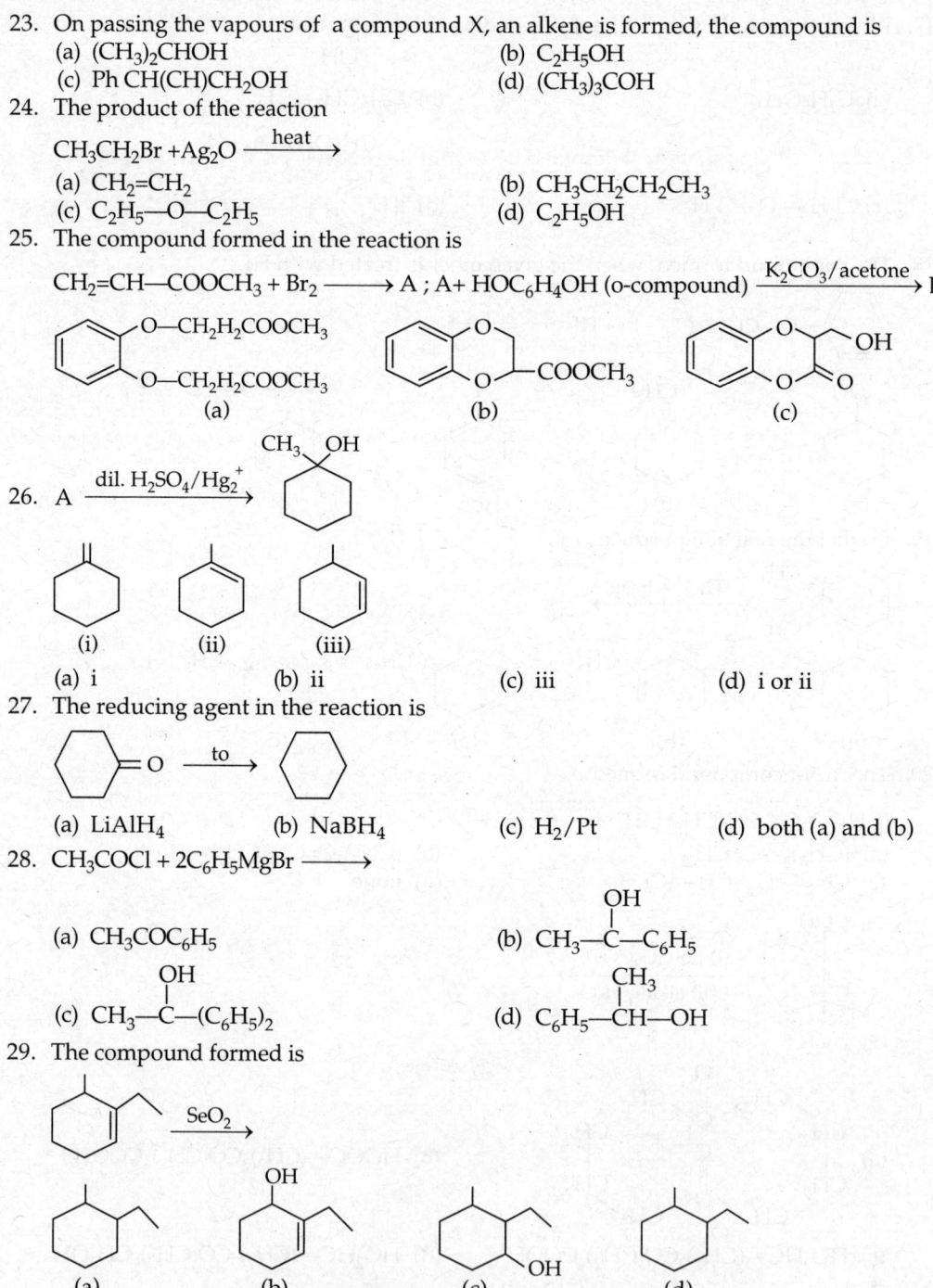

    (a)   (b)   (c)

26. A $\xrightarrow{\text{dil. } H_2SO_4/Hg_2^+}$

    (i)   (ii)   (iii)

    (a) i
    (b) ii
    (c) iii
    (d) i or ii

27. The reducing agent in the reaction is

    $\xrightarrow{\text{to}}$

    (a) $LiAlH_4$
    (b) $NaBH_4$
    (c) H_2/Pt
    (d) both (a) and (b)

28. $CH_3COCl + 2C_6H_5MgBr \longrightarrow$

    (a) $CH_3COC_6H_5$
    (b) $CH_3—\overset{OH}{\underset{|}{C}}—C_6H_5$
    (c) $CH_3—\overset{OH}{\underset{|}{C}}—(C_6H_5)_2$
    (d) $C_6H_5—\overset{CH_3}{\underset{|}{CH}}—OH$

29. The compound formed is

    $\xrightarrow{SeO_2}$

    (a)   (b)   (c)   (d)

30. $2C_2H_5OH \xrightarrow{Al_2O_3} C_2H_5O—C_2H_5 + H_2O$
    The role of Al_2O_3 in the above reaction is

(a) acts as absorbent of water molecule
(b) providing hot surface
(c) acts as a Lewis acid
(d) acts as a Lewis acid to facilitate the leaving of —OH group, by coordinating at the oxygen atom.

31. Which of the following compounds are soluble in $NaHCO_3$?

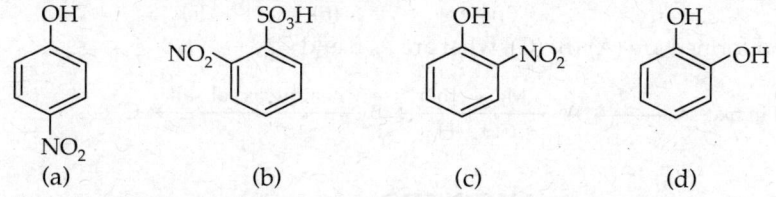

     (a)   (b)   (c)   (d)

32. Find out 'Z' in the given reaction?

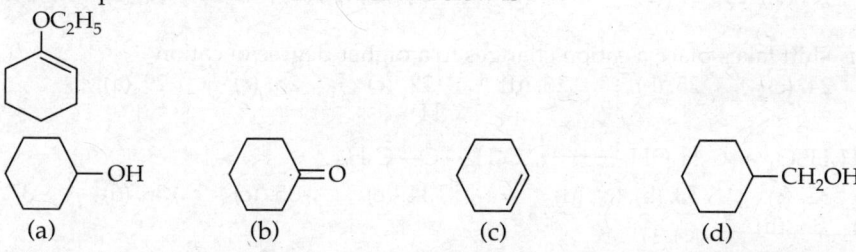

(a) Z in the reaction is identical to A   (b) a chain isomer of A
(c) position isomer of A   (d) reduced product of A

33. The compound formed in the reaction below is

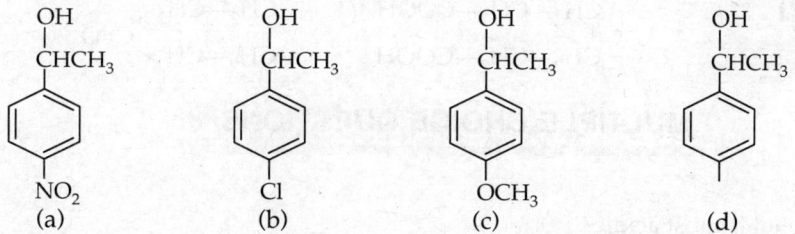

     (a)   (b)   (c)   (d)

34. Which of the compound will undergo dehydration easily?

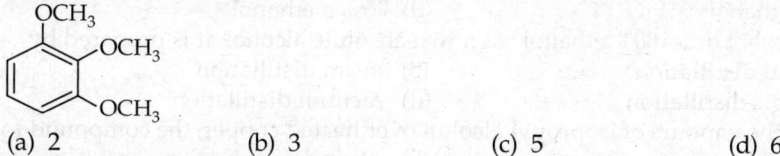

     (a)   (b)   (c)   (d)

35. The number of molecules of HIO_4 that will react with a mole of glucose.
(i) 4   (ii) 6   (iii) 5   (iv) 3

36. How many mole of HI (taken in excess) will react with the following compound namely 1,2,3-trimethoxy benzene

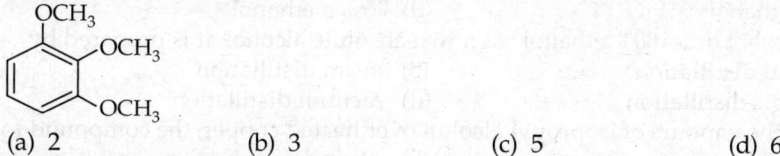

(a) 2   (b) 3   (c) 5   (d) 6

37. Diethyl ether and n-butanol can be distinguished by
(a) aqueous $FeCl_3$ (b) Na metal
(c) Tollen's reagent   (d) potassium dichromate

38. Find A, B formed in the reaction

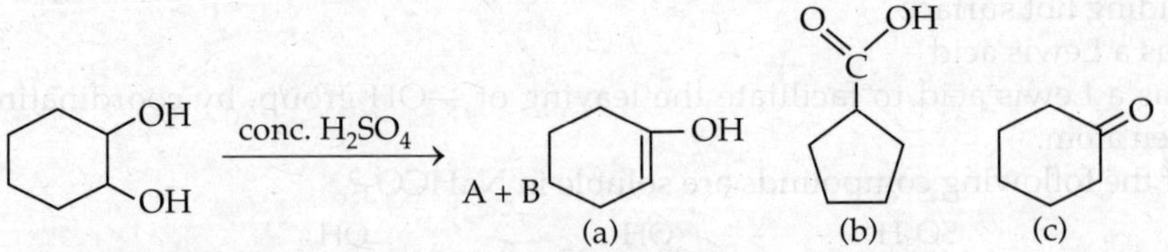

39. The compound formed are (A) (B) (C), what are A, B and C?

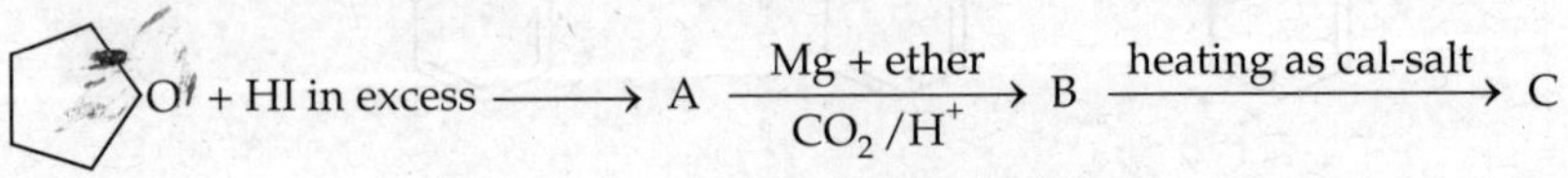

ANSWERS

(Group-C)

1. (a)   2. (b)   3. (c)   4. (c)
5. (A) cyclo hexyl-1-methyl-1,2-di-ol ; B = 1-keto-2-methyl 2-ol
6. (c)   7. (b)   8. (c)   9. (a)   10. (c)   11. (b)   12. (a)   13. (c)
14. (a)   15. (a), (b)   16. (b)   17. (c)   18. (a), (b), (d)   19. (a)   20. (b)
21. (c)
22. (c) *trans*-shift takes place a cation changes to a higher degree of cation
23. (d)   24. (c)   25. (b)   26. (d)   27. (c)   28. (c)   29. (a)

30. (d) $C_2H_5HSO_4$ + C_2H_5OH $\longrightarrow$ $C_2H_5\overset{\overset{H}{|}}{-}O-C_2H_5$

31. (b)   32. (c)   33. (b), (c), (d)   34. (c)   35. (c)   36. (6)   37. (a), (b)
38. A is (i)   B = (ii)
39. A = 1,4-di-iodo butane   B = adipic acid   C= cyclo pentanone

$$\begin{array}{l} CH_2CH_2I \\ | \\ CH_2CH_2I \end{array} \qquad \begin{array}{l} CH_2-CH_2-COOH \\ | \\ CH_2-CH_2-COOH \end{array} \qquad \begin{array}{l} CH_2-CH_2 \\ | \qquad\quad \diagdown \\ \qquad\qquad C=O \\ | \qquad\quad \diagup \\ CH_2-CH_2 \end{array}$$

MULTIPLE CHOICE QUESTIONS

(Group-D)

1. Which alcohol is most acidic?
   (a) CH_3OH (b) ethanol
   (c) propanol   (d) tertiary butyl alcohol
2. Methylated spirit is
   (a) methanol containing some pyridine   (b) ethanol containing some methanol
   (c) pure methanol   (d) 90% methanol
3. Absolute alcohol or a 100% ethanol is known absolute alcohol. It is prepared by
   (a) fractional distillation   (b) steam distillation
   (c) azeotropic distillation   (d) vacuum distillation
4. On passing the vapours of isopropyl alcohol over heated copper, the compound formed is
   (a) CH_3COCH_3 (b) ethanol
   (c) methanol   (d) acetaldehyde

5. Glycerol on heating with oxalic acid at a temperature $110\,^{\circ}C$, gives
   (a) formic acid
   (b) $CO_2 + CO$
   (c) allyl alcohol
   (d) glycol

6. The following reaction is known as

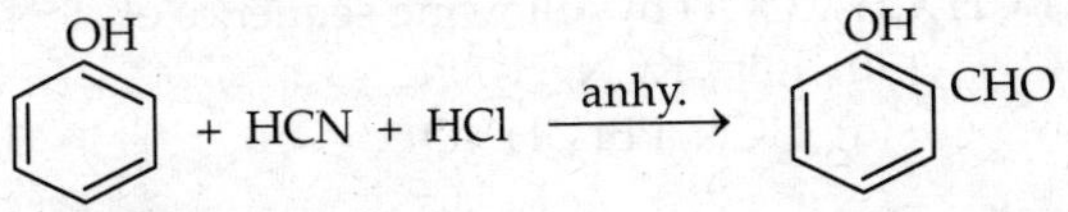

$$\text{(OH)} + HCN + HCl \xrightarrow{\text{anhy.}} \text{(OH)}\,CHO$$

   (a) Perkin's reaction
   (b) Gattermann
   (c) Kolbe's reaction
   (d) Gattermann–Koch synthesis

7. Primary, sec. and tertiary alcohols may be distinguished by employing
   (a) oxidation
   (b) Luca's reagent
   (c) Victor–Meyer's method
   (d) all

8. The product formed on heating phenyl methyl ether with HI is
   (a) phenol + methyl iodide
   (b) $C_6H_5I + CH_3OH$
   (c) $C_6H_5CH_3$
   (d) $C_6H_6 + CH_3OH$

9. 3-Methyl-2butanol on treatment with HCl, gives predominantly
   (a) 2-methyl-2-chlorobutane
   (b) 2-chloro-3-methylbutne
   (c) 2,2-dimethylpentane
   (d) none

10. On heating glycerol with conc.H_2SO_4, the compound is obtained, with bad odour
    (a) acrolein
    (b) formic acid
    (c) allyl alcohol
    (d) none

11. A reaction of tertiary butyl bromide and sod. methoxide produces
    (a) isobutane
    (b) isobutylene
    (c) sod.t-butoxide
    (d) t-butyl methyl ether

12. The intermediate produced in the preparation of ethylene from ethanol and H_2SO_4 is
    (a) diethyl ether
    (b) ethyl hydrogen sulphate
    (c) $(C_2H_5)_2SO_4$
    (d) none

13. An organic compound C_3H_6O, does not react with 2,4-DNP reagent and does not react with sodium metal, it could be
    (a) CH_3CH_2CHO
    (b) CH_3COCH_3
    (c) $CH_2=CH—CH_2OH$
    (d) $CH_2=CH—OCH_3$

14. The oxygen atom in ether is
    (a) very active
    (b) replaceable
    (c) comparably inert
    (d) active

15. Intramolecular hydrogen bonding are not present in
    (a) ethanol
    (b) ether
    (c) CH_3COOH
    (d) $C_2H_5NH_2$

16. Ethanol reacts with HCl but not with HCN
    (a) ethanol and HCN are weak bases
    (b) ethanol and HCN are strong and weak acids respectively
    (c) ethanol is a weak base and HCl is a strong acid
    (d) none

17. Diethyl ether reacts with Cl_2 in presence of sunlight to form
    (a) trichloro diethyl ether
    (b) perchloro di ethyl ether
    (c) trichloro acetaldehyde
    (d) 1,1-dichloro ether

18. The halogen acids on ether have the following order of reactivity
    (a) HCl > HBr > HI
    (b) HI > HCl > HBr
    (c) HI > HBr > HCl
    (d) none

19. $CH \equiv CH \xrightarrow{O_3/NaOH} X \xrightarrow{Zn/CH_3COOH} Y$
    (a) $CH_2OH—CH_2OH$
    (b) CH_3CH_2OH
    (c) CH_3OOH
    (d) CH_3OH (X=CHO—CHO)

20. RCH_2CH_2OH can be converted into RCH_2CH_2COOH by following sequence of steps
    (a) PBr_3, KCN, H^+_3O
    (b) PBr_3, KCN, H_2/Pt
    (c) KCN, H^+_3O
    (d) HCN, PBr_3, H^+_3O

21. The value of C—O—C angle is
    (a) 180°
    (b) 150°
    (c) 90°
    (d) 110°

22. Which among is Williamson's synthesis?
    (a) CH_3COCH_3—reduction $\longrightarrow CH_3CH(OH)CH_3$
    (b) $CH_3CHO \longrightarrow CH_3CH=CH—CHO$
    (c) $C_2H_5I + C_2H_5ONa \longrightarrow C_2H_5—O—C_2H_5$
    (d) $HCHO + NaOH \longrightarrow CH_3OH + HCOONa$

23. The products formed when glycerol is treated with Fenton's reagent
    (a) glyceric acid
    (b) dihydroxy acetone
    (c) glyceric aldehyde
    (d) both (b) and (c)

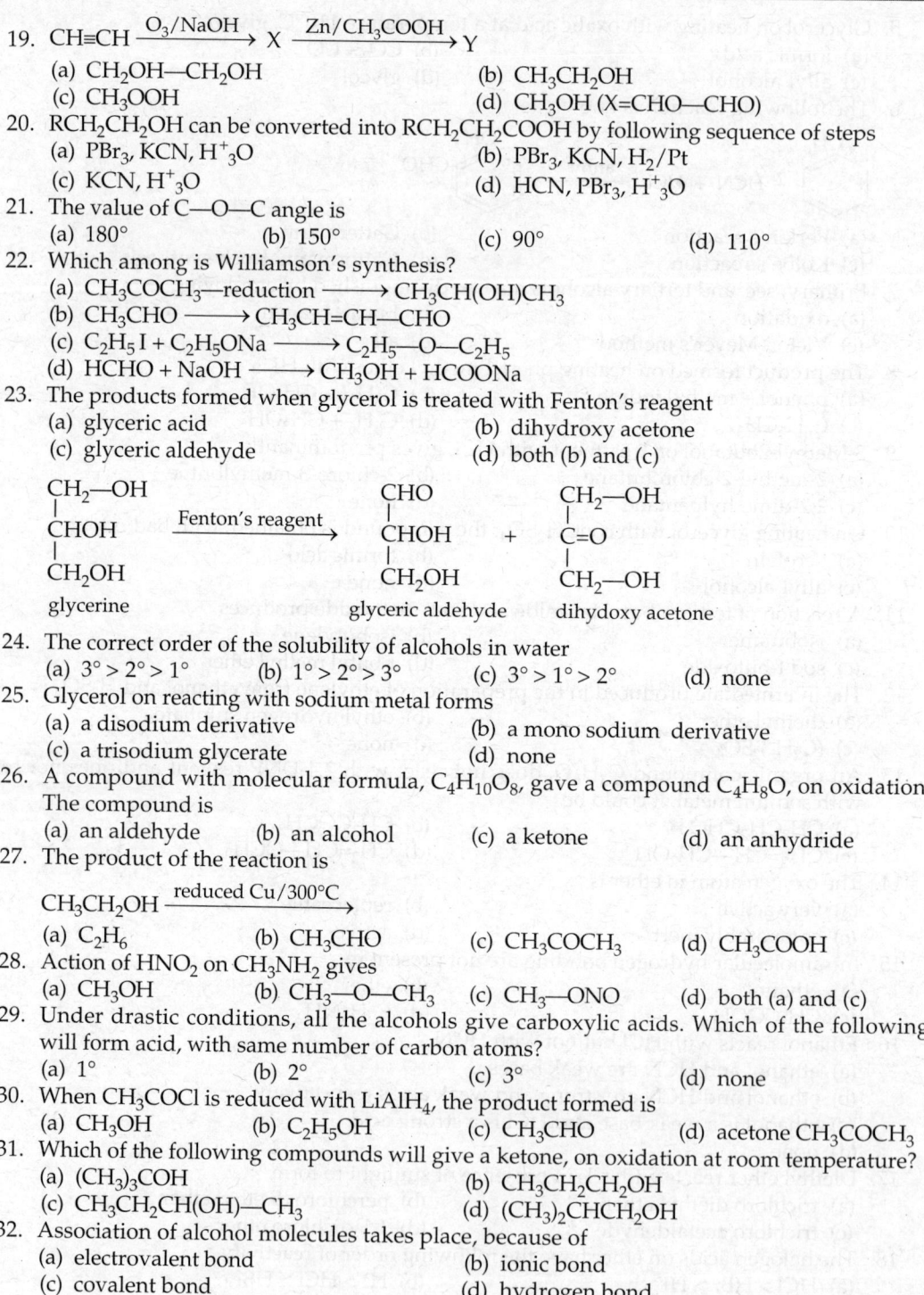

24. The correct order of the solubility of alcohols in water
    (a) 3° > 2° > 1°
    (b) 1° > 2° > 3°
    (c) 3° > 1° > 2°
    (d) none

25. Glycerol on reacting with sodium metal forms
    (a) a disodium derivative
    (b) a mono sodium derivative
    (c) a trisodium glycerate
    (d) none

26. A compound with molecular formula, $C_4H_{10}O_8$, gave a compound C_4H_8O, on oxidation. The compound is
    (a) an aldehyde
    (b) an alcohol
    (c) a ketone
    (d) an anhydride

27. The product of the reaction is
    $CH_3CH_2OH \xrightarrow{reduced\ Cu/300°C}$
    (a) C_2H_6
    (b) CH_3CHO
    (c) CH_3COCH_3
    (d) CH_3COOH

28. Action of HNO_2 on CH_3NH_2 gives
    (a) CH_3OH
    (b) $CH_3—O—CH_3$
    (c) $CH_3—ONO$
    (d) both (a) and (c)

29. Under drastic conditions, all the alcohols give carboxylic acids. Which of the following will form acid, with same number of carbon atoms?
    (a) 1°
    (b) 2°
    (c) 3°
    (d) none

30. When CH_3COCl is reduced with $LiAlH_4$, the product formed is
    (a) CH_3OH
    (b) C_2H_5OH
    (c) CH_3CHO
    (d) acetone CH_3COCH_3

31. Which of the following compounds will give a ketone, on oxidation at room temperature?
    (a) $(CH_3)_3COH$
    (b) $CH_3CH_2CH_2OH$
    (c) $CH_3CH_2CH(OH)—CH_3$
    (d) $(CH_3)_2CHCH_2OH$

32. Association of alcohol molecules takes place, because of
    (a) electrovalent bond
    (b) ionic bond
    (c) covalent bond
    (d) hydrogen bond

33. In esterification of an acid, the other reagents is
    (a) alcohol
    (b) aldehyde
    (c) amine
    (d) H_2O

34. Dunstan's test is done for identification of
    (a) acetone
    (b) alcohol
    (c) glycerine
    (d) carbonyl compound

35. Glycerin on warming with excess of HI gives
    (a) 2-iodo propane
    (b) 1-iodopropane
    (c) 1,2,3-tri iodo propane
    (d) none

36. The reaction of C_2H_5 —O—CH_3 with HI gives
    (a) C_2H_5OH
    (b) CH_3I
    (c) $CH_3I + C_2H5OH$
    (d) $CH_3OH + C_2H_5I$

37. In CH_3CH_2OH the bond which most easily undergoes heterolytic fission, during a reaction with CH_3COOH in presence of H_2SO_4, is
    (a) C—C
    (b) C—O
    (c) O—H
    (d) C—H

38. CH_3COOH reacts readily with
    (a) C_2H_5OH
    (b) $(CH_3)_2CHOH$
    (c) $(CH_3)_3C$—OH
    (d) all

39. In the reaction involving C—OH bond, the order of reactivity in the alcohols is
    (a) $1 > 2 > 3$
    (b) $3 > 2 > 1$
    (c) $2 > 3 > 1$
    (d) none

40. Which of the following gives a +ve iodoform test
    (a) pentanal
    (b) 1-phenyl ethanol
    (c) 2-phenyl ethanol
    (d) 3-pentanal

41. If methanol vapours are passed over heated copper, at 300 °C, formaldehyde is formed, by
    (a) hydrogenation
    (b) dehydrogenation
    (c) dehydration
    (d) oxidation

42. Ethers are quite stable towards
    (a) oxidation
    (b) reduction
    (c) sodium metal
    (d) all

43. Methanol and ethanol can be distinguished
    (a) by reaction with Na metal
    (b) reaction with NaOH
    (c) by heating with I_2 and washing soda
    (d) none

44. Dehydration of ethylene glycol with $ZnCl_2$ gives
    (a) ethylene oxide
    (b) acetaldehyde
    (c) ethylene
    (d) ethyl alcohol

45. Which of the following is an example of elimination reaction?
    (a) chlorination of methane
    (b) dehydration of ethanol
    (c) nitration of benzene
    (d) hydration of C_2H_4

46. Which of the reducing agent will reduce it to the corresponding alcohol
    $RCHO + H_2 \longrightarrow RCH_2OH$
    (a) Ni
    (b) Pd
    (c) Pt
    (d) any one

47. Which of the following reagents can convert acetic acid into ethanol?
    (a) Sn + HCl
    (b) $Pt + H_2$
    (c) $LiAl_4$ in ether
    (d) Na + alcohol

48. The reagent to convert ethylene glycol into HCHO
    (a) acidified $KMnO_4$
    (b) HIO_4
    (c) HNO_3
    (d) Chromic acid

49. The unsaturated aldehyde may be reduced into unsaturated alcohol.
    $RCH=CHCHO \longrightarrow RCH=CH—CH_2OH$
    (a) Na/Hg amalgam + H_2O
    (b) dil. H_2SO_4
    (c) Zn/HCl
    (d) $LiAlH_4$

50. To convert propene to 1-propanol, which set of reagents can be used
    (a) H_2O
    (b) B_2H_6/H_2O_2
    (c) H_2SO_4
    (d) none

51. The presence of an alcoholic group can be tested by
    (a) sodium metal   (b) ester test   (c) ceric ammo test   (d) all
52. Methanol and ethanol can be distinguished by
    (a) a reaction with Na metal   (b) heating with NaOH
    (c) haloform test   (d) none
53. Excessive solubility of lower alcohol is due to
    (a) covalent bond   (b) H-bonds   (c) ionic bond   (d) none
54. Methanol reacts with PCl_3 to form
    (a) methyl chloride   (b) acetyl chloride
    (c) methane   (d) dimethyl ether
55. 2-Methyl-2-butanol on treatment with HCl gives predominantly
    (a) 2-chloro-2-methyl butane   (b) 2-chloro-3-methyl butane
    (c) 2,2-dimethyl pentane   (d) none
56. $(CH_3)_3CONa$ with CH_3Br gives
    (a) $(CH3)_3$—C—OCH_3 (b) CH_3–O—CH_3
    (c) $(CH_3)_3C$—$OC(CH_3)_3$
57. When CH_3MgBr is made to react with acetone and the product formed after hydrolysis is
    (a) a 1° alcohol   (b) a 2° alcohol
    (c) a 3° alcohol   (d) an aldehyde
58. The central oxygen atom in ether is
    (a) sp^2 (b) sp^3 (c) sp (d) sp^3d^2
59. Which has the maximum bond angle among the followings?
    (a) H_2O (b) C_2H_5—O—C_2H_5
    (c) H_2S (d) NH_3
60. R—CH=CH_2 reacts with BH_3, in presence of H_2O_2 to form
    (a) R—CH(OH)CH_3 (b) RCH_2CH_2OH
    (c) $RCOCH_3$ (d) RCH_2CHO
61. An alcoholic group can be distinguished from a phenolic group by
    (a) methyl orange   (b) phenolphthalein
    (c) $FeCl_3$ (d) Na_2CO_3
62. CH_2Cl—CH_2OH is stronger acid than $ClCH_2OH$ as
    (a) the + inductive effect of Cl, disperses the –ve charge on oxygen atom, to produce a more stable anion
    (b) –IE of Cl increases –ve charge on oxygen atom, of alcohol
    (c) none
63. $CH≡CH \xrightarrow{O_3/NaOH} X \xrightarrow{Zn + CH_3COOH} Y$
    (a) X = OHC—CHO, CH_2OH—CH_2OH (b) X = CH_3CH_2OH, Y = CHO
    (c) Y = CH_3COOH (d) X = CH = CHO, Y = CH_3OH
64. RX + 2Na + XR ⟶ R—R + 2NaX; the reaction is termed
    (a) Friedel–Craft reaction   (b) Perkin's reaction
    (c) Wurtz reaction   (d) Gattermann's reaction
65. Which reagent cannot be used to differentiate between phenol and ethanol?
    (a) neutral $FeCl_3$ (b) Na metal   (c) oxidising agent   (d) I_2 + base
66. Which compound be oxidized to prepare methyl ethyl ketone?
    (a) 2-propanol   (b) 1-butanol   (c) 2-butanol   (d) tertiary butyl alcohol
67. Association of alcohol molecule takes place, because of
    (a) electrovalent bond   (b) ionic bonds
    (c) covalent bonds   (d) hydrogen bonds

68. What is the basis for the formation of organometallic compounds, which are quite stable and can be subjected to various type of reactions as in the case of organic compounds?
    (a) the halogen attached to the metal atom is quite reactive
    (b) the halogen atom is very firmly bound to the metal
    (c) the halogen atom is joined to the metal atom by a covalent bond
    (d) no other reason

69. In a reaction of an ether with HI, the H^+ goes to a higher alkyl group as in the case of ethyl methyl ether-HI $\longrightarrow H^+ + I^-$

    (a) $C_2H_5-O-CH_3 + H^+ \xrightarrow[H^+]{} C_2H_5-O^+-CH_3 \xrightarrow{I^-} C_2H_5OH + CH_3^+ \xrightarrow{H} CH_3I$

    (b) $C_6H_5-O-CH_3 + H^+ \longrightarrow C_6H_5-O^+-CH_3 \xrightarrow{I^-} C_6H_5OH + CH_3I$

70. Identify Z in the series $CH_2{=}CH_2 + HBr \xrightarrow{OH^-} X \xrightarrow{I_2/NaOH} Y \longrightarrow Z$
    (a) CHI_3 (b) C_2H_5OH (c) C_2H_5I (d) CH_3CHO

71. Reaction of sodium metal is faster in the case of
    (a) primary alcohol (b) a sec. alcohol
    (c) a tertiary alcohol (d) same with all type of alcohols

72. On conversion to Grignard's reagent, followed by treatment with absolute alcohol, how many isomeric alkyl chlorides would yield 2-methyl butane.
    (a) 2 (b) 3 (c) 4 (d) 8

73. The following compounds will not give an iodoform test
    (a) CH_3COCl (b) CH_3CONH_2
    (c) $CH_3COCH_2 \cdot COOC_2H_5$ (d) $C_2H_5CONH_2$

74. Primary and sec. alcohols on passing over hot reduced copper (300°C)
    (a) form aldehyde and a ketone respectively
    (b) give a ketone and aldehyde respectively
    (c) only aldehyde
    (d) ketone only

75. Reaction $CO + H_2 + H_2 \xrightarrow{400°C/300 \text{ atoms}/Cr_2O_3 + ZnO}$ may be used for the manufacturing of
    (a) HCHO (b) CH_3OH
    (c) HCOOH (d) CH_3OOH

76. In ethyl alcohol, the bond that undergoes heterolytic cleavage most easily is
    (a) C—C (b) C—O
    (c) C—H (d) O—H

77. Which reagent cannot be used to identify phenol and ethanol?
    (a) neutral $FeCl_3$ (b) Na metal
    (c) oxidising agent (d) I_2 in presence of a base

78. Formation of oxonium salt shows that ethers are
    (a) acidic (b) basic
    (c) neutral in nature (d) none

79. Arrange these in the decreasing order of their acidic nature
    (a) $H_2O > C_2H_5OH > NH_3 > CH{\equiv}CH$ (b) $NH_3 > CH{\equiv}CH > NH_2 > C_2H_5OH$
    (c) $C_2H_2 > H_2O > C_2H_5OH > CH{\equiv}CH$ (d) $C_2H_5OH > NH_3 > CH{\equiv}CH > H_2O$

80. Vinyl carbinol is
    (a) $CH_2{=}CH-CH_2OH$ (b) $CH_3-C(OH){=}CH_2$
    (c) $CH_3-CH{=}CH-OH$ (d) $CH_2{=}CHOH$

81. Reaction of $H_2C=CH_2$ with RMgX, followed by hydrolysis gives
    (a) RCH_2—CH_2OH
    (b) $RCH=CHOH$
    (c) $RCHOHCH_3H_2CrO_4$

82. Identify (X) in the sequence $C_3H_8O \xrightarrow[X]{} C_3H_6O \xrightarrow{I_2/NaOH} CHI_3$

83. Glycerol on oxidation with conc. HNO_3 gives
    (a) tartronic acid
    (b) glyceric acid
    (c) meso-oxalic acid
    (d) forms both (a) and (b)

84. Ethers are
    (a) neutral
    (b) Lewis acids
    (c) Lewis bases
    (d) cannot predict

85. Diethyl ether and dimethyl ether are
    (a) enantiomers
    (b) conformational isomer
    (c) metamers
    (d) geometrical isomers

86. Which is the best reagent to accomplish the following conversion?
    CH_3CH_2—O—$CH_2CH_3 \longrightarrow CH_3CH_2Br$
    (a) Br_2 in CCl_4
    (b) NaBr
    (c) Br_2 in water
    (d) conc. HBr

87. Which of the following pairs of compounds will not form H–bonding with other?
    (a) methanol and ethanol
    (b) CH_3SH and C_2H_5SH
    (c) dimethyl ether and diethyl ether
    (d) acetic acid and water

88. Ethylene oxide react with NH_3 to give
    (a) 1-amino ethanol
    (b) 2-aminoethanol
    (c) ethylene
    (d) acetamide

89. The compound formed when ethylene oxide is treated with acid methanol.
    (a) CH_3—O—CH_2—CH_2OH
    (b) $CH_3CH_2CH_2H_2OH$
    (c) HO—CH_2CH_2OH
    (d) CH_5—O—C_2H_5

90. Which of these is used as an anti-freeze?
    (a) ethylene glycol
    (b) glycerol
    (c) ether
    (d) none

91. Compound 'A' reacts with sodium metal to form one molecule of H_2. The compound can be
    (a) $CH_3CH_2CH=CH_2$
    (b) $HOCH_2H_2CH_2OH$
    (c) $CH_2=CH$—$CH=CH_2$
    (d) $CH_3CH_2CH_2OH$

92. Glycerol, on heating with oxalic acid, to a temperature $260°C$ it gives
    (a) 1,2-propane-diol
    (b) 1,3-propane diol
    (c) vinyl alcohol
    (d) allyl alcohol

93. Which is the best reagent for carrying out the following conversion?

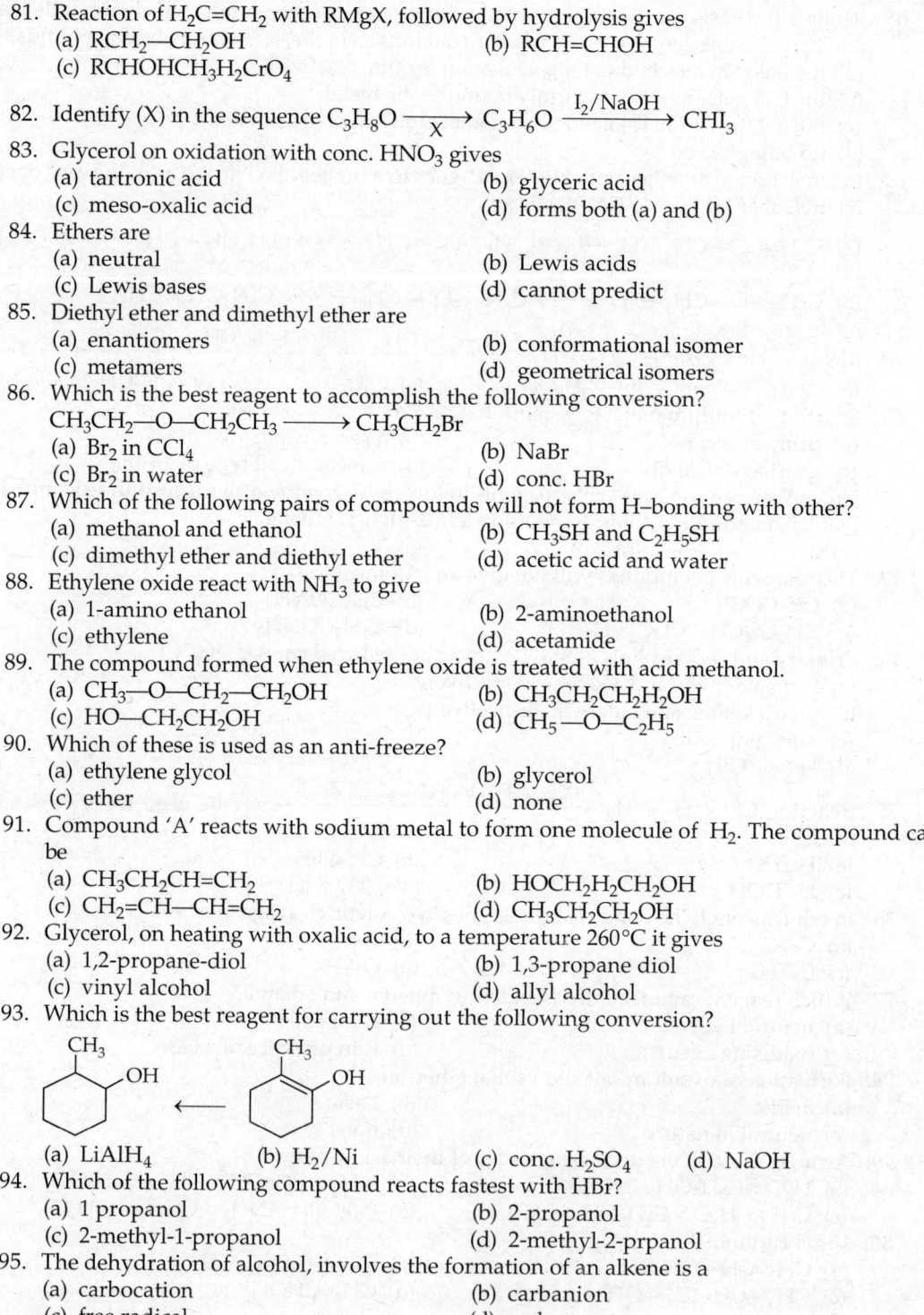

    (a) $LiAlH_4$ (b) H_2/Ni (c) conc. H_2SO_4 (d) NaOH

94. Which of the following compound reacts fastest with HBr?
    (a) 1 propanol
    (b) 2-propanol
    (c) 2-methyl-1-propanol
    (d) 2-methyl-2-prpanol

95. The dehydration of alcohol, involves the formation of an alkene is a
    (a) carbocation
    (b) carbanion
    (c) free radical
    (d) carbene

96. Which of the following is most resistant to oxidation?
    (a) 1° alcohol   (b) 2° alcohol
    (c) 3° alcohol   (d) none
97. Hydroboration–oxidation of 2-methyl propene gives
    (a) 2-methyl-2-propanol   (b) 1,2,3-propane triol
    (c) 2-methyl-1-propanol   (d) 1,2-propanediol
98. The acid catalyzed dehydration mechanism of alcohol is best described as
    (a) E1   (b) E2
    (c) S_N1 (d) S_N2
99. Which of the following reactions will form an ionic compound?
    (a) $C_2H_5OH + PCl_3$ (b) $C_2H_5OH + KBr + H_2SO_4$
    (c) $C_2H_5OH + Na$ (d) $C_2H_5OH + SOCl_2$
100. Why do alcohols boil at a higher temperature as compared to alkanes of similar mole. wt?
    (a) alcohols have greater van der Waals attractive forces
    (b) alcohol molecules have molecular symmetry
    (c) alcohols must break H-bonds during the process of volatilization
    (d) alcohols must overcome the greater ionic forces and the process of volatilization

ANSWERS

(Group-D)

1. (a)	2. (b)	3. (c)	4. (a)	5. (a)	6. (b)	7. (d) all	8. (a)
9. (a) 1, 2 shift	10. (a)	11. (b)	12. (b)	13. (d)	14. (c)	15. (b)	
16. (c)	17. (b)	18. (c)	19. (a)	20. (a)	21. (d)	22. (c)	23. (d)
24. (b)	25. (a)	26. (b) an alcohol	27. (b) acetaldehyde	28. (a)	29. (a)		
30. (b)	31. (c)	32. (d)	33. (a)	34. (c)	35. (a)		

36. (c) O goes with higher alkyl group
37. (b) C—O bond 38 (a) the order of reactivity of alcohols is 1° > 2° > 3°   39. (b)
40. (b), CH_3CH—OH this group gives a + iodoform test
41. (b)   42. (d)
43. (c) ethanol gives iodoform test
44. (b)   45. (b)   46. (d)   47. (c)
48. (b) periodic oxidation
49. (d)   50. (b) BH_3 and BH_2^+ and H^-
51. (d)   52. (c)   53. (b)   54. (a)   55. (a)   56. (a)   57. (c)   58. (b)
59. (b)
60. (b) the process is termed hydroboration
61. (c) no colourwith analcohol, but develops a distinct colour with phenols
62. (b) $Cl \leftarrow CH_2- \leftarrow CH_2^- \leftarrow O^- + H^+$; a –ve charge on 'o' dispersed, making more stable
63. (a) X = OHC—CHO, Y = CH_2OH—CH_2OH
64. (c) Wurtz reaction
65. (d)   66. (c)   67. (c)   68. (c)
69. ROH + CH_3I (R = higher alkyl group/a phenyl group) The phenyl group is stabilized by resonance
70. CHI_3 71. (a)
72. (d) $RMgBr + C_2H_5OH \longrightarrow RH + Mg(OC_2H_5)Br$; RH = $(CH_3)_2$ CH—CH_2—CH_3; RX should be $(CH_3)_2CHCH_2CH_2Cl$; $(CH_3)_2CHCHClCH_3$; $(CH_3)_2CClCH_2CH_3$; CH_2Cl—$C(CH_3)CH$—CH_2CH_3

73. (d)
74. (a) an aldehyde and a ketone respectively
75. (b) CH_3OH
76. (b) C—O has less bond energy than O—H bond
77. (b) both react with sodium metal
78. (b) basic in nature, as they donate a pair of electron
79. (a) $H_2O > C_2H_5OH > NH_3 > CH{\equiv}CH$
80. (d) carbinol means alcohol
81. (a)   82. $CH_3CH(OH)CH_3$

83. (d)	84. (c)	85. (c)	86. (d)	87. (c)	88. (c)	89. (a)	90. (a)
91. (b)	92. (d)	93. (b)	94. (d)	95. (a)	96. (c)	97. (c)	98. (a) E1
99. (c)	100. (c)						

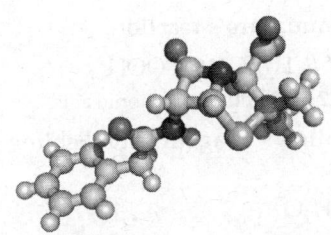

Chapter 4

Aldehydes and Ketones

Salient Features

1. Only aliphatic aldehydes reduce **Benedict's reagent or solution.** Aromatic aldehydes do not reduce Benedict's reagent.

2. Only **aliphatic aldehydes, methyl ketones and cyclic ketones** react with $NaHSO_3$ to form **sodium bisulphite an** addition compound.

$$>C=O \quad + \quad NaHSO_3 \quad \longrightarrow \quad >C \overset{HO}{\underset{SO_3Na}{\big<}}$$

3. **Halogen acids fail** to form addition compound with the carbonyl group, because the addition product being analogous to a **gem di-hydroxy product, which decomposes back to form ketone.**

4. Keto compound containing a CH_3CO-group or $CH_3CH(OH)$-group in the molecule, attached to a carbon or a hydrogen, undergo Haloform reaction, i.e. they react with iodine, in presence of a dilute alkali, to form iodoform, CHI_3.

5. **Aldehydes** react with **hydroxyl amine, NH_2OH,** to form **aldoxime** which exists in **two isomeric forms, as syn- and anti- forms.**

$$\overset{R}{\underset{R}{\big>}}C=O + H_2N{-}OH \quad \longrightarrow \quad \overset{R}{\underset{H}{\big>}}C=N{\underset{OH}{\diagdown}} \quad + \quad \overset{R}{\underset{H}{\big>}}C=N{\diagup}^{OH}$$

$$\qquad\qquad\qquad\qquad\qquad\qquad\qquad syn \qquad\qquad\qquad anti$$

Ketoximes (and some aldoxime), on treatment with P_2O_5, H_2SO_4, H_3PO_4 etc. rearrange to form acid amide derivatives, RCONHR. This arrangement is known as **Beckmann's rearrangement.**

$$\overset{R}{\underset{R}{\big>}}C=O + H_2N{-}OH \quad \longrightarrow \quad \overset{R}{\underset{H}{\big>}}C=N{-}OH \quad \longrightarrow \quad RCONHR$$

ketoxime

$$\overset{R}{\underset{R}{\big>}}C=O + H_2N{-}OH \quad \longrightarrow \quad \overset{R}{\underset{H}{\big>}}C=N{-}OH \quad \longrightarrow \quad RCONH_2 + HCONHR$$

aldoxime

6. Aldehydes that do not contain α-hydrogen atom, undergo **Cannizzaro's reaction.**

$$2\ (CH_3)_3C—CHO + NaOH \longrightarrow (CH_3)_3C—CH_2OH + (CH_3)_3C—C\ COOH$$

2,2-dimethyl propionaldehyde   2,2-dimethylpropylalcohol 2,2-dimethyl propionic acid

7. **α-Hydroxy ketones, RCO—CHOHR′, readily reduce Tollen's reagent and** Fehling solution.

$$RCOCH(OH)R + Ag_2O \longrightarrow RCO·COR + 2Ag + H_2O$$

Tollen's reagent

8. Aliphatic aldehydes can be prepared by the following methods.

a. **Wacker's reaction**

$$CH_2=CH_2 + PdCl_2 + H_2O \longrightarrow CH_3CHO + 2HCl + Pd$$

b. **Oxo process**

$Co(CO)_4/\Delta/$under pressure

$$CH_3—CH=CH_2 + CO + H_2 \longrightarrow CH_3CH_2CH_2CHO + (CH_3)_2—CHCHO$$

propene   butyraldehyde   iso butyraldehyde

9. Methyl ketones can be prepared from the corresponding esters and dimethyl sulphoxide.

$$RCOOCH_3 \xrightarrow[\text{(3) Al–Hg–THF + H}_2\text{O}]{\text{(1) NaH (2) (CH}_3)_2\text{SO}_4} RCOCH_3$$

10. Aliphatic as well as aromatic aldehydes can be prepared by oxidation of primary alcohols, by a Sarret-Collins reagent (CrO_3 + pyridine in CH_2Cl_2), or by PCC-pyridinium chlorochromate (CrO_3 + HCl + pyridine).

$$RCH_2OH \xrightarrow[\text{pyridinium chloro chromate}]{\text{Sarret-Collins reagent or}} RCHO$$

Aldehydes so formed are, not oxidised further, under these condition.

11. Benzaldehyde, C_6H_5CHO, reacts with Cl_2, under boiling conditions, to form benzoyl chloride, C_6H_5COCl.

$$C_6H_5CHO + Cl_2 \longrightarrow C_6H_5COCl + HCl$$

Aliphatic aldehydes do not form Cl_2, such compounds, when reacts with Cl_2 instead the α-hydrogen atoms, are substituted to give chloro derivatives.

12. **Benzoin condensation.** Benzaldehyde, forms a condensation product in the presence of KCN.

$$C_6H_5CHO + C_6H_5CHO \xrightarrow{\text{KCN/heat}} C_6H_5CHOH—COC_6H_5$$

benzoin

Aliphatic aldehydes do not undergo such reaction.

13. **Benzaldehyde does not reduce Fehling solution** while aliphatic aldehydes do reduce Fehling solution.

14. **Aldehydes can be converted into methyl ketones $RCOCH_3$, by their reaction with diazo** methane CH_2N_2.

$$RCHO + CH_2N_2 \longrightarrow RCOCH_3 + N_2$$

15. C_6H_5CHO, on reduction with zinc and HCl or with sodium amalgam, forms hydro benzoin, which on oxidation with chromic acid gives benzil

$$C_6H_5CHO + C_6H_5CHO + 2H \xrightarrow[\text{Na–Hg/H}_2\text{O}]{\text{Zn + HCl}} \underset{\substack{| \quad | \\ \text{H} \quad \text{H} \\ \text{hydrobenzoin}}}{C_6H_5-\overset{\text{HO OH}}{\underset{}{C}}-\overset{}{C}-C_6H_5} \xrightarrow[\text{(O)}]{\text{H}_2\text{CrO}_4}$$

$$\underset{\text{benzil}}{C_6H_5-\overset{O}{\overset{||}{C}}-\overset{O}{\overset{||}{C}}-C_6H_5}$$

Benzil on heating with KOH solution, undergoes rearrangement, to form benzilic acid.

$$\underset{\text{benzil}}{C_6H_5-\overset{O}{\overset{||}{C}}-\overset{O}{\overset{||}{C}}-C_6H_5} \xrightarrow{\text{KOH}} \underset{\text{benzilic acid}}{\begin{array}{c} C_6H_5 \\ \diagdown \\ C_6H_5 \end{array} C \begin{array}{c} OH \\ \diagup \\ COOH \end{array}}$$

16. **Acetone, on reduction with Mg-Hg (amalgam),** forms **pinacol,** which on warming with **dil. H$_2$SO$_4$** undergoes rearrangement **to form pinacolone.**

$$CH_3CO–CH_3 + CH_3COCH_3 \xrightarrow[\text{H}_2\text{O}]{\text{Mg–Hg}} \underset{\substack{| \quad | \\ \text{H} \quad \text{H} \\ \text{pinacol}}}{CH_3-\overset{\text{HO OH}}{\underset{}{C}}-\overset{}{C}-CH_3} \xrightarrow[\text{heat}]{\text{dil. H}_2\text{SO}_4} \underset{\substack{| \quad | \\ \text{H} \quad \text{H} \\ \text{pinacolone}}}{CH_3-\overset{\text{HO OH}}{\underset{}{C}}-\overset{}{C}-CH_3}$$

This rearrangement is known as **Pinacol–pinacolone rearrangement.**

Aldehyde and ketones are the keto compounds, containing a carbonyl group, >C=O, in the molecule. They are also isomeric compounds having the same molecular composition $–C_nH_{2n}O$. When one valency of the carbon of the >C=O group is occupied or joined by a hydrogen atom, the compounds are known as an aldehyde. When both the valences of the carbon of the >C=O group are joined by two alkyl groups, same or different alkyl groups, or aryl groups, the compounds are known as the ketones.

$$>C=O \qquad \underset{\text{H}}{\overset{\text{H/R}}{\diagdown}}C=O \qquad \underset{\text{R}}{\overset{\text{R}}{\diagdown}}C=O$$

a carbonyl group   an aldehyde   a ketone

Aldehydes and ketones may also be regarded as the oxidation products of the alkanes.

$$\underset{\text{alkane}}{R–CH_3} \longrightarrow \underset{\text{alcohol}}{RCH_2OH} \longrightarrow \underset{\text{aldehyde}}{RCHO} \longrightarrow \underset{\text{acid}}{RCOOH}$$

$$\underset{\text{alkane}}{\overset{\text{R}}{\underset{\text{H}}{\diagdown\diagup}}CH_2} \longrightarrow \underset{\text{alcohol}}{\overset{\text{R}}{\underset{\text{R}}{\diagdown\diagup}}CHOH} \longrightarrow \underset{\text{ketone}}{\overset{\text{R}}{\underset{\text{R}}{\diagdown\diagup}}C=O}$$

As an aldehydic group, –CHO, is a chain terminating group, a suffix "al" is added, to indicate its presence in the molecule. In IUPAC system of nomenclature, the aldehydes are known as "alkanals". Similarly in ketones, the presence of a keto groups is indicated by a suffix "one" in IUPAC system the ketones are known as "alkanones".

In aldehydes the carbon of –CHO group is always numbered one. In ketones the keto group is in the middle of the carbon chain is joined by two alkyl or aryl groups, it is given a lowest possible number. The suffix "one" used after the number of carbon of the keto group.

The carbon of carbonyl group in aldehydes and ketones is sp^2 hybridised and it is bonded to oxygen by a double bond, one bond σ and other π-bond. The molecule is triangular planar with an angle of 120°.

On count of high electronegativity of oxygen (3.5) as compare to that of carbon (2.5), the labile electron of bond are shifted towards oxygen. Due to inductive effect of oxygen (–I effect), the >C=O group becomes more polar in nature.

$$\overset{}{>C=O} \longrightarrow \overset{\delta+ \ \delta-}{>C-O}$$
$$2.5 \ \ 3.5 \longrightarrow \text{polarity}$$

As a result the carbonyl compounds, undergo nucleophilic addition reaction. The >C=O is more polar in aldehyde as compared to ketones, as one valency of carbon is satisfied by hydrogen. The >C=O is less polar in ketones because of +I effect of two alkyl groups attached to carbon.

The decreasing order of polarity in aldehydes and ketones

a. $HCHO > CH_3CHO > CH_3COCH_3$

$$\underset{H}{\overset{H}{>}}C=O \ \ > \ \ \underset{H}{\overset{CH_3}{>}}C=O \ \ > \ \ \underset{CH_3}{\overset{CH_3}{>}}C=O$$

b. $HCHO > RCHO > C_6H_5CHO > C_6H_5COCH_3 > C_6H_5CO \ C_6H_5$

Since an attack by an electrophile on the >C=O group may lead to formation of a less stable specie (>C=O$^+$H), than an attack by a nucleophile, the species formed is more stable

(>C—O—), A positive charge on oxygen is less stable than a negatively charged oxygen

$$>C=O \xrightarrow{\ H^+\ } >C=O^+H \quad \text{by}$$

$$>C=O \xrightarrow{\ Nu^-\ } C-O^- \xrightarrow{\ H^+\ } >C\underset{H^+}{\overset{Nu^-}{<}}$$

This is also the decreasing order of nucleophilic attack by CN— on aldehyde and ketones. Greater is the extent of positive charge on carbon of >C=O group, the easier will be for the nucleophile to add.

$$>C=O + CN^- \longrightarrow >C\underset{CN}{\overset{O^-}{<}} \longrightarrow >C\underset{CN}{\overset{OH}{<}}$$

On account of polarity in aldehyde and ketones, their boiling points (b.p.) and melting points (m.p.) are higher than alkanes of comparable molecular weight (mole. wt.). In general, the b.p. of the various organic compounds are in the following decreasing order. Ionic compounds > compounds with H-bonding > compounds with dipole-dipole attraction > compound with van der Waal forces.

For example: $RCOONa > RCOOH > ROH > RCHO > RCOR > RH$

On account of molecular weight

$$HCOONa > CH_3COOH > CH_3CH_2CH_2OH > CH_3CH_2CHO > CH_3COCH_3$$

Mole wt.   60/118°C   60/98°C   58/73°C   58/56°C

Nomenclature: In IUPAC system of nomenclature, 'al' is used as a suffix to indicate the presence of a —CHO group, in the molecules. Since an aldehydic group is a chain terminating group, in aldehydes it is always numbered as number 1.

Aldehydes	Common name	IUPAC system
HCHO	formaldehyde	methanal
CH_3COOH	acetaldehyde	ethanal
C_2H_5CHO	propionaldehyde	propanal
C_3H_7CHO	butyraldehyde	butanal
$(CH_3)_2CHCHO$	isobutyraldehyde	2-methylpropanal

Ketones or Alkanones, RCOR

In alkanones the keto group is joined by two alkyl or two aryl groups, either same or different, the common name of the specie is dialkyl ketone. In IUPAC system, the presence of a keto group and its position is, indicated by the number of its position in the chain.

Ketone	*Common name*	*IUPAC system*
CH_3COCH_3	Acetone	Propanone

Aldehydes

Preparation

1. **By oxidation of a primary alcohol (RCH_2OH)**

   i. Acidified $K_2Cr_2O_7$ –Primary alcohols are oxidized to give corresponding aldehyde. The b.p. of the alcohol is higher than the corresponding aldehyde formed

   $$K_2Cr_2O_7 + 4H_2SO_4 \longrightarrow K_2SO_4 + Cr_2(SO_4)_3 + 4H_2O + 3O$$

   a. $CH_3OH + O \longrightarrow HCHO + H_2O$

   b. $RCH_2OH + O \longrightarrow RCHO + H_2O$

   The aldehyde formed, distil over, are condense and collected. In this manner the aldehyde formed escapes further oxidation to the corresponding acids.

   ii. By oxidation of a primary alcohol by Sarret–Collins reagent (it consists of a mixture of CrO_3 and pyridine take in CH_2Cl_2 as a solvent) or by pyridinium chloro chromate PCC, (it consists of CrO_3 pyridine and dry HCl gas).

   $$RCH_2OH \xrightarrow{\text{Sarret-Collins reagent/PCC}} RCHO + H_2O$$

   iii. By Jones reagent–Jones reagent consists of CrO_3, taken in aqueous acetone. It oxidises a 1° and 2° alcohol to the corresponding aldehyde and a ketone, respectively.

   $$RCH_2OH \longrightarrow RCHO + H_2O$$
   $$R_2CHOH \longrightarrow RCOR + H_2O$$

   iv. By catalytic oxidation of 1° alcohol–The vapours of 1° alcohol mixed with air are passed over Ag catalyst heated to 250°C, when aldehydes formed.

2. **Dehydrogenation:** Primary alcohols are dehydrogenated by passing its vapours over hot Cu, kept at 300°C.

   $$RCH_2OH \xrightarrow{\text{Cu at 300°C}} RCHO + H_2$$
   $$R_2CHOH \longrightarrow RCOR + H_2$$

3. **By alkaline hydrolysis of gem-dihalides:**

   $$RCHCl_2 + 2NaOH \longrightarrow RHO + 2HCl$$
   $$CH_2Cl_2 + 2NaOH \longrightarrow HCHO + 2HCl$$
   $$CH_3CHCl_2 + 2NaOH \longrightarrow CH_3CHO + 2NaCl$$

4. **By Stephen's reaction:** Alkyl cyanides, when subjected to reductive hydrolysis form aldehydes.

   $$RCN + 2H \xrightarrow{\text{SnCl}_2 + 2HCl} RCH=NH \longrightarrow RCHO$$

5. **By Rosenmund reaction:** In boiling xylene solution, acid chlorides, RCOCl are subjected to reduction by H_2 in presence of a catalyst, consisting of Pd mounted on $BaSO_4$, poisoned by quinoline or sulphur, are reduced to corresponding aldehyde (quinoline or sulphur to check further reduction of the aldehyde to alcohol). This reaction is termed as Rosenmund reaction.

   $$\underset{\text{acid chloride}}{RCOCl} \xrightarrow{\text{H}_2/\text{Pd}/\text{BaSO}_4/ \text{ poisoned with quinoline}} \underset{\text{aldehyde}}{RCHO}$$

$$CH_3COCl \longrightarrow CH_3CHO$$

HCHO cannot be prepared as corresponding formyl chloride, HCOCl, is a unstable compound

6. **From alkenes:** Alkenes may be converted into aldehyde by any one of the following methods.

   a. By *Waker's method*: Alkenes are treated with Palladium chloride in the presence of $CuCl_2$ as catalyst to form an aldehyde

   $$CH_2{=}CH_2 + H_2O + PdCl_2 \longrightarrow CH_3CHO + Pd + 2HCl$$

   b. By *Oxo process*: Alkenes are treated with a mixture of $CO + H_2$, in the presence of Cobalt hydrogen tetra carbonyl, $CoH(CO)_4$, a catalyst, when alkenes are converted into the aldehydes, containing one carbon more.

   $$CH_2{=}CH_2 + CO + H_2 \longrightarrow CH_3CH_2CHO$$

   $$CH_2{=}CH{-}CH_3 + O + H_2 \longrightarrow CH_3CH_2CH_2CHO + (CH_3)_2{-}CHCHO$$

7. **From calcium salts of fatty acids:** On subjecting a mixture a salt of a fatty acid with calcium formate to dry distillation aldehydes are obtained.

$$\begin{array}{c} CH_3COO \\ CH_3COO \end{array}\!\!\!\Big\rangle Ca + \begin{array}{c} HCOO \\ HCOO \end{array}\!\!\!\Big\rangle Ca \longrightarrow 2CH_3CHO + 2CaCO_3$$

Physical properties

Formaldehyde is a gas at room temperature, boiling point $-21\,°C$, highly soluble in water. Its 40% aqueous solution is termed 'formalin', which is used for preservation of dead species, on account of its coagulating power that kills bacteria.

Acetaldehyde is a liquid, boiling point $21°C$, soluble in water, and miscible in organic solvents.

Lower members have a very offensive odour, miscible in organic solvents. Owing to dipole-dipole attractions, the boiling points of aldehyde and ketones, are higher than those of non-polar alkanes, but lower than the alcohols.

Chemical properties

1. **Reducing properties**

   Aldehydes are powerful reducing agents

   a. *Reduce Tollen's reagent*: Tollen's reagent is prepared by adding an excess of ammonia, to a solution of $AgNO_3$, till a clear solution is obtained.

   $$AgNO_3 + NH_3 \longrightarrow Ag(NH_3)_2NO_3 \longrightarrow Ag(NH_3)_2^+ + NO_3^-$$

   $Ag(NH_3)_2$ behaves as Ag_2O in the solution

   $$RCHO + Ag_2O \longrightarrow RCOOH + 2Ag$$
   $$\text{grey powder of silver}$$

   b. *Reduce Fehling solution*: It is prepared by mixing equal volumes of Fehling solution A, (an aqueous solution of 1 equivalent of $CuSO_4$ and Fehling solution B (an aqueous solution of two equivalent NaOH, containing sodium potassium tartarate) when a deep blue coloured solution is obtained, due to the formation of copper chelate compound.

   $$CuSO_4 + 2NaOH \longrightarrow Cu(OH)_2 + Na_2SO_4$$

$$\begin{array}{c} OH \\ Cu{<} \\ OH \end{array} + \begin{array}{c} CH(OH)COONa \\ | \\ CH(OH)COONa \end{array} \longrightarrow \begin{array}{c} OCH(COONa) \\ Cu{<} \\ OCHCOOK \end{array}$$

copper hydroxide Rochelle salt   copper chelate complex

Copper chelate in the solution behaves as CuO

$$RCHO + 2CuO \longrightarrow RCOOH + Cu_2O$$

Cu_2O is obtained as a brick red coloured powder. Formation of a red coloured powder confirms the reducing properties of the compound.

c. *Benedict's solution*: It Is a blue coloured aqueous solution containing $CuSO_4$, Na_2CO_3 and citric acid (as sodium citrate)

This reagent also behaves as CuO, in aqueous, solution

$$RCHO + 2CuO \longrightarrow RCOOH + Cu_2O$$

A red coloured powder is formed

Only aliphatic aldehydes reduce Benedict's solution. Aromatic aldehyde do not reduce Benedict's solution.

d. *Schiff's reagent*: Aldehydes restore the pink colour of the Schiff's reagent

An aqueous or alcoholic solution of magenta or fuschine is a pink coloured solution

On passing SO_2 through the solution, it becomes colourless,

On treating Schiff's reagent with aldehyde, it becomes pink again.

2. **Oxidation:** Aldehydes are oxidised easily to corresponding acids, containing the same number of carbon atoms, by oxidising agents like acidified $KMnO_4$ or $K_2Cr_2O_7$.

$$RCHO + O \longrightarrow RCOOH$$

3. **Reduction of aldehydes:** Aldehyde are reduced by various reducing agents, to form different reduced product, the >C=O group is reduced to $-CH_2$ group or to >CHOH group.

a. *Clemmensen's reduction or reaction*: Aldehydes are reduce to corresponding alkanes by Zn-Hg amalgam and conc. HCl.

$$Zn + 2HCl \longrightarrow ZnCl_2 + H_2$$
$$>C=O \longrightarrow >CH_2$$
$$RCHO + 4H \longrightarrow >RCH_3 + H_2O$$

b. *Wolf–Kishner reduction*: The aldehyde are reduced to corresponding alkane, by this method.

The reaction takes place in two stages:

i. Firstly the aldehyde is treated with hydrazine, NH_2-NH_2, thereby it is converted into a hydrazone.

ii. He hydrazone is then heated with either KOH or with potassium tertiary butoxide, $(CH_3)_3-COK$, in a high boiling solvent, like ethylene glycol B.P. 190°C, hydrazone is decomposed to form alkane. This method is termed as Millon's method.

$$RCHO + H_2N-NH_2 \xrightarrow{\overset{+}{KOH} \text{ ethylene glycol}} RCH=N-NH_2 \xrightarrow{190°C} RCH_3 + N_2$$

iii. By catalytic reduction: Aldehydes are reduced to alkanes by hydrogen in presence of Ni, heated to 300°C.

$$RCHO + 4H \xrightarrow{Ni/300°C} RCH_3 + H_2O$$

iv. Reduction with NaOH and ROH: Aldehydes are reduced to primary alcohol on reduction with Na + ROH.

$$C_2H_5OH + Na \longrightarrow C_2H_5ONa + H$$
$$RCHO + 2H \longrightarrow RCH_2OH$$

Reduction of an aldehyde to a primary alcohol is termed Bouveault-Blanc reaction.

v. Reduction of an aldehyde by $LiAlH_4$:

Aldehydes are reduced to a primary alcohol, on reduction by $LiAlH_4$

Mechanism: i. $LiAlH_4 \longrightarrow >LiH + AlH_3$

ii. $LiH \longrightarrow Li^+ + H^-$

iii. $RC\overset{\overset{\displaystyle H}{|}}{=}C \xrightarrow{Li^+} R-\overset{\overset{\displaystyle H}{|}}{\underset{}{C^+}}-OLi \xrightarrow{H^-} R-\overset{\overset{\displaystyle H}{}}{\underset{\underset{\displaystyle H}{|}}{C}}-OLi \xrightarrow{H_2O} R-\overset{\overset{\displaystyle H}{|}}{\underset{\underset{\displaystyle H}{|}}{C}}-OH$

It is the hydride ion, H^-, which is the reducing agent.

vi. Reduction with HI: Aldehydes are reduced to alkane, when heated with HI and red phosphorous.

$$RCHO + 4HI \longrightarrow RCH_3 + H_2O + I_2$$

4. **Reaction with PCl_5:** Aldehyde react with PCl_5 to form alkylidene chloride.

$$RCHO + PCl_5 \longrightarrow RCHCl_2 + POCl_3$$
$$3\ RCHO + PCl_3 \longrightarrow 3RCHCl_2 + H_3PO_3$$

5. **Addition reactions:**

a. With RMgX

HCHO reacts with formaldehyde to form a primary alcohol, the complex formed in between aldehyde and Grignard reagent is hydrolysed, by dil. acid, when the corresponding alcohols are obtained.

$$\overset{\displaystyle H}{\underset{\displaystyle H}{>}}C=O \longrightarrow R-\overset{\overset{\displaystyle H}{|}}{\underset{\underset{\displaystyle H}{|}}{C}}-OMgX \longrightarrow R-\overset{\overset{\displaystyle H}{|}}{\underset{\underset{\displaystyle H}{|}}{C}}-OH$$

Other aldehyde react with RMgX to form a secondary alcohol

$$\overset{\displaystyle CH_3}{\underset{\displaystyle H}{>}}C=O \longrightarrow R-\overset{\overset{\displaystyle CH_3}{|}}{\underset{\underset{\displaystyle H}{|}}{C}}-OMgX \longrightarrow R-\overset{\overset{\displaystyle CH_3}{|}}{\underset{\underset{\displaystyle H}{|}}{C}}-OH$$

Ketones react with R MgX to form a tertiary alcohol

$$\overset{\displaystyle CH_3}{\underset{\displaystyle CH_3}{>}}C=O \longrightarrow R-\overset{\overset{\displaystyle CH_3}{|}}{\underset{\underset{\displaystyle CH_3}{}}{C}}-OMgX \longrightarrow R-\overset{\overset{\displaystyle CH_3}{}}{\underset{\underset{\displaystyle CH_3}{|}}{C}}-OH$$

b. Addition compound with sodium bi sulphite

$$>C=O + NaHSO_3 \longrightarrow >C=O + NaHSO_3 \longrightarrow >C\overset{\displaystyle OH}{\underset{\displaystyle SO_3Na}{<}}$$

sodium bisulphites
addition compound

These addition compounds, are decomposed back with the liberation of the aldehyde. This may be use to purify the aldehydes.

c. Addition compound with HCN- Formation of a cyanohydrin-HCN adds to the keto group, to form a cyanohydrin

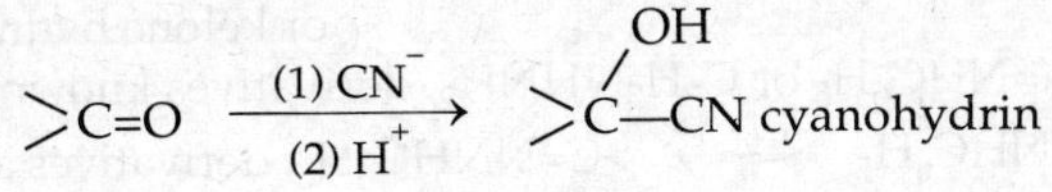

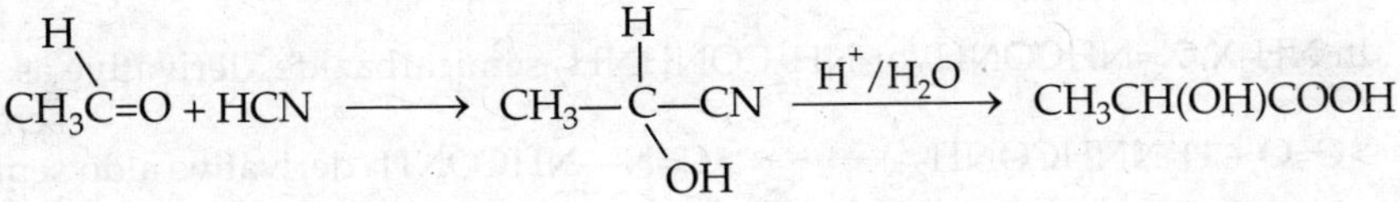

$$>C=O \xrightarrow[\text{(2) H}^+]{\text{(1) CN}^-} >\overset{\overset{\displaystyle OH}{|}}{C}-CN \text{ cyanohydrin}$$

$$\underset{H}{\overset{}{CH_3C}}=O + HCN \longrightarrow CH_3-\overset{\overset{\displaystyle H}{|}}{\underset{\underset{\displaystyle OH}{|}}{C}}-CN \xrightarrow{H^+/H_2O} CH_3CH(OH)COOH$$

lactic acid

Mechanism:

i. $HCN + H_2O \longrightarrow CN^- + H^+{}_3O$

ii. 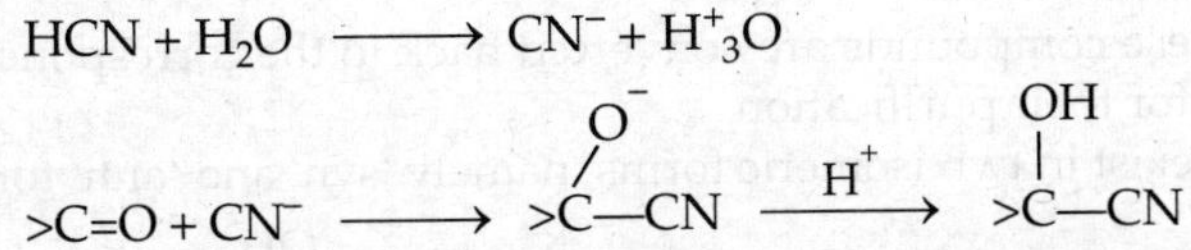 $>C=O + CN^- \longrightarrow >\overset{\overset{\displaystyle O^-}{}}{C}-CN \xrightarrow{H^+} >\overset{\overset{\displaystyle OH}{|}}{C}-CN$

These cynohydrins are converted into α-hydroxy acids on hydrolysis. Thus acetaldehyde cyanohydrin on hydrolysis forms lactic acid.

d. Addition of ammonia-All other aldehydes react with ammonia to form addition compounds, except HCHO, which forms a condensation product

$$\underset{H}{\overset{R}{>}}C=C + NH_3 \longrightarrow \underset{H}{\overset{R}{>}}C\underset{N^+H_2-H}{\overset{O^-}{<}} \longrightarrow \underset{H}{\overset{R}{>}}C\underset{NH_2}{\overset{OH}{<}}$$

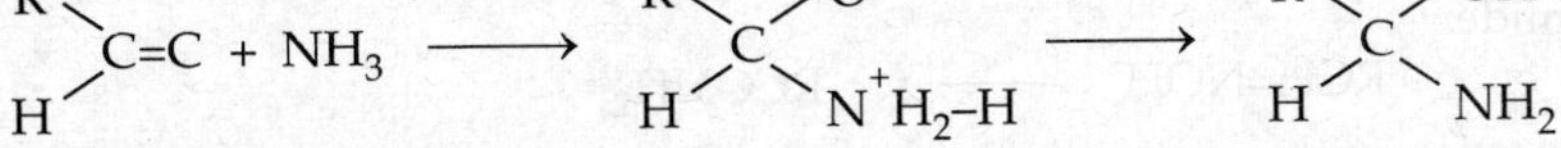

α-Hydroxy acid

HCHO forms a condensation product, when it reacts with NH_3–

$$6HCHO + 4NH_3 \longrightarrow (CH_2)_6N_4 + 6H_2O$$

Hexa methylene tetra amine
or urotropine

6. Addition compounds are formed by condensation, followed by loss of a molecule of water.
   a. Reaction with alcohols-formation of acetals-Aldehydes and ketones react with alcohols, in presence of dry HCl gas to form acetals

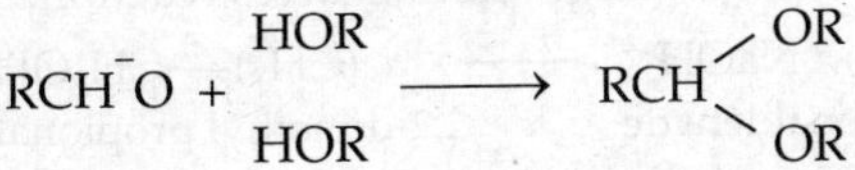

$$RCH\overset{-}{O} + \underset{HOR}{\overset{HOR}{}} \longrightarrow RCH\underset{OR}{\overset{OR}{<}}$$

acetal

   b. Reaction with ammonia derivatives-Aldehydes and ketones react with ammonia derivatives, to form initially an addition compound, which loses a molecule of water to give a condensation product.

$$>C=O + HNHX \longrightarrow >C\underset{\underset{\underset{\displaystyle H}{|}}{N^+HX}}{\overset{O^-}{<}} \longrightarrow >C\underset{H-N-X}{\overset{OH}{<}} \xrightarrow{-H_2O} >C=NX$$

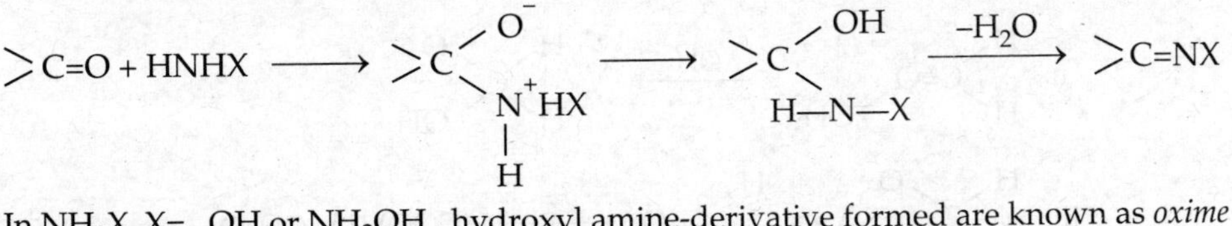

In NH_2X, X= OH or NH_2OH , hydroxyl amine-derivative formed are known as *oxime*

$$>C=O + H_2N-X \longrightarrow >C=NOH \text{ either aldoxime or ketoxime}$$

In NH_2X $X=NH_2$ or NH_2NH_2, hydrazine derivatives are known as aldo or

$>C=O + NH_2NH_2 \longrightarrow >C=NNH_2$ either aldehyde-hydrazone
or ketone hydrazone

In NH_2—X, $X=NHC_6H_5$ or $C_6H_5NHNH_2$, derivatives known as phenyl hydrazone

$>C=O + H_2NNHC_6H_5 \longrightarrow >C=NNHC_6H_5$, derivatives as aldo or ketophenyl
hydrazone

In NH_2X, $X=NHCONH_2$ or $NH_2CONHNH_2$ semicarbazide, derivative as
semicarbazone

$>C=O + H_2NNHCONH_2 \longrightarrow >C=N—NHCONH_2$ derivative aldo semicarbazone
Keto semicarbazone

All these derivatives are colourless crystalline solids with a sharp m.p. which can be used for their identification.

On hydrolysis these compounds are converted back to the corresponding aldehydes. This can be used for their purification

Aldehyde oxime exist in two isomeric forms, namely 'syn' and 'anti' forms.

'syn' form   'anti' form

On treating oxime with acidic reagents like conc. H_2SO_4, P_2O_5, PCl_5, acid chloride, CH_3COCl and acetic anhydride, $(CH_3CO)_2O$, these oxime rearrange to form acid amide.

$$RCH=NOH \longrightarrow RCONH_2$$

c. Reaction with caustic alkali solution—Cannizzaro's reaction

Formaldehyde, HCHO, reacts with conc. alkali, NaOH or KOH, formaldehyde undergoes self oxidation–reduction or disproportionation.

$$2\,HCHO + NaOH \longrightarrow CH_3OH + HCOOH$$
methanol   formic acid

In this reaction, one molecule of HCHO is oxidise to formic acid while the other molecule is reduced to methanol.

This reaction is termed Cannizzaro's reaction. All those aldehydes which do not contain α-hydrogen atom, undergo Cannizzaro's reaction.

$(CH_3)_3C$—CHO + NaOH $\longrightarrow$ $(CH_3)_3$—CH_2OH + $(CH_3)_3C$—COOH
2,2-dimethyl propionaldehyde   2,2-dimethyl propional-1   2,2-dimethyl propanoic acid

Benzaldehyde and other aromatic nuclear substituted aldehyde also undergo Cannizzaro's reaction.

$$2C_6H_5CHO + NaOH \longrightarrow C_6H_5CH_2OH + C_6H_5COOH$$

Mechanism:

i. NaOH $\longrightarrow$ $Na^+ + OH^-$

ii.

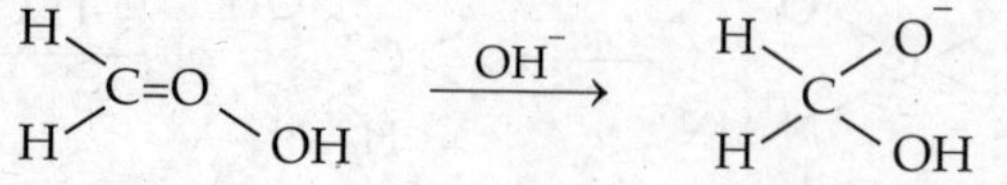

iii. 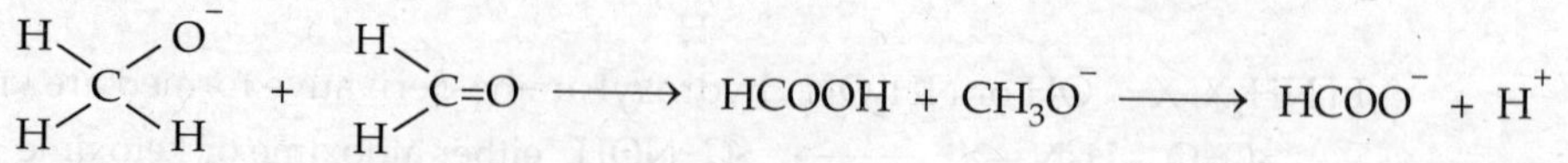 $\longrightarrow$ $HCOOH + CH_3O^- \longrightarrow HCOO^- + H^+$

iv. $CH_3O^- + H + \longrightarrow CH_3OH$

As both formaldehye and benzaldehyde undergo Cannizzaro's reaction individually, a mixture of the two when taken together should also undergo this type of oxidation-reduction reaction.

$$HCHO + C_6H_5CHO + NaOH \longrightarrow HCOOH + C_6H_5CH_2OH$$

This type of reaction with mixed nonhydrogen containing aldehyde is termed Cross-Cannizzaro's reaction. If one of the aldehyde happen to be formaldehyde, it is always oxidised to form formic acid and the other aldehyde is reduced to the corresponding alcohol.

As Cannizzaro's reaction involves two molecule of aldehydes, same or different, each providing one aldehydic groups, a compound containing two aldehydic groups (one in the form of aldehydic form and the other in the form of a keto group or a keto aldehyde, would also undergo this type of self oxidative-reductive reaction or internal Cannizzaro's reaction. Eg

i.
$$\begin{matrix} CHO \\ | \\ CHO \end{matrix} \xrightarrow{NaOH} \begin{matrix} CH_2OH \\ | \\ COOH \end{matrix}$$
glyoxal   glycolic acid

ii. $CH_3COCHO + NaOH \longrightarrow CH_3CH(OH)COOH$
pyruvic aldehyde   lactic acid

iii. $C_6H_5COCHO + NaOH \longrightarrow C_6H_5CH(OH)COOH$
phenyl glyoxal   mandelic acid

One of the keto group happen to be in the form of an aldehydic group and the other is in the form of a keto group, it is the keto group is always reduced and the aldehydic group is oxidise to a carboxylic hydrogen atom, in the presence of Exception-Iso butyraldehyde, $(CH_3)_2CH—CHO$, an α-hydrogen containing aldehyde undergoes Cannizzaro's reaction. With alkali it forms corresponding alcohol and acid.

$$(CH_3)_2CHCHO + NaOH \longrightarrow (CH_3)_2CHCH_2OH + (CH_3)_2CH—COOH$$
2- methyl-propanol-1   2-methylpropanoic acid

d. **Aldol condensation:** Aldehyde containing hydrogen condense with other molecules of aldehyde, same or different, also containing hydrogen atoms, in the presence of a very dilute alkali, and forms a product that contains both an aldehydic group as well as alcoholic group. This reaction is called as aldol condensation.

$$CH_3CHO + CH_3CHO \longrightarrow CH_3CH(OH)CH_2CHO$$
very dil alkali   aldol

Mechanism:

i. $NaOH \longrightarrow Na^+ + OH^-$

ii. $OH^- + H—CH_2CHO \longrightarrow {}^-CH_2CHO + H_2O$

iii. $\begin{matrix} CH_3C=O + {}^-CH_2CHO \longrightarrow CH_3CH—CH_2CHO \\ \quad\quad | \quad\quad\quad\quad\quad\quad\quad\quad\quad\quad | \\ \quad\quad H \quad\quad\quad\quad\quad\quad\quad\quad\quad\quad OH \end{matrix}$

iv. $\begin{matrix} CH_3CHCH_2CHO \xrightarrow{\Delta} CH_3CH=CH—CHO + H_2O \\ | \\ OH \quad\quad\quad\quad\quad\quad crotonaldehyde \end{matrix}$

When two different hydrogen containing aldehydes are subjected to undergo aldol condensation, four different products are formed.

Taking a mixture of acetaldehyde and propionaldehyde, the formation of four different product may be explained.

i. $CH_3CHO + CH_3CHO \longrightarrow CH_3CH(OH)CH_2CHO$
   acetaldehyde   aldol

ii. $CH_3CH_2CHO + CH_2CHO \longrightarrow CH_3CH(OH)CH_2CHO$
   propionaldehyde   acetaldehyde   3-hydoxy-2-methyl butanal

iii. $CH_3CH_2CHO + CH_2CHO \longrightarrow CH_3CH_2CH(OH)CHCHO$
               CH_3 CH_3
      propionaldehyde   2-hydrox 3-mehyl pentanal

iv. $CH_3CHO + CH_2CHO \longrightarrow CH_3CH(OH)CHCHO$
                     CH_3 CH_3
   acetaldehyde propionaldehyde   3-hydroxy-2-methylbutanal

REASONING TYPE QUESTIONS (RTQs)

(Group-A)

1. Both aldehydes and ketones contain a keto, group C=O in the compounds, yet aldehydes show reducing properties but ketones do not. Explain.
2. Both aldehydes and ketones undergo nucleophilic addition reactions, but these compounds do not show any electrophilic addition properties. Explain.
3. Alkenes can be converted into aldehydes or ketones. Discuss the processes used in these conversions.
4. Alkenes react with halogen acids to form addition compounds, but keto compounds, C=O, fail to form addition compounds with these halogen acids. Explain.
5. Discuss the basic difference between a compound containing a keto group and a compound containing an olefinic double bond.
6. Describe the following in decreasing order of reactivity of the following keto compounds, in nucleophilic addition, in terms of steric and electronic features.

   $HCHO > CH_3—CHO > C_6H_5CHO > CH_3—CO—CH_3 > C_6H_5COCH_3$

7. Explain, why chloral, CCl_3CHO, exists as chloral hydrate normally.
8. Benzaldehyde gives a positive test with Tollen's reagent, but fails with Fehling's solution and Benedict's solution. Why? Explain.
9. A pure and dry hydrogen cyanide HCN fails to react with aldehydes. Why? Explain.
10. Name the reagent, that is used to oxidise a primary alcohol to aldehyde, but this reagent does not oxidise aldehyde to a carboxylic acid. What are other methods used to convert a primary alcohol to an aldehyde? Discuss.
11. Formaldehyde differs from other aldehydes, in its reaction with ammonia, and aqueous NaOH solution. Explain.
12. Discuss 'Cannizzaro's and 'Cross Cannizzaro's reaction.
13. Acetaldehyde adds on a molecule of HCN, to form an addition compound, which on hydrolysis give another substance that exists in various isomeric forms. Discuss the chemistry involved.
14. Discuss the addition of HCl to the molecule of acrolein, $CH_2=CH=CHO$.
15. Write short notes on the following reactions.
    (a) Rosenmund reaction   (b) Benzoin condensation   (c) Perkin's reaction.
16. Which type of compound undergoes aldol condensation? Also discuss mixed aldol condensation.
17. Write short notes on the following reactions
    (i) Tischenko reaction   (ii) Knoevenegel's reaction   (iii) Claisen-Schmidt reaction.

18. The bond energy of the carbon-oxygen double bond (179 kcal/mole) is higher than that of carbon-carbon double bond (145.8 kcal/mole). Explain.

ANSWERS

1. Both aldehyde and ketones contain keto group, C=O, however the keto group in the case of a ketone is joined to two carbon atoms, whereas in the case of an aldehyde the keto group is joined to only one carbon atom and the hydrogen atom.

   The C—H bond in an aldehyde is easily oxidised to a C—OH group, but the bond between carbon to carbon is not oxidised at all, with the result, aldehydes show reducing properties, but ketones do not.

$$\underset{\text{H}}{\overset{\text{H}}{\underset{|}{\text{R—C=O}}}} \xrightarrow{\text{(O)}} \underset{\text{OH}}{\overset{\text{OH}}{\underset{|}{\text{R—C=O}}}}$$

2. The aldehyde and ketone undergo nucleophilic addition, reaction as the—C=O bond is polarized, on account or higher electronegativity of the oxygen atom. The pair of electrons shifts towards oxygen, carbon assuming a positive charge and oxygen acquiring a negative charge.

   Polarised $\qquad \overset{+}{-}\text{C} = \overset{-}{\text{O}} \longrightarrow \text{C} = \text{O}$

   Electronegativity $\qquad$ 2.5 $\quad$ 3.5

   A nucleophile adds to the positive charged carbon, shifts the pair of electrons to the oxygen atom. For example, addition of HCN

$$\text{HCN} \longrightarrow \text{H}^+ + \text{CN}^-$$

$$\overset{+}{>}\text{C} = \overset{-}{\text{O}} + \text{—CN} \longrightarrow >\text{C} = \text{O} \longrightarrow >\text{C—O—H}$$
$$\text{cyanohydrin}$$

   If, an electrophile adds to the negatively charged oxygen atom, the species formed, corresponds to being analogous to gem-dihydroxyl compound, it decomposes back into original substance.

   Therefore, the keto compounds undergo a nucleophilic addition reactions.

3. The processes used to prepare aldehydes and ketones from alkenes:

   a. Oxo process – The process is used to prepare aldehydes from alkenes. In this process the alkene is treated with carbon monoxide, CO, and hydrogen in the presence of cobalt hydrogen tetra carbonyl, $Co(CO)_4$, catalyst, at a high temperature and pressure.

$$\underset{\text{alkene}}{\text{R—CH=CH}_2} + \text{CO} + \text{H}_2 \xrightarrow{\text{CoH(CO)}_4} \underset{\text{aldehyde}}{\text{R—CH}_2\text{—CH}_2\text{—CHO}}$$

   b. Waker process – This method can be used for the preparation for both aldehydes and ketones. In this process, an alkene is treated with an acidified aqueous solution of palladium chloride and cupric chloride.

$$\underset{\text{ethene}}{\text{CH}_2\text{=CH}_2} + \text{PdCl}_2 + \text{H}_2\text{O} \longrightarrow \underset{\text{acetaldehyde}}{\text{CH}_3\text{—CHO}} + \text{Pd} + \text{HCl}$$

   Palladium is then converted into palladium chloride again by reacting with HCl and $CuCl_2$.

$$\text{Pd} + 2\text{HCl} \longrightarrow \text{PdCl}_2 + \text{H}_2$$
$$\text{Pd} + \text{CuCl}_2 \longrightarrow \text{PdCl}_2 + \text{CuCl}$$

Acetone may be prepared from propene.

$$CH_3—CH=CH_2 + PdCl_2 + H_2O \longrightarrow CH_3—CO—CH_3 + Pd + 2HCl$$

propene   acetone

$$Pd + 2HCl \longrightarrow PdCl_2 + H_2$$

4. Alkenes undergo electrophilic addition reaction with hydrogen halides, HX, to form addition compounds.

$$R—CH=CH_2 + \overset{+}{H} \longrightarrow R—\overset{}{CH}—CH_3 \xrightarrow{+X^-} R—CHX—CH_3$$

Keto compounds undergo nucleophilic addition reactions. The compound formed on the addition of HX, to the keto compound, is analogous to the gem-dihydroxy compound; therefore, it breaks up to form the starting material again.

$$C=O \xrightarrow{H^+} C=O—H \xrightarrow{X^-} C\overset{OH}{\underset{X}{<}} \longrightarrow C=O + HX$$

5. Alkenes as well as keto compounds both form addition products, when they react with the reactant.

   The basic difference between an alkene and a keto compound are–

   i. In the case of alkenes, electromeric change takes place at the time of the attack, by an electrophile.

$$\overset{}{C}=\overset{}{C} \xrightarrow{HX} -\overset{|}{\underset{H}{C}}-\overset{|}{\underset{X}{C}}-$$

   ii. In keto compounds, the C=O group is already polarized, facilitating a nucleophilic addition.

6. The order of reactivity of the keto compounds is in the given order, on account of the electron donating or electron repelling nature of the alkyl groups, attached to the keto group. The alkyl groups, on account of their electron donating nature, reduce electron deficiency on the carbon atom, of the keto group, making it less electropositive, thus decreasing the velocity of the attacking nucleophile.

   In formaldehyde, there are no alkyl groups, attached to the keto group; therefore it is most reactive towards a nucleophilic attack. Next comes the acetaldehyde, that contains one alkyl group, therefore the compound is less reactive towards nucleophilic attack. In the same way, acetone has two alkyl groups, attached the keto group, in comparison, it will be much less reactive towards a nucleophilic attack.

$$\underset{\text{formaldehyde}}{\overset{H}{\underset{H}{>}}C=O} > \underset{\text{acetaldehyde}}{\overset{CH_3}{\underset{H}{>}}C=O} > \underset{\text{benzaldehyde}}{\overset{C_6H_5}{\underset{H}{>}}C=O} > \underset{\text{acetone}}{\overset{CH_3}{\underset{CH_3}{>}}C=O} > \underset{\text{acetophenone}}{\overset{C_6H_5}{\underset{CH_3}{>}}C=O}$$

7. In the molecule of chloral, the inductive effect of —CCl_3 group, the carbon atom of the keto group acquires a little +ve charge, due to the inductive effect of the chlorine atoms. The positive charge on the carbon atom, prevents the release of —OH ion from the carbon atom, and therefore no water molecule is eliminated.

$$Cl \leftarrow \overset{Cl}{\underset{Cl}{\overset{\delta+}{C}}} \overset{\delta+}{—} \overset{\delta-}{C} = O \xrightarrow{H_2O} Cl \leftarrow \overset{Cl}{\underset{Cl}{C}} \overset{OH}{\underset{OH}{\overset{|}{—}}} H$$

8. Benzaldehyde reduces Tollen's reagent but it neither reduces Fehling solution nor Benedict's solution.

$$C_6H_5CHO \ + \ Ag_2O \ \longrightarrow \ C_6H_5COOH \ + \ 2Ag$$

Tollen's   grey precipitate of
reagent   metallic silver

The reduction of Fehling solution does not take place conclusively. The Fehling and Benedict's solution are weaker reducing agents, as compared to Tollen's reagent, they oxidise aliphatic aldehydes but not aromatic aldehydes.

9. As HCN is a weak acid, it does not ionize on its own. In the presence of moisture or a base, HCN ionizes to release a cyanide ion, which is then available for the reaction

$$\overline{B}: \ + \ H{-}C{\equiv}N \ \longrightarrow \ B{-}H \ + \ C{\equiv}\overline{N}$$

$$\overset{\delta+}{>}C = \overset{\delta-}{O} \ + \ >C{\equiv}N \ \longrightarrow \ >\underset{\underset{H^+}{|}}{C}{-}O^- \ \longrightarrow \ >\underset{\underset{CN}{|}}{C}{-}OH$$

CN   cyanohydrin

10. Primary alcohols may be oxidised to aldehydes, by PCC, (pyridinium chloro chromate), $C_5H_6NCrO_3Cl$, (pyridine + CrO_3 + HCl), in dichloromethane. The aldehyde formed, is not oxidised further.

$$RCH_2OH \ \xrightarrow{\text{PCC/CH}_2\text{Cl}_2} \ RCHO \quad \text{(No further oxidation)}$$

primary alcohol   aldehyde

11. a. Formaldehyde reacts with ammonia, to form a condensation product, known as hexa-methylene tetraamine, $(CH_2)_6N_4$, or urotropine.

$$6HCHO + 4NH_3 \ \longrightarrow$$

hexamethylene tetraamine

12. All aromatic aldehydes react with aqueous caustic alkali to give a mixture of benzoic acid and benzyl alcohol. This reaction is known as Cannizzaro's reaction. One molecule of the aldehyde is oxidised while the other molecule is reduced. The reaction is given by those aldehydes that, do not contain an α-hydrogen atom in the molecule.

$$2C_6H_5CHO + NaOH \ \longrightarrow \ C_6H_5COOH + C_6H_5CH_2{-}OH$$

benzaldehyde   benzoic acid   benzyl alcohol

Among aliphatic aldehydes, formaldehyde undergoes Cannizzaro's reaction.

$$2HCHO + NaOH \ \longrightarrow \ HCOOH + CH_3{-}OH$$

formaldehyde   formic acid   methanol

When a mixture of benzaldehyde and formaldehyde is treated with caustic alkali, benzal-dehyde is reduced to give benzyl alcohol, and formaldehyde is oxidised, to for formic acid. This reaction is termed 'cross Cannizzaro's' reaction.

$$C_6H_5CHO + HCHO + NaOH \ \longrightarrow \ C_6H_5CH_2OH + HCOOH$$

benzaldehyde   formaldehyde   benzyl alcohol   formic acid

In case, one of the aldehyde happen to be formaldehyde in the 'cross Cannizzaro's' reaction, formaldehyde is always oxidised to formic acid and the other aldehyde is reduced, to form a corresponding alcohol.

13. Acetaldehyde reacts with HCN to form acetaldehyde cyanohydrin. On hydrolysis, acetaldehyde cyanohydrin is converted lactic acid.

$$CH_3-CHO + HCN \longrightarrow CH_3-\overset{\overset{\displaystyle OH}{|}}{\underset{\underset{\displaystyle CN}{|}}{C}}-H \xrightarrow{H_2O} CH_3-CH(OH)COOH$$

$$\underset{C}{|}$$

acetaldehyde   lactic acid
cyanohydrin

As lactic acid contains one asymmetric carbon atom, the compound shows optical activity. Lactic acid exists in two optically isomeric forms. Lactic acid exist in dextro rotatory and a laevo rotatory form.

However, as lactic acid is synthesized in the laboratory, there is obtained an optically inactive form, i.e. a racemic mixture.

A racemic mixture, contains an equal amount of both, dextro and laevo rotatory forms, therefore the racemic form is optically inactive. However, it can be resolved. In other words the racemic form can be separated in dextro and laevo forms.

14. a. Rosenmund reaction: Acid chlorides, aliphatic as well as aromatic acid chlorides, R—COCl or ArCOCl, both, are reduced by hydrogen, in the presence of Pd, mounted on $BaSO_4$, as a catalyst. The catalyst is poisoned by mixing it with sulphur, as to check the further reduction of the aldehyde, to form alcohol.

$$RCOCl + H_2 \xrightarrow[\text{sulphur}]{Pd/BaSO_4} RCHO + HCl$$

acid chloride   aldehyde

b. Benzoin condensation: Aromatic aldehydes form benzoin, on treatment with aqueous NACN or KCN.

In this reaction, two molecules of the aromatic aldehyde, react with each other, in the presence of KCN or NaCN, to form an α-hydroxyl ketone, known as benzoin.

$$C_6H_5\overset{\overset{\displaystyle O}{\|}}{\underset{\underset{\displaystyle H}{|}}{C}} + \overset{\overset{\displaystyle O}{\|}}{\underset{\underset{\displaystyle H}{|}}{C}}-C_6H_5- \xrightarrow[H_2O + C_2H_5OH]{KCN} C_6H_5-\overset{\overset{\displaystyle H}{|}}{\underset{\underset{\displaystyle OH}{|}}{C}}-\overset{\overset{\displaystyle O}{\|}}{C}-C_6H_5$$

benzaldehyde   benzoin

Aliphatic aldehydes do not undergo such type of condensation.

c. Perkin's reaction: On heating, an aromatic aldehyde with anhydride of an aliphatic acid in the presence of the sodium salt of that acid, there is formed an α- β-unsaturated acid.

$$C_6H_5-CHO + (CH_3CO)_2O \xrightarrow[\text{sodium acetate}]{CH_3COONa} C_6H_5-CH=CH-COOH$$

benzaldehyde   acetic anhydride   cinnamic acid

Mechanism:

i. First step is the formation of a carbanion

$$CH_3COONa \rightleftharpoons CH_3CO\bar{O} + \overset{+}{Na}$$

acetate carbanion

$$CH_3CO\bar{O} + H-CH_2CO-O-CO-CH_3 \longrightarrow CH_2-CO-O-CO-CH_3 + \bar{C}H_3COOH$$

acetate carbanion   carbanion

ii. Carbanion then reacts with benzaldehyde

$$C_6H_5-\overset{\overset{O}{\|}}{C}-H + \bar{C}H_2-CO-O-CO-CH_3 \longrightarrow C_6H_5-\overset{\overset{O^-}{|}}{\underset{H}{C}}-CH_2-CO-O-CO-CH_3$$

$$\longrightarrow C_6H_5-\overset{\overset{H^+}{|}}{\underset{H}{C}}-\overset{\overset{OH}{|}}{C}H_2-CO-O-CO-CH_3 \longrightarrow C_6H_5-CH=CH-CO-O-CO-CH_3$$

iii. Hydrolysis: Formation of cinnamic acid

$$C_6H_5-CH=CH-CO-O-CO-CH_3 \xrightarrow{\ H_2O\ } C_6H_5-CH=CH-COOH + CH_3COOH$$
$$\qquad\qquad\qquad\qquad\qquad\qquad\qquad\qquad \text{cinnamic acid}$$

15. Aldol condensation: Two molecules of acetaldehyde condense together, in the presence of dilute alkali, potassium carbonate or HCl, to give a compound, known as aldol. Aldol contains both an aldehyde as well as a hydroxyl group.

$$CH_3-\overset{\overset{O}{\|}}{C}-H + H-CH_2-CHO \xrightarrow{\text{dilute alkali}} CH_3-\overset{\overset{OH}{|}}{\underset{H}{C}}-CH_2-CHO$$
$$\qquad\quad \text{acetaldehyde} \qquad\qquad\qquad\qquad\qquad\qquad \text{aldol}$$

On heating aldol loses a molecule of water, to form an unsaturated aldehyde.

$$CH_3-\overset{\overset{OH}{|}}{C}H-\overset{\overset{H}{|}}{C}H\,CHO \xrightarrow{\text{heat}} CH_3-CH=CH-CHO + H_2O$$
$$\qquad\quad \text{aldol} \qquad\qquad\qquad\qquad\qquad \text{crotonaldehyde}$$

Aldol condensation is a general reaction, and it takes place in between two aldehydes or ketones, which contains at least one hydrogen, attached to the carbon, α- to the —CO group.

Acetone condensation with another molecule of acetone, in the presence of Ba(OH)$_2$, to form diacetonyl alcohol. On heating it loses a water molecule to form mesityl oxide.

$$\overset{\overset{CH_3}{|}}{\underset{\underset{CH_3}{|}}{C}}=O + HCH_2\,CO-CH_3 \xrightarrow{Ba(OH)_2} \overset{\overset{CH_3}{|}}{\underset{\underset{CH_3}{|}}{C}}-CH_2-CO-CH_3 \xrightarrow[-H_2O]{\text{Heating}} \overset{\overset{CH_3}{|}}{\underset{\underset{CH_3}{|}}{C}}=CH-CO-CH_3$$
$$\text{acetone} \qquad\qquad\qquad\qquad\qquad \text{diacetonyl alcohol} \qquad\qquad\qquad \text{mesityloxide}$$

16. a. Tischenko reaction: Benzaldehyde on heating with aluminium alkoxide (ethoxide or isopropoxide) and anhydrous AlCl$_3$, undergoes an intramolecular oxidation and reduction, similar to Cannizzaro's reaction, instead of producing an acid and an alcohol, the two react to produce an ester, namely benzyl benzoate.

$$2C_6H_5CHO \xrightarrow{Al(OC_2H_5)_3} C_6H_5-CH_2-OCOC_6H_5$$
$$\text{Benzaldehyde} \qquad\qquad\qquad \text{benzyl benzoate}$$

Acetaldehyde also undergoes Tischenko's reaction and forms methyl propionate.

$$2CH_3-CHO \xrightarrow{Al(OC_2H_5)_3} CH_3-CH_2-OCOCH_3$$
$$\text{Acetaldehyde} \qquad\qquad\qquad \text{ethyl acetate}$$

b. Knoevenegel's reaction: Benzaldehyde condenses with the compounds containing a reactive methylene group, like malonic ester, in the presence of either pyridine or ethanolic ammonia, to form cinnamic acid.

$$C_6H_5CH(O+H_2)C\text{—}(COOC_2H_5)_2 \xrightarrow{\text{pyridine}} C_6H_5\text{—}CH\text{=}CH\text{—}(COOC_2H_5)_2 \xrightarrow{H^+/H_2O}$$
$$\underset{\text{malonic ester}}{} \qquad\qquad \underset{\text{benzylidine malonic ester}}{}$$

$$C_6H_5\text{—}CH\text{=}CH(COOC_2H_5)_2 \longrightarrow C_6H_5\text{—}CH\text{=}CH(COOH)_2$$

On hydrolysis, it is converted into the corresponding dibasic acid, being a gem-dibasic acid, loses a molecule of CO_2, on heating to form cinnamic acid finally.

$$\underset{\text{benzylidine malonic acid}}{C_6H_5CH\text{=}CH(COOH)_2} \xrightarrow{\rho\backslash\text{--}CO_2} \underset{\text{cinnamic acid}}{C_6H_5CH\text{=}CH\text{—}COOH}$$

c. Claisen–Schmidt reaction or Claisen's reaction: Benzaldehyde condenses with aldehydes or ketones, containing α-hydrogen atom, in the presence of dilute alkali, to form benzylidene derivatives.

$$\underset{\text{acetaldehyde}}{C_6H_5CHO + H_2CHCHO} \xrightarrow{\text{NaOH}} \underset{\text{cinnamaldehyde}}{C_6H_5\text{—}CH\text{=}CH\text{—}CHO + H_2O}$$

$$\underset{\text{acetone}}{C_6H_5CHO + H_2CH\text{—}CO\,CH_3} \xrightarrow{\text{NaOH}} \underset{\text{benzylidene acetone}}{C_6H_5\text{—}CH\text{=}CH\text{—}CO\text{—}CH_3}$$

17. The bond energy of the carbon-oxygen double bond, (—C=O—179 kcal/mol), is higher than the that of carbon to carbon double bond (—C=C—145.8 kcal) double bond, because of the type of the bonds between the atoms joined.

$$-\overset{..}{\underset{..}{C}}\text{=}O \longrightarrow -\overset{..}{\underset{..}{C}}\text{—}O^- :$$

A C=O bond can be represented by the following resonating structures.
In contrast to the carbon-oxygen double bond that can exist in a **number of resonating structures, the carbon-carbon double bond does not exist in resonating structures. In case, number of the resonating structures can be taken as a qualitative measure of stability, the C=O bond is more stable.**
Whereas the carbon-oxygen bond is polar and the carbon-carbon bond is nonpolar. A polar bond requires more energy to break. The carbon to carbon is symmetrical, may require less energy to break.

MULTIPLE CHOICE QUESTIONS

(Group-B)

1. Hybridisation of carbon in carbonyl group is
   (a) sp (b) sp^2
   (c) sp^3 (d) none
2. Aldehyde and ketones form addition products with
   (a) HCN (b) $NaHSO_3$
   (c) both (a) and (b) (d) none
3. Carbonyl group undergoes
   (a) electrophilic addition reactions (b) nucleophilic addition reactions
   (c) both (a) and (b) (d) none
4. Which carbonyl compound does not undergo aldol condensation?
   (a) HCHO (b) CH_3CHO
   (c) CH_3CH_2CHO (d) CH_3COCH_3

5. Most reactive compound carbonyl compound is
   (a) HCHO
   (b) CH_3CHO
   (c) CH_3COCH_3
   (d) C_2H_5CHO
6. Which will show haloform reaction?
   (a) HCHO
   (b) acetone
   (c) CH_3—O—CH_3
   (d) CH_3CH_2Cl
7. Aldol condensation between the following compounds, followed by dehydration, will give, methyl vinyl ketone
   (a) methanal and ethanal
   (b) two molecule of HCHO
   (c) methanal and propanone
   (d) two molecules ofethanol.
8. Acetaldehyde reacts with HCN followed by hydrolysis forms a compound, which shows
   (a) optical isomerism
   (b) geometrical isomerism
   (c) metamerism
   (d) diastereoisomerism
9. Hybridisation of carbon in carbonyl group is
   (a) sp
   (b) sp^2
   (c) sp^3
   (d) none
10. Aldehyde and ketones do not form addition products with
    (a) HCN
    (b) $NaHSO_3$
    (c) both (a) and (b)
    (d) none
11. Carbonyl group undergoes
    (a) electrophilic addition reactions
    (b) nucleophilic addition reactions
    (c) both (a) and (b)
    (d) none
12. Which carbonyl compound undergoes aldol condensation
    (a) propanal chloride
    (b) benzaldehyde
    (c) CH_3CH_2CHO
    (d) CH_3COCH_3
13. The reagent which does not react with both acetone and benzaldehyde
    (a) Sodium hydrogen sulphite
    (b) Phenyl hydrazine
    (c) Grignard reagent
    (d) Fehling's Solution
14. Through which of the following reactions number of carbon atoms can be increased in the chain?
    (a) Grignard reaction
    (b) Cannizaro's reaction
    (c) Aldol condensation
    (d) HVZ reaction
15. Aldol condensation between the following compounds, followed by dehydration, will give, methyl vinyl ketone
    (a) methanal and ethanal
    (b) two molecule of HCHO
    (c) methanal and propanone
    (d) two molecules of ethanol
16. Acetaldehyde reacts with HCN followed by hydrolysis forms a compound, which shows
    (a) optical isomerism
    (b) geometrical isomerism
    (c) metamerism
    (d) diastereoisomerism
17. Which compound is oxidised to prepare methyl ethyl ketone?
    (a) propanol-2
    (b) butanol-1
    (c) butanol-2
    (d) none
18. Which of the following has most acidic hydrogen?
    (a) acetone
    (b) $(CH_3)_2CH=CH_2$
    (c) $CH_3COCH_2COCH_3$
    (d) $CH_3CO.O.OCH_3$
19. When propyne is treated with aqueous H_2SO_4, in presence of $HgSO_4$, the major product formed is
    (a) propanol
    (b) acetone
    (c) propanal
    (d) propyl hydrogen sulphate

20. Methyl ketones are characterised by
    (a) Tollen's reagent
    (b) Haloform test
    (c) Schiff's reagent
    (d) Benedict's reagent

21. Acetone of saturation with dry HCl gas gives
    (a) diacetonyl alcohol
    (b) mesityl oxide
    (c) mesitylene
    (d) propane

22. Among the following the most susceptible to a nucleophilic attack at carbonyl group is
    (a) CH_3COCl
    (b) CH_3CHO
    (c) CH_3COOCH_3
    (d) $CH_3CO.O.COCH_3$

23. The following reaction is known by the name of
    $$CH_3COCl + H_2 \xrightarrow{Pd/BaSO_4} CH_3CHO$$
    (a) Stephen's reaction
    (b) Rosenmund reaction
    (c) Cannizzaro's reaction
    (d) none

16. CHO $\longrightarrow$ X; X =
    (a) $CH_2OH—COOH$
    (b) $CH_3OH + CH_3OH$
    (c) CH_3COO^-
    (d) $COO^- COO^-$

17. Fehling solution is
    (a) acidified $CuSO_4$ solution
    (b) amm. $CuSO_4$ solution
    (c) $CuSO_4$ + Rochelle salt + NaOH
    (d) one

18. In Cannizzaro's reaction, the intermediate is the best hydride donor is

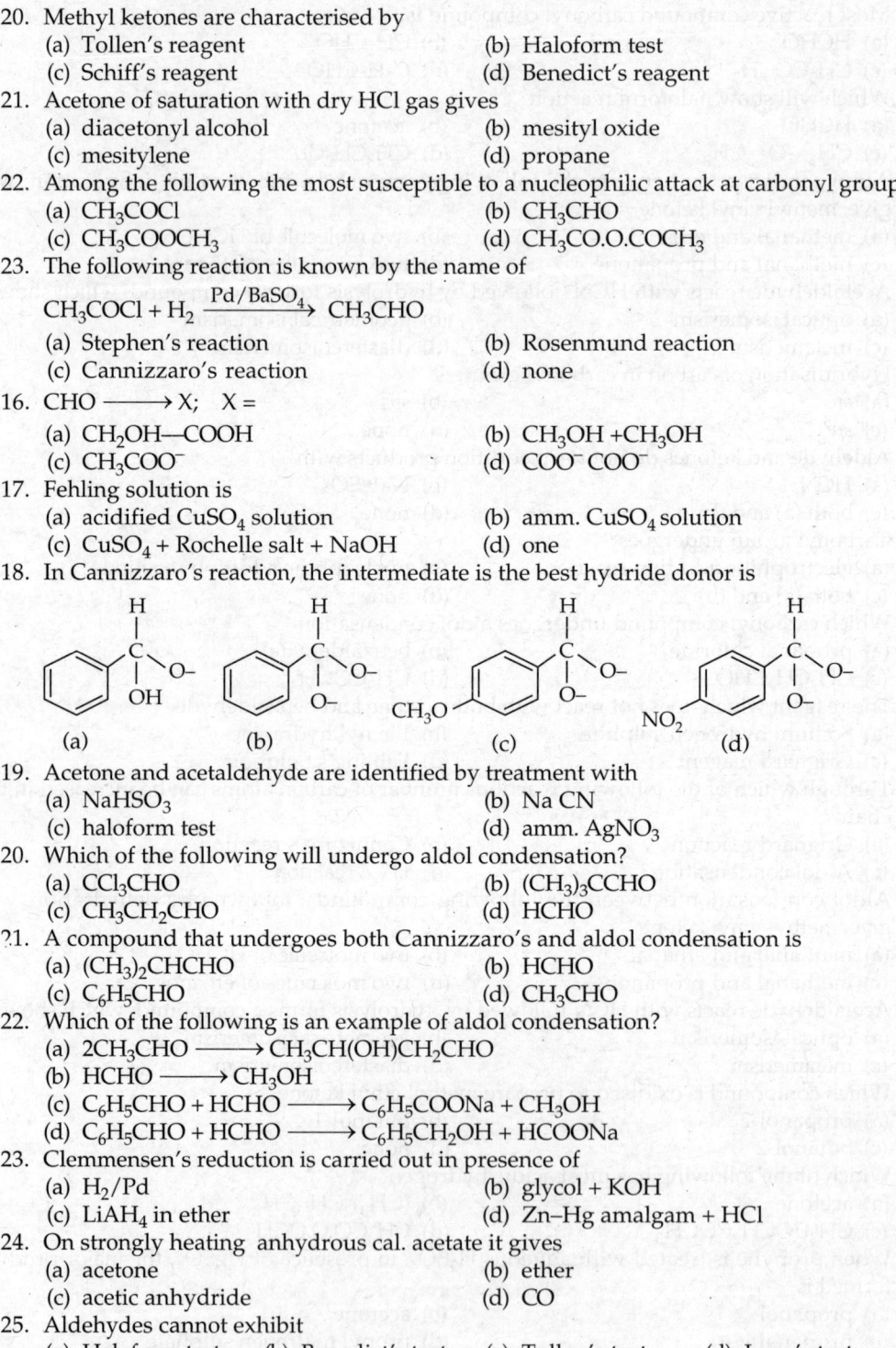

    (a)   (b)   (c)   (d)

19. Acetone and acetaldehyde are identified by treatment with
    (a) $NaHSO_3$
    (b) Na CN
    (c) haloform test
    (d) amm. $AgNO_3$

20. Which of the following will undergo aldol condensation?
    (a) CCl_3CHO
    (b) $(CH_3)_3CCHO$
    (c) CH_3CH_2CHO
    (d) HCHO

21. A compound that undergoes both Cannizzaro's and aldol condensation is
    (a) $(CH_3)_2CHCHO$
    (b) HCHO
    (c) C_6H_5CHO
    (d) CH_3CHO

22. Which of the following is an example of aldol condensation?
    (a) $2CH_3CHO \longrightarrow CH_3CH(OH)CH_2CHO$
    (b) $HCHO \longrightarrow CH_3OH$
    (c) $C_6H_5CHO + HCHO \longrightarrow C_6H_5COONa + CH_3OH$
    (d) $C_6H_5CHO + HCHO \longrightarrow C_6H_5CH_2OH + HCOONa$

23. Clemmensen's reduction is carried out in presence of
    (a) H_2/Pd
    (b) glycol + KOH
    (c) $LiAH_4$ in ether
    (d) Zn–Hg amalgam + HCl

24. On strongly heating anhydrous cal. acetate it gives
    (a) acetone
    (b) ether
    (c) acetic anhydride
    (d) CO

25. Aldehydes cannot exhibit
    (a) Haloform test   (b) Benedict's test   (c) Tollen's test   (d) Luca's test

26. Condensation of chloroform with acetone in the presence of alkali gives
    (a) chloropicrin
    (b) chloretone
    (c) chloral
    (d) chloro acetone

27. Acetaldehyde, on oxidation with SeO_2 gives
    (a) CH_3COOH
    (b) C_2H_5OH
    (c) CHO—CHO
    (d) none

28. Benzaldehyde does not reduce Fehling's solution, because
    (a) of a bulky ring and —CHO group acting as a hinder
    (b) owing to resonance oxidation of benzaldehyde is difficult
    (c) —CHO is present as a cyclic structure
    (d) of all the reason

29. IPUAC, name of acrolein is
    (a) propanal
    (b) prop-2-ene-1-al
    (c) propane-2-ol
    (d) pro-1-ene-2-al

30. The difference between Cannizzaro's reaction and aldol condensation is that
    (a) aldol condensation takes place I the presence of an α-H atom
    (b) Cannizzaro's reaction takes place in the absence of α-H atom
    (c) aldol condensation takes place in the presence of a β-H atom
    (d) none

31. Carbonyl compounds react with phenyl hydrazine to form
    (a) oxime
    (b) phenyl hydrazone
    (c) semicarbazone
    (d) hydrazone

32. Oppenauer oxidation is the reverse process of
    (a) Wolff's-Kishner's reduction
    (b) Rosenmond's reaction
    (c) Clemmensen's reduction
    (d) Meerwein-Pondrof-Verley reduction

33. Which of the following will not undergo aldol condensation?
    (a) CH_3CHO
    (b) propionaldehyde
    (c) benzaldehyde
    (d) CD_3CHO

34. Which of the following compounds will undergo self-condensation, in presence of an alkali?
    (a) C_6H_5CHO
    (b) $CH_2=CH$—CHO
    (c) CH_3CH_2CHO
    (d) none

35. The reaction $C_6H_5CHO + CH_3CHO \longrightarrow C_6H_5CH=CHCHO$, is called
    (a) benzoin condensation
    (b) Claisen's condensation
    (c) aldol condensation
    (d) condensation

36. Acetophenone and acetone can be identified by
    (a) by burning on a spatula
    (b) adding a saturated solution of $NaHSO_3$
    (c) HCN
    (d) all are correct

37. The reagent which can be u to distinguish between acetophenone and benzophenone is
    (a) aqueous $NaHSO_3$
    (b) 2,4-DNP
    (c) Benedict's solution
    (d) $I_2 + Na_2CO_3$

38. The IUPAC name of crotonaldeyde is
    (a) propanal
    (b) but-2-ene-1-al
    (c) but-1-ene-2-al
    (d) butanol

39. The correct order of reactivity of >C=O group, in the given compounds is
    (a) CH_3CHO (A) > CH_3COCH_3 (B) > $C_2H_5COCH_3$ (C)
    (b) C > B > A
    (c) C > A > B
    (d) B > C > A

40. Aldol condensation of acetaldehyde results in the formation of
    (a) $CH_3CO-C(OH)CHCH_3$
    (b) $CH_3CH(OH)CH_2CHO$
    (c) $CH_3CH_2CH(OH)$ CHO
    (d) $CH_3COH=CHCH_2OH$
41. On treating CH_3COCHO with NaOH solution, the compound formed is
    (a) $CH_3CH(OH)CHO$
    (b) $CH_3COCOOH$
    (c) $CH_3CH(OH)COOH$
    (d) C_2H_5COOH
42. Which one of the following aldehyde will not form aldol, on treatment with NaOH?
    (a) CH_3CHO
    (b) CH_3CH_2CHO
    (c) CH_3CCHO
    (d) $C_6H_5CH_2CHO$
43. On treating benzaldehyde with hydroxyl amine hydrochloride an oxime is formed, the oxime exits in how many forms
    (a) 1   (b) 2   (c) 3   (d) none
44. Most suitable reducing agent for reducing of a ketone without effecting double bond is
    (a) $LiAlH_4$
    (b) sodium borohydride
    (c) H_2/Pd
    (d) Na+ethanol
45. Schiff's reagent and Schiff's bases are
    (a) they are same compound
    (b) different compound
    (c) physically same but different chemically
    (d) none
46. Pick up the correct statement from the following
    (a) Secondary alcohols are oxidised to ketones, containing the same number of carbon atom
    (b) tetraethyl lead is a good anti-knock compound
    (d) both aldehydes and ketones have a sp^2 hybridised carbon atom
    (d) acetic acid neither reduces Fehling solution nor Tollen's reagent
    (e) all statements are correct
47. The reaction $CH_3CHO + NH_2-NH_2 \longrightarrow CH_3CH=N-NH_2$–is
    (a) elimination
    (b) addition
    (c) addition-elimination
    (d) none
48. Propanone does not undergo
    (a) oxime formation
    (b) hydrazone formation
    (c) cyanohydrin formation with HCN
    (d) does not reduce Fehling solution
49. When HCHO is heated with NH_3, the compound formed is
    (a) methyl amine
    (b) amino formaldehyde
    (c) hexa methylene tetramine, $(CH_2)_6N_4$
    (d) formalin
50. What is Schiff's reagent?
    (a) magenta solution decolourised by Cl_2 gas
    (b) magenta solution decolourised by H_2SO_3
    (c) decolourised by oxidation with chlorine gas
    (d) reduced by SO_2 gas
51. Mild oxidation of carboxylic acids occurs at —— positions
    (a) α (b) β (c) δ (d) both (a) and (b)
52. Among the give compounds, the most susceptible to nucleophilic attack at the carbonyl group is-
    (a) CH_3COCl
    (b) CH_3CHO
    (c) CH_3COOCH_3
    (d) $(CH_3CO)_2O$
53. The Cannizzaro's reaction is given below
    $$2C_6H_5CHO \xrightarrow{OH-} C_6H_5CH_2OH + C_6H_5COO^-$$

The slowest step is
(a) The attack of OH$^-$ group on the carbonyl group
(b) Transfer of hydride to the carbonyl group
(c) the abstraction of a proton from carboxylic group
(d) The disproportionation of $C_6H_5CH_2O$

54. Which can distinguish between aldehydes and ketones?
(a) Fehling solution
(b) H_2SO_4 solution
(c) $NaHSO_3$
(d) NH_3

55. Stephen's reaction is to prepare aldehydes from
(a) alcohols
(b) alkyl cyanide
(c) alkenes
(d) acid chloride

56. Acrolein, on complete reduction gives
(a) allyl alcohol
(b) propanol
(c) propanal
(d) none

57. Cannizzaro's reaction involves
(a) conversion of an aldehyde into an acid only
(b) Conversion of an aldehyde into an alcohol only
(c) redox system only
(d) aromatic transformation

58. Base catalyzed aldol condensation occurs in
(a) Propionaldehyde
(b) benzaldehyde
(c) 2, dimethyl propionaldehyde
(d) none

59. Which of the following shows electromeric effect?
(a) aldehydes
(b) alkenes
(c) alkyl amines
(d) alkyl halides

60. Acetaldehyde and benzaldehyde differ in their reaction with
(a) $NaHSO_3$
(b) NH_3
(c) PCl_5
(d) $C_6H_5NHNH_2$

61. Aldol condensation of aldehydes and ketones takes place through the formation of
(a) carbene
(b) nucleophile
(c) electrophile
(d) free radical

62. Which of the following combination give tertiary butyl alcohol, when treated with Grignard's reagent?
(a) $CH_3MgBr + CH_3COCH_3$
(b) $C_2H_5MgBr + CH_3COCH_3$
(c) $CH_3MgBr + CH_3CH_2CHO$
(d) $CH_3MgBr + (CH_3)_3C—OH$

63. Benedict's solution provides
(a) Ag^+
(b) Cu_2+
(c) Ba_{2+}
(d) $Li+$

64. Which of the carbonyl compound does not undergo aldol condensation?
(a) HCHO
(b) CH_3CHO
(c) C_2H_5CHO
(d) CH_3COCH_3

65. Which of the following compound is used as a preservative for dead species
(a) formalin a 40% aqueous solution of HCHO
(b) CH_3CHO
(c) CH_3CH_2CHO
(d) none of above

66. The enol form of acetone contains
(a) 9σ-bonds, 1π-bond and 2-lone pairs
(b) 8σ-bonds, 2π-bonds and 2-lone pairs
(c) 10σ-bonds, 1π-bond and 1-lone pair
(d) 9σ-bonds, 2π-bonds and 1-lone pair

67. The number of bonds present in CH_3COCH_3 and in its enol form (of acetone) are
    (a) different
    (b) same
    (c) number of bonds corresponds but number of σ-bonds differ in number of π-bonds
68. What property of aldehydes and ketones, having low mole. wt. account for the magnitude of their boing and melting points?
    (a) the ability to form H-bond, between the molecules
    (b) the ability of the carbonyl group to form H-bonding with other carbonyl groups
    (c) the ability of the polar carbonyl group to attract other polar groups
    (d) the ability of the carbonyl group to attract electrophiles and form a bond
69. A compound 'A' gave +v iodoform test, but did not reduces Tollen's reagent. The compound could be
    (a) CH_3COOH (b) $CH_3CH_2\,CO\,CH_2CH_3$
    (c) $C_6H_5COC_2H_5$ (d) $C_6H_5COCH_3$
70. Appearance of a silver mirror in a test indicate the presence of
    (a) —CHO group   (b) a keto group
    (c) a —OH group   (d) COOH group
71. An aldehyde can be distinguished from a ketone by using
    (a) amm. $AgNO_3$ (b) phenyl hydrazine
    (c) $NaHSO_3$ (d) thionyl chloride
72. Which will give acetone?
    (a) $CH_3CH_2CH_2OH + Na_2Cr_2O_7$
    (b) passing vapours of $CH_3CH(OH)CH_3$ over hot Cu at 300 C
    (c) oxidation of propane by HNO_3
    (d) heating propene with H_2SO_4
73. Write the IUPAC name of the given compound-OHC—COOH
    (a) glyoxalic acid   (b) oxalic aldehyde
    (c) 2-oxo ethanoic acid   (d) 2-oxo-methanoic acid

ANSWERS

(Group-B)

1. (b)   2. (c)   3. (c)   4. (a)   5. (a)   6. (b)   7. (c)   8. (a)
9. (c)   10. (b)   11. (b)   12. (b)   13. (c)   14. (a) and (c)   15. (b)
16. (a) it is known as internal Cannizzaro's reaction
17. (c)   18. (d)   19. (d)   20. (c)   21. (a); it contains an hydrogen
22. (a)   23. (d)   24. (a)   25. (d)   26. (b)   27. (c)   28. (d)   29. (b)
30. (b)   31. (b)   32. (d)   33. (c)   34. (c)   35. (b)   36. (a)   37. (d)
38. (b) $CH_3CH=CH—CHO$ 39.   (a) 40.   (a)
41. (c) internal Cannizzaro's reaction
42. (c)   43. 2 (b)   44. (b)   45. (d)   46. (d)   47. (c)   48. (d)   49. (c)
50. Schiff's reagent is an alcoholic solution, of magenta decolourise by SO_2 gas
51. β-position
52. (c) though it has an H, yet undergoes Cannizzaro's reaction
52. (a)

53. slowest step is the transfer of hydride ion to the carboxylic group

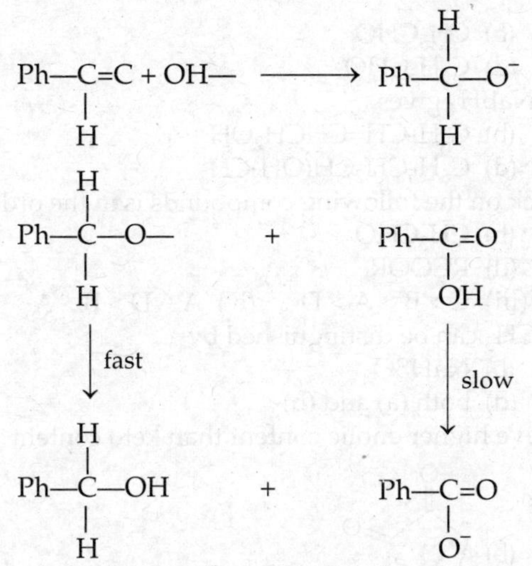

$$Ph-\underset{\underset{H}{|}}{C}=C + OH- \longrightarrow Ph-\underset{\underset{H}{|}}{\overset{\overset{H}{|}}{C}}-O^-$$

$$Ph-\underset{\underset{H}{|}}{\overset{\overset{H}{|}}{C}}-O- \quad + \quad Ph-\underset{\underset{OH}{|}}{C}=O$$

fast ↓   slow ↓

$$Ph-\underset{\underset{H}{|}}{\overset{\overset{H}{|}}{C}}-OH \quad + \quad Ph-\underset{\underset{O^-}{|}}{C}=O$$

54. (a) all aldehydes reduce Fehling solution
55. (b)   56. (b)   57. (c)   58. (a)
59. alkenes $C=O \longrightarrow C^- + -O$
60. (a) only α-H containing aldehydes give aldol condensation >)
60. (b) acetaldehyde forms an addition compound, but acetone forms (ab) product-diacetone amine
61. (b) nucleophile $CH_3-\underset{\underset{H}{|}}{C}=O \longrightarrow CH_3-\underset{\underset{H}{|}}{\overset{+}{C}}-O^- + H-CH_2-CHO \longrightarrow CH_3^-$
62. (a)   63. (b)   64. (a) HCHO –no α-hydrogen atom
65. (a)   66. (a)   67. (b) same   68. (a)   69. (d)   70. (a)   71. (ba)
72. (b)   73. (a)

MULTIPLE CHOICE QUESTIONS

(Group-C)

1. In Cannizzaro's reaction the slowest step is
   $2C_6H_5CHO + OH^- \longrightarrow C_6H_5COO^- + C_6H_5CH_2OH$
   (a) The attack of OH^- group on the carboxylic group
   (b) The transfer of the hydride ion (H^-) to the COOH group
   (c) The abstraction of a proton from the —COOH group
   (d) The deprotonation of the $C_6H_5CH_2OH$
2. Phenyl glyoxal, C_6H_5COCHO, on treatment with NaOH gives
   $C_6H_5COCHO + NaOH \longrightarrow$
   (a) $C_6H_5CH(OH)COO\,Na$
   (b) $C_6H_5CH_2OH + HCOONa$
   (c) $C_6H_5COONa + HCOOH$
   (d) both the compounds in the form of the sodium salt
3. Ketone on reaction with Peroxy acid gives an
   (a) ester   (b) amine   (c) acid   (d) aldehyde

4. Which among the following aldehydes, will undergo both aldol condensation as well as Cannizzaro's reaction
   (a) HCHO
   (b) CH_3CHO
   (c) $(CH_3)_2CH—CHO$
   (d) C_6H_5CHO

5. The reaction of $C_6H_5CH=CH—CHO$ with $NaBH_4$ gives
   (a) $C_6H_5CH_2CH_2CH_2CHO$
   (b) $C_6H_5CH=CHCH_2OH$
   (c) $C_6H_5CH_2CH_2CHO$
   (d) $C_6H_5CH_2CH(OH)CH_3$

6. The decreasing order of a nucleophilic attack on the following compounds is in the order of
   (a) CH_3COCl
   (b) CH_3CHO
   (c) RCOR
   (d) RCOOR′
   (i) A > B > C > D   (ii) D > C > B > A   (iii) C > B > A > D   (iv) A > D > B > A

7. $CH_3COCH_2CH_2CH_3$ and $CH_3CH_2COCH_2CH_3$ can be distinguished by
   (a) $I_2 + NaOH$
   (b) $NaHSO_3$
   (c) HCN
   (d) both (a) and (b)

8. Which of the following compounds will have higher enolic content than keto content.
   (a) $CH_3—CO—CO—CH_3$
   (b) 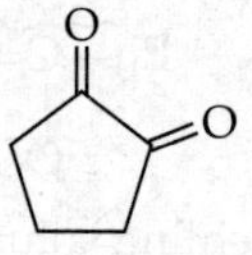
   (c) 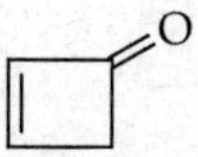
   (d)

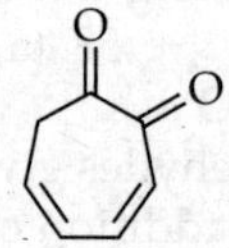

9. The tautomer of the compound is/are

   $C_6H_5—\overset{O}{\overset{||}{C}}—CH_2—\overset{O}{\overset{||}{C}}—C_6H_5$

   (a) $C_6H_5—\overset{OH}{\overset{|}{C}}=CH—\overset{O}{\overset{||}{C}}—C_6H_5$
   (b) $C_6H_5—\overset{O}{\overset{||}{C}}—CH=C—\overset{OH}{\overset{|}{}}C_6H_5$
   (c) $Ph—\overset{OH}{\overset{|}{C}}=C=\overset{OH}{\overset{|}{C}}—Ph$
   (d) $Ph—\overset{OH}{\overset{|}{C}}=CH—\overset{OH}{\overset{|}{C}}=C_6H_4$

10. Which of the following will undergo aldol condensation
    (a) CCl_3CHO
    (b) $(CH_3)_3C/CHO$
    (c) CH_3CH_2CHO
    (d) HCHO

11. The correct order of the dipole moment of the following
    (i) HCHO   (ii) CH_3CHO (iii) CH_3COCH_3
    (a) (i) > (ii) > (iii)
    (b) (iii) > (ii) > (i)
    (c) (ii) > (iii) > (i)
    (d) (ii) > (i) > (iii)

12. The decreasing order of the nucleophilic addition reaction
    HCHO CH_3CHO CH_3COCH_3 $C_6H_5COCH_3$
    A   B   C   D
    (a) B > C > D > A   (b) A > B > C > D   (c) D > C > B > A   (d) B > C > A > D

13. The decreasing order (D.O.) of nucleophilic addition reaction of the following compounds is
    CH_3COCH_3 CH_3COCl CH_3CONH_2 $(CH_3CO)_2O$
    A   B   C   D
    (a) B > D > A > C   (b) A > B > C > D   (c) D > C > B > A   (d) C > A > B > D

14. The decreasing order of nucleophilic addition of

CH_3COCH_2Cl CH_2ClCHO $HCHO$ CH_3COCH_3
     A   B   C   D

(a) $A > B > C > D$ (b) $A > B > D > C$ (c) $B > C > A > D$ (d) $B > A > C > D$

15. In the given compound which carbonyl group will be most reactive?

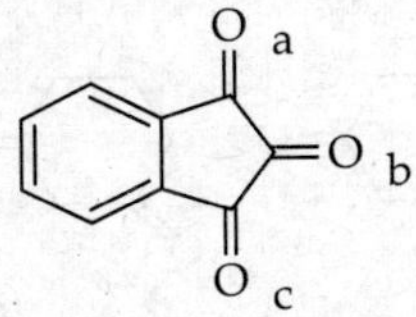

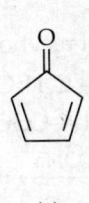

(a) a   (b) b
(c) c   (d) all have same reactivity

16. Which carbonyl compound will have a maximum dipole moment?

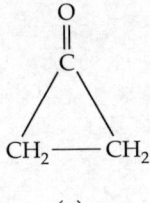 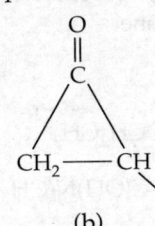 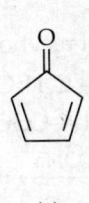 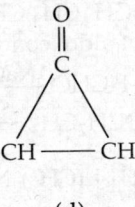

   (a)   (b)   (c)   (d)

17. On treating which of the following compound with dil. acid, will form stable cation

CH_3COCH_3 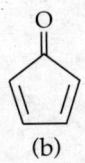   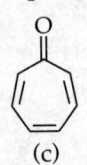

   (a)   (b)   (c) 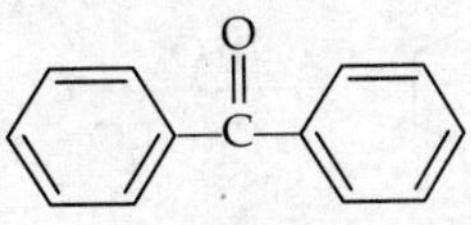   (d)

18. Predict the formation of compound in the following reaction.

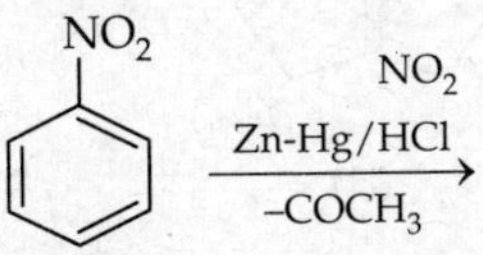

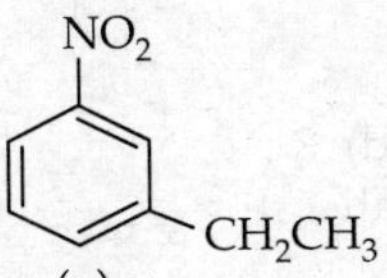

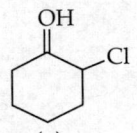

   (a) 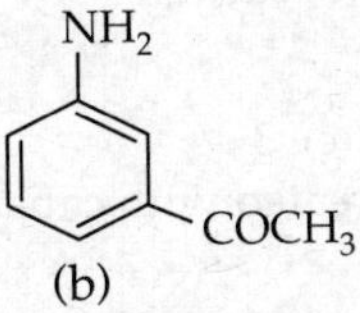   (b) 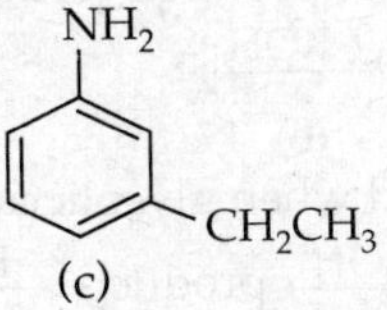   (c)

19.

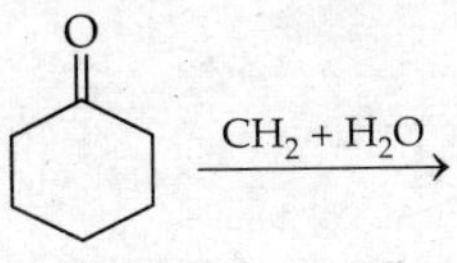

   (a) 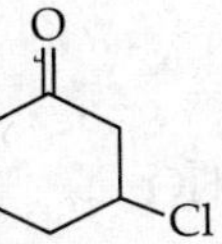   (b) 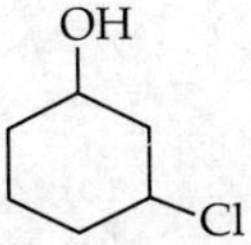   (c) 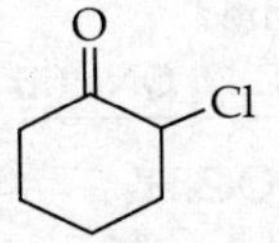   (d)

20. On treating cyclohexanone with NaOH, the product is

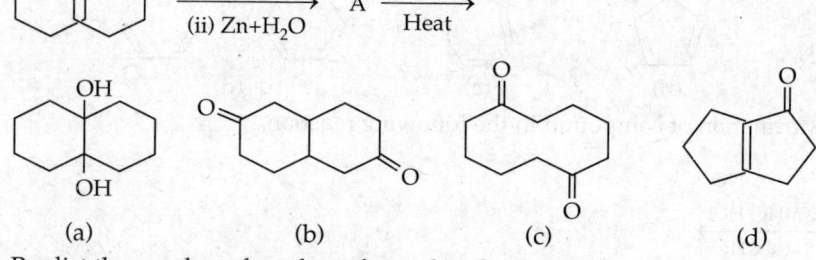

    (a)   (b)   (c)   (d)

21. A reaction between $C_6H_5CHO + CH_3CH_2CHO \xrightarrow{Ca(OH)_2}$
    (a) $C_6H_5CH=CH—CH_2CHO$
    (b) $C_6H_5CH(OH)CH_2CH_2CHO$
    (c) $C_6H_5CH=CH(CH_3)CHO$
    (d) none

22. An example of aldol condensation reaction
    (a) $C_6H_5HO+HCHO \xrightarrow{NaOH/heat} C_6H_5CH_2OH$
    (b) $2CH_3COOCH_2CH_3 \longrightarrow CH_3—COCH_2COOCH_2CH_3$
    (c) $CH_3COCH_3 + (CH_3)_2NH \xrightarrow{C_2H_5ONa} (CH_3)_2—C(OH)N(CH_3)_2$
    (d) $2CH_3COCH_3 \xrightarrow{Ba(OH)_2} CH_3COCH=C(CH_3)_2$

23. On oxidising, the compounds formed are

    $\xrightarrow[\text{(ii) Zn+H}_2\text{O}]{\text{(i) O}_3} A \xrightarrow[\text{Heat}]{\text{NaOH}}$

    (a)   (b)   (c)   (d)

24. Predict the number of products formed in the reaction
    $C_6H_5CHO + HCHO \xrightarrow{KOH}$
    (a) 1   (b) 4   (c) 3   (d) 2

25. The product formed when glycolic acid is treated with caustic alkali
    $CHO—COOH \xrightarrow{NaOH} \text{products} \xrightarrow{HCl}$
    (a) CH_2OH
        $|$
        $COOH$
    (b) $COOH$
        $|$
        $COOH$
    (c) both (a) and (b)
    (d) none

26. The compound formed
    $CH_3COOC_2H_5 \xrightarrow{\text{(i) C}_2\text{H}_5\text{ONa (ii) HCl}}$
    (a) $CH_3COCH_2COOC_2H_5$
    (b) $2CH_3CHO$
    (c) $CH_3COOCH_2COOC_2H_5$
    (d) none

27. The compound fomed is

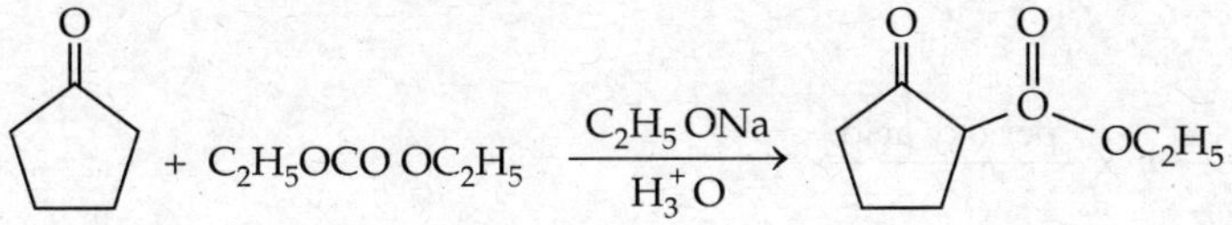

28.

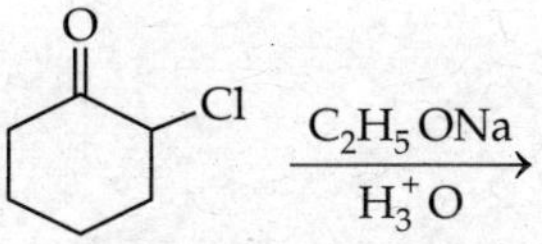

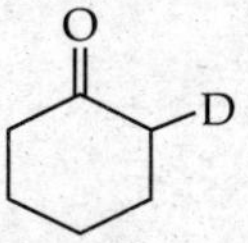

(a) 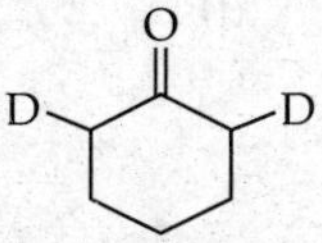   (b) 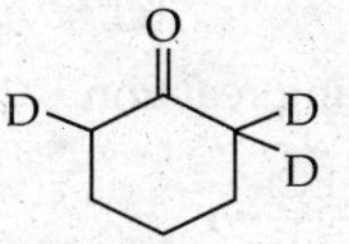   (c) 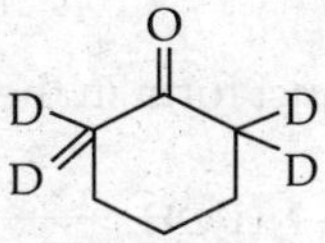   (d)

29.

$COCH_3 + I_2 + NaOH \longrightarrow X + Y \xrightarrow{Ag/heat} Z$

(a) CHI_3 (b) $CH{\equiv}CH$
(c) $CH_2{=}CH_2$ (d) C_6H_5COONa

30. $CH_3CHO \xrightarrow{CH_3COOOH}$

(a) CH_3COOH (b) CH_3OCOOH
(c) $HCOOH$ (d) $HO{-}CH_2COOH$

31. $CH_3CO{-}CO{-}C_2H_5 \xrightarrow{CH_3CO_3H \text{ (per oxy acetic acid)}}$

(a) $CH_3CO.O.COC_2H_5$ (b) $CH_3{-}CO{-}COOC_2H_5$
(c) $HOCH_2{-}CO{-}O{-}COC_2H_5$ (d)

32. Which hydrogen is most acidic and which hydrogen is least acidic

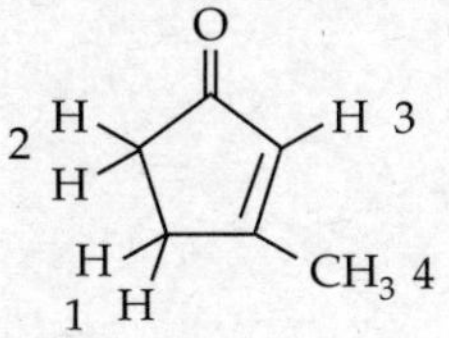

(a) 1 and 4   (b) 3 and 4   (c) 2 and 4   (d) 4 and 2

33. The compound formed is

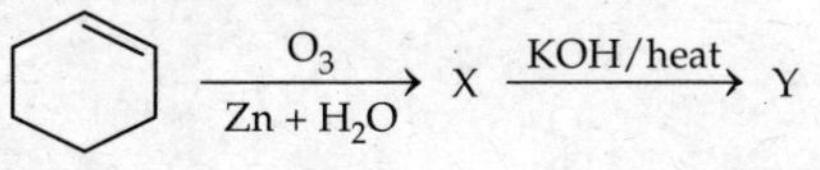

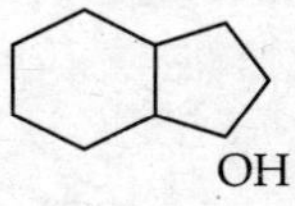

$\xrightarrow[Zn + H_2O]{O_3}$ X $\xrightarrow{KOH/heat}$ Y

(a) 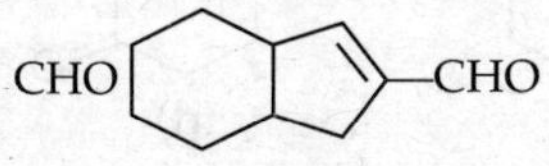   (b) 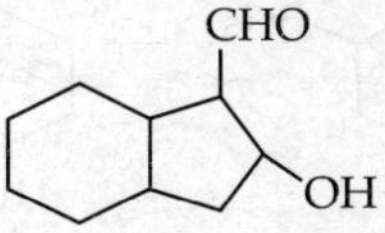   (c) 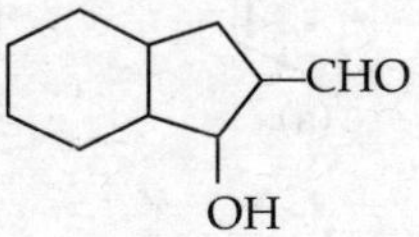   (d)

34. The compound formed is

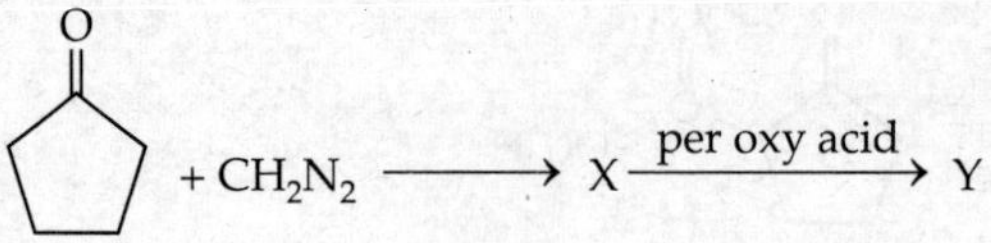

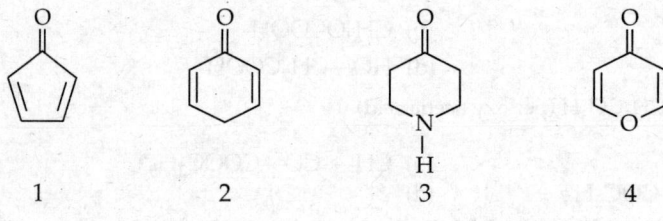

Y =

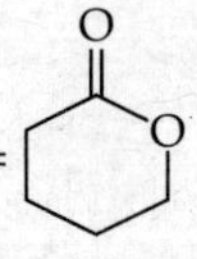

(a) 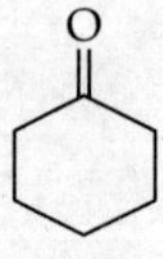　　　(b) 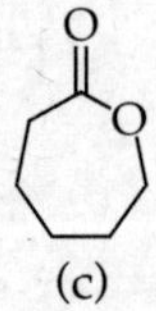　　　(c) 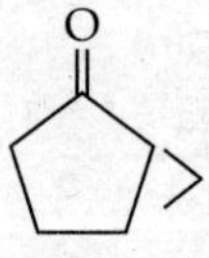　　　(d)

35. The product form in the following reaction

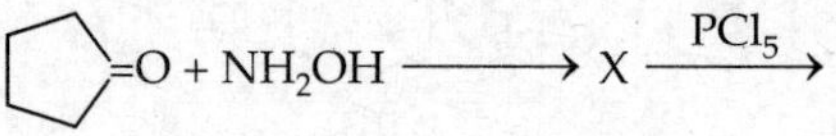

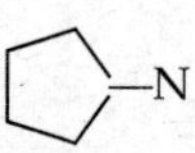

(a) 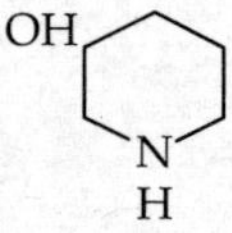　　　(b) 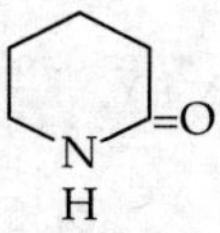　　　(c) 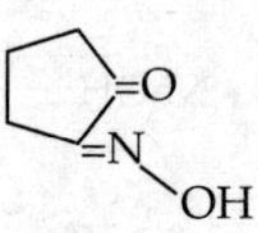　　　(d)

36. Arrange the following keto compounds in the decreasing order their dipole moment

1　　　2　　　3　　　4

(a) 1 > 2 > 3 > 4 　　　(b) 4 > 3 > 2 > 1
(c) 4 > 2 > 3 > 1 　　　(d) 2 > 4 > 3 > 1

37. CH_3CHO reacts most readily with
(a) NH_2NH_2 　　　(b) $NH_2NHCONH_2$
(c) $C_6H_5NHNH_2$ 　　　(d) NH_2OH

38.

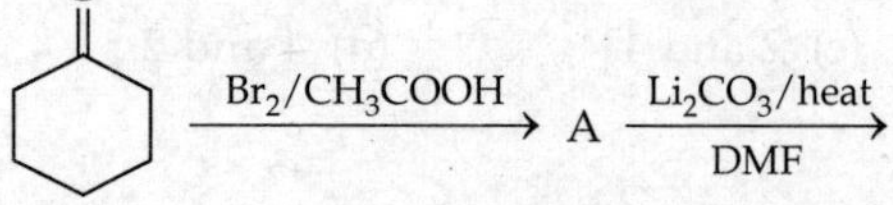

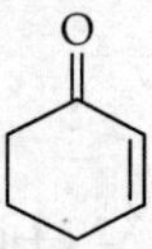

(a) 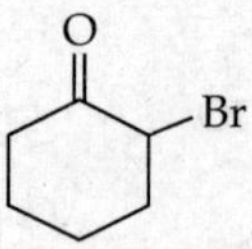　　　(b) 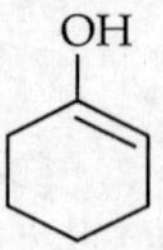　　　(c) 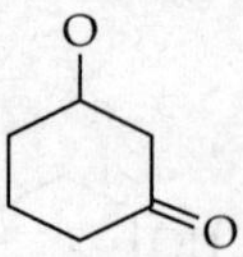　　　(d)

ANSWERS

(Group-C)

1. (b)
2. (a) an example of internal Cannizzaro's reaction

3. (a)	4. (c)	5. (b)	6. (a)	7. (d)	8. (b)	9. (a, b)	10. (c)
11. (c)	12. (b)	13. (i)	14. (iii)	15. (b)			

16. (d)

17. (c)	18. (c)	19. (d)	20. (c)	21. (c)	22. (d)	23. (b)	24. (c) (2)
25. (c)	26. (a)	27. (a)	28. (d)	29. (b)	30. (a)	31. (a)	32. (a)
33. (b)	34. (b)	35. (c)	36. (c)	37. (b)	38. (a)		

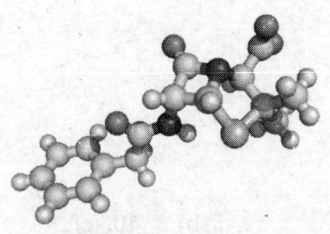

Chapter 5

Aliphatic and Aromatic Carboxylic Acids

Salient Features

1. The strength of hydrogen bonds in carboxylic acids, is even more than the alcohols, as the boiling points are higher than the boiling points of alcohols.
2. The m.ps of the carboxylic acids, containing even number of carbon atoms, is always higher than those containing odd number of carbon atoms.
3. Smaller the value of pKa, stronger is the acid.
4. Both carboxylic acids and carboxylate ions undergo resonance.
5. Due to resonance, oxygen acquires a +ve charge, which facilitates the release of proton.

$$R-C\underset{\ddot{O}-H}{\overset{CH_3}{\lesssim}} \longrightarrow R-C\underset{\overset{+}{O}-H}{\overset{O^-}{\lesssim}} \longrightarrow R-C\underset{O^-}{\overset{O^-}{\lesssim}} + H^+$$

6. Carboxylate ion is more stable as compared to the carboxylic acid (in aqueous solution) owing to a greater dispersal of the negative charge.
7. +I effect (electron donating inductive effect). Effect of various groups is in the following order:
$$(CH_3)_3-C > (CH_3)_2-CH > CH_3CH_2 > CH_3-$$
8. Phenyl group has –I effect (electron withdrawing inductive effect).
9. In benzoic acid, phenyl group release electrons towards the —COOH group, as the resonance effect is stronger than the –I effect, due to which deceasing order of acidic character of the following acids is as follows:
$$HCOOH > C_6H_5COOH > CH_3COOH$$
10. In sodium formate, both the carbon-oxygen bonds have the same value i.e 1.27Å

$$H-C\underset{\underset{1.27A}{O1/2-}}{\overset{\overset{1.27A}{O1/2-}}{\lesssim}} + Na^+$$

11. A, —C≡C—, triple bond is a better electron with drawing group than a >C=C<
$$C≡C—COOH > CH_2=CH—COOH$$
pKa value = 3.32   4.35
12. The b.p.s of acid chloride, RCOCl and esters RCOOR', are lower than those of corresponding alcohols, because of absence of hydrogen bonding, in their molecules.

Decreasing order of the melting points and boiling points is as follows:
$$RCONH_2 > RCOOH > RCOCl$$

13. The boiling point's of anhydride are higher than of corresponding acids because of those of their large size and van der Waal forces, of inter molecular attraction.

14. Electron with drawing groups, when attached to the benzoic acid, make benzoic acid stronger from all positions.
On the other hand, electron donating groups, make benzoic acid weaker, when attached in the m-position or p-position, but make acid stronger when attached in the o-position. This is known as **ortho-effect**, in the case of benzoic acid.

ALIPHATIC ACIDS

Formic and Acetic Acids

General Methods of Preparation

1. By oxidation of a primary alcohol by oxidation with chromic acid leads to the formation of carboxylic acid.
$$RCH_2OH + H_2SO_4 + K_2Cr_2O_7 \longrightarrow RCHO \longrightarrow RCOOH$$

2. By the hydrolysis of an alkyl cyanide. By boiling an alkyl cyanide with dil. H_2SO_4, the cyanide is converted to give an acid.
$$RX + Na\ CN\text{---}RCN \longrightarrow RCOOH + NH_3$$

3. By hydrolyzing an ester-An ester is warmed with conc. NaOH. After this the, contents are acidified with HCl, to isolate the acid.

4. By Grignard's reagent with CO_2
$$RX + Mg \longrightarrow RMgX \xrightarrow{CO_2} R\text{---}COOMg \xrightarrow{H_3^+O} R\text{---}COOH$$

5. From alkenes-By Koch synthesis
$$CH_2=CH_2 + CO + H_2O \xrightarrow{H_3PO_4/400°C} CH_3CH_2COOH$$

6. On heating malonic acid, a geminal or a gem-dibasic it loses a molecule of CO_2 to give
$$CH_2(COOH)_2 \longrightarrow CH_3COOH + CO_2$$

7. By reducing an aldehyde
$$RCHO + 2H \longrightarrow RCH_2OH$$

The aldehydes may be formed from alkenes either by Waker's process or by oxo-process
$$CH_2=CH_2 + H_2O + PdCl_2 \longrightarrow CH_3CHO + Pd + 2HCl \text{ (Waker's process)}$$
By Oxo-process-   under pressure
$$CH_3\text{---}CH=CH_2 + CO + H_2O \longrightarrow CH_3\text{---}CH_2\text{---}CH_2CHO + (CH_3)_2\text{---}CH\ CHO$$

Physical properties

a. Lower members are pungent smelling liquids but higher members are colour less solids.

b. The boiling points and melting points increase with rise in molecular weight. First five members are miscible with water (the carbonyl group is able to form bonds with water).This solubility decrease with increase in the carbon chain.
The molecular weight of first few members has been found to be the double of their actual molecular weight, because the lower member form hydrogen bonds in between the two molecular.
For ex-The molecular weight of HCOOH has been found to be equal to 92 in benzene. Similarly the molecular weight of acetic acid has been determined to be 120. It means that formic acid and acetic acid exist as dimers in benzene.

$$CH_3-C \overset{O-HO}{\underset{OH-O}{<}} C-CH_3$$

The strong acids like HCl or H_2SO_4 completely ionise in aqueous solution. But weak organic acids, like HCOOH and CH_3COOH, ionise partially and there exists an equilibrium in between ionized and unionized forms. This is expressed in terms of Ka, known as acidity constant.

The Ka may be defined as the concentration of the product of ionisation, in moles per liter divided by concentration of unionised acid also in moles per liter.

$$Ka \xrightarrow{\quad} \frac{[RCOO-][H_3O]}{[RCOOH]}$$

Therefore, the higher acids will have a higher numerical value of Ka.

Why acids show acidity

Since being acids they lose a proton easily because the carboxylate ion is stabilised by resonance.

On the basis of X-ray analysis, the COOH has two types of CO bonds C=O (bond distance = 1.20Å and a C—O bond (bond distance = 1.42Å. Where as in the sodium formate the two carbon-oxygen bonds are of identical bond length which is in between that of carbon-oxygen single and double bonds.

$$H-C \overset{\overset{1.23A}{O}}{\underset{\underset{1.36Å}{OH}}{<}} \qquad\qquad H-C \overset{\overset{1.27A}{O}}{\underset{\underset{1.27Å}{O}}{<}}$$

Stability of carboxylate ion may also be explained on the basis of molecular orbital theory.

The carbon is in sp^2 hybridised form. It is bound to both oxygens by a σ-bond. The unused carbon p-orbital overlaps the two oxygens by p-orbital to form a stable delocalized molecular orbitals.

As the 4 electrons are bound to three atoms, this delocalisation is responsible for extra stability of the carboxylate on.

Effect of EWG on the acidity: EWG like Cl, Br, OH CN, increase acidity, as these groups decrease the negative charge on the carboxylate ion, the carboxylate ion is stabilised further.

Chemical properties

1. Salt formation-It reacts with alkali and bases to form salts.
   $$RCOOH + NaHCO_3 \longrightarrow RCOO\,Na + CO_2 + H_2O$$

2. Reaction with PCl_5, PCl_3 and $SOCl_2$
   Acids react with PCl_3 to form acid chlorides-
   $$3\,RCOOH + 3\,PCl_3 \longrightarrow 3\,RCOCl + H_3PO_3$$

3. Reaction with NH_3-
   $$RCOOH + NH_3 \xrightarrow{heat} RCOONH_4 \xrightarrow{\;\Delta\;} RCONH_2;$$
   carboxylic acids react with NH_3 to form amides

4. Reaction with alcohols-Acids react with alcohols to form esters, in presence of a little conc. H_2SO_4 or HCl. The process is termed as 'esterification'
   $$RCOOH + R'OH \longrightarrow RCOOR' + H_2O$$

Mechanism:

i. Protonation of carboxylation

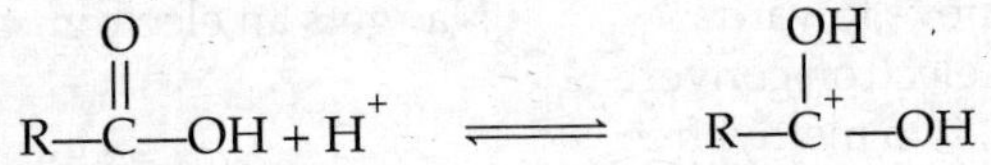

$$R-\overset{\overset{\displaystyle O}{\|}}{C}-OH + H^+ \quad \rightleftharpoons \quad R-\overset{\overset{\displaystyle OH}{|}}{\overset{+}{C}}-OH$$

ii. Attack by a nucleophile

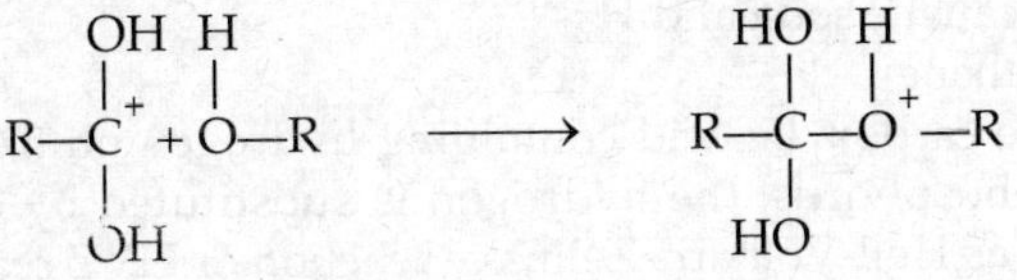

$$R-\overset{\overset{\displaystyle OH}{|}}{\underset{\underset{\displaystyle OH}{|}}{\overset{+}{C}}} + \overset{\displaystyle H}{\overset{|}{O}}-R \quad \longrightarrow \quad R-\overset{\overset{\displaystyle HO}{|}}{\underset{\underset{\displaystyle HO}{|}}{C}}-\overset{\overset{\displaystyle H}{|}}{\overset{+}{O}}-R$$

iii. Transfer of proton

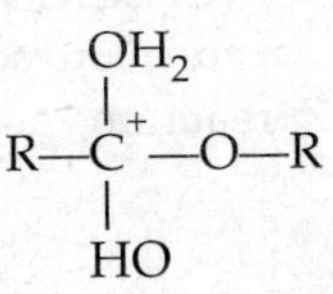

$$R-\overset{\overset{\displaystyle OH_2}{|}}{\underset{\underset{\displaystyle HO}{|}}{\overset{+}{C}}}-O-R$$

iv. Elimination of H_2O

$$R-C-O-R \quad \longrightarrow \quad R-C-OR + H_2O$$

$$R-\overset{\overset{\displaystyle H_2O^+}{\diagdown}}{\underset{\underset{\displaystyle O-H}{\diagup}}{C}}-O-R \quad \longrightarrow \quad R-\overset{\overset{\displaystyle O}{\|}}{C}-OR + H_2O$$

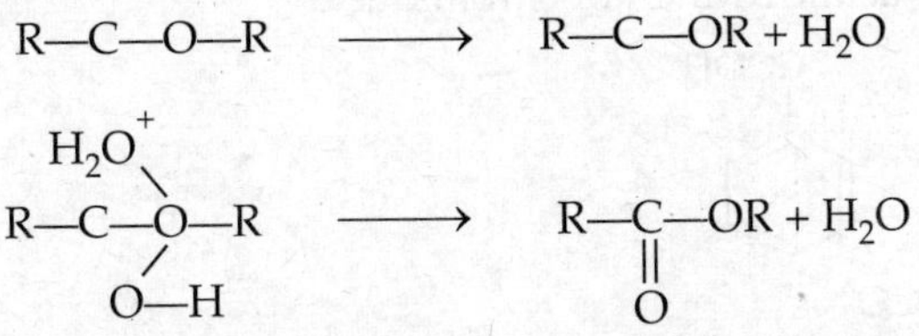

On heating the acid with P_2O_5, two molecules react to form an anhydride.

$$\begin{matrix} RCOOH \\ RCOOH \end{matrix} \quad \xrightarrow{P_2O_5} \quad (RCO)_2 + HPO_2$$

6. On reduction with $LiAlH_4$-Acids are converted to form alcohol

$$RCOOH \quad \longrightarrow \quad RCH_2OH$$

7. On heating the sodium salt of the acid with soda lime, alkanes are formed

$$R\,COO\,Na + NaOH \quad \xrightarrow[\Delta]{NaOH} \quad RH + Na_2CO_3$$

8. On electrolysis an aqueous solution of sodium acetate, a higher alkane is formed containing just the double the number of carbon. The reaction is termed as Kolbe's electrolytic synthesis.
   Mechanism: The reaction is a free radical reaction

   i. Being an electrolyte sodium acetate dissociate of its own-

$$CH_3COONa \quad \longrightarrow \quad CH_3COO^- + Na^+$$

   ii. On switching the current, the ions move toward the opposite electrodes, get discharged

At anode	At cathode
CH_3COO-	Na^+
$\downarrow e^-$	$\downarrow + e^-$
$CH_3\text{-}COO$	Na
$\downarrow H_2O$	$\downarrow H_2O$
$CH_3\bullet + CO_2\downarrow$	$NaOH + H_2\uparrow$

Reaction at anode
$$CH_3\bullet + CH_3\bullet = C_2H_6$$
CH_3COO- ion move towards anode, loses its electron converts to a free radical, two methyl free radicals combine of metal, reacts with water to caustic soda and H_2 form to form ethane

Reaction at cathode
$$Na^+ + e \rightarrow Na$$
Na^+ gets an electron, converts into NaOH

9. **Halogenation:** The carboxylic acid containing hydrogen on treating with Cl_2 and Br_2 in presence of red phosphorus, the hydrogen is substituted by chlorine or bromine. This reaction is known as Hell-Volhard-Zelinsky reaction or HVZ reaction.

$$CH_3COOH + Cl_{12} \longrightarrow CH_2ClCOOH$$
α-chloro acetic acid

HCOOH does not undergo HVZ reaction, as compound does not contain hydrogen
Benzoic acid (C_6H_5COOH)

Methods of Preparation

1. **Oxidation of toluene:** by acidic $KMnO_4$ or chromic acid

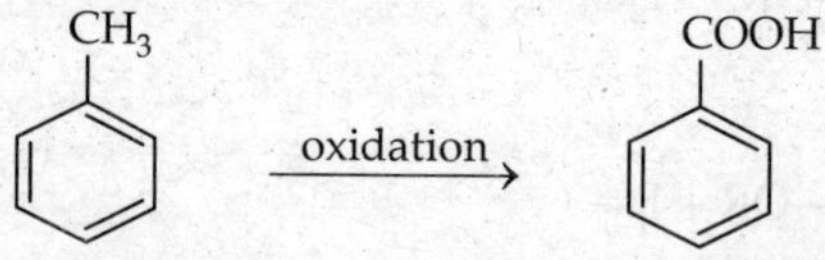

2. **Grignard's reagent and CO_2**

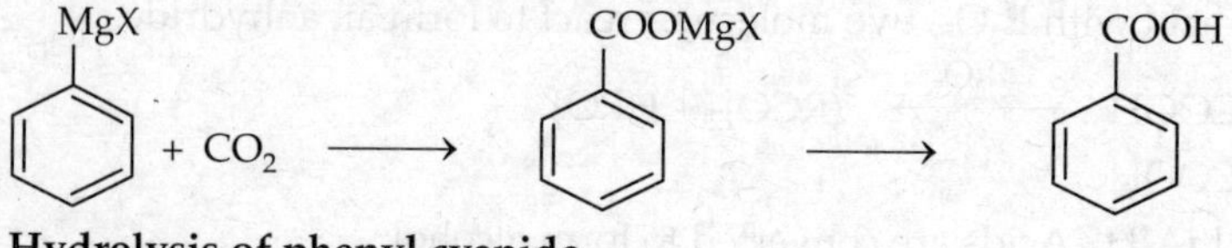

3. **Hydrolysis of phenyl cyanide**

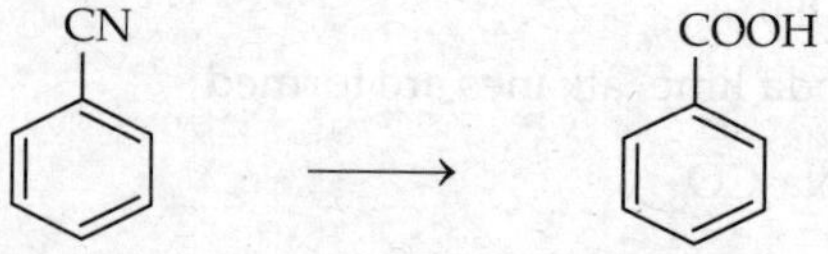

4. **By oxidation of benzaldehyde**

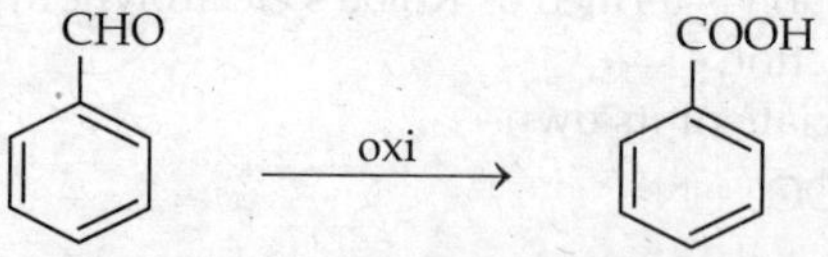

5. **By oxidation of aromatic compound containing a side chain.**

Ortho effect:

It is specific to only o-substitute benzoic acids

According to this effect, the o-substituted benzoic acids are stronger than benzoic acid, irrespective of whether the group is o and p-directing or m-directing.

Electron withdrawing groups will make the acid stronger, yet on account of the- ortho effect the m-directing groups, shows increased acidic behavior. For ex-salicylic acid or o-hydroxy benzoic acid.

Properties:
1. It is a colourless crystalline needle shaped compound, soluble in hot water and soluble in organic solvent
2. There is no hydrogen atom present in the molecule there fore it does not undergo HVZ reaction
3. Benzoic acid also shows, the properties due to the benzene ring

Ortho effect:
It is specific to only o-substitute benzoic acids
According to this effect, the o-substituted benzoic acids are stronger than benzoic acid, irrespective of whether the group is o p-directing or m-directing.
Electron withdrawing groups will make the acid stronger, (the —OH group is electron donating; it is expected to make benzoic acid weaker) yet on account of the ortho effect the m-directing groups, shows increased acidic behavior. For example
Salicylic acid

Preparation:
1. By heating sodium phenoxide with CO_2 under 10 atom pressure and at a temperature 120-140°C, sodium salt of salicylic aid is formed. On acidification salicylic acid is obtained.
2. The other method is to prepare salicylaldehyde by Riemer-Teimann reaction, followed by a selective oxidation by ammo. $AgNO_3$ to finally give salicylic acid.
   It is a colourless crystalline compound mp. 159°C

As the molecule contains both —OH group and a COOH group, therefore it shows the properties of a phenol and a monocarboxylic aid.

The combined properties due to —OH as well as COOH group
1. On slow distillation, it lose CO_2 and it is converted to form benzoic acid.
   On rapid distillation it is converted to form an ester, phenyl salicylate.
2. On treatment with bromine, it forms 2,4,6-tribromo phenol. A bond between C to C is broken. This reaction is termed as an IPSO reaction.
3. On reaction with HNO_3 it is converted to give 2,4,6-tri nitro phenol also known as picric acid.
4. Acetylation-Salicylic acid on acetylation gives an acetyl salicylic acid. It is used in medicine as anti-septic, also used as a pain killer.

REASONING TYPE QUESTIONS (RTQs)

(Group-A)
1. What is the cause of acidity of fatty acids? Explain.
2. What happens when a sodium salt of acetic acid is electrolyzed? Discuss the mechanism involved in the formation of the products.
3. What happens when a silver salt of a fatty acid is heated with chlorine or bromine? Write the equation involved. Does the reaction with iodine, also give the same product? Discuss.
4. In what ways formic acid differs from the other fatty acids? Give reasons.
5. Write the reactions of formic with
   a. conc. H_2SO_4 b. acidified $KMnO_4$.
6. What are acidity constants? In a dibasic acid, what is the relation between Ka1 and Ka2. Discuss the relation between pKa and Ka values?
7. What is the effect of substitution on the acidity of the fatty acids? Discuss.
8. Explain: Fluoroacetic acid is a stronger than chloroacetic acid.
9. Explain acetic acid is a stronger acid as compared to ethyl alcohol.

10. Arrange the following acids in the decreasing order of acidity. Give the reasons.
$(CH_3)_2CHCOOHCH_3COOH, (CH_3)_3 C—COOH$

11. How will you synthesize crotonic acid from acetaldehyde? Give the equations involved.

12. How will you prepare oxalic acid from formic acid in two step reaction? Write the steps involved.

13. What do you under stand by the word NBS? What are its uses?

14. Discuss the compound formed, when adipic acid reacts with hexamethylenediamine.

15. Discuss the relation between maleic acid and fumaric acid. How will you distinguish two acids? Explain

16. Predict the compound formed, when bromine reacts with salicylic acid.

17. a. The b.p. of first two fatty acids are higher than can be explained on the basis of their molecular wts. Explain
    b. Also explain why benzoic acid is stronger than acetic acid.

18. What is the action of heat on α-, β-, γ-, δ- hydroxyl acids? Discuss.

19. Discuss Hofmann's bromamide reaction, along with the mechanism involved.

20. Write notes on the following reactions:
    i. HVZ Reaction   ii. Koch Reaction

ANSWERS

(Group-A)

1. The carboxylic acids readily lose a proton, either by a reaction with a base or by ionisation, and the carboxylate ion formed is stabilised by resonance.

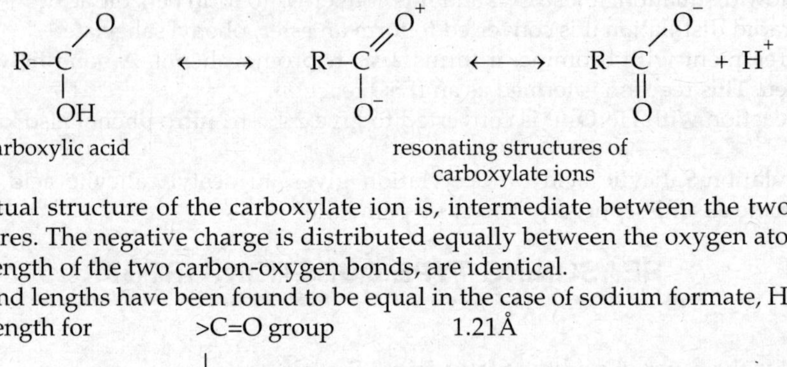

carboxylic acid   resonating structures of carboxylate ions

The actual structure of the carboxylate ion is, intermediate between the two resonating structures. The negative charge is distributed equally between the oxygen atoms, and the bond length of the two carbon-oxygen bonds, are identical.

The bond lengths have been found to be equal in the case of sodium formate, HCOONa.

Bond length for   >C=O group   1.21Å

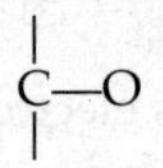

$$\underset{|}{\overset{|}{C}}{-}O \qquad 1.43\text{Å}$$

sodium formate   both C—O bonds are equal and have a bond length equal to 1.27Å

The structure of a carboxylate ion is often represented as follows:

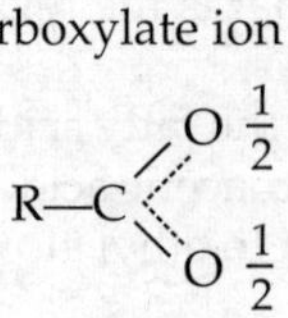

resonance hybrid structure

When a charge is spread over several atoms, rather than concentrating over only one atom, the species becomes more stable. In this case the negative charge is spread over three atoms, two oxygen atoms and one carbon atom; the carboxylate ion becomes more stable. Since each of the oxygen has a half negative charge on either oxygen atom, the proton has much less chance of joining on either of the oxygen atom, and thus remain ionized. Hence carboxylic acids show acidity.

2. The electrolysis of an aqueous solution of a sodium salt of acetic acid, CH_3COONa, leads to the formation of ethane. The reaction follows a free radical mechanism.

   The various steps involved are as follows:

   i. Sodium acetate, dissociates of its own, being an electrolyte, to form an acetate ion and a sodium ion.

   $$\underset{\text{sodium acetate}}{CH_3\text{—}COO^- Na^+} \rightleftharpoons \underset{\text{acetate ion}}{CH_3\text{—}COO^-} + \underset{\text{sodium ion}}{Na^+}$$

   ii. On switching the current, sodium ions move towards cathode, gains an electron, gets converted into sodium metal, attacks water, to form NaOH and hydrogen.
   At *cathode*-

   $$Na^+ + e \longrightarrow Na$$
   $$2\,Na + 2\,H_2O \longrightarrow 2\,NaOH + H_2\uparrow$$

   iii. Acetate ion goes towards the anode, loses an electron, gets converted into an acetate free radical.
   At *anode*-

   $$\underset{\text{acetate anion}}{CH_3\text{—}COO^-} \longrightarrow \underset{\text{acetate free radical}}{CH_3COO\bullet} + e$$

   Acetate free radical, then decomposes into a methyl free radical and carbon dioxide.

   $$\underset{\text{acetate free radical}}{CH_3 : COO^\bullet} \longrightarrow \underset{\text{methyl free radical}}{CH_3\bullet} + CO_2$$

   Two methyl free radicals formed, then combine to form ethane.

   $$CH_3\bullet + CH_3\bullet \longrightarrow CH_3 : CH_3$$

   This method of preparation of an alkane, on electrolysis of an aqueous solution of a sodium salt of fatty acid, is known as Kolbe's electrolytic synthesis.

3. When a silver salt of a fatty acid, RCOOAg, is treated with chlorine or bromine, the silver salt of fatty acid decomposes to form an alkyl halide.

   $$\underset{\text{silver salt of a fattyacid}}{RCOOAg} + X_2 \longrightarrow \underset{\text{alkyl halide}}{RX} + CO_2 + AgX \quad (X = Cl_2 \text{ or } Br_2)$$

   This reaction is known as Borodus Hunsdiecker's reaction. The alkyl halide formed, contains one carbon less.

   On the other hand when a silver salt of a fatty acid, is treated with iodine, an ester is formed.

   The reaction is called Birnbaum-Simonini reaction

   $$2\,R\text{—}COOAg + I_2 \longrightarrow \underset{\text{ester}}{R\text{—}COOR} + AgI_2$$

4. Formic acid differs from other fatty acids in being a powerful reducing agent. It reduces Tollen's reagent, reduces mercuric chloride, and shows other reducing properties like formaldehyde.

$$\begin{matrix} H \\ \quad \diagdown \\ \qquad C{=}O \\ \quad \diagup \\ HO \end{matrix} + Ag_2O \longrightarrow H_2O + 2\,Ag$$

Tollen's reagent   grey ppt of metallic silver

$$\begin{matrix} H \\ \quad \diagdown \\ \qquad C{=}O \\ \quad \diagup \\ OH \end{matrix} + Ag_2O \longrightarrow H_2O + 2\,Ag$$

mercuric chloride   mercurous chloride

It is because of the fact that formic acid contains a –CHO group in the molecule.

$$\begin{matrix} H \\ \quad \diagdown \qquad \text{aldehydic group} \\ \qquad C{=}O \quad \text{Formic acid} \\ \quad \diagup \\ HO \end{matrix}$$

5. **Reactions of formic acid with:**
   a. Formic acid is dehydrated on reacting with conc. H_2SO_4, and forms carbon monoxide, CO.

   $$HCOOH \xrightarrow{\;H_2SO_4\;} H_2O + CO$$

   It is considered a good method for the preparation of carbon oxide.
   b. Formic acid decolourises the pink colour of $KMnO_4$, when treated with acidified $KMnO_4$ solution. Formic is oxidised to CO_2 and H_2O.

   $$2\,KMnO_4 + 3\,H_2SO_4 \longrightarrow K_2SO_4 + 2\,MnSO_4 + 3\,H_2O + 5\,O$$
   $$HCOOH + O \longrightarrow H_2O + CO_2$$

6. Fatty acids are weak acids and these ionize only partially in water. There exists an equilibrium between the ionized and unionized forms of the acid.

$$\begin{matrix} O & & O \\ \parallel & & \parallel \\ R{-}C{-}O{-}(H + H_2O) & \rightleftharpoons & R{-}C{-}O^- + H_3O \end{matrix}$$

This extent of dissociation of the fatty acid is described by an equilibrium constant, pKa, which is known as acidity constant.

Acidity constant, Ka is defined as concentration of the products of ionisation in moles per litre, divided by the concentration of the unionized acid.

$$Ka = \frac{[RCOO^-][H_3^+O]}{[RCOOH]}$$

Acidity constant describes the relative strength of a weak acid. It can also be described in terms of Ka values or pKa values. The pKa value is the negative log of the Ka value.

$$Ka = 10^{-pKa}$$ The smaller the pKa value, stronger is the acid.

7. Effect of the substituents on the acidity of the carboxylic acids as follows:
   The most important factor affecting the acidity is the inductive effect of the substituents on the α-carbon atom.
   a. Electron releasing alkyl groups decrease the acidity. The alkyl groups when present, increase the electron density on the carboxylate ion. As a result of which, the carboxylate ion is destabilised and the release of proton becomes difficult. As the size of alkyl groups increases, the acidity decreases.

$$HCOOH > CH_3COOH > CH_3\text{---}CH_2\text{---}COOH > CH_3\text{---}CH_2\text{---}CH_2\text{---}COOH$$

$Ka = 17.7 \times 10^{-5}$ 1.76×10^{-5} 1.34×10^{-5} 0.63×10^{-5}

  formic acid   acetic acid   propionic acid   butyric acid

Thus formic acid is the strongest acid among the fatty acids.

b. Effect of electron withdrawing groups increase the acidity: Some of these atoms and groups are $(F, Cl, Br, OH, C\equiv N)$.

The electron withdrawing groups, decrease the electron density on the carboxylate ion. Thus the decrease of a negative charge on the carboxylate ion, stabilises it. As a result the release of a proton becomes relatively easy.

The electronegativity of the various halogen atoms are

$$F \quad > \quad Cl \quad > \quad Br \quad > \quad I$$
$$4 \qquad\quad 3 \qquad\quad 2.8 \qquad 2.6$$

$$F\text{---}CH_2COOH > Cl\text{---}CH_2COOH > Br\text{---}CH_2COOH > I\text{---}CH_2COOH$$

$Ka = 260 \times 10^{-5}$ 136×10^{-5} 125×10^{-5} 67×10^{-5}

fluoro acetic acid   chloroaceticacid   bromo aceticacid   iodoaceticacid

$C\equiv N\text{---}CH_2COOH$ $(HO)\text{---}CH_2COOH$

$Ka = 4 \times 10^{-3}$

cyanoacetic acid   glycolic acid

The number of electron-withdrawing groups also increases the acidity of the acids.

$$Cl_3\text{---}C\text{---}COOH > Cl_2\text{---}CH\text{---}COOH > ClCH_2\text{---}COOH$$

$Ka = 23200 \times 10^{-5}$ 5530×10^{-5} 136×10^{-5}

trichloro acetic acid   dichloro acetic acid   chloroacetic acid

Similarly the effect of the electron withdrawing group in the α-position is stronger than in the β-position. The r-substituted compound is still weaker in the series. As the distance between the electron withdrawing group and the carboxylic group increases the acidity decreases.

For example:

$$CH_3\text{---}CH_2\text{---}CHCl\text{---}COOH > CH_3\text{}^-CHCl\text{---}CH_2\text{}^-COOH > ClCH_2\text{---}CH_2\text{---}CH_2\,COOH$$

$Ka = 139 \times 10^{-5}$ 8.9×10^{-5} 2.96×10^{-5}

α-chloro butyric acid   β-chlorobutyric acid   γ-chlorobutyric acid

2-chloro butyric acid   3-chloro butyric acid   4-chloro butyric acid

8. Fluoroacetic acid is stronger than chloroacetic acid, on account of the presence of a stronger electron withdrawing group, on the α-carbon atom.

Fluorine is a stronger electron withdrawing atom, than chlorine, owing to the higher electronegativity.

                     F   Cl

Electronegativity   4   3

$$F\text{---}CH_2\text{---}COOH > Cl\text{---}CH_2\text{---}COOH$$

9. The acidities of the various hydrogen containing organic compounds is as follows:

$$RCOOH > C_6H_5OH > H_2O > ROH > CH\equiv CH > NH_3 > CH_2{=}CH_2 > CH_3\text{---}CH_3$$

1×10^{-5} 1×10^{-11} 1×10^{-14} 1×10^{-18} 1×10^{-22} 1×10^{-35} 1×10^{-36} 1×10^{-40}

The carboxylic group in the acids is stabilised by resonance, but in the case of the alcohol, resonance is not possible. It is the higher inductive effect of the oxygen atom that withdraws the bonding electrons forming the bond between hydrogen and itself, that makes the alcohol slightly acidic in nature.

10. The decreasing order of the acidic character of the acids is as follows:

$$CH_3COOH > CH_3—CH_2COOH > (CH_3)_2—CH—COOH > (CH_3)_3—C—COOH$$

    acetic acid   >   propionic acid   isobutyric acid   tertiary butyric acid

   The methyl groups are electron donating or electron repelling, they do not stabilize the carboxylate ion as a result of which the release of proton becomes more difficult, the acids become weaker. Hence is the given order.

11. Crotonaldehyde and crotonic acids from acetaldehyde

   a. Crotonaldehyde $CH_3—CH=CH—CHO$, can be prepared from acetaldehyde by subjecting acetaldehyde to aldol condensation.

$$CH_3CHO + H—CH_2—CHO \xrightarrow{\text{Dil. alkali}} CH_3—CH(OH)—CH_2CHO$$

      acetaldehyde   aldol

$$\Delta \downarrow H_2O$$

$$CH_3—CH=CH—CHO$$

                                              crotonaldehyde

   b. Crotonic acid: $CH_3—CH=CH—COOH$

      It can be obtained from crotonaldehyde on

      oxidation by Tollen's reagent $\quad \downarrow Ag_2O$

$$CH_3—CH=CH—COOH$$

12. Oxalic acid can be obtained from formic acid, by heating its sodium salt to 360°C.

$$\begin{array}{ccc} H—COONa & COONa & COOH \\ | & | & | \\ H—COONa & COONa & COOH \end{array}$$

   $\qquad\qquad \longrightarrow \qquad +H_2 \longrightarrow$

      sodium formate   sodium oxalate   oxalic acid

13. N-bromosuccinimide

$$\begin{array}{c} CH_2—CO \\ | \qquad\qquad N—Br \\ CH_2—CO \end{array}$$

   The compound is prepared from succinic acid, by heating it with ammonium carbonate, when succinic acid is converted into succinimide. On treating the imide formed with bromine, it gives N-bromosuccinimide.

$$\begin{array}{c} CH_2—COOH \\ | \\ CH_2—COOH \end{array} \xrightarrow{NH_3} \begin{array}{c} CH_2CO \\ \diagdown NH \\ CH_2CO \diagup \end{array} \xrightarrow{Br_2} \begin{array}{c} CH_2—CO \\ \diagdown N—Br \\ CH_2—CO \diagup \end{array}$$

   sodium formate   succinimide   N-bromosuccinimide or NBS

   The reagent NBS is used to brominate alkenes. A free radical reaction takes place between NBS and alkenes, commonly initiated by light, peroxide, or other catalyst, and yields allylic bromide.

$$\begin{array}{c} CH_2—CO \\ \diagdown NBr \\ CH_2—CO \diagup \end{array} + CH_3CH=CH_2 \longrightarrow CH_3BrCH=CH_2 + \begin{array}{c} CH_2—CO \\ \diagdown NH \\ CH_2—CO \diagup \end{array}$$

   NBS   propene   allyl bromide

14. A reaction between adipic acid and hexamethylenediamine, leads to the formation of a polymer, known as nylon 66.

$(HO-)OC-(CH_2)_4-CO(-OH + H)-NH-(CH_2)_6-NH-(H + HO-)OC-(CH_2)_4CO(-OH)$

$\longrightarrow OC-(CH_2)_4-CO-NH-(CH_2)_6-NH-CO-(CH_2)_4-CO \longrightarrow$

nylon 66

15. Distinction between maleic acid and fumaric acid

Maleic acid is a *cis*-dibasic acid. On heating, it loses a molecule of water and is converted into maleic anhydride.

$$\begin{array}{c} HC-COOH \\ \| \\ HC-COOH \end{array} \xrightarrow[150°C]{\text{heating}} \begin{array}{c} HC-CO \\ \| \qquad \rangle O \\ HC-CO \end{array} \quad \text{m.p.} = 130.5°C$$

maleic acid   maleic anhydride

Fumaric acid is a *trans*-dibasic acid. On heating it does not form anhydride.

$$\begin{array}{c} H-C-COOH \\ \| \\ HOOC-C-H \end{array} \xrightarrow{\Delta} \text{No anhydride formation m.p.} = 287°C$$

fumaric acid

16. On a treating salicylic acid with bromine it forms 2, 4, 6-tribromophenol. The carboxylic group is knocked off and this position is also substituted by bromine, besides 2- and 4-positions.

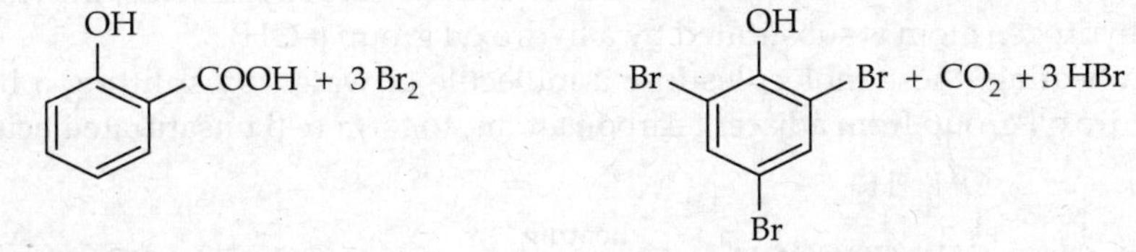

salicylic acid   2, 4, 6-tribromophenol

17. a. The b.p.s of formic acid and acetic acids are higher than can be predicted on the basis of their molecular wts. It is on account of the fact that formic acid and acetic acids exist as dimers.

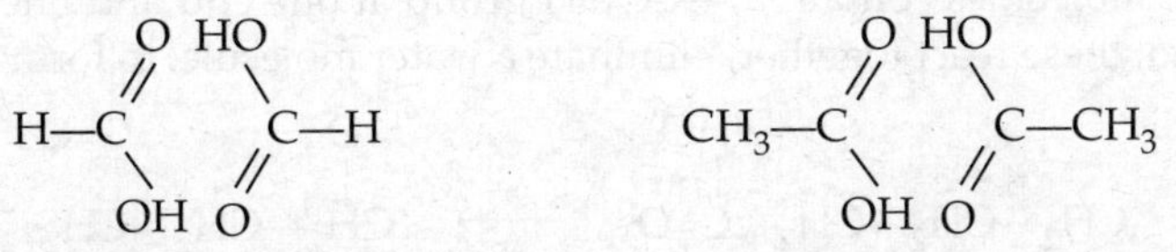

mole. wt = 92   mole. wt = 120

formic acid as dimier   acetic acid as dimer

The molecular wts of formic and acetic acids in benzene solution, have found to be just double of their molecular wt., indicating that these acids exist as dimer in benzene.

b. Benzoic acid (pKa = 4.2) ; Acetic acid (pKa = 4.76)

The carboxylate formed from benzoic acid, is stabilised by resonance, whereas in the case of acetate ion formed from acetic acid, does not form resonating structures.

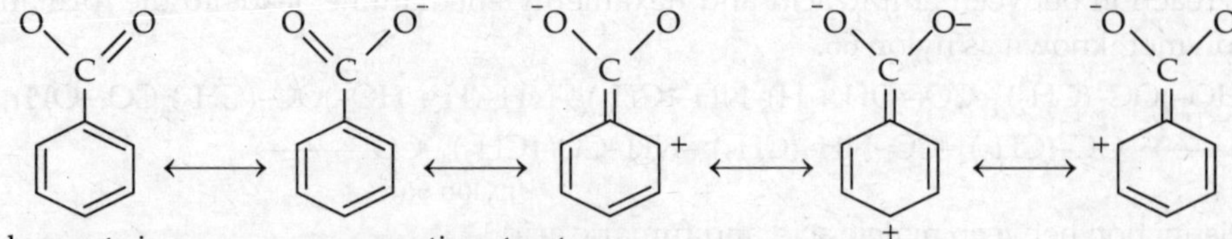

benzoate ion resonating structures

18. Action of heat on the hydroxy acids, in the presence of either HCl or H_2SO_4

   i. **On α-hydroxy acids:** These are the monobasic acids, which contain one of the α-hydrogen atom, substituted by a hydroxyl, (–OH), group. On heating, two molecules of these acids they react in such a way, that the hydroxyl group of one acid molecule react with the carboxylic group of the other molecule, to form what is known as lactide formation.

   Lactides are the intermolecular cyclic diesters. Two molecules of glyollic acid, $CH_2(OH)—COOH$, react together, to form glycollide.

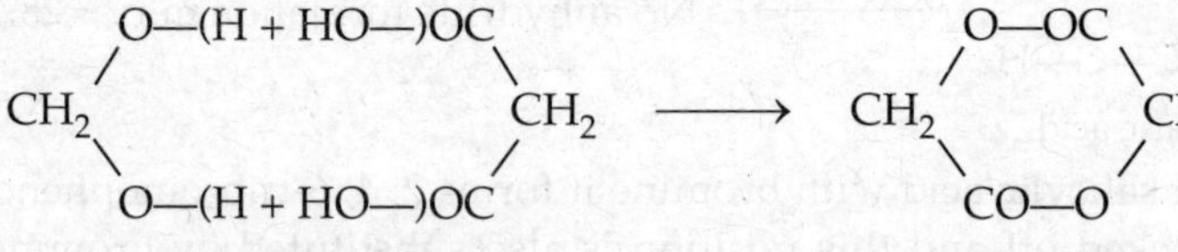

glycollic acid glycollide

   ii. **Action of heat on α-hydroxyl acids:** The mono carboxylic acids, in which one of the β-hydrogen atom is substituted by a hydroxyl group (–OH).

   On heating, these molecules lose a molecule of water, by splitting a hydrogen and hydroxyl group from adjacent carbon atoms, to form α–β **unsaturated acids**.

$$\underset{\text{β-hydroxy butyric acid}}{CH_3—\overset{\overset{OH}{|}}{CH}—\overset{\overset{H}{|}}{C}HCOOH} \xrightarrow[-H_2O]{\text{heating}} \underset{\text{crotonic acid}}{CH_3—CH{=}CH—COOH}$$

   iii. **Action of heat on γ- and δ-acids:** On heating these hydroxy acids form internal esters, known as lactones.

   As these molecules contain a —COOH group at one end and one —OH group at the other end, these react together, eliminate a water molecule, to form α- and δ-lactones.

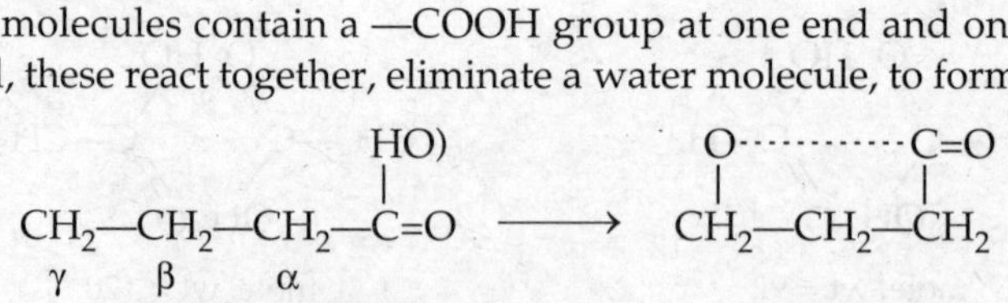

γ-hydroxy acid γ-lactone

   iv. **Action of heat on espilon and other longer hydroxyl acids**

   γ-Hydroxy acids undergo intermolecular esterification to form liner polymers.

$$H)—O—\underset{\delta}{CH_2}—\underset{\gamma}{CH_2}—\underset{\beta}{CH_2}—\underset{\alpha}{CH_2}—CH_2—CO\,(—OH + H)—O—(CH_2)_5\,CO\,(OH \longrightarrow$$

$$O—(CH_2)_5—CO—O—(CH_2)_5—CO—$$

a linear polymer

19. **Hofmann's Bromamide reaction or Hofmann's degradation of amides**
    The reaction is termed as Hofmann's rearrangment.
    It is a very good method to convert an amide into a primary amine.
    $$RCONH_2 + Br_2 + 4NaOH \longrightarrow RNH_2 + 2NaBr + Na_2CO_3 + 2H_2O$$
    The amine formed has one carbon less than the original amide.
    Mechanism:
    Step 1: $RCONH_2 + Br_2 + 4NaOH \longrightarrow Na^+O^-Br + NaBr + H_2O$

    Step 2: $R-\overset{\overset{O}{\|}}{C}-\overset{\overset{H}{|}}{N}-H + NaOBr \longrightarrow R-\overset{\overset{O}{\|}}{C}-\overset{\overset{H}{|}}{N}-Br + NaOH$
    N-bromamide

    Step 3: $R-\overset{\overset{O}{\|}}{C}-\overset{\overset{H}{|}}{N}-Br + OH^- \longrightarrow R-\overset{\overset{O}{\|}}{C}-\ddot{N}: \longrightarrow R-N=C=O$
    Alkyl isocyanate

    Step 4: $R-N=C=O + 2NaO \longrightarrow R-NH_2 + Na_2CO_3$

20 a. **HVZ reaction or a-halogenation:** Carboxylic acids that contain α-hydrogen, on treated with Cl_2 or Br_2, in the presence of phosphorous, the α-hydrogen is substituted by chlorine or bromine atoms. This reaction is termed as Hell-Volhard-Zelinsky reaction.
    $$R-CH_2-COOH + X_2 \longrightarrow RCHXCOOH + HX \ (X = Cl, Br)$$
    b. **Koch reaction or carboxylation of alkenes:** When alkenes are heated to 400°C, with carbon mono-oxide, CO, and steam, under pressure in the presence of phosphoric acid H_3PO_4, carboxylic acids are formed.
    $$CH_2=CH_2 + CO + H_2O \longrightarrow CH_3-CH_2-COOH$$
    Ethene   propionic acid

MULTIPLE CHOICE QUESTIONS

(Group-B)

1. Pick out the reactions in which HCOOH differ from acetic acid
   (a) formic acid reduces Tollen's reagent, and reduces Fehling solution
   (b) HCOOH possess reducing properties
   (c) both do not possess any reducing property
   (d) none
2. HCOOH is
   (a) more acidic than acetic acid   (b) less acidic than HCOOH
   (c) both are equally acidic
3. Which of the following acid is strongest acid
   (a) CH_3COOH (b) HCOOH
   (c) $ClCH_2COOH$ (d) $CHCl_2COOH$
4. When sodium formate is heated with soda lime, the compound formed is
   (a) H_2 (b) CH_4 (c) C_2H_4 (d) CHCH
5. Lactic acid on oxidation with Fenton's reagent gives
   (a) CH_3COOH (b) oxalic acid
   (c) $CH_3CO.COOH$ (d) none

6. Which of the carboxylic acids undergoes decarboxylation easily
   (a) $C_6H_5CO.COOH$
   (b) $C_6H_5COCH_2OH$
   (c) $C_6H_5CH(OH)COOH$
   (d) $C_6H_5CH(NH_2) COOH$

7. The main products of the following reaction
   $CH_3CONH_2 + HNO_2 \longrightarrow$
   (a) CH_3COOH
   (b) $CH_3CH_2NH_2$
   (c) CH_3NH_2
   (d) CH_3OONH_4

8. Hydrolysis of an ester gives acid an "A" and an alcohol "B". The acid reduces Fehling solution. Oxidation of B gives acid A. The ester is
   (a) methyl formate
   (b) ethyl formate
   (c) methyl acetate
   (d) ethyl acetate

9. What are the organic products formed in the following reaction

   $$C_6H_5COOCH_3 \xrightarrow{\text{(1) LiAlH}_4 \text{ (2) } H_3^+ O}$$

   (a) $C_6H_5CH_2$—OH and CH_3OH
   (b) $C_6H_5OH + CH_3OH$
   (c) $C_6H_5CH_3 + CH_3OH$
   (d) $C_6H_5CH_2OH + CH_4$

10. Which of the products formed, on heating adipic acid

    (a) $\begin{array}{l} CH_2\text{—}CH_2 \\ | \qquad\qquad\;\; \rangle O \\ CH_2\text{—}CH_2 \end{array}$

    (b) $\begin{array}{l} CH_2\text{—}CH_2\text{—}CO \\ | \qquad\qquad\qquad\;\; \rangle O \\ CH_2\text{—}CH_2\text{—}CO \end{array}$

    (c) $\begin{array}{l} CH_2\text{—}CH_2 \\ | \qquad\qquad\;\; \rangle C\text{—}O \\ CH_2\text{—}CH_2 \end{array}$

    (d) $\begin{array}{l} CH_2\text{—}CH_2\text{—}COOH \\ | \qquad\qquad\qquad\qquad \rangle O \\ CH_2\text{—}CH_2\text{—}COOH \end{array}$

11. In esterification the reactivity of alcohol is
    (a) $3 > 2 > 1$
    (b) $1 > 2 > 3$
    (c) same in all cases

12. Which acid would give wine red colour with $FeCl_3$?
    (a) Propaonic acid   (b) CH_3COOH (c) HCOOH   (d) none

13. The reaction of HCOOH with conc. H_2SO_4 gives
    (a) CO_2 (b) CO   (c) oxalic acid   (d) acetic acid

14. Two molecules of acetic acid are heated woth P_2O_5, the products formed are
    (a) 2-molesof ethanol
    (b) formic anhydride
    (c) acetic anhydride
    (d) none

15. CH_3COCl reacts with Grignard 's reagent to form
    (a) esters
    (b) ethers
    (c) carbonyl compound
    (d) none

16. Acetic anhydride reacts with ammonia to form
    (a) acetamide
    (b) formamide
    (c) ethyl amine
    (d) methyl amine

17. The main reason for the acids to ionise is
    (a) carboxylic acids do not have a-H atom
    (b) the resonance stabilization of the carboxylate ion
    (c) high reactivity of-hydrogen
    (d) h-bonding

18. Maleic acid and fumaric acids
    (a) have identical m.p.
    (b) have identicalsolubility in water
    (c) form same anhydride on heating

19. An organic halide was treated with KCN and the compound was boiled with dil. HCl, to give a compound B, B can be
    (a) an alkane
    (b) an alkyl halide
    (c) a carboxylic acid
    (d) a ketone

20. The product formed when acetic acid is treated with PCl_3
    (a) CH_3CO—O—PCl_2
    (b) CH_3COOCl
    (c) CH_3COCl
    (d) Cl—CH_2COOH

21. Carboxylic acids react with diazomethane to give
    (a) amines
    (b) alcohols
    (c) esters
    (d) amides ether $HCHOH_2O$

22. Consider the following reactions

    $CH_3Br + Mg \xrightarrow{\quad} A \xrightarrow{HCHO} B \xrightarrow{H_2O} C$; C is

    (a) acetic acid   (b) CH_3CHO (c) ethanol   (d) ethers

23. For hydrolysis of the following functional groups, decreasing order of the reactivity is RCOOR (A), RCOCl (B), $RCONH_2$ (C)
    (a) A > B > C   (b) B > A > C   (c) B > C > A   (d) C > B > A

24. Reactivity of carboxylic acids in esterification is
    HCOOH (A), CH_3COOH (B), RCH_2COOH (C), $R_2CHCOOH$ (D) R_3—CCOOH
    (a) A > B > C > D > E
    (b) B > A > E > D > C
    (c) A > B > D
    d) D > C > B > A

25. $CH_3CH_2COOH \xrightarrow{Cl_2/Fe} X \xrightarrow{alco. KOH} Y$
    (a) CH_3CH_2OH
    (b) CH_3CH_2CN
    (c) CH_2=CH—COOH
    (d) $CH_3CHClCOOH$

26. Ammonolysis of an ester gives
    (a) amine   (b) amide   (c) uried   (d) none

27. IUPAC name of caproic acid is
    (a) pentanoic acid
    (b) hexanoicacid
    (c) heptanoic acid
    (d) octanoic acid

28. The reaction of Br_2 and NaOH or NaOBr, on acetamide gives
    (a) ethyl amine   (b) methyl amine   (c) ethane   (d) ethanol

29. The irritation caused by red ants bite is due to
    (a) lactic acid
    (b) formic acid
    (c) ceric acid
    (d) acetic acid

30. Which of the following, cannot be used for replacing-OH group in organic compound
    (a) S_2Cl_2 (b) $SOCl_2$ (c) PCl_5 (d) PCl_3

31. IUPAC name for tartaric acid is
    (a) 2,3-dihydroxy butane-1,4-dioic acid
    (b) 1,4-dihdroxybutane-2,3-dioic acid
    (c) butane-1,4-dicaboxylic acid

32. The correct order of acid strength is CH_3COOH(A), $CH_2Cl COOH$(B), $CHCl_2COOH$(C)
    (a) A > B > C   (b) C > B > A   (c) B > C > A   (d) D > A > C

33. The end product "B" in the sequence of reactions R—X $\xrightarrow{CN^-}$ A $\xrightarrow{NaOH}$ B
    (a) an alkane
    (b) a carboxylic acid
    (c) sodium saltof acarboxylic acid
    (d) a ketone

34. The identical C—O bond length in carboxylate ion is due to
    (a) resonance
    (b) presence of alkyl groups
    (c) presenceof aphenyl ring
    (d) none

35. Hydrolysis of HCN gives
    (a) CH_3COOH (b) HCOOH   (c) CH_3CHO (d) HCHO

36. Waker's process is used to convert alkene into by using $PdCl_2$
    (a) an alcohol   (b) ketone   (c) aldehyde   (d) an acid

37. In glycine, the basic group is
    (a) NH_2 (b) $-NH_3^+$
    (b) COOH group   (d) COO—

38. The reaction
    $$RCOOAg + Br_2 \longrightarrow RBr + AgBr + CO_2$$
    is called
    (a) HVZ reaction   (b) Borodine Hunsdiecker reaction
    (c) Hofmann's reaction   (d) carbylamine reaction

39. Oxalic acid may be distinguished from tartaric acid by
    (a) $NaHCO_3$ (b) amm. $AgNO_3$ (c) litmus paper   (d) phenolphthalein

40. Weakest acid among the following is
    (a) CH_3COOH (b) $ClCH_2COOH$ (c) $(CH_3)_2CHCOOH$ (d) CCl_3COOH

41. The conversion of –COOH to NH_2 may be carried by
    (a) HVZ Reaction   (b) Hunsdiecker reaction
    (c) Schmidt reaction   (d) decarboxylation

42. IUPAC name for $OCH—(CH)_4—COOH—$
    (a) 1-aldo-hexan-6-oic acid   (b) Hexan-1-al-6-oic acid
    (c) 6-oxo-hexanoic acid   (d) 6-aldo-hexan-1-oic acid

43. The reaction-2 $RCOOAg + I_2 \longrightarrow RCOOR + 2AgI + CO_2$
    The reaction is known as
    (a) Borodine Hunsdiecker reaction   (b) Simonini–Birnbaum reaction
    (c) Hofmann's reaction   (d) Wacker reaction

44. The strongest acid among the following is
    (a) CCl_3COOH (b) CF_3COOH (c) CBr_3COOH (d) CH_3COOH

45. Weakest acid among the following:
    (a) H_2O (b) CH_3COOH (c) C_6H_5OH (d) C_2H_2

46. Which reduces carboxylic acid directly to primary alcohol?
    (a) $LiAlH_4$ (b) Na + ethanol   (c) $NaBH_4$ (d) all

47. Which part of —COOH group, is involved in the reaction of the acid with the metals?
    (a) only H atom   (b) only –OH part
    (c) both (a) and (b)   (d) none

48. Which is a polyprotic acid?
    (a) acetic acid   (b) oxalic acid   (c) benzoic acid   (d) salicylic aid

49. Which of the following groups will increase the acidity of acids?
    (a) electron withdrawing groups   (b) electron donating groups
    (c) both (a) and (b)   (d) none

50. Identify X in the sequence
    $$C_4H_7OCl \xrightarrow{NH_3} C_4H_9ON \xrightarrow{Br_2/KOH} CH_3CH_2CH_2NH_2$$
    (a) $(CH_3)_2CHCOCl$ (b) $CH_3CH_2CH(OH)CH_2—Cl$
    (c) $CH_3CH_2CH_2COCl$ (d) $OHC—CH_2CH_2CH_2—Cl$

51. The principle behind the acidity order of $CCl_3COOH > CHI_2COOH > CH_2ClCOOH > CH_3COOH$, is the withdrawal of electrons and liberation of proton is due to
    (a) inductive effect   (b) resonance
    (c) electromeric effect   (d) mesomeric effect

52. Acetic acid on heating with urea, gives
    (a) $CH_3CONH_2 + CO_2 + NH_3$
    (b) $(NH_4)_2CO_3 + CO_2$
    (c) $CH_3ONH_2 + CH_3COONH_4 + CO_2$
    (d) none of these

53. Iso electric point is the pH at which
    (a) an amino acid becomes acidic
    (b) an amino acid becomes basic
    (c) zwitterion has a +ve charge
    (d) zwitterion has no charge

54. At isoelectric point for amino acids, the species present are

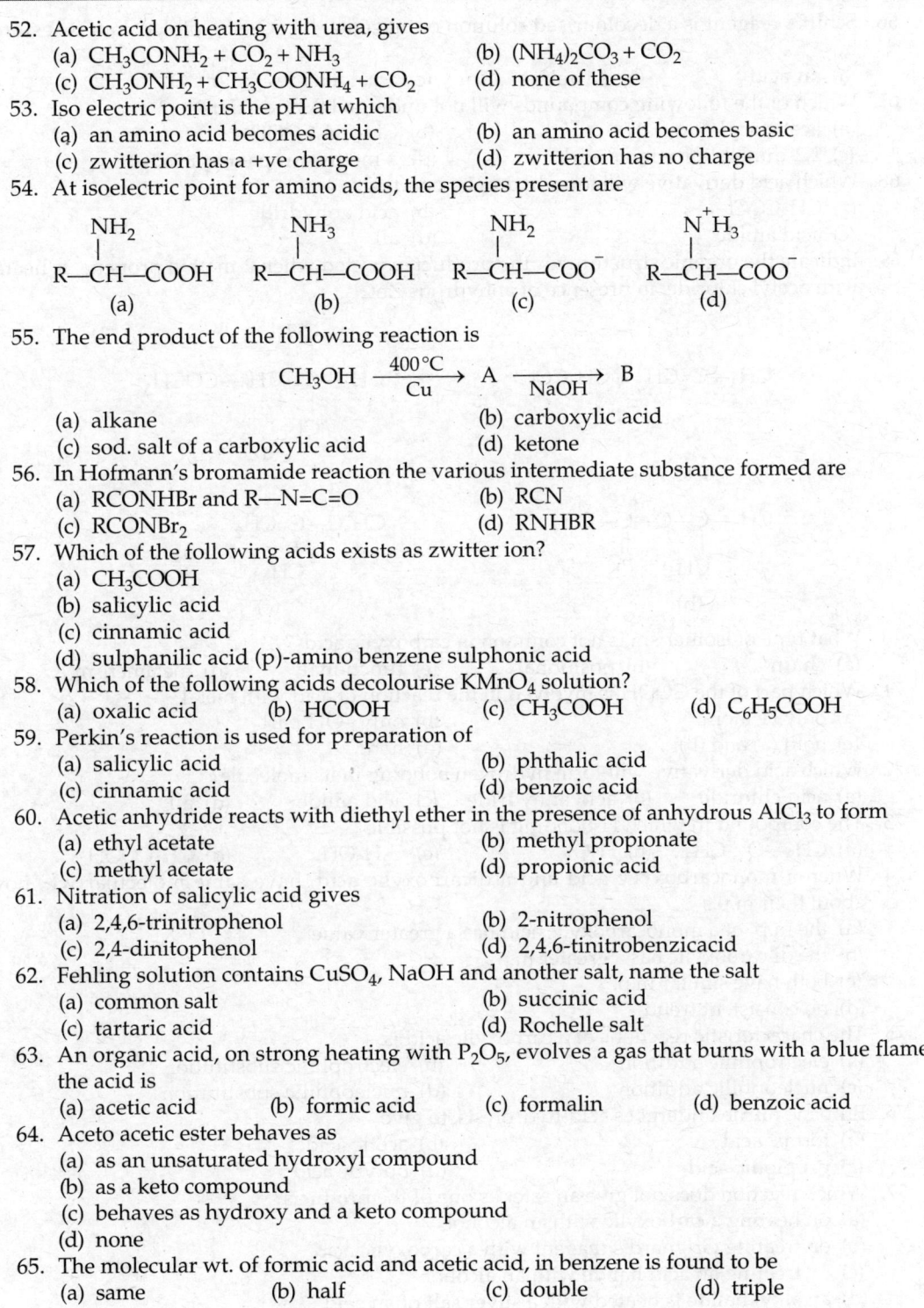

55. The end product of the following reaction is
$$CH_3OH \xrightarrow[Cu]{400\,°C} A \xrightarrow{NaOH} B$$
    (a) alkane
    (b) carboxylic acid
    (c) sod. salt of a carboxylic acid
    (d) ketone

56. In Hofmann's bromamide reaction the various intermediate substance formed are
    (a) $RCONHBr$ and $R—N=C=O$
    (b) RCN
    (c) $RCONBr_2$
    (d) $RNHBR$

57. Which of the following acids exists as zwitter ion?
    (a) CH_3COOH
    (b) salicylic acid
    (c) cinnamic acid
    (d) sulphanilic acid (p)-amino benzene sulphonic acid

58. Which of the following acids decolourise $KMnO_4$ solution?
    (a) oxalic acid
    (b) $HCOOH$
    (c) CH_3COOH
    (d) C_6H_5COOH

59. Perkin's reaction is used for preparation of
    (a) salicylic acid
    (b) phthalic acid
    (c) cinnamic acid
    (d) benzoic acid

60. Acetic anhydride reacts with diethyl ether in the presence of anhydrous $AlCl_3$ to form
    (a) ethyl acetate
    (b) methyl propionate
    (c) methyl acetate
    (d) propionic acid

61. Nitration of salicylic acid gives
    (a) 2,4,6-trinitrophenol
    (b) 2-nitrophenol
    (c) 2,4-dinitrophenol
    (d) 2,4,6-tinitrobenzicacid

62. Fehling solution contains $CuSO_4$, NaOH and another salt, name the salt
    (a) common salt
    (b) succinic acid
    (c) tartaric acid
    (d) Rochelle salt

63. An organic acid, on strong heating with P_2O_5, evolves a gas that burns with a blue flame, the acid is
    (a) acetic acid
    (b) formic acid
    (c) formalin
    (d) benzoic acid

64. Aceto acetic ester behaves as
    (a) as an unsaturated hydroxyl compound
    (b) as a keto compound
    (c) behaves as hydroxy and a keto compound
    (d) none

65. The molecular wt. of formic acid and acetic acid, in benzene is found to be
    (a) same
    (b) half
    (c) double
    (d) triple

66. Schiff's reagent is a decolourised solution of magenta, by SO_2, the pink colour is restored by
    (a) an acid   (b) an amine   (c) an aldehyde   (d) O_3

67. Which of the following compounds will not undergo HVZ reaction?
    (a) acetic acid   (b) CH_3CH_2COOH
    (c) 2,2-dimethl propionic acid   (d) 2-methyl propanoic acid

68. Which acid derivative will form hydrogen bonding?
    (a) CH_3COCl (b) acid anhydride
    (c) acid amide   (d) all

69. Indicate the organic structure for the product expected when 2-methyl propene, is heated with acetyl chloride, in presence of anhydrous $ZnCl_2$.

$$CH_3-\underset{\underset{CH_3}{|}}{C}=CH_2 + CH_3COCl \longrightarrow CH_3-\underset{\underset{Cl}{|}}{\overset{\overset{CH_3}{|}}{C}}-CH_2-COCH_3$$
(a)

$$CH_3-\underset{\underset{CH_3}{|}}{\overset{\overset{CH_3}{|}}{C}}-O-\underset{\underset{O}{||}}{C}-CH_3$$
(b)

$$CH_3\overset{\overset{O}{||}}{C}-\underset{\underset{CH_3}{|}}{C}=CH_2$$
(c)

70. What type of isomerism is not common in carboxylic acids?
    (a) chain   (b) positional   (c) functional   (d) metamerism

71. Which part of the COOH is involved in the reaction of acid with metal?
    (a) only H atom   (b) only –OH part
    (c) both (a) and (b)   (d) none

72. Which acid derivative will form hydrogen bonding in its molecule?
    (a) acid chloride   (b) acid anhydride   (c) acid amide   (d) all

73. The compound in which H-bonding is not possible
    (a) CH_3-O-CH_3 (b) H_2O (c) C_2H_5OH (d) CH_3COOH

74. When a monocarboxylic acid and a dicarboxylic acid, have same molecular wt., how about their m.p.s
    (a) the m.p. of a monocarboxylic acid has a greater value
    (b) the di carboxylic has a greater m.p.
    (c) both have similar m.p.
    (d) no consistent trend

75. The characteristic reactions of a carboxylic acid is
    (a) electrophilic addition   (b) electrophilic substitution
    (c) nucleophilic addition   (d) nucleophilic substitution

76. Propane nitrile undergoes acid-hydrolysis, to give
    (a) formic acid   (b) acetic acid
    (c) propionic acid   (d) butyric acid

77. Which reaction does not give an ester as one of the products?
    (a) on heating a carboxylic with an alcohol
    (b) on treating Grignard's reagent with a carboxylic acid
    (c) by treating an acid halide with an alcohol
    (d) an alkyl halide is heated with a silver salt of an acid

78. A compound is subjected to reduction with $LiAlH_4$, forms an alcohol which on oxidation forms the compound again. The compound is soluble in alkali. The compound is
    (a) an aldehyde   (b) an ester
    (c) a carboxylic acid   (d) a anhydride

ANSWERS

(Group-B)

1. (a)	2. (a)	3. (d)	4. H_2	5. (d)
6. (b)	7. (a)	8. (a)	9. (b)	10. (b)
11. (a)	12. (b)	13. (b) CO	14. (c)	15. (c)
16. (a)	17. (a)	18. (c)	19. (c)	20. (c)
21. (c) methyl esters		22. (c)	23. (b)	24. (a)
25. (c)	26. (b)			

27. (b) $CH_3(CH_2)_4 COOH$

28. (b)	29. (b)	30. (a)	31. (a)	32. (b)
33. (c)	34. (a)	35. (b)	36. (c) aldehyde	37. (a) $-NH_2$
38. (b)				

39. (b) tartaric acid reduces Tollen's reagent

40. (c)	41. (c)	42. (d)	43. (b)	44. (b)
45. (d) C_2H_2	46. (d)	47. (a)		

48. (b) poly protic acid contains more than one proton

49. (a)	50. (c)	51. (a)	52. (a)	53. (a)
54. (d)	55. (c)	56. (a)	57. (d)	58. (a)
59. (c)	60. (a)	61. (a)	62. (d)	

63. (b) HCOOH, CO is evolved, which burns with a blue flame
64. (c) due to tautomerism
65. (c) owing to association
66. (c)
67. (c) no α-H is present

68. (c)	69. (a)	70. (d)	71. (a)

72. (c) H is attached to N atom will form H-bonding

73. (a)	74. (b)	75. (d)	76. (c)	77. (b)

78. (c)

REASONING TYPE QUESTIONS (RTQs)

(Group-C)

1. Explain the acid strength
   i. HCOOH >> CH_3COOH;   ii. $C_6H_5COOH > CH_3COOH$
2. Chloro acetic acid is stronger than glycolic acid.
3. The b.p.s of carboxylic acids are higher than the acid chloride or acid anhydrides.
4. Benzene sulphonic acid is a very strong acid.
5. o-Nitro benzoic acid is stronger than p-nitro benzoic acid.
6. Ka1 value of maleic acid is higher than Ka1 value of fumaric acid, but Ka2 value of fumaric acid is higher than that of maleic acid.
   Ka1 maleic acid > Ka1 fumaric acid
   Ka2 fumaric acid > Ka2 of maleic acid

7. Anthranilic acid does not exists as a zwitter ion.
8. The both C—O bond length in sodium formate are equal.
9. It is difficult to esterify 2,4,6-tri methyl benzoic acid.
10. The acid strength o-HO—C_6H_4COOH > C_6H_5COOH > p-HO—C_6H_4COOH.
11. CH≡C—COOH, propynoic acid is stronger than CH_2=CH—COOH propenoic acid.

ANSWERS

(Group-C)

1. i. Formic acid is stronger than acetic acid as acetic acid contains a —CH_3 group attached to a COOH groups, on account of +I effect (electron donating or electron releasing inductive effect) of the —CH_3 group, increases the negative charge on the COOH group, destabilises the negative charge on it, with the result the loss of proton becomes more difficult.
    As the size of alkyl groups increases the acidity decreases.
    ii. Benzoic acid is stronger than acetic acid because the benzoate ion is stabilised on account of the resonance in the benzene ring. On the other hand in acetic acid, the +I effect makes the acid weaker.
    Electron withdrawing groups (NO_2, Cl, Br, OH, CN) increase the acidity. This is because the electron releasing groups decrease the –ve charge on the carboxylate ion, and stabilise it.
2. Chloro acetic acid is stronger than glycolic acid because Cl negative effect has aquatic –I-effect than –OH group.
    $$CH_2ClCOOH > CH_2(OH)COOH$$
3. Carboxylic acids have extensive H-bonding, where as acid chlorides or anhydrides do not have any H-bonding.
    $$RCOOH > RCOCl \quad or \quad (RCO)_2O$$
4. H_2SO_4 is a mineral acid and all acids are very strong acids. Sulphuric acid is a dibasic acid. In forming a benzene sulphonic acid, only one hydrogen is substituted by a phenyl group, it still contains one replaceable H-atom.
    $$H_2SO_4 \longrightarrow C_6H_5SO_3H$$
5. This is on account of ortho-effect. All ortho-substituted benzoic acids are stronger than benzoic acids even if the group happen to be electron donating.
6. Maleic acid ionises easily on account of the two COOH groups being closer to each other, the resulting in the formation of a more stable maleate ion. On the other hand fumaric acid being a *trans* acid ionises to form a fumerate ion.

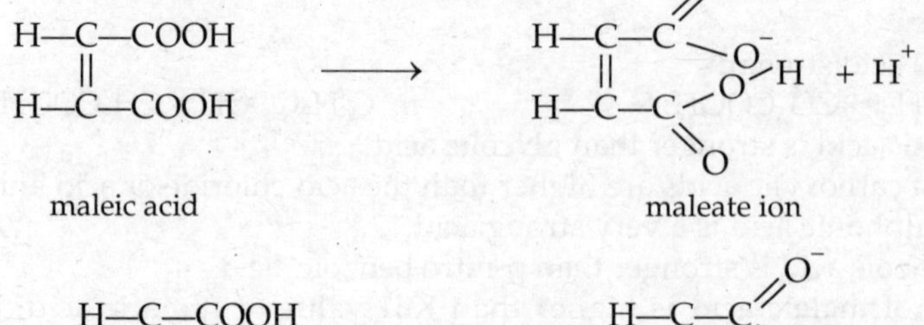

maleic acid   maleate ion

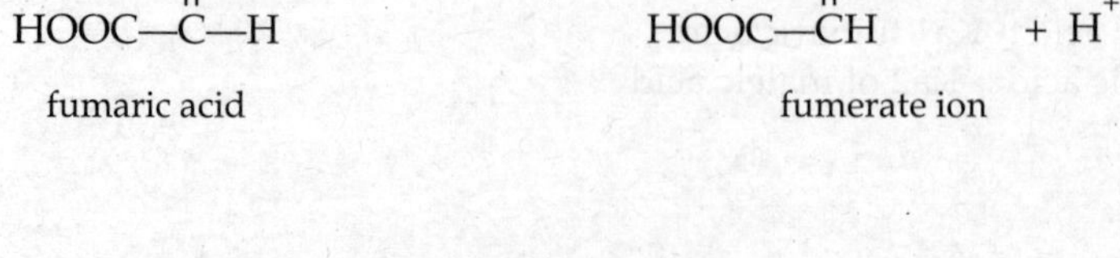

fumaric acid   fumerate ion

Ka1 of maleic acid > Ka1 of fumaric acid
In the second ionisation, the maleate ion does not ionise easily as the proton is held by one oxygen through a bond and by the oxygen through a hydrogen bonding. Fumerate ion ionises as usual. This is why Ka1 of maleic acid > Ka1 of fumaric acid, and in the same way Ka2 of fumerate ion > Ka2 of maleate ion.

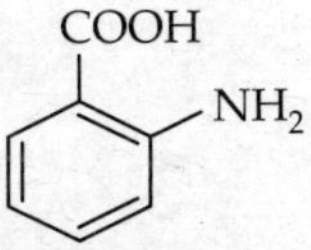

COOH
NH$_2$

7. Anthranilic acid is a o-amino benzoic acid.
   The –COOH is an electron withdrawing group, therefore it deactivates the ring. Since –NH$_2$ happen to be in the ortho position to the –COOH group, the –NH$_2$ group become electron deficient, with the result amino group hold its protons tightly.

8. In sodium formate, both the C—O bond have same bond length, i.e. 1.27Å, instead of having a C=O bond length = 1.23Å and the C—O bond = 1.36Å, proves that the carboxylate ion exists as a resonance hybrid

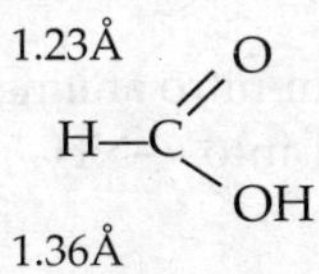

1.23Å   O
H—C
        OH
1.36Å

formic acid

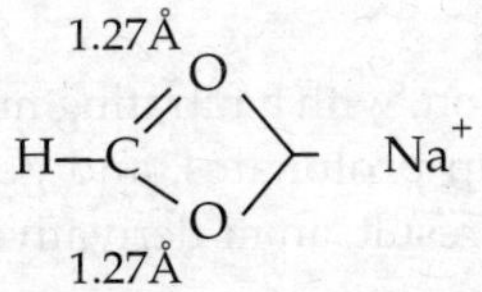

1.27Å   O
H—C   Na$^+$
        O
1.27Å

sodium formate

9. It is difficult to esterify 2,4,6-trimethyl benzoic acid, because of the steric hindrance.
   Owing to the presence of two methyl groups in the o-positions to the COOH group, the alcohol finds it difficult to reach the —COOH group. This is known as steric hindrance or space problem.

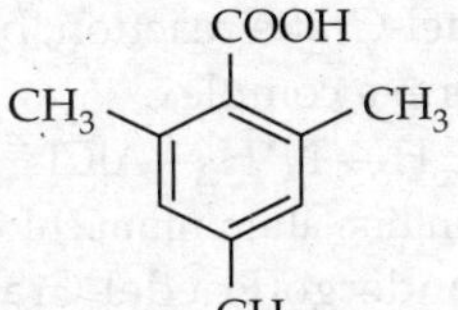

COOH
CH$_3$ CH$_3$

CH$_3$

10. Owing to the ortho effect, the o-hydroxy benzoic acid is strongest.
    The p-hydroxy benzoic acid as a weaker acid an even than benzoic acid, because of the –OH group, which is electron releasing group, destabilises the negative charge on the –COOH group, and makes it weaker.

11. CH≡C—COOH, Propynoic acid is stronger than propenoic acid, CH$_2$=CH—COOH, because the CH≡C group is more electron withdrawing than a CH$_2$=CH group.

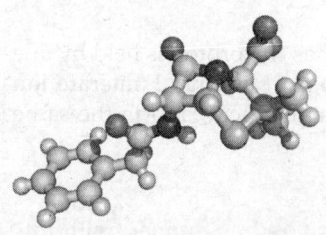

Chapter 6

Amines

Salient Features

1. Aniline on nitration, with a nitrating mixture gives m-nitro aniline. In the acidic medium. The —NH_2 group protonates, and gets converted into —NH_3^+ group, which becomes deactivating as a result, an m-derivative is formed.

aniline   + HNO_3 + H_2SO_4 ⟶   anilinium ion

        nitrating mixture

2. Aniline does not undergo Friedel-Crafts reaction, because aniline combines with electron deficient, anhydrous $AlCl_3$ to form a complex.

$$C_6H_5NH_2 \ + \ AlCl_3 \longrightarrow C_6H_5\text{—}N^+H_2\text{—}AlCl_3^-$$
   aniline   aniline-aluminium chloride complex

3. Nitro benzene also does not undergo Friedel-Crafts reaction. It is because, the ring is deactivated by —NO_2 group, to such an extent that the benzene ring, does not form a π-complex with the electrophile. Hence no reaction takes place.

   Moreover, nitrobenzene is used as a solvent for carrying out, Friedel-Crafts reaction.

4. Alkyl amines are more basic than aryl amines (nuclear substituted amines).

   R—NH_2 $ArNH_2$
   alkyl amine   aryl amine
   Kb = 4×10^{-4} 4.2×10^{-10} (value for aniline)

5. Nitro alkanes, (R—NO_2), and nitro arenes, (Ar—NO_2), are highly polar molecules, due to very strong dipole-dipole molecular attractions.

6. 1°, (RNH_2) and 2° amines, (R_2NH), form intermolecular hydrogen bonding.

7. The bond angle, $CH_3 \diagup \overset{N}{\underset{CH_3}{|}} \diagdown CH_3$ in trimethyl amine, $(CH_3)_3N$, is 108°. The molecule is 'chiral'.

8. 'Ortho effect' in aniline–All o-substituted anilines, are weaker bases, as compared to aniline.

9. Amines, having a chiral N—atom, cannot be resolved, into enantiomers, because of rapid flipping of one enantiomer into the other.

10. CH_2—CH_2 –The compound is named as aziridine or azocyclopropane.
    $\underset{NH}{\diagdown\diagup}$

11. Boron hydride, B_2H_6 in THF, can reduce amides to the corresponding amines, with the same number of carbon atoms.

12. Alkyl cyanides, RCN, and alkyl isocyanides, RNC, are more polar than RX; they have high boilng point, than RX, because of intermolecular dipole-dipole attractive interactions.

13. Alkyl cyanides, RCN, are more polar than alkyl isocyanides, RNC, because the inter-molecular forces are weaker in RNC.

14. RCN are soluble in water (R = lower alkyl groups).

15. Alkyl isocyanides, RNC, do not undergo basic hydrolysis, because of their inability to undergo attack by OH^-.

16. On account of the +I effect of the alkyl groups, the basicity of the alkyl amines should be, in the decreasing order:

    $$R_3N \quad > \quad R_2NH \quad > \quad RNH_2$$
    $$3° \quad > \quad 2° \quad > \quad 1°$$

    This is observed only in the gaseous phase.

    However in the aqueous medium, the decreasing order, (D.O.) of the alkyl amines, is as follows:

    $$R_2N \quad > \quad RNH_2 \quad > \quad R_3N \quad > \quad NH_3$$
    $$2° \quad > \quad 1° \quad > \quad 3°$$

    In the aqueous medium, the amines react with H_2O to form corresponding ammonium hydroxides.

    On account of the association, as well crowding of various groups around the nitrogen atom, the H^+ is not able to approach the nitrogen of the tertiary amine, with the result, the tertiary amine is much weaker than even a primary amine.

17. The decreasing order of basicity's of various amines.

    a. The D.O. order of the basicity, of various alkyls as well aryl amines is as follows:
       $R_2NH > RNH_2 > R_3NH > ArCH_2NH_2 > NH_3 > ArNHR > ArNH_2 > Ar_2NH > Ar_3N$

    b. The D.O. order of basicity of the various carbanion is as follows:

       $H:^- > R:^- > H_2N:^- > CH\equiv C:^- > OR:^- > HO:^- > RCOO:^-$

    c. The D.O. of the basicity of the amines is as follows:
       $$RNH_2 \quad > \quad RCH=NR \quad > \quad RC\equiv N:$$
       $$sp^3 \qquad\quad sp^2 \qquad\qquad sp$$

       More is the s-character in the hybridised orbitals, of the nitrogen with the unshared pair of electrons; the less basic is the molecule.

    d. The D.O. of the amines, showing –I effect of the —OH group:
       CH_3—CH_2—NH_2 > HO—CH_2—CH_2—CH_2—NH_2 > HO—CH_2—CH_2—NH_2

    e. The D.O. of the following amines with the unsaturation in the chain
       i. '–I' effect of the various group is in the D.O.

          $$HC \equiv C—CH_2— \quad > \quad CH_2=CH—CH_2— \quad > \quad CH_3—CH_2—CH_2—$$
          $$\text{propargyl} \qquad\qquad\qquad \text{allyl} \qquad\qquad\qquad \text{propyl}$$

       ii. CH_3—CH_2—CH_2—NH_2 > CH_2=CH—CH_2—NH_2 > $HC\equiv C$—CH_2—NH_2

6. Basicity in cyclic compounds

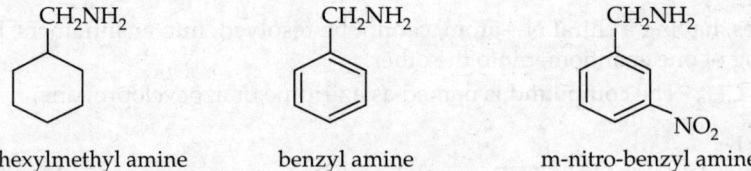

| cyclohexylmethyl amine | benzyl amine | m-nitro-benzyl amine |

7. Aniline is a weaker base as compared to the cyclohexylamine

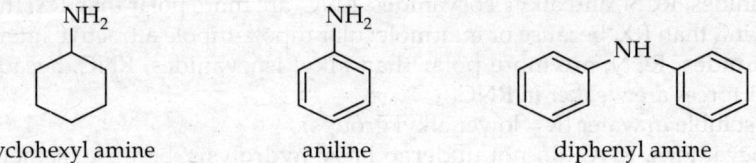

| cyclohexyl amine | aniline | diphenyl amine |

8. The basicity of the nitro-anilines is as follows

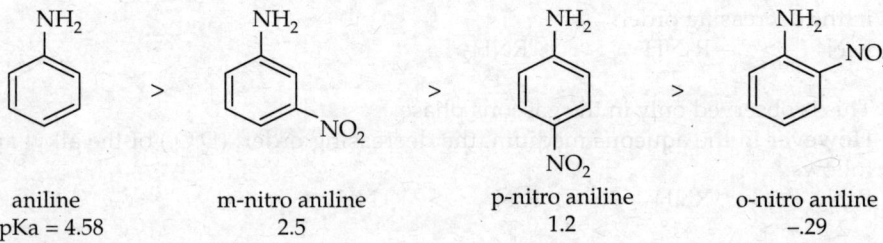

| aniline | m-nitro aniline | p-nitro aniline | o-nitro aniline |
| pKa = 4.58 | 2.5 | 1.2 | −.29 |

9. The basicity of methoxy-anilines is as follows

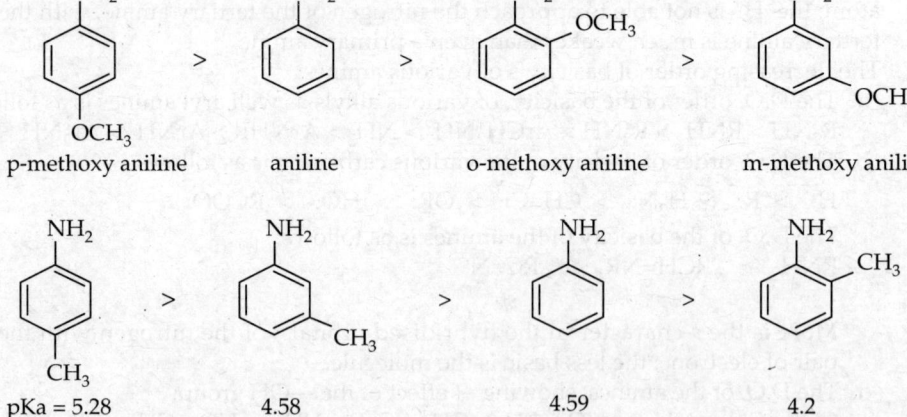

| p-methoxy aniline | aniline | o-methoxy aniline | m-methoxy aniline |

| pKa = 5.28 | 4.58 | 4.59 | 4.2 |

Methoxy group —OCH_3, is electron donating from o and p-positions, but acts as electron withdrawing group.

(Note: In amino derivatives, NH_2— is assigned the lowest number on the longest chain)

10. The basicity of methyl anilines is as follows

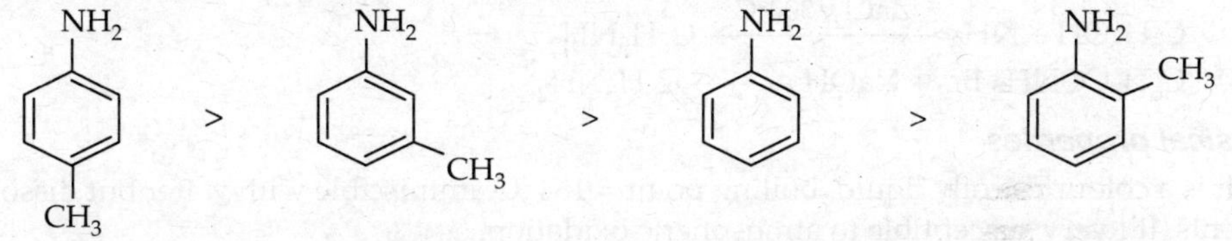

| p-methyl aniline pKa = 5.12 | m-methyl aniline 4.69 | aniline 4.58 | o-methyl aniline 4.39 |

11. The D.O. of basicity of alkyl amine as compared with that of aryl amide and acetamide.

$$CH_3—CH_2—NH_2 \quad > \quad C_6H_5CONH_2 \quad > \quad CH_3CONH_2$$
$$\text{ethyl amine} \qquad\qquad \text{benzamide} \qquad\qquad \text{acetamid}$$

Benzamide is weaker than ethyl amine but stronger than acetamide.

The amides are weaker bases than corresponding amines, because the ability of >C=O to participate in the base weakening extended p-bonding with —NH$_2$ group

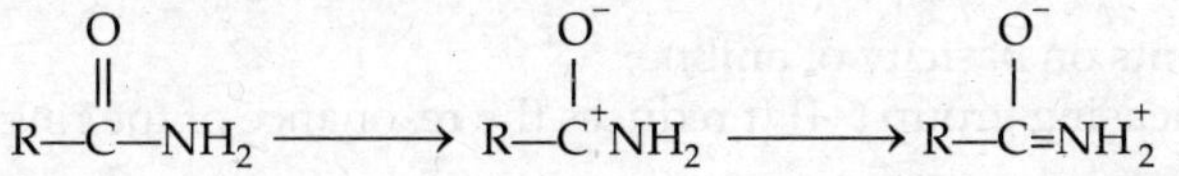

The amido group —CONH$_2$, enters into conjugation with benzene ring.

12. (a) The D.O. of basicity of some cyclic nitrogen compounds

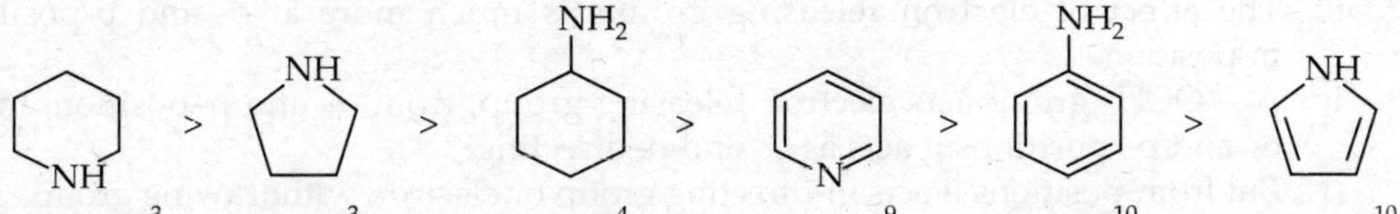

| piperidine Kb = 2×10^{-3} | pyrrolidine 1×10^{-3} | cyclohexyl amine 5×10^{-4} | pyridine 2.3×10^{-9} | aniline 11×10^{-10} | pyrrole 2.5×10^{-10} |

(b) Aniline and N— alkyl substituted derivatives

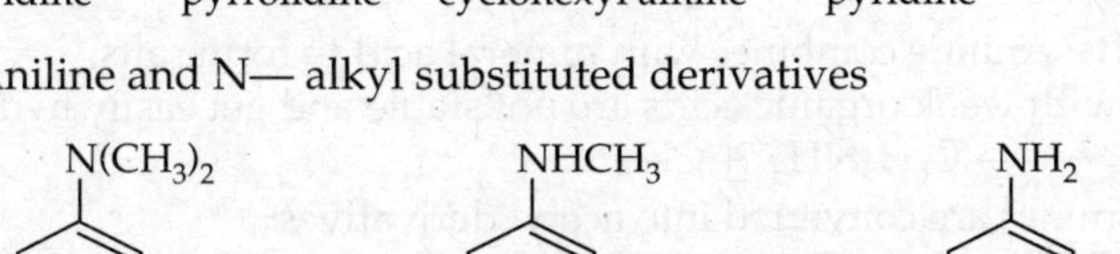

| dimethyl aniline Kb = 2×10^{-10} pKa = 5.06 | methyl aniline 7×10^{-10} 4.85 | aniline 4.2×10^{-10} 4.58 |

(Note: A higher value for pKa for amines mean, the amine is more basic).

Aniline, C$_6$H$_5$NH$_2$

Method of Preparation

1. From nitro benzene: nitro benzene on reduction with metal and HCl forms aniline

$$C_6H_5NO_2 + 6H \longrightarrow C_6H_5NH_2 + 2H_2O$$

2. From chloro benzene

$$C_6H_5Cl + NH_3 \xrightarrow{\text{Cu}_2\text{O}} C_6H_5NH_2$$

3. From phenol

$$C_6H_5OH + NH_3 \xrightarrow{ZnCl_2/300°C} C_6H_5NH_2$$

4. $C_6H_5CONH_2\ Br_2 + NaOH \longrightarrow C_6H_5NH_2$

Physical properties

It is a colourless oily liquid, boiling point = 184°C, immiscible with water but dissolve organic solvents. It I very susceptible to atmospheric oxidation.

Chemical properties

It forms salts with mineral acids and the salt are quite stable 20. It is a weaker base than aliphatic primary amine. Owing to resonance, the lone pair of electron is involved in resonance, with the result, the lone pair of electrons is much less available on aniline

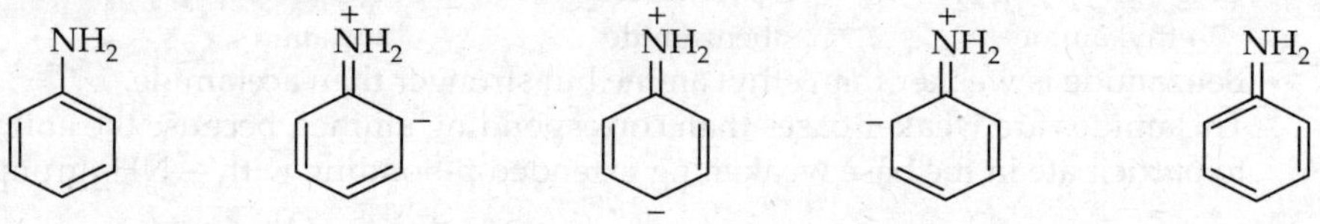

Effect of substituents on basicity of aniline,

  i. An electron releasing group (+I) It reduces the resonance of the ring, therefore increase the basicity.
 ii. The electron with drawing group (–I effect) withdraws the electron towards itself self, therefore these groups decrease the basicity.
iii. The effect of electron releasing groups is much more at o- and p-positons, than at m-position.
 iv. A —OCH$_3$ group acts electron releasing group, from o– and p–position, therefore from o– and p– positions it acts as o– and p–directing.
  v. But from-positions it acts m-directing group or electron withdrawing group.

Method of Preparation

1. **Formation of salts:** Aniline combines with mineral acid to form salts.
   The salt formed with weak organic acids are not stable and get easily hydrolysed.
   $C_6H_5NH_2 + HCl \longrightarrow C_6H_5NH_3^+ + Cl—$
2. **Acylation:** The amines are converted into acetyl derivatives.
   $C_6H_5NH_2 + CH_3COCl \longrightarrow CH_3NHCOCH_3 + HC$
3. **Alkylation:** On treating aniline withdimethyl sulphate $(CH_3)_2SO_4$, aniline is converted to form a mono and a dimethyl derivatives. Halogens have both –I and +I effect. However +R effect is very small. The presence of halogen decreses the basicity.
4. **Carbyl amine reaction or isocyanide test:** On heating any primary amine (aliphatic or aromatic primary amine) with CHCl$_3$ and KOH, a very foul smelling compound, the isocyanide is formed
   $C_6H_5NH_2 + CHCl_3 + 3KOH \longrightarrow C_6H_5NC + 3KCl + 3H_2O$

   Mechanism: i. $CHCl_3 + OH \longrightarrow C^-Cl_3 + H_2O$

   ii. $CCl_3^- \longrightarrow :CCl_2 + Cl–$

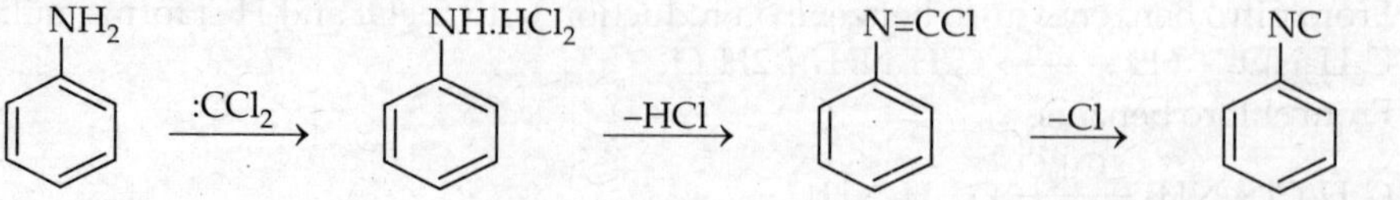

5. Condensation with benzaldehyde

$C_6H_5NH_2 + OHC\ C_6H_5 \longrightarrow C_6H_5—N=CH\ C_6H_5$ –Schiff's base

benzal aniline

6. On heating aniline with conc. H_2SO_4 to 180°C, it forms finally sulphanilic acid.

7.

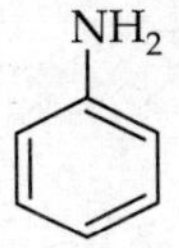

(C$_6$H$_5$—NH$_2$)$_2$ H$_2$SO$_4$ $\xrightarrow{\text{aniline}}$ C$_6$H$_5$NH$_2$ H$_2$SO$_4$ $\longrightarrow$

aniline sulphate   aniline hydrogen
                        sulphate

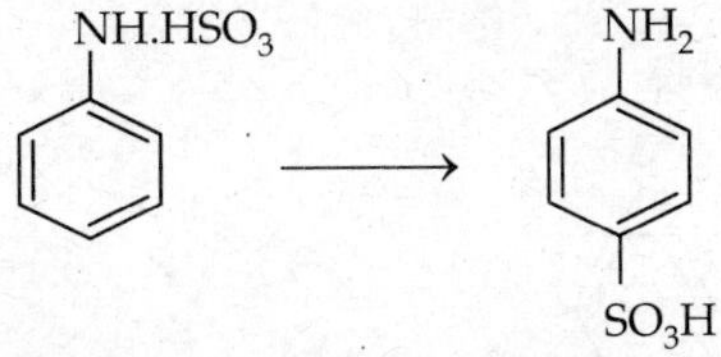

sulphaminic acid   sulphanilic acid

8. On treating aniline with Br_2, it forms a 2, 4, 6-tribromo aniline.
9. Sulphanilic acid with Br_2, it is convrteto form 2, 4, 6-tribromo aniline

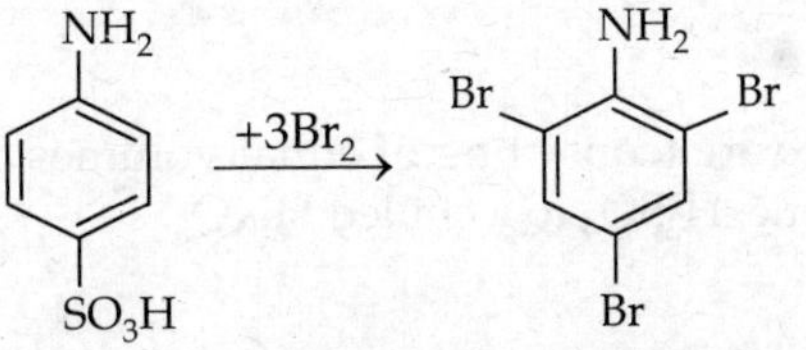

This reaction is termed as IPSO reaction.

10. On treating aniline with conc. HCl and $NaNO_2$ below a temprature 10°C benzene is converted to form benzene diazonium chloride. This process is known as *diazotisation*.

$C_6H_5NH_2 + HCl \longrightarrow C_6H_5NH_2.HCl$

$NaNO_2 + HCl \longrightarrow HNO_2 + Na\ Cl$

$C_6H_5NH_2.HCl^- + HNO_2 \xrightarrow{-5°C} C_6H_5N_2Cl$

Benzene Diazonium Chloride

Benzene diazonium chloride is a crystalline compound, soluble in water. As the compound is stable only below 10°C, it cannot be preserved. Therefore as soon it is prepared, it is used immediately. HNO_2 so formed and is consumed, without isolating. This is termed as using HNO_2' in situ.

Properties

Benzene diazonium chloride is highly reactive substance. On keeping it reacts with water, and is hydrolysed to form phenol.

$C_6H_5N_2Cl + H_2O \longrightarrow C_6H_5OH + HCl + N_2$

1. Reduction: On treating it with hypo phosphorous acid, H_3PO_2 it is reduced to form benzene

$C_6H_5N_2Cl + H_3PO_2\ H_2O \longrightarrow C_6H_6 + N_2 + H_3PO_3$

2. —N_2Cl group can be replaced by Cl. Br, I or CN groups,

3. $C_6H_5N_2Cl\ HCl \xrightarrow{Cu_2Cl_2} C_6H_5Cl + N_2$

This reaction is known as **Sandmeyer's reaction**

4. Similarly, bromo an iodo compounds can be obtained by warming it with corresponding bromo or cyano compounds of copper.
5. Sandmeyer's reaction
   Mechanism: It is a free radical reaction
   i. $C_6H_5N_2Cl \longrightarrow C_6H_5 N_2^+ + Cl^-$
   ii. $C_6H_5 N_2^+ + Cu^+ \longrightarrow C_6H_5 \bullet + Cu_2^+ + N_2$
   iii. $C_6H_5 \bullet + Cl^- \longrightarrow C_6H_5Cl + e$
   iv. $Cu_2^+ + e \longrightarrow Cu^+$
6. Cupling

$$C_6H_5N_2Cl + \text{aniline} \longrightarrow C_6H_5N= N—NH_6H_5$$

Phenol

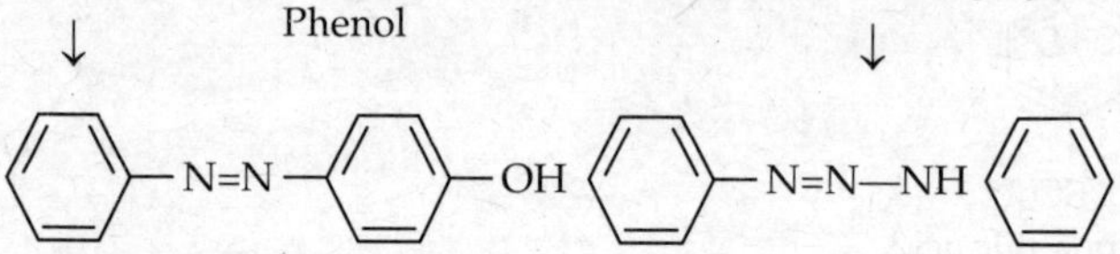

p-hydroxy azo benzene   diazo amino benzene

Nitro Benzene, $C_6H_5NO_2$

Method of Preparation

It prepared by nitrating benzene, by nitrating mixture (consisting of equal volumes of conc. HNO_3 and conc. H_2SO_4, prepared by adding slowly conc. H_2SO_4 to a chilled HNO_3.

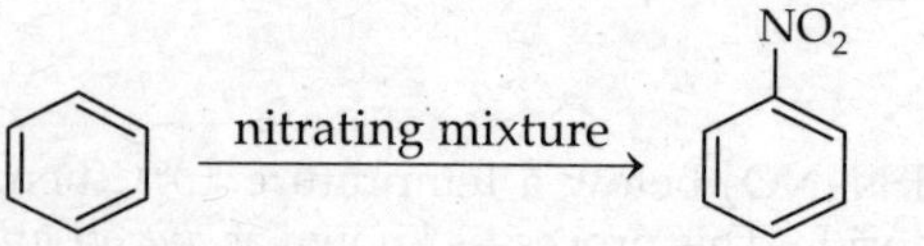

Physical properties

It is a slightly yellow coloured liquid, b.p. 210 °C it is affected only by reducing agents
i. Reduction *in acidic medium*: it is reduced in presence metal and HCl, to form aniline

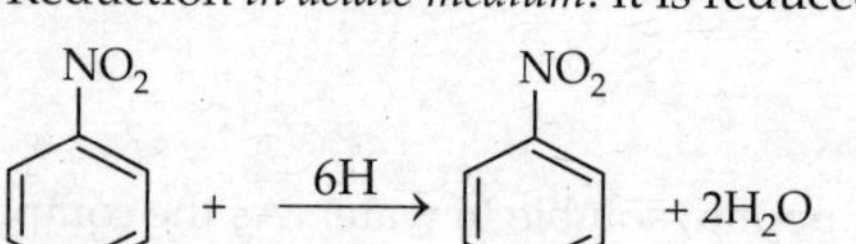

ii. Reduction *in alkaline medium* with metal like tin, aluminium or zinc, in presence of caustic alkali, two molecules are reduced to form a mixture of three products, azoxy benzene, azo benzene and hydrazo benzene.

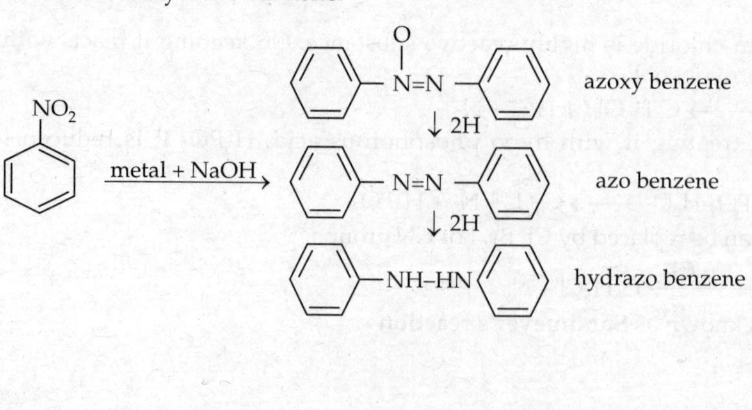

azoxy benzene

$\downarrow$ 2H

azo benzene

$\downarrow$ 2H

hydrazo benzene

On fusing with solid KOH nitrobenzene form mixture of o- and p- nitro phenols

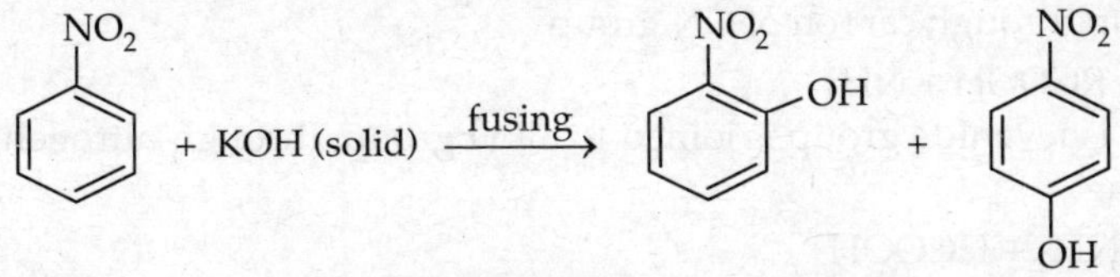

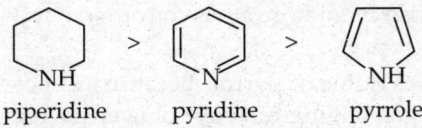

iii. It is a rare example of nucleophilic substitution.

REASONING TYPE QUESTIONS (RTQs)

(Group-A)

1. Nitrobenzene is neutral.
2. Acetamide is less basic than aniline.
3. Alkyl cyanides give acids on hydrolysis whereas alkyl isocyanides give amines.
4. Nitro alkanes on reduction give an amine but alkyl nitrite give an alcohol.
5. Phenyl isocyanide reacts with sulphur to form phenyl iso thio cyanate.
6. The basicity of the following bases is in the order

piperidine > pyridine > pyrrole

7. An aqueous solution of alkyl amines precipitates the metal as their hydroxide form their salt solution.
8. Salts formed by amine with HCl or with H_2SO_4 are quite stable but the salt formed by aniline with HCl or H_2SO_4 are easily hydrolysed.
9. Glycine exists as a zwitterion.

ANSWERS

1. The NO_2 group does not have any replaceable H atom. It is joined to a sp^2 hybridised carbon, which also does not have any H-atom.

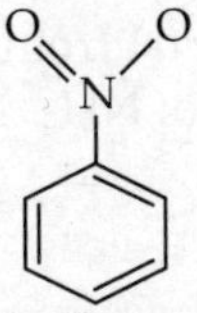

nitro benzene

2. Acetamide has $CO-NH_2$ group in the molecule. A lone par of electron is available for protonation. However, the lone pair of electrons enters in conjugation with the C=O group and forms a peptide type of linkage and not available for protonation.

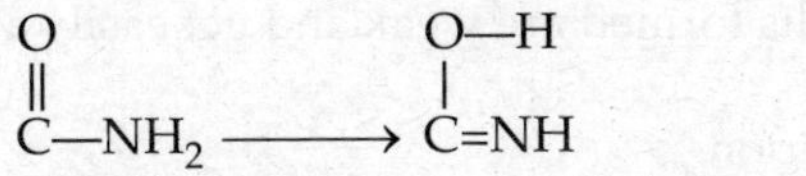

$$\underset{\text{amide}}{\overset{O}{\underset{\|}{C}}-NH_2} \longrightarrow \underset{\text{a peptide linkage}}{\overset{O-H}{\underset{|}{C}=NH}}$$

In aniline the lone pair is available much more than in the case of an amide. Thus amide is less basic than aniline.

3. On hydrolysis alkyl cyanides form acids on hydrolysis because the cyanide group is joined to the alkyl group through carbon of CN group.

$$R\text{—}CN + 2H_2O \longrightarrow RCOOH + NH_3$$

Alkyl isocyanide have isocyanide group is joined to alkyl group, through nitrogen, hence the formation of n amine

$$RNC + 2H_2O \longrightarrow RNH_2 + HCOOH$$

4. The nitro alkanes have the NO_2 group joined through nitrogen

$$R\text{—}N{\overset{\diagup O}{\diagdown O}} \longrightarrow R\text{—}NH_2$$

   nitro alkane   amine

Alkyl nitrite have nitrogen joined through oxygen

$$R\text{—}O\text{—}N{=}O \longrightarrow ROH$$

   alkyl nitrite   alcohol

5. Alkyl isocyanide have a pair of electrons on the carbon of –NC: group. Therefore on reaction with sulphur it is easily converted into iso thiocyanate.

$$R\text{—}NC: + S \longrightarrow R\text{—}NCS$$

In the same way on treating it with HgO, the iso cyanide group is converted into —NCO.

$$R\text{—}NC: + Hg\,O \longrightarrow RNCO + Hg$$

6. Piperidine is a stronger base as compared to pyridine or pyrrole because the molecule has $+I$ effect of $5\text{–}CH_2$, in ring, besides the nitrogen having full control over the loan pair of electrons.

Pyridine is stronger because the loan pair of nitrogen is available on the nitrogen to form salts.

The loan pair of electrons in the case of pyrrole to nitrogen is involved in the formation of a sextet by Huckel's rule for the compound to behave as aromatic compound. As result, pyrrole does not behave as a base but is slightly acidic in nature.

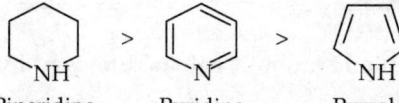

   Piperidine   Pyridine   Pyrrole

7. Aqueous solution of an alkyl amine precipitates the hydroxides of metal from their salt solution. In aqueous solution the amine form hydroxides, as in the case of NH_3.

$$NH_3 + H_2O \longrightarrow NH_4OH \;;\; NH_4OH \longrightarrow NH_4^+ + OH^-$$
$$M^{2+} + 2OH \longrightarrow M(OH)_2$$
$$\qquad\qquad\qquad\text{precipitate}$$
$$RNH_2 + H_2O \longrightarrow RNH_3OH \longrightarrow RNH_3^+ + OH^-$$
$$M^{2+} + 2OH \longrightarrow M(OH)_2$$

8. Owing to resonance, the lone pair responsible for aniline behaving as a base are not always located on nitrogen, with the result, salts formed are weak and get easily attacked by water and get hydrolysed.

9. Glycine is amino acetic acid, it exits as a zwitterion.

$$NH_2\text{—}CH_2COOH \longrightarrow NH^+{}_3\text{—}CH_2COO^-$$

MULTIPLE CHOICE QUESTIONS

(Group-B)

1. Which amine gives, the characteristic smell of an isocyanide on heating it with KOH and CHCl$_3$.
   (a) C$_6$H$_5$NH$_2$ (b) (CH$_3$)$_2$NH   (c) C$_6$H$_5$NHCH$_3$ (d) (C$_6$H$_5$)$_2$NH

2. A reaction between RCONH$_2$ and Br$_2$/KOH, the main product is RNH$_2$, the intermediate involved in the reaction are
   (a) RCONHBr   (b) RNHBr   (c) R—N=C=O   (d) RCOBr$_2$

3. The reaction RCONH$_2$ $\xrightarrow{\text{NaOBr}}$ gives
   (a) RBr   (b) RH   (c) ROBr   (d) RNH$_2$

4. CH$_3$CONH$_2$ $\xrightarrow{\text{P}_2\text{O}_5/\text{heat}}$ gives
   (a) CH$_3$CN
   (c) CH$_3$CHO
   (b) CH$_3$NH$_2$
   (d) CH$_3$—CH$_3$ + CO + NH$_3$

5. Which one of the following gives a carbyl amine test, on heating with KOH and CHCl$_3$
   (a) RCONH$_2$ (b) CCl$_4$ (c) RNH$_2$ (d) R$_2$NH

6. C$_2$H$_5$NH$_2$ can be prepared by the action of Br$_2$ and KOH on
   (a) acetamide   (b) propionamide   (c) CH$_3$CN   (d) CH$_3$NH$_2$

7. Which one of the following is the correct statement?
   (a) an aqueous solution of C$_2$H$_5$NH$_2$ is basic
   (b) C$_2$H$_5$NH$_2$ is less basic than NH$_3$
   (c) ethyl amine is a stronger base ammonia
   (d) ethyl amine forms salt with bases

8. Among the following which compound will react with acetone to form a product containing >N=C< group
   (a) aniline   (b) tri methyl amine (c) C$_2$H$_5$OH   (d) phenyl hydrazine

9. A positive carbylamine test is given by
   (a) N, N-Dimethyl aniline
   (c) N-methyl O-methyl aniline
   (b) 2, 4-dimethyl-N-methyl aniline
   (d) p-methyl benzyl amine

10. The common feature among the species is CN$^-$, CO and NO$^+$ are
   (a) each species has 14 electrons and the bond order is same
   (b) no relation
   (c) with 18 electrons
   (d) with 8 electrons in outer orbit

11. The correct order of the basicities of the following compound
   (a) 2 > 1 > 3 > 4   (b) 1 > 3 > 2 > 4   (c) 3 > 1 > 2 > 4   (d) 1 > 2 > 3 > 4

$$CH_3-C\begin{matrix} \nearrow NH \\ \searrow NH_2 \end{matrix} \qquad CH_3-CH_2NH_2 \qquad (CH_3)_2-NH \qquad CH_3-\overset{\overset{\displaystyle O}{\|}}{C}-NH_2$$

   1   2   3   4

12. Allyl isocyanide contains
   (a) 9σ ; 3π   (b) 9σ ; 9π   (c) 3σ ; 4π   (d) 5σ ; 7π bonds

13. CH$_3$CONH$_2$ $\xrightarrow{\text{P}_2\text{O}_5}$ CH$_3$CN, discuss the change in the state of hybridisation of carbon atom
   (a) sp^3 to sp^2 (b) sp^3 to sp (c) sp to sp^3 (d) sp^2 to sp

14. Tertiary nitro compounds cannot show tautomerism because
    (a) they are stable
    (b) isomerise to form sec. nitro compound
    (c) do not have labile hydrogen atom
    (d) they are highly reactive
15. Acetone oxime on catalytic hydrogenation gives
    (a) 1-propanamide
    (b) iso propyl amine
    (c) ethyl methyl amine
    (d) CH_4 + ethanamine

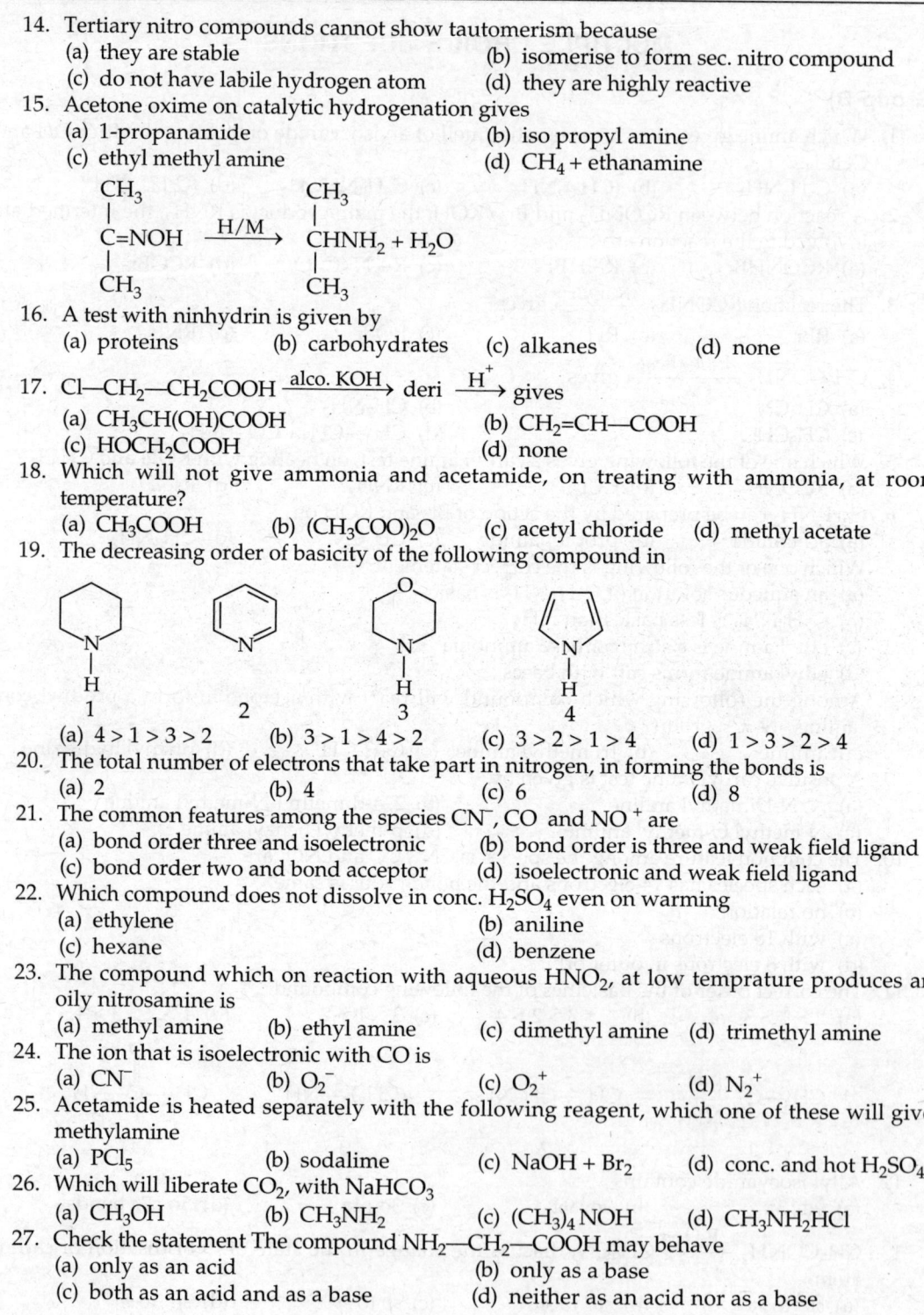

$$\begin{array}{c} CH_3 \\ | \\ C=NOH \\ | \\ CH_3 \end{array} \xrightarrow{H/M} \begin{array}{c} CH_3 \\ | \\ CHNH_2 + H_2O \\ | \\ CH_3 \end{array}$$

16. A test with ninhydrin is given by
    (a) proteins   (b) carbohydrates   (c) alkanes   (d) none

17. $Cl—CH_2—CH_2COOH \xrightarrow{alco.\ KOH}$ deri $\xrightarrow{H^+}$ gives
    (a) $CH_3CH(OH)COOH$
    (b) $CH_2=CH—COOH$
    (c) $HOCH_2COOH$
    (d) none
18. Which will not give ammonia and acetamide, on treating with ammonia, at room temperature?
    (a) CH_3COOH (b) $(CH_3COO)_2O$ (c) acetyl chloride   (d) methyl acetate
19. The decreasing order of basicity of the following compound in

<!-- structures 1: piperidine, 2: pyridine, 3: morpholine, 4: pyrrole -->
1   2   3   4

    (a) $4 > 1 > 3 > 2$ (b) $3 > 1 > 4 > 2$ (c) $3 > 2 > 1 > 4$ (d) $1 > 3 > 2 > 4$
20. The total number of electrons that take part in nitrogen, in forming the bonds is
    (a) 2   (b) 4   (c) 6   (d) 8
21. The common features among the species CN^-, CO and NO^+ are
    (a) bond order three and isoelectronic
    (b) bond order is three and weak field ligand
    (c) bond order two and bond acceptor
    (d) isoelectronic and weak field ligand
22. Which compound does not dissolve in conc. H_2SO_4 even on warming
    (a) ethylene
    (b) aniline
    (c) hexane
    (d) benzene
23. The compound which on reaction with aqueous HNO_2, at low temprature produces an oily nitrosamine is
    (a) methyl amine   (b) ethyl amine   (c) dimethyl amine   (d) trimethyl amine
24. The ion that is isoelectronic with CO is
    (a) CN^- (b) O_2^- (c) O_2^+ (d) N_2^+
25. Acetamide is heated separately with the following reagent, which one of these will give methylamine
    (a) PCl_5 (b) sodalime   (c) $NaOH + Br_2$ (d) conc. and hot H_2SO_4
26. Which will liberate CO_2, with $NaHCO_3$
    (a) CH_3OH (b) CH_3NH_2 (c) $(CH_3)_4NOH$ (d) CH_3NH_2HCl
27. Check the statement The compound $NH_2—CH_2—COOH$ may behave
    (a) only as an acid
    (b) only as a base
    (c) both as an acid and as a base
    (d) neither as an acid nor as a base

28. Urea on heating slowly forms the product
    (a) N_2 (b) CO_2 (c) biuret   (d) amm. carbamate
29. Benzaldehyde reacts with methyl amine to give
    (a) $C_6H_5NH_2$ (b) $C_6H_5CH_2NH_2$ (c) $C_6H_5CH=NCH_3$ (d) $C_6H_5CONH_2$
30. Which of the following, when heated with KOH and a primary amine, will give a carbylamine test?
    (a) $CHCl_3$ (b) CH_3Cl (c) CCl_4 (d) CH_3CN
31. Primary and secondary amines are distinguished by
    (a) $Br_2 + KOH$ (b) HClO   (c) HNO_2 (d) NH_3
32. The primary, sec and tertiary amines are distinguished by
    (a) Hinsberg's reagent   (b) Grignard's reagent
    (c) Fehling's Solution   (d) Tollen's reagent
33. Ethyl amine undergoes oxidation, in the presence of $KMnO_4$, to form
    (a) an acid   (b) an alcohol   (c) an aldehyde   (d) an N-oxide
34. Which of the following does not yield an amine
    (a) $RX + NH_3$ (b) $RCH=NOH \xrightarrow{Na + C_2H_5OH}$
    (c) $RCN + H_2O$ (d) $RCONH_2 + 4H—(LiAlH_4) \longrightarrow$
35. The decreasing order of the basic character of the amines and ammonia is
    (a) $NH_3 > CH_3NH_2 > C_2H_5NH_2 > C_6H_5NH_2$
    (b) $CH_3NH_2 > C_2H_5NH_2 > C_6H_5NH_2 > NH_3$
    (c) $C_2H_5NH_2 > CH_3NH_2 > NH_3 > C_6H_5NH_2$
    (d) $C_6H_5NH_2 > C_2H_5NH_2 > CH_3NH_2 > NH_3$
36. Hinsberg's method of separating the amines is based on the use of
    (a) $C_6H_5SO_2Cl$ (b) benzene sulphonic acid
    (c) ethyl oxalate   (d) CH_3COCl
37. An amine reacts with benzene sulphonyl chloride is soluble in alkali, the amine is
    (a) $1°$ (b) $2°$ (c) $3°$ (d) all
38. Which is the weakest amine?
    (a) NH_3 (b) CH_3NH_2 (c) $(CH_3)_2NH$ (d) $(CH_3)_3N$
39. The action of HNO_2 on a $1°$ amine, gives
    (a) nitro alkane   (b) alkyl nitrite   (c) alcohol   (d) $2°$amine
40. Identify 'X' in the sequence
    $$C_3H_9N \xrightarrow{HNO_2} C_3H_8O \xrightarrow{CrO_3 + H_2SO_4} C_3H_6O$$
    (a) $CH_3NHCH_2H_5$ (b) $C_2H_5CH_2NH_2$ (c) $(CH_3)_3N$ (d) none
41. In diazotisation of aniline, a large quantity of acid is used
    (a) to suppress the concentration
    (b) to suppress the hydrolysis to phenol
    (c) ensure stoichiometric amount of nitrous acid
    (d) to neutralise the base
42. Sulphonation of nitrobenzene gives
    (a) p-nitro benzene sulphonic acid
    (b) o-nitrobenzene sulphonic acid
    (c) m-nitrobenzene sulphonic acid
    (d) a mixture of o- and p-nitro benzene sulphonic acid
43. Diazomethane reacts with carboxylic acid to produce
    (a) an ester   (b) alcohol   (c) amine   (d) imines

44. When alcohol is mixed with ammonia, and the mixture is passed over catalyst, the compound formed is
    (a) $C_2H_5NH_2$ (b) C_2H_4
    (c) $C_2H_5—O—C_2H_5$ (d) $CH_3—O—CH_3$

45. The basic character of amine can be explained
    (a) in terms of Lewis and Arrhenius concept
    (b) only in terms of Lowry-Bronsted concept
    (c) in terms of Lowry Bronsted concept
    (d) only in terms of Lewis concept

46. The hydrolysis of methyl cyanide in presence of acid gives
    (a) CH_3COOH (b) ethanoic acid (c) methylamine (d) methanol

47. Isocyanide test is done to detect
    (a) primary alcohol (b) primary amine (c) sec. amine (d) sec. alcohol

48. Aniline on oxidation with acidified $K_2Cr_2O_7$, forms
    (a) p-aminophenol (b) p-benzoquinone
    (c) benzoic acid (d) phenyl hydroxyl amine

49. When benzene diazonium chloride is heated with methanol, the compound formed is
    (a) methyl benzene (b) phenyl methyl ether
    (c) methyl benzoate (d) toluene

50. A primary amine is heated with CS_2, in presence of $HgCl_2$, gives iso thiocyanate. The reaction is called as
    (a) Hofmann's bromamide reaction (b) Hofmann's mustard oil reaction
    (c) Perkin's reaction (d) Hofmann's elimination

51. The reaction of aniline with benzoyl chloride is known as
    (a) Freidel-Craft reaction (b) Claisen's condensation
    (c) Schotten-Baumann reaction (d) none

52. The compound formed when malonic acid reacts with urea
    (a) barbituric acid (b) butyric acid (c) crotonic acid (d) cinnamic acid

53. The compound formed when a carboxylic acid reacts with hydrazoic acid is
    (a) an amine (b) an amide (c) an aldehyde (d) none

54. The name of the reaction between a carboxylic acid and N_3H is known as
    (a) Schmidt reaction (b) Sanderen's reaction
    (c) Sandmeyer's reaction (d) none

55. $CH{\equiv}CH + CH_2N_2 \longrightarrow$
    (a) $CH_3—C{\equiv}CH$ (b) $CH_3—C{\equiv}C—CH_3$
    (c) a cyclic compound (d) none

56. What is the decreasing order of basicities of CH_3NH_2, methyl amine, $(CH_3)_2NH$ dimethyl amine and $\{(CH_3)_3N\}$ trimethylamine
    (a) tert > sec > pri amine (b) sec > tert > pri
    (c) pri > sec > pri (d) none

57. The amines are soluble in water and the aqueous solution is basic. What is the decreasing order of basicities of amines in aqueous solution.
    (a) tert > sec > pri (b) pri > sec > tert (c) sec > pri > tert (d) one

58. The correct order of basicities of methyl amine, (A) ammonia (B) and aniline (C)
    (a) A < B < C (b) C < A < B (c) C < A < B (d) C < B < A

59. The following reaction is termed
    $CH_3CN + 4H \longrightarrow CH_3CH_2NH_2$
    (a) Hofmann's bromamide reaction (b) Mendius reaction
    (c) Sabatier reaction (d) one

60. The no of σ bonds in the given structure are

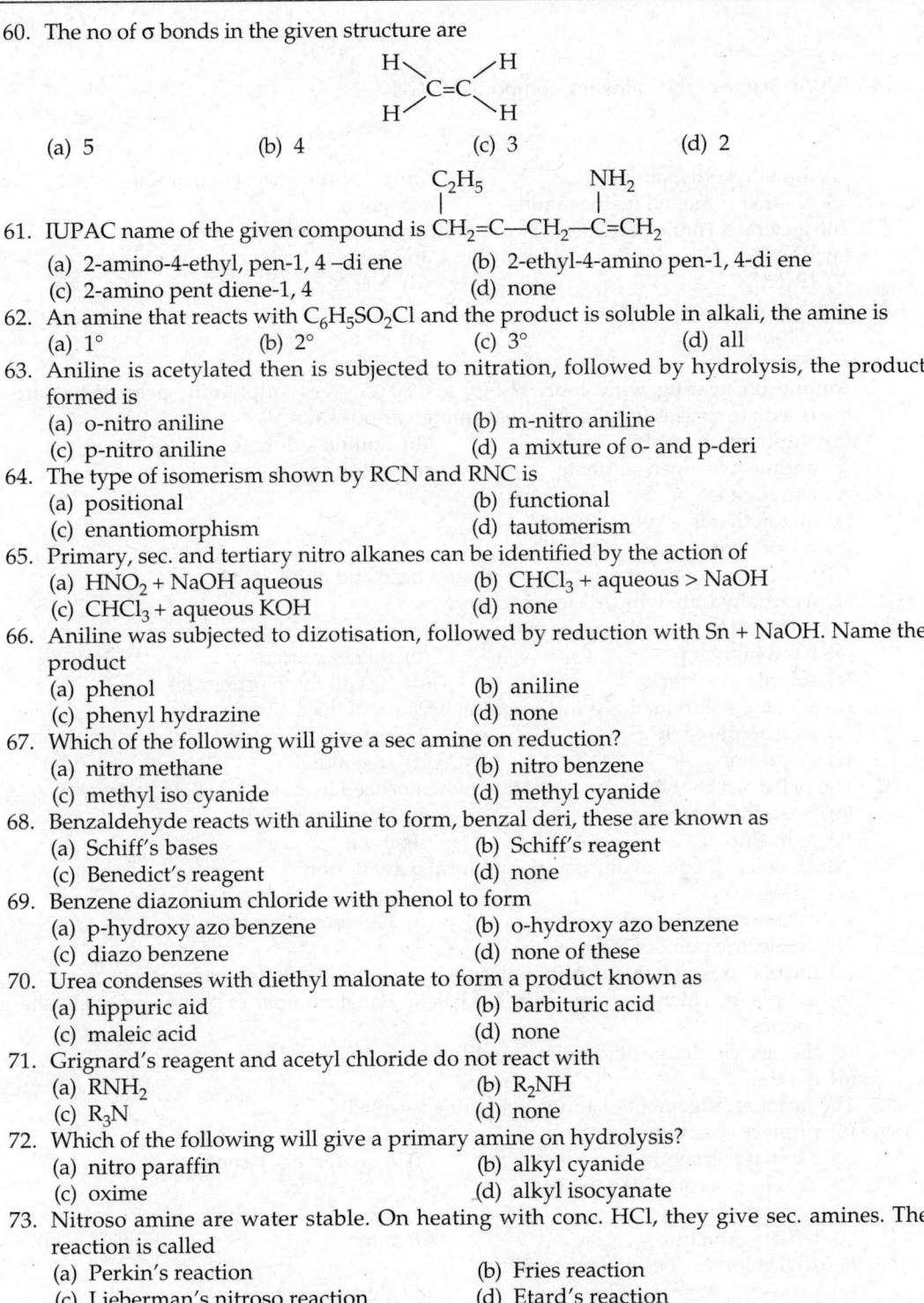

 (a) 5   (b) 4   (c) 3   (d) 2

61. IUPAC name of the given compound is $CH_2=C-CH_2-C=CH_2$ (with C_2H_5 and NH_2 substituents)
 (a) 2-amino-4-ethyl, pen-1, 4 –di ene   (b) 2-ethyl-4-amino pen-1, 4-di ene
 (c) 2-amino pent diene-1, 4   (d) none

62. An amine that reacts with $C_6H_5SO_2Cl$ and the product is soluble in alkali, the amine is
 (a) 1°   (b) 2°   (c) 3°   (d) all

63. Aniline is acetylated then is subjected to nitration, followed by hydrolysis, the product formed is
 (a) o-nitro aniline   (b) m-nitro aniline
 (c) p-nitro aniline   (d) a mixture of o- and p-deri

64. The type of isomerism shown by RCN and RNC is
 (a) positional   (b) functional
 (c) enantiomorphism   (d) tautomerism

65. Primary, sec. and tertiary nitro alkanes can be identified by the action of
 (a) HNO_2 + NaOH aqueous   (b) $CHCl_3$ + aqueous > NaOH
 (c) $CHCl_3$ + aqueous KOH   (d) none

66. Aniline was subjected to dizotisation, followed by reduction with Sn + NaOH. Name the product
 (a) phenol   (b) aniline
 (c) phenyl hydrazine   (d) none

67. Which of the following will give a sec amine on reduction?
 (a) nitro methane   (b) nitro benzene
 (c) methyl iso cyanide   (d) methyl cyanide

68. Benzaldehyde reacts with aniline to form, benzal deri, these are known as
 (a) Schiff's bases   (b) Schiff's reagent
 (c) Benedict's reagent   (d) none

69. Benzene diazonium chloride with phenol to form
 (a) p-hydroxy azo benzene   (b) o-hydroxy azo benzene
 (c) diazo benzene   (d) none of these

70. Urea condenses with diethyl malonate to form a product known as
 (a) hippuric aid   (b) barbituric acid
 (c) maleic acid   (d) none

71. Grignard's reagent and acetyl chloride do not react with
 (a) RNH_2 (b) R_2NH
 (c) R_3N (d) none

72. Which of the following will give a primary amine on hydrolysis?
 (a) nitro paraffin   (b) alkyl cyanide
 (c) oxime   (d) alkyl isocyanate

73. Nitroso amine are water stable. On heating with conc. HCl, they give sec. amines. The reaction is called
 (a) Perkin's reaction   (b) Fries reaction
 (c) Lieberman's nitroso reaction   (d) Etard's reaction

74. IUPAC name of the following compound is

$$CH_3—N \begin{matrix} C_2H_5 \\ | \\ | \\ CH_3 \end{matrix}$$

   (a) dimethyl ethyl amine
   (b) N, N-dimethyl ethanamine
   (c) N-ethyl-n-methyl methanamine
   (d) none

75. The general formula of quaternary ammonium compound is
   (a) RNH_2
   (b) R_3N
   (c) $[R_4N]^+X^-$
   (d) NH_4X

76. Amines are more basic than
   (a) alcohols
   (b) ethers
   (c) esters
   (d) all

77. Aniline on heating with conc. H_2SO_4 at 180°C gives sulphanilic acid. What are the intermediate species formed, beside forming various salts
   (a) sulphaminic acid
   (b) aniline sulphate
   (c) aniline hydrogen sulphate
   (d) sulphanilic acid

78. A zwitterion is
   (a) an ion that is + vely charged
   (b) an ion that is – vely charged
   (c) is a compound that can ionise both as a base and as an acid
   (d) a carbohydrate with an electric charge

79. A zwitterion that has the properties
   (a) has a high m.p.
   (b) no net charge
   (c) soluble in water
   (d) has all these properties

80. An aqueous solution of glycine is neutral because of the formation of
   (a) a carbonium ion
   (b) carbanion
   (c) zwitterion
   (d) free radial

81. The pH at which there is no migration, when peaced in an electric field, is known as
   (a) isoelectric point
   (b) dipole moment
   (c) iodine no
   (d) none

82. Which one of the following compound form a zwitterion?
   (a) phenols
   (b) amino acids
   (c) carbonyl compounds
   (d) heterocyclic compounds

83. The isoelectric point of a protein is
   (a) the pH, at which the protein molecule has no charge on its surface
   (b) the pH at which a protein solution has an equal number of +vely and –vely charged species
   (c) the electric charge at identical conditions
   (d) none

84. The linear arrangement of amino acid units is called
   (a) primary structure
   (b) a secondary structure
   (c) a tertiary structure
   (d) a quaternary structure

85. The α-helix is a common form of
   (a) primary structure
   (b) secondary structure
   (c) tertiary structure
   (d) none

86. Benzoyl chloride does not react with
   (a) primary and sec. amines
   (b) aliphatic compound
   (c) aromatic compound
   (d) carboxylic acid

87. Give IUPAC name for the following compound: $(C_2H_5)_2N$—$CH_2CHClCOOH$
    (a) 2-chloro-4-N-diethyl pentanoic acid
    (b) 2-chloro-3-N, N diethyl amino propanoic acid
    (c) 2-chloro-3-N, N diethyl amino propanoic acid
    (d) none of the above

ANSWERS

(Group-B)

   1. (a)   2. (a)   3. (d)   4. (a)   5. (c)   6. (b)   7. (c)   8. (d)
   9. (d)
   10. (a) Each species has 14 electrons and the bond order is same.
   11. (b) a 2° amine is more basic as compared to a 1° amine; amides are less basic owing to less availability of the pair of electrons; the –ve charge developed on, make it more basic.
   12. (a)   13. (a)
   14. (c) R_3NO_2 : does not have labile hydrogen
   15. (b)   16. (a)
   17. An elimination reaction.
   18. (a) $CH_3COOH + NH_3 \longrightarrow CH_3COONH_4$
   19. (d)
   20. (c) 6 electrons.
   21. (a) bond order three and isoelectronic.
   22. (c) alkanes are insoluble in conc. H_2SO_4.
   23. (c) sec amines form nitrosamine.
   24. (a)   25. (a)   26. (c)   27. (c)   28. (c)   29. (c)   30. (a)   31. (a)
   32. (a)
   33. (c) an aldehyde, $CH_3CH=NH \longrightarrow CH_3CHO$.
   34. (c)   35. (c)   36. (a)   37. (a)   38. (a)   39. (c)   40. (b)   41. (b)
   42. (c)   43. (a)   44. (a)   45. (a)   46. (a)   47. (b)   48. (b)   49. (b)
   50. (b)   51. (c)   52. (a)   53. (a)   54. (a)
   55. (c) pyrazolidine.
   56. (b)   57. (c)   58. (d)   59. (b)   60. (a)   61. (b)   62. (a)   63. (b)
   64. (b)
   65. (a) this method is known a Victor Meyer's method.
   66. (c)   67. (c)   68. (a)   69. (a)
   70. (b) barbituric acid.
   71. (c)   72. (c)   73. (c)   74. (b)   75. (c)
   76. (d) these compounds do not infulence the pH.
   77. (a) sulphaminic acid $C_6H_5NHSO_3H$.
   78. (c)   79. (d)   80. (b)   81. (a)   82. (b)   83. (b)   84. (b)   85. (b)
   86. (d)   87. (c)

MULTIPLE CHOICE QUESTIONS

(Group-C)

   1. The intermediate formed in **Hofmann's bromamide** reaction
      (a) RCONHBr   (b) RCONBr$_2$
      (c) R—N=C=O   (d) RNHBr

2. In Gabriel synthesis the amine formed is always
   (a) a primary amine   (b) a sec amine
   (c) a tertiary amine   (d) a mixture of all amines

3. What is the order of basicity of the amines in gaseous phase or in chloro benzene (where is no association with the solvent) is
   (a) $R_3N > R_2NH > RNH_2 > NH_3$ (b) $NH_3 > RNH_2 > R_2NH > R_3N$
   (c) $R_2NH > R_3N > RNH_2 > NH_3$ (d) none

4. Which amine is most volatile?
   (a) RNH_2 (b) R_2NH
   (c) R_3N (d) None

5. Which amine is most basic in aqueous solution? Arrange these in the decreasing order of basicity.
   (i) RNH_2 (ii) R_2NH (iii) R_3N (iv) NH_3

6. The strongest base among the following compounds is

$$
\begin{array}{ccc}
\overset{\displaystyle NH}{\underset{\|}{}} & \overset{\displaystyle O}{\underset{\|}{}} & \overset{\displaystyle \overset{+}{N}H_2}{\underset{|}{}} \\
\end{array}
$$

   (a) $NH_2—C—NH_2$ (b) $NH_2—C—NH_2$ (c) $NH_2—C—NH_2$ (d) None

7. Which among the following is most basic? Arrange these in the decreasing order of the basic nature.

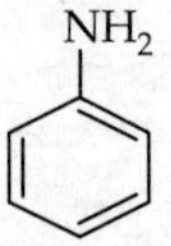

aniline   pyridine
   (a)   (b)   (c) 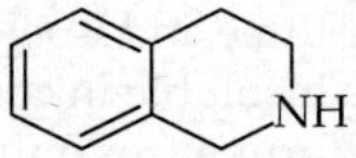   (d)

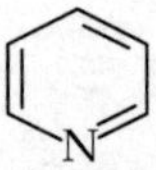

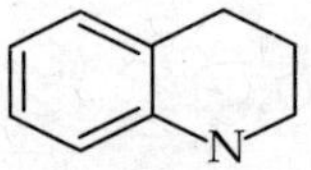

8. For the separation of amines the reagent used is
   (a) Hinsberg's reagent-benzene sulphonyl chloride
   (b) phenyl isocyanate
   (c) p-toluene sulphonic acid
   (d) none

9. Which of the following compounds will not form an acetyl derivative?

   (a) RNH_2 (b) $Ar—NH—R$ (c) R_3N (d)

10. Which one will form nitroso derivative when treated with HNO_2?

   (a) RNH_2 (b) $C_6H_5NH_2$ (c) C_6H_5NHR (d)

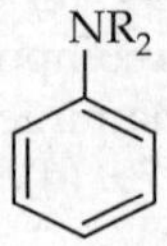

11. The compounds formed on heating
    $C_6H_5 N^+ (CH_3)_3OH^- \longrightarrow$
    (a) $C_6H_5N (CH_3)_2 + CH_3OH$ (b) $C_6H_5N (CH_3)_2 + H_2O$
    (c) $p\text{-}CH_3C_6H_4 NHCH_3 + CH_3OH$ (d) none

12. The compound formed on treating $(CH_3)_3C—CH_2NH_2$ with HNO_2 is
    (a) $(CH_3)_3C—CH_2OH$ (b) $(CH_3)_3—C—CH=CH_2$
    (c) $(CH_3)_2C(OH)CH_2CH_3$ (d) none

13. In Hinsberg's separation of amines, an organic compound $C_4H_{11}N$ forms a solid with NaOH, the composition of the compound is
    (a) $CH_3CH_2CH_2CH_2NH_2$
    (b) $(CH_3)_2CHCH_2NH_2$
    (c) $CH_3CHNHCH_2CH_3$
    (d) $CH_3CH_2N(CH_3)_2$

14. The product in the reaction is

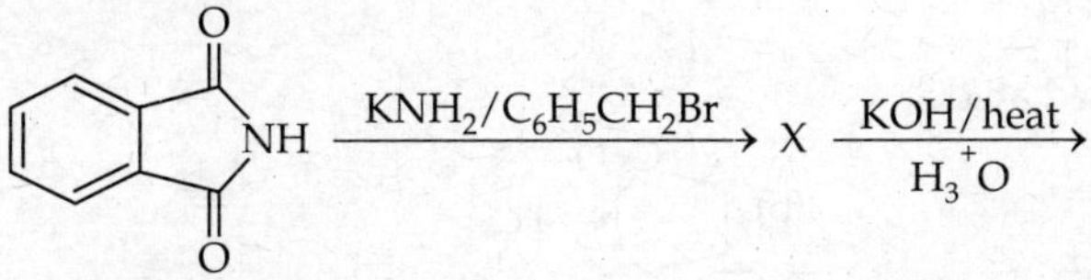

$\xrightarrow{KNH_2/C_6H_5CH_2Br} X \xrightarrow[H_3\overset{+}{O}]{KOH/heat}$

    (a) $C_6H_5CH_2NH_2$
    (b) $C_6H_5NHC_6H_5$
    (c) $C_6H_5CONHC_6H_5$
    (d) none

15. The product formed in the reaction

$CH_2=CH—CH=CH_2 + Br_2$ (one mole) $\longrightarrow A \xrightarrow{KCN(excess)/H_2, Ni} B$

    (a) $NH_2—(CH_2)_6NH_2$
    (b) $NH_2(CH_2)_4 CN$
    (c) $NH_2—CH_2—CH=CH—CH_2CN$
    (d) none

16. $HOOC—C(CH_3)_2—CN \xrightarrow{200°C/heat} A \xrightarrow{\text{reduction with } LiAlH_4} B$; the compound 'B' is

    (a) $HO—CH_2—C(CH_3)_2CN$
    (b) $(CH_3)_3C—CH_2NH_2$
    (c) $(CH_3)_2CHCH_2NH_2$
    (d) $(CH_3)_2 CONHCH_3$

17. Which will give a sec. alcohol, on reaction with HNO_2?
    (a) $(CH_3)_2—CH—CO—NH_2$
    (b) $(CH_3)_2CHCH_2NH_2$
    (c) $CH_3—CH (CH_3)NH_2$
    (d) 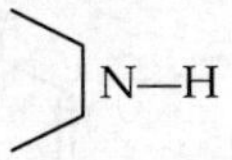 image N—H

18. 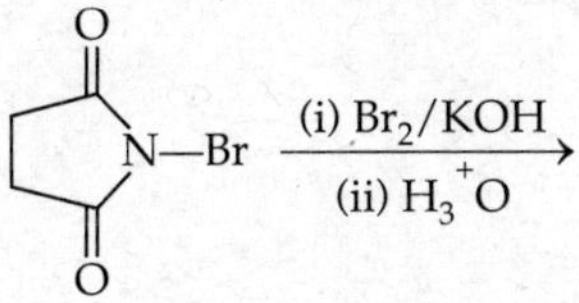 image $\xrightarrow[\text{(ii) } H_3\overset{+}{O}]{\text{(i) } Br_2/KOH}$

    (a) 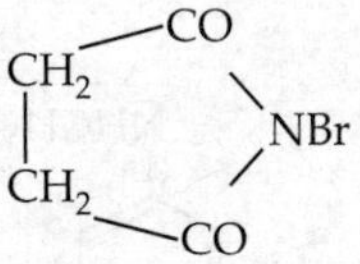 image
    (b) $CH_2—NH_2$ / CH_2NH_2
    (c) 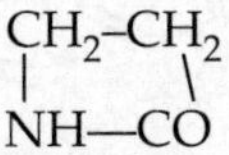 image
    (d) $HN_2—CH_2—CH_2—CO—OH$

19. How will you will convert cyclo pentyl cyanide into cyclo pentyl methylene
    image —CN $\longrightarrow$ image =CH_2

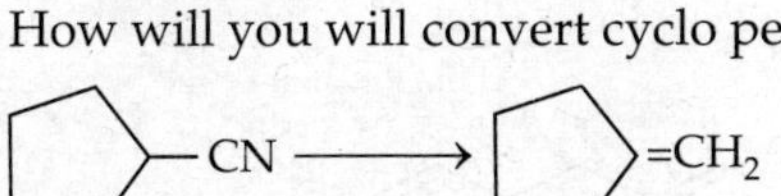

20. Which is the strongest base?
    (a) $(CH_3)_3N$
    (b) 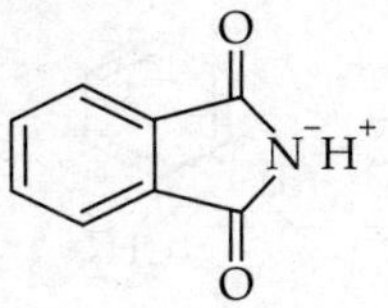 image
    (c) 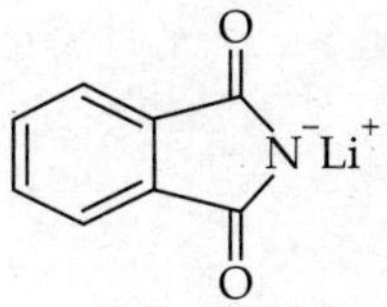 image
    (d) 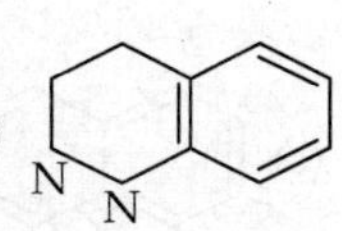 image

21. How to open a cyclic base, to form an open chain compound?
    (a) Hofmann's elimination reaction
    (b) exhaustive methylation followed by a reaction with AgOH, followed by heat
    (c) only by A or B methods
    (d) both methods

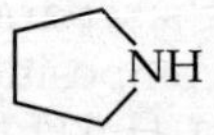

22. Which is the strongest Bronsted acid?

    (a) 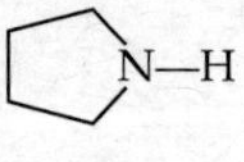 [N—H ring]   (b) [C=O N—H ring]

    (c) [NH₂ benzene]   (d) 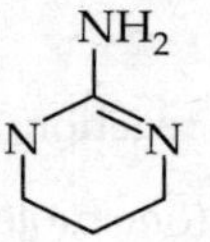 [NH₂ / N N ring]

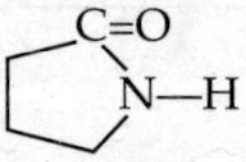

23. Which is the strongest base in Q. No. 22?

24. The best method for the preparation of a sec amine is
    (a) $R_2CHBr + NaNH_2$
    (b) $RNH_2 + OHCR, H_2/Pt$
    (c) $RBr + NH_3$
    (d) by Gabriel's method

25. Which of the following will react fast with an ester ($C_6H_5COOCH_3$)?

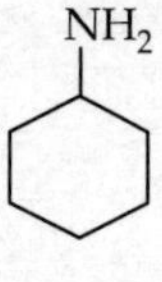

    (a) 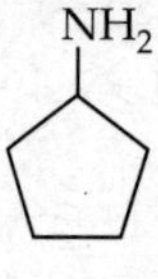   (b) 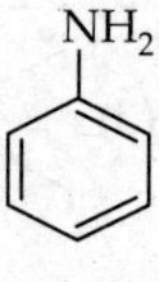   (c) 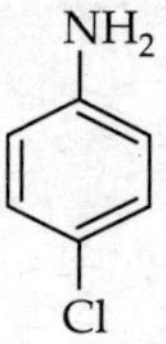   (d)

26. The compound formed in the reaction

    $$C_6H_5CONH_2 \xrightarrow{\text{(i) Br}_2 \text{ (ii) CH}_3O^- Na^+ \text{ in CH}_3OH}$$

    (a) [NH₂]   (b) [NHCOOCH₃]   (c) [NHOCH₃]   (d) [NHCH₃]

27.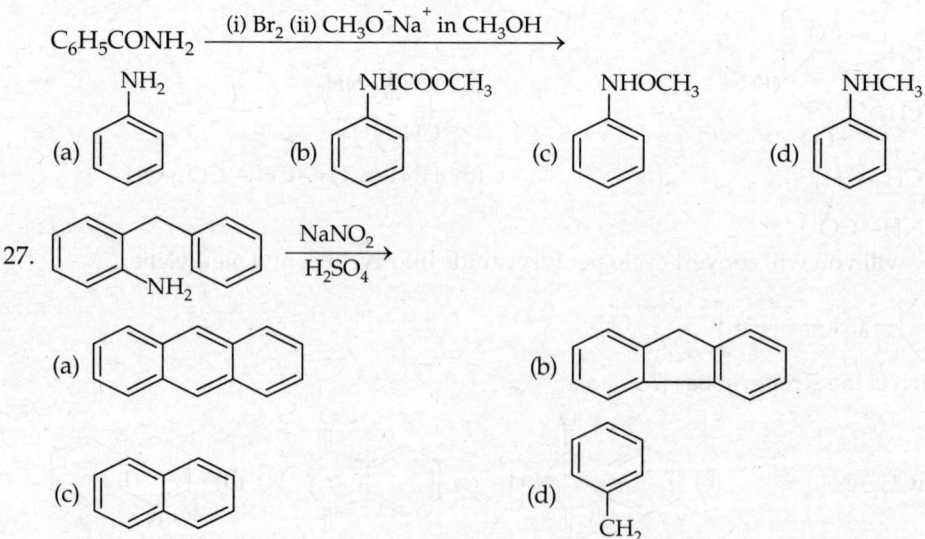

    NaNO₂ / H₂SO₄

    (a)   (b)

    (c)   (d)

28.

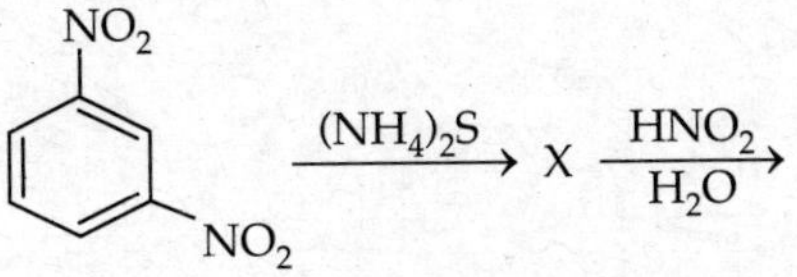

$\xrightarrow{(NH_4)_2S}$ X $\xrightarrow[H_2O]{HNO_2}$

(a)

(b)

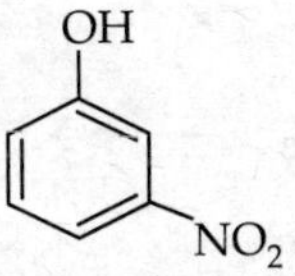

(c)

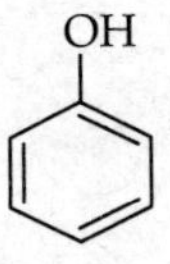

(d)

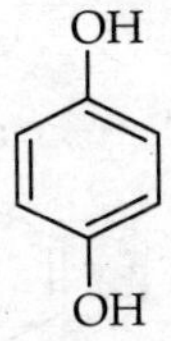

29. The compound formed in the reaction is

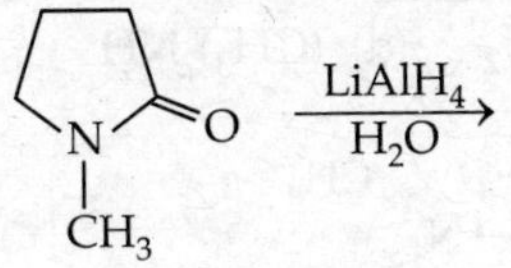

$\xrightarrow[H_2O]{LiAlH_4}$

(a)

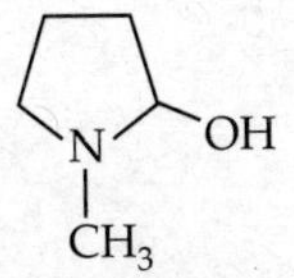

(b)

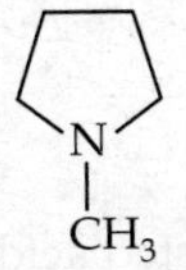

(c) $CH_3\text{—}N\text{—}(CH_2)_2\text{—}CHO$
       $|$
       CH_3

(d) $CH_3\text{—}N\text{—}(CH_3)_3\text{—}CH_2OH$

30. The best leaving group
   (a) OH—
   (c) N_2
   (b) NH_2—
   (d) CH_3COO—

31. Which is the Hinsberg's reagent?
   (a) Ph CN
   (c) $C_6H_5SO_2Cl$
   (b) $p\text{-}CH_3C_6H_4SO_2Cl$
   (d) $o\text{-}C_6H_4Cl_2$

32. The reaction is termed
   (a) Cope reaction
   (c) carbylamine reaction
   (b) Saytzeff' reaction
   (d) E2 elimination

33. The main product formed in the reaction is
   $Me_3 C\text{—}CH_2NH_2 + HNO_2 \longrightarrow$
   (a) $Me_3C\text{—}CH_2OH$
   (c) $Me_3CCHNHNO$
   (b) $Me_3C\text{—}CH=CH_2$
   (d) $Me_3C\text{—}C(OH)C_2H_5$

34. The compound formed in the following reaction

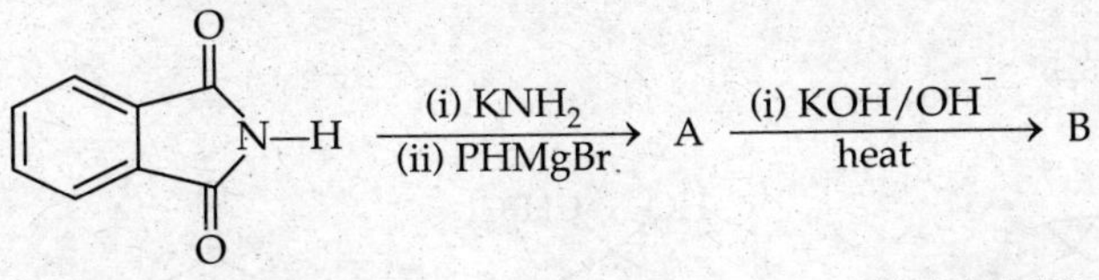

$\xrightarrow[\text{(ii) PHMgBr}]{\text{(i) KNH}_2}$ A $\xrightarrow[\text{heat}]{\text{(i) KOH/OH}^-}$ B

   (a) $PhCH_2NH_2$
   (c) Ph CONH Ph
   (b) Ph_2NH
   (d) $PhNH_2NHPh$

35. The major product formed in the reaction is

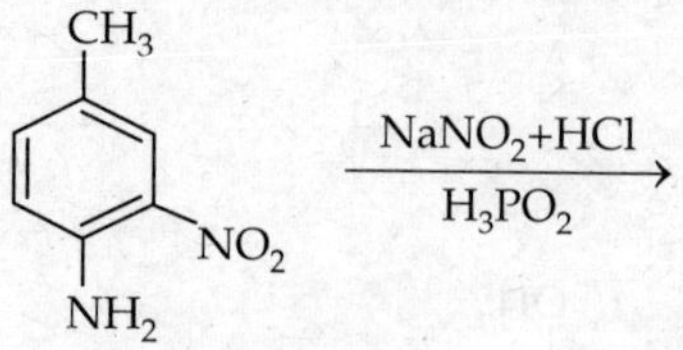

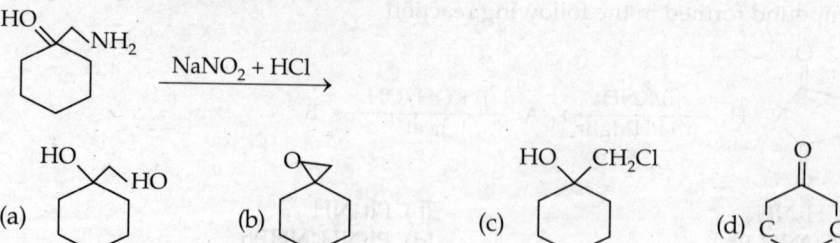

(structure: benzene ring with CH₃ at top, NO₂ and NH₂ at bottom) → NaNO₂+HCl / H₃PO₂

(a) 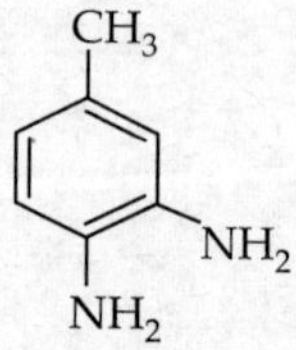 CH₃ ring with NH₂ and NH₂

(b) 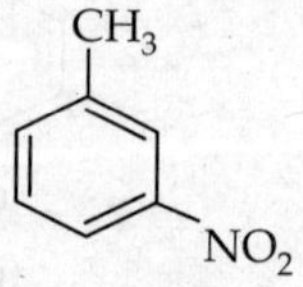 CH₃ ring with NO₂

(c) 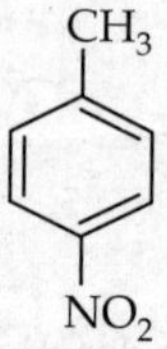 CH₃ ring with NO₂ (para)

(d) 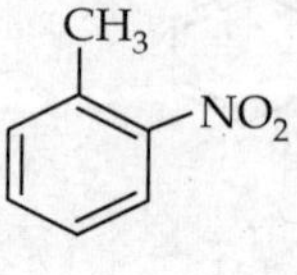 CH₃ ring with NO₂

36. Which of the following will not liberate N₂ on treatment with HNO₂?
    (a) CH_3CONH_2 (b) NH_2CONH_2 (c) $(CH_3)_2CHNH_2$ (d) $(CH_3)_2NH$

37. Which is the strongest Bronsted base?

(a) 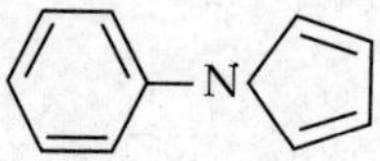 (ring—N linked to cyclopentadiene)

(b) 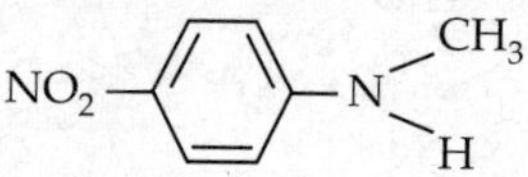 NO_2—(ring)—N(CH₃)(H)

(c) 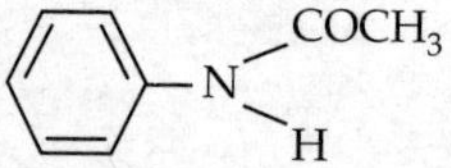 (ring)—N(COCH₃)(H)

(d) 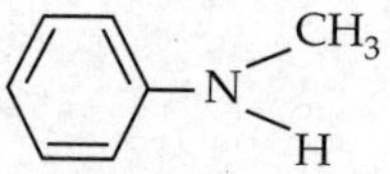 (ring)—N(CH₃)(H)

38. Which is the strongest Bronsted acid?

(a) 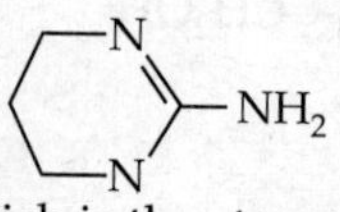 (cyclic amidine)—NH₂   (b) 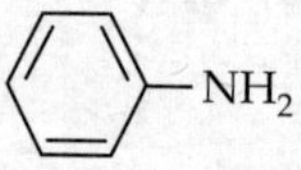 (ring)—NH₂   (c) 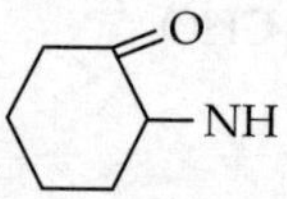 (cyclohexanone ring)—NH   (d) 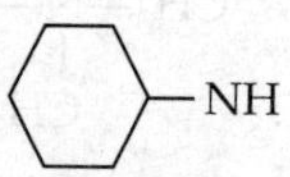 (cyclohexane ring)—NH

39. Which is the strongest base and strongest acid?

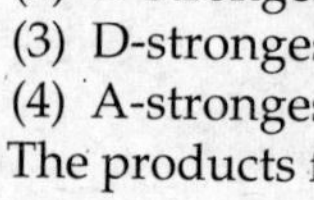

A: (ring)—NH₂   B: (cyclohexanone)—NH   C: (ring)—S(=O)(=O)—NH₂   D: (imidazole ring with N—H)

A 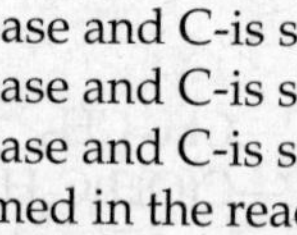   B 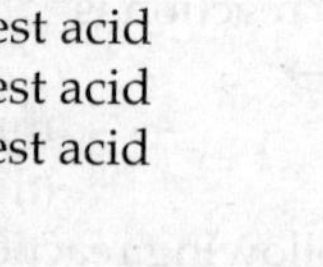   C   D

(1) B-strongest base and C-strongest acid
(2) D-strongest base and C-is strongest acid
(3) D-strongest base and C-is strongest acid
(4) A-strongest base and C-is strongest acid

40. The products formed in the reaction

HO, NH₂ (on cyclohexane ring) → NaNO₂ + HCl

(a) HO, HO (on cyclohexane ring)

(b) O (epoxide on cyclohexane)

(c) HO, CH₂Cl (on cyclohexane ring)

(d) O (cyclohexanone with C—C)

41. Which of the following is an enamine?

(a)

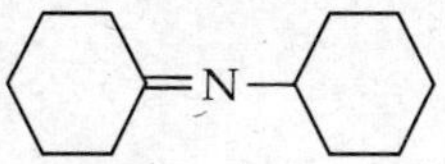

(b)

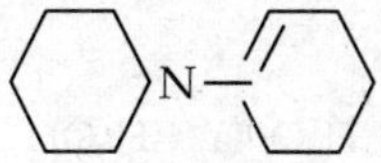

(c)

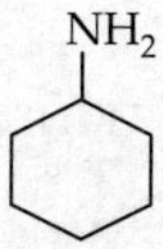

(d)

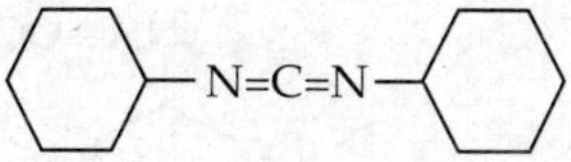

42. The product formed in the following reaction is

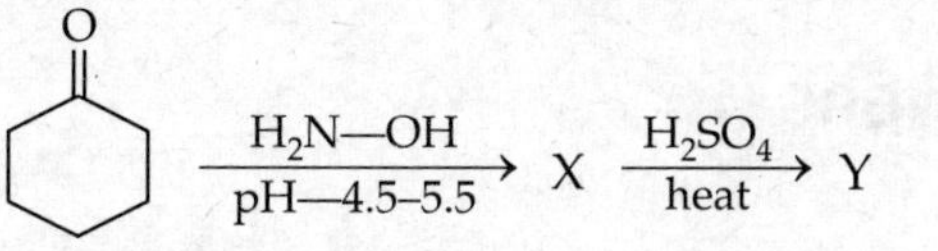

$$\xrightarrow[\text{pH—4.5–5.5}]{H_2N—OH} X \xrightarrow[\text{heat}]{H_2SO_4} Y$$

(a) 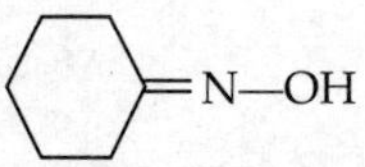 (b)

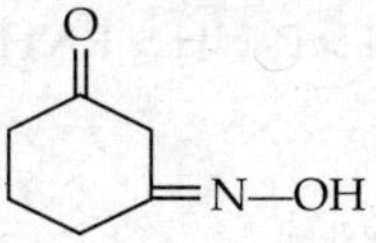

(c)

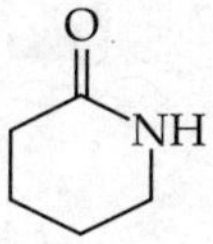

(d)

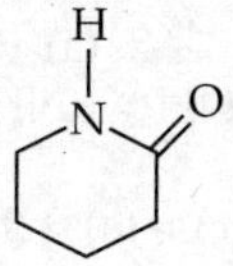

43. Which of the compound is most basic?

(a)

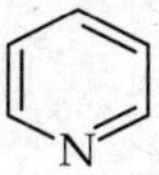

(b)

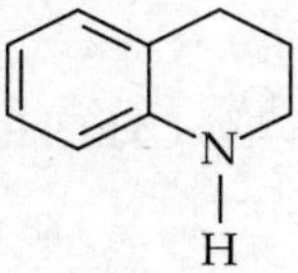

(c)

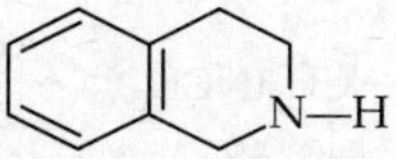

(d)

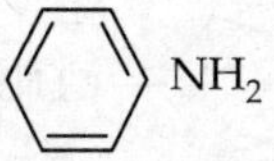

44. Which of the following compounds will react with cyclo pentanone to form enamine?

(a)

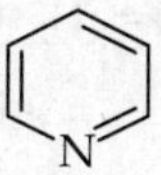

(b)

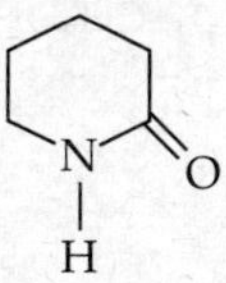

(c)

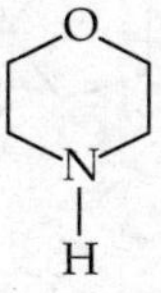

(d)

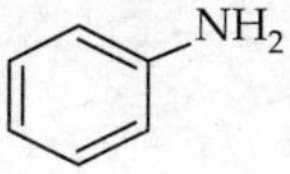

45. The main product formed in the reaction

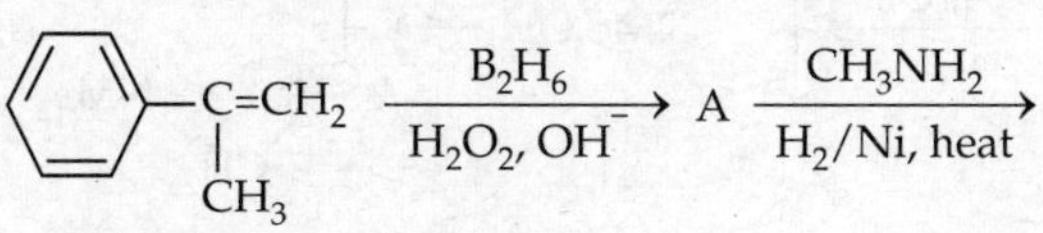

$\xrightarrow[\text{H}_2\text{O}_2, \text{OH}^-]{\text{B}_2\text{H}_6}$ A $\xrightarrow[\text{H}_2/\text{Ni, heat}]{\text{CH}_3\text{NH}_2}$

(a) Ph—CH—NHCH$_3$
(b) Ph—CH$_2$—CH$_2$NHCH$_3$
(c) PhCH=N—CH$_3$
(d) Ph—CH$_2$—CH=N—CH$_3$

46. What product are formed?

Propionic acid $\xrightarrow{\text{SOCl}_2}$ X $\xrightarrow{\text{NH}_3}$ Y $\xrightarrow{\text{KOH/Br}_2}$ Z

(a) CH$_3$—CH$_2$NH—CH$_3$
(b) CH$_3$—CH$_2$NH$_2$
(c) CH$_3$CH$_2$CH$_2$NH$_2$
(d) CH$_3$CH$_2$CONH$_2$

47. The products formed in the reaction

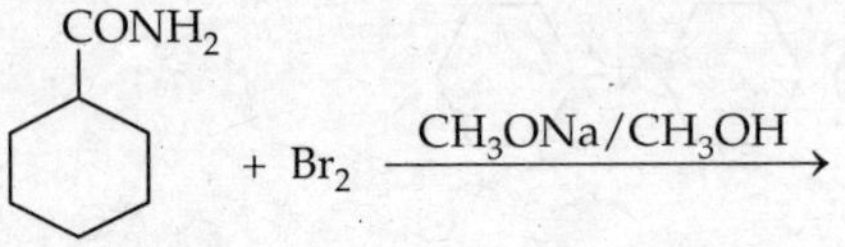

(a) 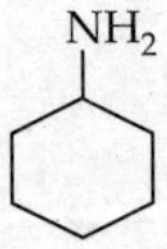   (b) 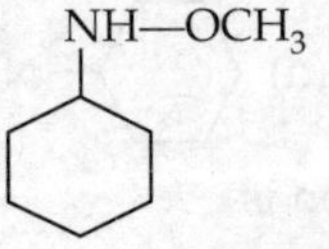   (c) 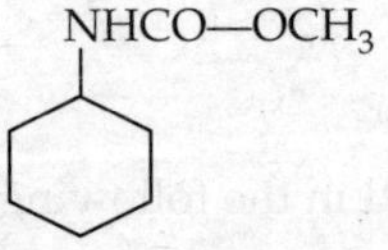   (d)

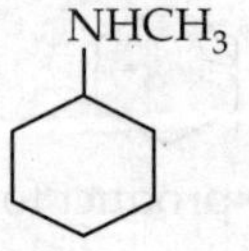

ANSWERS

(Group-C)

1. (c)   2. (a)   3. (a)   4. (a) $R_3N > R_2NH > RNH_2$
5. (b) $R_2NH > RNH_2 > R_3N$
6. (a)
7. (d) > (c) > (a) > (b)
8. (a)   9. (c)   10. (d)   11. (a)   12. (c)
13. (d) (compound should be a tertiary amine)
14. (a)   15. (a)

16. (c) $HOOC(CH_3)_2CN \xrightarrow{-CO_2} CH(CH_3)_2-CN \longrightarrow (CH_3)_2-CHCH_2NH_2$
17. (c)
18. (d)

19.

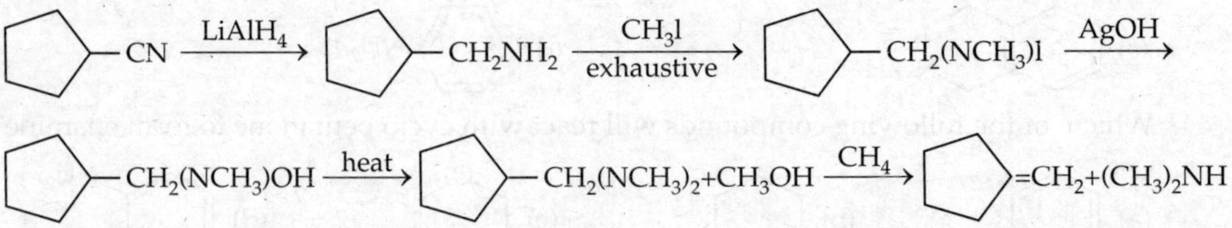

20. (b)

21. (b)

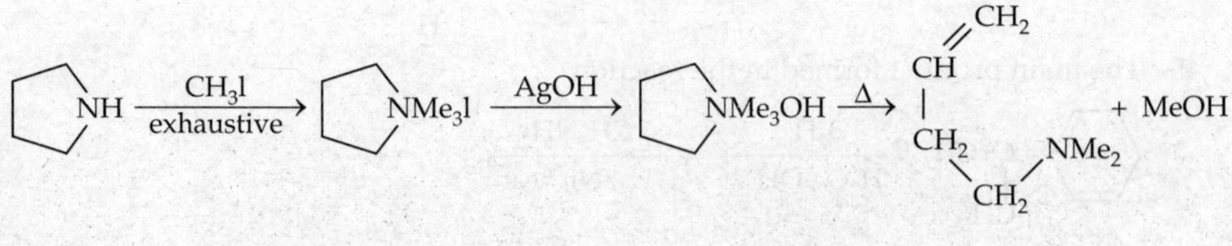

22. (b)
23. (d)
24. (b) $RNH_2 + OHCR \longrightarrow RN=CHR \longrightarrow RNHCH_2R$
25. (a) being a strongest amine.
26. (b)   27. (b)   28. (b)   29. (b), (a)   30. (c)   31. (c)   32. (a)   33. (c)
34. (b)   35. (c)   36. (a)   37. (c)   38. (a)   39. (a)   40. (d)
41. (b) a condensation product of keto compound with a sec amine.
42. (c)   43. (c)   44. (c)   45. (b)   46. (b)   47. (c)

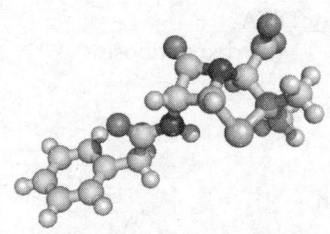

Chapter 7

Biomolecules—Proteins and Nucleic Acid

MULTIPLE CHOICE QUESTIONS

1. In DNA, the complementary base are
   (a) adenine, thymine, guanine, cytosine
   (b) uracil, adenine, cytosine, guanine
   (c) adenine, guanine, thymine and cytosine
   (d) adenine, thymine guanine and uracil

2. Carbohydrates contain
   (a) C—OH   (b) >C=O   (c) —CHO   (d) all groups

3. An aqueous solution of a carbohydrate, with 2 drops of alcohol solution of naphthol and H_2SO_4 produces a ring, the colour of the ring is
   (a) yellow   (b) green   (c) violet   (d) red

4. Which reagent is used to detect sugar in urine?
   (a) Baeyer's reagent   (b) Fehling solution   (c) ozonolysis   (d) none

5. α-glucose and β-glucose are
   (a) epimers   (b) anomers   (c) isomers   (d) tautomers

6. Glucose shows mutarotation, when solvent is
   (a) neutral   (b) amphoteric   (c) acidic   (d) acidic

7. Which are called biomolecules?
   (a) carbohydrates   (b) proteins   (c) lipids   (d) all

8. Cellulose is a linear polymer of
   (a) α-glucose   (b) β-glucose   (c) α-fructose   (d) none

9. Glycogen is a branched polymer of
   (a) α-glucose   (b) β-glucose   (c) α-fructose   (d) none

10. Number of atoms in a pyranose ring are
    (a) 5 + 1   (b) 4 + 1   (c) 3 + 2   (d) 4 + 2

11. Methylation of glucose with dimethyl sulphate indicates the presence of
    (a) —CHO   (b) >C=O   (c) —OH   (d) none

12. Starch is made up of
    (a) glucose and fructose unit   (b) amylose and amylopectin units
    (c) amylose and amylopectin   (d) amylopectin and glycogen

13. Which of the following sugar units are present in genetic factor DNA?
    (a) glucose   (b) maltose   (c) ribose   (d) deoxy ribose
14. The main structural feature of the proteins are
    (a) the presence of an ester linkage   (b) the ether linkage
    (c) the peptide linkage   (d) none
15. The function of DNA is
    (a) proteins synthesis   (b) self-replication
    (c) storage of here ditary information   (d) none
16. Which of the following have a coiled structure?
    (a) carbohydrates   (b) proteins   (c) lipids   (d) vitamins
17. Blood protein is
    (a) albumin   (b) hemoglobin   (c) both (a) and (b)   (d) none
18. DNA is a polymer of
    (a) sugars   (b) ribose   (c) amino acids   (d) pyridoxine
19. Nucleotide and nucleosides differ from each other
    (a) presence of a phosphate unit   (b) presence of a base unit
    (c) presence of a nucleic unit   (d) none
20. What are enzymes? Enzymes are
    (a) catalyst   (b) fatty acid   (c) proteins   (d) carbohydrates
21. Albumin has
    (a) H-bonding   (b) intermolecular and intramolecular bonding
    (c) both   (d) none
22. Deoxyribonucleic acid consists of the following units.
    (a) peptides   (b) sugar   (c) base   (d) H_3PO_4
23. The storehouse all biological information is
    (a) RNA   (b) DNA   (c) m-RNA   (d) none
24. The process of transformation RNA from DNA is termed
    (a) translation   (b) transcription   (c) replication   (d) mutation
25. An antigen develops antibodies, which protects the body from their harmful effect. The antibodies are
    (a) immune globulins   (b) phospholipids
    (c) albumins   (d) lymphocytes
26. Hormones function as
    (a) chemical messenger   (b) coenzyme
    (c) pro vitamins   (d) none
27. Which is the sweetest sugar?
    (a) sucrose   (b) glucose   (c) fructose   (d) lactose
28. Nucleic acid contains
    (a) 4 purine bases
    (b) 2 purine and 3-pyrimidine bases
    (c) 4 pyrimidine bases
    (d) 4 pyrimidine bases and one purine base
29. Increased blood pressure may be caused by excess secretion of
    (a) thyroxin   (b) testosterone   (c) adrenalin   (d) estradiol
30. The hormone used as a contraceptive is
    (a) aldosterone   (b) cortisone   (c) progesterone   (d) testosterone
31. Starch can be used as an indicator for the detection of traces of
    (a) glucose in aqueous solution   (b) protein in blood
    (c) iodine in aqueous solution   (d) urea in blood

32. Glucose give a silver mirror test, because it contains
    (a) —CHO group   (b) a keto group   (c) an ester group   (d) an amide group
33. Glucose reacts with methanol to form
    (a) α-methyl glucoside   (b) β-methyl glucoside
    (c) forms both (a) and (b)   (d) none
34. Glucose shows mutarotation, when the solvent is
    (a) acidic   (b) basic   (c) neutral   (d) amphoteric
35. Glucose forms osazone involving only two carbon atoms, because of
    (a) chelation   (b) oxidation   (c) reduction   (d) hydrolysis
36. Amylose is a polymer of
    (a) α-d-glucopyranose   (b) β-d-fructose
    (c) fructose   (d) none
37. The number of asymmetric carbons in fructose is
    (a) 2   (b) 3   (c) 4   (d) 5
38. Fructose also occurs as
    (a) furanose   (b) contain 5-member ring
    (c) 4-carbon and one oxygen atom   (d) all
39. On what basis, fructose has been also a ring structure?
    (a) fructose occurs in two forms   (b) it also shows mutarotation
    (c) possesses reducing properties (d) none
40. The letter 'D' and 'L' are substituents of 'd' and 'l'
    (a) yes   (b) no
    (c) can't be explained   (d) all

ANSWERS

1. (a) DNA contains these four bases; adenine, thymine, guanine, cytosine.
   RNA contains adenine, uracil, guanine and cytosine.
2. (d) all.
3. (c) colour is violet, known as Molisch test.
4. (b)   5. (a) epimers   6. (b)   7. (d) all
8. (b) a polymer of glucose.
9. (a)   10. (a)   11. (c) —OH group.
12. (b)   13. (d)
14. (c) The peptide   15. (c)
16. (b) proteins   17. (c)
18. (d) a polymer of nucleotide.
19. (a) nucleotide have a phosphate units
20. (c) are proteinous molecules.
21. (b)   22. (b)   23. (b)   24. (a)
25. (a) immunoglobulins are gamma globulins, responsible for immune response.
26. (a)   27. (c)   28. (c)
29. (c) it functions as a heart stimulant.
30. (c) use for protecting embryo.
31. (c) a blue colour is formed when iodine reacts with starch.
32. (a)   33. (c)   34. (d)   35. (a)   36. (a)
37. (b) 3 only   38. (d)   39. (b)
40. (a) Capital 'D' and 'L' stand for the arbitrary assignment, while writing the structures of monosaccharides. The small letters 'd' and 'l' stand for dextro and laevo rotatory sugars.

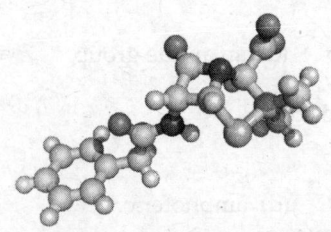

Chapter 8

Chemistry in Action

MULTIPLE CHOICE QUESTIONS

1. Penicillin was discovered by
   (a) A Fleming   (b) Tence and Salke   (c) SA Wanska   (d) Lewis Pasteur
2. Aspirin is chemically
   (a) methyl salicylate   (b) ethyl salicylate
   (c) acetyl salicylate   (d) o-hydroxy benzoic acid
3. An example of psychedelic is termed
   (a) DNA   (b) LSD   (c) DDT   (d) TNT
4. The full name of DDT is
   (a) dichloro diphenyl toluene   (b) dichloro diphenyl tri chloro ethane
   (c) dichloro deri   (d) dichloro diphenyl thiocarbazide
5. Which is of the following is known as a broad spectrum drug?
   (a) streptomycin   (b) chloramphenicol
   (c) ampicillin   (d) penicillin G
6. One of the widely used drug 'Iodex' is
   (a) methyl salicylate   (b) ethyl salicylate
   (c) salicylic acid   (d) acetyl salicylic acid
7. Which one is a hypnotic drug?
   (a) luminal   (b) salol   (c) catechol   (d) none
8. The drug used as an antidepressant is
   (a) luminol   (b) tofranil   (c) mescaline   (d) sulphadiazine
9. Which one is being used as a refrigerant?
   (a) CCl_4 (b) acetone
   (c) CF_4 (d) dichloro difluoro methane (CF_2Cl_2)
10. Morphine is
    (a) an alkaloid   (b) an enzyme   (c) a carbohydrate (d) a protein-a
11. Which statement is false?
    (a) some disinfectant can be used in low concentration
    (b) sulphadiazine is a synthetic antiseptic compound
    (c) ampicillin is a natural antiseptic
    (d) aspirin is analgestic and antipyretic

12. Phenacetin is used as
   - (a) antipyretic
   - (b) antiseptic
   - (c) antimalarial drug
   - (d) analgesic
13. Arsenic drugs are used in the treatment of
   - (a) syphilis
   - (b) jaundice
   - (c) typhoid
   - (d) cholera
14. Which is correct about saccharine?

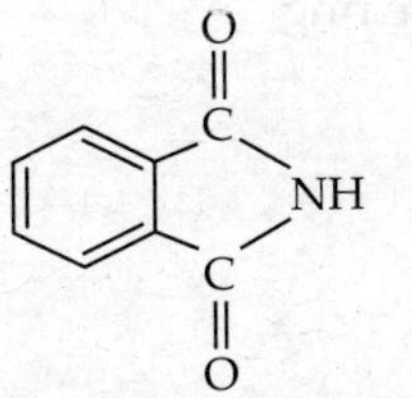

   - (a) it has the above structure
   - (b) it is 600 times sweeter than sugar
   - (c) it is used as a sweetening agent
   - (d) all
15. The following compound is used as

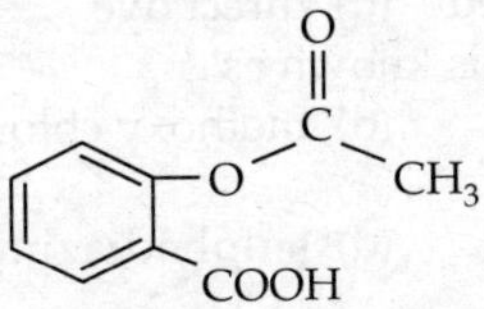

   - (a) an anti-inflammable compound
   - (b) analgesic
   - (c) antipyritic
   - (d) antiseptic
16. Which of the following is antipyritic?
   - (a) quinine
   - (b) paracetamol
   - (c) luminol
   - (d) paprazine
17. Which of the following can be used as an analgesic without any addition or modification?
   - (a) morphine
   - (b) acetyl p-amino phenol
   - (c) diazepam
   - (d) tetrahydro catechol
18. Which is the correct structure of paracetamol? What other name can be given to this compound?

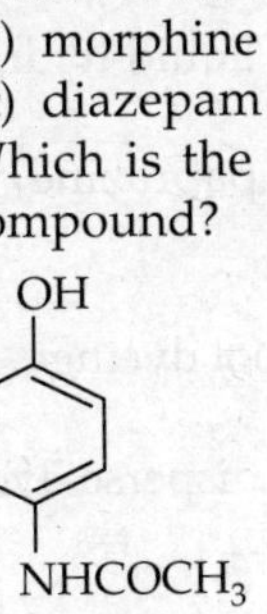

   (a) 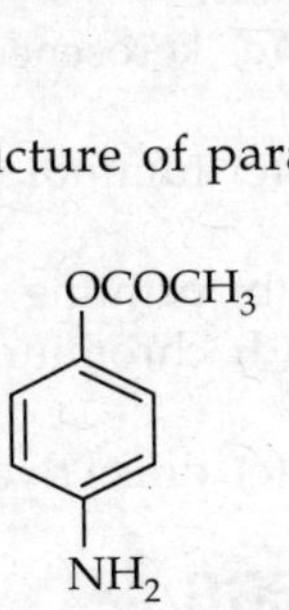   (b)
19. Which one among the following is not a chromophore?
   - (a) —N=N—
   - (b) —NO
   - (c) NO_2
   - (d) —NH_2
20. Which one of the following is not an auxochrome?
   - (a) —$N(CH_3)_2$
   - (b) —OH
   - (c) —OCH_3
   - (d) >C=N
21. An azo dye is formed by the interaction of a diazonium salt with
   - (a) phenol
   - (b) an aliphatic primary amine
   - (c) benzene
   - (d) HNO_2
22. An insoluble compound formed by the action of a metallic salt on the dye is known as
   - (a) lake
   - (b) mordant
   - (c) dye intermediate
   - (d) none of these
23. Alizarin belongs to the class of
   - (a) vat dyes
   - (b) mordant dye
   - (c) substantive dye
   - (d) reactive dye

24. To which class the phenolphthalein and methyl orange belong to
    (a) a basic dye and an acid dye   (b) an basic dye
    (c) a neutral dye
25. Which one of the following is not an antipyretic?
    (a) aspirin   (b) paracetamol
    (c) barbituric acid   (d) phenacitin
26. An antiseptic that contains a nitro group, attached to the benzene ring
    (a) penicillin   (b) streptomycin
    (c) chloramphenicol   (d) tetracycline
27. Further growth of cancerous cells are arrested by the body
    (a) psychotherapy   (b) electrotherapy
    (c) chemotherapy   (d) physiotherapy
28. Chloramphenicol is an effective drug in the
    (a) treatment of typhoid   (b) malaria
    (c) cholera   (d) tuberculosis
29. Indigo belongs to the class of
    (a) a mordant dye   (b) vat dye   (c) dispersed dyed   (d) direct dye
30. The insecticide, containing isomer of benzene hexa chloride, is known as
    (a) lindane   (b) malathion   (c) TNT   (d) methoxy chlor
31. Anti-depressant drug is
    (a) luminol   (b) tofranil   (c) mescaline   (d) sulphadiazine
32. Which of the following drug is a hallucinogenic drug?
    (a) methedrine   (b) calmpose   (c) LSD   (d) seconal
33. A space is propelled by
    (a) an automobile engine   (b) rocket propellant
    (c) steam engine   (d) nuclear submarine
34. Which of these is not used as liquid propellant?
    (a) liquid He   (b) liquid hydrogen (c) kerosene   (d) liquid hydrazine
35. Which of the following is antipyretic?
    (a) quinine   (b) paracetamol   (c) luminol   (d) paprazine
36. Alizarin is a mordant dye, it is not used in
    (a) cotton printing   (b) printing
    (c) painting   (d) chromium lake for wool dyeing
37. Indigo belongs to the class of
    (a) mordant dye   (b) vat dyes   (c) direct dye   (d) disperse dye

ANSWERS

1. (a)   2. (c)
3. (b) a drug is termed psychedelic, the consumption of which produces hallucination (for ex-
    LSD lysergic acid dimethyl amide) and mascaline.
4. (b)   5. (c)
6. (a) methyl salicylate.
7. (a)
8. (a) luminal.
9. (d) dichloro difluoro methane-(CCl_2F_2).
10. (a) morphine is an alkaloid, it is a mixture of several alkaloids.
11. (c) ampicillin is a synthetic antiseptic compound.
12. (a) Phenacitin is used as antipyretic or a drug for lowering the body temperature.

The structure of phenacitin is p-ethoxy acetanilide

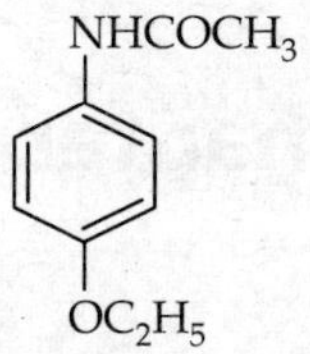

NHCOCH$_3$

OC$_2$H$_5$

13. (a) the arsenic drug is used in the treatment of syphilis.
14. (d) all.
15. (b) the compound is acetyl salicylic acid.
16. (b) these compound cause sleep and are termed as hypnotic-ex-luminal and seconal.
17. (b) acetyl p-aminophenol.
18. (a) the compound can be called as p-acetamide phenol.
19. (d) a compound bearing a chromophore may be coloured or may not be, but the presence of anauxochrome make the colour dark.
20. (d) dye is made up of two parts:
    1. chromophore-colour bearing; usually these are unsaturated groups.
    2. colour enhancing-acidic, (COOH); basic groups and neural groups.
21. (a) phenols undergo coupling with diazonium salt to form dyes.
22. (b) mordant dye.
23. (b) it is a mordant dye-different colours are formed in the presence of different metal ions.
24. (a) it belongs to phthalein dye.
25. (c) barbituric is a hypnotic.
26. (c) Chloramphenicol

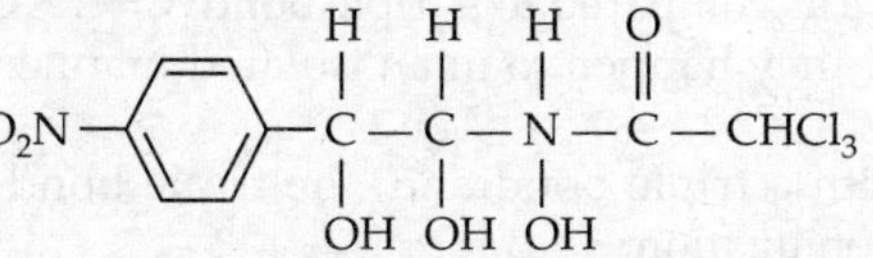

$$O_2N-\underset{}{\bigcirc}-\overset{\overset{H}{|}}{C}-\overset{\overset{H}{|}}{C}-\overset{\overset{H}{|}}{N}-\overset{\overset{O}{\|}}{C}-CHCl_3$$
$$\quad\quad\quad\quad\; OH\; OH\; OH$$

27. (c) Chemotherapy; Paul Ehlirich-is known as the father of chemotherapy.
28. (a) Chloramphenicol (or chloromycin) is an effective drug in the treatment of typhoid.
29. (c)   30. (a)   31. (b)
32. (c) lysergic diethyl amide (LSD).
33. (b)   34. (a)
35. Paracetamol at both as analgesic as well as antipyretic.
36. (c)
37. (b) vat dye.

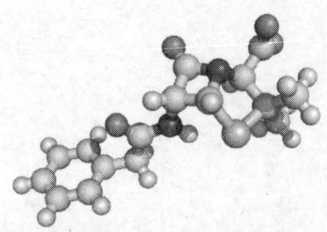

Chapter 9

Important Points to Remember

1. In aliphatic hydrocarbons
   a. Alkanes only undergo free radical reactions
   b. If O_2 is added in a reaction undergoing a free radical reaction, the reaction is retarded.
2. The stability of alkenes in the following order
   R_2—C=CR_2 > R_2C=CHR > RHC=CHR > R_2C=CH_2 > RHC=CH > CH_2=CH_2
   Reactivity of alkenes is in the reverse order.
3. The reactivity of alkenes and alkynes is in the following order:
   i. In general a double bond is more reactive as compared to a triple bond; C=C > C≡C
   ii. If in a compound both are present, and they happen to in an isolated manner, then a double bond will be attacked first.
   iii. When a double bond is conjugated with a triple bond and, the triple bond will be attacked first, in order to maintain the conjugation.
4. In alkyl halides, the iodides are more reactive than the corresponding alkyl bromides or alkyl chlorides.
   i. A 1° alkyl halides undergoes S_N2 reaction where as a 3° alkyl halide undergoes S_N1 reaction in nucleophilic substitution reactions.
   ii. A 3° alkyl halides undergoes elimination, when attacked by a strong nucleophile and results in the formation of an alkene.
   iii. In a S_N2 substitution reactions, an inversion must take place; in S_N1 substitution reactions, racemisation takes place.
5. The alcohols are slightly acidic in nature. The acidity of various hydrogen compounds is in the following deceasing order:
   HCl > RCOOH > C_6H_5OH > H_2O > ROH > C_2H_2 > NH_3 > C_2H_4 > C_2H_6
   $1×10^8$ $1×10^{-5}$ $1×10^{-110}$ $1×10^{-14}$ $1×10^{-18}$ $1×10^{-22}$ $1×10^{-35}$ $1×10^{-36}$ $1×10^{42}$
   These are also termed as Lewis acids.
   The decreasing order of the basic character of their corresponding or the reverse is the basic strength of the Lewis bases:
   C_2H_5 > CH_2=CH^- > NH_2^- > CH≡C^- > RO^- > OH^- > $C_6H_5O^-$ > $RCOO^-$
6. CH_3OH is a stronger acid as compare to other alcohols.
7. A 3° alcohol undergoes dehydration more easily as compared to a 2° or to a 1° alcohohls
   The bond between a C—O break up faster as compared to a O—H bond in a 3° alcohol
   In a 1° alcohol a O—H bond breaks up faster as compared to a 3° alcohol.

8. Formaldehyde is the strongest reducing aldehyde.

9. Aldehydes are easily oxidised, as compared to the keto compounds.

10. Aliphatic keto compounds form sodium bisulphites addition compounds, but aromatic to compounds do not.

11. All aliphatic aldehydes reduce Fehling solution but benzaldehyde and other aromatic aldehydes do not.

12. HCOOH, formic acid gives the properties of an aldehyde besides behaving as a mono carboxylic acid.

$$\underset{\text{H—C—H}}{\overset{\overset{\displaystyle O}{\|}}{}} \qquad \underset{\text{H—C—OH}}{\overset{\overset{\displaystyle O}{\|}}{}}$$

13. Aliphatic amines are stronger bases than aromatic nuclear substituted amines.
    $RNH_2 > ArNH_2$

14. In benzene and other aromatic rings ii is the electrophilic substitution takes place.

15. In nitrobenzene, there is a rare example of a nucleophilic substitution; on fusing solid KOH with $C_6H_5NO_2$ there is formed a mixture of o-nitro and p-nitro phenols are formed.

16. In hybridisation, if sp^3-sp^3 bond changes to a sp^2-sp^2 bond, the total number bonds decreases by only one.

17. When sp^2-sp^2 bond changes to sp-sp bond, there is decrease in total number of bonds by only one.

18. In case the sp^3-sp^3 bond change to a sp-sp bond, the total number of bonds changes by only two.

19. In tautomers the total number of bond remains the same.

20. The groups with –I effect (electron withdrawing) when attached to a benzene ring, are

    $—N^+(CH_3)_3$; NO_2 ; $—CN$; SO_3H ; $—COOH$; $—CHO$; $—COR$;

    Order of the directing group

    $Me_3N^+ > NO_2 > CN > SO_3H > CHO > COCH_3 > COOH$

    The substituents which have a positive or a partial charge on the adjacent to the ring are m-directing groups. These groups deactivate the ring and act as m-directing groups.

    The groups with +I effect (electron donating or electron releasing) when attached to a benzene ring.

    $—NH_2$; $—OH$; $—OCH_3$; $—NHCOCH_3$; $—C_6H_5$; $—CH_3$; X

    The substituents which have one or more non-bonding pair of electron (loan pair) on the atom, adjacent to the ring are o- and p-directing groups.

    These groups activate the ring and therefore act as o- and p-directing groups.

21. In the case of open chain compounds, the groups with –I effect are (any atom having electronegativity higher than the carbon, to which it is joined, will attract electrons from the carbon and such a group will act as having –I effect.

    $—NO_2$; F ; Cl ; Br ; I ; OH ; C_6H_5

    In open chain compounds, the groups with +I effect are

    $(CH_3)_3C—$; $(CH_3)_2CH$; $CH_3CH_2—$; $CH_3—$

22. Compounds which rotate the plane of polarised light to the right or in a clockwise direction are known as dextro rotatory and the compounds which rotate the plane polarised light to the left they are known as laevo rotatory.

23. The optical activities of (+) tartaric acid is named as d(D) tartaric acid because it has a +ve rotation.

IMPORTANT POINTS

1. The number of bonds (both σ- and π-bonds) remains same in a compound and its tautomer.
2. When a sp^3-sp^3 bond changes to a sp^2-sp^2 bond, the total number of bonds decreases by one only.
3. Similarly when a sp^2-sp^2 bond changes to a sp-sp bond, the total number of bonds decreases by one only.
4. In other words, bond sp^3-sp^3 change to a sp-sp bond the total number bonds decreases by two.

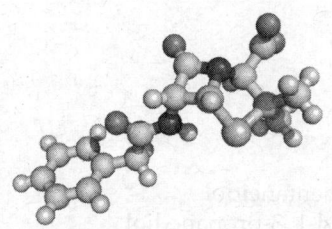

Chapter 10

IUPAC Names

1. The correct name of the compound is

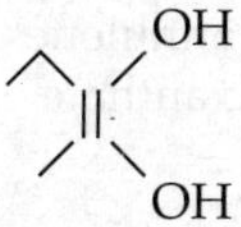

   (a) but-2-ene-2, 3-diol
   (b) pen-2-ene-2, 3-diol
   (c) 2-methyl but-2-ene-2, 3 diol
   (d) hex-2-ene-2, 3di-ol

2. Give IUPAC name

   $$OHC—CH=CH—CH—CH=CH_2$$
   $$|$$
   $$CH_2CH_2CH_2CH_3$$

   (a) 4-butyl-2, 5-hexadiene-1-al
   (b) 5-vinyl oct-3-en-1-al
   (c) 4-n-butyl2, 5- hexa-1-diene-1-al
   (d) none of the above

3. IUPAC name of the compound is

   $$CH_3$$

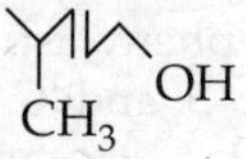

   (a) 5-methyl-hexanol
   (b) 2-methyl hexanol
   (c) 2-methylhex-3-ol
   (d) 4-methylpent-2-ene-1-ol

4. IUPAC name of the compound is

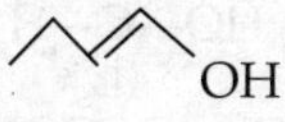

   (a) But-2-ene-1-ol
   (b) 1-hydroxy but-1-ol
   (c) 4-hydroxybutane-1-ol
   (d) but-1-ene-1-ol

5. IUPAC name of the compound is

   (a) bicyclo (2, 1, 0) pentane
   (b) 1, 2-cyclo propyl cyclo butane
   (c) cyclopentane-3, 4-annulene
   (d) 1, 2 methylene cyclo butane

6. The IUPAC name for

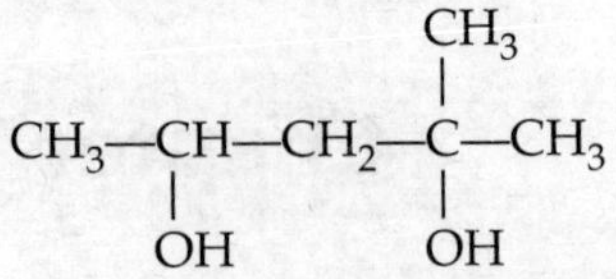

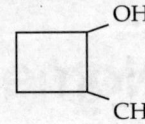

   (a) 1, 1-dimethyl-1, 3-butanediol   (b) 4-methyl-2, 4-pentaneidol
   (c) 2-methyl-2, 4-pentanediol   (d) 1, 3, 3-trimethyl-1, 3-propanediol

7. IUPAC name of the compound is

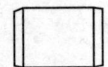

   (a) 3-methyl cyclo-1-but-2-ol   (b) 4-methylcyclo-2-butene -1 al
   (c) 4-methyl cyclo-1-butene-1ol   (d) 2-methylcyclo-3-butene-2-ol

8. The compound is an example of

   (a) aromatic compound   (b) annulene
   (c) heterocyclic compound   (d) xanthate

9. Correct name of the compound is

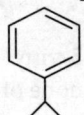

   (a) 1, 2-dimethyl-2-butanol   (b) 2, 3-dimethyl-4-pentene-1-ol
   (c) 3, 4-dimethyl-2-butene-4-ol   (d) 3-methyl pent-3-ene-2-ol

10. The correct IUPAC name of the compound is

   (a) isopropyl benzene   (b) cumeme
   (c) 2-phenyl iso propane   (d) phenyl iso propane

11. The correct statement about the compound 'A', 'B' and 'C

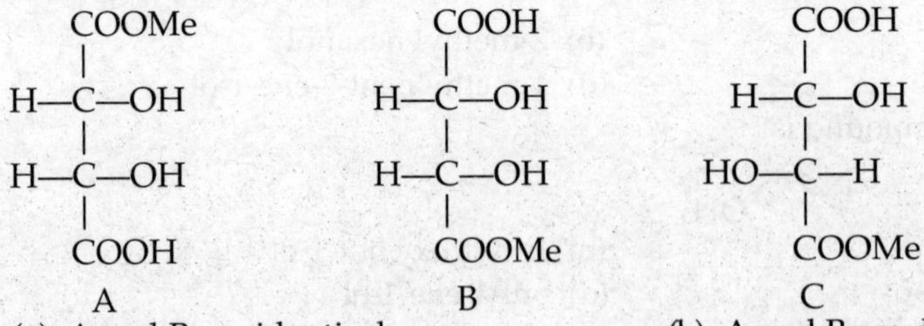

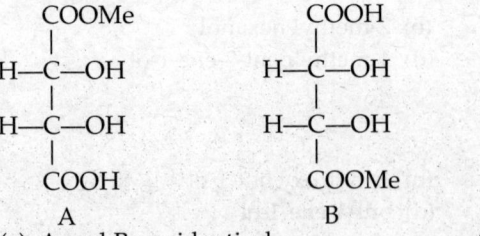

   (a) A and B are identical   (b) A and B are diastereomers
   (c) B and C are enantiomers   (d) A and B are enantiomers

12. IUPAC name of

   (a) 4, 5-dimethyl oct-4-ene   (b) 3, 4-dimthyl oct-5-ene
   (c) 4, 5-dimethyloct-5-ene   (d) none

13. The structure of *cis*-bis (propenyl) ethene

(a)

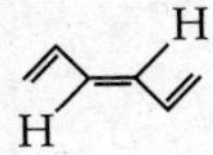

(b)

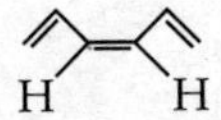

(c)

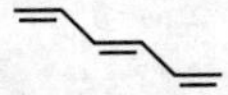

(d)

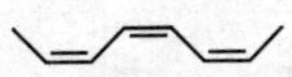

14. Hydroenation of the compound, in presence of poisoned Pd, gives you
   (a) an optically active compound
   (b) an optically inactive compound
   (c) a racemic mixture
   (d) a single diastereomeric mixture

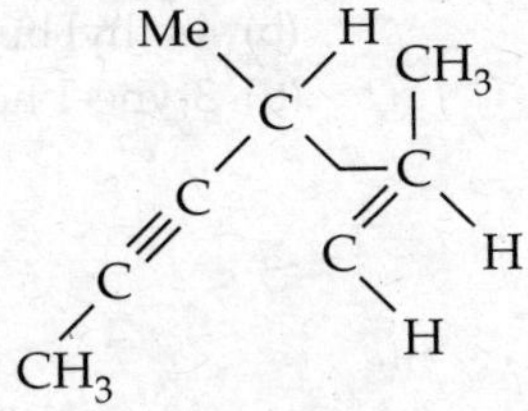

15. $CH_3—(CH_2)_2\ CH—\overset{\overset{\displaystyle CH_3}{|}}{\underset{\underset{\displaystyle (CH_2)_2—CH_3}{|}}{C}}—CH_2CH_3$

   (a) 4-ethyl-4, 5-dimethyl octane
   (b) 5-ethyl-5, 6-dimethyl octane
   (c) 4-ethyl-3, 4-di methyl octane
   (d) none of these

16. $CH_3—\overset{\overset{\displaystyle }{}}{\underset{\underset{\displaystyle CH_3}{|}}{CH}}—CH_2—\overset{\overset{\displaystyle }{}}{\underset{\underset{\displaystyle OH}{|}}{CH}}—CH_2—Cl$

   (a) 1-chloro-4-methyl-2-hexanol
   (b) 1-chloro-4-methyl-2-hydroxy pentane
   (c) 1-chloromethyl-3-methyl butanol-1
   (d) none

17. $CH_3—\overset{\overset{\displaystyle CH_3}{|}}{CH}—CH_2—CH_2—Cl$

   (a) 1-chloroisopentane
   (b) 1-isopentyl chloride
   (c) 1-chloro-3-methyl butane
   (d) none

18. $C_2H_5—\overset{\overset{\displaystyle CH_2}{|}}{C}—CH_2—\overset{\overset{\displaystyle CH_3}{|}}{CH}—NH_2$

   (a) 2 amino-4-ethyl pentene-4
   (b) 2-amino pentane
   (c) 2-ethyl-4-amino-1-pentene
   (d) 4 amino-2-ethyl-1-pentene

19. $CH_2{=}\overset{\overset{\displaystyle C_2H_5}{|}}{C}—CH_2C{\equiv}CH$

   (a) 2-methyl pent-1-ene-4-yne
   (b) 4-ethyl-pent-1-ene-4yne
   (c) 2-methylene-4-amino-pentene
   (d) 2-methyl-4 pent-4yne

20. $(CH_3)_2—N—C_2H_5$
   (a) N-ethyl dimethyl amine
   (b) N-dimethyl ethylamine
   (c) dimethyl ethyl amine
   (d) none of these

21. Which of the following is the correct name?

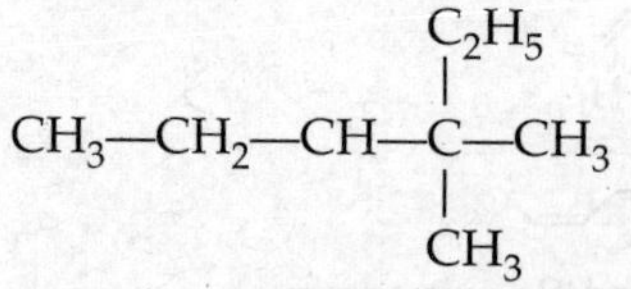

$$CH_3-CH_2-CH-\underset{\underset{CH_3}{|}}{\overset{\overset{C_2H_5}{|}}{C}}-CH_3$$

   (a) 2-methyl-3-ethylpentane
   (b) 2-ethtyl-3-methylpentane
   (c) 3-ethyl-2-methylpentane
   (d) 2,2-dimethyl hexane

22. $CH_3-CH_2-C\equiv C-CH=CH_2$
   (a) 1-ethyl-3-ene-1-yne
   (b) 4-ethyl but-1-ene-3-yne
   (c) ethylvinyl acetylene
   (d) 3-yne-1 hexene

23. 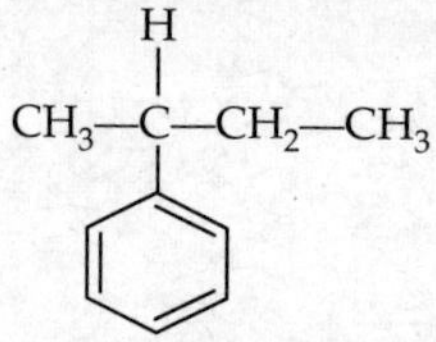 $CH_3-\underset{\underset{\text{(phenyl ring)}}{|}}{\overset{\overset{H}{|}}{C}}-CH_2-CH_3$

   (a) 2-phenyl n-butane
   (b) cumene
   (c) iso-butyl benzene
   (d) none

24. Among the following, the name, acceptable, as per IUPAC rules
   $CH_2=CH-CH_2-CH_2-C\equiv CH$
   (a) 4-pentene-1-yne
   (b) 1-pentene-4-yne
   (c) 2, 2-dimethyl-4-ethyl pentane
   (d) 3-hydroxy butane-1-ol

25. $Cl-CH_2-CH_2-COOH$
   (a) 3-choro propanoic acid
   (b) 3-chloro propionic acid
   (c) 2-chloro propionic acid
   (d) 2-chloro acetic acid

26. CH_3CHCl_2
   (a) 1, 1 dichloroethane
   (b) ethylidine chloride
   (c) ethylene chloride
   (d) dichloro ethane

27. $CH_2=CH-CH_2Cl$
   (a) 3-chloro-1-propene
   (b) allyl chloride
   (c) vinyl chloride
   (d) propyl chloride

28. 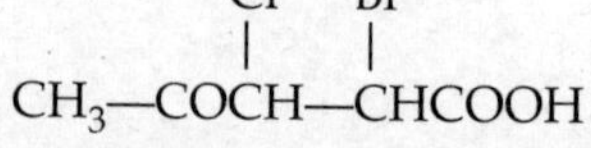 $CH_3-\overset{}{C}O\underset{\underset{Cl}{|}}{C}H-\underset{\underset{Br}{|}}{C}HCOOH$
   (a) 4-oxo-2-bromo-3-chloro pentanoic acid
   (b) 4-oxo2-bromo-3-chloro valeric acid
   (c) 2-bromo-3-chloro-1-one pentanoic acid
   (d) none

29. 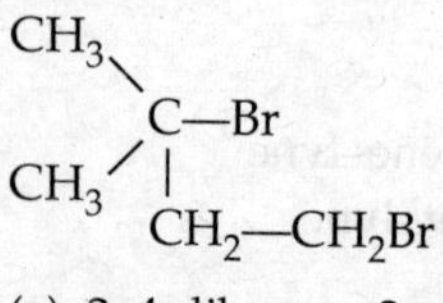 $\underset{CH_3}{\overset{CH_3}{>}}\underset{\underset{CH_2-CH_2Br}{|}}{C}-Br$

   (a) 2, 4-dibromo-2-methyl butane
   (b) 1, 3-dibomo-3-methyl butane
   (c) 3-methyl-1, 3-dibromopropane
   (d) none

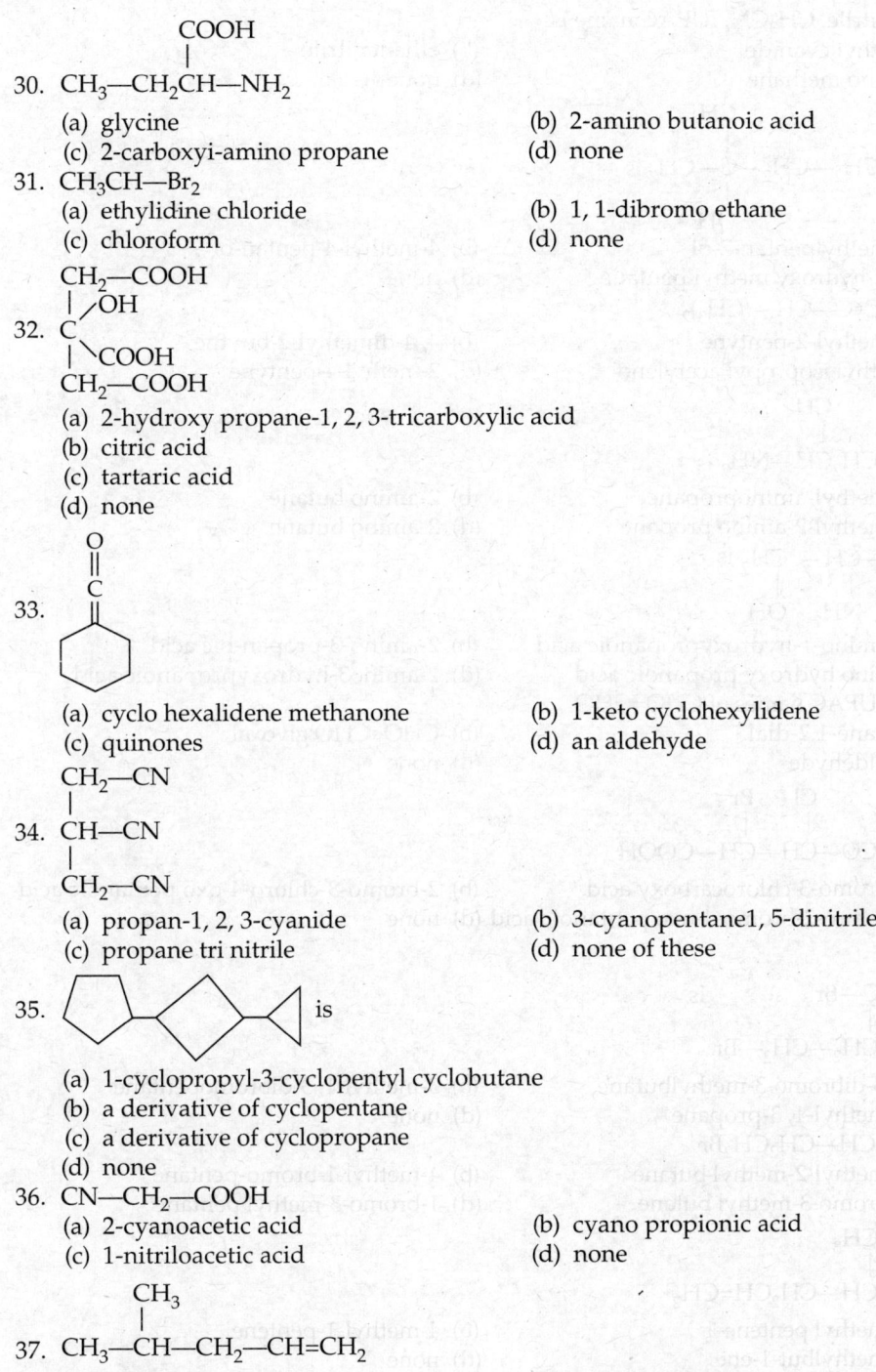

30. $CH_3—CH_2CH—NH_2$ (with COOH substituent)
    (a) glycine
    (b) 2-amino butanoic acid
    (c) 2-carboxyi-amino propane
    (d) none

31. $CH_3CH—Br_2$
    (a) ethylidine chloride
    (b) 1, 1-dibromo ethane
    (c) chloroform
    (d) none

32. C with $CH_2—COOH$, OH, COOH, $CH_2—COOH$
    (a) 2-hydroxy propane-1, 2, 3-tricarboxylic acid
    (b) citric acid
    (c) tartaric acid
    (d) none

33.
    (a) cyclo hexalidene methanone
    (b) 1-keto cyclohexylidene
    (c) quinones
    (d) an aldehyde

34. $CH_2—CN$ / $CH—CN$ / $CH_2—CN$
    (a) propan-1, 2, 3-cyanide
    (b) 3-cyanopentane1, 5-dinitrile
    (c) propane tri nitrile
    (d) none of these

35. is
    (a) 1-cyclopropyl-3-cyclopentyl cyclobutane
    (b) a derivative of cyclopentane
    (c) a derivative of cyclopropane
    (d) none

36. $CN—CH_2—COOH$
    (a) 2-cyanoacetic acid
    (b) cyano propionic acid
    (c) 1-nitriloacetic acid
    (d) none

37. $CH_3—CH—CH_2—CH=CH_2$ (with CH_3 substituent)
    (a) 4-methylpent-1-ene
    (b) 2-methyl-1-ene
    (c) methyl hexane
    (d) 2-methyl pent-4-ene

38. Acetonitrile, CH_3CN, IUPAC name is
    (a) methyl cyanide
    (b) ethanenitrile
    (c) cyano methane
    (d) none

39. 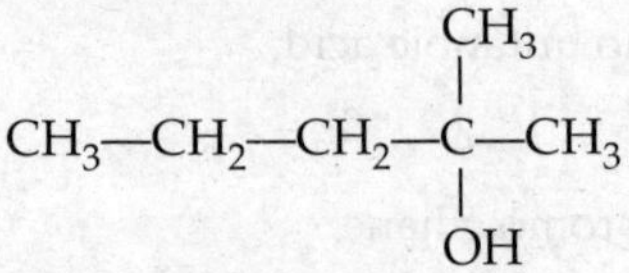 $CH_3-CH_2-CH_2-\underset{\underset{OH}{|}}{\overset{\overset{CH_3}{|}}{C}}-CH_3$ is
    (a) 2-methylpentan-2-ol
    (b) 4-methyl-4-pentan-ol
    (c) 2, 2-hydroxy methyl pentane
    (d) none

40. $CH_3-C\equiv C-CH-(CH_3)_2$
    (a) 4-methyl-2-pentyne
    (b) 4, 4-dimethyl-2-buytne
    (c) methyl isopropyl acetylene
    (d) 2-methyl-4-pentyne

41. 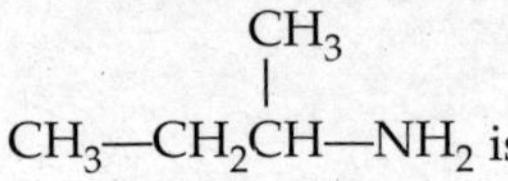 $CH_3-CH_2CH-NH_2$ (with CH_3 above) is
    (a) 1-methyl-aminopropane
    (b) 2-amino butane
    (c) 2-methyl-2-amino propane
    (d) 2-amino butane

42. 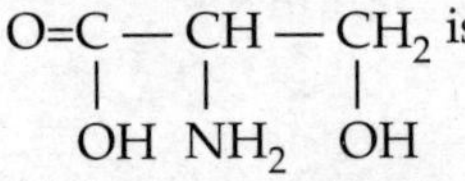 $O{=}C-CH-CH_2$ is (with OH, NH_2, OH below)
    (a) 3-amino-1-hydroxypropanoic acid
    (b) 2-amino-3-propan-oic acid
    (c) amino hydroxy propanoic acid
    (d) 2-amino3-hydroxypropanoic acid

43. Write IUPAC names of CHO=CHO
    (a) ethane-1,2-dial
    (b) CHO≡CHO glyoxal
    (c) dialdehyde
    (d) none

44. 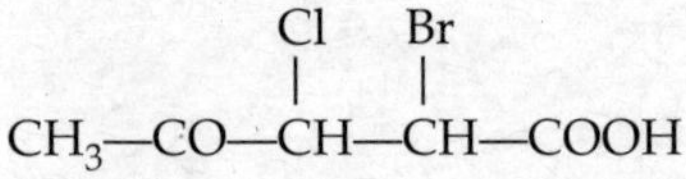 $CH_3-CO-CH-CH-COOH$ (with Cl, Br above)
    (a) 4-bromo-3-chlorocarboxy acid
    (b) 2-bromo-3-chloro-4-oxo pentanoic acid
    (c) 3 chloro-2-bromo-4-oxo pentanoic acid
    (d) none

45. 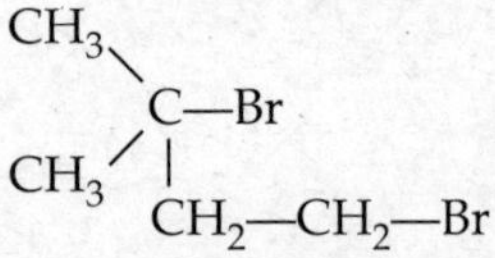 CH_3 and CH_3 attached to $C-Br$, with CH_2-CH_2-Br below, is
    (a) 1, 3-dibromo-3-methylbutane
    (b) 3-methyl-1, 3-dibromopentane
    (c) 3-methyl-1, 3-propane
    (d) none

46. $(CH_3)_2 CH-CH_2CH_2Br$
    (a) 3-methyl-2-methyl butane
    (b) 4-methyl-1-bromo-pentane
    (c) 1-bromo-3-methyl butane
    (d) 1-bromo-3-methyl pentane

47. 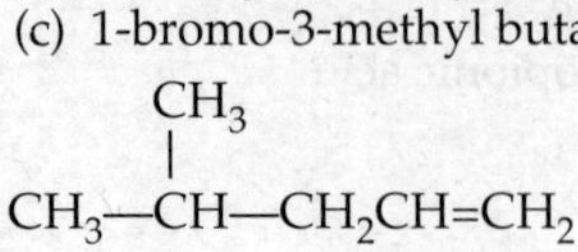 $CH_3-CH-CH_2CH=CH_2$ (with CH_3 above)
    (a) 2-methyl pentene-1
    (b) 4-methyl-1-pentene
    (c) 2-methylbut-1-ene
    (d) none

48. $(CH_3CO)_2O$
    (a) acetic anhydride
    (b) ethanoic anhydride
    (c) acid anhydride
    (d) none

49. CH_2=CH—$CH(CH_3)_2$ is
   (a) 2-methyl-but-3-ene
   (b) 3-methyl-1-butene
   (c) 1, 1-dimethyl Pentene-1
   (d) none

50. CH_3O—$CH(CH_3)_2$ is
   (a) 2-methoxypropane
   (b) 1-methoxy propane
   (c) 3-methoxy propane
   (d) methyl-isopropyl ether

51. The IUPAC name of picric acid is
   (a) 2, 4, 6-tri nitro phenol
   (b) 2, 4, 6-rinitro-1-hydroxy benzene
   (c) 1, 3 ,5-trinitro-6-hydroxybenzene
   (d) none

52. CH(OH)—COOH is
    |
    CH(OH) COOH
   (a) Tartaric acid
   (b) 2,3-dihydroxy1,4-butane dioic acid
   (c) Malic acid
   (d) none

53. $(C_2H_5)_2N$ CH_2CHClCOOH is
   (a) 2-chloro-4-N-ethyl pentanoic acid
   (b) 2-chloro-4-N,N diethyl-amino pentanoic acid
   (c) 2-chloro-4-(N,N)-diethyl amino propanoic acid
   (d) none

54. OHC—CO—COOH is
   (a) aldo, keto propionic acid
   (b) 3-aldehyde-2-keto-propionic acid
   (c) 2, 3 di oxo-1-propanoic acid
   (d) none of these

ANSWERS

1. (b)   2. (a)   3. (d)   4. (d)   5. (a)   6. (c)   7. (b)   8. (b)
9. All are correct IUPA name-3, -dimethyl-2-but -4-ol.
10. (c) All are correct names.
11. (a) Both are identical; A and B are identical. 12. (a)
13. (d) the two propenyl groups attached to 1, 2-positions on carbon in *cis*-form.
14. (b)   15. (a)   16. (b)   17. (c)   18. (c)   19. (a)   20. (a)   21. (d)
22. (d)   23. (a)   24. (b)   25. (a)   26. (a)   27. (a)   28. (a)   29. (b)
30. (b)   31. (b)   32. (a)   33. (a)   34. (b)   35. (a)   36. (a)   37. (a)
38. (b)   39. (a)   40. (a)   41. (b)   42. (d)   43. (a)   44. (b)   45. (b)
46. (d)   47. (b)   48. (b)   49. (b)   50. (a)   51. (a)   52. (c)   53. (c)
54. (c)

MCQs on Synthetic Polymers

Synthetic and Natural Polymers

1. The fibre obtained by the condensation of hexa methylene diamine and adipic acid is
   (a) dacron
   (b) nylon-55
   (c) rayon
   (d) teflon

2. Orlon is a polymer of
   (a) styrene
   (b) acrylonitrile
   (c) vinyl chloride
   (d) F_2—C=C—F

3. The compound that is used in the manufacturing of terylene are
   (a) phthalic acid
   (b) caprolactum
   (c) p-benzene di carboxylic acid
   (d) m-phthalic acid
4. Which of the following belong to the class of natural polymer?
   (a) proteins
   (b) cellulose
   (c) rubber
   (d) all of these
5. Toluene di-isocyanate is used for the preparation of
   (a) polyesters
   (b) polyamides
   (c) polycarbons
   (d) polyurethanes
6. Acetate rayon is prepared form
   (a) acetic acid
   (b) glycerol
   (c) starch
   (d) cellulose
7. The catalyst used in the manufacture of polythene is Zeigler Natta
   (a) $TiCl_4 + (C_6H_5)_3Al$
   (b) $TiCl_4 + (C_2H_5)_3Al$
   (c) TiO_2
   (d) titanium iso peroxide
8. Nylon-6 and nylon-66 are the same compound or
   (a) no they are different compounds compounds
   (b) nylon-66 is a condensation product between adipic acid and diethyl terphthalate; nylon-6 is an addition polymerisation
   (c) both (a) and (b)
   (d) none
9. Terylene is a
   (a) polyamide
   (b) polyester
   (c) poly ether
   (d) long chain polymer
10. Nylon-6,10-is a polymer of
    (a) hexa methylene and adipic acid
    (b) hexa methylene + sebacic acid
    (c) caprolactum
    (d) none
11. Which one can be remelted again and again without any change?
    (a) thermo setting polymer
    (b) thermo plastic polymer
    (c) Bakelite
    (d) melamine
12. Which one is a protein fibre?
    (a) cotton
    (b) rayon
    (c) silk
    (d) polyester
13. The thermopolastic are
    (a) linear polymers
    (b) can be soft on heating
    (c) molten polymer can be moulded
    (d) all the above
14. Polymers obtained by condensation are
    (a) polythene
    (b) teflon
    (c) PVC
    (d) formaldehye-phenol resin
15. The example of addition polymerisation
    (a) buna-S
    (b) bakelite
    (c) nylon-6
    (d) malamac
16. Thermosetting plastic are
    (a) cross-linked polymers
    (b) do not melt or soften
    (c) cross-linking takes place duing the time of heating
    (d) all

17. Melamine is
    (a) 2,4,6-triamine- 1,3,5 triazine
    (b) 1,3,5 triazine
    (c) 2,4,6-triamine
    (d) none
18. Condensation of caprolactum gives
    (a) nylon-66
    (b) nlon-6
    (c) nylon-6
    (d) a nitrile rubber
19. Which of the following is not natural polymer
    (a) glycogen
    (b) cellulose
    (c) pepsin
    (d) polybutadiene
20. Strongest inter particle forces are in
    (a) elastomer
    (b) thermoplastic
    (c) fibres
    (d) thermosetting plastic

ANSWERS

1. (b) nylon-66   2. (b) it is a polymer of acrylonitile (CH_2—CHCN)
3. (c)   4. (d)   5. (d)   6. (d)   7. (b)   8. (b)   9. (d)   10. (b)
11. (b)   12. (c)   13. (d)   14. (d)   15. (a)   16. (c)   17. (a)
18. nylon-6
19. (d)   20. (d)

MCQs on Oils and Fats

1. What are edible oils?
    (a) esters of higher fatty acid and gycerol
    (b) solid and liquid hyrocarbons
    (c) lipids
    (d) weak carboxylic acids
2. Lipids are
    (a) long chain fatty acid esters
    (b) long chain sulphonic acid esters
    (c) polymeric hydrocarbons
    (d) polymeric aldehydes
3. When oils and fats are heated with caustic soda the substances formed are
    (a) soap and glycerine
    (b) soap and oil
    (c) sodium carbonate and oil
    (d) oil, H_2O and soap
4. Detergent is a
    (a) cleansing agent
    (b) catalyst
    (c) drug
    (d) soap
5. Commercial detergent contains mainly
    (a) RONa
    (b) RCOONa
    (c) $ROSO_3Na$
    (d) $ROCH_2CH(OH)CH_2OR$
6. Toilet soap is
    (a) a mixture of calcium and potasium salt of higher fatty acids
    (b) a mixture of calcium stearate and glycerine
    (c) a mixture of sodium salts f higher fatty acids
    (d) a mixture of potassium salt of higher fatty acids
7. Sodium dodecyl benzene sulphonated is used a
    (a) pesticide
    (b) soap
    (c) fertilizer
    (d) detergent

8. Essential oils are
   - (a) mixture of various hydrocarbons
   - (b) pleasant smelling liquids from paints
   - (c)
   - (d)

9. Formation of soap involves
   - (a) saponification
   - (b) polymerization
   - (c) fermentation
   - (d) none

10. Rancidity of oils and fats is due to
   - (a) partial hydrolysis by atmospheric oxidation of fatty acids
   - (b) absorption of foul smelling substances
   - (c) fermentation, caused by micro-organism
   - (d) slow decomposition of fatty acids

11. What is the difference between oils an fats?
   - (a) oils are liquids
   - (b) fats are solids
   - (c) oils are esters of unsaturated acids and fats are saturated compounds
   - (d) steroids

12. Lipids are
   - (a) long chain fatty acid esters
   - (b) long chain sulphonic ester
   - (c) polymeric hydrocarbons
   - (d) polymeric aldehydes

13. Hydrolytic fission of fat is known as
   - (a) acetylation
   - (b) saponification
   - (c) esterification
   - (d) carboxylation

14. Metal lauryl sulphate acts as
   - (a) soap
   - (b) disinfectant
   - (c) antiseptic
   - (d) detergent

15. Oils and fats serve as
   - (a) reserve food for body
   - (b) immediate source of energy
   - (c) nitrogenous food
   - (d) control metabolism

16. Triglycerides, which is the composition of oils and fats, can be broken to form glycerine and fatty acids by
   - (a) diastase
   - (b) lipase
   - (c) trypsin
   - (d) pepsin

17. Purity of butter is determined in terms of
   - (a) saponification value
   - (b) iodine value
   - (c) acetyl value
   - (d) Reichert-Meissl value

18. Waxes belong to which one of the following class?
   - (a) acids
   - (b) alcohols
   - (c) esters
   - (d) ethers

19. The number of milligram of KOH required to nutralise 1 g of oil or fat, is called
   - (a) saponification value
   - (b) iodine value
   - (c) acetyl value
   - (d) acid value

20. Detergents are better cleansing agents than soaps, because
   - (a) they wash clothes better
   - (b) absorb the hardness of water
   - (c) they are less affected by hard water
   - (d) they are less soapy

ANSWERS

1. (a) Edible oils are esters of higher fatty acids with glycerol

2. (a)   3. (a)   4. (a)   5. (c)   6. (d)   7. (d)   8. (b)   9. (a)

10. (a)   11. (c)   12. (a)   13. (b)   14. (d)   15. (a)   16. (b)

17. (d) It is defined as the number of milliliters of N/10 KOH solution required to neutralise the distillate acid of 5 g of fat.

18. (c) esters; waxes are long chain esters of mono hydric alcohols with higher fatty acids.

19. (a) it is a definition of saponification.

20. (a) characteristic of detergent.

ANSWERS

Conversions and Interconversions

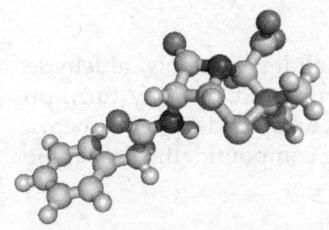

Conversions and Interconversions

There are five routes following which a lower member of any class of compounds may be converted into a higher member. In the same manner, a higher member may be converted to a lower one, by following **two** routes. These may be applied to any class of compounds. Given are:

- An alkane to either higher alkane or to a lower alkane.
- An alkene to either higher alkene or to a lower alkene.
- An alkyne to either higher alkyne or to a lower alkyne.

Route for Converting Aromatic Compounds

- An aromatic compound maybe converted into either a higher one or to a lower one, provided in all aromatic compounds the benzene ring has to remain intact.

  Therefore aromatic compounds must contain a minimum of "seven" carbon atoms, either in the form of a side chain or the fuctional group should be attached to a side chain. Only then an aromatic compound may be converted into a lower one.

Route for Converting Cycloalkane

- The size of a cycloalkane ring, either to increase or decrease.
- An alkyl halide either to a higher alkyl halide or to a lower alkyl halide.
- An aryl halide to a higher or to a lower one.
- An alcohol to a higher or to a lower one.
- A phenol may be converted either to a higher one or to a lower one.
- An aldehyde to higher one or to a lower one.
- An aromatic aldehyde to a higher or to lower one.
- A carboxylic acid to a higher or to a lower one.
- An aromatic acid to either a higher one or to a lower one.
- Alkyl amines to higher or to lower one.
- An amide to a higher or to a lower one.

Carbohydrates

- Glucose to fructose–two methods
- Fructose to glucose–two methods
- Interconversions

Discussed

Treating methane, ethane, propane, alkenes, alkynes, alkyl halides, alcohols, aldehydes, ketones, carboxylic acids, amines and amides, as the starting compound, turn by turn, preparation of all other compounds. In a similar way benzene, toluene, aryl halide, phenol, benzaldehyde, benzoic acid, aniline, etc. taking turn by turn, as the starting compound, the preparation of all aromatic compounds discussed.

Hydrocarbons

1. **CH_3—CH_2—CH_3 to acetone $CH_3\ COCH_3$**

$$CH_3CH_2CH_3 \xrightarrow{Cl_2/UV} \underset{\text{2-chloro propane}}{CH_3CHClCH_3} \xrightarrow{\text{aq. KOH}} \underset{\text{iso propyl alcohol}}{CH_3CH(OH)CH_3} \xrightarrow{\text{oxi}} CH_3COCH_3$$

2. **2-Methyl hexane from propene**

$$CH_3CH=CH_2 \xrightarrow{HBr} CH_3CHBrCH_3 \xrightarrow{2Li} (CH_3)_2CHLi \xrightarrow{CuI} \underset{\text{di iso propyl lithium cuperate}}{(CH_3)_2CH_2CuLi}$$

$$\xrightarrow{CH_3CH_2CH_2CH_2Br} \underset{\text{2-methyl hexane}}{(CH_3)_2CH\ CH_2CH_2CH_2CH_3}$$

3. **n-Hexane from propane**

$$CH_3CH_2CH_3 \xrightarrow{Cl_2/UV} CH_3CHClCH_3 \xrightarrow{\text{alco. KOH}} CH_3CH=CH_2 \xrightarrow{HBr/peroxide}$$

$$CH_3CH_2CH_2Br \xrightarrow{2Li} CH_3CH_2CH_2Li \xrightarrow{CuI} (CH_3CH_2CH_2)CuLi \xrightarrow{CH_3CH_2CH_2Br}$$
$$CH_3CH_2CH_2{-}CH_2CH_2CH_3$$

4. **Acetylene from ethane**

$$CH_3{-}CH_3 \xrightarrow{Cl_2/UV} CH_3CH_2Cl \xrightarrow{\text{alco. KOH}} CH_2=CH_2 \xrightarrow{Br_2} CH_2Br{-}CH_2Br \xrightarrow[\text{(ii) NaNH}_2]{\text{(i) alco. KOH}}$$
$$CH{\equiv}CH$$

Alkyl Halides

1. **Allyl chloride from propane**

$$CH_3{-}CH_2{-}CH_3 \xrightarrow{Cl_2/UV} CH_3CH(Cl)CH_3 + CH_3CH_2CH_2Cl \xrightarrow{\text{alco. KOH}} CH_3CH=CH_2$$
$$\xrightarrow{Cl_2/500°C} ClCH_2CH=CH_2$$

2. **1-Bromopropane from 1-chloropropane**

$$CH_3CH_2CH_2Cl \xrightarrow{\text{alco. KOH}} CH_3CH=CH_2 \xrightarrow{HBr/peroxide} CH_3CH_2CH_2Br$$

3. **Isopropyl bromide from n-propyl bromide**

$$CH_3{-}CH_2{-}CH_2{-}Br \xrightarrow{\text{alco. KOH}} CH_3{-}CH=CH_2 \xrightarrow{HBr} CH_3CHBr{-}CH_3$$

Alcohols

1. **Propane to isopropyl alcohol**

$$\underset{\text{propane}}{CH_3{-}CH_2{-}CH_3} \xrightarrow{Cl_2/\Delta} \underset{\text{2-chloropropane}}{CH_3{-}CH(Cl){-}CH_3} \xrightarrow{\text{aq. KOH}} \underset{\text{isopropyl alcohol}}{CH_3{-}CH(OH){-}CH_3}$$

2. **Acetylene to n-butyl alcohol $CH_3CH_2CH_2CH_2OH$**

$$CH{\equiv}CH \xrightarrow{H_2 \text{ (1 mole)}} CH_2=CH_2 \xrightarrow{HBr} CH_3CH_2Br \xrightarrow{CH{\equiv}CNa} CH{\equiv}C{-}CH_2CH_3 \longrightarrow CH_2=CHCH_2CH_3 \xrightarrow{BH_3/H_2O_2}$$
$$CH_3CH_2CH_2CH_2OH$$

3. **1-butene to 1-butane**

$$CH_3CH_2CH=CH_2 \xrightarrow{BH_3/H_2O_2} CH_3CH_2CH_2CH_2OH$$

4. **Acetophenone to 2-hydroxy-2-phenyl hexane**

$$C_6H_5COCH_3 + CH_3CH_2CH_2MgBr \longrightarrow C_6H_5\overset{\overset{\displaystyle OMgBr}{|}}{\underset{\underset{\displaystyle CH_3}{|}}{C}}-CH_2CH_2CH_3 \longrightarrow C_6H_5\overset{\overset{\displaystyle OH}{|}}{\underset{\underset{\displaystyle CH_3}{|}}{C}}-CH_2CH_2CH_3$$

5. $CH_3CH_2OH \longrightarrow$ **2-hydroxy-2-methyl-propane**

$$CH_3-CH_2OH \xrightarrow{oxidation} CH_3COOH \xrightarrow[\Delta]{distill\ as\ Ca\ salt} CH_3COCH_3 \xrightarrow{CH_2MgBr}$$

$$CH_3-\overset{\overset{\displaystyle OH}{|}}{\underset{\underset{\displaystyle CH_3}{|}}{C}}CH_3$$

6. **Cumene to phenol**

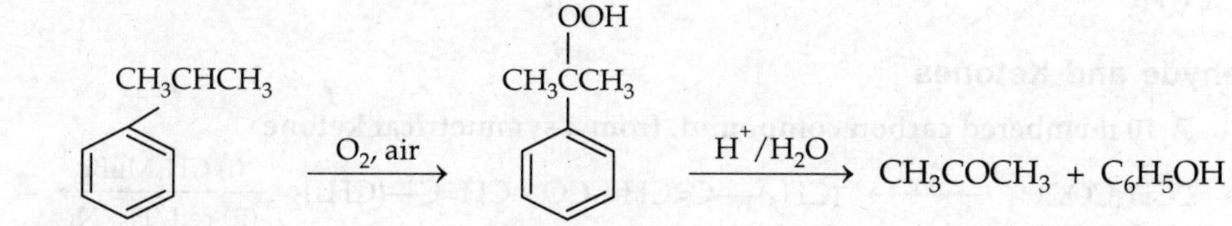

isopropyl benzene   peroxide

7. **But-1-ene and but-2-ene to butane-2-ol**

$$CH_2=CHCH_2CH_3 \xrightarrow{H_2SO_4} CH_3-\underset{\underset{\displaystyle HSO_4}{|}}{C}HCH_2CH_3 \xrightarrow{H_2O} CH_3-\underset{\underset{\displaystyle OH}{|}}{C}HCH_2CH_3$$

But-1-ene

$$CH_3CH=HCCH_3 \xrightarrow{H_2SO_4} CH_3CH_2-\underset{\underset{\displaystyle HSO_4}{|}}{H}CCH_3 \xrightarrow{H_2O} CH_3-\underset{\underset{\displaystyle OH}{|}}{C}HCH_2CH_3$$

But-2-ene

8. **1, 1-diphenyl prop-1-ene, $(C_6H_5)_2$—C=CH—CH$_3$ to 3, 3-di phenyl propane-2-ol**

$$(C_6H_5)_2C=CHCH_3 \xrightarrow{BH_3/H_2O_2} (C_6H_5)_2CHCH\underset{\underset{\displaystyle CH_3}{|}}{\overset{\overset{\displaystyle OH}{|}}{C}}$$

9. **A 1° alcohol to a 3° alcohol**

1° alcohol $\xrightarrow{oxi}$ an acid $\xrightarrow{distillation\ as\ cal.\ salt}$ ketone $\xrightarrow{RMgX}$ a 3° alcohol.

10. **A 2° alcohol to a 3° alcohol**

$$RCH(OH)R \xrightarrow{oxi} RCOR \xrightarrow{RMgX} \underset{\underset{R}{|}}{\overset{\overset{R}{|}}{R-C-OH}}$$

11. **A 1° alcohol to a 2° alcohol**

$$CH_3CH_2CH_2OH \xrightarrow{conc.\ H_2SO_4} CH_3CH=CH_2 \xrightarrow{HBr} \underset{\underset{Br}{|}}{CH_3-CHCH_3} \xrightarrow{aq.\ KOH} \underset{\underset{OH}{|}}{CH_3CHCH_3}$$

12. **A 3° alcohol to a 2° alcohol**

Tertiary butyl alcohol to iso butyl alcohol

$$CH_3-\underset{\underset{CH_3}{|}}{\overset{\overset{CH_3}{|}}{C}}-OH \xrightarrow{conc.\ H_2SO_4} \underset{CH_3}{\overset{CH_3}{\diagdown}}C=CH_2 \xrightarrow[peroxide]{HBr} \underset{CH_3}{\overset{CH_3}{\diagdown}}CH-CH_2Br \xrightarrow{aq.\ KOH}$$

$$\underset{CH_3}{\overset{CH_3}{\diagdown}}CHCH_2OH$$

Aldehyde and Ketones

1. **A 10 membered carbon compound, from a symmetrical ketone**

$$2\underset{acetone}{CH_3COCH_3} \xrightarrow{HCl} \underset{phorone}{(CH_3)_2-C=CH-CO-CH=C-(CH_3)_2} \xrightarrow[(ii)\ red/H_2/Ni]{(i)\ CH_3MgBr}$$

$$(CH_3)_2-CH-CH_2-\underset{\underset{CH_3}{|}}{\overset{\overset{OH}{|}}{C}}-CH_2-CH-(CH_3)_2$$

2,4,6-trimethyl-4-hydroxyheptane

2. **A 9-membered carbon compound from acetone**

$$2CH_3COCH_3 + 6\ HCHO \longrightarrow (CH_2OH)_3-C-CO-C\ (CH_2OH)_3 \xrightarrow{(1)\ HI\ (2)\ Zn/Hg + HCl}$$

1,5-dihdroxy-2,2,4,4-tetra hydroxyl methyl-
3-ketopentane

$$(CH_3)_3-C-CH_2-C(CH_3)_3$$

2,2,4,4-tetramethylpentane

3. **From benzal acetone, $C_6H_5-CH=CHCOCH_3$ to benzal pyruvic aldehyde**

$$C_6H_5-CH=CH-COCH_3 \xrightarrow{SeO_2} C_6H_5CH=CH-CO-CHO$$

4. **Benzal acetone to 4-phenyl-2-butanone**

$$C_6H_5-CH=CH-COCH_3 \xrightarrow{Li/NH_3,\ ether} C_6H_5CH_2-CH_2-CO-CH_3\ (Birch\ reaction)$$

5. **Di phenyl ketoxime to benzoyl aniline**

$$(C_6H_5)_2C=NOH \xrightarrow{C_6H_5COCl} (C_6H_5)_2NHCOC_6H_5$$

6. **Ethyl cyanide to 3-keto-*n*-hexane**

$$CH_3CH_2CN + CH_3—CH_2—CH_2—Li \longrightarrow CH_3CH_2COCH_2CH_2CH_3$$

7. **3–Chlorocyclohexyl-1-methanol—to 3-chloro-1-cyclohexyl-aldehyde**

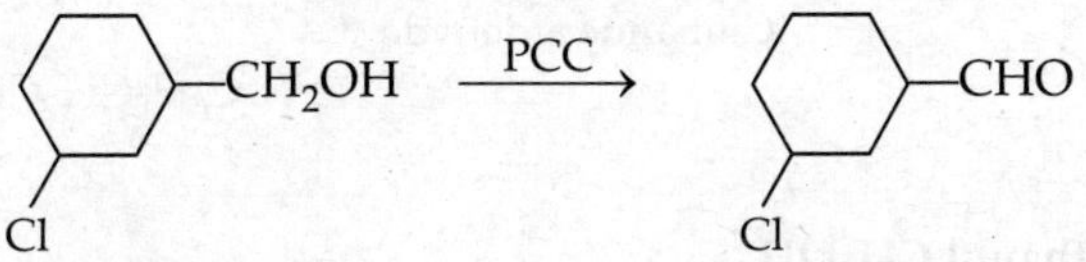

8. **4–Bromo-1-cyclohexanone to 4-ethenyl-1-cyclohexanone**

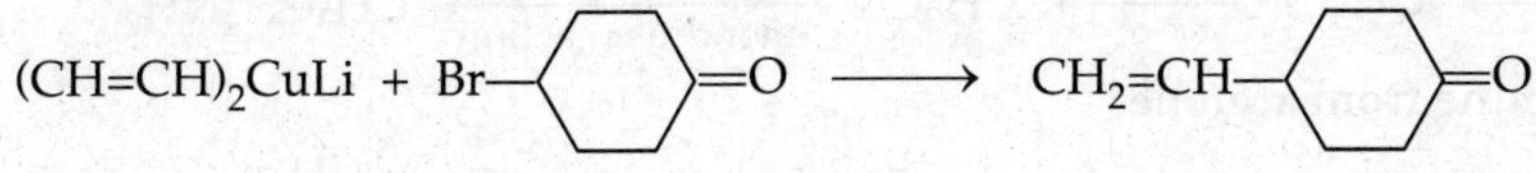

4-vinyl cyclohexanone

Carboxylic Acids

1. **Succinic anhydride to tetralone**

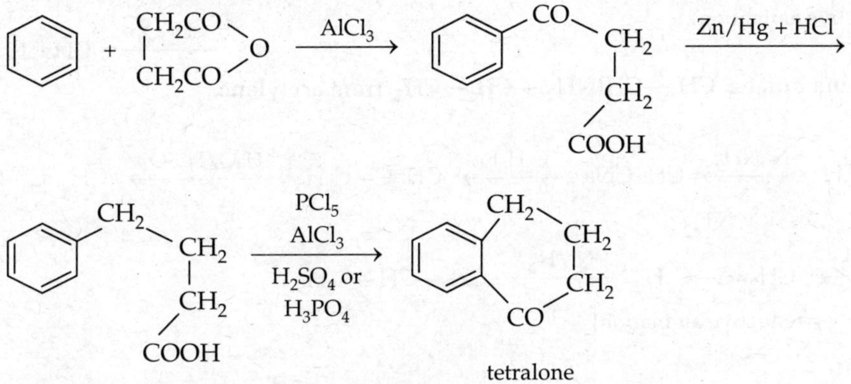

tetralone

2. **2-Methyl-1-butene to di methyl ethyl acetic acid**

$$CH_3CH_2\overset{\overset{\displaystyle CH_3}{|}}{C}=CH_2 \xrightarrow[H_2PO_4/300°400°C]{CO/H_2O} CH_3CH_2—\underset{\underset{\displaystyle CH_3}{|}}{\overset{\overset{\displaystyle CH_3}{|}}{C}}—COOH$$

3. **Acetic acid to malonic acid**

$$CH_3COOH \xrightarrow{Cl_2/UV} CH_2ClCOOH \xrightarrow{KCN} CH_2CNCOOH \xrightarrow{H^+/H_2O} CH_2\begin{matrix} ⁄ COOH \\ \backslash COOH \end{matrix}$$

4. **An amide to an acid**

$$RCONH_2 \xrightarrow{\text{by boiling with dil. alkali}} —RCOOH + NH_3$$

5. **A 1° alcohol to a 3° alcohol**

$$RCH_2OH \xrightarrow{oxi} RCOOH \xrightarrow{\text{disti. as cal salt}} RCOR \xrightarrow{RMgX} (R)_3C—OH$$

6. **Acetaldehde, CH_3CHO, to lactic acid, $CH_3CH(OH)COOH$**

$$CH_3CHO + HCN \longrightarrow CH_3CH(OH)CN \xrightarrow{H^+/H_2O} CH_3CH(OH)\ COOH$$

7. **Benzaldehyde to 3-phenyl-propionaldehyde**

$$C_6H_5{-}CHO + (CH_3CO)_2O \xrightarrow[\text{CH}_3\text{COONa}]{\text{Perkin's reaction}} C_6H_5{-}CH{=}CH{-}CHO \xrightarrow{\dot{B}_2H_6/CH_3COOH}$$

$$\underset{\text{Cinnamic aldehyde}}{} $$

$$C_6H_5CH_2CH_2CHO$$

Nitrogen Compounds

1. **Ethyl amine, $C_2H_5NH_2$, from methanol CH_3OH**

$$CH_3OH \xrightarrow{PCl_5} CH_3Cl \xrightarrow{KCN} CH_3CN \xrightarrow[\text{mendius reaction}]{\text{red. Na/alcohol}} CH_3CH_2NH_2$$

2. **Iso propyl amine from acetone**

$$CH_3COH_3 + NH_2OH \longrightarrow (CH_3)_2C{=}NOH \xrightarrow{LiAlH_4} CH_3{-}\overset{\overset{\displaystyle CH_3}{|}}{C}HNH_2$$

3. **Diethyl amine from methyl bromide**

$$CH_3Br \xrightarrow{KCN} CH_3CN \xrightarrow{red} CH_3CH_2NH_2 \xrightarrow{CH_3CHO} CH_3CH_2\,N{=}CHCH_3$$
(reductive amination)

$$\xrightarrow{H_2/Ni} CH_3CH_2NC_2H_5$$

4. **2-Amino butane, $CH_3{-}CH(NH_2){-}CH_2{-}CH_3$, from acetylene**

$$CH{\equiv}CH \xrightarrow{NaNH_2} CH{\equiv}CNa \xrightarrow{C_2H_5Br} CH{\equiv}C{-}C_2H_5 \xrightarrow{H_2O/H_2SO_4} CH_2{=}\overset{\overset{\displaystyle OH}{|}}{C}{-}C_2H_5$$

$$\xrightarrow{NH_3} CH_2{=}\overset{\overset{\displaystyle NH_2}{|}}{C}{-}C_2H_5 \xrightarrow{Ni/H_2} CH_3{-}\overset{\overset{\displaystyle NH_2}{|}}{C}H{-}C_2H_5$$

(reductive amination)

Conversions

An alkane to a higher alkane. To **increase the chain** by only one carbon
Method

i. A reaction with diazomethane CH_2N_2

$$CH_3{-}CH_2{-}CH_2{-}CH_2 \longrightarrow CH_3{-}CH_2{-}CH_2{-}CH_2{-}CH_3 + CH_3{-}CH_2{-}\overset{\overset{\displaystyle CH_3}{|}}{C}H{-}CH_3$$

The addition may take place either lengthwise or it may take place by forming iso structure. The reaction may increase length of the chain by one carbon, to form the isomers of the alkane.

All other methods are based on reaction of the corresponding alkyl halides.

ii. The alkane is first converted into the corresponding alkyl halide, with which KCN will form alkyl cyanide. On reduction with Na + ROH, the cyanide is converted into a primary amine. This with HNO_2 will be converted to form an alcohol, which may be reduced, by phosphorus and red HI, to an alkane containing one carbon more.

Alternatively, the alkyl halide may be treated with sodium metal in the presence of dry ether, when two molecules of the alkyl halides, react together, to form an alkane, containing the double of carbon atoms in the molecule. Only 1° and 2° alkyl halides, undergo

Wurtz reaction. This method is applicable only for the preparation of linear alkanes, containing even number of carbon atoms.

$$RX + RX \xrightarrow{2Na} R—R + 2NaX$$

iii. For the preparation of a branched chain alkane, Corey–House synthesis is used.

$$CH_3\text{-}CH_2\text{-}CH_3 \longrightarrow \underset{\underset{Cl}{|}}{CH_3CHCH_3} \xrightarrow{\Delta} CH_2=CH\text{-}CH_3 \xrightarrow{peroxide} CH_3\text{-}CH_2\text{-}CH_2Br$$

$$\xrightarrow{2Li} CH_3CH_2CH_2Li \xrightarrow{CuI} (CH_3CH_2CH_2)_2CuLi \xrightarrow{RX} CH_3CH_2CH_2R$$

iv. A reaction with Grignard's reagent can also increase the number of carbon atoms in the molecule.

$$an\ alkane \longrightarrow \underset{R}{\overset{R}{\diagdown}}CO \xrightarrow{RMgX} \underset{R}{\overset{R}{\diagdown}}\underset{R'}{\overset{OMgX}{\diagup}}C \longrightarrow \underset{R}{\overset{R}{\diagdown}}\underset{R'}{\overset{OH}{\diagup}}C$$

Decreasing one carbon in the molecule on conversion of a higher alkane to a lower alkane

To convert a higher member to a lower one, i.e. to decrease the carbon chain by one carbon, there are two methods

i. First method is to convert a compound into the corresponding carboxylic acid, using the conventional methods.
The carboxylic acid in the form of a sodium salt is then heated with soda lime, to decarboxylate to get the corresponding compound with one carbon less.

$$R—COONa + NaOH \longrightarrow RH + Na_2CO_3$$

ii. The other method is based on the acid formation.
The acid formed is then treated with ammonia, followed by heating the ammonium salt it is converted in to an amide.

$$RCOOH + NH_3 \longrightarrow RCOONH_4 \longrightarrow RCONH_2$$

The amide is then treated with bromine/KOH or the Hoffmann's bromamide reaction, when the amide is converted to an amine

$$RONH_2 + Br_2 + 4KOH \longrightarrow RNNH_2 + 2KBr + K_2CO_3$$

The amine, so obtained, may be converted into the desired compound.

ALKENES. A higher alkene may be converted to a **lower alkene.**
In alkenes since the double bond is to remain intact, any alkyl group attached to the doubly bonded carbon atoms, may be chlorinated by Cl_2 at an elevated temperature. The chlorinated compound may be converted to a carboxylic acid followed by oxidation.

$$CH_3—CH=CH_2 \xrightarrow{Cl_2/UV} ClCCH_2CH=CH_2 \xrightarrow{aq.\ KOH} CH_2=CH—CH_2OH \xrightarrow{oxi}$$

$$CH_2=CH—COOH \xrightarrow{soda\ lime} CH_2=CH_2$$

The double bond can be protected by converting it to a dibromide. This dibromide on treatment with alcoholic KOH or with $NaNH_2$, gives back the double bond.

Conversion of a lower alkene to a higher one
An alkene, other than ethylene, may be converted into a chlorinated compound, with another alkyl halide, in the presence of sodium metal and dry ether (Wurtz reaction or by Corey–House synthesis) may be converted to give a higher alkene.

$$CH_3CH=CH_2 \longrightarrow CH_2=CH—CH_2Cl \longrightarrow +RX \xrightarrow{Wurtz\ reaction} CH_2=CH—CH_2R$$

By Corey–House synthesis: $(CH_2=CH—CH_2)_2CuLi \longrightarrow CH_2=CH—CH_2R'$

The double bonds in alkene may be protected by treating it with bromine solution (bromine dissolved in $CHCl_3$). These addition compounds formed may be debrominated by treating with Zn dust and methanol, to bring back the double bond.

$$CH_2=CH—CH_3 \xrightarrow{Br_2/CHCl_3} CH_2Br—CHBr—CH_3 \xrightarrow{Zn + CH_3OH} CH_2=CH—CH_3$$

ALKYNES. A lower alkyne to a higher alkyne

The terminal alkyne are acidic in nature. Therefore they may be treated with sodium metal to form a mono-sodio (or a di-sodio) acetylide derivatives. As these derivatives are ionic in nature, they on treatment with any alkyl halide, will very easily be converted into higher alkyl derivatives.

$$RC≡CH \xrightarrow{Na} RC≡CNa \xrightarrow{R'X} RC≡CR$$

The triple bond in alkynes may be protected by treating it with Br_2 to convert it into a dibromo derivative, which may be restored by treating the derivative with the alcoholic KOH debromination

$$RC≡CH \xrightarrow{HBr} R—CBr=CH_2 \xrightarrow{HBr} R—CBr_2—CH_3 \xrightarrow{NaNH_2} R—C≡CH$$

A higher alkyne to a lower alkyne

A higher alkyne may be converted to a lower alkyne, by protecting the triple bond, as already suggested, following the same route as in the case of alkenes.

Cyclo Alkanes

To increase the size of the ring:

i. The cyclo alkanes, when treated with diazomethane, CH_2N_2, ring may in corporate di azo molecule to increase the size of the ring.

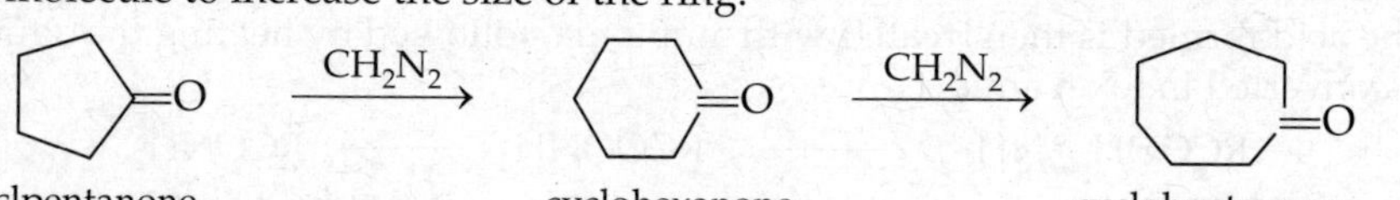

cyclpentanone   cyclohexanone   cycloheptanone

ii. Sometimes some ring may be expanded by subjecting these rings to Beckmann's rearrangement.

iii. Cyclo hexanone may be converted into caprolactum, a compound containing an even membered rings, by any of the following methods.

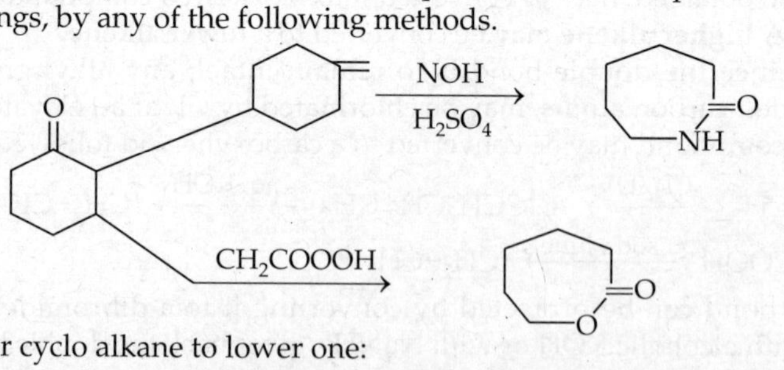

From a higher cyclo alkane to lower one:

Cyclo alkanes may be converted into a lower one, by using the same manner as in the case of alkanes, because the cyclo alkanes behave in the same as alkanes. The higher member is treated with in the presence of Cl_2/UV, when it form cyclo alkane chloro derivative. On treatment of the cyclo alkyl chloride with aqueous NaOH it is converted to form an alcohol. The alcohol on further oxidation gives an open chain dibasic acid.

On distilling a calcium salt of the dibasic acid, there is formed a cyclic ketones, containing one carbon atom less. This on reduction may give the required cyclo alkanes.

Taking the case of cyclohexanes

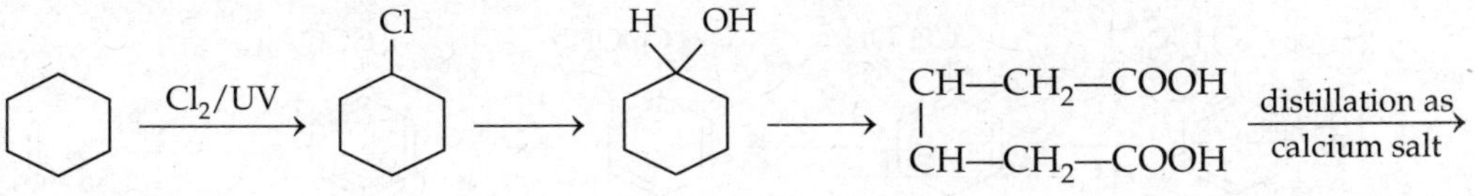

adipic acid

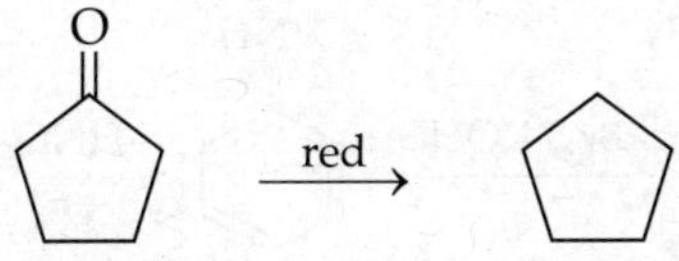

cyclopentnone   cyclopentane

Aromatic hydrocarbons

The aromatic compound fall in two categories

a. Nuclear substituted derivatives: Excepting alkyl substituted derivatives of benzene, all other nuclear substituted derivatives cannot be converted to the lower one.

b. All other type derivatives are side chain subsituted derivative, having functional groups attached to the side chain. All these compound may be regarded as phenyl subsituted aliphatic compounds. All these compounds may be converted into a lower ones.

For example:

Toluene may be converted into benzene

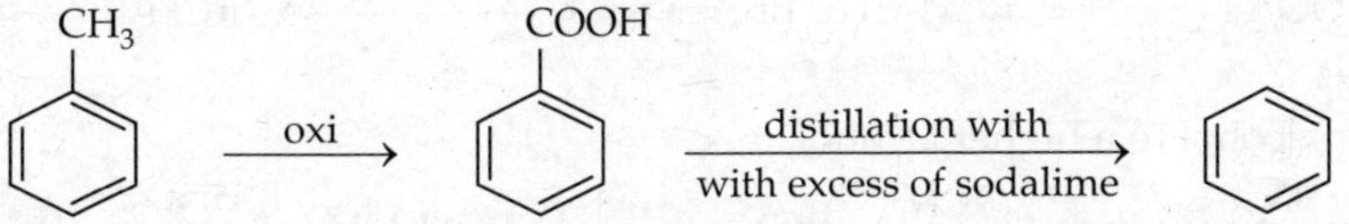

In the case of aromatic compounds the higher member would be the compound, a benzene ring containing an alkyl group attached to ring

Conversions

1. **Alkyl halides**
   A lower alkyl halide to higher one

   a. $RX \xrightarrow{KCN} RCN \xrightarrow{red} RCH_2NH_2 \xrightarrow{HNO_2} RCH_2OH \xrightarrow{PX_3} RCH_2X$

   b. $RX \xrightarrow{KCN} RCN \xrightarrow{H_2O/H^+} RCOOH \xrightarrow{PCl_5} RCOCl \xrightarrow{Ag_2O} R{-}COCHN_2$

   $\xrightarrow{CH_2N_2} RCH=C=O \longrightarrow RCH_2COOH \longrightarrow RCH_2COOAg \xrightarrow{Cl_2 \text{ or } Br_2} RCH_2X$

   A higher alkyl halide to a lower alkyl halide

   $RCH_2X \xrightarrow{aq. KOH} RCH_2OH \xrightarrow{oxi} RCOOH \longrightarrow RCOOAg \xrightarrow{Cl_2 \text{ or } Br_2} RCl$

   or $RBr \xrightarrow{\text{Finklestein reaction}} RI$

   A lower aryl halide to a higher one

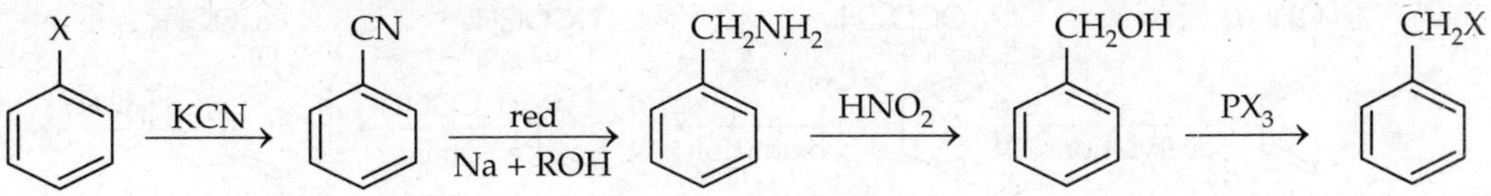

A higher aryl halide to a lower one

a.

$$C_6H_5CH_2X \longrightarrow C_6H_5CH_2OH \longrightarrow C_6H_5COOH \longrightarrow C_6H_5COOAg \xrightarrow{Cl_2 \text{ or } Br_2} C_6H_5(Cl \text{ or } Br)$$

(rings labelled: CH_2X, CH_2OH, $COOH$, $COOAg$, Cl or Br)

b.

$$C_6H_5CH_2X \xrightarrow{oxi} C_6H_5COOH \xrightarrow[heat]{NH_3} C_6H_5CONH_2 \xrightarrow{Br_3/KOH} C_6H_5NH_2 \xrightarrow[OC]{HNO_2} C_6H_5X$$

(rings labelled: CH_2X, $COOH$, $CONH_2$, NH_2, X)

(1) diazotisation
(2) CuX/HX

Increasing one carbon only

$$RX \longrightarrow RCN \xrightarrow{red} RCH_2NH_2 \xrightarrow{HNO_2} RCH_2OH \xrightarrow{PX_3} RCH_2X$$

By reaction with diazomethane, CH_2N_2

$$RX \xrightarrow{AgO} RCN \longrightarrow RCOCl \xrightarrow{CH_2N_2} RCOCHN_2 \longrightarrow R\,CH=C=O \longrightarrow$$

$$RCH_2COOH \xrightarrow{Ag}$$

$$RCH_2COOAg \longrightarrow RCH_2Cl \text{ or } Br \xrightarrow{Finklestein's \ reaction} RCH_2I$$

2. **Alcohols**

   A lower alcohol to a higher alcohol

   $$ROH \xrightarrow{PCl_5} RCl \xrightarrow{KCN} RCN \xrightarrow{red} RCH_2NH_2 \xrightarrow{HNO_2} RCH_2OH$$

   A higher alcohol to a lower alcohol

   $$RCH_2OH \xrightarrow{oxi} RCOOH \xrightarrow{NH_3/\Delta} RCONH_2 \xrightarrow{Br_2/NaOH} RNH_2 \xrightarrow{HNO_2} ROH$$

3. **Ethers**

   Two alkyl groups are attached to a –O–group
   To convert a lower ether to a higher ether.
   The alkyl group is chlorinated, in presence of UV radiations and the chloro compound so obtained may be converted into a higher alkyl group by Wurtz reaction
   Aromatic ethers are of two type
   i. **Ar—O—R**
   ii. **Ar—O—Ar.** These are difficult to prepare and cannot be converted into a higher or lower homologue

3. **Phenols**

   Phenols may contain an alkyl group in the molecules, which are termed as cresols. The cresols may be converted to phenol, by side chain oxidation, after protecting the —OH group.

   $$C_6H_5OH \xrightarrow{acetylation} \xrightarrow{oxidation} \xrightarrow[heat]{} \xrightarrow{H_2O/H^+} C_6H_5OH$$

   (rings labelled: OH; OCOCH$_3$ with CH$_3$; OCOCH$_3$ with COOH; OCOCH$_3$; OH)

Side chain containing OH group may also be converted to phenols (as in the case of aryl halides)

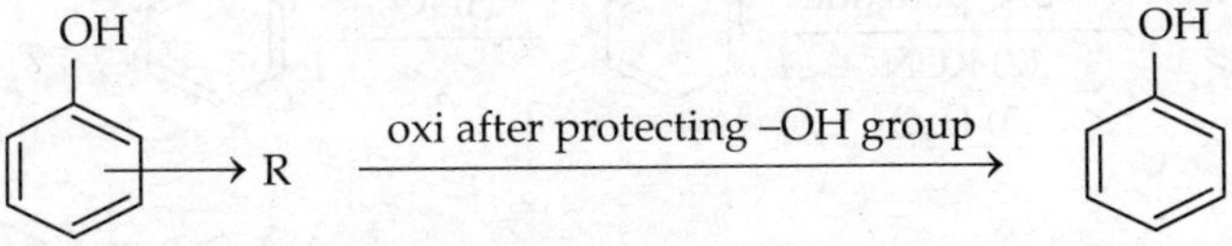

A higher phenol to a lower phenol
Note: In all aromatic compounds, the meaning of converting it to lower member is to increase the number of carbon atom in the side chain.

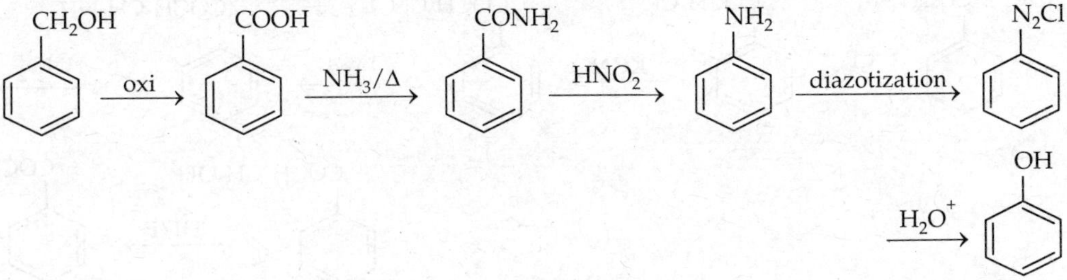

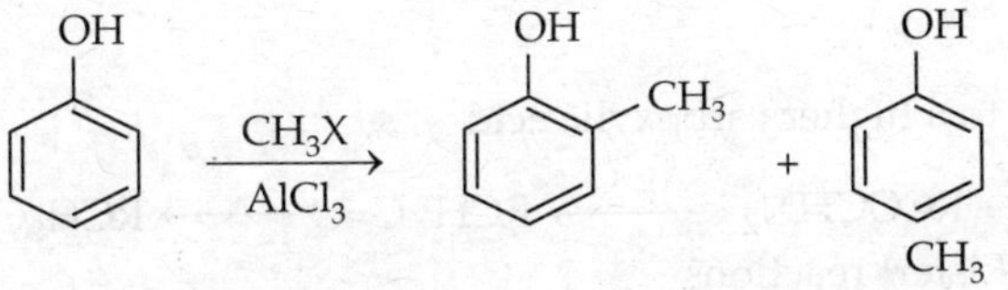

5. **Aldehyde and Ketones**
   A higher aldehyde to a lower aldehyde

   $$C_2H_5CHO \xrightarrow{oxi} C_2H_5COOH \xrightarrow{NH_3/\Delta} C_2H_5COCNH_2 \xrightarrow{Br_2/NaOH} C_2H_5NH_2$$

   $$\longrightarrow C_2H_5OH \xrightarrow{PCC} CH_3CHO$$

   A lower aldehyde to a higher one

   $$HCHO \xrightarrow{red} CH_3OH \longrightarrow CH_3X \xrightarrow{Wurtz\ reaction} C_2H_6 \xrightarrow{X_2} C_2H_5X \longrightarrow$$

   $$C_2H_5OH \longrightarrow CH_3COOH \longrightarrow CH_3CHO$$

   A romatic aldehyde
   As already stated, these also fall in two categories:
   Nuclear substituted aldehydes and side chain substituted aldehydes

Converting a Higher Aldehyde into a Lower Aldehyde

In this case, if the nuclear substituted compounds, by side chain oxidation
Converting a lower aromatic aldehyde, into a higher aldehyde.
By alkylating the benzene ring, a lower aldehyde may be converted into a higher one

Ketones

1. $R-CO-R + CH_2N_2 \longrightarrow RCOCH_2R$

   Aromatic ketones. These may be aryl alkyl ketones and di aryl ketone
   a. Only alkyl group in alkyl aryl ketone maybe converted to a higher or a lower alkyl group.

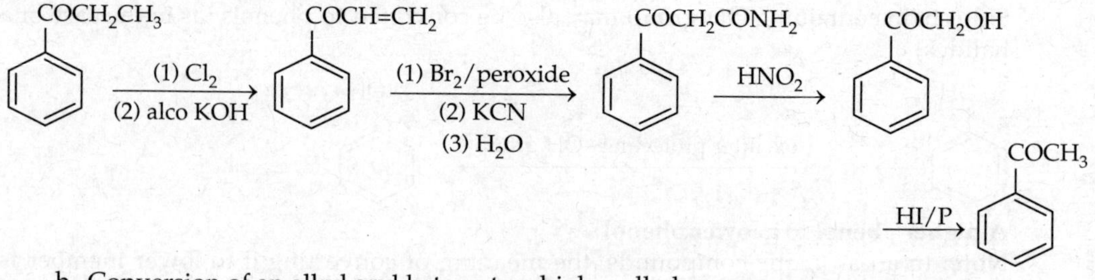

$$\underset{\text{(benzene)}}{C_6H_5}COCH_2CH_3 \xrightarrow[\text{(2) alco KOH}]{\text{(1) Cl}_2} C_6H_5COCH=CH_2 \xrightarrow[\substack{\text{(2) KCN} \\ \text{(3) H}_2\text{O}}]{\text{(1) Br}_2/\text{peroxide}} C_6H_5COCH_2CONH_2 \xrightarrow{\text{HNO}_2} C_6H_5COCH_2OH$$

$$\xrightarrow{\text{HI/P}} C_6H_5COCH_3$$

b. Conversion of an alkyl aryl ketone to a higher alkyl group

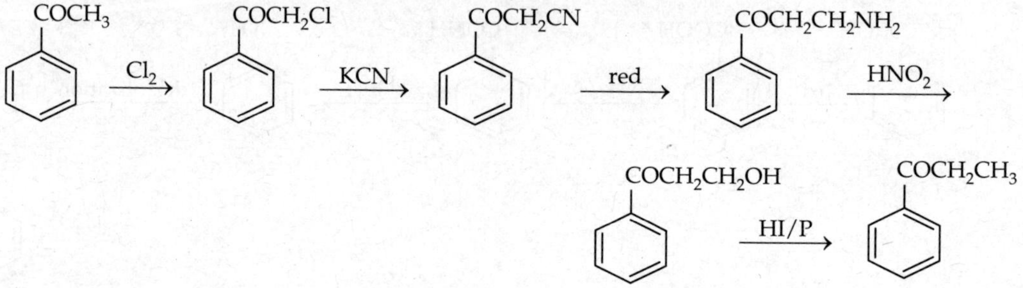

$$C_6H_5COCH_3 \xrightarrow{\text{Cl}_2} C_6H_5COCH_2Cl \xrightarrow{\text{KCN}} C_6H_5COCH_2CN \xrightarrow{\text{red}} C_6H_5COCH_2CH_2NH_2 \xrightarrow{\text{HNO}_2}$$

$$C_6H_5COCH_2CH_2OH \xrightarrow{\text{HI/P}} C_6H_5COCH_2CH_3$$

6. **Carboxylic acid**
   a. **Aliphatic acids**
   From a lower mono basic acid to a higher carboxylic acid

$$RCOOH \xrightarrow{PX_3} ROCl \xrightarrow{CH_2N_2} RCOCHN_2 \xrightarrow{Ag_2O} RCH=C=O \longrightarrow RCH_2COOH$$

   This reaction is termed Arndt-Eistert reactions
   From a higher acid to a lower one.

$$RCH_2COOH \longrightarrow RCH_2CONH_2 \longrightarrow RCH_2NH_2 \longrightarrow RCH_2OH \longrightarrow RCOOH$$

   All aromatic carboxylic acids compounds, fall in two categories
   b. **Nuclear subsituted acids. When in addition to the —COOH, the ring contains an alkyl group, only such compounds may be converted to lower ones.**

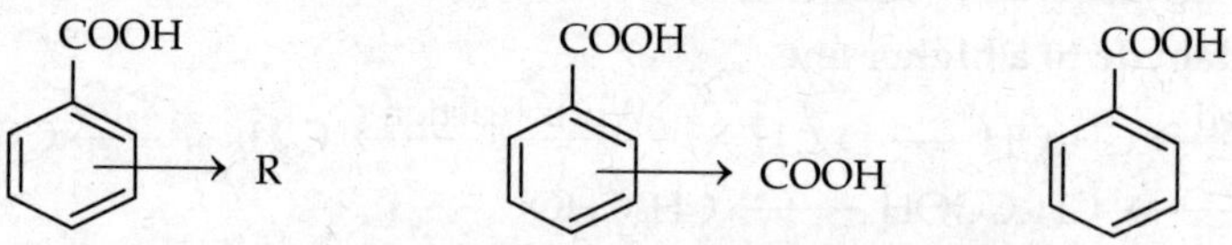

$$C_6H_4(COOH)(R) \longrightarrow R \qquad C_6H_4(COOH) \longrightarrow COOH \qquad C_6H_5COOH$$

The side chain is oxidised to —COOH, attached to the ring. On further decarboxylation of the dibasic acids, it is easily converted a lower acid.
A lower acid may be converted into a higher one.

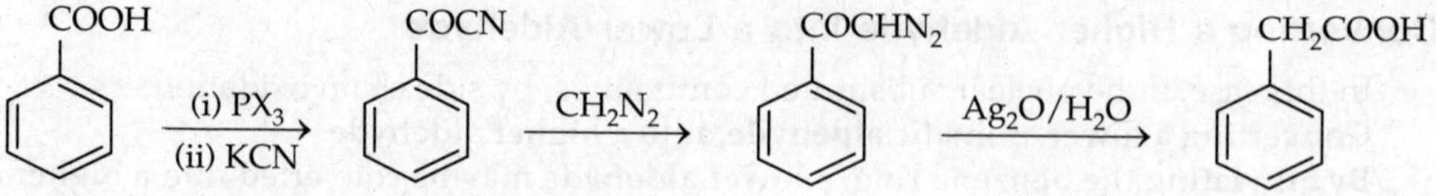

$$C_6H_5COOH \xrightarrow[\text{(ii) KCN}]{\text{(i) PX}_3} C_6H_5COCN \xrightarrow{CH_2N_2} C_6H_5COCHN_2 \xrightarrow{Ag_2O/H_2O} C_6H_5CH_2COOH$$

Another method is to alkylate the ring.

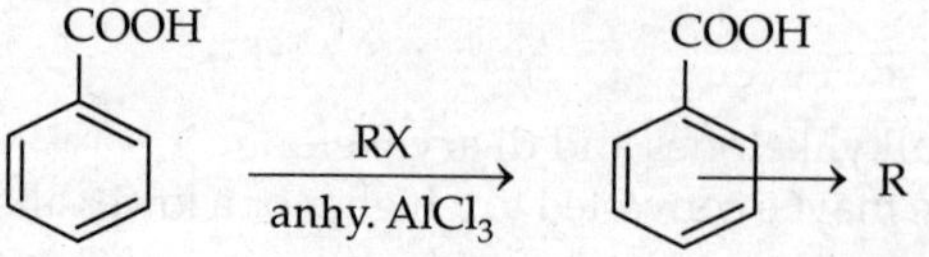

$$C_6H_5COOH \xrightarrow[\text{anhy. AlCl}_3]{RX} C_6H_4(COOH)(R)$$

7. **Amines: Aliphatic amines**

A lower amine to a higher amine:

$$RNH_2 \xrightarrow{HNO_2} ROH \xrightarrow{PX_3} RX \xrightarrow{KCN} RCN \xrightarrow{red} RCH_2NH_2$$

A higher amine to a lower amine

$$RCH_2NH_2 \xrightarrow{HNO_2} RCH_2OH \xrightarrow{oxi} RCOOH \xrightarrow{NH_3/\Delta} RCONH_2 \xrightarrow{Br_2/KOH} RNH_2$$

Amines—Aromatic amines

A lower amine to higher amine

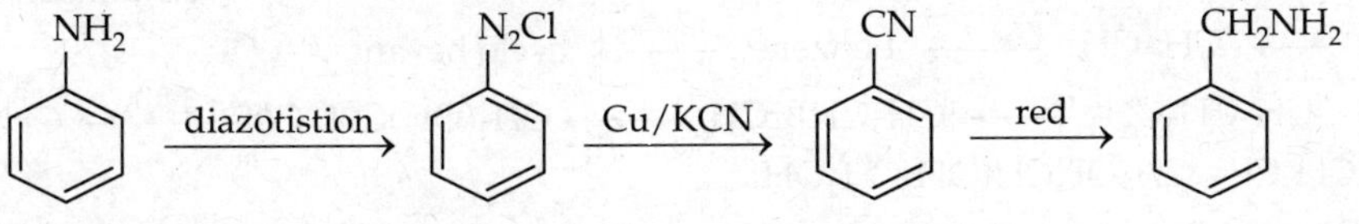

A higher amine to a lower amine

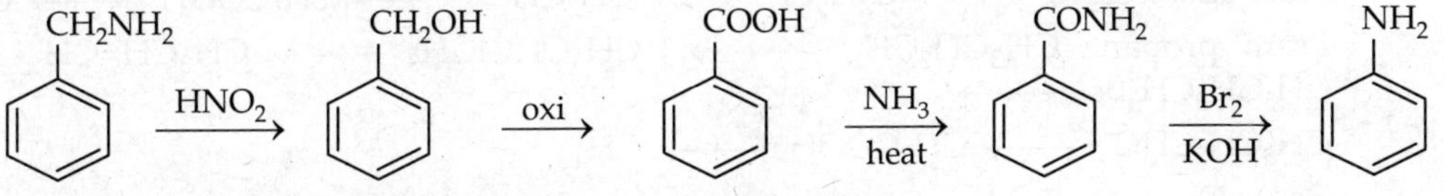

8. **Amides: Aliphatic and aromatic amides behave similarly**

A higher amide to a lower amide

$$RCH_2CONH_2 \xrightarrow{Br_2/KOH} RCH_2NH_2 \xrightarrow{HNO_2} RCH_2OH \xrightarrow{oxi} RCOOH \xrightarrow{NH_3/heat} RCONH_2$$

A lower amide to a higher amide

$$RCONH_2 \xrightarrow{Red/Na + ROH} RCH_2NH_2 \xrightarrow{HNO_2} >RCH_2OH \xrightarrow{oxi} RCH_2Cl \xrightarrow{KCN}$$
$$RCH_2CN \longrightarrow RCH_2CONH_2$$

$$RCH_2COOH \xrightarrow{NH_3/\Delta} RCH_2CONH_2$$

An important way is to treat every organic compound as a starting compound, is to prepare or synthesize all other compound (that are in the syllabus).

Taking methane CH_4, as the starting compound, here are the synthesis or preparation of all other compounds.

Route A

$$CH_4 \xrightarrow{Cl_2/UV} CH_3Cl \xrightarrow{KCN} \underset{\text{methyl cyanide}}{CH_3CN} \xrightarrow{red} \underset{\text{ethylamine}}{CH_3CH_2NH_2} \xrightarrow{HNO_2} CH_3CH_2OH$$

$$\xrightarrow{oxi} \underset{\text{acetaldehyde}}{CH_3CHO} \xrightarrow{oxi} \underset{\text{acetic acid}}{CH_3COOH}$$

$$CH_4 \xrightarrow{Cl_2/UV} CH_3Cl \xrightarrow{Na/ether} C_2H_6 \xrightarrow{Cl_2/UV} C_2H_5Cl \xrightarrow{alco. KOH} \underset{\text{ethylene}}{CH_2{=}CH_2}$$

$$\xrightarrow{Br_2} \underset{\text{dibromo ethane}}{CH_2BrCH_2Bt} \xrightarrow{alco. KOH} \underset{\text{acetylene}}{CH{\equiv}CH}$$

$$CH_4 \longrightarrow CH_3CH_2Cl \xrightarrow{Na/ether} CH_3{-}CH_2{-}CH_2{-}CH_3 \text{ } n\text{-butane}$$

$$CH_3Br + 2Na + C_2H_5Br \longrightarrow CH_3CH_2CH_3 \xrightarrow{Br_2} CH_3CHBrCH_3 \xrightarrow[Cu_2I_2]{(1) Li}$$
$$+$$
$$C_2H_6 \ C_4H_{10}$$

$$\{(CH_3)_2CH\}_2CuLi$$
by Corey-House synthesiss

$$\underset{\underset{CH_3}{|}}{CH_3CH}-\underset{\underset{CH_3}{|}}{CHCH_3}$$

2,3-dimethyl butane

$$CH_4 \longrightarrow CH{\equiv}CH \longrightarrow benzene \longrightarrow cyclo\ hexane$$

$$CH_4\ to\ {-}CH_3CH_2CH_3 \longrightarrow CH_3CHBrCH_3 \longrightarrow CH_2(OH)CH{=}CH_2 \xrightarrow{HOCl} CH_3OH$$
$$CHOHCH_2Cl{-}\ CH_2OHCH(OH)CH_2OH$$

Route B

From ethane $C_2H_6 \longrightarrow C_2H_5Cl \longrightarrow CH_3CHO \longrightarrow CH_3COOH \longrightarrow CH_4$

From propane $CH_3CH_2CH_3 \longrightarrow CH_3CHClCH_3 \longrightarrow CH_3CH{=}CH_2 \longrightarrow$
$CH_3CH_2CH_2Br \longrightarrow$

$$CH_3CH_2CHO \longrightarrow CH_3COOH \longrightarrow CH_4$$

Route C

$$CCH_3Cl \longrightarrow CH_3CN \longrightarrow CH_3CH_2NH_2 \longrightarrow CH_3CH_2OH \longrightarrow CH_3CHO \longrightarrow$$
$$CH_3COOH \longrightarrow$$

$$CH_3COOH \longrightarrow CH_3CONH_2 \longrightarrow CH_3NH_2 \longrightarrow CH_3OH \quad amines$$

$$CH_3COONH_4 \longrightarrow CH_3CONH_2 \longrightarrow CH_3NH_2 \longrightarrow CH_3OH \quad \underset{1}{RNH_2} \ \underset{2}{R_3NH_2} \ \underset{3}{R_3NH_2}$$

Route D

$$CH_3COCH_3 \longrightarrow CH_3$$

$$\underset{CH_3}{\overset{CH_3}{\diagdown}}CO + RMgBr \longrightarrow \underset{CH_3}{\overset{CH_3}{\diagdown}}\underset{R'}{\overset{OMgBr}{C\diagup}} \longrightarrow \underset{CH_3}{\overset{CH_3}{\diagdown}}\underset{R'}{\overset{OH}{C\diagup}}$$

$$CH_4 \longrightarrow C_2H_6 \longrightarrow CH_3CHO \longrightarrow CH_3CH(OH)CHCHO \xrightarrow{\Delta} CH_3CH{=}CH{-}CHO$$

Buy using any of the A, B, C and D routes, the following compounds may be prepared, by treating each one as a starting material,
$C_2H_5ClCH_3CH_3OH, C_2H_5OH,\ HCHO,$
$CH_3CHO, HCOOH, CH_3COOH, CH_3COCH_3, CH_3NH_2, C_2H_5NH_2$ isopropyl alcohol

Conversion in Carbohydrate

Glucose to Fructose

1. Glucose isomerises in presence of warm dil. aqueous NaOH solution and it forms, two compounds
   a. mannose and .
   b. fructose.
   This reaction is termed as **LOBRY de Bryun von Ekenstein reaction.**

$$
\begin{array}{ccc}
\text{CHO} & \text{CHO} & \text{CH}_2\text{OH} \\
\text{H—C—OH} & \text{HC—C—H} & \text{C=O} \\
\text{HO—C—H} & \text{HO—C—H} & \text{HO—C—H} \\
\text{H—C—OH} & \text{H—C—OH} & \text{H—C—OH} \\
\text{H—C—OH} & \text{H—C—OH} & \text{H—C—OH} \\
\text{CH}_2\text{OH} & \text{CH}_2\text{OH} & \text{CH}_2\text{OH} \\
\text{glucose} & \text{mannose} & \text{fructose}
\end{array}
$$

(**Mannose and glucose** are **epimers** as they differ in configuration at **carbon number 2 only**)

2. The other method, consists of forming an osazone in a reaction in between glucose and excess of phenyl hydrazine. On hydrolysis osazone decomposes to form a compound known as **ozone**. On reduction with Zn/CH_3COOH ozone forms **fructose**.

$$
\begin{array}{cccc}
\text{CHO} & \text{CH=NNHC}_6\text{H}_5 & \text{CHO} & \text{CH}_2\text{OH} \\
\text{H—C—OH} & \text{C=NNHC}_6\text{H}_5 & \text{C=O} & \text{C=O} \\
\text{HO—C—H} & \text{HO—C—H} & \text{HO—C—H} & \text{HO—C—H} \\
\text{H—C—OH} & \text{H—C—OH} & \text{H—C—OH} & \text{H—C—OH} \\
\text{H—C—OH} & \text{H—C—OH} & \text{H—C—OH} & \text{H—C—OH} \\
\text{CH}_2\text{OH} & \text{CH}_2\text{OH} & \text{CH}_2\text{OH} & \text{CH}_2\text{OH} \\
\text{glucose} & \text{glucosazone} & \text{ozone} & \text{fructose}
\end{array}
$$

(reactions: glucose $\longrightarrow$ glucosazone $\xrightarrow[H_2O/H^+]{H_2O}$ ozone $\longrightarrow$ fructose)

Fructose to Glucose

There are two methods.

1. On warming **fructose** with dil. aqueous NaOH solution fructose rearranges to form **glucose** and **mannose**.

$$
\begin{array}{ccc}
\text{CH}_2\text{OH} & \text{CHO} & \text{CHO} \\
\text{C=O} & \text{HO—C—H} & \text{H—C—OH} \\
\text{HO—C—H} & \text{HO—C—H} & \text{HO—C—H} \\
\text{H—C—OH} & \text{H—C—OH} & \text{H—C—OH} \\
\text{H—C—OH} & \text{H—C—OH} & \text{H—C—OH} \\
\text{CH}_2\text{OH} & \text{CH}_2\text{OH} & \text{CH}_2\text{OH} \\
\text{fructose} & \text{mannose} & \text{glucose}
\end{array}
$$

(reaction: fructose $\xrightarrow{\text{dil. NaOH}}$ mannose + glucose)

2. Fructose is reduced with H_2, in presence of Ni to form hexitol, which on oxidation with HNO_3, gives a mixture of gluconic and mannonic acid (hexanoic acid). This on heating forms γ-lactone which on reduction with Na + Hg amalgam forms glucose.

```
  CH₂OH                CH₂OH                CH₂OH                COOH
   |                    |                    |                    |
  C=O                HO—C—H               H—C—OH               H—C—OH
   |                    |                    |                    |
HO—C—H      red     HO—C—H               HO—C—H     HNO₃     HO—C—H
   |        ──→        |           +        |        ──→         |        +
 H—C—OH              H—C—OH               H—C—OH      oxi     H—C—OH
   |                    |                    |                    |
 H—C—OH              H—C—OH               H—C—OH               H—C—OH
   |                    |                    |                    |
  CH₂OH                CH₂OH                CH₂OH                COOH
 fructose                        hexitol                      hexanoic acid
```

```
         COOH                      CO                      CHO
          |                        |  \                     |
       HO—C—H                    H—C—OH \                  H—C—OH
          |                        |      \O                 |
       HO—C—H                    HO—C—H   |               HO—C—H
          |        heating         |      /      red         |
        H—C—OH      ──→           C—H   /      ──→          H—C—OH
          |                        |   /                      |
        H—C—OH                    H—C—OH                    H—C—OH
          |                        |                          |
         COOH                     CH₂OH                      COOH
                                  lactone                   glucose
```

Hexose to Pentose

1. **Wohl's method:** This method consists in treating hexose with hydroxyl amine, to form an oxime. Oxime is then treated with acetic anhydride, in presence of zinc chloride or sodium acetate, to dehydrate, to convert it into acetylated aldononitrile. It is then with ammoniacal $AgNO_3(Ag_2O)$ to hydrolyse the acetyl derivative and also to eliminate one molecule of HCN. The aldose, so obtained, contains one carbon less than in the original compound. Thus an aldo hexose, D-glucose, is converted into aldo pentose or arabinose.

```
  CH₂OH                CH=NOH               CN                   COOH
   |                    |                    |                    |
  C=O                 H—C—OH               H—C—OAC              H—C—OH
   |                    |                    |                    |
HO—C—H     N₂HOH     HO—C—H     Ac₂O     AcO—C—H     Ag₂O     HO—C—H
   |        ──→         |        ──→         |        ──→         |
 H—C—OH              H—C—OH      ZnCl₂    H—C—OAC              H—C—OH
   |                    |                    |                    |
 H—C—OH              H—C—OH               H—C—OAC              H—C—OH
   |                    |                    |                    |
  CH₂OH                CH₂OH                CH₂OAc               CH₂OH
 aldohexose            oxime           D-glucose acetonitrile   D-glucose
 D-glucose                               penta-acetate          acetonitrile
```

```
              CHO
               |
            HO—C—H
               |
  –HCN       H—C—OH
   ──→         |
             H—C—OH
               |
              CH₂OH
   aldopentose (D-arabinose)
```

   i. Hexose is treated with hydroxylamine to form oxime.

  ii. Oxime on treatment with acetic anhydride, in the presence of $ZnCl_2$ or sodium acetate, to dehydrate and convert it into acetylated aldono nitrile.

 iii. Acetylated aldono nitrile is then treated with ammo. $AgNO_3$ to hydrolyse and also to eliminate one molecule of HCN. An aldose is obtained, containing one carbon less, than present in glucose.

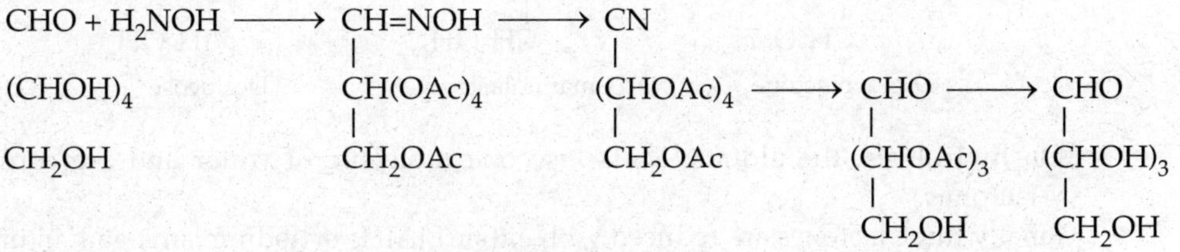

$$
\begin{array}{ccccccc}
\text{CHO} + \text{H}_2\text{NOH} & \longrightarrow & \text{CH=NOH} & \longrightarrow & \text{CN} & & \\
| & & | & & | & & \\
\text{(CHOH)}_4 & & \text{CH(OAc)}_4 & & \text{(CHOAc)}_4 & \longrightarrow \text{CHO} & \longrightarrow \text{CHO} \\
| & & | & & | & | & | \\
\text{CH}_2\text{OH} & & \text{CH}_2\text{OAc} & & \text{CH}_2\text{OAc} & \text{(CHOAc)}_3 & \text{(CHOH)}_3 \\
& & & & & | & | \\
& & & & & \text{CH}_2\text{OH} & \text{CH}_2\text{OH}
\end{array}
$$

Pentose to Hexose

1. **Killiani-Fischer cyanohydrin synthesis:** In this method, the aldo pentose is treated HCN, with the result that the aldo pentose forms two cyanohydrins or two aldo nitrilies. These aldo nitriles differ only at configuration at carbon number 2(epimers).

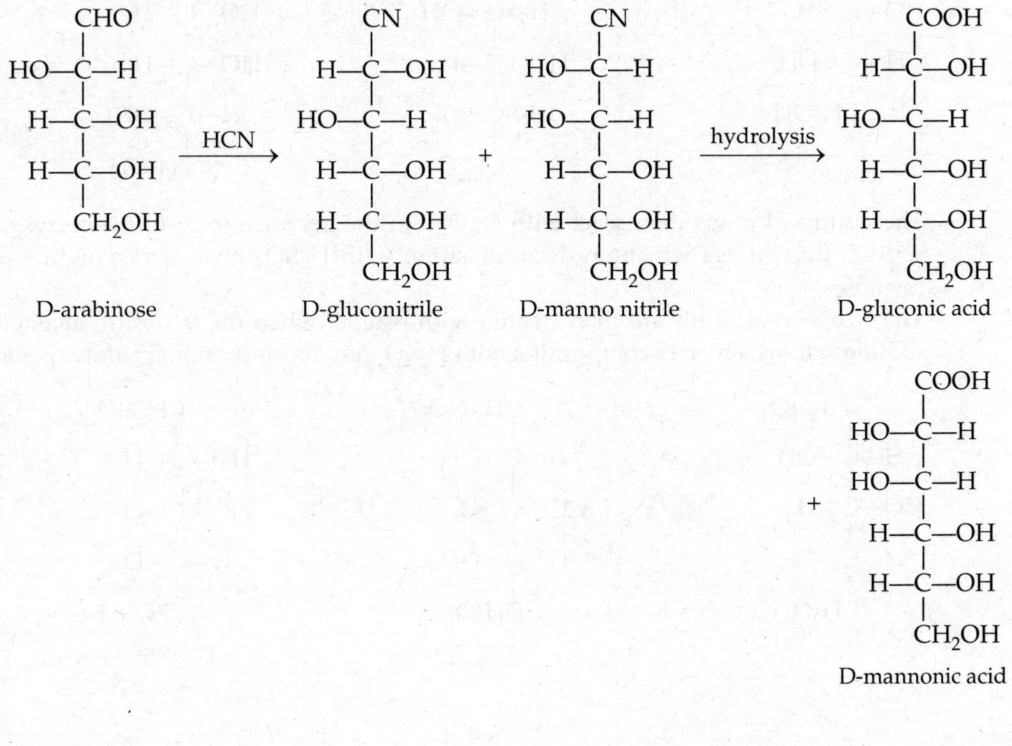

D-arabinose   D-gluconitrile   D-manno nitrile   D-gluconic acid

D-mannonic acid

$$
\underset{\text{heat}}{\longrightarrow}
\quad
\begin{array}{c}
\text{C=O} \\
| \\
\text{H—C—OH} \\
| \\
\text{HO—C—H} \\
| \\
\text{H—C} \\
| \\
\text{H—C—OH} \\
| \\
\text{CH}_2\text{O–H}
\end{array}
\text{O}
\qquad
\begin{array}{c}
\text{C=O} \\
| \\
\text{HO—C—H} \\
| \\
\text{HO—C—H} \\
| \\
\text{H—C} \\
| \\
\text{H—C—OH} \\
| \\
\text{CH}_2\text{OH}
\end{array}
\text{O}
\quad\longrightarrow\quad
\begin{array}{c}
\text{CHO} \\
| \\
\text{H—C—OH} \\
| \\
\text{HO—C—H} \\
| \\
\text{H—C—OH} \\
| \\
\text{H—C—OH} \\
| \\
\text{CH}_2\text{OH}
\end{array}
\qquad
\begin{array}{c}
\text{CHO} \\
| \\
\text{HO—C—H} \\
| \\
\text{HO—C—H} \\
| \\
\text{H—C—OH} \\
| \\
\text{H—C—OH} \\
| \\
\text{CH}_2\text{OH}
\end{array}
$$

D-glucolactone   D-mannolactone   D-glucose   D-mannose

On hydrolysis, the aldonic acids lose one molecule of water and are converted into γ-lactones.

Finally these lactones are reduced with either $LiAlH_4$ or sodium amalgam in presence of a weakly acidic solution, to form D-glucose and D-mannose respectively.

2. **Snowden-Fischer nitro methane synthesis:** As aldose is condensed with CH_3NO_2 ,in the presence of a base, with the result there are formed two different nitro alcohols, which are then separated by fractional distillation.

$$
\begin{array}{c}
\text{CHO} \\
| \\
\text{HO—C—H} \\
| \\
\text{H—C—OH} \\
| \\
\text{CH}_2\text{OH}
\end{array}
\quad\underset{}{\xrightarrow{\text{CH}_3\text{Na}}}\quad
\begin{array}{c}
\text{CH}_2\text{NO}_2 \\
| \\
\text{H—C—OH} \\
| \\
\text{HO—C—H} \\
| \\
\text{H—C—OH} \\
| \\
\text{CH}_2\text{OH}
\end{array}
\qquad
\begin{array}{c}
\text{CH}_2\text{NO}_2 \\
| \\
\text{HO—C—H} \\
| \\
\text{HO—C—H} \\
| \\
\text{H—C—OH} \\
| \\
\text{CH}_2\text{OH}
\end{array}
$$

tetrose

These nitro alcohols then react with NaOH, in the aci form to form the corresponding sodium derivatives which on decomposition with H_2SO_4 give corresponding aldoses (pentoses).

The two nitroalcohols are than treated with NaOH, when the two nitro alcohols form sodium salt, which on decomposition with H_2SO_4, give corresponding aldose (pentoses).

$$
\begin{array}{c}
\text{CH}_2\text{NO}_2 \\
| \\
\text{H—C—OH} \\
| \\
\text{HO—C—H} \\
| \\
\text{H—C—OH} \\
| \\
\text{CH}_2\text{OH}
\end{array}
\quad\xrightarrow{\text{NaOH}}\quad
\begin{array}{c}
\text{CH=NO}_2\text{Na} \\
| \\
\text{H—C—OH} \\
| \\
\text{HO—C—H} \\
| \\
\text{H—C—OH} \\
| \\
\text{CH}_2\text{OH}
\end{array}
\quad\xrightarrow{\text{H}_2\text{SO}_4}\quad
\begin{array}{c}
\text{CH}_2\text{NO}_2 \\
| \\
\text{HO—C—H} \\
| \\
\text{HO—C—H} \\
| \\
\text{H—C—OH} \\
| \\
\text{CH}_2\text{OH}
\end{array}
$$

pentose

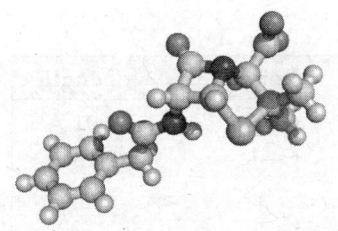

Appendix

Bond Energy (for homolytic fission) A : B $\longrightarrow$ A· + B·

	kJ	kcal		kJ	kcal
C—C	347.3	83	Cl—Cl	242.7	58
C=C	606.7	145	Br—Br	188.3	45
C≡C	803.3	192	I—I	105.6	36
C—O	334.7	80	H—Cl	426.8	102
C=O	94.5	166	H—Br	364.0	87
C—N	284.5	68	H—I	297.1	71
C=N	615.1	147	O—H	464.4	111
C≡N	866.1	207	O=C=O	803.3	192
C—F	326.4	107	H—H	435	104
C—Cl	284.5	78			
C—Br	284.5	68			
C—I	213.4	51			
C—H	414.2	99			

Bond Energy (heterolytic fission) A : B $\longrightarrow$ A$^+$ + B:$^-$

	kJ	kcal		kJ	kcal
H : H	1678	401	CH$_3${ : H	1310	313
H : F	1548	370	CH$_3$: F	1071	256
H : Cl	1397	334	CH$_3$: Cl	950	227
H : Br	1356	324	CH$_3$: Br	916	219
H : I	1318	315	CH$_3$: I	887	212
H : OH	1632	390	CH$_3$: {OH	1146	274

Bond Lengths

Bond	Length	Bond	Length	Bond	Length
C—C	1.54 Å	C—O	1.43 Å	C—Br	1.91 Å
C=C	1.40 Å	C=O	1.20 Å	C—I	2.13 Å
C≡C	1.20 Å	O—H	0.97 Å		
C—H	1.12 Å	N—H	1.03 Å		
C—N	1.47 Å	C—F	1.42 Å		
C—S	1.82 Å	C—Cl	1.77 Å		